Lecture Notes in Bioinformatics   6840

Subseries of Lecture Notes in Computer Science

W0038567

De-Shuang Huang   Yong Gan
Prashan Premaratne   Kyungsook Han (Eds.)

Bio-Inspired Computing and Applications

7th International Conference
on Intelligent Computing, ICIC 2011
Zhengzhou, China, August 11-14, 2011
Revised Selected Papers

 Springer

Series Editors

Sorin Istrail, Brown University, Providence, RI, USA
Pavel Pevzner, University of California, San Diego, CA, USA
Michael Waterman, University of Southern California, Los Angeles, CA, USA

Volume Editors

De-Shuang Huang
Tongji University, Shanghai, China
E-mail: dshuang@tongji.edu.cn

Yong Gan
Zhengzhou University of Light Industry, Zhengzhou Henan, China
E-mail: ganyong@zzuli.edu.cn

Prashan Premaratne
University of Wollongong, North Wollongong, NSW, Australia
E-mail: prashan@uow.edu.au

Kyungsook Han
Inha University, Inchon, Korea
E-mail: khan@inha.ac.kr

ISSN 0302-9743   e-ISSN 1611-3349
ISBN 978-3-642-24552-7   ISBN 978-3-642-24553-4 (eBook)
DOI 10.1007/978-3-642-24553-4
Springer Heidelberg Dordrecht London New York

Library of Congress Control Number: 2011938157

CR Subject Classification (1998): I.2, I.4, I.5, F.1, H.3, H.4

LNCS Sublibrary: SL 8 – Bioinformatics

Typesetting: Camera-ready by author, data conversion by Scientific Publishing Services, Chennai, India

Printed on acid-free paper

Springer is part of Springer Science+Business Media (www.springer.com)

Preface

The International Conference on Intelligent Computing (ICIC) was formed to provide an annual forum dedicated to the emerging and challenging topics in artificial intelligence, machine learning, pattern recognition, image processing, bioinformatics, and computational biology. It aims to bring together researchers and practitioners from both academia and industry to share ideas, problems, and solutions related to the multifaceted aspects of intelligent computing.

ICIC 2011, held in Zhengzhou, China, August 11–14, 2011, constituted the 7th International Conference on Intelligent Computing. It built upon the success of ICIC 2010, ICIC 2009, ICIC 2008, ICIC 2007, ICIC 2006, and ICIC 2005 that were held in Changsha, Ulsan/Korea, Shanghai, Qingdao, Kunming, and Hefei, China, respectively.

This year, the conference concentrated mainly on the theories and methodologies as well as the emerging applications of intelligent computing. Its aim was to unify the picture of contemporary intelligent computing techniques as an integral concept that highlights the trends in advanced computational intelligence and bridges theoretical research with applications. Therefore, the theme for this conference was "Advanced Intelligent Computing Technology and Applications". Papers focusing on this theme were solicited, addressing theories, methodologies, and applications in science and technology.

ICIC 2011 received 832 submissions from 28 countries and regions. All papers went through a rigorous peer-review procedure and each paper received at least three review reports. Based on the review reports, the Program Committee finally selected 281 high-quality papers for presentation at ICIC 2011, which are included in three volumes of proceedings published by Springer: one volume of *Lecture Notes in Computer Science* (LNCS), one volume of *Lecture Notes in Artificial Intelligence* (LNAI), and one volume of *Lecture Notes in Bioinformatics* (LNBI). In addition, among them, the 10 and 44 high-quality papers have also, respectively, been recommended to *BMC Bioinformatics* and *Neurocomputing*. This volume of *Lecture Notes in Bioinformatics* (LNBI) includes 93 papers.

The organizers of ICIC 2011, including Zhengzhou University of Light Industry, Institute of Intelligent Machines of Chinese Academy of Sciences, made an enormous effort to ensure the success of ICIC 2011. We hereby would like to thank the members of the Program Committee and the referees for their collective effort in reviewing and soliciting the papers. We would like to thank Alfred Hofmann, from Springer, for his frank and helpful advice and guidance throughout and for his continuous support in publishing the proceedings. In particular, we would like to thank all the authors for contributing their papers. Without the high-quality submissions from the authors, the success of the conference would not have been possible. Finally, we are especially grateful to the IEEE Computational Intelligence Society, the International Neural Network Society, and the National Science Foundation of China for their sponsorship.

July 2011

De-Shuang Huang
Yong Gan
Prashan Premaratne
Kyungsook Han

ICIC 2011 Organization

General Co-chairs

De-Shuang Huang, China
DeLiang Wang, USA
Yanli Lv, China

Program Committee Co-chairs

Zhongming Zhao, USA
Kang-Hyun Jo, Korea
Jianhua Ma, Japan

Organizing Committee Co-chairs

Yong Gan, China
Sushi Zhang, China
Hong-Qiang Wang, China
Wei Jia, China

Award Committee Chair

Laurent Heutte, France

Publication Chair

Juan Carlos Figueroa, Colombia

Special Session Chair

Phalguni Gupta, India

Tutorial Chair

Vitoantonio Bevilacqua, Italy

International Liaison Chair

Prashan Premaratne, Australia

Publicity Co-chairs

Xiang Zhang, USA
Kyungsook Han, Korea
Lei Zhang, Hong Kong, China

Exhibition Chair

Xueling Li, China

Organizing Committee Members

Xunlin Zhu, China
Shengli Song, China
Haodong Zhu, China
Xiaoke Su, China
Xueling Li, China
Jie Gui, China

Conference Secretary

Zhi-Yang Chen, China

Program Committee Members

Andrea Francesco Abate, Italy
Vasily Aristarkhov, Russian Federation
Costin Badica, Romania
Shuhui Bi, Japan
David B. Bracewell, USA
Martin Brown, UK
Zhiming Cai, Macau, China
Chin-chih Chang, Taiwan, China
Pei-Chann Chang, China
Guanling Chen, USA
Jack Chen, Canada
Shih-Hsin Chen, China
Wen-Sheng Chen, China
Xiyuan Chen, China
Yang Chen, China
Yuehui Chen, China
Ziping Chiang, China
Michal Choras, Poland

Angelo Ciaramella, Italy
Jose Alfredo F. Costa, Brazil
Youping Deng, USA
Eng. Salvatore Distefano, Italy
Mariagrazia Dotoli, Italy
Meng Joo Er, Singapore
Ahmed Fadiel, USA
Karim Faez, Iran
Jianbo Fan, China
Minrui Fei, China
Wai-Keung Fung, Canada
Jun-Ying Gan, China
Liang Gao, China
Xiao-Zhi Gao, Finland
Carlos Alberto Reyes Garcia, Mexico
Dunwei Gong, China
Valeriya Gribova, Russia
M. Michael Gromiha, Japan

Kayhan Gulez, Turkey
Anyuan Guo, China
Phalguni Gupta, India
Sung Ho Ha, Korea
Fei Han, China
Kyungsook Han, Korea
Nojeong Heo, Korea
Laurent Heutte, France
Wei-Chiang Hong, Taiwan, China
Zeng-Guang Hou, China
Yuexian Hou, China
Kun Huang, USA
Peter Hung, Ireland
Sajid Hussain, USA
Peilin Jia, USA
Minghui Jiang, China
Zhenran Jiang, China
Kang-Hyun Jo, Korea
Yoshiaki Kakuda, Japan
Sanggil Kang, Korea
Muhammad Khurram Khan, Saudi Arabia
Sungshin Kim, Korea
In-Soo Koo, Korea
Bora Kumova, Turkey
Yoshinori Kuno, Japan
Wen-Chung Kuo, Taiwan, China
Takashi Kuremoto, Japan
Vincent C.S. Lee, Australia
Guo-Zheng Li, China
Jing Li, USA
Kang Li, UK
Peihua Li, China
Ruidong Li, Japan
Shutao Li, China
Xiaoou Li, Mexico
Hualou Liang, USA
Honghuang Lin, USA
Chunmei Liu, USA
Liu Chun-Yu Liu, USA
Ju Liu, China
Van-Tsai Liu, Taiwan, China
Jinwen Ma, China
Tarik Veli Mumcu, Turkey
Igor V. Maslov, Japan

Filippo Menolascina, Italy
Primiano Di Nauta, Italy
Roman Neruda, Czech Republic
Ben Niu, China
Sim-Heng Ong, Singapore
Ali Özen, Turkey
Vincenzo Pacelli, Italy
Francesco Pappalardo, Italy
Witold Pedrycz, Canada
Caroline Petitjean, France
Pedro Melo-Pinto, Portugal
Susanna Pirttikangas, Finland
Prashan Premaratne, Australia
Daowen Qiu, China
Yuhua Qian, China
Seeja K.R., India
Marylyn Ritchie, USA
Ivan Vladimir Meza Ruiz, Mexico
Fariba Salehi, Iran
Angel Sappa, Spain
Jiatao Song, China
Stefano Squartini, Italy
Hao Tang, China
Antonio E. Uva, Italy
Jun Wan, USA
Bing Wang, USA
Ling Wang, China
Xue Wang, China
Xuesong Wang, China
Yong Wang, Japan
Yufeng Wang, Japan
Zhong Wang, USA
Wei Wei, Norway
Zhi Wei, China
Ling-Yun Wu, China
Junfeng Xia, USA
Shunren Xia, China
Hua Xu, USA
Jianhua Xu, China
Shao Xu, Singapore
Ching-Nung Yang, Taiwan, China
Wen Yu, Mexico
Zhi-Gang Zeng, China
Jun Zhang, China

Xiang Zhang, USA
Yanqing Zhang, USA
Zhaolei Zhang, Canada
Lei Zhang, Hong Kong, China
Xing-Ming Zhao, China
Zhongming Zhao, USA

Chun-Hou Zheng, China
Huiru Zheng, UK
Bo-Jin Zheng, China
Fengfeng Zhou, USA
Mianlai Zhou, China
Li Zhuo, China

Reviewers

Ibrahim Sahin
Bora Kumova
Birol Soysal
Yang Xiang
Gang Feng
Francesco Camastra
Antonino Staiano
Alessio Ferone
Surya Prakash
Badrinath Srinivas
Dakshina Ranjan Kisku
Zilu Ying
Guohui He
Vincenzo Pacelli
Pasqualedi Biase
Federica Miglietta
Junying Zeng
Yibin Yu
Kaili Zhou
Yikui Zhai
WenQiang Yang
WenJu Zhou
Dae-Nyeon Kim
Ilmari Juutilainen
Alessandro Cincotti
Marzio Alfio Pennisi
Carme Julià
Santo Motta
Nestor Arana-Arexolaleiba
Myriam Delgado
Giuliana Rotunno
Agostino Marcello Mangini
Carson K. Leung
Gabriella Stecco
Yaser Maddahi
Jun Wan
Jiajun Bracewell

Jing Huang
Kunikazu Kobayashi
Feng Liangbing
JoaquinTorres-Sospedra
Takashi Kuremoto
Fabio Sciancalepore
Valentina Boschian
Chuang Ma
Juan Xiao
Lihua Jiang
Changan Jiang
Ni Bu
Shengjun Wen
Aihui Wang
Peng Wang
Myriam Delgado
Wei Ding
Kurosh Zarei-nia
Li Zhu
Hoang-HonTrinh
Alessia Albanese
Song Zhu
Lei Liu
Feng Jiang
Bo Liu
Ye Xu
Gang Zhou
ShengyaoWang
Yehu Shen
Liya Ding
Hongjun Jia
Hong Fu
Tiantai Guo
Liangxu Liu
Dawen Xu
Zhongjie Zhu
Jayasuha J.S.

Aravindan Chandrabose
Shanthi K.J.
Shih-Hsin Chen
Wei-Hsiu Huang
Antonio Maratea
Sandra Venske
Carolina Almeida
Richard Goncalves
Ming Gao
Feng Li
Yu Xue
Qin Ma
Ming Gao
Gang Xu
Yandong Zhang
Benhuai Xie
Ran Zhang
Mingkun Li
Zhide Fang
Xiaodong Yang
Lein Harn
Wu-Chuan Yang
Bin Qian
Quan-ke Pan
Junqing Li
Qiao Wei
Xinli Xu
Hongjun Song
Michael Gromiha
Xueling Li
Y-h. Taguchi
Yu-Yen Ou
Hong-Bin Shen
Ximo Torres
Weidong Yang
Quanming Zhao
Chong Shen

Xianfeng Rui
Phalguni Gupta
Yuan Xu
Yuefang Zhao
Custiana Cucu
Xiaojuan Wang
Guihui Zhang
Xinyu Li
Yang Shi
Hongcheng Liu
Lijun Xu
Xiaomin Liu
Tonghua Su
Junbiao Pang
Chun Nie
Saihua Lin
Alfredo Pulvirenti
Melo-Pinto Pedro
Armando Fernandes
Atsushi Yamashita
Kazunori Onoguchi
Liping Zhang
Qiong Zhu
Chi Zhou
Qirong Mao
Lingling Wang
WenYong Dong
Wenwen Shen
Gang Bao
Shiping Wen
Giorgio Iacobellis
Paolo Lino
Qi Jiang
Yan-Jie Li
Gurkan Tuna
Tomoyuki Ohta
Jianfei Hu
Xueping Yu
Shinji Inoue
Eitaro Kohno
Rui-Wei Zhao
Shixing Yan
Jiaming Liu
Wen-Chung Kuo
Jukka Riekki

Jinhu Lu
Qinglai Wei
Michele Scarpiniti
Simone Bassis
Zhigang Liu
Pei Wang
Qianyu Feng
Jingyi Qu
Mario Foglia
Michele Fiorentino
Luciano Lamberti
Lein Harn
Kai Ye
Zhenyu Xuan
Francesco Napolitano
Raphael Isokpehi
Vincent Agboto
Ryan Delahanty
Shaohui Liu
Ching-Jung Ting
Chuan-Kang Ting
Chien-Lung Chan
Jyun-Jie Lin
Liang-Chih Yu
Richard Tzong-Han Tsai
Chin-Sheng Yang
Jheng-Long Wu
Jun-Lin Lin
Chia-Yu Hsu
Wen-Jia Kuo
Yi-Kuei Lin
K. Robert Lai
Sumedha Gunewardena
Qian Xiang
Joe Song
Ryuzo Okada
Handel Cheng
Chin-Huang Sun
Tung-Chen Huang
Bin Yang
Changyan Xiao
Mingkui Tan
Zhigang Ling
Lei Zhou
Hung-Chi Su

Chyuan-Huei Yang
Rey-Sern Lin
Cheng-Hsiung Chiang
Chrisil Arackaparambil
Valerio Bianchi
Zhi Xie
Ka-Chun Wong
Zhou Yong
Aimin Zhou
Yong Zhang
Yan Zhang
Jihui Zhang
Xiangjuan Yao
Jing Sun
Jianyong Sun
Yi-Nan Guo
Yongbin Zhang
Vasily Aristarkhov
Hongyan Sang
Aboubekeur Hamdi-Cherif
Chen Bo
Min Li
Linlin Shen
Jianwei Yang
Lihua Guo
Manikandan Narayanan
Masoumeh Esfandiari
Amin Yazdanpanah
Ran Tao
Weiming Yu
Aditya Nigam
Kamlesh Tiwari
Maria De Marsico
Stefano R.
Wei Wei
Lvzhou Li
Haozhen Situ
Bian Wu
Linhua Zhou
Shaojing Fan
Qingfeng Li
Rina Su
Hongjun Song
Bin Ye
Jun Zhao

Yindi Zhao
Kun Tan
Chen Wei
Yuequan Yang
Qian Zhang
Zhigang Yan
Jianhua Xu
Ju-Yin Cheng
Yu Gu
Guang Zeng
Xuezheng Liu
Weirong Yuan
Ren Xinjun
Futian Yu
Mingjing Yang
Chunjiang Zhang
Yinzhi Zhou
William Carswell
Andrey Vavilin
Sang-Hee Lee
Yan Fan
Hong Wang
Fangmin Yao
Angelo Ciaramella
Eric Hsu
Xiao-Feng Wang
Jing Deng
Wanqing Zhao
Weihua Deng
Xueqin Liu
Sung Shin Kim
Gyeongdong Baek
Seongpyo Cheon
Bilal Khan
Maqsood Mahmud
Pei-Wei Tsai
Lin Zhang
Bo Peng
Jifeng Ning
Yongsheng Dong
Chonglun Fang
Yan Yang
Hongyan Wang
Min Wang
Rong-Xiang Hu

Xiaoguang Li
Jing Zhang
Yue Jiao
Hui Jing
Ruidong Li
Wei Xiong
Toshiaki Kondo
Suresh Sundaram
Hai Min
Donghui Hu
Xiaobin Tan
Stefano Dell'Atti
Rafal Kozik
Michal Choras
R. Phan
Yuan-Fang Li
Tsung-Che Chiang
Ming Xia
Weimin Huang
Xinguo Yu
Sabooh Ajaz
ZhengMao Zou
Prashan Premaratne
Ibrahim Aliskan
Yusuf Altun
Ali Ahmed Adam
Janset Dasdemir
Turker Turker
Ibrahim Kucukdemiral
JunSheng Zhou
Yue Wang
Yoshiaki Kakuda
Daqiang Zhang
Min-Chih Chen
Aimin Zhou
Shihong Ding
Ziping Chiang
Xiaoyu Wen
Gao Liang
Orion Reyes-Galaviz
Miguel Mora-Gonzalez
Pilar Gomez-Gil
Miguel Mora-Gonzalez
Jida Huang
Insoo Koo

Nhan Nguyen-Thanh
ThucKieu Xuan
Yang Zhao
Andreas Konstantinidis
Canyi Lu
Nobuo Funabiki
Yukikazu Nakamoto
Xin Zhou
Qian Wang
Xiaoyan Yin
Juan Cui
Francesco Polese
Sen Jia
Crescenzio Gallo
Yu Sun
Xuewen Xia
Chuan Peng
Chen Jing-Yuan
Edison Yu
Petra Vidnerová
Klara Peskova
Martin Pilat
Liu Zhaochen
Jun Du
Ning Lv
Yoko Kamidoi
Meng Wang
Hao Xin
Dingfei Ge
Xin Gao
Ivan Vladimir Meza Ruiz
Tsang-Yi Wang
Sangyoon Oh
Li Ruichang
Fan Jing
Lin Wang
Chunlu Lai
Hamide Cheraghchi
Wen-Tsai Sung
Theanh Bui
Zhong Qishui
Duyu Liu
Keliang Jun
Ying Qiu
Huisen Wang

Maria Elena Valcher
Alex Muscar
SorinIlie
Amelia Badica
Guanghai Liu
Changbin Du
Jianqing Li
Hao Wang
Yurong Cheng
Mingyi Wang
Claudio Franca
Jose Alfredo Ferreira Costa
Tomasz Andrysiak
Ajay Kumar
Lei Zhang
Zhoumian Wang
Ji-Xiang Du
Xibei Yang
Junhong Wang
Wei Wei
Guoping Lin
Dun Liu
Changzhong Wang
Xiaoxiao Ma
Xueyang Xiao
Wei Yu
Ming Yang
Francesca Nardone
Kok-Leong Ong
David Taniar
Nali Zhu
Hailei Zhang
My HaLe
Haozhen Situ
Lvzhou Li
Mianlai Zhou
Chin-Chih Chang

Carlos A. Reyes-Garcia
Jack Chen
Wankou Yang
Qijun Zhao
Jin Xie
Xian Chen
Gustavo Fontoura
Xiaoling Zhang
Ondrej Kazik
Bo Yan
Yun Zhu
B.Y. Lee
Jianwen Hu
Keling Chang
Jianbo Fan
Chunming Tang
Hongwei Ma
Valeria Gribova
Ailong Wu
William-Chandra Tjhi
Gongqing Wu
Yaohong Liang
Bingjing Cai
Lin Zhu
Li Shang
Bo Li
Jun Zhang
Peng Chen
Wenlong Sun
Xiaoli Wei
Bing Wang
Jun Zhang
Peng Chen
Karim Faez
Xiaoyan Wang
Wei-Chiang Hong
Chien-Yuan Lai

Sugang Xu
Junfeng Xia
Yi Xiong
Xuanfang Fei
Jingyan Wang
Zhongming Zhao
Yonghui Wu
Samir Abdelrahman
Mei Liu
Fusheng Wang
Shao-Lun Lee
Wen Zhang
Zhi-Ping Liu
Qiang Huang
Jiguang Wang
Rui Xue
Xiao Wang
Jibin Qu
Bojin Zheng
Susanna Pirttikangas
Ukasz Saganowski
Chunhou Zheng
Zheng Chunho
Mei Jun
Geir Solskinnsbakk
Satu Tamminen
Laurent Heutte
Mikko Perttunen
Renqiang Min
Rong-Gui Wang
Xinping Xie
Horace Wang
Hong-Jie Yu
Wei Jia
Huqing Wang

Table of Contents

Neural Networks

Systems Biology and Computational Biology

Computational Genomics and Proteomics

Knowledge Discovery and Data Mining

Evolutionary Learning and Genetic Algorithms

Biological and Quantum Computing

Machine Learning Theory and Methods

Biomedical Informatics Theory and Methods

Complex Systems Theory and Methods

Intelligent Computing Theory and Methods

Natural Language Processing and Computational Linguistics

Granular Computing and Rough Sets

Fuzzy Theory and Models

Fuzzy Systems and Soft Computing

Particle Swarm Optimization and Niche Technology

Swarm Intelligence and Optimization

Supervised and Semi-supervised Learning

Intelligent Computing in Bioinformatics

Intelligent Computing in Brain Imaging and Bio-medical Engineering

Intelligent Computing in Computational Biology and Drug Design

Intelligent Computing in Pattern Recognition

Special Session on Sparse Representation-Based Machine Learning

Special Session on Bio-inspired Computing

Special Session on Molecular Diagnosis of Tumor and Tumor Biomarker Discovery

Special Session on Computer Aided Early Diagnosis in Medicine - CAEDiM

Special Session on Sparse Representation in Bioinformatics

Special Session on Sparse Representation in Bioinformatics

Special Session on Advances in Particle Swarm Optimization

Special Session on Advances in Intelligent Information Processing

Application of Neural Network on Solid Boronizing

YuXi Liu and ZhiFeng Zhang

Software College
ZhengZhou University of Light Industry
ZhengZhou, China
yuxiliu@yeah.net

Abstract. This paper discusses an application of neural network system on the performance prediction of solid boronizing. To build the mathematics model between the solid boronizing and the prediction of boronizing performance, a neural network approach is adopted. This approach overcomes a lot of problems in the traditional approaches and provides a stable and effective approach.

Keywords: solid boronizing, mathematics model, neural network.

1   Introduction

In chemistry, a boride is a chemical compound between boron and a less electronegative element (transition metal, alkaline earth metal, rare earth metal), such as Ni_2B, NiB_2, MnB, MnB_2, Cu_3B_2 etc. Some borides exhibit very useful physical properties. They generally have high melting (2000-3000°C) and high hardness(Mohs' scale number 8-10), and they are not ionic in nature. They have high electrical conductivity and chemical stability(not affected by hydrochloric acid and hydrofluoric acid, but could be attacked by hot alkali metal hydroxides). They were widely used in high temperature material and abrasive material.

Boronizing could be divided into solid boronizing, gas boronizing, liquid boronizing (electric salt bath and salt bath). For solid boronizing, granular and powdered medium was used for the entire boronizing for small and medium parts; paste was used for large parts. These mediums were composed by boron supplier agent (B4C, B-Fe, amorphous boron powder); energizer (KBF4, NH4C1, NH4F); filler of activity regulation (A12O3, SiC, SiO2). Boronizing could be operated in the temperature 650~1000°C, usually be 850~950°C; Heat preservation for 2~6 hours, 50~200 µm boronizing layer was acquired for different steel. For salt bath, one is on basis of borax, $5\% \sim 15\%$ chlorizated salts was added; The other is on basis of chlorizated salts, which BoC or SiC was added [6].

A lot of research has been done for boronizing [1-2]. The relationship between the boronizing technics coefficient and the performance indicator of diffusion layers is obscure. It is affected by many of factors. In the practical application, there are several of technics coefficients and diffusion layers indicators that are affected by each other. Data fitting is an optional approach, but it is very difficult to be expressed by function relation [3].

D.-S. Huang et al. (Eds.): ICIC 2011, LNBI 6840, pp. 1–7, 2012.
© Springer-Verlag Berlin Heidelberg 2012

Neural network is the technology that is inspired by biology. That will be used in the modeling in sophisticated nonlinear function. For its structure, a neural network can be divided into the input layer, output layer, hidden layer. Each node of input layer is corresponding to a prediction variant; the node of the output layer is corresponding to the objective variant. Between the input layer and the output layer is the hidden layer. Each node of the neural network is linked to the front node, and each link corresponds to a weight. The job of construction of the neural network is to adjust the weight of the nodes [1].

Neural network has the ability of self-structured, high fault-tolerant and non-linear function approximation. It also has the ability of self-studied. So it has been widely used in field of mental, alloy, inorganic nonmetallic materials. This paper is focus on the application of neural network in the solid boronizing to predict the performance of diffusion layer.

2   The Design and the Implementation of the Neural Network

2.1   The Collection of the Sample Data

Steel Q235, which size was $\varphi 15mm \times 10mm$ is the test sample, and borax, B_4C were selected as the boron supply agent; Na3AlF, MgCl2, CaF2 were selected as the activating agent. The temperature of boronizing, insulation time, and boron supply agent purity were selected as the technics coefficients. The boron supply agent packaged the sample and it was prepared proportionately. That were putted in crucible, and circled by bulking agent. Then they were putted into the box resistance furnace, keeping warm in the experiment temperature for some time. At last they were taken from the crucible, cooled to the room temperature.

Considering that there were only three technics coefficients, three level factors were selected in table 1. And 125 groups of experiment were tested. The object indicator was thickness of diffusion layers, micro hardness of surface and the micro hardness that was 0.1mm away from the surface.

Table 1. Level factors in solid boronizing

Boronizing Time(hour)	2	3	4	5	6
Boronizing Temperature(°C)	850	870	900	930	950
Boron supply agent purity (%)	60	65	70	75	80

2.2   Neural Network Topology and Algorithms

2.2.1   Neural Network Topology

Three layers of 3×6×3 BP neural network was adopted, its structure was illustrated in figure I. Error back propagation model of neural network included: input layer, hidden layer, and output layer.

1) The number of the hidden layer

Firstly the single hidden layer architecture was adopted. If the number of the nodes in the hidden layer were too many, two hidden layer could be adopted. In this experiment, single hidden layer could get a satisfied result.

2) The number of the nodes in the hidden layer

If the number of the nodes in the hidden layer were too less, the capacity of the learning would be limited; on the other hand, too many nodes would increase the training time significantly. According to the empirical formula:

$$m = \sqrt{n+l} + \alpha \tag{1}$$

Here m is the number of the nodes in the hidden layer, n is the number of the nodes in the input layer, l is the number of the nodes in the output layer, α is the regulation constant, usually between 1~10. Adjusting the value of m and using the single sample set for training, when the network error is minimum, the number of the nodes in the hidden layer could be selected. Six nodes were selected in the hidden layer finally.

3) The inputs and outputs of the network

Each neural between the upper and lower layer was fully linked together. That meant each neural in the lower layer would be linked to each neutral in the upper layer with a weight, and there was no link in the neural which in the same layer. The input of the neural network was corresponding to the technics coefficients of the solid boronizing; the output was corresponding to the indicators of the diffusion layers. In the neural network, there were three nodes in the input layers, corresponding to the boronizing temperature, insulation time, and boron supply agent purity that could be adjusted. There were three nodes in the output layer, corresponding to the three indicators of the diffusion layers: thickness of diffusion layers, micro hardness of surface and the micro hardness that was 0.1mm under the surface.

2.2.2   The Construction of the Neural Network

The weight of the neural link between the input layer and the hidden layer was W_{hi}, and the weight of the neural link between the hidden layer and the output layer was W_{ij}; which were feed by the minor random values. The nodes of the input layer were n; the nodes of hidden layer were p; the nodes of output layer were q.

Given the input/output sample couple, the output of network was calculated: if the p-th input, output of the sample of group was:

$$u_p = (u_{1p}, u_{2p}, \ldots\ldots, u_{np}). \tag{2}$$

$$dp = (d_{1p}, d_{2p}, \ldots\ldots d_{qp}). \tag{3}$$

$$p=1,2\ldots\ldots L. \tag{4}$$

L was the maxim number of the samples group.

p-th group sample couple was inputted to node I, the output was:

$$y = f[x_{ip}(t)] = f[\sum_{j} w_{ij}(t)I_{jp}]. \tag{5}$$

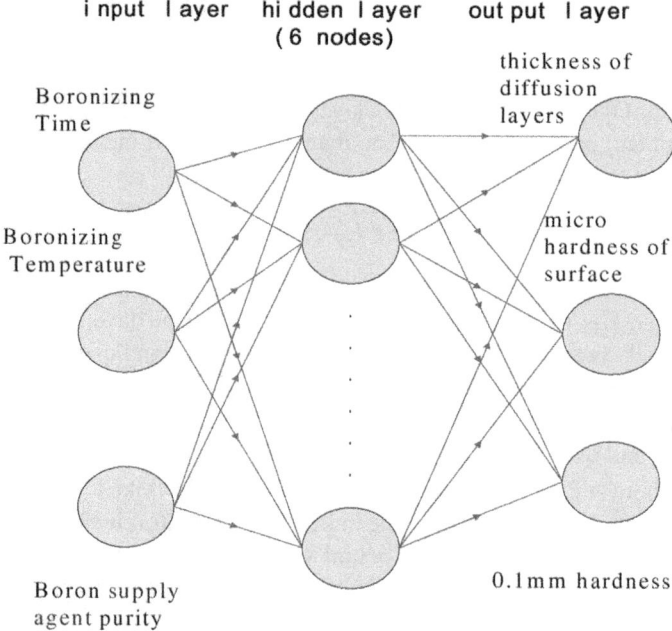

input layer   hidden layer   output layer
                 (6 nodes)

Boronizing Time

Boronizing Temperature

Boron supply agent purity

thickness of diffusion layers

micro hardness of surface

0.1mm hardness

Fig. 1. Neural network model of property prediction

In the formula, I_{jp} was the *j*-th input for the node *I*, when *p*-th sample group was inputted; W_{ij} was the weight of the neural link between the hidden layer and the output layer; *f* was the activation function, sigmoid model adopted:

$$f(x) = \frac{1}{1+e^x}. \tag{6}$$

The node's input in the output layer could be calculated from the input layer to output layer via the hidden layer.

Unit deviation of output layer was:

$$E(\omega) = \sum_{i=1}^{q} \left\| y_i - y_i^k \right\|^2. \tag{7}$$

In the formula, y_i was the expected output for the output unit I. y_i^k was the actual output value for unit I. If $E(\omega)$ was less than the expected error value by agreement, the training of network would be ended; otherwise back propagation process would be started.

The backpropagation algorithm trains a given feed-forward multilayer neural network for a given set of input patterns with known classifications [5]. Outline of the backpropagation learning algorithm can be gotten in reference 5 written by Beale, R. and Jackson.

2.3   Study Training of Neural Network

If the data were inputted which were different greatly, that would reduce the effect of radial function (when small data affected on it). To overcome the problem, the raw data would be treated by normalization. In this paper, the input data would be normalized linearly to section [0,1], formula was:

$$t_i = 0.1 + 0.9 \times \left(\frac{t - t_{min}}{t_{max} - t_{min}} \right).$$ (8)

In this formula, t was the input parameter of network; t_{max} and t_{min} are the maxim and minimum value of correspondent data.

According to the BP network process, the model training would be programmed by Matalab 6.5; the sample data were selected from the experiment result. 64 group data were selected for network training. System converged after 9000 times of iteration in training. From the result, the network output was very closely to the actual result, which verified the neural network could reflect the relationship of input and output.

3   The Result and the Comparison

In order to verify the reliability and the practicability of the program, the technics coefficients of the solid boronizing and the performance indicators prediction of diffusion layers of solid boronizing will be tested by a confirmatory experiment. Setting a group of technics coefficients values (temperature of boronizing, insulation time, and boron supply agent purity), that were inputted into the performance prediction network model, a group of values about the prediction of diffusion layers performance would be got; doing a solid boronizing experiment according to the setting technics, the value of diffusion layer performance could be determined. Five group of performance prediction and the reality experiment result were listed in the table 2. From the result, the performance prediction value of the diffusion layer and the experiment measurement value was less than 10% in deviation. Therefore, that verified the reliability of performance prediction neural network model.

Table 2. Comparison of measuring results and property prediction values

Measuring group	diffusion layer thickness (µm)		surface hardness (HV0.1)		hardness of surface under 0,1mm(HV0.1)	
	measuring	prediction	measuring	prediction	measuring	prediction
1	129	121	1965	1860	1180	1100
2	131	122	1970	1870	1190	1129
3	133	125	2010	1930	1195	1160
4	135	129	2060	1940	1270	1165
5	137	135	2063	1990	1280	1176

When the boronizing time(3 hours), temperature(850°C) and the boron supply agent purity(65%) selected, the number of nodes in the hidden layer(1~10) were changed for experiment. Such experiments had been tested for 15 groups. One group result is in table 3(only part of digit in one group). The deviation also was shown in percentage in the table. These experiments determined the number of nodes in the hidden layer.

Table 3. Result comparison of different number of hidden nodes

Hidden nodes	diffusion layer thickness (µm)	surface hardness (HV0.1)	hardness of surface under 0,1mm(HV0.1)
5	116 (>10%)	1765 (>10%)	1009 (>10%)
6	121 (<10%)	1860 (<10%)	1100 (<10%)
7	114 (>10%)	1870 (<10%)	1098 (<10%)

4   Conclusion

Via neural network, we set up the mapping of technics coefficients and the performance indicator of solid boronizing. That fulfilled the prediction of the diffusion layers. The experiment showed the mapping model was reliable. Neural network provided an advanced and reasonable approach for performance prediction of diffusion layer in solid boronizing.

References

1. Liu, X.: Orthogonal Analysis of Boriding Process for 45 Steel. Hot Working Technology 37(4), 36–37 (2004)
2. Yuan, X.B., Yang, R.C., Chen, H.: The optimum Technologies and Prospects of Solid Boriding. China Surface Engineering 27(5), 5–9 (2003)
3. Zong, B., Su, X.K., Li, H.E.: Application of Artificial Neural Network in Prediction of Properties and Process Parameters Optimization of Ion Nitriding. Heat Treatment Of Metals 34(6), 3–4 (2001)
4. Zhu, D.Q., Shi, H.: The theory and application of artificial neural network. The science publishing house, Beijing (2006)
5. Beale, R., Jackson, T.: Neural Computing: An Introduction. Hilger, Philadelphia, PA (1991)
6. Chen, S.W., Cheng, H.W., Chen, W.D.: The application and research of boronizing. Hteat Treatment Of Mentals Abroad 33(5), 8–9 (2003)

Improved Stability Criteria for Discrete-Time Stochastic Neural Networks with Randomly Time-Varying Delays*

Mengzhuo Luo[1] and Shouming Zhong[1,2]

[1] School of Mathematical Sciences, University of Electronic Science and Technology of China, 611731, Chengdu, P.R. China
zhuozhuohuahua@163.com, zhongsm@uestc.edu.cn
[2] Key Laboratory for Neuroinformation of Ministry of Education, University of Electronic Science and Technology of China, 610054, Chengdu, Sichuan, P.R. China

Abstract. This paper investigates the problem of exponential stability for a class of discrete-time stochastic neural network with randomly time-varying delays. Compared with some previous results, the new conditions obtained in this paper are less conservative. Finally, a numerical example is exploited to show the usefulness of the results derived.

Keywords: Delay-dependent stability, Neural networks, Time-varying delay, Linear matrix inequality(LMI), Free-weighting matrix.

1 Introduction

Neural networks have great potential applications in various areas such as signal processing, pattern recoganization, static image processing, associative memory and combinatorial optimization. Therefore, the stability problem of time-delay neural networks have become a topic of great theoretic and practical importance [1–3].

It is worth noting that the synaptic transmission is a noisy process brought on by random fluctuations from the release of neurotransmitters and other probabilistic causes in real nerves systems. So the stochastic disturbance is probably the main resource of the performance degradation of the implemented neural networks. Therefore, the stability for stochastic neural networks with delay have attracted increasing interests and some results related to stochastic disturbances have been published [4–7].

In this paper, the problem of stability analysis for discrete-time stochastic neural networks with delays is investigated. By using the discrete-time Jensen inequality, free-weighting matrix method and the probability distribution of the time delays, some sufficient conditions are established to ensure the stochastic neural networks are globally exponential stability in the mean square, which

---

* This work was supported by the National Natural Science Foundation of China (Grant No. 60736029) and the National Basic Research Program of China(2010CB732501).

D.-S. Huang et al. (Eds.): ICIC 2011, LNBI 6840, pp. 8–13, 2012.

proved to be less conservative than previous results. Finally, a numerical example is given to demonstrate the effectiveness of the proposed results.

2   Preliminaries

Consider the following discrete-time stochastic neural networks (DSNNs) with time-varying delays described by

$$x(k+1) = Cx(k) + Ag(x(k)) + Bg(x(k - \tau(k))) + \delta(k, x(k))\omega(k) \quad (1)$$

where $x(t) \in \mathbb{R}^n$ is the neuron state vector, $g(x(\cdot)) \in \mathbb{R}^n$, denotes the neuron activation function, $A, B \in \mathbb{R}^{n \times n}$ are the connection weight matrix and the delayed connection weight matrix,respectively. $C = diag(c_1, c_2, \cdots, c_n)$ with $|c_i| < 1, i = 1, 2, \cdots, n$. $\tau(k)$ is time-varying and satisfies $0 < \tau_1 \le \tau(k) \le \tau_2$.

For further discussion, we introduce the following assumption and definition.

Assumption 1: For any $x, y \in R, x \ne y$,

$$l_i^- \le \frac{g_i(x) - g_i(y)}{x - y} \le l_i^+ \quad (2)$$

where l_i^-, l_i^+, are some constants.

Assumption 2: There exist a positive constant ε such that

$$\delta^T(k, x(k))\delta(k, x(k)) \le \varepsilon x^T(k)x(k) \quad (3)$$

Assumption 3: The time-varying delay $\tau(k)$ is bounded and its probability distribution can be established, for example, assume that $\tau(k)$ takes values in $[\tau_1, \tau_0]$ or $(\tau_0, \tau_2]$, and $\text{Pro}\{\tau(k) \in [\tau_1, \tau_0]\} = \rho$, where τ_0 is integer satisfying $\tau_1 \le \tau_0 < \tau_2, 0 \le \rho \le 1$.

At first, we define the following two sets

$$H_1 = \{k \mid \tau(k) \in [\tau_1, \tau_0]\} \quad H_2 = \{k \mid \tau(k) \in (\tau_0, \tau_2]\} \quad (4)$$

Define two mapping functions

$$\tau_1(k) = \begin{cases} \tau(k) & k \in H_1 \\ \tau_1 & \text{else} \end{cases} \quad \tau_2(k) = \begin{cases} \tau(k) & k \in H_2 \\ \tau_0 & \text{else} \end{cases} \quad (5)$$

so from the above definitions, it is easy to derive that $k \in H_1$ implies the event $\tau(k) \in [\tau_1, \tau_0]$ takes place and $k \in H_2$ means $\tau(k) \in (\tau_0, \tau_2]$ happens. To develop our results, we define a stochastic variable as

$$\rho(k) = \begin{cases} 1 & k \in H_1 \\ 0 & k \in H_2 \end{cases} \quad (6)$$

then the DSNNs (1) can be rewritten as

$$\begin{aligned} x(k+1) = &Cx(k) + Ag(x(k)) + \rho(k)Bg(x(k - \tau_1(k))) \\ &+ (1 - \rho(k))Bg(x(k - \tau_2(k))) + \delta(k, x(k))\omega(k) \end{aligned} \quad (7)$$

Remark 1: From the assumption 3 and the definition of $\rho(k)$, it is obviously that the stochastic variable $\rho(k)$ is Bernouli distributed white sequences with $\text{Pro}\{\rho(k) = 1\} = E[\rho(k)] = \rho$, $\text{Pro}\{\rho(k) = 0\} = 1 - \rho = \bar{\rho}$, further it is can be seen that $E[\rho(k) - \rho] = 0$, $E\left[(\rho(k) - \rho)^2\right] = \rho\bar{\rho}$, meantime, we know that ρ is dependent on the values of τ_1, τ_0, τ_2 and we can easy to derive that $\rho = \frac{\tau_0 - \tau_1}{\tau_2 - \tau_1}$ if the time delay $\tau(k)$ satisfies uniform distribution in $[\tau_1, \tau_2]$.

3   Main Results

Theorem 1. Suppose that Assumption 1-3 hold. Then the DSNNs (7) is globally exponential stable in the mean square if there exists positive definite matrices $P, Q_i, T_i, i = 1, 2, 3, 4$, $Z_j, j = 1, 2$, diagonal matrices D_1, D_2, D_3, four positive scalars $\varepsilon, \rho_1^*, \rho_2^*, \rho_3^*$ and for any matrices $P_i, i = 1, \cdots, 12$ such that the following LMIs hold

$$P < \rho_1^* I \qquad T_2 < \rho_2^* I \qquad T_4 < \rho_3^* I \tag{8}$$

$$\Xi = \begin{bmatrix} \Xi_{11} & \cdots & \Xi_{1,12} \\ * & \ddots & \vdots \\ * & * & \Xi_{12,12} \end{bmatrix} < 0 \tag{9}$$

Where
$\Xi_{11} = 2C^T PC - 2P + 2\rho_1^* \varepsilon I + (\tau_1 + 1)Q_1 + (\tau_2 + 1)Q_2 + (\tau_2 - \tau_1)Q_3 + Q_4$
$\quad + (\tau_2 - \tau_0)R_2 + T_1 + (\tau_2 - \tau_0)^2 (C - I)^T T_2 (C - I) + (\tau_2 - \tau_0)^2 \rho_2^* \varepsilon I + T_3$
$\quad + (\tau_2 - \tau_0)^2 (C - I)^T T_4 (C - I) + (\tau_2 - \tau_0)^2 \rho_3^* \varepsilon I - D_1 L_1 + (\tau_0 - \tau_1 + 1)R_1$
$\Xi_{12} = P_1$ $\Xi_{14} = -P_1$ $\Xi_{17} = 2C^T PA + (\tau_2 - \tau_0)^2 (C - I)^T T_2 A + (\tau_0 - \tau_1)^2 (C - I)^T$
$T_4 A + D_1 L_2$ $\Xi_{18} = 2\rho C^T PB + (\tau_2 - \tau_0)^2 \rho (C - I)^T T_2 B + (\tau_0 - \tau_1)^2 \rho (C - I)^T T_4 B$
$\Xi_{19} = 2\bar{\rho} C^T PB + (\tau_2 - \tau_0)^2 \bar{\rho} (C - I)^T T_2 B + (\tau_0 - \tau_1)^2 \bar{\rho} (C - I)^T T_4 B$
$\Xi_{1,10} = -P_1$ $\Xi_{1,11} = P_1$ $\Xi_{1,12} = -P_1$ $\Xi_{22} = P_2 + P_2^T - T_3 - \frac{\tau_0 - \tau_1}{\tau_2 - \tau_1} T_4$ $\Xi_{23} = P_3^T$
$\Xi_{24} = P_4^T - P_2$ $\Xi_{25} = P_5^T + \frac{\tau_0 - \tau_1}{\tau_2 - \tau_1} T_4$ $\Xi_{26} = P_6^T$ $\Xi_{27} = P_7^T$ $\Xi_{28} = P_8^T$ $\Xi_{29} = P_9^T$
$\Xi_{2,10} = P_{10}^T - P_2$ $\Xi_{2,11} = P_{11}^T + P_2$ $\Xi_{2,12} = P_{12}^T - P_2$ $\Xi_{33} = -\frac{\tau_0 - \tau_1}{\tau_2 - \tau_1} T_4 - Q_4 -$
$\frac{\tau_2 - \tau_0}{\tau_2 - \tau_1} T_2$ $\Xi_{34} = -P_3$ $\Xi_{35} = \frac{\tau_0 - \tau_1}{\tau_2 - \tau_1} T_4$ $\Xi_{36} = \frac{\tau_2 - \tau_0}{\tau_2 - \tau_1} T_2$ $\Xi_{3,10} = -P_3$ $\Xi_{3,11} =$
P_3 $\Xi_{3,12} = -P_3$ $\Xi_{44} = -\frac{\tau_2 - \tau_0}{\tau_2 - \tau_1} T_2 - T_1 - P_4 - P_4^T$ $\Xi_{45} = -P_5^T$ $\Xi_{46} = \frac{\tau_2 - \tau_0}{\tau_2 - \tau_1} T_2 - P_6^T$
$\Xi_{47} = -P_7^T$ $\Xi_{48} = -P_8^T$ $\Xi_{49} = -P_9^T$ $\Xi_{4,10} = -P_{10}^T - P_4$ $\Xi_{4,11} = -P_{11}^T + P_4$
$\Xi_{4,12} = -P_{12}^T - P_4$ $\Xi_{55} = -2\frac{\tau_0 - \tau_1}{\tau_2 - \tau_0} T_4 - R_1 - D_2 L_1$ $\Xi_{58} = D_2 L_2$ $\Xi_{5,10} = -P_5$
$\Xi_{5,11} = P_5$ $\Xi_{5,12} = -P_5$ $\Xi_{66} = -R_2 - D_3 L_1 - 2\frac{\tau_2 - \tau_0}{\tau_2 - \tau_1} T_2$ $\Xi_{69} = D_3 L_2$
$\Xi_{6,10} = -P_6$ $\Xi_{6,11} = P_6$ $\Xi_{6,12} = -P_6$ $\Xi_{7,10} = -P_7$ $\Xi_{7,11} = P_7$ $\Xi_{7,12} = -P_7$
$\Xi_{77} = 2A^T PA + (\tau_0 - \tau_1 + 1)Z_1 - D_1 + (\tau_0 - \tau_1)^2 A^T T_4 A + (\tau_2 - \tau_0)Z_2 +$
$(\tau_2 - \tau_0)^2 A^T T_2 A$ $\Xi_{78} = 2\rho A^T PB + (\tau_2 - \tau_0)^2 \rho A^T T_2 B + (\tau_0 - \tau_1)^2 \rho A^T T_4 B$
$\Xi_{79} = 2\bar{\rho} A^T PB + (\tau_2 - \tau_0)^2 \bar{\rho} A^T T_2 B + (\tau_0 - \tau_1)^2 \bar{\rho} A^T T_4 B$
$\Xi_{88} = 2\rho B^T PB - Z_1 + (\tau_2 - \tau_0)^2 \rho B^T T_2 B + (\tau_0 - \tau_1)^2 \rho B^T T_4 B - D_2$
$\Xi_{8,10} = -P_8$ $\Xi_{8,11} = P_8$ $\Xi_{8,12} = -P_8$ $\Xi_{9,10} = -P_9$ $\Xi_{9,11} = P_9$ $\Xi_{9,12} = -P_9$
$\Xi_{99} = 2\bar{\rho} B^T PB - Z_2 + (\tau_2 - \tau_0)^2 \bar{\rho} B^T T_2 B + (\tau_0 - \tau_1)^2 \bar{\rho} B^T T_4 B - D_3$
$\Xi_{10,10} = -P_{10} - P_{10}^T - \frac{1}{\tau_1} Q_1$ $\Xi_{10,11} = P_{10} - P_{11}^T$ $\Xi_{10,12} = -P_{10} - P_{12}^T$

$$\Xi_{11,11} = P_{11} + P_{11}^T - \frac{1}{\tau_2} Q_2 \quad \Xi_{11,12} = -P_{11} + P_{12}^T \quad \Xi_{12,12} = -P_{12} - P_{12}^T - \frac{1}{\tau_2 - \tau_1} Q_3$$

$$L_1 = diag\left(l_1^- l_1^+, l_2^- l_2^+, \cdots, l_n^- l_n^+\right) \qquad L_2 = diag\left(\frac{l_1^- + l_1^+}{2}, \frac{l_2^- + l_2^+}{2}, \cdots, \frac{l_n^- + l_n^+}{2}\right)$$

Proof: Consider the Lyapunov-Krasovskii functional as follows:

$$v_1(k) = 2\alpha^T(k) \begin{pmatrix} P & 0 & \cdots & 0 \\ 0 & 0 & \cdots & 0 \\ \vdots & \vdots & \ddots & 0 \\ 0 & 0 & \cdots & 0 \end{pmatrix}_{12n \times 12n} \alpha(k) \qquad v_2(k) = \sum_{j=k-\tau_1}^{k-1} \sum_{i=j}^{k-1} x^T(i) Q_1 x(i)$$

$$v_3(k) = \sum_{j=k-\tau_2}^{k-1} \sum_{i=j}^{k-1} x^T(i) Q_2 x(i) \qquad v_4(k) = \sum_{j=k-\tau_2+1}^{k-\tau_1} \sum_{i=j}^{k-1} x^T(i) Q_3 x(i)$$

$$v_5(k) = \sum_{i=k-\tau_0}^{k-1} x^T(i) Q_4 x(i)$$

$$v_6(k) = \sum_{i=k-\tau_1(k)}^{k-1} x^T(i) R_1 x(i) + \sum_{i=-\tau_0+1}^{-\tau_1} \sum_{j=k+i}^{k-1} x^T(j) R_1 x(j)$$

$$v_7(k) = \sum_{i=k-\tau_2(k)}^{k-1} x^T(i) R_2 x(i) + \sum_{i=-\tau_2+1}^{-\tau_0-1} \sum_{j=k+i}^{k-1} x^T(j) R_2 x(j)$$

$$v_8(k) = \sum_{i=k-\tau_1(k)}^{k-1} g^T(x(i)) Z_1 g(x(i)) + \sum_{i=-\tau_0+1}^{-\tau_1} \sum_{j=k+i}^{k-1} g^T(x(j)) Z_1 g(x(j))$$
$$+ \sum_{i=k-\tau_2(k)}^{k-1} g^T(x(i)) Z_2 g(x(i)) + \sum_{i=-\tau_2+1}^{-\tau_0+1} \sum_{j=k+i}^{k-1} g^T(x(j)) Z_2 g(x(j))$$

$$v_9(k) = \sum_{i=k-\tau_2}^{k-1} x^T(i) T_1 x(i) + (\tau_2 - \tau_0) \sum_{j=-\tau_2}^{-\tau_0-1} \sum_{i=k+j}^{k-1} \eta^T(i) T_2 \eta(i)$$

$$v_{10}(k) = \sum_{i=k-\tau_1}^{k-1} x^T(i) T_3 x(i) + (\tau_0 - \tau_1) \sum_{j=-\tau_0}^{-\tau_1-1} \sum_{i=k+j}^{k-1} \eta^T(i) T_4 \eta(i)$$

$$\alpha^T(k) = \left[x^T(k), x^T(k-\tau_1), x^T(k-\tau_0), x^T(k-\tau_2), x^T(k-\tau_1(k)), x^T(k-\tau_2(k)), \right.$$
$$\left. g^T(x(k)), g^T(x(k-\tau_1(k))), g^T(x(k-\tau_2(k))), \sum_{i=k-\tau_1}^{k-1} x^T(i), \right.$$
$$\left. \sum_{i=k-\tau_2}^{k-1} x^T(i), \sum_{i=k-\tau_2+1}^{k-\tau_1} x^T(i) \right]$$

Defining $\eta(k) = x(k+1) - x(k)$, then along the solution of DSNNs (7), we have

$$E[\Delta v_1(k)] = E\left(2x^T(k+1) P x(k+1) - 2\alpha^T(k) \begin{pmatrix} P & P_1 \\ 0 & P_2 \\ \vdots & \vdots \\ 0 & P_{12} \end{pmatrix}\right.$$
$$\left. \times \begin{pmatrix} I & 0 & 0 & 0 & 0 & 0 & 0 & 0 & 0 & 0 & 0 & 0 \\ 0 & -I & 0 & I & 0 & 0 & 0 & 0 & 0 & I & -I & I \end{pmatrix} \alpha(k)\right)$$

$$E[\Delta v_2(k)] \le E\left(\tau_1 x^T(k) Q_1 x(k) - \frac{1}{\tau_1} \sum_{i=k-\tau_1}^{k-1} x^T(i) Q_1 \sum_{i=k-\tau_1}^{k-1} x(i)\right)$$

$$E[\Delta v_3(k)] \le E\left(\tau_2 x^T(k) Q_2 x(k) - \frac{1}{\tau_2} \sum_{i=k-\tau_2}^{k-1} x^T(i) Q_2 \sum_{i=k-\tau_2}^{k-1} x(i)\right)$$

$$E\left[\Delta v_4\left(k\right)\right] \le E\left(\left(\tau_2 - \tau_1\right) x^T\left(k\right) Q_3 x\left(k\right) - \frac{1}{\tau_2-\tau_1} \sum_{i=k-\tau_2+1}^{k-\tau_1} x^T\left(i\right) Q_3 \sum_{i=k-\tau_2+1}^{k-\tau_1} x\left(i\right)\right)$$

$$E\left[\Delta v_5\left(k\right)\right] = E\left(x^T\left(k\right) Q_4 x\left(k\right) - x^T\left(k - \tau_0\right) Q_4 x\left(k - \tau_0\right)\right)$$

$$E\left[\Delta v_6\left(k\right)\right] \le E\left(\left(\tau_0 - \tau_1 + 1\right) x^T\left(k\right) R_1 x\left(k\right) - x^T\left(k - \tau_1\left(k\right)\right) R_1 x\left(k - \tau_1\left(k\right)\right)\right)$$

$$E\left(\Delta v_7\left(k\right)\right) \le E\left(\left(\tau_2 - \tau_0\right) x^T\left(k\right) R_2 x\left(k\right) - x^T\left(k - \tau_2\left(k\right)\right) R_2 x\left(k - \tau_2\left(k\right)\right)\right)$$

$$E\left(\Delta v_8\left(k\right)\right) \le E\left(\left(\tau_0 - \tau_1 + 1\right) g^T\left(x\left(k\right)\right) Z_1 g\left(x\left(k\right)\right) - g^T\left(x\left(k - \tau_1\left(k\right)\right)\right) Z_1 g\left(x\left(k - \tau_1\left(k\right)\right)\right)\right.$$
$$\left. + \left(\tau_2 - \tau_0\right) g^T\left(x\left(k\right)\right) Z_2 g\left(x\left(k\right)\right) - g^T\left(x\left(k - \tau_2\left(k\right)\right)\right) Z_2 g\left(x\left(k - \tau_2\left(k\right)\right)\right)\right)$$

$$E\left(\Delta v_9\left(k\right)\right) \le E\left(x^T\left(k\right) T_1 x\left(k\right) - x^T\left(k - \tau_2\right) T_1 x\left(k - \tau_2\right) + \left(\tau_2 - \tau_0\right)^2 \eta^T\left(k\right) T_2 \eta\left(k\right)\right.$$

$$+ \tfrac{\tau_0-\tau_1}{\tau_2-\tau_1} \begin{bmatrix} x\left(k - \tau_0\right) \\ x\left(k - \tau_2\left(k\right)\right) \end{bmatrix}^T \begin{bmatrix} T_2 & -T_2 \\ -T_2 & T_2 \end{bmatrix} \begin{bmatrix} x\left(k - \tau_0\right) \\ x\left(k - \tau_2\left(k\right)\right) \end{bmatrix}$$

$$+ \begin{bmatrix} x\left(k - \tau_2\left(k\right)\right) \\ x\left(k - \tau_0\right) \\ x\left(k - \tau_2\right) \end{bmatrix}^T \begin{bmatrix} -2T_2 & T_2 & T_2 \\ * & -T_2 & 0 \\ * & * & -T_2 \end{bmatrix} \begin{bmatrix} x\left(k - \tau_2\left(k\right)\right) \\ x\left(k - \tau_0\right) \\ x\left(k - \tau_2\right) \end{bmatrix}$$

$$\left. + \tfrac{\tau_0-\tau_1}{\tau_2-\tau_1} \begin{bmatrix} x\left(k - \tau_2\left(k\right)\right) \\ x\left(k - \tau_2\right) \end{bmatrix}^T \begin{bmatrix} T_2 & -T_2 \\ -T_2 & T_2 \end{bmatrix} \begin{bmatrix} x\left(k - \tau_2\left(k\right)\right) \\ x\left(k - \tau_2\right) \end{bmatrix}\right)$$

$$E\left(\Delta v_{10}\left(k\right)\right) \le E\left(x^T\left(k\right) T_3 x\left(k\right) - x^T\left(k - \tau_1\right) T_3 x\left(k - \tau_1\right) + \left(\tau_0 - \tau_1\right)^2 \eta^T\left(k\right) T_4 \eta\left(k\right)\right.$$

$$+ \tfrac{\tau_2-\tau_0}{\tau_2-\tau_1} \begin{bmatrix} x\left(k - \tau_1\right) \\ x\left(k - \tau_1\left(k\right)\right) \end{bmatrix}^T \begin{bmatrix} T_4 & -T_4 \\ -T_4 & T_4 \end{bmatrix} \begin{bmatrix} x\left(k - \tau_1\right) \\ x\left(k - \tau_1\left(k\right)\right) \end{bmatrix}$$

$$+ \begin{bmatrix} x\left(k - \tau_1\left(k\right)\right) \\ x\left(k - \tau_0\right) \\ x\left(k - \tau_1\right) \end{bmatrix}^T \begin{bmatrix} -2T_4 & T_4 & T_4 \\ * & -T_4 & 0 \\ * & * & -T_4 \end{bmatrix} \begin{bmatrix} x\left(k - \tau_1\left(k\right)\right) \\ x\left(k - \tau_0\right) \\ x\left(k - \tau_1\right) \end{bmatrix}$$

$$\left. + \tfrac{\tau_2-\tau_0}{\tau_2-\tau_1} \begin{bmatrix} x\left(k - \tau_1\left(k\right)\right) \\ x\left(k - \tau_0\right) \end{bmatrix}^T \begin{bmatrix} T_4 & -T_4 \\ -T_4 & T_4 \end{bmatrix} \begin{bmatrix} x\left(k - \tau_1\left(k\right)\right) \\ x\left(k - \tau_0\right) \end{bmatrix}\right)$$

We denote that

$$D_1 = diag\left(\lambda_1, \lambda_2, \cdots, \lambda_n\right) > 0 \qquad D_2 = diag\left(\gamma_1, \gamma_2, \cdots, \gamma_n\right) > 0$$
$$D_3 = diag\left(\beta_1, \beta_2, \cdots, \beta_n\right) > 0$$

Combining above discussion , we can easily obtain the following inequality from the Assumption 1 and the (9)

$$E\left(\Delta v\left(k\right)\right) \le E\left(x^T\left(k\right) \Xi x\left(k\right)\right) \le \lambda_{\max}\left(\Xi\right) E\left\|x\left(k\right)\right\|^2$$

Now use the similar proof as [6] we can easily derive that the system (7) is globally exponential stable in the mean square. This completes the proof of the Theorem 1.

4   Numerical Examples

Consider discrete-time stochastic neural network (7) with

$$C = \begin{bmatrix} 0.8 & 0 \\ 0 & 0.9 \end{bmatrix} \qquad A = \begin{bmatrix} 0.001 & 0 \\ 0 & 0.005 \end{bmatrix} \qquad B = \begin{bmatrix} -0.1 & -0.01 \\ -0.2 & -0.1 \end{bmatrix}$$

The activation function satisfy Assumption 1 with $L_1 = diag\left(-0.64, 0\right)$, $L_2 = diag\left(0, -0.2\right)$. Choosing $\varepsilon = 0.001 I$, setting $\tau_1 = 1$ and $\tau_0 = 2$ in Theorem 1. Table 1 show the corresponding maximum allowable value of τ_2 for the case of ρ

Table 1. Calculated the maximum τ_2 with different probability distribution of time delay

ρ	0.05	0.1	0.15	0.2	0.3	0.4	0.5	0.6	0.7	0.8	0.9	1
[7]Th1	5	5	5	6	6	7	9	11	16	27	64	$+\infty$
Theorem 1	13	14	14	15	16	18	20	23	28	38	66	$+\infty$

varying from 0.05 to 1, one can see that stability criteria propose in this paper significantly improve the existing results of Theorem 1 in [7].

5   Conclusions

In this paper,a improved delay-distribution-dependent exponential stability criterion for stochastic discrete-time neural networks with time-varying delay is proposed. Finally, a numerical example has been given to demonstrate the effectiveness of the proposed method.

References

1. Xiong, W., Song, L., Cao, J.: Adaptive robust convergence of neural networks with time-varying delays. Nonlinear Analysis 9, 1283–1291 (2008)
2. Song, Q., Cao, J.: Global robust stability of interval neural networks with multiple time-varying delays. Mathematics and Computers in Simulation 74, 38–46 (2007)
3. Zhang, B., Xu, S., Li, Y.: Delay-dependent robust exponential stability for uncertain recurrent neural networks with time-varying delays. International Journal of Neural Systems 17, 207–218 (2007)
4. Liu, Y., Wang, Z., Liu, X.: Robust stability of discrete-time stochastic neural networks with time-varying delays. Neurocomputing 71, 823–833 (2008)
5. Ou, Y., Liu, H., Si, Y., Feng, Z.: Stability analysis of discrete-time stochastic neural networks with time-varying delay. Neurocomputing. 773, 740-748 (2010)
6. Luo, M., Zhong, S., Wang, R., Kang, W.: Robust stability nanlysis of discrete-time stochastic neural networks with time-varying delays. Applied Mathematics and Computation 209, 305–313 (2009)
7. Zhang, Y., Yue, D., Tian, E.: Robust delay-distribution-dependent stability of discrete-time stochastic neural networks with time-vaying delay. Neurocomputing 72, 1265–1273 (2009)

A Partially Connected Neural Evolutionary Network for Stock Price Index Forecasting

Didi Wang[2], Pei-Chann Chang[1], Jheng-Long Wu[1], and Changle Zhou[2]

[1] Department of Information Management, Yuan Ze University, Taoyuan 32026, Taiwan
{iepchang,jlwu}@saturn.yzu.edu.tw, didiw7@gmail.com
[2] Cognitive Science Department, Fujian Key Laboratory of the Brain-like Intelligent Systems,
Xiamen University, Xiamen, China
dozero@xmu.edu.cn

Abstract. This paper proposes a novel partially connected neural evolutionary model (Parcone) architecture to simulate the relationship of stock and technical indicators to predict the stock price index. Different from artificial neural networks, the architecture has corrected three drawbacks: (1) connection between neurons of is random; (2) there can be more than one hidden layer; (3) evolutionary algorithm is employed to improve the learning algorithm and train weights. The more hidden knowledge stored within the historic time series data are needed in order to improve expressive ability of network. The genetically evolved weights mitigate the well-known limitations of gradient descent algorithm. In addition, the activation function is not defined by sigmoid function but $\sin(x)$. The experimental results show that Parcone can make the progress concerning the stock price index and it's very promising to calculate the predictive percentage by simulation results of proposed evolutionary system.

1 Introduction

Recently, more and more people pay attention to stock market. A central issue lies in the prediction of share prices. This is especially important for investment analysis and risk management. Many researchers have predicted stock market returns using artificial intelligence (AI) approaches during the past decades. Neural network [1] combines with Case Based Reasoning (CBR) to be applied to predict stock trading. They conclude that an NN can be used to make stock trading point. Two neural network architectures which include multi-layer perception (MLP) neural networks and generalized regression neural networks are used to predict the Kuwait stock exchange closing price movements [2]. Chang [3] also used an ensemble of neural networks to make a decision concerning stock trading. The results demonstrate the hybrid system can make a significant and constant amount of profit. Levenberg-Marquardt BP algorithm [4] has been adopted for stock market prediction in order to avoid local extra-mum and promote convergence speed. Back-propagation neural network is contrasted with Genetic Algorithms based Back-propagation neural

D.-S. Huang et al. (Eds.): ICIC 2011, LNBI 6840, pp. 14–19, 2012.

network for stock rates prediction accuracy [5-7].The experimental results show that genetically evolved weights mitigate the well-known limitations of the gradient descent algorithm. A piecewise linear representation method and a neural network model is integrated for stock trading points prediction [8-9]. Probabilistic neural networks [10] are used to make an array of bipolar predictions concerning stock.

This paper is to propose a novel partially connected neural evolutionary model to improve expressive ability and predictive percentage. As there can be more than one hidden layers, the more hidden knowledge stored within the historic time series data are provided for improving expressive ability of network. As gradient descent algorithm searches the local solution the genetically evolved weights mitigate limitations of gradient descent algorithm and search the global solution. The input and output of neural network respectively is technical indexes and stock price index. Finally, financial time series data from S&P 500 are applied to demonstrate the effectiveness of the system.

2   Partially Connected Neural Evolutionary Model (PARCONE)

The connectionism of Parcone [11] is quite different from other neural networks and the detail architecture of Parcone can be found in Fig. 1. Different to Backpropagation Network, there are three specific features in Parcon which are listed as follows:

1) Connection between neurons is random. Neuron i and j is in the form of probability p_{ij} in Parcone. Here, we define

$$X_{ij} = \begin{cases} 1, & neuron \quad i \quad and \quad j \quad is \quad connected, \\ 0, & neuron \quad i \quad and \quad j \quad is \quad unconnected. \end{cases} \tag{1}$$

where i, j=1, ..., N, $i \neq j$.

$$p(x_{ij} = 1) = p_{ij,} \\ p(x_{ij} = 0) = 1 - p_{ij} = \overline{p_{ij}}. \tag{2}$$

2) There can be more than one hidden layer.
3) Weights are trained by not back-propagation weight training step but evolutionary algorithm.

In addition, the structure of Parcone [12] consists of many layers: input layer, many hidden layers, and output layer. Each layer contains N_1,.., N_n nodes denoted, respectively, by circles. The node is also named after neuron or unit. The circles are connected by links, denoted by arrows, each of which represents a numerical weight. And connection of them is probability. w_{ij} and p_{ij} respectively denote numerical weights and probability between layers which contain input layer, hidden layers, and output layer.

2.1   Applying Genetic Algorithms to Parone

The Genetic Algorithm (GA) is based on the principle of "survival of the fittest", which imitates the way that biological creatures have adapted to their environment over billions of years [13]. GA has been used successfully to solve weight computational optimization problems. GA contains selection and reproduction, crossover, and mutation [14]. GA often works with a form of binary coding which is characterized as chromosomes. The operations of the GA work follows: 1). The selection chooses the fittest individuals according to the fittest function. The best are selected by fittest function for further iteration. The rest is fittest population and the best chromosomes are reproduced. 2). The cross-over is the method for combining those selected individuals into new individuals. The cross-over splits up the parent individuals and recombines them. 3). The mutation simply adds some noise to "genes" of the individual. Mutation arbitrarily alters one or more components of a selected chromosome. GA are applied to optimize weight problems. At first weights is denoted by binary coding. Bad chromosomes are washed out by the fittest function and good ones are selected. Then they are dealt with by cross-over and mutation. Finally, weight values are defined by maximizing the fitness function.

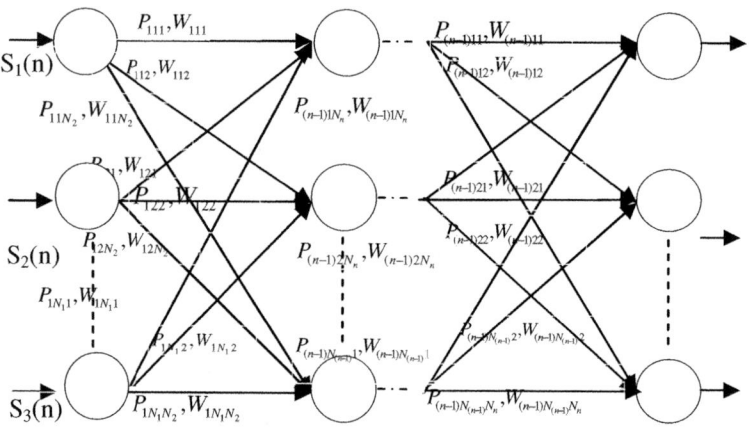

Fig. 1. Architectures of partially connected model (Canning A.& Gardner E, 1988)

Table 1. Variables included in analysis

Input variables in our system	Technical
5MA,6MA,10MA,20MA	Moving average(MA)
5BIAS,	Bias(Bias)
10BIAS	
6RSI, 12RSI	Relative strength index(RSI)
K,D	Nine days Stochastic line(K,D)
9MACD	Moving average convergence and divergence(MACD)
12W%R	Williams %R (W%R)
K-D	Differences of technical index between K and D (K-D).
△5MA, △6MA, △10MA, △5BIAS, △10BIAS, △6RSI, △12RSI, △9k, △9D, △12W%R, △MACD,	Differences of technical index(△)

3   Research Method

The paper uses a new neural network-Parone to predict stock price index. Different to other neural network, connection between neurons are probability and there can be more than one hidden layer. More hidden knowledge is stored in hidden layers. GA searches global optimized solution to train weights. Of course, GA has a self-disadvantage. It will take more time to train weights for global search

The input variables of PARCONE based on economic knowledge, technical indexes are required. In general, techniques used in economic forecasting are the fundamental and technical analysis. Table I gives 24 original technical indicators. Technical indicators attempt to determine the strength and direction of the trend. There are advantage and applying limitation in these technical indicators. Technical indicators are calculated according to original component which includes opening price, closing price, the highest price, and the lowest price.

3.1   The Implement of Parcone

It is very interesting to implement Parcone. It will have a close relation with many factors. The target of Parcone is to optimize weights. At first, it is important to define number of neurons and layers. Then, input variability need to be normalized since input value required to be [0, 1] in neural network. Weight value needs to be initialized. Also, it will make sure whether between neurons are connected. After that, input variability enters into neuron network. Weights will be optimized by GA. The overall framework of Parcone is shown in Fig. 2. The algorithms of Parcone consist of six phases.

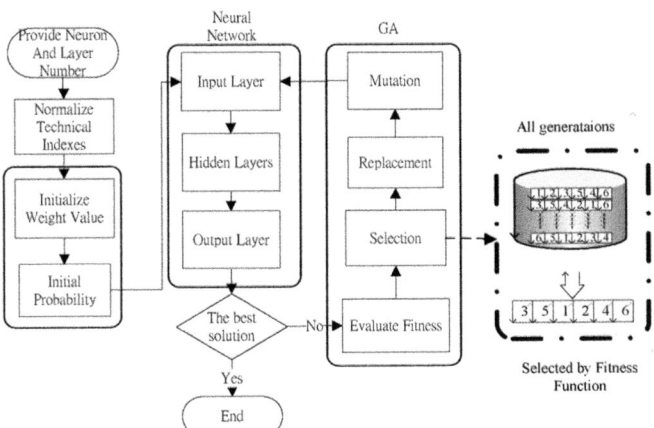

Fig. 2. The algorithm of Parcone consists of six phases

Phase 1. The number of neurons and layers is close relative with training time. When there are a lot of neurons and layers it will take lots of time to train weights.

Phase 2 The neural network requires input value to be [0, 1]. Therefore, 24 technical indexes and objective value called stock price are normalized to meet property of neural network.

Phase 3. In the third phase, weights are initialized randomly by Random number generator. Pseudo-random number generators (PRNG) are applied to create long runs of numbers with good random properties.

Phase 4. The fourth phase states how a partially connected neural network is implemented in detail. Neural network contains input layer, hidden layers, and output layer. We can make sure number of neurons in the input layer is 24 since there are 24 technical indicators. As stock price index is close contacted with technical indicators number of neurons in out layer is 1.

Phase 5. GA searches optimal or near-optimal connection weights. The parameters for searching must be encoded on chromosomes. That is, weights are transformed into binary code to change it by mutation. In fittest function, weights are filtrated.

Phase 6. The above process states how GA is applied in partially connected neural network. It will close combine with neural network to optimize weights.

4   Research Data and Experiments

In this section, two different stocks are selected. There are Citigroup and Motors Liquidation Company which come from S&P 500. The historic data covers the financial time-series data from 2008/01/01 to 2009/6/31. The training data is based on the data from 2008/01/01 to 2008/12/31 and the test data is based on the data from 2009/01/01 to 2009/06/31. For Citigroup the size of the training and test data respectively is 254 and 124. For Motors Liquidation Company the size of the training and test data respectively is 253 and 124.

In this paper we try to use various numbers of layers and nodes for testing our model of PARCONE. Figure 3 show as our model predicted the stock price index and actual value of original stock price index. The predication performance was well.

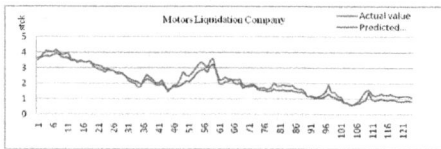 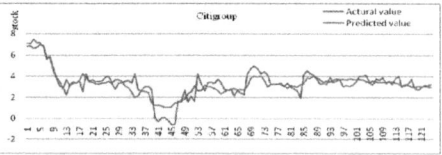

Fig. 3. The predicted result from propose algorithm of Parcone

5   Conclusions

As mentioned earlier, previous studies tried to optimize the controlling parameters of ANN using local search algorithms. However, this paper combined global search algorithms with a partially connected neural network to forecast stock price index trend. The expressive ability of Parcone is stronger than BPN since connection between neurons is probabilistic. The deposed technical indexes are inputted into neural networks to train the connection weight of the model. Then, a new set of input data can trigger the model when GA searches optimized weights solution. Since number of neurons and layers impacts predictive percentage, Parcone is tested on different neurons and layers. Experiment results show training percentage reaches as high as at least 94% and predictive percentage reaches as high as at least 97%. In summary, the proposed system is very effective and encouraging in its predictions with respect to stock price index trend forecasting. In the future, data preprocessing is one of the features that can be applied in financial time series data. Times series data is indirectly inputted into Parcone. Effective approaches such as wavelet analysis and clustering of data improve the predictive accuracy of the forecasting system.

References

1. Chang, P.C., Liu, C.H., Lin, J.L., Fan, C.Y., Ng, C.S.P.: A Neural Network with a Case Based Dynamic Window for Stock Trading Prediction. Expert Systems with Applications 36(3 PART 2), 6889–6898 (2009)
2. Mostafa, M.M.: Forecasting Stock Exchange Movements Using Neural Networks: Empirical Evidence from Kuwait. Expert Systems with Applications 37(9), 6302–6309 (2010)
3. Chang, P.-C., Liu, C.-H., Fan, C.-Y., Lin, J.-L., Lai, C.-M.: An ensemble of neural networks for stock trading decision making. In: Huang, D.-S., Jo, K.-H., Lee, H.-H., Kang, H.-J., Bevilacqua, V. (eds.) ICIC 2009. LNCS, vol. 5755, pp. 1–10. Springer, Heidelberg (2009)
4. Li, F., Liu, C.: Application Study of BP Neural Network on Stock Market Prediction. In: Ninth International Conference on Hybrid Intelligent Systems, Shgenyang, China, pp. 174–178 (2009)
5. Kim, K.J., Han, I.: Genetic Algorithms Approach to Feature Discretization in Artificial Neural Networks for the Prediction of Stock Price Index. Expert Systems with Applications 19, 125–132 (2000)
6. Mandziuk, J.: Jaruszewicz, m.: Neuro-evolutionary approach to stock market prediction. In: Proceedings of International Joint Conference on Neural Networks, Orlando, Florida, USA, pp. 12–17 (August 2007)
7. Kim, S.H., Chun, H.S.: Graded Forecasting Using an Array of Bipolar Predictions: Application of Probabilistic Neural Networks to a Stock Market Index. International Journal of Forecasting 14, 323–337 (1998)
8. Chang, P.C., Fan, C.Y., Liu, C.H.: Integrating a Piecewise Linear Presentation Method and a Neural Network Model for Stock Trading Points Prediction. IEEE Transactions on Systems, Man and Cybernetics Part C: Applications and Reviews 39(1), 80–92 (2009)
9. Chang, P.C., Liu, C.H.: A TSK type Fuzzy Rule Based System for Stock Price Prediction. Expert Systems with Applications 34(1), 135–144 (2008)
10. Montana, D.: A Weighted Probabilistic Neural Network. In: Advances in Neural Information Processing Systems, pp. 1110–1117 (1992)
11. Canning, A., Gardner, E.: Partially Connected Models of Neural Networks. Journal of Physics A 21, 3275–3284 (1998)
12. Hubert, C.: Design of Fully and Partially Connected Random Neural Networks for Pattern Completion. In: Mira, J., Cabestany, J., Prieto, A.G. (eds.) IWANN 1993. LNCS, vol. 686, pp. 137–142. Springer, Heidelberg (1993)
13. Goldberg, D.E.: Genetic Algorithms in Search, Optimization, and Machine Learning. Addison-Wesley, Reading (1989)
14. Adeli, H., Hung, S.: Machine Learning: Neural Networks, Genetic Algorithms, and Fuzzy Systems. Wiley, New York (1995)

Finite Precision Extended Alternating Projection Neural Network (FPEAP)

Yanfei Wang and Jingen Wang

New Star Research Institute of Applied Technology,
451 Huangshan Road, Hefei, Anhui, China
wangjingen@126.com

Abstract. The paper studies finite precision Extended Alternating Projection Neural Network (FPEAP) and its related problems. An improved training method of FPEAP has been present after considering the finite precision influence on the training method of EAP. Then the mathematical relation among the factors influencing the association times has been studied. Finally simulation experiments have been designed and simulation results demonstrate validity of theoretical analyses.

Keywords: Alternating projection, neural network, signal processing, association times, finite precision.

1 Introduction

Alternating Projection Neural Network(APNN)[1] is firstly proposed by Marks II etc. The literature [2] studies APNN thoroughly and proposes a new neural network— Extended Alternating Projection Neural Network (EAP) which functions in the complex domain. The topology architecture and association process of EAP are the same as those of APNN. In the literature [2] the stability of EAP has been studied and strict mathematical proofs to its stability has also been given. The literature [3] obtains the mathematical expression to the steady state value of EAP and gives the sufficient and necessary condition of EAP used for Content Addressable Memory (CAM).

EAP neural network is prone to parallel computation and VLSI design due to its simplicity, consequently has a bright future under the real time processing situations. It has been applied to the signal processing such as band-limited signal extrapolation[4], notch filters[5] and weak signal seperation[6]. In order to expand its application scope and apply it better in other field such as pattern recognition and sensor network, the literature [7] has made further research on the EAP.

While studying EAP the above literatures have made an assumption that each data is of unlimited accuracy and no errors exist during their operations. However, artificial neural networks in practical application are implemented by adopting ASCI or microprocessors. Hence its data is of limited accuracy and error must exist during the operations of data. The paper will study finite precision EAP (FPEAP) and its related problems.

D.-S. Huang et al. (Eds.): ICIC 2011, LNBI 6840, pp. 20–25, 2012.

2   EAP and FPEAP

EAP is full-interconnection neural network and its topology architecture is shown in Figure 1. Suppose EAP is made up of L neurons, the arbitrary neuron i and the neuron j are bidirectional connection. Weight t_{ij} equals weight t_{ji}. Neurons of EAP can be classified into clamped neurons or floating neurons according to their states. The state $s_i(m)$ of arbitrary floating neuron i at time m equals $\sum_{p=1}^{L} t_{pi} s_p(m-1)$, the state $s_j(m)$ of arbitrary clamped neuron j at time m is equal to $s_j(m-1)=\cdots=s_j(0)$. The weight-value of network can be obtained by the training method in the literature [2].

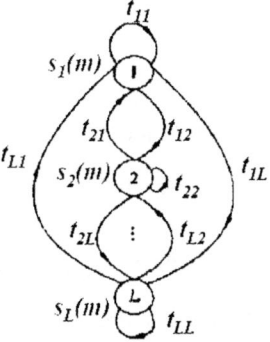

Fig. 1. Illustration of EAP

After EAP has been trained, it is time to decide which neurons are clamped neurons and which neurons are floating neurons. We can assume without loss of generality that neurons 1 through P are clamped and the remaining $Q=L-P$ neurons are floating. At time $m=0$ clamped neurons are initialized. If the state vector of EAP is $S(m)=[s_1(m) \quad s_2(m) \quad \cdots \quad s_L(m)]^T$ at time m, the state vector of clamped neurons is $S^P(m)$ and the state vector of floating neurons is $S^Q(m)$, then the state vector of EAP at time $m+1$ is

$$S(m+1)=\eta TS(m)=\begin{bmatrix} S^P(m+1) \\ S^Q(m+1) \end{bmatrix}=\eta \cdot \begin{bmatrix} T_2 & T_1 \\ T_3 & T_4 \end{bmatrix} \cdot \begin{bmatrix} S^P(m) \\ S^Q(m) \end{bmatrix}=\begin{bmatrix} S^P(m) \\ T_3 S^P(m)+T_4 S^Q(m) \end{bmatrix} \quad (1)$$

Where T is the interconnection matrix, operator η clamps the states of clamped neurons whose states will be altered by T. Thus the whole operating process of EAP is that operator T and operator η function by turns.

The entire operating process of EAP includes two stages, namely training stage and association stage. There exist addition, subtraction, multiplication and division etc at the training stage; there just exist addition and multiplication at the association stage.

If data of FPEAP are all of double precision, then we just discuss relative error. For FPEAP we find that relative error of subtraction of two approximate numbers which are very close or equal is remarkable.

For example, Let L=5, library pattern number $N=3$, $f_1 = \begin{bmatrix} 1 & 2 & 6 & 9 & 8 \end{bmatrix}^T$, $f_2 = \begin{bmatrix} 2 & 4 & 9 & 8 & 7 \end{bmatrix}^T$, $f_3 = \begin{bmatrix} 3 & 6 & 15 & 17 & 15 \end{bmatrix}^T$, $F = \begin{bmatrix} f_1 & f_2 & f_3 \end{bmatrix}$. Since $f_3 = f_1 + f_2$, $\|\varepsilon_3\|$ is equal to zero theoretically, but computation errors caused by finite precision make $\|\varepsilon_3\| = 4.6044 \times 10^{-15}$ (Matlab result) to be unequal to zero. Thus the rank of matrix T obtained by EAP's weight-learning method is 3, for $P=2$ the spectral radius of matrix T_4 is 1.4192>1, for $P=3$ the spectral radius of T_4 is 1.1498>1. The network is no more stable. Therefore weight-learning method for EAP does not suit FPEAP.

Herein we will provide the following improved weight-learning method for FPEAP:

(a) Let the interconnection matrix T equal $\mathbf{0}$, $i \leftarrow 1$;

(b) $\varepsilon_i = (I - T)f_i$, where I is $L \times L$ identity matrix, f_i is library pattern;

(c) if $\|\varepsilon_i\| \leq \lambda_\alpha$, then f_i is already in the subspace T and goto step (d)⬛ else $T \leftarrow T + \varepsilon_i \varepsilon_i^H / \varepsilon_i^H \varepsilon_i$;

(d) $i \leftarrow i+1$, if $i > N$ (N is number of library pattern) then end, else goto step (b).

In the 3th step of the above method λ_α is threshold value which can be determined by using different method according to practical application.

3   Factors Influencing Association Times

Theoretically the FPEAP will reach the steady state value only after infinite times association. However in practical application it is impossible and unnecessary to make infinite times association since proper error between finite times association value and the steady state value may be permitted. Then what are the factors that influence the association times when the FPEAP is used for CAM?

Suppose FPEAP has L neurons, and has learned N library patterns f_i $(i=1,2,...,N)$. We can assume without loss of generality that neurons 1 through P are clamped and the remaining $Q=L-P$ neurons are floating.

From the formula (1) in the above section we can deduce the following conclusions:

$$S^P(m) = S^P(m-1) = \cdots = S^P(0) \tag{2}$$

$$S^Q(m) = \sum_{i=0}^{m-1} T_4^i T_3 S^P(0) + T_4^m S^Q(0) \tag{3}$$

Where the matrix T_4 is $Q \times Q$ matrix, and T_4^0 is $Q \times Q$ identity matrix.

According to the formula (3) we can deduce $S^Q(\infty)$ as follows:

$$S^Q(\infty) = \sum_{i=0}^{\infty} T_4^i T_3 S^P(0) + T_4^\infty S^Q(0) \tag{4}$$

From the literature [7] we can learn that $T_4^\infty = 0$ because the FPEAP is used for CAM. Thus the above formula (4) can be rewritten as follows:

$$S^Q(\infty) = \sum_{i=0}^{\infty} T_4^i T_3 S^P(0) \tag{5}$$

The error between m-times association value $S^Q(m)$ and the steady state value $S^Q(\infty)$ can be defined as follows:

$$Er(m) = \left\| S^Q(m) - S^Q(\infty) \right\| \tag{6}$$

$$= \left\| \sum_{i=0}^{m-1} T_4^i T_3 S^P(0) + T_4^m S^Q(0) - \sum_{i=0}^{\infty} T_4^i T_3 S^P(0) \right\|$$

$$= \left\| T_4^m S^Q(0) - \sum_{i=m}^{\infty} T_4^i T_3 S^P(0) \right\| = \left\| T_4^m \left(S^Q(0) - S^Q(\infty) \right) \right\|$$

While $\left\| S^Q(\infty) \right\| \neq 0$ we can also define the relative error as follows:

$$REr(m) = \left\| S^Q(m) - S^Q(\infty) \right\| / \left\| S^Q(\infty) \right\| \tag{7}$$

$$= \left\| T_4^m \left(S^Q(0) - S^Q(\infty) \right) \right\| / \left\| S^Q(\infty) \right\| \leq \gamma$$

From the formula (7) it is obvious that association times m will be closely related with the permitted relative association error γ, the inital state vector $S^Q(0)$ of all floating neurons and the spectral radius $\rho(T_4)$ of the interconnection matrix T_4 formed by weights of all floating neurons when the FPEAP is used for CAM.

4   Simulation Experiment and Result Analysis

In this section simulation experiments are designed to verify the theoretical analyses in section 3. Suppose the first P neurons are clamped and the remaining $Q=L-P$ neurons are floating. The following simulation experiments will be made on MATLAB R2006a.

Let $L=200$, $N=17$, $P=50$ and $Q=150$. $F=randn(L,N)$, $S^P(0)=randn(P,1)$, $S^Q(0)=5*randn(Q,1)$, $\gamma=0.1*rand(1,1)+10^{-4}$, where $randn(\ )$ and $rand(\ )$ are both MATLAB function(see MATLAB handbook).

Case 1: we will make experiments for 500 times. For each experiment F, $S^P(0)$ and γ will remain invariant while $S^Q(0)$ varies randomly according to the above given expression. The 500-times variation of the minimum association times m_m satisfying $REr(m_m) \leq \gamma$ with $\left\| S^Q(0) \right\|$ will be illustrated in Figure 2.

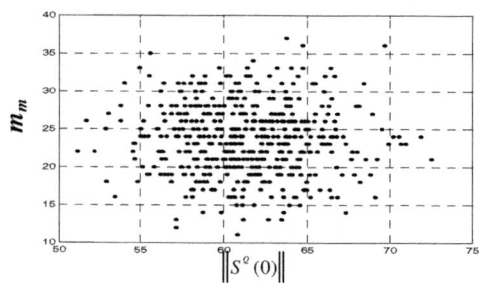

Fig. 2. 500-times experimental results for case 1

Case 2: for each experiment $S^Q(0)$, $S^P(0)$ and γ will remain invariant while F varies randomly according to the above given expression. The 500-times variation of the minimum association times $\boldsymbol{m}_m$ satisfying $REr(m_m) \leq \gamma$ with $\rho(T_4)$ will be illustrated in Figure 3.

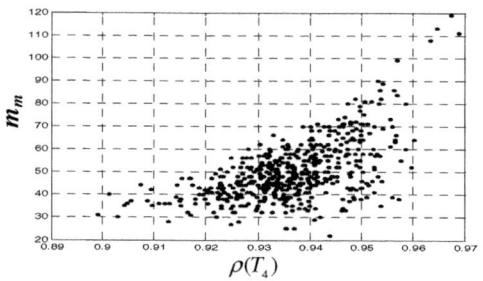

Fig. 3. 500-times experimental results for case 2

Case 3: for each experiment $S^Q(0)$, F and $S^P(0)$ will remain invariant while γ varies randomly according to the above given expression. The variation of the minimum association times $\boldsymbol{m}_m$ satisfying $REr(m_m) \leq \gamma$ with $|\ln(\gamma)|$ for 500 times will be illustrated in Figure 4.

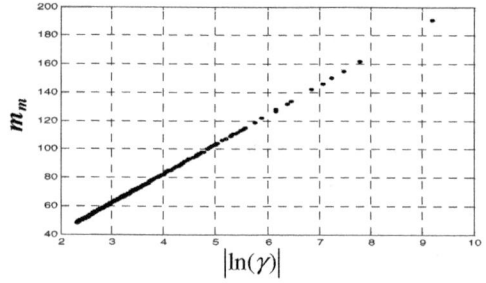

Fig. 4. 500-times experimental results for case 3

Figure 2 shows that $S^Q(0)$ influences the association times m_m and its multi-dimensional property causes the complex variation relation between m_m and $S^Q(0)$.

Figure 3 shows that $\rho(T_4)$ influences the association times m_m and m_m will increase with the increase of $\rho(T_4)$ on the whole.

Figure 4 shows that the association times m_m will almost linearly vary with $|\ln(\gamma)|$. This result can be theoretically analyzed and proved.

5   Peroration

The paper has studied finite precision EAP (FPEAP) and its related problems. Firstly the finite precision influence on the training method of EAP has been considered and an improved training method for FPEAP is present. Then the mathematical relation among association times of FPEAP used for CAM, the permitted relative association error, the inital state vector of all floating neurons and the spectral radius of the interconnection matrix formed by weights of all floating neurons has been studied. Finally simulation experiments have been designed and simulation results demonstrate validity of theoretical analyses.

Acknowledgments. This work was partially supported by the National Natural Science Foundation of China under grant No. 61040042.

References

1. Marks, R.J., Oh, S., Atlas, L.E.: Alternating Projection Neural Networks. IEEE Trans. CAS. 36(6), 846–857 (1989)
2. Wang, J.: Target signal detection and target localization [Doctoral dissertation]. Naval University of Engineering, Wuhan (May 2001)
3. Wang, J.-G., Gong, S.-G., Chen, S.-F.: Sufficient and necessary condition of the extended alternating projection neural network configured as a content addressable memory. Tien Tzu Hsueh Pao/Acta Electronica Sinica 32(4), 596–600 (2004)
4. Wang, J., Lin, C., Gong, S.: An extrapolation algorithm for band-limited signals based on Alternating Projection Neural Networks. Tien Tzu Hsueh Pao/Acta Electronica Sinica 28(10), 52–55 (2000)
5. Wang, J., Gong, S., Lin, C., Tang, J., Liu, S.: Notch filter based on Complex Alternating Projection Neural Networks. Journal of Data Acquisition & Processing 16(4), 440–445 (2001)
6. Wang, J.-g., Chen, S.-f., Gong, S.-g., Chen, Z.-q.: An extended alternating projection neural networks based weak-signal separation algorithm. In: IEEE International Conference on RISSP, October 8-13, vol. 1, pp. 554–558 (2003)
7. Wang, J., Wang, Y., Cui, X.: Further Research on Extended Alternating Projection Neural Network. In: Huang, D.-S., Zhao, Z., Bevilacqua, V., Figueroa, J.C. (eds.) ICIC 2010. LNCS, vol. 6215, pp. 33–40. Springer, Heidelberg (2010)
8. Zhang, M.: Matrix Theory for Engineering. Southeast University Press, Nanjing (1995)

Simulation of Visual Attention Using Hierarchical Spiking Neural Networks

QingXiang Wu[1,2], T. Martin McGinnity[1], Liam Maguire[1],
Rongtai Cai[2], and Meigui Chen[2]

[1] Intelligent Systems Research Center, School of Computing and Intelligent Systems
University of Ulster at Magee, Londonderry, BT48 7JL, Northern Ireland, UK
{q.wu,tm.mcginnity,lp.maguire}@ulster.ac.uk
[2] School of Physics and OptoElectronics Technology, Fujian Normal University
Fuzhou, 350007, China
{qxwu,rtcai,mgchen}@fjnu.edu.cn

Abstract. Based on the information processing functionalities of spiking neurons, a hierarchical spiking neural network model is proposed to simulate visual attention. The network is constructed with a conductance-based integrate-and-fire neuron model and a set of specific receptive fields in different levels. The simulation algorithm and properties of the network are detailed in this paper. Simulation results show that the network is able to perform visual attention to extract objects based on specific image features. Using extraction of horizontal and vertical lines, a demonstration shows how the network can detect a house in a visual image. Using this visual attention principle, many other objects can be extracted by analogy.

Keywords: Visual attention, spiking neural network, receptive field, visual system.

1 Introduction

The biological brain, with its huge number of neurons, displays powerful functionality in information processing, vision, reasoning, and other intelligent behaviours. Spiking neurons are regarded as essential components in the neural networks in the brain. Neurobiologists have found that various receptive fields exist in the visual cortex and play different roles [1, 2]. Visual Attention enables the visual system to process potentially important objects by selectively increasing the activity of sensory neurons that represent the relevant locations and features of the environment [3]. Given the complexity of the visual environment, the ability to selectively attend to certain locations, while ignoring others, is crucial for reducing the amount of visual information to manageable levels and for optimizing behavioral performance and response times. However, there are relatively few explanations of visual attention using spiking neural networks in the literature. Based on a spiking neuron model, a simulation of visual attention is demonstrated in this paper. The principles and simulation algorithms are presented in detail.

D.-S. Huang et al. (Eds.): ICIC 2011, LNBI 6840, pp. 26–31, 2012.
© Springer-Verlag Berlin Heidelberg 2012

2   Spiking Neural Network for Simulation of Visual Attention

Biological findings show that the visual system can use feedback signals to highlight the relevant locations[4][5]. In this paper, a spiking neural network, as shown in Fig.1, is proposed to simulate such visual attention. Suppose that a visual image is presented to the retina and an edge firing rate map [6] is obtained as the input for the network. The line detection layer contains two pathways. The horizontal pathway contains a neuron array N_h with the same size as the input neuron array. Each neuron has a receptive field corresponding to a horizontal synapse strength matrix W^h. The vertical pathway contains a neuron array N_v with the same size as input neuron array. Each neuron has a receptive field corresponding to a vertical synapse strength matrix W^v. Therefore, the spike rate map of the neuron arrays N_h and N_v represent horizontal and vertical lines respectively.

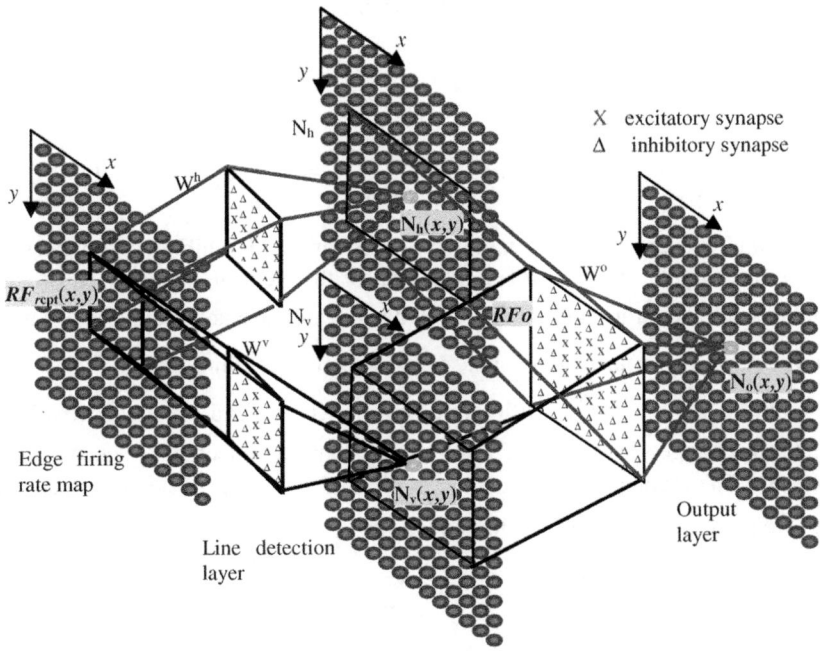

Fig. 1. Spiking Neural Network Model for Attention on Vertical and Horizontal Lines

The output layer is a neuron array with the same size as the edge firing rate map. Each neuron has two receptive fields on the horizontal and vertical neuron arrays respectively. The two receptive fields have the same synapse strength matrix W^o. If there is any horizontal line or vertical line around Neuron $N_o(x,y)$, the neuron will be activated, and a horizontal or vertical lines area can be obtained from the output layer. If these signals are regarded as feedback to an earlier layer of the visual system, objects in the area can be extracted and objects in other areas can be ignored. The simulation algorithms are shown in following sections.

3   Simulation Algorithms of the Spiking Neural Network

In this work a conductance-based integrate-and-fire model has been used to simulate the spiking neural networks since its behaviour is very close to the Hodgkin and Huxley neuron model but yet its computational complexity is much less than that neuron model [7]. Let $S_{x,y}(t)$ represent a spike train that produced by neuron (x,y). If neuron (x,y) fires at time t, $S_{x,y}(t)=1$, otherwise $S_{x,y}(t)=0$. In order to detect edges, the spike train array $S_{x,y}(t)$ can be obtained using the spiking neural network in [6]. Let q^{ex} and q^{ih} represent the peak conductance for excitatory synapse and inhibitory synapse respectively, $g_{x,y}^{hex}(t)$ and $g_{x,y}^{hih}(t)$ represent conductance for excitatory and inhibitory synapses respectively for neuron $N_h(x,y)$. $g_{x,y}^{vex}(t)$ and $g_{x,y}^{vih}(t)$ represent conductance for excitatory and inhibitory synapses respectively for neuron $N_v(x,y)$. In the conductance-based integrate-and-fire neuron model, we have

$$\frac{g_{x,y}^{hex}(t)}{dt} = -\frac{1}{\tau_{ex}}g_{x,y}^{hex}(t) + S_{x,y}(t)q^{ex}, \qquad \frac{g_{x,y}^{hih}(t)}{dt} = -\frac{1}{\tau_{ih}}g_{x,y}^{hih}(t) + S_{x,y}(t)q^{ih}, \qquad (1)$$

$$\frac{g_{x,y}^{vex}(t)}{dt} = -\frac{1}{\tau_{ex}}g_{x,y}^{vex}(t) + S_{x,y}(t)q^{ex}, \qquad \frac{g_{x,y}^{vih}(t)}{dt} = -\frac{1}{\tau_{ih}}g_{x,y}^{vih}(t) + S_{x,y}(t)q^{ih}, \qquad (2)$$

where τ_{ex} and τ_{ih} are time constants for excitatory and inhibitory synapses, ex is short for excitatory, and ih for inhibitory. If the neuron generates a spike, the conductance of excitatory and inhibitory synapses increase an amount of q^{ex} and q^{ih} respectively. If the neuron does not generate spike at time t, $g_{x,y}^{hex}(t)$, $g_{x,y}^{hih}(t)$, $g_{x,y}^{vex}(t)$ and $g_{x,y}^{vih}(t)$ decay with time constants τ_{ex} and τ_{ih} respectively. The conductance changes lead to different currents that are injected to neurons $N_h(x, y)$ and $N_v(x,y)$. The membrane potential $v_{h(x,y)}(t)$ and $v_{v(x,y)}(t)$ of neurons $N_h(x, y)$ and $N_v(x,y)$ are governed by the following equations.

$$c_m\frac{dv_{h(x,y)}(t)}{dt} = g_l(E_l - v_{h(x,y)}(t)) + \sum_{(x',y')\in RF_{rcpt}(x,y)}\frac{w_{x',y'}^{hex}g_{x',y'}^{hex}(t)}{A_{ex}}(E_{ex} - v_{h(x,y)}(t))$$
$$+ \sum_{(x',y')\in RF_{rcpt}(x,y)}\frac{w_{x',y'}^{hih}g_{x',y'}^{hih}(t)}{A_{ih}}(E_{ih} - v_{h(x,y)}(t)), \qquad (3)$$

$$c_m\frac{dv_{v(x,y)}(t)}{dt} = g_l(E_l - v_{v(x,y)}(t)) + \sum_{(x',y')\in RF_{rcpt}(x,y)}\frac{w_{x',y'}^{vex}g_{x',y'}^{vex}(t)}{A_{ex}}(E_{ex} - v_{v(x,y)}(t))$$
$$+ \sum_{(x',y')\in RF_{rcpt}(x,y)}\frac{w_{x',y'}^{vih}g_{x',y'}^{vih}(t)}{A_{ih}}(E_{ih} - v_{v(x,y)}(t)), \qquad (4)$$

where E_{ex} and E_{ih} are the reverse potential for excitatory and inhibitory synapses respectively, c_m represents the capacitance of the membrane, g_l represents the conductance of the membrane, A_{ex} is the membrane surface area connected to a excitatory synapse, A_{ih} is the membrane surface area connected to an inhibitory synapse, $w_{x',y'}^{hex}$ represents excitatory synapse strength in matrix $\mathbf{W}^h$, and $w_{x',y'}^{hih}$ represents inhibitory synapse strength in matrix $\mathbf{W}^h$. If the edge in the receptive field matches the pattern in $\mathbf{W}^h$, the neuron $N_h(x, y)$ receives a strong input current through the excitatory synapses and the membrane potential $v_{h(x,y)}(t)$ increases. When the potential reaches a threshold v_{th}, the neuron $N_h(x,y)$ generates a spike and moves into the refractory period for a time τ_{ref}. After the refractory period, the neuron receives inputs and prepares to generate another spike. By analogy, $w_{x',y'}^{vex}$ represents excitatory synapse strength in matrix $\mathbf{W}^v$, and $w_{x',y'}^{vih}$ represents synapse inhibitory strength in matrix $\mathbf{W}^v$. If the edge in the receptive field matches the pattern in $\mathbf{W}^v$, the neuron $N_v(x, y)$ receives a strong input current through the excitatory synapses and generates spikes. The matrix pair of synapse strength distribution for $\mathbf{W}^h$ and $\mathbf{W}^v$ is determined by the following equations.

$$
w_{x',y'}^{hex} = \begin{cases} e^{-(\frac{(x'-x)^2}{\delta_{hx}^2}+\frac{(y-y')^2}{\delta_{hy}^2})} & if \sqrt{(x'-x)^2+(y-y')^2} \le R_{RF} \ , \\ 0 & if \sqrt{(x'x)^2+(y'y)^2} > R_{RF} \end{cases} \tag{5}
$$

$$
w_{x',y'}^{hih} = \begin{cases} (e^{-(\frac{(x'-x)^2}{\delta_{hx}^2}+\frac{(y'-y-\Delta)^2}{\delta_{hy}^2})}+e^{-(\frac{(x'-x)^2}{\delta_{hx}^2}+\frac{(y'-y+\Delta)^2}{\delta_{hy}^2})}) & if \sqrt{(x'-x)^2+(y'-y)^2} \le R_{RF} \\ 0 & if \sqrt{(x'-x)^2+(y'-y)^2} > R_{RF} \end{cases} \tag{6}
$$

$w_{x,y}^{vex}$ and $w_{x,y}^{vih}$ have the same distribution as $w_{x,y}^{hex}$ and $w_{x,y}^{hih}$ replacing δ_{hx} and δ_{hy} with δ_{vx} and δ_{vy}, where (x,y) is the centre of receptive field, (x',y') is a neuron position within the receptive field, R_{RF} is a radius of the receptive field, for horizontal line detection $\delta_{hx} \gg \delta_{hy}$ and for vertical line detection $\delta_{vx} \ll \delta_{vy}$. The output spike train from the horizontal neuron array is represented by $S_{h(x,y)}(t)$ and $S_{v(x,y)}(t)$. The neuron $N_o(x,y)$ in output layer receives spike trains from both the horizontal and vertical pathways, and the behaviours are governed by the following equations.

$$
\frac{g_{x,y}^{oex}(t)}{dt} = -\frac{1}{\tau_{ex}}g_{x,y}^{oex}(t)+\sum_{(x',y')\in RF_o}w_{x',y'}^{oex}S_{h(x',y')}(t)q^{oex}+\sum_{(x',y')\in RF_o}w_{x',y'}^{oex}S_{v(x',y')}(t)q^{vex},
$$

$$
\frac{g_{x,y}^{oih}(t)}{dt} = -\frac{1}{\tau_{ih}}g_{x,y}^{oih}(t)+\sum_{(x',y')\in RF_o}w_{x',y'}^{oih}S_{h(x',y')}(t)q^{hih}+\sum_{(x',y')\in RF_o}w_{x',y'}^{oih}S_{v(x',y')}(t)q^{vih},
$$

$$
c_m\frac{dv_{o(x,y)}(t)}{dt} = g_l(E_l-v_{h(x,y)}(t))+\frac{g_{x,y}^{oex}(t)}{A_{ex}}(E_{ex}-v_{o(x,y)}(t))+\frac{g_{x,y}^{oih}(t)}{A_{ih}}(E_{ih}-v_{o(x,y)}(t)), \tag{7}
$$

where the synapse strength matrix is determined by the following equations. If $(x'-x)^2+(y-y')^2 \le R_{RFo}^2$, $w_{x',y'}^{oex}=\exp(-((x'-x)^2+(y-y')^2)/\delta^2)$ and $w_{x',y'}^{oih}=\exp(-(root((x'-x)^2+(y-y')^2)-\lambda)^2/\delta^2)$, otherwise $w_{x',y'}^{oex}=0$ and

$w_{x',y'}^{oih} = 0$. Let $S_{o(x,y)}$ (t) represent a spike train generated by Neuron $N_o(x,y)$ in the output layer. The firing rate for neuron $N_o(x,y)$ is calculated by the following expression.

$$r_{o(x,y)} = \frac{1}{T} \sum_{t-T}^{t} S_{o(x,y)}(t) \tag{8}$$

By plotting this firing rate as an image with a colour bar, areas of horizontal and vertical lines are obtained. Suppose that pixel values of the input image are represented by $I(x,y)$. Suppose that the output signals are regarded as feedback to filter out the attention area as follows.

$T(x,y)=I(x,y)r_{o(x,y)}/r_{max}$ ($r_{max}=max(r_{o(x,y)} \mid (x,y) \in$ all pixels in the image)) (9)

The attention area image is obtained by a plot of image $T(x,y)$.

4   Simulation Results

The network model was implemented in Matlab using a set of parameters for the network: v_{th} = -60 mv. v_{reset} = -70 mv. E_{ex}= 0 mv. E_{ih}= -75 mv. E_l= -70 mv. g_l =1.0 $\mu s/mm^2$. c_m=10 nF/mm^2. τ_{ex}=4 ms. τ_{ih}=10 ms. A_{ih}=0.028953 mm^2. A_{ex}=0.014103 mm^2. τ_{ref} = 6 ms for neurons in line detection layer. τ_{ref} = 25 ms for output layer. A_{ih}, A_{ex}, q^{ex} and q^{ih} can be adjudged from a large range for example [0.0001,1]. These parameters are consistent with biological neurons [8]. Parameters for the receptive fields are set as follows. δ_{hx} =20. δ_{hy} = 2. δ_{vx} = 2. δ_{vy} = 20. Δ=2. R_{RF} =15. δ = 30. R_{RFo} =40. λ=24.

The proposed spiking neural network is combined with the edge detection network model in [6]. If a visual image as shown in Fig.2 (A) presents to the edge detection network, edges of the image are reflected in the output neuron array in [6]. The spike trains from this neuron array are regarded as inputs for the proposed network in this paper. In our simulation, the firing rate map is recorded as shown in Fig.2 (B).

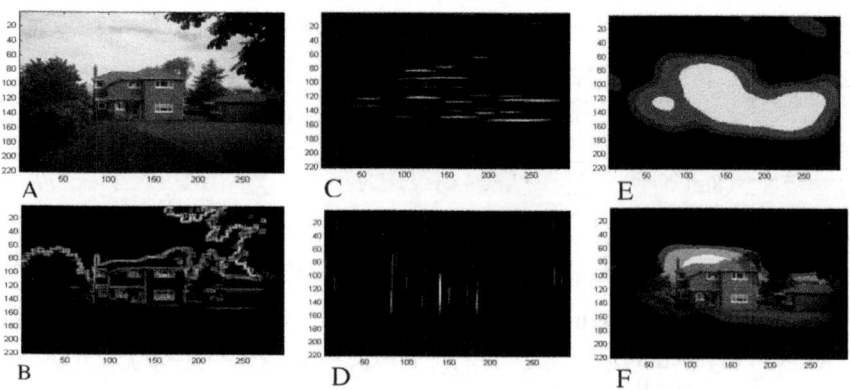

Fig. 2. Simulation results for extraction of attention area

The spike trains are transferred to horizontal and vertical line pathways according to the network architecture in Fig.1, the firing rates of the horizontal and vertical line neuron arrays are recorded as in Fig.2 (C) and (D). The spike trains from horizontal and vertical line neuron arrays are transferred to the output layer. The firing rates of the output neuron array are shown in Fig.2 (E). The output firing rates can be regarded as feedback signals to obtain the attention areas that are based on the key features of horizontal and vertical lines. The image corresponding to the attention area is obtained in Fig. 2 (F). It can be seen that the region around the house has been strengthened and other areas are ignored.

5   Discussion

Various receptive fields and hierarchical structures of spiking neurons enable a spiking neural network to perform very complicated computation tasks, learning tasks and intelligent behaviours in the human brain. This paper proposed a spiking neural network model that can perform visual attention based on image features of horizontal and vertical lines. Simulations show that the proposed network is able to obtain attention area based on the inherent horizontal and vertical lines, for example a house. Based on this principle, other more complicated image features can also be used in this attention mechanism to focus on more complicated objects. We will address this topic in further study.

References

1. Hosoya, T., Baccus, S.A., Meister, M.: Dynamic predictive coding by the retina. Nature 436, 71–77 (2005)
2. Kandel, E.R., Shwartz, J.H.: Principles of neural science. Edward Amold (Publishers) Ltd., London (1981)
3. Anderson, J.R.: Cognitive psychology and its implications, 6th edn. Worth Publishers, Belmont (2004)
4. Saalmann, Y.B., Pigarev, I.N., Vidyasagar, T.R.: Neural Mechanisms of Visual Attention: How Top-Down Feedback Highlights Relevant Locations. Science 316(5831), 1612–1615 (2007)
5. Lauritzen, T.Z., D'Esposito, M., Heeger, D.J., Silver, M.A.: Top-down flow of visual spatial attention signals from parietal to occipital cortex. Journal of Vision 9(13), 18, 1–14 (2009)
6. Wu, Q., McGinnity, M., Maguire, L.P., Belatreche, A., Glackin, B.: Edge Detection Based on Spiking Neural Network Model. In: Huang, D.-S., Heutte, L., Loog, M. (eds.) ICIC 2007. LNCS (LNAI), vol. 4682, pp. 26–34. Springer, Heidelberg (2007)
7. Wu, Q.X., McGinnity, T.M., Maguire, L.P., Belatreche, A., Glackin, B.: Adaptive co-ordinate transformation based on a spike timing-dependent plasticity learning paradigm. In: Wang, L., Chen, K., S. Ong, Y. (eds.) ICNC 2005. LNCS, vol. 3610, pp. 420–428. Springer, Heidelberg (2005)
8. Gerstner, W., Kistler, W.: Spiking Neuron Models: Single Neurons, pulations, Plasticity. Cambridge University Press, Cambridge (2002)

A Class of Chaotic Neural Network with Morlet Wavelet Function Self-feedback

Yqoqun Xu[1] and Xueling Yang[2]

[1] Institute of System Engineering, Harbin University of Commerce,
Harbin, 150028
[2] Department of mathematic, Harbin Engineering University,
Harbin, 150001
xuyq@hrbcu.edu.cn

Abstract. A Chaotic neural network model with Morlet wavelet function self-feedback is proposed by introducing Morlet wavelet function into self-feedback of chaotic neural network. The analyses of the optimization mechanism of the networks suggest that Morlet wavelet function self-feedback affects the original Hopfield energy function in the manner of the sum of the multiplications of Morlet wavelet function to the state, avoiding the network being trapped into the local minima. The energy function is constructed, and the sufficient condition for the networks to achieve asymptotical stability is analyzed and is used to instruct the parameter set of the networks for solving traveling salesman problem (TSP). Simulation researches on 10-city TSP indicate that the proposed networks can find the optimal solutions of combinatorial optimization problems.

Keywords: Self-feedback, Chaotic neural network, Energy function, Morlet wavelet function.

1 Introduction

Chaotic neural networks (CNNs) can acquire the ability to escape from the local minima of the energy function by introducing chaotic search mechanism into the original Hopfield neural network (HNN) [1-7]. Different from external chaos, chaotic search mechanism in CNN is generated by the self-feedback item of CNN. Besides, CNN possesses abundant dynamics characteristics and can traverse every point of the system by chaotic search. However, CNN cannot be easy to converge to a point steadily. Hence, Chen and Aihara have proposed a transient chaotic neural network (TCNN) with chaotic simulated annealing (CSA) by introducing a linear self-feedback into the original HNN and reducing the self-feedback connection weight exponentially. It is ensured that the TCNN has transient chaotic search behavior and can converge to a point steadily. In addition, it overcomes the limitation of HNN which is not enough for the network to escape from the local minima. In the theory, Chen and Aihara have proven that the TCNN is asymptotical stability. This paper proposes a novel TCNN model with Morlet wavelet function self-feedback which has new characteristics different from linear self-feedback network. This paper analyzes the effect of energy

D.-S. Huang et al. (Eds.): ICIC 2011, LNBI 6840, pp. 32–40, 2012.

modifier item by Kwok unified framework theory, constructs energy function of the network, and further analyzes stability of the proposed network. The analysis of the energy function suggests that the proposed network model is asymptotical stability under the given conditions. The simulations of 10-city TSP suggest that the proposed chaotic neural network has a good optimal performance.

2 TCNN With Morlet Wavelet Function Self-feedback

The proposed TCNN with Morlet wavelet function self-feedback can be described as follows.

$$x_i(t) = \frac{1}{1+\exp(-y_i(t)/\varepsilon)} \tag{1}$$

$$y_i(t+1) = ky_i(t) + \alpha\left[\sum_{j=1,j\neq i}^{n} w_{ij}x_j(t) + I_i\right] - z_i(t)\phi(s_i(t),u_i(t),x_i(t)-I_0) \tag{2}$$

$$s_i(t+1) = (1-\beta)s_i(t) \tag{3}$$

$$u_i(t+1) = 4u_i(t)(1-u_i(t)) \tag{4}$$

$$z_i(t+1) = (1-\beta)z_i(t) \tag{5}$$

$$\phi(s,u,x) = \exp\{-[(x-u)/s]^2/2\}\cdot\cos[5(x-u)/s] \quad (s>0,-1<u<1) \tag{6}$$

where x_i is the output of neuron i; y_i is the internal state of neuron i; w_{ij} is the connection weight from neuron j to neuron i, $w_{ij} = w_{ji}$; I_i is an input bias of neuron i; I_0 is a positive parameter; k is a damping factor of nerve membrane ($0<k<1$); z_i is the self-feedback connection weight; β is the damping factor of z_i.

The single neuron of the proposed TCNN model can be described as follows:

$$x(t) = \frac{1}{1+\exp(-y(t)/\varepsilon)} \tag{7}$$

$$y(t+1) = ky(t) - z(t)\phi(s(t),u(t),x(t)-I_0) \tag{8}$$

$$s(t+1) = (1-\beta)s(t) \tag{9}$$

$$u(t+1) = 4u(t)(1-u(t)) \tag{10}$$

$$z(t+1) = (1-\beta)z(t) \tag{11}$$

$$\phi(s,u,x) = \exp\{-[(x-u)/s]^2/2\}\cdot\cos[5(x-u)/s] \quad (s>0,-1<u<1) \tag{12}$$

In order to make the neuron behave transient chaotic behavior, the parameters are set as follows:

$\varepsilon = =1/11$, $y(1) = 0.283$, $k = 0.3$, $I_0 = 0.55$, $\beta = 0.0005$, $s(1) = 1.5$, $u(1) = 0.5$, $z(1) = 0.35$

The state bifurcation figures and the time evolution figures of the maximal Lyapunov exponent are respectively shown as Fig.1.

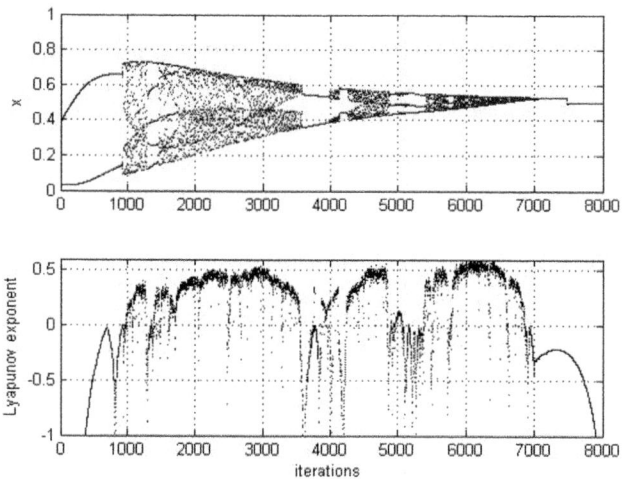

Fig. 1. State bifurcation figure and the maximal Lyapunov exponents of the single neuron

Seen from the above state bifurcation figures, the neuron behaves a transient chaotic dynamic behavior. The single neural unit first behaves the global chaotic search, and with the decrease of the value of z_i, the reversed bifurcation gradually converges to a stable equilibrium state. After the chaotic dynamic behavior disappears, the dynamic behavior of the single neural unit is controlled by the gradient descent dynamics. When the behavior of the single neural unit is similar to that of Hopfield, the network tends to converge to a stable equilibrium point.

3   Energy Function

Analysis of the energy function is very important for the study of nonlinear dynamics systems, and through such an analysis, we can easily observe how the sigmoid function self-feedback affects the optimization performance of the proposed CNN. In this section, One is the unified framework theory that allows the construction and comparison of various models from the basic HNN by the introduction of an energy modifier. The other is Lyapunov stability analysis that we use to investigate the asymptotical stability of the proposed CNN model. Moreover, the unified framework theory is applied to construct the energy function for the Lyapunov stability analysis in this section.

3.1   Energy Modifier of CNN

Based on the unified framework, the energy function can be described as follows:

$$E = E_{Hop} + H \tag{13}$$

$$E_{Hop} = -\frac{1}{2}\sum_{i,j}^{n} w_{i,j} x_i x_j - \sum_{i}^{n} I_i x_i + \frac{\varepsilon_1}{\tau}\sum_{i}^{n}\int_0^{x_i} \ln\frac{x}{1-x}dx \tag{14}$$

Where E_{Hop} is the energy function of HNN, H is the energy modifier, the connection weight matrix W is symmetric, $W = W^T$ and $w_{ii} = 0$. For the proposed CNN model, the energy modifier can be described as

$$H = \lambda\sum_{i}^{n}\int_0^{x_i} z_i\phi(s_i, u_i, x - I_0)dx \tag{15}$$

In the following, we apply the unified framework to verify (15) is the energy modifier of the proposed CNN:

$$\frac{dy_i}{dt} = -\frac{\partial E}{\partial x_i} = -\frac{\partial(E_{Hop} + H)}{\partial x_i}$$
$$= -\frac{y_i}{\tau} + \left[\sum_{j\neq i}^{n} w_{ij} x_j + I_i\right] - \lambda z_i\phi(s_i, u_i, x_i - I_0) \tag{16}$$

Where $-\dfrac{\partial E_{Hop}}{\partial x_i} = -\dfrac{y_i}{\tau} + \sum_j^n w_{ij} x_j + I_i$ and $\dfrac{dy_i}{dt} = -\dfrac{\partial E}{\partial x_i}$, the relationship are given by Hopfied. Applying Euler discretization, (12) can be rewritten as

$$y_i(t+1) = (1-\frac{\Delta t}{\tau})y_i(t) + \Delta t\left[\sum_{j\neq i}^{n} w_{ij} x_j(t) + I_i\right] - \Delta t\lambda z_i(t)\phi(s_i(t), u_i(t), x_i(t) - I_0) \tag{17}$$

where Δt is the time step. If the following relationships are satisfied:

$$k = 1-\frac{\Delta t}{\tau}, \quad \alpha = \Delta t, \quad \lambda = \frac{1}{\Delta t} \tag{18}$$

Then (17) is equal to (2). Therefore, (17) is a reasonable energy modifier of the proposed CNN.

Using the mean value theorem, (11) can be rewritten as

$$H = \lambda\sum_{i}^{n}\int_0^{x_i} z_i\phi(s_i, u_i, x - I_0)dx = \lambda\sum_{i}^{n} z_i x_i\phi(s_i, u_i, \bar{x}_i - I_0) \tag{19}$$

where $0 < \bar{x}_i < x_i(t)$. As seen from (19), the equation form of the energy modifier is a linear combination of sigmoid function and states.

3.2   Asymptotical Stability

In the following, a computational energy function of the proposed CNN is constructed by applying the unified framework theory [11] and is proven to be asymptotically stable by applying Lyapunov stability theory.

Applying the unified framework discussed, the energy function can be described as

$$E(X) = -\frac{1}{2}\sum_{i,j}^{n} w_{i,j} x_i x_j - \sum_{i}^{n} I_i x_i - \frac{(k-1)}{\alpha} \varepsilon_1 \sum_{i}^{n} \int_{0}^{x_i} \ln\frac{x}{1-x} dx + \frac{1}{\alpha}\sum_{i}^{n} \int_{0}^{x_i} z_i g(x-I_0) dx$$

then, the energy function can be described as

$$\alpha E(X) = -\frac{\alpha}{2}\sum_{i,j}^{n} w_{i,j} x_i x_j - \alpha\sum_{i}^{n} I_i x_i - (k-1)\varepsilon_1 \sum_{i}^{n} \int_{0}^{x_i} \ln\frac{x}{1-x} dx + \sum_{i}^{n} \int_{0}^{x_i} z_i g(x-I_0) dx \tag{20}$$

where $W = W^T$, $w_{ii} = 0$. Before analyzing the asymptotical stability of the discrete system containing (1) and (2), (1) and (2) are first combined into one equation as follows [1]:

$$\alpha\left[\sum_{j\neq i}^{n} w_{ij} x_j(t) + I_i\right] - z_i(t)g(x_i(t)-I_0) = \varepsilon_1 \ln\frac{x_i(t+1)}{1-x_i(t+1)} - k\varepsilon_1 \ln\frac{x_i(t)}{1-x_i(t)} \tag{21}$$

Applying the mean value theory, several equations can be described as follows:

$$\int_{x_i(t)}^{x_i(t+1)} z_i g(x-I_0) dx = \Delta x_i z_i g(\overline{x}_i - I_0) \tag{22}$$

where $\overline{x}_i(t) = x_i(t) + \overline{\theta}_i \Delta x_i$, $0 \leq \overline{\theta}_i \leq 1$

$$\Delta x_i z_i\left[g(\overline{x}_i - I_0) - g(x_i - I_0)\right] \leq \overline{\theta}_i z_i \left[\Delta x_i\right]^2 / \varepsilon_2 \tag{23}$$

where $\xi_i = x_i(t) + \tilde{\theta}(\overline{\theta}_i \Delta x_i)$, $0 \leq \tilde{\theta} \leq 1$ and the equality holds only for $\xi_i = I_0$

$$\int_{x_i(t)}^{x_i(t+1)} \ln\frac{x}{1-x} dx = \left[x_i(t+1)-x_i(t)\right]\ln\frac{\varphi_i}{1-\varphi_i} = \Delta x_i \ln\frac{\varphi_i}{1-\varphi_i} \tag{24}$$

where $\varphi_i = x_i(t) + \theta_i \Delta x_i$, $0 \leq \theta_i \leq 1$

$$\Delta x_i\left(\ln\frac{x_i(t+1)}{1-x_i(t+1)} - \ln\frac{\varphi_i}{1-\varphi_i}\right) \geq 4(1-\theta_i)\left[\Delta x_i\right]^2 \tag{25}$$

where $\eta_{i1} = \varphi_i + \eta_1\left[(1-\theta_i)\Delta x_i\right]$, $0 \leq \eta_1 \leq 1$

$$\Delta x_i\left(\ln\frac{\varphi_i}{1-\varphi_i} - \ln\frac{x_i(t)}{1-x_i(t)}\right) \geq 4\theta_i\left[\Delta x_i\right]^2 \tag{26}$$

where $\eta_{i2} = x_i(t) + \eta_2\left[\theta_i \Delta x_i\right]$, $0 \leq \eta_2 \leq 1$。

The detailed analysis of the asymptotical stability is shown as follows:

$$\alpha E(X(t+1)) - \alpha E(X(t))$$

$$= -\frac{\alpha}{2}\sum_{i,j}^n w_{i,j}\Delta x_i \Delta x_j - \alpha \sum_{i,j}^n w_{i,j} x_j \Delta x_i - \alpha \sum_i^n I_i x_i - (k-1)\varepsilon_1 \sum_i^n \int_{x_i(t)}^{x_i(t+1)} \ln\frac{x}{1-x}dx$$

$$+\sum_i^n \int_{x_i(t)}^{x_i(t+1)} z_i g(x-I_0)dx$$

where $\Delta x_i = x_i(t+1) - x_i(t)$, $\Delta x_j = x_j(t+1) - x_j(t)$. Substituting (26) into last formula, we have

$$\alpha E(X(t+1)) - \alpha E(X(t))$$

$$= -\frac{\alpha}{2}\sum_{i,j}^n w_{i,j}\Delta x_i \Delta x_j - \sum_i^n \Delta x_i \left\{ \alpha \left[\sum_j^n w_{i,j} x_j + I_i \right] - z_i g(x_i - I_0) \right\}$$

$$-(k-1)\varepsilon_1 \sum_i^n \int_{x_i(t)}^{x_i(t+1)} \ln\frac{x}{1-x}dx + \sum_i^n \Delta x_i z_i \left[g(\overline{x}_i - I_0) - g(x_i - I_0) \right]$$

Substituting (17) and (19)-(22) into $\alpha E(X(t+1)) - \alpha E(X(t))$, we have

$$\alpha E(X(t+1)) - \alpha E(X(t))$$

$$= -\frac{\alpha}{2}\sum_{i,j}^n w_{i,j}\Delta x_i \Delta x_j - \sum_i^n \Delta x_i \left\{ \varepsilon_1 \ln\frac{x_i(t+1)}{1-x_i(t+1)} - k\varepsilon_1 \ln\frac{x_i(t)}{1-x_i(t)} \right\}$$

$$+ (k-1)\varepsilon_1 \sum_i^n \Delta x_i \ln\frac{\varphi_i}{1-\varphi_i} + \sum_i^n \Delta x_i z_i \left[g(\overline{x}_i - I_0) - g(x_i - I_0) \right]$$

$$\leq -\frac{\alpha}{2}\sum_{i,j}^n w_{ij}\Delta x_i \Delta x_j - \varepsilon_1 \sum_i^n [\Delta x_i]^2 [4(1-\theta_i)+4k\theta_i] + \sum_i^n \overline{\theta}_i z_i [\Delta x_i]^2 / \varepsilon_2$$

$$= -\frac{1}{2}\Delta X(t)^T \left\{ \alpha W' + 2\varepsilon_1 \left[4(1-\theta_i)+4k\theta_i \right] I_n \right\} \Delta X(t)$$

If the minimal eigenvalue λ' of the matrix W' satisfies

$$-\lambda' < \frac{2\varepsilon_1}{\alpha}[4(1-\theta_i)+4k\theta_i] = \frac{8\varepsilon_1}{\alpha}[1-(1-k)\theta_i] \tag{27}$$

Then the matrix $\alpha W' + 2\varepsilon_1 \left[4(1-\theta_i)+4k\theta_i \right] I_n$ is positive definite, then we can obtain $E(X(t+1)) - E(X(t)) \leq 0$ and the proposed CNN model can achieve asymptotical stability.

4   Application to 10-City TSP

A solution of TSP with N cities is represented by $N \times N$-permutation matrix, where each entry corresponds to the output of a neuron in a network with $N \times N$ lattice structure. Assume x_{ij} to be the neuron output that represents city i in visiting order j.

A computational energy function can be described as:

$$E = \frac{W_1}{2} \left\{ \sum_{i=1}^{n} \left[\sum_{j=1}^{n} x_{ij} - 1 \right]^2 + \sum_{j=1}^{n} \left[\sum_{i=1}^{n} x_{ij} - 1 \right]^2 \right\} + \frac{W_2}{2} \sum_{i=1}^{n} \sum_{j=1}^{n} \sum_{k=1}^{n} (x_{k,j+1} + x_{k,j-1}) x_{i,j} d_{ik} \qquad (28)$$

where $x_{i0} = x_{in}$ and $x_{i,n+1} = x_{i1}$. W_1 and W_2 are the coupling parameters corresponding to the constraints and the cost function of the tour length, respectively. d_{xy} is the distance between city x and city y .

This paper adopts the following 10-city unitary coordinates:
(0.4, 0.4439), (0.2439, 0.1463), (0.1707, 0.2293), (0.2293, 0.716), (0.5171, 0.9414), (0.8732, 0.6536), (0.6878, 0.5219), (0.8488, 0.3609), (0.6683, 0.2536), (0.6195, 0.2634).
The shortest distance of the 10-city is 2.6776.
The parameters of the network are set as follows:

$$W_1 = 1, \ W_2 = 1, \ k = 1, \ I_0 = 0.75, \ z(1) = 0.1, \ \varepsilon = 0.2, \ \alpha = 0.8, \ u(1) = 0.5,$$
$$s(1) = 1.8.$$

Table 1. Results of 1000 different initial conditions for each value β on 10-city TSP

β	RGM(%)	RLM(%)	RIS(%)	AIC
0.01	68.9	31.1	0.0	86
0.005	76.9	23.1	0.0	126
0.004	80.0	20.0	0.0	147
0.003	84.9	15.1	0.0	172
0.002	89.8	10.2	0.0	212
0.001	100	0.0	0.0	252
0.0009	96.9	3.1	0.0	282
0.0008	96.4	3.6	0.0	299
0.0007	95.9	4.1	0.0	317
0.0006	95.5	4.5	0.0	352
0.0005	94.5	5.5	0.0	420
0.0004	89.7	10.3	0.0	564

1000 different initial conditions of y_i are generated randomly in the region [0, 1] for different β. The results are summarized in Table1, the column 'RGM', 'RLM', 'RIS' and 'AIC' respectively represents the rate of global minima, the rate of local minima, the rate of infeasible solutions and average iterations for convergence.

As seen from Table1, the CNN with sigmoid function self-feedback can solve the 10-city TSP effectively and find the global optimal solution with 1000 different initial conditions when β is 0.008. With the increase of the value of β when $\beta > 0.008$, the value of 'RGM' becomes small although it has a better constringency speed. At the same time, with the decrease of the value of β when $\beta < 0.008$, the value of 'AIC' becomes large although it has a better optimization effect.

5   Conclusion

The proposed CNN with sigmoid function self-feedback can acquire the ability to enhance the optimization and avoid trapping into local minimum. The analysis of the energy function of the proposed CNN with sigmoid function self-feedback indicates that there exists an energy modifier affecting the localized characterization ability of the proposed CNN. In addition, the stability analysis suggests that the proposed CNN with sigmoid function self-feedback can achieve asymptotical stability. The simulations on **continuous function optimization problems** and TSP indicate that the sigmoid function self-feedback has a better optimization ability.

References

1. Chen, L.N., Aihara, K.: Chaotic Simulated Annealing by a Neural Network Model with Transient Chaos. Neural Networks 8(6), 915–930 (1995)
2. Wang, L., Smith, K.: On Chaotic Simulated Annealing. IEEE Trans. on Neural Networks 9(4), 716–718 (1998)
3. Xu, Y.Q., Sun, M., Duan, G.R.: Wavelet Chaotic Neural Networks and Their Application to Optimization Problems. In: Adi, A., Stoutenburg, S., Tabet, S. (eds.) RuleML 2005. LNCS, vol. 3791, pp. 379–384. Springer, Heidelberg (2005)
4. Xu, Y.Q., Sun, M., Zhang, J.H.: A Model of Wavelet Chaotic Neural Network with Applications in Optimization. In: Proc. 6th World Congr. Intell. Control Autom., China, vol. 1, pp. 2901–2905 (2006)
5. Wang, L.: Intelligence Optimization Algorithm and its Application. Press of TUP (2001)
6. Xu, Y.-q., Sun, M., Guo, M.-s.: Improved Transiently Chaotic Neural Network and Its Application to Optimization. In: King, I., Wang, J., Chan, L.-W., Wang, D. (eds.) ICONIP 2006. LNCS, vol. 4233, pp. 1032–1041. Springer, Heidelberg (2006)
7. Lin, J.S.: Annealed Chaotic Neural Network with nonlinear Self-feedback and its Application to Clustering Problem. Pattern Recognition 34, 1093–1104 (2001)
8. Chen, L., Aihara, K.: Chaos and Asymptotical Stability in Discrete-time Neural Networks. Physica D 104, 286–325 (1998)

9. Kwok, T., Smith, K.: Experimental Analysis of Chaotic Neural Network Models for Combinatorial Optimization under a Unifying Framework. Neural Networks 13, 731–744 (2002)
10. Zhao, L., Sun, M., Cheng, J.H., Xu, Y.Q.: A Novel Chaotic Neural Network with the Ability to Characterize Local Features and its Application. IEEE Trans. On Neural Networks 20(4), 737–738 (2009)
11. Kwok, T., Smith, K.: A Unified Framework for Chaotic Neural-network Approaches to Combinatorial Optimization. IEEE Trans. Neural Network 10(4), 978–981 (1999)

A Novel Nonlinear Neural Network Ensemble Model Using K-PLSR for Rainfall Forecasting

Chun Meng and Jiansheng Wu

Department of Mathematics and Computer, Liuzhou Teacher College,
Liuzhou, Guangxi, 545004, China
biaji2003@yahoo.com.cn,
wjsh2002168@163.com

Abstract. In this paper, a novel hybrid Radial Basis Function Neural Network (RBF–NN) ensemble model is proposed for rainfall forecasting based on Kernel Partial Least Squares Regression (K–PLSR). In the process of ensemble modeling, the first stage the initial data set is divided into different training sets by used Bagging and Boosting technology. In the second stage, these training sets are input to the RBF–NN models of different kernel function, and then various single RBF–NN predictors are produced. Finally, K–PLSR is used for ensemble of the prediction purpose. Our findings reveal that the K–PLSR ensemble model can be used as an alternative forecasting tool for a Meteorological application in achieving greater forecasting accuracy.

Keywords: Neural network ensemble, kernel partial least square regression, rainfall forecasting.

1 Introduction

Accurate and timely rainfall forecasting is a major challenge for the scientific community because it can help prevent casualties and damages caused by natural disasters [1,2]. However, neural network are a kind of unstable learning methods, i.e., small changes in the training set and/or parameter selection can produce large changes in the predicted output. This diversity of neural networks is a naturally by-product of the randomness of the inherent data and training process, and also of the intrinsic non–identifiability of the model. For example, the results of many experiments have shown that the generalization of single neural network is not unique [3,4].

In order to overcome the main limitations of neural network, recently a novel ensemble forecasting model, i.e. artificial neural network ensemble (NNE), has been developed. Because of combining multiple neural networks learned from the same training samples, NNE can remarkably enhance the forecasting ability and outperform any individual neural network. It is an effective approach to the development of a high performance forecasting system [5].

In general, a neural network ensemble is constructed in two steps, i.e. training a number of component neural networks and then combining the component

D.-S. Huang et al. (Eds.): ICIC 2011, LNBI 6840, pp. 41–48, 2012.

predictions. Different from the previous work, this paper proposes a novel hybrid Radial Basis Function Neural Network (RBF-NN) ensemble forecasting method in terms of Kernel Partial Least Squares Regression (K–PLSR) principle, namely RBF–K–PLSR. The rainfall data of Guangxi is predicted as a case study for our proposed method. An actual case of forecasting monthly rainfall is illustrated to show the improvement in predictive accuracy and capability of generalization achieved by our proposed RBF–K–PLSR model.

The rest of this study is organized as follows. Section 2 describes the proposed RBF–LS–SVR, ideas and procedures. For further illustration, this work employs the method set up a prediction model for rainfall forecasting in Section 3. Discussions are presented in Section 4 and conclusions are drawn in the final Section.

2   The Building Process of the Neural Network Ensemble Model

In this section, a triple–phase nonlinear RBF–NN ensemble model is proposed for rainfall forecasting. First of all, many individual RBF–NN predictors are generated in terms of different kernel function. Then RBF–NN predictors are ensemble into an aggregated predictor by K–PLSR.

2.1   Radial Basis Function Neural Network

Radial basis function was introduced into the neural network literature by Broomhead and Lowe [6]. RBF–NN are widely used for approximating functions from given input–output patterns. The performance of a trained RBF network depends on the number and locations of the radial basis functions, their shape and the method used for learning the input–output mapping. The network is generally composed of three layers, the architecture of RBF–NN is presented in Figure 1.

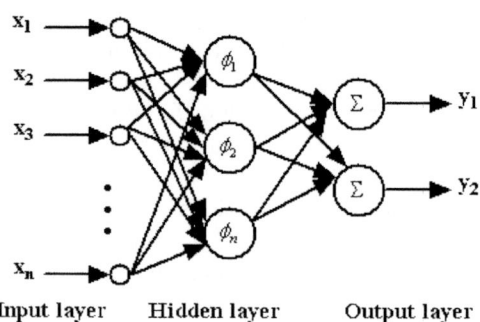

Fig. 1. The RBF–NN architecture

The output of the RBF–NN is calculated according to

$$y_i = f_i(x) = \sum_{k=1}^{N} w_{ik}\phi_k(\|x - c_k\|), i = 1, 2, \cdots, m \tag{1}$$

where $x \in \Re^{n \times 1}$ is an input vector, $\phi_k(\cdot)$ is a function from $\Re^+$ to $\Re$, $\| \cdot \|_2$ denotes the Euclidean norm, w_{ik} are the weights in the output layer, n is the number of neurons in the hidden layer, and $c_k \in \Re^{n \times 1}$ are the centers in the input vector space. The functional form of $\phi_k(\cdot)$ is assumed to have been given, and some typical choices are shown in Table 1.

Table 1. Types of function name and Function formula

Denote	Functional name	Function formula
RBF1	Cubic approximation	$\phi(x) = x^3$
RBF2	Reflected sigmoid	$\phi(x) = (1 + exp(x^2/\sigma^2))^{-1}$
RBF3	Thin-plate-spline function	$\phi(x) = x^2 ln x$
RBF4	Guassian function	$\phi(x) = exp(-x^2/\sigma^2)$
RBF5	Multi-quadratic function	$\phi(x) = \sqrt{x^2 + \sigma^2}$
RBF6	Inverse multi-quadratic function	$\phi(x) = \dfrac{1}{\sqrt{x^2+\sigma^2}}$

The training procedure of the RBF networks is a complex process, this procedure requires the training of all parameters including the centers of the hidden layer units ($c_i, i = 1, 2, \cdots, m$), the widths (σ_i) of the corresponding Gaussian functions, and the weights ($\omega_i, i = 0, 1, \cdots, m$) between the hidden layer and output layer. In this paper, the the orthogonal least squares algorithm (OLS) is used to training RBF based on the minimizing of SSE. The more detailed about algorithm is described by the related literature [7,8].

2.2 Kernel PLS Regression

Partial least squares (PLS) regression analysis was developed in the late seventies by Herman O. A. Wold [9]. It has been a popular modeling, regression, discrimination and classification technique in its domain of origin–chemometrics. It is a statistical tool that has been specifically designed to deal with multiple regression problems where the number of observations is limited, missing data are numerous and the correlations between the predictor variables are high.

PLS regression is is a technique for modeling a linear relationship between a set of output variables (responses) $\{Y_i, i = 1, 2, \cdots, N, Y \in R^L\}$ and a set of input variables (regressors) $\{X_i, i = 1, 2, \cdots, N, Y \in R^N\}$. The K–PLSR methodology was proposed by Roman Rosipal [10]. In kernel PLS regression a linear PLS regression model in a feature space F is considered. The data set y variables into a feature F_1 space. This simply means that $K_1 = YY^T$ and F_1 is the original Euclidian R^L space. In agreement with the standard linear PLS model it is assumed that the score variables $\{t_i\}_{i=1}^p$ are good predictors of Y. Further, a linear inner relation between the scores of t and u is also assumed; that is,

$$\hat{g}^m(x, d^m) = \sum_{i=1}^{N} d_i^m K(x, x_i) \qquad (2)$$

which agrees with the solution of the regularized form of regression in RKHS given by the Representer theorem [11,12]. Using equation (10) the kernel PLS model can also be interpreted as a linear regression model of the form

$$\hat{g}^m(x, c^m) = c_1^m t_1(x) + c_2^m t_2(x) + \cdots + c_N^m t_N(x) = \sum_{i=1}^{N} c_i^m t_i(x) \qquad (3)$$

where $\{t_i(x)\}_{i=1}^p$ are the projections of the data point x onto the extracted p score vectors and $c^m = T^T Y^m$ is the vector of weights for the m-th regression model. The more detailed about K–PLSR algorithm is described by the related literature [10,12].

2.3   Our Proposed Novel Hybrid RBF–K–PLS

To summarize, the proposed hybrid RBF–NN ensemble model consists of three main steps. Generally speaking, firstly, the initial data set is divided into different training sets by used Bagging and Boosting technology. Secondly, these training sets are input to the different individual RBF–NN models, and then various single RBF–NN predictors are produced based on different kernel function. Thirdly, K–PLSR is used to aggregate the ensemble results. The basic flow diagram can be shown in Figure 2.

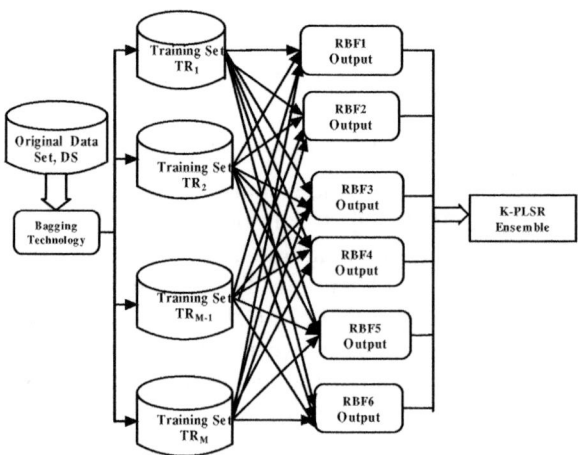

Fig. 2. A Flow Diagram of The Proposed Ensemble Forecasting Model

3   Experiments Analysis

3.1   Empirical Data

This study has investigated Modeling RBF–K–PLSR to predict average monthly precipitation from 1965 to 2009 on July in Guangxi (including Ziyuan, Guilin, Liuzhou, Bose, Wuzhou, Nanning, Yulin and Beihai). Thus the data set contained 529 data points in time series, 489 samples were used to train and the other 40 samples were used to test for generalization ability.

Firstly, the candidate forecasting factors are selected from the numerical forecast products based on 48h forecast field, which includes: the 23 conventional meteorological elements, physical elements from the T213 numerical products of China Meteorological Administration, the 500–hPa monthly mean geopotential height field of the Northern Hemisphere data and the sea surface temperature anomalies in the Pacific Ocean data. We get 10 variables as the predictors, which the original rainfall data is used as the predicted variables.

Method of modeling is one-step ahead prediction, that is, the forecast is only one sample each time and the training samples is an additional one each time on the base of the previous training.

3.2   Performance Evaluation of Model

In order to measure the effectiveness of the proposed method, three types of errors are used in this paper, such as, Normalized Mean Squared Error (NMSE), the Mean Absolute Percentage Error (MAPE) and Pearson Relative Coefficient (PRC), which be found in many paper [4]. The minimum values of NMSE indicate that the deviations between original values and forecast values are very small. The minimum values of MAPE indicate the smallest variability from sample to sample, which it is expressed in generic percentage terms that can be understandable to a wide range of users. The accurate efficiency of the proposed model is measured as PRC, The higher values of PRC (maximum value is 1) indicate that the forecasting performance of the proposed model is effective, which can capture the average change tendency of the cumulative rainfall data.

According to the previous literature, there are a variety of methods for rainfall forecasting model in the past studies. For the purpose of comparison, we have also built other three rainfall forecasting models: ARIMA and stacked regression (SR) ensemble [13] method based on RBF–NN.

The authors used Eviews statistical packages to formulate the ARIMA model. Akaike information criterion (AIC) was used to determine the best model. The model generated from the data set is AR(4). The equation used is presented in Equation 5.

$$x_t = 1 + 0.7326x_{t-1} - 0.6118x_{t-2} + 0.0231x_{t-3} + 0.0173x_{t-4} \qquad (4)$$

The standard RBF–NN with Gaussian-type activation functions in hidden layer were trained for each training set, then tested as an ensemble for each method for the testing set. Each network was trained using the neural network toolbox provided by Matlab software package. In addition, the best single RBF neural network using cross–validation method [14] (i.e., select the individual RBF network by minimizing the MSE on cross–validation) is chosen as a benchmark model for comparison.

3.3   Analysis of the Results

Table 2 illustrates the fitting accuracy and efficiency of the model in terms of various evaluation indices for 489 training samples. From the table, we can generally see that learning ability of RBF–K–PLSR outperforms the other two models under the same network input. The more important factor to measure performance of a method is to check its forecasting ability of testing samples in order for actual rainfall application.

Table 2. Fitting result of three different models about training samples

Moel	NMSE	MAPE	PRC
AR(4)	0.0146	0.2451	0.8765
SR Ensemble	0.0120	0.1890	0.9431
RBF-K-PLSR	0.0113	0.1209	0.9756

Figure 3 shows the forecasting results of three different models for 30 testing samples, we can see that the forecasting results of RBF–K–PLSR model are best in all models. Table 3 shows that the forecasting performance of three different models from different perspectives in terms of various evaluation indices. From the graphs and table, we can generally see that the forecasting results are very promising in the rainfall forecasting under the research where either the measurement of fitting performance is goodness or where the forecasting performance is effectiveness.

From more details, the NMSE of the AR(4) model is 0.2164. Similarly, the NMSE of the SR ensemble model is 0.1762; however the NMSE of the RBF–K–PLSR model reaches 0.0235. The NMSE result of the model has obvious advantages over two other models. Subsequently, for MAPE efficiency index, the proposed RBF–K–PLSR model is also the smallest.

Similarly, for PRC efficiency index, the proposed RBF–K–PLSR model is also deserved to be confident. As shown in Table 3, we can see that the forecasting rainfall values from RBF–K–PLSR model have higher correlative relationship with actual rainfall values; As for the testing samples, the PRC for the AR(4) model is only 0.8971, for the SR ensemble model PRC is 0.9032; while for the RBF–K–PLSR forecasting models, the PRC reaches 0.9765. It show that the PRC of RBF–K–PLSR model is close to their real values in different models and the model is capable to capture the average change tendency of the daily rainfall data.

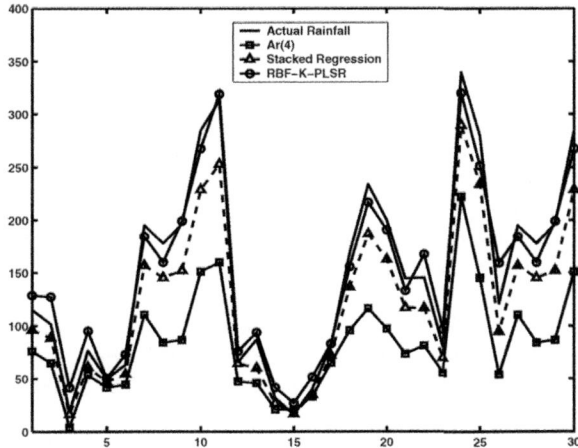

Fig. 3. The Forecasting Results of Testing Samples

Table 3. Forecasting result of three different models about testing samples

Moel	NMSE	MAPE	PRC
AR(4)	0.2164	0.3876	0.8971
SR Ensemble	0.1762	0.2912	0.9032
RBF-K-PLSR	0.0235	0.1400	0.9765

4   Conclusion

Accurate rainfall forecasting is crucial for a frequent unanticipated flash flood region to avoid life losing and economic loses. This paper proposes a novel hybrid Radial Basis Function Neural Network ensemble forecasting method in terms of Kernel Partial Least Squares Regression principle. This model was applied to the forecasting fields of monthly rainfall in Guangxi. The results show that using different the kernel function of RBF can establish the effective nonlinear mapping for rainfall forecasting. It demonstrated that K–PLSR is used to combine the selected individual forecasting results into a nonlinear ensemble model, which keeps the flexibility of the nonlinear model. So the proposed nonlinear ensemble model can be used as a feasible approach to rainfall forecasting.

Acknowledgments. The authors would like to express their sincere thanks to the editor and anonymous reviewer's comments and suggestions for the improvement of this paper. This work was supported in part by the Natural Science Foundation of China under Grant No.41065002, and in part by the Guangxi Natural Science Foundation under Grant No.0832019Z.

References

1. Nasseri, M., Asghari, K., Abedini, M.J.: Optimized Scenario for Rainfall Forecasting Using Genetic Algorithm Coupled with Artificial Neural Network. Expert Systems with Application 35, 1414–1421 (2008)
2. Yingni, J.: Prediction of Monthly Mean Daily Diffuse Solar Radiation Using Artificial Neural Networks and Comparison with Other Empirical Models. Energy Policy 36, 3833–3837 (2008)
3. Kannan, M., Prabhakaran, S., Ramachandran, P.: Rainfall Forecasting Using Data Mining Technique. International Journal of Engineering and Technology 2(6), 397–401 (2010)
4. Wu, J.S., Chen, E.h.: A Novel Nonparametric Regression Ensemble for Rainfall Forecasting Using Particle Swarm Optimization Technique Coupled with Artificial Neural Network. In: Yu, W., He, H., Zhang, N. (eds.) ISNN 2009. LNCS, vol. 5553, pp. 49–58. Springer, Heidelberg (2009)
5. French, M.N., Krajewski, W.F., Cuykendal, R.R.: Rainfall Forecasting in Space and Time Using a Neural Network. Journal of Hydrology 137, 1–37 (1992)
6. Broomhead, D.S., Lowe, D.: Multivariable Functional Interpolation and Adaptive Networks. Complex Systems 26, 321–355 (1988)
7. Moravej, Z., Vishwakarma, D.N., Singh, S.P.: Application of Radial Basis Function Neural Network for Differential Relaying of a Power Transformer. Computers and Electrical Engineering 29, 421–434 (2003)
8. Ham, F.M., Ivica, K.: Principles of Neurocomputing for Science & Engineering. The McGraw-Hill Companies, New York (2001)
9. Wold, S., Ruhe, A., Wold, H., Dunn, W.J.: The Collinearity Problem in Linear Regression: Rhe Partial Least Squares Aapproach To Generalized Inverses. Journal on Scientific and Statistical Computing 5(3), 735–743 (1984)
10. Rosipal, R., Trejo, L.J.: Kernel Partial Least Squares Regression in Reproducing Kernel Hilbert Space. Journal of Machine Learning Research 2, 97–123 (2001)
11. Wahba, G.: Splines Models of Observational Data. Series in Applied Mathematics. SIAM, Philadelphia (1990)
12. Rosipal, R., Trejo, L.J., Matthews, B.: Kernel PLS-SVC for Linear and Nonlinear Classication. In: Proceedings of the Twentieth International Conference on Machine Learning (ICML 2003), Washington DC (2003)
13. Yu, L., Wang, S.Y., Lai, K.K.: A Novel Nonlinear Ensemble Forecasting Model Incorporating GLAR and ANN for Foreign Exchange Rates. Computers & Operations Research 32, 2523–2541 (2005)
14. Krogh, A., Vedelsby, J.: Neural Network Ensembles, Cross Validation, and Active Learning. In: Tesauro, G., Touretzky, D., Leen, T. (eds.) Advances in Neural Information Processing Systems, vol. 7, pp. 231–238. The MIT Press, Cambridge (1995)

On-Line Extreme Learning Machine for Training Time-Varying Neural Networks

Yibin Ye, Stefano Squartini, and Francesco Piazza

A3LAB, Department of Biomedics, Electronics and Telecommunications,
Università Politecnica delle Marche, Via Brecce Bianche 1, 60131 Ancona, Italy
{yibin.ye,s.squartini,f.piazza}@univpm.it
http://www.a3lab.dibet.univpm.it

Abstract. Time-Varying Neural Networks(TV-NN) represent a power-ful tool for nonstationary systems identification tasks, as shown in some recent works of the authors. Extreme Learning Machine approach can train TV-NNs efficiently: the reference algorithm is named ELM-TV and is of batch-learning type. In this paper, we generalize an online sequential version of ELM to TV-NN and evaluate its performances in two nonstationary systems identification tasks. The results show that our proposed algorithm produces comparable generalization performances to ELM-TV with certain benefits to those applications with sequential arrival or large number of training data.

1 Introduction

Time-Varying Neural Networks(TV-NN) are a relevant example in neural architectures working properly in nonstationary environments. Such networks implement time-varying weights, each being a linear combination of a certain set of basis functions. An extended Back-Propagation algorithm has been proposed to train these networks (BP-TV) [1]. Later on, an Extreme Learning Machine approach is also developed for TV-NN, accelerating the training procedure significantly (ELM-TV) [2,3]. Recently, two variants are proposed based on ELM-TV, referred as EM-ELM-TV [4] and EM-OB [5], which are able to automatically determine the number of hidden neurons or output bases functions, respectively.

All the algorithms for TV-NN mentioned above belong to batch-learning type. BP-TV is a time consuming algorithm as it involve many epochs (or iterations). Meanwhile, learning parameters such as learning rate, number of learning epochs have to be properly chosen to ensure convergence. As for ELM-based algorithms, whenever a new data arrived, the past data along with the new one have to be used for retraining, thus consuming lots of time or/and memory. With the advantage of no requirement for retraining whenever receiving a new data, on-line learning algorithms are preferred over batch ones in several engineering applications.

Based on [6], in this paper, we introduced an online extreme learning machine for time-varying neural networks referred as OL-ELM-TV, which can learn the training data one-by-one or chunk-by-chunk. The data which have already been

D.-S. Huang et al. (Eds.): ICIC 2011, LNBI 6840, pp. 49–54, 2012.

used in the training can be discarded, so as to save more memory and computational load to process the new coming data.

2   Extreme Learning Machine for Time-Varying Neural Networks

The time-varying version of Extreme Learning Machine has been studied in [3]. In a time-varying neural network, the input weights, or output weights, or both are changing through training and testing phases. Each weight can be expressed as a linear combination of a certain set of basis function: $w[n] = \sum_{b=1}^{B} f_b[n]w_b$, in which $f_b[n]$ is the known orthogonal function at n-th time instant of b-th order, w_b is the b-th order coefficient of the basic function to construct time-varying weight w_n, while B is the total number of the bases preset by user.

If time-varying input weights are introduced in a SLFN, hidden neuron output function can be written as: $h_k[n] = g\left(\sum_{i=0}^{I} x_i[n]w_{ik}[n]\right)$, where $w_{ik}[n] = \sum_{b=1}^{B} f_b[n]w_{b,ik}$. Similarly, if time-varying output weights are introduced, the standard output equation can be written as: $\sum_{k=1}^{K} h_k[n]\beta_{kl}[n] = t_l[n]$, where $\beta_{kl}[n] = \sum_{b=1}^{B} f_b[n]\beta_{b,kl}$.

To train a TV-NN, we randomly generate input weights $\{w_{b,ik}\}$ and compute hidden-layer output matrix $\mathbf{H}$. Further assuming $\mathbf{f}[n] = [f_1[n], f_2[n], \ldots, f_B[n]]^T \in \mathbb{R}^B$, $\mathbf{h}[n] = [h_1[n], h_2[n], \ldots, h_K[n]]^T \in \mathbb{R}^K$, $\boldsymbol{\beta}_{kl} = [\beta_{1,kl}, \beta_{2,kl}, \ldots, \beta_{B,kl}]^T \in \mathbb{R}^B$, $\boldsymbol{\omega}_{(l)} = [\boldsymbol{\beta}_{1l}^T, \boldsymbol{\beta}_{2l}^T, \ldots, \boldsymbol{\beta}_{Kl}^T] \in \mathbb{R}^{B\cdot K\times 1}$, $\mathbf{F} = [\mathbf{f}[1], \mathbf{f}[2], \ldots, \mathbf{f}[N]]^T \in \mathbb{R}^{N\times B}$, $\mathbf{H} = [\mathbf{h}[1], \mathbf{h}[2], \ldots, \mathbf{h}[N]]^T \in \mathbb{R}^{N\times K}$, $\boldsymbol{\Omega} = [\boldsymbol{\omega}_{(1)}, \boldsymbol{\omega}_{(2)}, \ldots, \boldsymbol{\omega}_{(L)}] \in \mathbb{R}^{B\cdot K\times L}$, and desire output matrix $\mathbf{T} = \{t_l[n]\} \in \mathbb{R}^{N\times L}$, according to [4], we have: $\mathbf{G} \cdot \boldsymbol{\Omega} = \mathbf{T}$, where $\mathbf{G} = \mathbf{H} * \mathbf{F}$, in which $*$ denotes the *Khatri-Rao product* of matrices $\mathbf{H}$ and $\mathbf{F}$, with $\mathbf{h}[n]^T$ and $\mathbf{f}[n]^T$ as their submatrices, respectively. Hence, the time-varying output weight matrix $\boldsymbol{\Omega}$ can be computed by:

$$\hat{\boldsymbol{\Omega}} = \mathbf{G}^\dagger \cdot \mathbf{T} \tag{1}$$

where $\mathbf{G}^\dagger$ is the MP inverse of matrix $\mathbf{G}$, and consequently, $\hat{\boldsymbol{\Omega}}$ is a set of optimal output weight parameters minimizing the training error.

3   On-Line ELM-TV

All the training data(N samples) have to be available before using the batch algorithm ELM-TV stated in previous section. However, in some applications, the neural networks may receive the taining data sequentially, or large amount of data to process in limited memory at the same time. Therefore, it is necessary to develop an online algorithm for TV-NN.

We assume here the number of samples N is large and the rank of time-varying hidden matrix $R(G) = K \cdot B$, the number of hidden nodes by the number of output bases, $\mathbf{G}^\dagger$ in (1) can be expressed as:

$$\mathbf{G}^\dagger = (\mathbf{G}^T\mathbf{G})^{-1}\mathbf{G}^T \tag{2}$$

Given the initial training set $\{(\mathbf{x}[n], \mathbf{t}[n])\}_{n=1}^{N_0}$ and the number of samples is larger than the number of hidden nodes by the number of output bases, e.g. $N_0 > K \cdot B$. Using batch ELM-TV, according to (1) we will have optimal output weight matrix:

$$\boldsymbol{\Omega}^{(0)} = \mathbf{G}_0^{\dagger} \cdot \mathbf{T}_0 = (\mathbf{G}_0^T \mathbf{G}_0)^{-1} \mathbf{G}_0^T \mathbf{T}_0 = \mathbf{C}_0^{-1} \mathbf{A}_0 \tag{3}$$

where $\mathbf{C}_0 = \mathbf{G}_0^T \mathbf{G}_0; \mathbf{A}_0 = \mathbf{G}_0^T \mathbf{T}_0;$

$$\mathbf{G}_0 = \begin{bmatrix} \mathbf{h}[1]^T \otimes \mathbf{f}[1]^T \\ \vdots \\ \mathbf{h}[N_0]^T \otimes \mathbf{f}[N_0]^T \end{bmatrix}_{N_0 \times K \cdot B} ; and \; \mathbf{T}_0 = \begin{bmatrix} \mathbf{t}[1]^T \\ \vdots \\ \mathbf{t}[N_0]^T \end{bmatrix}_{N_0 \times L} \tag{4}$$

Suppose another chunk of data $\{(\mathbf{x}[n], \mathbf{t}[n])\}_{n=N_0+1}^{N_0+N_1}$, the optimal output weight matrix has to be modified as:

$$\boldsymbol{\Omega}^{(1)} = \mathbf{C}_1^{-1} \mathbf{A}_1 \tag{5}$$

where

$$\mathbf{G}_1 = \begin{bmatrix} \mathbf{h}[N_0+1]^T \otimes \mathbf{f}[N_0+1]^T \\ \vdots \\ \mathbf{h}[N_0+N_1]^T \otimes \mathbf{f}[N_0+N_1]^T \end{bmatrix}_{N_1 \times K \cdot B} ; \mathbf{T}_1 = \begin{bmatrix} \mathbf{t}[N_0+1]^T \\ \vdots \\ \mathbf{t}[N_0+N_1]^T \end{bmatrix}_{N_1 \times L} \tag{6}$$

$$\mathbf{C}_1 = \begin{bmatrix} \mathbf{G}_0 \\ \mathbf{G}_1 \end{bmatrix}^T \begin{bmatrix} \mathbf{G}_0 \\ \mathbf{G}_1 \end{bmatrix} = \begin{bmatrix} \mathbf{G}_0^T \; \mathbf{G}_1^T \end{bmatrix} \begin{bmatrix} \mathbf{G}_0 \\ \mathbf{G}_1 \end{bmatrix} = \mathbf{C}_0 + \mathbf{G}_1^T \mathbf{G}_1 \tag{7}$$

$$\mathbf{A}_1 = \begin{bmatrix} \mathbf{G}_0 \\ \mathbf{G}_1 \end{bmatrix}^T \begin{bmatrix} \mathbf{T}_0 \\ \mathbf{T}_1 \end{bmatrix} = \begin{bmatrix} \mathbf{G}_0^T \; \mathbf{G}_1^T \end{bmatrix} \begin{bmatrix} \mathbf{T}_0 \\ \mathbf{T}_1 \end{bmatrix} = \mathbf{A}_0 + \mathbf{G}_1^T \mathbf{T}_1 \tag{8}$$

The aim here is to express $\boldsymbol{\Omega}^{(1)}$ as a function of $\boldsymbol{\Omega}^{(0)}, \mathbf{C}_1, \mathbf{G}_1$ and $\mathbf{T}_1$. According to (3) and (7), $\mathbf{A}_0$ can be written as:

$$\mathbf{A}_0 = \mathbf{C}_0 \mathbf{C}_0^{-1} \mathbf{A}_0 = (\mathbf{C}_1 - \mathbf{G}_1^T \mathbf{G}_1) \boldsymbol{\Omega}^{(0)} = \mathbf{C}_1 \boldsymbol{\Omega}^{(0)} - \mathbf{G}_1^T \mathbf{G}_1 \boldsymbol{\Omega}^{(0)} \tag{9}$$

Combining (5),(7),(8), and (9), we have:

$$\boldsymbol{\Omega}^{(1)} = \mathbf{C}_1^{-1} (\mathbf{C}_1 \boldsymbol{\Omega}^{(0)} - \mathbf{G}_1^T \mathbf{G}_1 \boldsymbol{\Omega}^{(0)} + \mathbf{G}_1^T \mathbf{T}_1) = \boldsymbol{\Omega}^{(0)} + \mathbf{C}_1^{-1} \mathbf{G}_1^T (\mathbf{T}_1 - \mathbf{G}_1 \boldsymbol{\Omega}^{(0)}) \tag{10}$$

Since $\mathbf{C}_1^{-1}$ is used for computing $\boldsymbol{\Omega}^{(1)}$. Setting $\mathbf{P}_0 = \mathbf{C}_0^{-1}$ and $\mathbf{P}_1 = \mathbf{C}_1^{-1}$, based on Woodbury formula, we have:

$$\mathbf{P}_1 = (\mathbf{C}_0 + \mathbf{G}_1^T \mathbf{G}_1)^{-1} = \mathbf{P}_0 - \mathbf{P}_0 \mathbf{G}_1^T (\mathbf{I} + \mathbf{G}_1 \mathbf{P}_0 \mathbf{G}_1^T)^{-1} \mathbf{G}_1 \mathbf{P}_0 \tag{11}$$

Generalizing the previous arguments, when m-th chunk of data set arrives, we have:

$$\mathbf{P}_m = \mathbf{P}_{m-1} - \mathbf{P}_{m-1} \mathbf{G}_m^T (\mathbf{I} + \mathbf{G}_m \mathbf{P}_{m-1} \mathbf{G}_m^T)^{-1} \mathbf{G}_m \mathbf{P}_{m-1} \tag{12}$$

$$\boldsymbol{\Omega}^{(m)} = \boldsymbol{\Omega}^{(m-1)} + \mathbf{P}_m \mathbf{G}_m^T (\mathbf{T}_m - \mathbf{G}_m \boldsymbol{\Omega}^{(m-1)}) \tag{13}$$

When the training data is received one-by-one instead of chunk-by-chunk, e.g. $N_m = 1$, the above fomula have the following simple format:

$$\mathbf{P}_m = \mathbf{P}_{m-1} - \frac{\mathbf{P}_{m-1}\mathbf{g}_m\mathbf{g}_m^T\mathbf{P}_{m-1}}{1 + \mathbf{g}_m^T\mathbf{P}_{m-1}\mathbf{g}_m} \tag{14}$$

$$\mathbf{\Omega}^{(m)} = \mathbf{\Omega}^{(m-1)} + \mathbf{P}_m\mathbf{g}_m(\mathbf{t}_m^T - \mathbf{g}_m^T\mathbf{\Omega}^{(m-1)}) \tag{15}$$

where $\mathbf{g}_m = (\mathbf{h}_m^T \otimes \mathbf{f}_m^T)^T$

Now, the OL-ELM can be summarized as follows. Assume that we have a single hidden layer time-varying neural network to be trained, with K hidden nodes and B output bases functions, and receiving the training data set $\{(\mathbf{x}[n], \mathbf{t}[n])\}$ sequentially. We thus attain Algorithm 1.

Algorithm 1. OL-ELM

1: Randomly generate the input weights set $\{\omega_{b,ik}\}$ and choose a set of basis funtions.

2: Accumulate N_0 samples of training data (make sure $N_0 > K \cdot B$).
3: Calculate the time-varying hidden layer output matrix $\mathbf{G}_0$ by (4)
4: Calculate $\mathbf{P}_0 = (\mathbf{G}_0^T\mathbf{G}_0)^{-1}$
5: Calculate the initial output weight matrix $\mathbf{\Omega}^{(0)}$ by (3)
6: **for** $m = 1$ to M **do**
7:     When the m-th chunk of data arrive, calculate the time-varying partial hidden layer output matrix $\mathbf{G}_m$.
8:     Update the output weight matrix $\mathbf{\Omega}^{(m)}$ by,

$$\mathbf{P}_m = \mathbf{P}_{m-1} - \mathbf{P}_{m-1}\mathbf{G}_m^T(\mathbf{I} + \mathbf{G}_m\mathbf{P}_{m-1}\mathbf{G}_m^T)^{-1}\mathbf{G}_m\mathbf{P}_{m-1} \tag{16}$$

$$\mathbf{\Omega}^{(m)} = \mathbf{\Omega}^{(m-1)} + \mathbf{P}_m\mathbf{G}_m^T(\mathbf{T}_m - \mathbf{G}_m\mathbf{\Omega}^{(m-1)}) \tag{17}$$

9:     $m \leftarrow m + 1$
10: **end for**

---

4   Simulation Results

In this section, we compare OL-ELM-TV (in one-by-one learning mode) with ELM-TV in three identification problems. The polynomial activation function $(g(x) = x^2 + x)$ is used in TV-NN networks. All simulations have been accomplished by means of the MATLAB 7.8.0 programming environment, running on an Intel Core2 Duo CPU P8400 2.26GHz, with Windows Vista OS.

4.1   Time-Varying MLP System Identification

The system to be identified here is the same to that in [1] and [3]: a time-varying IIR-buffered MLP with 11 input lines, one 5 neurons hidden layer, one neuron output layer. The input weights and output weights are combination of 3 Chebyshev bases functions; the lengths of the input and output Time Delay

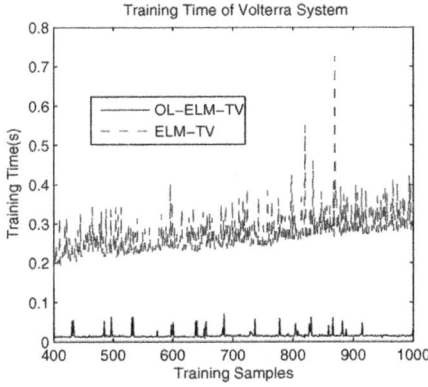 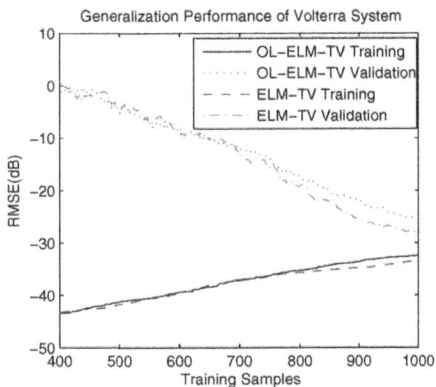

Fig. 1. Comparison performances of OL-ELM-TV and ELM-TV for MLP system with parameter settings as: Legendre basis functions, 3 input bases, 3 output bases and 100 hidden nodes

Lines (TDLs) are equal to 6 and 5 respectively. Note that the output neuron of this system is not linear, both the hidden neurons and output neuron use tangent sigmoid activation function.

4.2   Time-Varying Narendra System Identification

The next test identification system is a modified version of the one addressed in [7], by adding the coefficients $a[n]$ and $b[n]$, which are low pass filtered versions of a random sequence, to form a time-varying system, as done in [1] and [3]:

$$y[n] = a[n] \cdot \frac{y[n-1] \cdot y[n-2] \cdot y[n-3] \cdot x[n-1] \cdot (y[n-3]-1) + x[n]}{1 + b[n] \cdot (y[n-3]^2 + y[n-2]^2)} \quad (18)$$

The model used for this identification task is a time-varying IIR-buffered neural network with 5 input lines, where 3 of them are produced by the input TDL and the remaining 2 by the output feedback TDL.

The results of these identification tasks consistently show that OL-ELM-TV has comparable generalization performance to ELM-TV. Due to the space constraint, only the result of MLP identification task is depicted here in Fig. 1. Note that as number of training samples increases, the training RMSE become worse, because more training samples means more rows in $\mathbf{G}$ and $\mathbf{T}$, hence more equations to be satisfied in $\mathbf{G} \cdot \mathbf{\Omega} = \mathbf{T}$, and thus results in increasing dimensions of the term $\|\mathbf{G\Omega} - \mathbf{T}\|$. However, the validation RMSE turns out to be better as expected: the more the network is trained (if no overfitting occurs), the better it can identifies the target systems.

On the other hand, OL-ELM consumes much less training time when updating the output weights in the applications in which the training data set arrived sequentially. As shown in Fig. 1, once the TV-NN receives a new training sample, OL-ELM-TV takes less than 0.01 seconds to update its output weights, while ELM-TV takes more than 0.06 seconds to retrain its network.

5 Conclusions

In this paper, we have extended the online version of extreme learning machine to time-varying neural networks, namely OL-ELM-TV. The proposed algorithm is oriented to the applications with sequential arrival or large number of training data set, the main advantage of the online approach consists in updating the output weights with much less time when new training data arrived, or consuming less memory if large number of samples have to be trained. Simulation results showed that with the benefits discussed above, our OL-ELM-TV achieved comparable generalization performances to ELM-TV. It has also to be observed that the OL-ELM-TV behavior perfectly matches with that one of the original OS-ELM [6], working in stationary environments, allowing us to positively conclude about the effectiveness of the proposed generalization to the time-varying neural network case study. Future works are intended to apply the promising ELM-based approach for TV-NNs to real-life problems.

References

1. Titti, A., Squartini, S., Piazza, F.: A New Time-variant Neural Based Approach for Nonstationary and Non-linear System Identification. In: Proc. IEEE International Symposium on Circuits and Systems, ISCAS, pp. 5134–5137 (2005)
2. Huang, G.B., Zhu, Q.Y., Siew, C.K.: Extreme Learning Machine: Theory and Applications. Neurocomputing 70(1-3), 489–501 (2006); Neural Networks - Selected Papers from the 7th Brazilian Symposium on Neural Networks (SBRN 2004) (2006)
3. Cingolani, C., Squartini, S., Piazza, F.: An Extreme Learning Machine Approach for Training Time Variant Neural Networks. In: Proc. IEEE Asia Pacific Conference on Circuits and Systems APCCAS, pp. 384–387 (2008)
4. Ye, Y., Squartini, S., Piazza, F.: Incremental-based Extreme Learning Machine Algorithms for Time-variant Neural Networks. In: Advanced Intelligent Computing Theories and Applications, pp. 9–16 (2010)
5. Ye, Y., Squartini, S., Piazza, F.: ELM-Based Time-Variant Neural Networks with Incremental Number of Output Basis Functions. In: Liu, D., Zhang, H., Polycarpou, M., Alippi, C., He, H. (eds.) ISNN 2011, Part I. LNCS, vol. 6675, pp. 403–410. Springer, Heidelberg (2011)
6. Liang, N.Y., Huang, G.B., Saratchandran, P., Sundararajan, N.: A Fast and Accurate Online Sequential Learning Algorithm for Feedforward Networks 17(6), 1411–1423 (2006)
7. Narendra, K., Parthasarathy, K.: Identification and Control of Dynamical Systems Using Neural Networks. IEEE Transactions on Neural Networks 1(1), 4–27 (1990)

Oscillatory Behavior for a Class of Recurrent Neural Networks with Time-Varying Input and Delays[*]

Chunhua Feng[1] and Zhenkun Huang[2]

[1] College of Mathematical Science, Guangxi Normal University,
Guilin, Guangxi, 541004, P.R. China
[2] Department of Mathematics, Jimei University, Xiamen, 350211, P.R. China
chfeng@mailbox.gxnu.edu.cn

Abstract. In this paper, the existence of oscillations for a recurrent neural network with time delays between neural interconnections is investigated. Several simple and practical criteria to determine the oscillatory behavior are obtained.

Keywords: Recurrent neural network, time-varying input, delay, oscillation.

1 Introduction

In [9], Hu and Liu have studied the stability of a class of continuous-time recurrent neural networks (RNNs) defined by the following model:

$$x_i'(t) = \sum_{j=1}^{n} w_{ij} g_j(x_j(t)) + u_i(t), \quad x_i(0) = x_{i0}, \quad i = 1, 2, \cdots, n. \tag{1}$$

The equivalent form of (1) in matrix format is given by

$$x'(t) = Wg(x(t)) + u(t), \quad x(0) = x_0. \tag{2}$$

where $x = [x_1, x_2, \cdots x_n]^T$ is the state vector, $W = (w_{ij})_{n \times n}$ is a constant connection weight matrix, $u(t) = [u_1(t), u_2(t), \cdots, u_n(t)]^T$ is a continuous input vector function which is called the time-varying input, $g(x) = [g_1(x_1), g_2(x_2), \cdots, g_n(x_n)]^T$ is a nonlinear vector-valued activation function. Assume that time-varying input $u_i(t)$ tend to constants u_i as t tends to infinity, and $g(\cdot)$ belongs to the class of globally Lipschitz continuous and monotone increasing activation functions. The authors established two sufficient conditions for global output convergence of this class of neural networks.

It is known that time-varying inputs $u(t)$ can drive quickly $x(t)$ to some desired region of the activation space [7]. Apart from discussing of convergence of neural

---

[*] Supported by NNSF of China (10961005).

D.-S. Huang et al. (Eds.): ICIC 2011, LNBI 6840, pp. 55–63, 2012.

network models, their oscillatory behaviors have also been exploited in many applications. For example, combustion-instability control [4], robotic control [10], the design of associative memory [13], sleep and walking oscillation modeling [16], wall oscillation in wall-bounded turbulent flows [15], oscillations of brain activity [14], mixed-mode oscillations in chemistry, physics and neuroscience [2], heat flow oscillation [17], oscillation in a network model of neocortex [3], neuronal population oscillations during epileptic seizures [11], oscillation in population model [18]. biochemical oscillations [5], oscillations in power systems [12]. In reality, time-delay neural networks are frequently encountered in various areas, and a time delay is often a source of instability and oscillations in a system. This is due to the finite switching speed of amplifiers in electronic neural networks or to the finite signal propagation time in biological networks. Therefore, in this paper, we discuss the oscillatory behavior of the time delay neural network with the time-varying input as follows:

$$x_i'(t) = \sum_{j=1}^{n} w_{ij} g_j(x_j(t - \tau_j)) + u_i(t), \ x_i(0) = x_{i0}, \ i = 1, 2, \cdots, n. \tag{3}$$

where w_{ij}, τ_j are constants, in which n corresponds to the number of units in the networks, $x_i(t) \, (i = 1, 2, \cdots, n)$ correspond to the state vectors of the ith neural unit at time t, w_{ij} are the synaptic connection strengths, $\tau_j > 0 \, (j = 1, 2, \cdots, n)$ represent delays. The equivalent form of (3) in matrix format is given by

$$x'(t) = W g(x(t - \tau)) + u(t), \quad x(0) = x_0 . x + y = z . \tag{4}$$

where $x(t - \tau) = [x_1(t - \tau_1), x_2(t - \tau_2), \cdots, x_n(t - \tau_n)]^T$. We assume that the function $g_j(x(t - \tau_j)) \, (j = 1, 2, \cdots, n)$ in (3) belong to the class of Lipschitz continuous and monotone activation functions; that is, for each g_j, there exist $k_j > 0$ such that

$$0 < [g_j(x_j) - g_j(y_j)]/(x_j - y_j) \le k_j \quad (j = 1, 2, \cdots, n). \tag{5}$$

It should be note that such activation functions may be unbounded. There are many frequently used activation functions that satisfy this condition, for example, tanh(x), $0.5(|x+1| - |x-1|)$, max(0, x), and so on. We also assume that the time-varying input u_i satisfy the following conditions

$$\lim_{t \to \infty} u_i(t) = u_i . \tag{6}$$

where $u_i \, (i = 1, 2, \cdots, n)$ are constants, i.e. we assume that $\lim_{t \to \infty} u(t) = u$.

Definition 1. A solution of system (3) is called oscillatory if the solution is neither eventually positive nor eventually negative. If $x(t) = [x_1(t), x_2(t), \cdots x_n(t)]^T$ is an oscillatory solution of system (3), then each component of $x(t)$ is oscillatory.

2   Preliminary

Lemma 1 (Lemma 2 in [9]). If (6) is satisfied and there exists a constant vector $x^* = [x_1^*, x_2^*, \cdots, x_n^*]^T$ such that $Wg(x^*) + u = 0$, then given any $x_0 \in R^n$, system (3) has a unique solution $x = (t, x_0)$ defined on $[0, +\infty)$, where $W = (w_{ij})_{n \times n}$.

Thus, under the conditions of Lemma 1 hold, if we set

$$\tilde{u}(t) = u(t) - u; \quad z(t) = [z_1(t), z_2(t), \cdots, z_n(t)]^T = x - x^*. \tag{7}$$

then system (3) can be transformed to the following equivalent system

$$z'(t) = Wf(z(t - \tau)) + \tilde{u}(t), \quad z(0) = x_0 - x^*. \tag{8}$$

where $f(z) = g(z + x^*) - g(x^*)$ satisfies that $f(0) = 0$.

Lemma 2. If the matrix W is a nonsingular matrix, then system (8) has a unique equilibrium point.

Proof. An equilibrium point $z^* = [z_1^*, z_2^*, \cdots, z_n^*]^T$ is a constant solution of the following algebraic equation

$$Wf(z^*) + \tilde{u}^* = 0. \tag{9}$$

Noting that (9) holds for any t, from (6) and the definition of $\tilde{u}(t) = u(t) - u$, if t is sufficiently large, it implies that $\tilde{u}^* = 0$. Meanwhile W is a nonsingular matrix, from (9) we get. $f(z^*) = 0$. Noting that $f(z)$ is a continuous monotone activation function, there exists only one z^* such that $f(z^*) = 0$. Since $f(0) = 0$, so the unique equilibrium point z^* is exactly 0. Since system (8) is an equivalent system of (3), hence, the oscillatory behavior of system (8) implies that the oscillatory behavior of system (3). Therefore, in the following we only deal with the oscillatory behavior of the unique equilibrium point of system (8). Noting that $f(0) = 0$. and is a continuous monotone activation function, then in a sufficiently small neighborhood of zero point, there exist $0 < \beta_j \leq k_j$ $(j = 1, 2, \cdots, n)$, a sufficiently small constant $\varepsilon (0 < \varepsilon \ll 1)$ such that

$$(\beta_j - \varepsilon)z_j(t - \tau_j) \leq f_j(z_j(t - \tau_j)) \leq (\beta_j + \varepsilon)z_j(t - \tau_j) \ (j = 1, 2, \cdots, n)$$

hold. According to [1, 6], the oscillatory behavior of system (8) about equilibrium point equals to the oscillation of the following system about zero point:

$$z_i'(t) = \sum_{j=1}^{n} (\beta_j \pm \varepsilon)w_{ij}z_j(t - \tau_j) + \tilde{u}_i(t), \quad i = 1, 2, \cdots, n. \tag{10}$$

3   Oscillation Analysis

Theorem 1. Suppose that the matrix W is a nonsingular matrix and system (8) has a unique equilibrium point. Let $\rho_1, \rho_2, \cdots, \rho_n$ represent the characteristic values of matrix $B = (\beta_j w_{ij})_{n \times n}$. Assume that there exists at least an ρ_j such that $\rho_j > 0$, $j \in \{1, 2, \cdots, n\}$. Then the unique equilibrium point of system (8) is unstable, and system (8) has a oscillatory solution.

Proof. Since ε is a sufficiently small constant, the oscillatory behavior of system (10) is the same as the following

$$z_i'(t) = \sum_{j=1}^{n} \beta_j w_{ij} z_j (t - \tau_j) + \tilde{u}_i(t), \quad i = 1, 2, \cdots, n. \tag{11}$$

From condition (6), when t is sufficiently large, $u_i(t) = u_i \pm \delta_i$, where δ_i are sufficiently small positive constants. This means that when t is sufficiently large, $\tilde{u}_i(t) = u_i \pm \delta_i - u_i = \pm \delta_i$. Therefore, if we neglect $\tilde{u}_i(t)$ in system (11), and get the following system:

$$z_i'(t) = \sum_{j=1}^{n} \beta_j w_{ij} z_j (t - \tau_j), \quad i = 1, 2, \cdots, n. \tag{12}$$

We claim that system (12) has an oscillatory solution implies that system (11) has an oscillatory solution. Because the unique equilibrium point $z^* = [z_1^*, z_2^*, \cdots, z_n^*]^T$ in system (12) is unstable. Then for the sufficiently small positive constant ε, $z^* \pm \varepsilon$ is still neither eventually positive nor eventually negative. Since $\rho_1, \rho_2, \cdots, \rho_n$ are characteristic values of B, the characteristic equation of (12) can be written as $\prod_{i=1}^{n} (\lambda - \rho_i e^{-\lambda \tau_i}) = 0$. This means that for each i we get

$$\lambda - \rho_i e^{-\lambda \tau_i} = 0, i = 1, 2, \cdots, n. \tag{13}$$

There is at least one characteristic value of (13) larger than zero. Noting that the characteristic equation (13) is a transcendental equation, the characteristic values may be complex numbers. However, there still exists a real positive root under our assumptions. Since for some $j \in \{1, 2, \cdots, n\}$, $\rho_j > 0$, we consider function of λ for this j as follows:

$$h(\lambda) = \lambda - \rho_j e^{-\lambda \tau_j}. \tag{14}$$

Obviously, $h(\lambda)$ is a continuous function of λ, and $h(0) = -\rho_j < 0$. Noting that delay $\tau_j > 0$. Then $e^{-\lambda \tau_j}$ tends to zero as λ tends to positive infinity. Therefore, there exists a suitable large $\lambda^* > 0$ such that $h(\lambda^*) = \lambda^* - \rho_j e^{-\lambda^* \tau_j} > 0$. By means of the mean value theorem, there exists an $\overline{\lambda} \in (0, \lambda^*)$ such that $h(\overline{\lambda}) = 0$. In other words, $\overline{\lambda}$ is a positive characteristic value of system (14). This means that equation (13) has a positive characteristic value. Therefore, the unique equilibrium point of (12) is unstable, which implies that system (8) has an oscillatory solution.

Theorem 2. If the matrix W is a nonsingular matrix and system (8) has a unique equilibrium point. Suppose that the following conditions hold:

$$\beta_i w_{ii} < 0, \quad \max_{1 \le i \le n}(\beta_i w_{ii} + \sum_{j=1, j \ne i} |\beta_j w_{ij}|) = -\delta < 0, \quad \delta \tau e > 1 \tag{15}$$

where $\tau = \min\{\tau_1, \tau_2, \cdots \tau_n\}$. Then system (8) has an oscillatory solution.

Proof. From system (12), assume, for the sake of contradiction, there exists a t^* such that for $t > t^*$, we always have $|z_i(t)| > 0 \ (i = 1, 2, \cdots, n)$. Noting that $\beta_i w_{ii} < 0$, then for

$t > t^*$ and $i = 1, 2, \cdots, n$, we get

$$|z_i'(t)| \le -\beta_i w_{ii} |z_i(t - \tau_i)| + \sum_{j=1, j \ne i} |\beta_j w_{ij}| \cdot |z_j(t - \tau_j)| \tag{16}$$

therefore,

$$\sum_{i=1}^n |z_i'(t)| \le -\delta \sum_{i=1}^n |z_i(t - \tau_i)| \tag{17}$$

It is easily to see from (17) that $\lim_{t \to \infty} z_i(t) = 0$. Otherwise, suppose that $\lim_{t \to \infty} z_i(t) = c > 0$. Then there exists a sufficiently large $t_0 (> \tau_i)$ such that when $t > t_0 - \tau_i$ we have $z_i(t) > 0.5c$. By integrating both sides of (17) from t_0 to t, we get

$$\sum_{i=1}^n (|z_i(t)| - |z_i(t_0)|) \le -\delta \int_{t_0}^t \sum_{i=1}^n |z_i(s - \tau_i)| ds = -\delta \int_{t_0-\tau_i}^{t-\tau_i} \sum_{i=1}^n |z_i(s)| ds \tag{18}$$

namely,

$$\sum_{i=1}^n |z_i(t)| + \delta \int_{t_0-\tau_i}^{t-\tau_i} n \cdot \frac{c}{2} ds \le \sum_{i=1}^n |z_i(t_0)| \tag{19}$$

Noting that $\delta \int_{t_0-\tau_i}^{t-\tau_i} n \cdot \frac{c}{2} ds \to \infty$ as $t \to \infty$, and the right hand of (19) is a constant. This contradiction implies that $\lim_{t\to\infty} z_i(t) = 0$. Again integrating both sides of (17) from $t_0(>\tau_i)$ to $+\infty$, we get

$$0 - \sum_{i=1}^{n} |z_i(t)| \le -\delta \int_{t}^{+\infty} \sum_{i=1}^{n} |z_i(s-\tau_i)| ds \tag{20}$$

Namely, $\sum_{i=1}^{n} |z_i(t)| \ge \delta \int_{t}^{+\infty} \sum_{i=1}^{n} |z_i(s-\tau_i)| ds$. Set $\tau = \min\{\tau_1, \tau_2, \cdots \tau_n\}$, then for each τ_i we have $t - \tau \ge t - \tau_i$, hence

$$\sum_{i=1}^{n} |z_i(t)| \ge \delta \int_{t}^{+\infty} \sum_{i=1}^{n} |z_i(s-\tau_i)| ds \ge \delta \int_{t-\tau}^{+\infty} \sum_{i=1}^{n} |z_i(s)| ds \tag{21}$$

Set $y(t) = \sum_{i=1}^{n} |z_i(t)|$, then $y(t) \ge 0$ and from (21) we get $y(t) \ge \delta \int_{t-\tau}^{+\infty} y(s)ds$.

Define a sequence as follows:

$$\xi_0(t) = y(t), \quad \xi_{k+1}(t) = \begin{cases} y(t) - y(T) + \delta \int_{t-\tau}^{+\infty} \xi_k(s)ds, & t \le T, \\[2ex] \delta \int_{t-\tau}^{+\infty} \xi_k(s)ds, & t > T. \end{cases} \tag{22}$$

For $t > T$ by induction we can easily to see that $\xi_0(t) \ge \xi_1(t) \ge \cdots \ge \xi_k(t) \ge \cdots \ge 0$. Therefore, $\lim_{k\to\infty} \xi_k(t) = \xi$ exists, and ξ is an eventually positive bounded solution of the following equation:

$$y'(t) = -\delta y(t-\tau), \ t > T. \tag{23}$$

Thus, the characteristic equation of (23) has a real root which is negative. From $\lambda = -\delta e^{-\lambda\tau}$ has a negative root, so that $-\lambda > 0$, then using formula of $e^x \ge ex$, $(x > 0)$, we obtain that $1 = \frac{\delta \tau e^{-\lambda\tau}}{-\lambda\tau} \ge \frac{\delta \tau e(-\lambda\tau)}{-\lambda\tau} = \delta \tau e$, which contradicts assumption (15). The result follows.

4   Simulation Results

Example 1. Consider the following three-neuron system

$$\begin{cases} x_1' = -18g(x_1) + 0.2g(x_2) - 1.2g(x_3) + 0.5t/(1+t), \\ x_2' = -2g(x_1) - 6.4g(x_2) + 0.1g(x_3) + 1.5t/(1+t), \\ x_3' = -0.24g(x_1) + 6g(x_2) - 12g(x_3) + 2.5t/(1+t). \end{cases} \qquad (24)$$

where $g(x_i) = g(x_i(t - \tau_i)), i = 1, 2, 3.$ Let $g(x) = 0.5 \times (|x+1| - |x-1|).$
For function $g(x),$ we can select $\beta_1 = \beta_2 = \beta_3 = 1,$ then $\beta_1 w_{11} = -18,$
$\beta_2 w_{22} = -6.4,$ $\beta_3 w_{33} = -12.$ $\max\limits_{1 \le i \le 3}(\beta_i w_{ii} + \sum\limits_{j=1, j \ne i} |\beta_j w_{ij}|) = -\delta = -4.3.$
From (24), $u = (0.5,\ 1.5,\ 2.5).$ When we select $\tau_1 = 1,\ \tau_2 = 2,\ \tau_3 = 4,$ and
$\tau_1 = 3,\ \tau_2 = 6,\ \tau_3 = 12,$ respectively, the conditions of Theorem 2 are satisfied.
System (24) generates oscillations (Fig.1A and Fig.1B).

Example 2. Consider the following four-neuron system

$$\begin{cases} x_1' = -6.8g(x_1) + 0.2g(x_2) - 0.2f(x_3) + 0.5f(x_4) + u_1(t) \\ x_2' = 2.4g(x_1) - 8.5g(x_2) + 0.2f(x_3) - 0.1f(x_4) + u_2(t) \\ x_3' = 0.25g(x_1) + 0.3g(x_2) - 3.8f(x_3) - 1.2f(x_4) + u_3(t) \\ x_4' = 1.6g(x_1) + 0.2g(x_2) - 0.2f(x_3) - 4.2f(x_4) + u_4(t) \end{cases} \qquad (25)$$

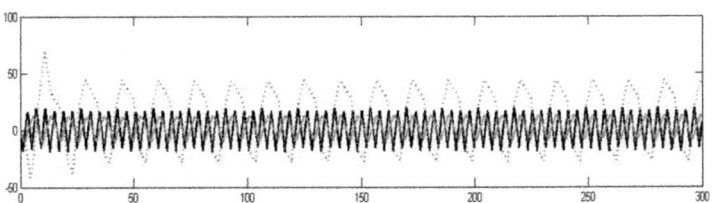

Fig. 1A. Oscillation of the equilibrium point, u=(0.5, 1.5, 2.5),
Solid line $X_1(t)$, dashed line: $X_2(t)$, dotted line: $X_3(t)$, delays: 1, 2, 4.

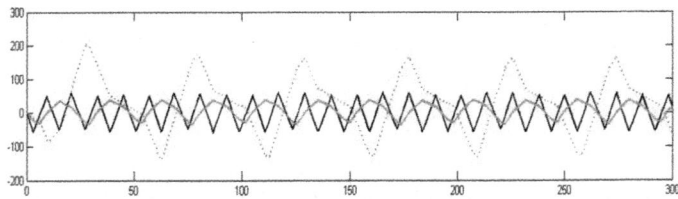

Fig. 1B. Oscillation of the equilibrium point, u=(0.5, 1.5, 2.5),
Solid line $X_1(t)$, dashed line: $X_2(t)$, dotted line: $X_3(t)$, delays: 3, 6, 12.

where $g(x) = \arctan(x), f(x) = \tanh(x)$, we can select $\beta_1 = \beta_2 = \beta_3 = \beta_4 = 1$, then the characteristic values of $B = (\beta_j w_{ij})_{4 \times 4}$ are 0.2020, -3.9043, -6.8447, -8.7530. It is known that a characteristic value 0.2020>0. When $u_1(t), u_2(t), u_3(t), u_4(t)$ are taken the values and $(2t^2/(1+t^2), 4t^2/(1+t^2), 6t^2/(1+t^2), 4t^2/(1+t^2))$ respectively, we select $\tau_1 = 1, \tau_2 = 2, \tau_3 = 2, \tau_4 = 1$, the conditions of Theorem 1 are satisfied. System (25) generates oscillations (Fig.2A and Fig.2B).

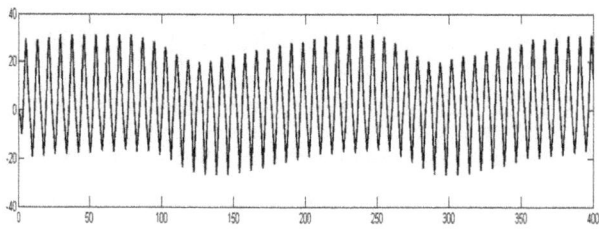

Fig. 2A. Oscillation of the second neuron output,
u= (2, 0, 2, 0), delays: 1, 2, 2, 1.

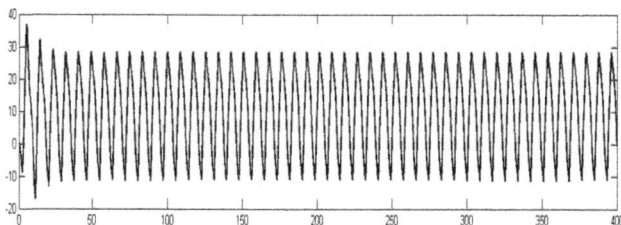

Fig. 2B. Oscillation of the second neuron output,
u= (2, 4, 6, 5), delays: 1, 2, 2, 1.

5   Conclusion

This paper discusses the oscillatory behavior for a recurrent neural network with time delays between neural interconnections and time-varying input. By using continuous and monotone increasing activation function and continuous time-varying inputs, two criteria to guarantee oscillations of global output have been proposed.

References

1. Arino, O., Gyori, I.: Necessary and Sufficient Condition for Oscillation of a Neutral Differential System with Several Delays. J. Diff. Eqns. 81, 98–105 (1989)
2. Curtu, R.: Singular Hopf Bifurcations and Mixed-mode Oscillations in a Two-cell Inhibitory Neural Network. Physica D 239, 504–514 (2010)

3. Dwyer, J., Lee, H., Martell, A., Stevens, R., Hereld, M., Drongelen, W.W.: Oscillation in a Network Model of Neocortex. Neurocomputing 73, 1051–1056 (2010)

4. Fichera, A., Pagano, A.: Application of Neural Dynamic Optimization to Combustion-instability Control. Applied Energy 83, 253–264 (2006)

5. Goldstein, B.N., Aksirov, A.M., Zakrjevskaya, D.T.: A New Kinetic Model for Biochemical Oscillations: Graph-theoretical Analysis. Biophysical Chemistry 145, 111–115 (2009)

6. Gyori, I., Ladas, G.: Oscillation Theory of Delay Differential Equations with Applications. Clarendon Press, Oxford (1991)

7. Hirsch, M.W.: Convergent Activation Dynamics in Continuous Time Networks. Neural Networks 2, 331–349 (1989)

8. Horn, R.C., Johnson, C.R.: Matrix Analysis, pp. 345–348. Cambridge University Press, Cambridge (1985)

9. Hu, S.Q., Liu, D.R.: On the Global Output Convergence of a Class of Recurrent Neural Networks with Time-varying Inputs. Neural Networks 18, 171–178 (2005)

10. Jin, H., Zacksenhouse, M.: Oscillatory Neural Networks for Robotic yo-yo Controll. IEEE Trans. Neural Networks 14, 317–325 (2003)

11. Li, X.X., Jefferys, J.G., Fox, J., Yao, X.: Neuronal Population Oscillations of Rat Hippocampus during Epileptic Seizures. Neural Networks 21, 1105–1111 (2008)

12. Naggar, K.M.: Online Measurement of Low-frequency Oscillations in Power Systems. Measurement 42, 716–721 (2009)

13. Nishikawa, T., Lai, Y.C., Hoppensteadt, F.C.: Capacity of Oscillatory Associative-memory Networks with Error-free Retrieval. Phy. Rev. Lett. 92, 108101 (2004)

14. Picchioni, D., Horovitz, S.G., Fukunag, M., Carr, W.S., Meltzer, J.A., Balkin, T.J., Duyn, H., Braun, A.R.: Infraslow EEG Oscillations Organize Large-scale Cortical–subcortical Interactions during Sleep: A Combined EEG/fMRI Study. Brainresearch 1374, 63–72 (2011)

15. Ricco, P., Quadrio, M.: Wall-oscillation Conditions for Drag Reduction in Turbulent Channel Flow. Inter. J. of Heat and Fluid Flow 29, 601–612 (2008)

16. Steriade, M., Timofeev, I.: Neuronal Plasticity in Thalamocortical Networks during Sleep and Waking Oscillations. Neuron 37, 563–576 (2003)

17. Su, G.H., Morita, K., Fukuda, K., Pidduck, M., Jia, D.N., Miettinen, J.: Analysis of the Critical Heat Flux in Round Vertical Tubes under Low Pressure and Flow Oscillation Conditions. Nuclear Engin. Design 220, 17–35 (2003)

18. Wang, C., Wang, S.S., Yan, X.P., Li, L.R.: Oscillation of a Cass of Partial Functional Population Model. J. Math. Anal. Appl. 368, 32–42 (2010)

A Saturation Binary Neural Network for Bipartite Subgraph Problem

Cui Zhang[1], Li-Qing Zhao[2], and Rong-Long Wang[2]

[1] Department of Autocontrol, Liaoning Institute of Science and Technology,
Benxi, China
[2] Graduate School of Engineering, University of Fukui, Bunkyo 3-9-1,
Fukui-shi, Japan
bxlkyzhangcui@163.com, nkzlq@hotmail.com, wang@u-fukui.ac.jp

Abstract. In this paper, we propose a saturation binary neuron model and use it to construct a Hopfield-type neural network called saturation binary neural network to solve the bipartite sub-graph problem. A large number of instances have been simulated to verify the proposed algorithm, with the simulation result showing that our algorithm finds the solution quality is superior to the compared algorithms.

Keywords: Saturation binary neuron model, Saturation binary neural network, Combinatorial optimization problems, Bipartite sub-graph problem.

1 Introduction

The objective of the bipartite subgraph problem[1] is to find a bipartite subgraph with maximum number of edges of a given graph. It was proved by Garey, Johnson, and Stockmeyer [1], Karp [2], and Even and Shiloach [3] that the problem of finding a bipartite subgraph of a given graph with the maximum number of edges is NP-complete. Thus, the efficient determination of "maximum" bipartite subgraphs is a question of both practical and theoretical interest. Because efficient algorithms for this NP-complete combinatorial optimal problem are unlikely to exist, the bipartite subgraph problem has been widely studied by many researchers on some special classes of graphs. An algorithm for solving the largest bipartite subgraphs in triangle-free graph with maximum degree three has been propossed for practical purpose [4]. Grotschel and Pulleyblank [5] defined a class of weakly bipartite graphs. Barahona [6] characterized another class of weakly bipartite graphs. For solving such combinatorial optimal problems, Hopfield neural networks [7]-[11] constitute an important avenue. These networks contain many simple computing elements (or artificial neurons), which cooperatively traverse the energy surface to find a local or global minimum. The simplicity of the neurons makes it promising to build them in large numbers to achieve high computing speeds by way of massive parallelism. Using the Hopfield neural network, Lee, Funabiki and Takefuji [12] presented a parallel algorithm for bipartite subgraph problem. However with the Hopfield network, the state of the system is forced to converge to local minimum and the rate

D.-S. Huang et al. (Eds.): ICIC 2011, LNBI 6840, pp. 64–70, 2012.
© Springer-Verlag Berlin Heidelberg 2012

to get the maximum bipartite subgraph is very low. Global search methods such as simulated annealing can be applied to such problem [13], but they are generally very slow [14]. No tractable algorithm is known for solving the bipartite subgraph problem.

In this paper we propose a saturation binary neuron model and use it to construct a Hopfield-type neural network for efficiently solving the bipartite sub-graph problems. In the proposed saturation binary neuron model, once the neuron is in excitatory state, then its input potential is in positive saturation where the input potential can only be reduced but cannot be increased, and once the neuron is in inhibitory state, then its input potential is in negative saturation where the input potential can only be increased but cannot be reduced. Using the saturation binary neuron model, a saturation binary neural network is constructed to solve the bipartite subgraph problems. The simulation results show that the saturation binary neural network can find better solutions than the other method, for example the Lee et al.'s algorithm.

2   Saturation Binary Neural Network

In the proposed algorithm, a novel neuron updating rule is proposed. To avoid the neural network entering a local minimum stagnation, the neuron network is updated according to the state of the neuron. According to different states, the updating rule is different. In this section, the proposed saturation binary neural network is introduced.

For solving the bipartite subgraph problem, the state of the neuron is binary, which is called binary neuron model. Note that the standard energy function of Hopfield network can be written as follows:

$$E = -\frac{1}{2}\sum_{i=1}^{N}\sum_{j=1}^{N}\omega_{ij}V_iV_j - \sum_{i=1}^{N}\theta_iV_i \qquad (1)$$

Where ω_{ij} is weight of a synaptic connection from j_{th} neuron to the i_{th} neuron, θ_i is the external input of neuron #i and is also called threshold. V_i, V_j is the output of the neuron which also called the state of the neuron.

The Hopfield neural network can find the solution by seeking the local minimum of the energy function E using the motion equations of neurons.

$$\frac{dU_i(t)}{dt} = \sum_{j\neq i}^{N}\omega_{ij}V_j + \theta_i \qquad (2)$$

In this paper, the output Vi is updated from ui using sigmoid function:

$$V_i = 1/(1 + e^{(-U_i/T)}) \qquad (3)$$

In the Hopfield neural network, the updating method of input potential U_i is especially important. In conventional neuron model, the input potential U_i is updated from the Eq. 4 or Eq. 5.

$$U_i(t+1) = \frac{dU_i(t)}{dt} \qquad (4)$$

$$U_i(t+1) = U_i(t) + \frac{dU_i(t)}{dt} \tag{5}$$

Where $dU_i(t)/dt$ derives from the energy function E based on the gradient descent method:

$$\frac{dU_i(t)}{dt} = -\frac{\partial E(V_1, V_2, ..., V_n)}{\partial V_i} \tag{6}$$

The Hopfield-type binary neural network is usually constructed using the above neuron model. In order to improve the global convergence quality and shorten the convergence time, we propose a new neuron model called saturation binary neuron model (SBNM) which consists of the following important ideas.

(1). Once the neuron is in excitatory state (the output $V_i=1$), then the input potential is assumed to be in positive saturation. In the positive saturation, the input potential U_i can only be reduced but cannot be increased.

For the case of $V_i=1$:
if $dU_i(t)/dt < 0$

$$U_i(t+1) = U_i(t) + \frac{dU_i(t)}{dt} \tag{7}$$

else

$$U_i(t+1) = U_i(t) \tag{8}$$

(2). Once the neuron is in inhibitory state (the output $V_i=0$), then the input potential is assumed to be in negative saturation. Then the input potential can only be increased but cannot be reduced.

For the case of $V_i=0$
if $dU_i(t)/dt > 0$

$$U_i(t+1) = U_i(t) + \frac{dU_i(t)}{dt} \tag{9}$$

else

$$U_i(t+1) = U_i(t) \tag{20}$$

The above process can be presented in Fig. 1. The neuron update process can be set to three parts according to Eq. 7~Eq.10.

Note that the input/output function in the McCulloch-Pitts neuron [15] or hysteresis McCulloch-Pitts neuron [16] can be used to update the output V_i. Using the above neuron model, we construct a Hopfield-type neural network called saturation binary neuron network (SBNN). The following procedure describes the synchronous parallel algorithm using the SBNN to solve a COP. Note that N is the number of

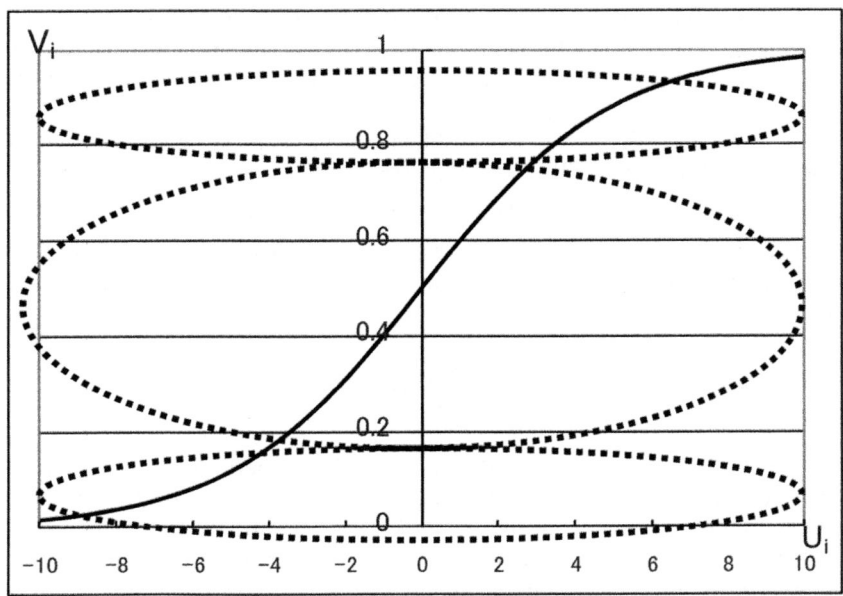

Fig. 1. The neuron state of the saturation binary neuron model

neuron, *targ_cost* is the target total cost set by a user as an expected total cost and *t_limit* is maximum number of iteration step allowed by user.

1. Set $t=0$ and set *targ_cost, t_limit,* and other constants.
2. The initial value of U_i for $i=1,...,N$ is randomized.
3. Evaluate the current output $V_i(t)$ for $i=1,...,N$.
4. Check the network, if *targ_cost* is reached, then terminate this procedure.
5. Increment t by 1. if $t> t\_limit$, then terminate this procedure.
6. For $i=1,...,N$
   a. Compute Eq. 6 to obtain $dU_i(t)/dt$
   b. Update $U_i(t+1)$, using the proposed saturation binary neuron model(Eq. 7~Eq.10).
7. Go to the step 3.

3   Solving Biprtite Sub-graph Problems Using SBNN

For an *N*-vertex *M*-edge graph G=(*V, E*), if the vertex set V can be partitioned into two subsets *V1* and *V2*, such that each edge of G is incident to a vertex in *V1* and to a vertex in *V2*, then the graph G is called bipartite graph, and be denoted by G=(*V1, V2, E*). Given a graph G=(*V, E*) with a vertex set *V* and an edge set *E*, the goal of bipartite sub-graph problem is to find a bipartite sub-graph with the maximum number of edges. In other words, the goal of bipartite sub-graph problem is to remove the minimum number of edges from a given graph such that the remaining graph is a bipartite graph. Consider a simple undirected graph composed of four vertices and four edges as shown in Fig.2 (a). The graph is bipartite as long as one edge is removed. Fig.2 (b) shows a bipartite graph. A bipartite graph is usually shown with the two subsets as top and bottom rows of vertices, as in Fig.2 (b), or with the two subsets as left and right columns of vertices.

An *N*-vertex bipartite sub-graph problem can be mapped onto a Hopfield neural network with *N* neurons [17]. Neuron #*i* (*i=1,…,N*) expresses the *i*-vertex, and the output ($y_i = 1$ or $y_i = 0$) of neuron #*i* (*i=1,…,N*) expresses that the *i*-vertex is partitioned into the subset *V1* or *V2*, respectively. Thus the *N*-vertex bipartite sub-graph problem can be mathematically transformed into the following optimization problem.

$$\text{Min} \left[\sum_{i=1}^{N} \sum_{j \neq i}^{N} d_{ij} y_i y_j + \sum_{i=1}^{N} \sum_{j=1}^{N} d_{ij} (1 - y_i)(1 - y_j)) \right] \tag{11}$$

where d_{ij} is 1 if edge (i, j) exists in the given graph, 0 otherwise. The first term of Eq. 8 is the number of edges connecting two vertices in the subset *V1*, and the second term is the number of edges in the subset *V2*. The number of the edges both in the subset *V1* and *V2* is smaller the number of the edges between the *V1* and *V2* is larger. Thus, we can remove the minimum number of edges defined by Eq. 8 to obtain a bipartite sub-graph from a given graph.

When we follow the mapping procedure by Hopfield and Tank, the energy function for the bipartite sub-graph problem is given by

$$E = A \sum_{i=1}^{N} \sum_{j \neq i}^{N} d_{ij} y_i y_j + B \sum_{i=1}^{N} \sum_{j=1}^{N} d_{ij} (1 - y_i)(1 - y_j)) \tag{12}$$

where A, B are parameters which are used to balance the two terms of Eq. 12.

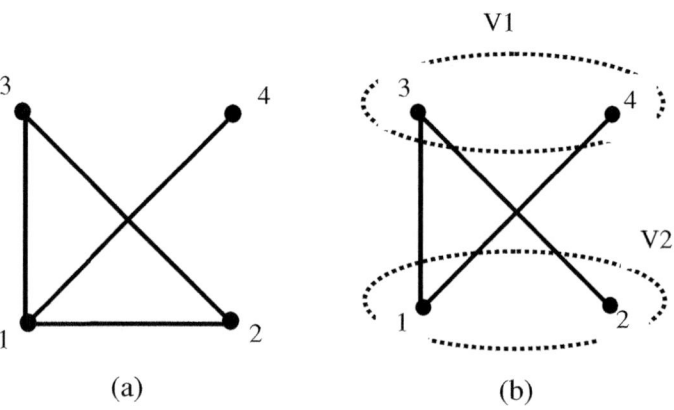

Fig. 2. (a) Graph is not bipartite. (b) Graph is bipartite.

Table 1. The simulation results

No.vertex	No.edges	Lee et al.[12]	Proposed algorithm
50	61	52	53
50	183	133	136
50	305	203	205
80	158	127	134
80	474	325	329
80	790	504	513
100	247	196	207
100	742	492	502
100	1235	761	779
150	558	402	423
150	1676	1062	1076
150	2790	1645	1693
200	995	685	719
200	2985	1838	1862
200	4975	2886	2954
250	1556	1060	1104
250	4668	2809	2854
250	7778	4435	4516
300	2242	1486	1532
300	6727	3987	4051
300	11212	6393	6451

4   Simulation Results

To widely verify the proposed algorithm, we have tested the algorithm with a large number of randomly generated graphs [18] defined in terms of two parameters, n and p. The parameter n specifies the number of vertices in the graph; the parameter p, $0<p<1$, specifies the probability that any given pair of vertices constitutes an edge. In the experiment, up to 300-vertex graphs with different probability are used.

To evaluate our results, we compared the results of Lee et al. with our results. For each of instances, 100 simulation runs were performed. Information on the test graphs as well as all results is shown in Table.1. The results that we recorded for each graph are the solutions in number of embedded edges, produced by the algorithm of Lee et al., and by the proposed algorithm. Table.1 shows that the proposed method can find better solutions than Lee et al.'s algorithm in all problems, although the computing time becomes longer.

5   Conclusion

We have proposed a saturation binary neuron model and used it to construct a Hopfield-type neural network which is called saturation binary neural network (SBNN). The SBNN is used to solve the bipartite sub-graph problems. The simulation results show that SBNN is capable of finding better solutions than other methods. Also, it can be seen that the SBNN is problem independent and can be used to solve other combinatorial optimization problems.

References

1. Garey, M.R., Johnson, D.S., Stockmeyer, L.J.: Some simplified NP-complete graph problem. Theor. Comput. Sci., 237–267 (1976)
2. Karp, R.M.: Reducibility among combinatorial problems. In: Complexity of Computer Computations, pp. 85–104. Plenum, New York (1972)
3. Even, S., Shiloach, Y.: NP-completeness of several arrangement problems. Technical Report 43, Department of computer Science, Technion, Haifa Israel (1975)
4. Bondy, J.A., Locke, S.C.: Largest bipartite subgraph in triangle-free graphs with maximum degree three. J. Graph Theory, 477–504 (1986)
5. Grotscheland, M., Pulleyblank, W.R.: Weakly bipartite graphs and the max-cut problem. Oper. Res. Lett., 23–27 (1981)
6. Barahona, F.: On some weakly bipartite graph. Oper. Res. Lett. 2(5), 239–242 (1983)
7. Hopfield, J.J.: Neural networks and physical systems with emergent collective computational abilities. Proc. Nat. Acad. Sci. U.S. 79, 2554–2558 (1982)
8. Hopfield, J.J.: Neurons with graded response have collective computation properties like those of two-state neurons. Proc. Nat. Acad. Sci. U.S. 81, 3088–3092 (1982)
9. Hopfield, J.J., Tank, D.W.: 'Neural' computation of decisions in optimization problems. Bio. Cybern. 52, 141–152 (1985)
10. Hopfield, J.J., Tank, D.W.: Simple 'Neural' optimization networks: An a/d converter, signal decision circuit, and a linear programming circuit. IEEE Trans. Circuits Syst. 33(5), 533–541 (1986)
11. Hopfield, J.J., Tank, D.W.: Computing with neural circuits: A model. Science 233, 625–633 (1986)
12. Lee, K.C., Funabiki, N., Takefuji, Y.: A parallel improvement algorithm for the bipartite subgraph problem. IEEE Trans. Neural Networks 3(1), 139–145 (1992)
13. Garey, M.R., Johnson, D.S.: Computers and Intractability: a Guide to the Theory of NP-Completeness. W.H. Freeman, New York (1979)
14. Ackley, D.H., Hinton, G.E., Sejnowski, T.J.: A learning algorithm for Boltzman Machines. Cognitive Science 9, 147–169 (1985)
15. McCulloch, W.S., Pitts, W.H.: A logical calculus of ideas immanent in nervous activity. Bull. Math. Biophys 5, 115–133 (1943)
16. Takefuji, Y., Lee, K.C.: An artificial hysteresis binary neuron: A model suppressing the oscillatory behaviors of neural dynamics. Biol. Cybern. 64, 353–356 (1991)
17. Wang, R.L., Tang, Z., Cao, Q.P.: A Hopfield Network Learning Method for Bipartite Subgraph Problem. IEEE Trans. Neural Networks 15(6), 1458–1465 (2004)
18. Johnson, D.S., Aragon, C.R., McGeoch, L.A., Schevon, C.: Optimization by simulated annealing: An experimental evaluation; Part 1, graph partitioning. Operations Research 37(6), 865–892 (1989)

Modeling and Analysis of Network Security Situation Prediction Based on Covariance Likelihood Neural

Chenghua Tang[1], Xin Wang[1], Reixia Zhang[1], and Yi Xie[2]

[1] School of Computer Science and Engineering, Guilin University of Electronic Technology,
Guilin 541004, China
{tch,wxin,rxzhang}@guet.edu.cn
[2] Department of Information Science and Technology, Sun Yat-Sen University,
Guangzhou 510275, China
Xieyi5@mail.sysu.edu.cn

Abstract. Security situation is the premise of network security warning. For lack of self-learning on situation data processing in existing complex network, a modeling and analysis of network security situation prediction based on covariance likelihood neural is presented. With the introduction of the error covariance likelihood function, and considering the impact of sample noise, the network security situation prediction model using the situation sequences as input sequences, and in the back-propagation to achieve the parameters adjustment. Results show that the model can take advantage of the relationship characteristics between the complexity and efficiency in complex neural networks, and the method has good performance of situation prediction.

Keywords: network security, situation prediction, covariance, neural.

1 Introduction

Most of the current situation prediction techniques are a part of situation assessment, which make reference on whether to the early warning after the situation evaluation. Main methods used in Multi-sensor data fusion [1], grey correlation analysis [2], AHP [3], etc., these methods rely on specialists than those given the initial elements of the security situation beginning weights, required during operation of man-made changes to the weights, some algorithm do not have the self-learning ability. Because of the uncertainty, ambiguity and variability characteristics, of the attack information, situation prediction involves computer science, military strategy, political science, and other disciplines, its importance has been related to people's lives and national security. After "9.11"incident, the European Union accelerated the implementation pace of establishment of electronic information security program, requiring strict inspection early warning and emergency response capacity of information network infrastructure and network system. British Institute of King's College London researched the information warfare threat assessment and early warning of attacks, and proposed the decision-making intelligent early warning system [4]. Kijewski studied for early warning and attack of a prototype system identification framework [5]. NetSA Working Group developed the SILK to carry out large-scale real-time

D.-S. Huang et al. (Eds.): ICIC 2011, LNBI 6840, pp. 71–78, 2012.
© Springer-Verlag Berlin Heidelberg 2012

monitoring of network security situation, the potentially malicious network behavior before they become unable to control the identification, response and early warning [6]. Honeywell Laboratories proposed using the theory of plan recognition for intrusion prediction that the behavior from the observed sequence of reasoning, called the plan recognition or task tracking in the field of artificial intelligence [7]. Honeynet Project proposed to predict the invasion of hacker's intent and possible future theoretical methods using moving average models and other statistical theory [8]. U.S. Department of Defense Advanced Research also heavily subsidized project development such as the purpose of hacking prediction, but due to various reasons, the progress of their projects have not received specific information. In China, Hu Huaping proposed for large-scale network intrusion detection and early warning system architecture [9]. Hu Wei proposed an improved model for prediction of Grey Verhulst [10]. For periodic attacks, Zhang Feng proposed method of network security warning based on intrusion events [11]. An Xifeng analyzed the characteristics of security incidents, security incidents to establish a distributed network of early warning model [12].

These results are mainly stay in the theoretical framework, the traditional situation prediction methods rely on experts to give too much weight, and lack of self-learning. This paper presents a security situation prediction method based on covariance likelihood neural, introduced the concept of state sequences, and the back propagation process to achieve the right value for the specified parameter self-learning adjustment. In order to avoid the situation of all elements of algorithms to interpolate the sample points, abandoning the traditional minimum variance error function, the introduction of maximum likelihood estimation, re-defined error function, considered the impact of sample covariance and noise on the network training, and from the global Point of the network to achieve unsupervised learning.

2   Principle of Situation Prediction Control

Network security situation prediction model based on the past situation input and situation output predicts the future trend of the situation. The past input and output values as the neural network training signal, namely, the scalar adjustable parameters of the transfer function, and output the result, that is, the future of security situation. The basic structure of network security situation prediction mode based on neural network is shown in Fig. 1.

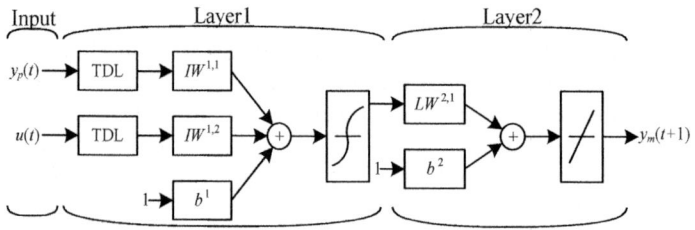

Fig. 1. Network security situation prediction model

The curve (Layer1) and linear (Layer2) transfer function in the basic structure model, respectively, for the network back propagation and linear regression. The prediction model can be trained offline, and the training objects can be a large number of off-line historical security situation data.

Considering a M layers feed forward network, the number of neurons of each layer is N_m, ($m=1,2,...,M$), then the basic network equation can be expressed as:

$$y_i^m = f((W_i^m)^T Y^{m-1})$$ (1)

Where $i=1,2,...,N_m$, represents the serial number of neurons m layer, y_i^m indicates that the network's first i-m layer of output neurons, $f(.)$ indicates the S-type nonlinear function $f(x) = 1/(1+e^{-x})$, $(W_i^m)^T$ is the vector transpose composed of connecting weights about the first i-m layer of neurons with the first m-1 layer all the neurons, and Y^{m-1} is the first m-1 layer composed of all the output neurons in the vector.

Let the network output is $\hat{y}_i^m = f(X_i, W)$, real output is $\overline{y}_i^m = f(X_i)$, and sample output is $y_i^m = f(X_i) + \xi_i$, then the output layer neuron fitting error is $e_i = y_i^m - \hat{y}_i^m$, and the real bias is $\overline{e} = \overline{y}_i^m - \hat{y}_i^m$, so according to the traditional neural algorithm to obtain the minimum variance method LS type network error function is the following function:

$$E_{LS} = \sum_{i=1}^{N} \phi(e_i) = \sum_{i=1}^{N} e_i^2$$ (2)

Successful application of system identification using neural algorithm depends on the quality of samples, because the training algorithm does not consider the training samples error, using the least square error to get the best value of function E_{LS} only when the deviation submit to the Gaussian distribution. E_{LS} guides the learning process, and the ultimate goal is to make all the opportunity to sample the fitting error tends to zero balance, that approximate the network output for each sample by the noise pollution output, rather than real output, so in the sample case with noise, it interpolates all the training samples in the identification model, rather than approaching the real object model, which will lead to the practical application of the more iterations, the smaller the training error, and generalization capability is worse, but significantly lower efficiency.

3   Neural with Covariance Likelihood

Learning sample of data set (X_i, Y_i^m), ($i=1,2,..., N_m$), given a set of weights W, because of the existence of the error, the conditional probability density of the network output vector $\hat{y}_i^m$ relative to the weight W is:

$$l(W) = P(\hat{y}_i^m | W)$$ (3)

If the sample is independent between the statistics, the continuous application of the Bayesian formula to get all the joint probability density of the sample, the sample likelihood function is:

$$L(W) = P(\hat{y}_1^m, \hat{y}_2^m, \ldots, \hat{y}_{Nm}^m \mid W) = \prod_{i=1}^{Nm} p(\hat{y}_i^m \mid W) \tag{4}$$

If the network model is correct, then all the differences between the samples output and the actual output mainly derive from the samples noise, therefore the fitting error must to be considered, then assume the sample output error obeys the Gaussian distribution, the density function of the network output vector with respect to weight W Conditions and Gaussian distribution parameters:

$$P(\hat{y}_i^m \mid W, \mu, \sigma) = (2\pi \mid Cy_i \mid)^{-1/2} \exp\{-\frac{(y_i^m - \hat{y}_i^m)^T (y_i^m - \hat{y}_i^m)}{2Cy_i}\} \tag{5}$$

Where $Cy_i = \text{cov}[\hat{y}_i^m \mid y_{i-1}^m]$, is the output condition of the sample covariance matrix reflects the error structure of the sample output, and fitting deviation ($y_i^m - \hat{y}_i^m$) is the mean of sample error.

According to maximum likelihood theory, the parameters μ, σ determining conditions, so that the $W*$ for the largest value of likelihood function $L(W)$ is the most optimal estimation of W.

$$L(W*, \mu, \sigma) = \max L(W, \mu, \sigma) \tag{6}$$

Apparently equivalent to the following error function is minimized:

$$E(W, \mu, \sigma) = -2\ln L(W, \mu, \sigma) \tag{7}$$

Therefore, the new network error function is:

$$E(W, \mu, \sigma) = \sum_{i=1}^{Nm} \frac{(y_i^m - \hat{y}_i^m)^T (y_i^m - \hat{y}_i^m)}{Cy_i} + 2\sum_{i=1}^{Nm} \ln \mid Cy_i \mid + 2N_m \ln \sqrt{2\pi} \tag{8}$$

Constant term omitted, the final error function is:

$$E(W, \mu, \sigma) = \sum_{i=1}^{Nm} \frac{(y_i^m - \hat{y}_i^m)^T (y_i^m - \hat{y}_i^m)}{Cy_i} \tag{9}$$

According to this building method the network is called covariance likelihood neural in this paper.

Compared with the conventional error function, where taking the sample covariance into account. In addition, when $L(W*)$ to take the maximum, that is, the deviation of training samples tends to the expected error in the center of the surface, the learning intensity maximum deviation tends to zero faster, its resistance to errors and noise samples interference.

4   Security Situation Prediction Model

According to the service, host and network security system in the threat situation provided by the target network, the situation assessment model can be established. Because of the space limitations, we only assess the security situation on the service as example.

Definition 1. The function F_S said the security situation in the target network service status, denoted as:

$$F_S(S,C,N,D,t) = N(t) \cdot 10^{D(t)} \qquad (10)$$

Where S represents a service the target network provided; C indicates the type of service attacks on the; N is said that services by the number of attacks; D is said that the severity of the attack; $N(t)$ is said that the severity of attacks in t time; $D(t)$ is said that the number of attacks occurred in t time.

Definition 2. The function of F_H said the security situation in the host status of the network, denoted as:

$$F_H(H,V,Fs,t) = V \cdot Fs(t) \qquad (11)$$

Where H represents the target hosts on the network; V indicates that the service's weight of all opened services.

Definition 3. Assuming in t (as small as possible) time period, select a state sequence from the state database, as the future network security situation prediction model of the input sequence, denoted by $X^{(0)} = (x^{(0)}_1, x^{(0)}_2, ..., x^{(0)}_n)$, where $x^{(0)}_t \geq 0$, t = 1,2, ..., n; $X^{(1)}$ is $X^{(0)}$ of 1-AGO sequence, denoted as $X^{(1)} = (x^{(1)}_1, x^{(1)}_2, ..., x^{(1)}_n)$, where:

$$x^{(1)}_t = \sum_{i=1}^{t} x^{(0)}i \,, \ t{=}1,2,...,n. \qquad (12)$$

By definition 1 and definition 3, the services available for the target network security situation prediction function model:

$$F_S(t+(n+1))=f(\sum_{i=1}^{n} F_S(t+i)) \qquad (13)$$

Thus, parameters $F_S(t+(n+1))$ and $\sum_{i=1}^{n} F_S(t+i)$ can be recognized the existence of a nonlinear relationship.

According to Kolmogorov mapping existence theorem of multi-layer neural network, the nonlinear mapping relationship can be three layers feed-forward artificial neural network approximation achieved. Then the known time situation series act as network input, and the network output is to identify the situation prediction. This paper requires $F_S(t+(n+1))$ and time $(t+(n+1))$ within the range, by definition given in 1 and 2, network services, host situation assessment methods to obtain network training samples required for the application covariance likelihood neural algorithm to train the network, follow these steps:

1. According to the history and current situation of network security information data, the service and the host of multi-input single-output prediction of artificial neural network model and corresponding network error function $E(W)$ are established:

$$P(\hat{y}_i^m \mid W) = (2\pi \mid C_{yi} \mid)^{-1/2} \exp\{-\frac{(y_i^m - \hat{y}_i^m)^T (y_i^m - \hat{y}_i^m)}{2C_{yi}}\} \tag{14}$$

$$E(W) = \sum_{i=1}^{Nm} \frac{(y_i^m - \hat{y}_i^m)^T (y_i^m - \hat{y}_i^m)}{C_{yi}} \tag{15}$$

Where $\hat{y}_i^m$ and y_i^m, respectively, for the actual output and expected output of the m layer i neurons, corresponding to the situation prediction value; for every single point of communication during the process input parameters, where W can represent the situation assessment model Severity of attacks D, the importance of services V parameters.

2. Training the neural network for the fit error (y_i^m - $\hat{y}_i^m$) tending to zero, the weights of self-learning adjustment, to find the optimal parameters, and output the prediction model finally. The network output is what we want recognition the next moment on the service or host situation prediction values. In the real world, real-time training the network, the fitting curve of situation prediction can be available.

5   Experiments and Results

Purpose of the experiment is to verify the effective and reasonable of the security situation prediction model.

Experimental environment is configured with Ubuntu 10.04 LTS / Inter Core 2 Duo E7200/2G/250G host, SUN, IBM and other large servers, and multi-layer routers, Gigabit switches, IDS, firewall, and fiber optic cable to construct more complex Network. Using Domain 3.5, Namp3.5 and Trinity V3 to attack a protected server, collect IDS, firewall and system log information, and assess the service or host security situation every 10 hours. The value of each assessment with the previous values are to be established the time series, input into the covariance likelihood neural, finally, output the situation prediction value each moment in turn, and calculate the actual situation value in the next time point for comparison.

Firstly, Experiment gets the number of attacks on the ftp, telnet, rpc, dns, socks, www, etc., and then assesses each service's security situation by definition 1. Trial of 60 consecutive made the security situation in the value of rpc services, as with the previous 45 training samples, and after 15 testing samples, after pretreatment, input into covariance likelihood neural to train. Pre-set training speed factor = 0.6, target error goal = 0.0001, after running about 32.116s, achieve the target error for the 6107 iterations. At the same time using traditional neural to test security situation prediction for comparative experiments, found that after iteration 10989 times the error has not reached the goal, and time has been occupied 149.973 seconds. Fig. 2 shows the situation prediction results about rpc service security situation based on covariance likelihood neural.

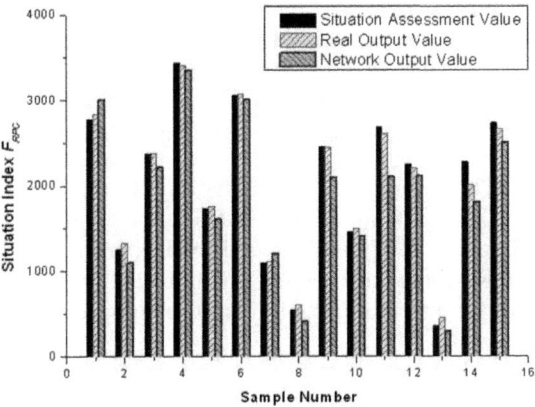

Fig. 2. Prediction of rpc service security situation

Finally, considering the various services on the server, by definition 2, evaluate the value of the server host's security situation, the same method draw neural prediction on the host in Fig. 3.The figure shows that the value of the neural with covariance likelihood prediction can better approximate the true assessed value, with a good prediction effect.

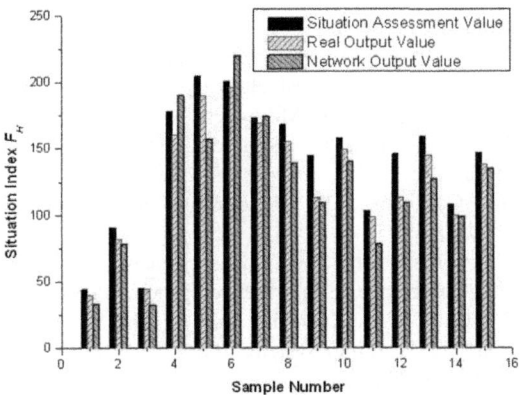

Fig. 3. Prediction of host security situation

6   Conclusions

This paper introduced neural and studied its improvement to establish a security situation prediction model based on covariance likelihood neural. The traditional error function is replaced by the maximum likelihood error function. The impact of sample covariance and noise on the network training is considered. The situation sequences established through the situation assessment model are used as the training input sequences, and the self-learning adjustment of the appointed parameters' values is implemented in the process of back propagation training. The new method can make full use of the characteristics of the network more complex, finer grain size, the higher the efficiency, and results show it can effectively predict security situation and provides an effective way of network security strategic early warning.

Acknowledgments. This work was supported by the National Natural Science Foundation of China under Grant No.60970146, the China Postdoctoral Science Foundation under Grant No.20070420793, and the Department of Education research project in Guangxi, P.R. China under Grant No.201012MS088. The helpful comments from anonymous reviewers are also gratefully acknowledged.

References

1. Onwubiko, C.: Functional requirements of Situational Awareness in Computer Network Security. In: Proc of the IEEE International Conference on Intelligence and Security Informatics (ISI 2009), Dallas, Texas, USA (2009)
2. Zhao, G.S., Wang, H.Q., Wang, J.: Study on Situation Evaluation for Network Survivability Based on Grey Relation Analysis. Journal of Chinese Computer Systems 27(10), 1861–1864 (2006)
3. Chen, X.Z., Zheng, Q.H., Guan, X.H.: Quantitative Hierarchical Threat Evaluation Model for Network Security. Journal of Software 17(4), 885–897 (2006)
4. US Infrastructure Assurance Strategic Roadmaps. Strategies for Preserving Our National Security. Sandia National Laborato-ries, Sand Report, 98-1496 (1998)
5. Kijewski, P.: ARAKIS-An early warning and attack identification system. In: Proc. of the 16th Annual First Conference, Dudapest, Hungary (2004)
6. Carrie, G., Michael, C., Michael, D.: More Netflow Tools: for Performance and Security. In: Proc. of the 18th Large Installation Systems Administration Conference (LISA 2004), Atlanta, GA, USA (2004)
7. Christopher, W.G., Goldman, R.P.: Honeywell Labs. Plan Recognition in Intrusion Detection Systems (2001)
8. Das, S., Lawless, D.: Trustworthy Situation Assessment via Belief Networks. In: Proc. of the 5th International Conference on Information Fusion, USA (2002)
9. Hu, H., Zhang, Y., Chen, H.T., Xuan, L., Sun, P.: The Study of Large Scale Networks Intrusion Detection and Warning System. Journal of National University of Defense Technology 25(1), 21–25 (2003)
10. Hu, W., Li, J.H., Chen, X.Z.: Network Security Situation Prediction Based on Improved Adaptive Grey Verhulst Model. Journal of Shanghai Jiaotong University (Science) 15(4), 408–413 (2010)
11. Zhang, F., Qin, Z.g., Liu, J.d.: Intrusion Event Based Early Warning Method for Network Securiyt. Computer Science 31(11), 79–81, 131 (2004)
12. An, X.f., Li, W.H., Liu, Z.: Research on early-alert, orientation and rapid isolation control system for large-scale networks. Computer Engineering and Design 29(8), 78–81 (2008)

Local Meta-models for ASM-MOMA

Martin Pilát[1,2] and Roman Neruda[2]

[1] Department of Theoretical Computer Science and Mathematical Logic,
Faculty of Mathematics and Physics, Charles University in Prague,
Malostranské náměstí 25, Prague, Czech Republic
[2] Institute of Computer Science, Academy of Sciences of the Czech Republic,
Pod Vodárenskou věží 2, Prague 8, Czech Republic
Martin.Pilat@mff.cuni.cz, roman@cs.cas.cz

Abstract. Evolutionary algorithms generally require a large number of objective function evaluations which can be costly in practice. These evaluations can be replaced by evaluations of a cheaper meta-model of the objective functions. In this paper we describe a multiobjective memetic algorithm utilizing local distance based meta-models. This algorithm is evaluated and compared to standard multiobjective evolutionary algorithms as well as a similar algorithm with a global meta-model. The number of objective function evaluations is considered, and also the conditions under which the algorithm actually helps to reduce the time needed to find a solution are analyzed.

Keywords: Multiobjective optimization, meta-model, evolutionary algorithm.

1 Introduction

Many real life optimization tasks require optimizing multiple conflicting objectives at once. It has been shown and widely accepted that multiobjective evolutionary algorithms (MOEA) are among the best methods for multiobjective optimization. In the past years several multiobjective evolutionary algorithms [3,12,1] were proposed and used to deal with these problems. However, most of them require lots of evaluations of each objective function, which makes them problematic to use for solving real life problems. These problems may have complex objective functions whose evaluations are expensive (either in terms of time or money).

The use of the meta-models aims at lowering the number of objective function evaluations which are needed to obtain the final solution. The meta-model is a simplified and cheaper approximation of the real objective function. Meta-models can be used in several ways to augment the multiobjective evolutionary algorithms. In one of the first approaches [10] its authors used the NSGA-II [3] and replaced the objective functions with their meta-models. In [7] and [8] authors describe an aggregate meta-model based on various SVM architectures. Although the memetic variant is also possible in multiobjective setting, only a few references were found in the literature which deal with meta-model assisted multiobjective memetic algorithms [5].

The paper is organized as follows: In the next section 2. The tests and their results are described in sections 3 and 4. Section 5 concludes the paper and provides ideas for future research.

D.-S. Huang et al. (Eds.): ICIC 2011, LNBI 6840, pp. 79–84, 2012.

2   Algorithm Description

In one of our previous papers [9] we proposed a multiobjective memetic algorithm with aggregate meta-model (ASM-MOMA). This algorithm was able to reduce the number of required evaluations of the objective functions by the factor of 5 to 10 on most problems. ASM-MOMA uses a single global meta-model trained after each generation as a fitness function during the local search.

In this paper, we propose a new variant of ASM-MOMA with local models instead of a single global one. We call this variant LAMM-MMA. The main difference between LAMM-MMA and other multiobjective evolutionary algorithms is the addition of a special memetic operator, which performs local search on some of the newly generated individuals (the generation of the new individuals is handled by the respective MOEA to which is this operator added). The operator uses the meta-model constructed based on previously evaluated points in the decision space, for which the values of objective functions are known. The meta-model is trained to predict the distance to the currently known Pareto front. Moreover, as an addition to ASM-MOMA, in LAMM-MMA the points do not not have the same weight, as those that are closer to the locally optimized one are considered more important during the model building phase, see Equation 1 for details.

The main idea is that points closer to the known Pareto front are more interesting during the run of the algorithm and the memetic operator moves the individuals closer to the Pareto front. The purpose of the meta-model is not to precisely predict the value but rather provide a general direction in which the memetic search should proceed. To obtain a training set for the meta-models we also added an external archive of individuals with known objective values. This archive is updated after each generation when new individuals are added and at the same time the archive is truncated to ensure it does not grow indefinitely – random individuals are removed to match the limit on the number of individuals, see [9] for analysis of this approach.

The following sections detail the important parts of the algorithm. The main loop is essentially a generic MOEA with added memetic operator. We train a dedicated model for each individual I which shall be locally optimized by the memetic operator. For such an individual I we create a weighted training set

$$T_I = \left\{ \langle (x_i, y_i), w_i \rangle | y_i = -d(x_i, P), w_i = \frac{1}{1 + \lambda d(x_i, I)} \right\} \tag{1}$$

where $d(x, y)$ is the Euclidean distance of individuals x and y in the decision space, P is the set of non-dominated individuals in the archive and $d(x, P)$ is the distance of individual x to the closest point in the set P. λ is a parameter which controls the locality of the model, larger values of λ lead to more local model, whereas lower values lead to more global one.

The points which are closer to the individual I are more important during the training of the model. This distance weighting adds some locality to the models trained for each individual. The training set is constructed in such a way, that for the individuals closer to the currently known Pareto front the meta-model should return larger values. This fact is used during the local search phase (which uses the meta-model as a fitness function).

Table 1. Times needed for training and evaluation of selected meta-models, in seconds

Model	Training (T_t)	Evaluation (T_m)
Linear regression	0.142	8.46×10^{-7}
Support vector regression	0.328	7.14×10^{-7}
Multilayer perceptron	3.75	1.80×10^{-5}

In the local search phase we use another evolutionary algorithm (this time it is only a single objective one) to find better points in the surroundings of each individual. The algorithm runs only for a few generations and it uses only meta-model evaluations. The newly found individuals are placed back to the population. During the initialization of the local search the individual which should be optimized is inserted in the initial population and its variables are perturbed to create the rest of the initial population.

The algorithm uses quite large number of meta-model evaluations and even trainings. This might lead to significant overhead. To find out how large this overhead is, we run a few benchmarks (archive size/training set size of 400 individuals, Intel Core i7 920 (2.87Ghz) processor and 6GB RAM). Table 2 shows the results. We can see that the evaluations are faster than the training by several orders of magnitude and that each training takes only a fraction of a second. Even if there are 100 trainings per generation, it would mean an overhead of roughly 15 to 30 seconds per generation, which still might be faster than a single evaluation of the real objective function. Therefore the overhead of the training and evaluation is easily compensated by the reduced evaluations of the objective functions.

3   Test Setup

We tested our approach on the widely used ZDT [11] benchmark problems. These problems are all two dimensional, and we used 15 variables for each of them. In the local search phase we used various meta-models: namely multilayer perceptron, support vector regression, and linear regression. All the models use default parameters from the Weka framework [6] (which we used to run the experiments).

Table 2. Parameters of the multiobjective algorithm

Parameter	MOEA value	Local search value
Stopping criterion	50,000 objective evaluations	30 generations
Population size	50	50
Crossover operator	SBX	SBX
Crossover probability	0.8	0.8
Mutation operator	Polynomial	Polynomial
Mutation probability	0.1	0.2
Archive size	400	–
Memetic operator probability	0.25	–
Meta-model locality parameter λ	–	1

See Table 2 for the parameters of the main multiobjective algorithm and the internal single-objective algorithm.

As the base multiobjective evolutionary algorithm we used the NSGA-II and ϵ-IBEA with Simulated Binary Crossover [2] and Polynomial Mutation [4]. In the local search phase we used a simple single objective evolutionary algorithm with the same operators and the meta model as the fitness function.

To compare the results we use a measure we call H_{ratio}, it is defined as the $H_{ratio} = \frac{H_{real}}{H_{optimal}}$, where H_{real} is the hypervolume of the dominated space attained by the algorithm and $H_{optimal}$ is the hypervolume of the real Pareto set of the solutions. As the Pareto set is known for all the ZDT problems, we can compute this number directly. We use the vector $\mathbf{2} = (2, 2)$ as the reference point in the hypervolume computation. All points that do not dominate the reference point are excluded from the hypervolume computation. We compare the median number of function evaluations needed to attain the H_{ratio} of $0.5, 0.75, 0.9, 0.95$, and 0.99 respectively.

4   Results

Table 3 shows the results of our algorithm compared to original ϵ-IBEA and ASM-MOMA. IBEA denotes the original ϵ-IBEA. LR, SVM, and MLP stands for the model used: linear regression, support vector regression and multilayer perceptron respectively. G denotes the single global model of ASM-MOMA and L stands for the local models as described in this paper.

The numbers in the table represent the median number of objective function evaluations needed to reach the specified H_{ratio} value. Twenty runs for each configuration were made.

From the results, we can see that the global models generally significantly decrease the number of required function evaluations, and the local models are even better than the global ones. Generally, linear regression gives better results than support vector regression and multilayer perceptrons. It probably creates simpler models which indicate the right general direction in which the local search should proceed. Moreover, we can see that the results of local models are almost always better than those of a single global model, thus we recommend using the faster models, i.e. linear regression or support vector regression instead of multilayer perceptrons.

On ZDT1 the global model was able to reduce the number of function evaluations to reach the $H_{ratio} = 0.95$ by a factor of 6.8 (LR) and the local model reduced it further, yielding the reduction factor of 7.3 for LR and even 7.7 with the SVM. For the $H_{ratio} = 0.99$ the reductions are not that large, but we can still see the number decreased by the factor of almost 4. In this case, local models did not improve the result.

On ZDT2 the results improved largely even for the $H_{ratio} = 0.99$. ASM-MOMA reduced the required number of objective function evaluations 8.4 times (SVM), while LAMM-MMA was able to reduce it almost 9 times with the same model and 9.6 times with the LR as the model.

Reaching the $H_{ratio} = 0.99$ was a problem for the original ϵ-IBEA on ZDT3, but both ASM-MOMA and LAMM-MMA were much more successful, both reducing the

Table 3. Median number of function evaluations needed to reach the specified H_{ratio} on ZDT1 test problem

	ZDT1					ZDT2				
H_{ratio}	0.5	0.75	0.9	0.95	0.99	0.5	0.75	0.9	0.95	0.99
IBEA	7400	13750	18200	20000	25550	750	2050	5150	7800	13000
IBEA-LR-G	1450	2500	2800	2950	7450	350	550	750	900	1650
IBEA-SVM-G	1400	2050	2700	3100	**6850**	350	550	850	1050	1550
IBEA-MLP-G	1800	2550	4000	4600	10100	450	650	950	1200	2700
IBEA-LR-L	**1300**	1900	2400	2750	7500	**300**	**500**	**700**	**850**	**1350**
IBEA-SVM-L	1350	1900	**2350**	**2600**	7100	350	550	800	1000	1450
IBEA-MLP-L	1400	**1850**	2450	3250	9650	350	550	750	900	1400

	ZDT3					ZDT6				
H_{ratio}	0.5	0.75	0.9	0.95	0.99	0.5	0.75	0.9	0.95	0.99
IBEA	650	1550	5400	8150	33350	10300	13650	18400	23150	34050
IBEA-LR-G	**350**	550	850	950	**1300**	3050	**6500**	13400	**17600**	32100
IBEA-SVM-G	**350**	550	850	1000	**1300**	**3000**	7250	14100	19250	34150
IBEA-MLP-G	450	800	1100	1250	1800	3500	7250	13250	18900	32450
IBEA-LR-L	**350**	**450**	**750**	**900**	**1300**	3050	6850	13050	18750	**31400**
IBEA-SVM-L	400	650	850	1050	1450	**3000**	**6500**	**12650**	17850	32550
IBEA-MLP-L	400	650	950	1150	1600	3400	7050	13300	18200	32950

number of evaluations over 25 times. For the $H_{ratio} = 0.95$ (which is not that difficult for ϵ-IBEA) we can see again reduction factors of 8.5 and 9 for ASM-MOMA and LAMM-MMA respectively (LR in both cases).

ZDT6, as in our previous paper [9], proved to be the most difficult problem. Although we can see reductions by the factor of 3.5 for the $H_{ratio} = 0.5$, this factor drops and there are only slight reductions of 6% for the $H_{ratio} = 0.99$. Note that LAMM-MMA again provided better reductions for this value, even though, the difference is not very large. Dealing with the difficulty of this problem is a motivation for further research.

5   Conclusions

In this paper we presented a memetic evolutionary algorithm for multiobjective optimization with local meta-models. We showed that the local models give better results than a single global model, usually reducing the number of needed function evaluations by another 10%. Although this difference may seem rather small it may greatly reduce the associated costs in practical tasks. We also showed that the algorithm is usable even for problems with quite simple objective functions, which take only milliseconds to evaluate, thus making it more widely usable. However, some problems are still difficult to solve with LAMM-MMA, and these provide the motivation for further research. Another open question is whether real life problems are among those easily solvable, or not.

We will continue the work on memetic multiobjective algorithms with aggregate meta-models. One of the goals is the reduction of the number of times the model is trained which is a problem especially for more expensive local models. These are trained multiple times in each generation. One possibility could be to cluster the individuals before the model is constructed and create a single local model for all the individuals in the cluster. Another open question is the effect of the degree of locality (represented by the λ parameter) on the evolution convergence speed and the possibility to change this parameter adaptively.

Acknowledgments. This research has been supported by the Ministry of Education of the Czech Republic under project no. OC10047, and by Czech Science Foundation project no. 201/09/H057.

References

1. Bader, J., Zitzler, E.: HypE: An Algorithm for Fast Hypervolume-Based Many-Objective Optimization. TIK Report 286, Computer Engineering and Networks Laboratory (TIK), ETH Zurich (November 2008)
2. Deb, K., Agrawal, R.B.: Simulated binary crossover for continuous search space. Complex Systems 9, 115–148 (1995)
3. Deb, K., Agrawal, S., Pratap, A., Meyarivan, T.: A fast elitist non-dominated sorting genetic algorithm for multi-objective optimisation: NSGA-II. In: Deb, K., Rudolph, G., Lutton, E., Merelo, J.J., Schoenauer, M., Schwefel, H.-P., Yao, X. (eds.) PPSN 2000. LNCS, vol. 1917, pp. 849–858. Springer, Heidelberg (2000)
4. Deb, K., Goyal, M.: A combined genetic adaptive search (geneas) for engineering design. Computer Science and Informatics 26, 30–45 (1996)
5. Georgopoulou, C., Giannakoglou, K.: Multiobjective metamodel-assisted memetic algorithms. Multiobjective Memetic Algorithms, 153–181 (2009)
6. Hall, M., Frank, E., Holmes, G., Pfahringer, B., Reutemann, P., Witten, I.H.: The WEKA data mining software: an update. SIGKDD Explorations 11(1), 10–18 (2009), http://dx.doi.org/10.1145/1656274.1656278
7. Loshchilov, I., Schoenauer, M., Sebag, M.: A mono surrogate for multiobjective optimization. In: Pelikan, M., Branke, J. (eds.) GECCO, pp. 471–478. ACM, New York (2010)
8. Loshchilov, I., Schoenauer, M., Sebag, M.: Dominance-based pareto-surrogate for multiobjective optimization. In: Deb, K., Bhattacharya, A., Chakraborti, N., Chakroborty, P., Das, S., Dutta, J., Gupta, S.K., Jain, A., Aggarwal, V., Branke, J., Louis, S.J., Tan, K.C. (eds.) SEAL 2010. LNCS, vol. 6457, pp. 230–239. Springer, Heidelberg (2010)
9. Pilát, M., Neruda, R.: ASM-MOMA: Multiobjective memetic algorithm with aggregate surrogate model. In: IEEE Congress on Evolutionary Computation. IEEE, Los Alamitos (to appear, 2011)
10. Voutchkov, I., Keane, A.: Multiobjective optimization using surrogates. Presented on Adaptive Computing in Design and Manufacture, ACDM 2006 (2006)
11. Zitzler, E., Deb, K., Thiele, L.: Comparison of Multiobjective Evolutionary Algorithms: Empirical Results. Evolutionary Computation 8(2), 173–195 (2000)
12. Zitzler, E., Künzli, S.: Indicator-Based Selection in Multiobjective Search. In: Yao, X., Burke, E.K., Lozano, J.A., Smith, J., Merelo-Guervós, J.J., Bullinaria, J.A., Rowe, J.E., Tiňo, P., Kabán, A., Schwefel, H.-P. (eds.) PPSN 2004. LNCS, vol. 3242, pp. 832–842. Springer, Heidelberg (2004)

Stock Market Trend Prediction Model for the Egyptian Stock Market Using Neural Networks and Fuzzy Logic

Maha Mahmoud Abd ElAal[1], Gamal Selim[1], and Waleed Fakhr[2]

[1] Arab Academy for Science, Technology and Maritime Transport,
Computer Engineering, Cairo, Egypt
[2] Arab Academy for Science, Technology and Maritime Transport,
Computer Science, Cairo, Egypt
maha.elaal@staff.aast.edu, dgamal55@yahoo.com, waleedf@aast.edu

Abstract. This paper presents a study and implementation of a stock trend prediction system based on Artificial Neural Network (ANN) and fuzzy logic rules. Technical analysis tools such as technical indicators and Elliott's wave theory were deployed in the presented prediction system. In this approach the neural network functions as a classifier, where the technical analysis indicators are its input features. The multilayer perceptron (MLP), Support Vector Machine (SVM) and Radial Bases Function (RBF) are tested as classification tools. Also, a fuzzy rule based system based on the Elliott's wave theory is developed to predict the short term stock trend. Finally, integration between these two modules is established using neural network. The System was trained and tested with real data from the Egyptian stock market. The obtained results are encouraging.

Keywords: Stock Prediction, Neural Networks, Fuzzy logic.

1 Introduction

Prediction of financial time-series such as stock market is considered a complicated and daring task [1]. This is because of the uncertainties in the market movements since political events, economic conditions, and traders' expectations are all factors affecting the market [2].

Since the Egyptian Stock Market is a rising market it shows a highly volatile return which is difficult to predict. Most of the previous work in stock market prediction has aimed to predict the price levels of the stock market. However, recently the studies have proposed that prediction of the direction of price change may be more efficient and lead to higher profits [3]. The aim of this study is to attempt to predict the short-term future trend of the Stock Market by developing predictor system, consisting of three main phases. The first phase is a trend predictor model based on neural network. The second phase is a fuzzy rule based system to predict the short term stock trend based on the Elliott's wave theory [4]. The third phase is integration between the first two phases.

D.-S. Huang et al. (Eds.): ICIC 2011, LNBI 6840, pp. 85–90, 2012.
© Springer-Verlag Berlin Heidelberg 2012

In the following sections an overview of technical analysis will be discussed. Then the main features of the proposed system will be explained. Finally the system test results will be represented with a conclusion about the current state and future work.

2   Technical Analysis

One of the main techniques for stock price evolution analysis is Technical Analysis. Technical Analysis depends on many tools to analyze stock price such as technical indicators. In Technical Analysis the information needed is limited to stock price history of the value to be studied [5]. Two of technical analysis tools are:

• Technical Indicators: A technical indicator is a series of data points derived by applying a formula to the price data series. A technical indicator works to understand the behavior and psychology of the investor. It provides additional information to discover the future trending of price activity [5].
• Elliott Wave Theory: The Elliott wave theory has been developed from the observation that rhythmic regularity has been observed in the stock market over an eighty year period. It has been further observed that the market moved forward in a series of five (5) waves and declined in a series of three (3) waves [6].

3   The Proposed System

This paper presents an efficient stock market prediction system based on machine learning techniques. Fig. 1 shows a block diagram of the proposed system. In the following, each part is thoroughly discussed.

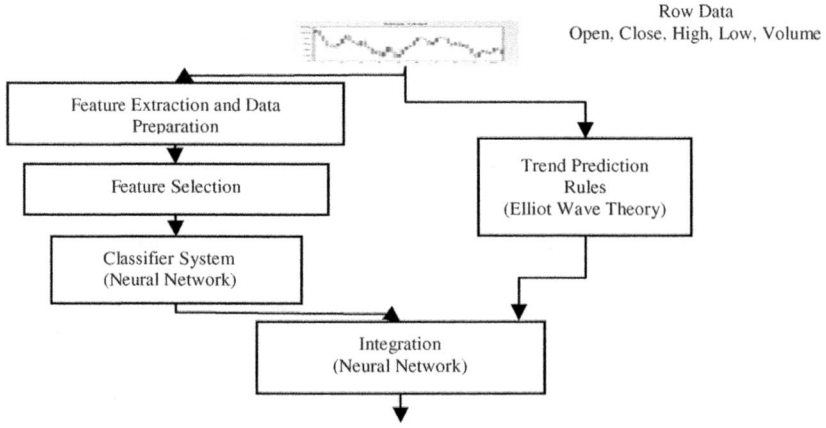

Fig. 1. Proposed System

3.1   Predictor Model Based on Neural Network

In this model the neural network predicts the future trend by classifying the historical data of the stock. This model is constructed of three stages:

The first stage is the Feature Extraction and Data Preparation stage, where the features used in the system are being calculated from the raw price values. The input features used are Technical indicators, as they analyze and give more information about the studied stock price. Out of the many technical indicators used by traders, 11 indicators have been chosen. These indicators are Moving Average, Relative Strength Index (RSI), Stochastic Oscillator, Williams %R, Exponential Moving Average, Bollinger Bands, Moving Average Convergence/divergence (MACD), Momentum, On Balance Volume (OBV), Commodity Channel Index (CCI), and Price Rate of Change (ROC). Beside these indicators the stock price is also used as an input feature.

Also in the Feature Extraction and Data Preparation stage data normalization is performed. As the stock price moves in a totally different price ranges over time. In our system we applied normalization per input window to keep the prices in it within a certain range. In the proposed system the statistical normalization technique is deployed, as it reduces the effect of outlier in the input data [7]. This is done by equation (1) which is applied for each price in the window.

$$N = \frac{M - C}{\sigma} \tag{1}$$

Where N is the normalized value of the price, M is the mean value of the closing price over the same window, C is the closing price that needs to be normalized and σ is the standard deviation of price for the same window.

The second stage is Feature selection, where a subset of original features is selected. One of the most famously used techniques is Sequential Selection. Sequential Selection uses hill climber search algorithm to optimize feature selection. In the proposed system Backward Sequential Selection is used in the feature selection phase.

The third stage is the classifier stage, where prediction is performed based on classifying the input vector as one of three classes; Up, Down and Sideways. The time frame for the input vector is 5 days as it represents a whole trading week [8]. The time frame window used is a sliding window so that in any given point the system can predict the trend. The system was trained for each stock separately to react according to the stock characteristics and to make use of the similar historical attitude of the price movement. The data used was annotated manually by technical analysis experts.

MLP, RBF and SVM were examined as classifiers. MLP was trained using a Powel-Beale conjugate gradient back-propagation method: (CGB in Matlab [9]), RBF was implemented by the Matlab radial basis network [9], while for SVM we used the package SVM-km for Matlab [10].

3.2 Fuzzy Rule Based Model

The main idea of this model is to learn how to correctly detect the Elliott wave patterns that tend to occur repeatedly in the market. The main challenge is to count the waves, and mark the current position of the stock on an Elliott wave pattern. The Elliott wave oscillator (EWO) emerged to help track this pattern. EWO have the following behavior: a higher value on the third wave, lower but positive values on the first and fifth waves, and negative values on biggest corrections, or downtrend impulse waves [11].

Trend Prediction rules is a fuzzy implementation for Elliot wave theory rules given by Kaufman [12] to predict the stock trend. The first step of this model is to calculate the Elliott wave oscillator (EWO) as shown in equation (3).

$$\text{mean} = \frac{\text{High} + \text{Low}}{2} \tag{2}$$

$$\text{EWO} = \text{MA(mean, 5)} - \text{MA(mean, 35)} \tag{3}$$

The second step is to implement the set of rules given by Kaufman to predict the trend based on Elliott wave theory. These rules are:

If EOW is highest(EOW,n) and trend is constant then Elliot Wave trend is up.

If EOW is lowest(EOW,n) and trend is constant then Elliot Wave trend is down.

If lowest(EOW,n) is less than zero and trend is down and EOW is greater than $-1 \times trigger \times lowest(EOW, n)$ then Elliot Wave trend is up.

If highest(EOW,n) is greater than zero and trend is up and EOW is less than $-1 \times trigger \times highest(EOW, n)$ then Elliot Wave trend is down.

According to our experiments n was chosen to be 50 and trigger 0.70. The presented model is a Mamdani type fuzzy system that was implemented using Matlab Fuzzy Logic Toolbox [9]. The member ship functions used are triangular, Z shaped and S shaped functions. The output of this model is the prediction of the next days' trend, where +1 indicating an uptrend, -1 for a down trend, and 0 for constant price movement.

3.3   Integration

To obtain the final system result, the output of the two predictor models are integrated using ANN. ANN's ability to learn from scratch makes it particularly suitable to perform this integration. The Integration phase works on predicting the future trend by classifying the output of the two predictor models. MLP is employed for the decision integration. The integrated system output represents the predicted trend as one of three classes; Up, Down and Sideways.

4   Experimental Results

First an experiment is conducted to evaluate the most relevant classifier system for the presented model. The data used in our system is the stock daily updates for the period from 1998 to 2009. The examined stocks are: National Societe Generale Bank (NSGB), Ezz Steel (ESRS), Egyptian Chemical Industries (EGCH), Suez Cement (SUCE), Housing and Development Bank (HDBK, Extracted Oils (ZEOT, and Arab Pharmaceuticals (ADCI). Prices from January, 1998 to January, 2004 are used for training while those from January, 2006 to January, 2009 are for testing. Table 1 shows the accuracy results of each classifier with and without the Backward Sequential Selection feature selection results. As it is shown from this experiment, the MLP gives better performance accuracy than the other tested classifiers and adding a feature selection method increases the prediction accuracy. Therefore the feature selection method for data preprocessing and the MLP will be used in the system.

Second, an experiment was performed to evaluate the efficiency of the proposed system by comparing its results with the two integrated models. The integration phase is trained using the output of the classifier model and the fuzzy rule based model for the period from January, 2004 to January, 2006.The performance of the integrated system is shown in table 2. It is shown that integrating the neural network and the fuzzy rule based models gives a better performance than each of the two models.

In order to quantify the presented work another trading test was performed to calculate the profit gained by the presented system and compare it to the gain that is obtained by buy and hold strategy. In this test virtual profile was made that starts trading by January, 2006 with only 10000 L.E and goes on for 3 years. When the system indicated a predicted up trend we buy 10 stocks and if it indicated a predicted down trend we sell 10 stocks. Fig. 2 shows a graphical representation of the results for one of the stocks.

5   Conclusion

The proposed system introduced in this paper was tested on different stocks in the Egyptian Stock Market for a 3 years period. The system is simulated using Matlab[9] and the results show a good accuracy near 90%. The proposed system has the ability to predict the future trend at any trading day.

Future work includes detailed analysis of the errors and more advanced methods to overcome these errors. Also, the proposed system was only applied and examined on the Egyptian market stocks. The system must explore other stock market.

References

1. Coupelon, O.: Neural Network Modeling for Stock Movement Prediction - A State of the Art (2007)
2. Choudhry, R., Garg, K.: A Hybrid Machine Learning System for Stock Market Forecasting. World Academy of Science, Engineering and Technology 29, 29–59 (2008)
3. Tilakaratne, C.D., Morris, S.A., Mammadov, M.A., Hurst, C.P.: Predicting Stock Market Index Trading Signals Using Neural Networks. In: Proceedings of the 14th Annual Global Finance Conference (GFC 2007), Melbourne, Australia, pp. 171–179 (September 2007)
4. Holter, J.T.: The Markets' Hidden Order. Futures 25(13), 66–71 (1997)
5. Fernández-Blanco, P., Bodas-Sagi, D., Solterol, F., Hidalgo, J.I.: Technical Market Indicators Optimization using Evolutionary Algorithms. In: GECCO 2008, pp. 1851–1857 (2008)
6. Chatterjee, A., Felix Ayadi, O., Maniam, B.: The Applications Of The Fibonacci Sequence And Elliott Wave Theory In Predicting The Security Price Movements: A Survey. Journal of Commercial Banking and Finance 1, 65–76 (2002)
7. Priddy, K.L., Keller, P.E.: Artificial Neural Networks: An Introduction. SPIE Press, Washington (2005)
8. Vanstone, B.J.: Trading in the Australian Stock market using Artificial Neural Networks. Thesis (P.Hd). Bond University (2005)
9. MATLAB, MATLAB Reference Guide, The Math Works Inc. (version 7.5.0.342) (2007)

10. Canu, S., Grandvalet, Y., Guigue, V., Rakotomamonjy, A.: SVM and Kernel Methods Matlab Toolbox, Perception Systèmes et Information. In: INSA de Rouen, Rouen, France (2005)
11. Atsalakis, G.S., Dimitrakakis, E.M., Zopounidis, C.D.: Elliott Wave Theory and neuro-fuzzy systems. In: Stock Market Prediction: The WASP system, Expert Systems with Applications (2011), doi:10.1016/j.eswa.2011.01.068
12. Kaufman, P.J.: Trading Systems and Methods, 3rd edn. John Wiley & Sons, Inc., USA (1998)

Table 1. Feature selection prediction accuracy results

| Stocks | SVM | | MLP | | RBF | |
	all features	selected features	all features	selected features	all features	selected features
NSGB	71.57%	75.29%	77.29%	79.86%	57.14%	61.63%
ESRS	68.23%	72.33%	70.40%	73.61%	56.42%	61.88%
EGCH	54.42%	56.72%	64.83%	67.67%	46.55%	51.25%
SUCE	75.23%	78.55%	79.56%	81.92%	59.09%	63.38%
HDBK	71.66%	75.37%	71.42%	74.54%	52.03%	57.95%
ZEOT	73.17%	76.15%	75.98%	78.67%	62.99%	66.87%
ADCI	69.43%	73.38%	71.26%	75.30%	58.86%	61.38%
Average	69.10%	72.54%	72.96%	75.94%	56.15%	60.62%

Table 2. Prediction accuracy of the two integrated models and the overall system

Stock	Predictor Model Based on Neural Network Results	Fuzzy Rule Based Model Results	integrated System Results
NSGB	79.86%	69.27%	90.40%
ESRS	73.61%	69.83%	88.51%
EGCH	67.67%	69.76%	91.69%
SUCE	81.92%	75.35%	93.02%
HDBK	74.54%	75.30%	86.79%
ZEOT	78.67%	72.72%	92.04%
ADCI	75.30%	73.68%	91.02%
Average	75.94%	72.27%	90.50%

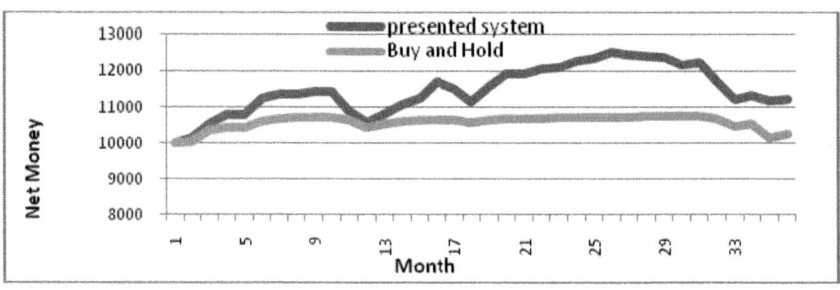

Fig. 2. The return from trading in the NSGB

Inferring Protein-Protein Interactions Based on Sequences and Interologs in Mycobacterium Tuberculosis

Zhi-Ping Liu[1], Jiguang Wang[2], Yu-Qing Qiu[3], Ross K.K. Leung[4],
Xiang-Sun Zhang[3], and Stephen K.W. Tsui[4], and Luonan Chen[1]

[1] Key Laboratory of Systems Biology, Shanghai Institutes for Biological Sciences,
Chinese Academy of Sciences, Shanghai 200031, China
[2] Beijing Institute of Genomics, Chinese Academy of Sciences, Beijing 100029, China
[3] Academy of Mathematics and Systems Science, Chinese Academy of Sciences,
Beijing 100190, China
[4] Hong Kong Bioinformatics Centre, The Chinese University of Hong Kong,
Shatin N. T., Hong Kong, China
lnchen@sibs.ac.cn

Abstract. *Mycobacterium tuberculosis* is a pathogenic bacterium that poses serious threat to human health. Inference of the protein interactions of *M. tuberculosis* will provide cues to understand the biological processes in this pathogen. In this paper, we constructed an integrated *M. tuberculosis* H37Rv protein interaction network by machine learning and ortholog-based methods. Firstly, we developed a support vector machine (SVM) method to infer the protein interactions by gene sequence information. We tested our predictors in *Escherichia coli* and mapped the genetic codon features underlying protein interactions to *M. tuberculosis*. Moreover, the documented interactions of other 14 species were mapped to the proteome of *M. tuberculosis* by the interolog method. The ensemble protein interactions were then validated by various functional linkages i.e., gene coexpression, evolutionary relationship and functional similarity, extracted from heterogeneous data sources.

1 Introduction

M. tuberculosis is the causative agent that causes tuberculosis and leads to lesions in lungs and other organs. Tuberculosis is the second leading cause of death in infectious diseases. An extensive protein-protein interaction (PPI) network of *M. tuberculosis* can lead to more comprehensive screens of its cellular operations. To date, genome-wide experimental and computational systems for studying PPIs in *M. tuberculosis* is not available [1]. It is urgently necessary to develop approaches capable of converting available genomic data into functional information for *M. tuberculosis*. *E. coli* is one of the best model systems to study bacterial physiology, with well-characterized interactome, genome and transcriptome [2]. Interaction features can be learned by machine learning methods, such as support

D.-S. Huang et al. (Eds.): ICIC 2011, LNBI 6840, pp. 91–96, 2012.

vector machines (SVMs) [3], and also it is common to predict protein interactions from known interactions of other organisms by interolog method [4].

Genetic information in the form of codons, i.e. tri-nucleotide sequences, specifies amino acid sequence in the polypeptide during the synthesis of proteins. It is well known that codon usage is correlated with expression level [5]. Genetic codons will be selected as the sequence features in the learning of interaction patterns. Moreover, the corresponding orthologs of interacting proteins in other organisms will provide more information about the potential interaction mappings by comparative genomics.

In this work, we developed a systematic method combining heterogeneous data sources to infer a comprehensive protein interaction network in *M. tuberculosis*. The codon features of interacting protein pairs are detected and used to train an SVM classifier. Moreover, the interactions from other 14 species are mapped to *M. tuberculosis* by the interolog method. The available data from multiple levels including gene coexpression and evolutionary relationship to functional similarity are implemented to assess these predicted interactions by confidence significance. The predicted protein interaction network as well as the proposed method provide a framework for the functional specificities study of *M. tuberculosis*.

2   Methods

2.1   Framework of Prediction

The protein interactions were predicted by two main pipelines. Firstly, we built the protein interaction network of *M. tuberculosis* from codon features of interacting proteins in *E. coli* by machine learning approach. The integrated interaction maps and gene sequences of *E. coli* were retrieved from EcID [2]. The ORFs of *M. tuberculosis* were derived from the laboratory strain H37Rv. We used the information of protein interaction network of *E. coli* to train an SVM classifier to get the genetic codon features underlying the interacting pairs. The interactions in *M. tuberculosis* were then predicted by the trained SVM predictor with the genetic codons of ORFs in gene sequences of *M. tuberculosis*. Secondly, we inferred the protein interactions of *M. tuberculosis* by interolog method from the documented protein interactions in 14 species. We collected these interactions from IntAct [6] and DIP [7] and the *M. tuberculosis* orthologs of these interologs were identified by BLAST [8]. The homologs of two interacting proteins will be identified as the predicted interactors. As for the validation of predicted results, we tested our method in *E. coli*. Three pieces of available information of *M. tuberculosis*, i.e., gene expression profiling, evolutionary relationship from ortholog database and functional similarity, were used to evaluate the prediction results.

2.2   Validation from Multiple Resources

We implemented multiple available resources to access the constructed PPI network in *M. tuberculosis*. The confidence of interactions was evaluated by three extra data sources, namely, gene expression, evolutionary relationship and functional similarity. Firstly, we identified the Pearson correlation coefficients (PCC) of gene coexpression of pairwise proteins in the predicted network. We downloaded the gene expression data of *M. tuberculosis* from NCBI GEO (ID: GSE9776). Secondly, we presented the evaluation of evolutionary relationship between the predicted interacting proteins. Clusters of orthologous groups (COGs) were delineated by comparing protein sequences encoded in complete genomes, representing major phylogenetic lineages. Each COG consists of individual proteins or groups paralogs from at least 3 lineages and thus corresponds to an ancient conserved domain. The maximum of COG value between two groups in which the interacting proteins located were regarded as the value representing their evolutionary relationship. Thirdly, Gene Ontology (GO) similarity between the predicted pairs were identified to evaluate their functional relationship. We used semantic similarity measures [9] to evaluate the similarity of GO term lists corresponding to the interacting proteins.

3   Results

3.1   Performance of Predictor

E. coli is one of the best characterized organisms and has been served as a model system to study many aspects of bacterial physiology [2]. The positive and negative sets of protein interactions in *E. coli* were designed to test the performance of our codon-based prediction methods. The genome and proteome of *E. coli* were downloaded and prepared for the interacting sets as well as all known opening reading frames (ORFs). The distance of two ORFs in terms of usage of codon c is defined as $d_{ij}(c) = |f_i(c) - f_j(c)|$, where $f_i(c)$ and $f_j(c)$ are relative frequencies of codon c in ORF i and ORF j. By codon definition, $\sum_k f_i(c_k) = 1$ and $\sum_k f_j(c_k) = 1$ for $k = 1, 2, ..., 64$ in all codons. There are 14058 pairs of interactions and 27882 pairs of non-interactions in 4227 proteins of *E. coli*. A five-fold cross validation process is implemented in these pairs. Figure 1 shows the performance of prediction results by the SVM predictor using genetic codon features. There are several codons corresponding to the same amino acid in genetic code. The prediction performance of merging the frequency of these degenerate codons ('codon-mer') is also shown in Figure 1. The details of prediction precision and accuracy are listed in Table 1. The results provide evidences for the effectiveness and efficiency of predicting protein interactions from the genetic codons by machine learning method.

Table 1. Prediction performances of the codon-based SVM predictor in *E. coli*.

Feature	ACC	SN	SP	PRE	AUC
Codon	0.9003	0.7576	0.9486	0.8327	0.9507
Codon-mer	0.9595	0.8986	0.9801	0.9386	0.9835

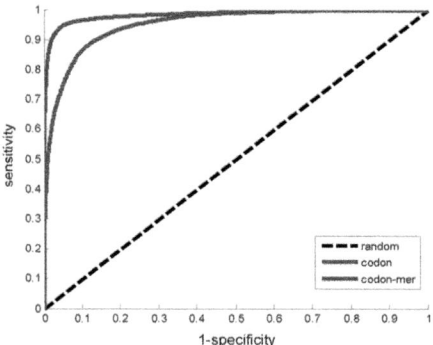

Fig. 1. ROC curves of the five-fold cross validation predictions in *E. coli*.

3.2 Protein Interactions in *M. tuberculosis*

To explore protein interactions in *M. tuberculosis*, we used the formerly trained SVM classifier to infer the interactions of *M. tuberculosis* by the codon message of ORFs in gene sequence level. Based on the genetic codons of *M. tuberculosis* H37Rv, we predicted 12,899 interactions in 3,266 proteins. Furthermore, the known protein interactions of other species were mapped to the proteome of *M. tuberculosis* by interolog method. We collected the documented interactions of 14 species from PPI databases, IntAct and DIP, and the sequence features of interacting proteins were transferred into the *M. tuberculosis* proteome by ortholog detection. Table 2 lists the detailed prediction results by interolog method. The known protein interactions were also included in our inferred interactome of *M. tuberculosis*. So far, we also found 530 pairs of protein interactions of *M. tuberculosis* from various databases, such as BIND [10] and Reactome [11]. Combining with these known interactions, we built a comprehensive protein interaction map totally with 46,119 interactions of 3,465 proteins in *M. tuberculosis*.

3.3 Validation Results

Protein interacting pairs are identified with close relationship with gene coexpression, coevolution, similar GO annotations. To every predicted interacting pairs of *M. tuberculosis*, we collected these available heterogeneous data sources to annotate them. Firstly, we annotated the predicted pairs by their corresponding PCC of gene coexpression. For comparison, we calculated the corresponding correlation values of these same-size random selected protein pairs. Every prediction was then annotated by a coexpression value in gene expression profiling. Figure 2 (a) shows the boxplot of coexpression values in the predictions. From Figure 2, we identified that the coexpression values in the predicted interacting pairs tend to be more correlated when compared to that of random selected ones. Secondly, we identified the evolutionary relationship of the interacting proteins by COG information. The interacting proteins were detected in their own COG individually. Figure 2 (b) shows the boxplot of evolutionary relationship values in

Table 2. Details of predicted protein interactions in *M. tuberculosis*

Species	Database	Original PPI	Predicted PPI	Percentage (%)
By machine Learning				
E. coli	ECID	14,058(positive)+ 27,882(negative)	12,899	27.97
By interolog				
Escherichia coli	IntAct	14,158	16,468	35.71
Campylobacter jejuni	IntAct	11,870	7,674	16.64
Treponema pallidum	IntAct	3,744	324	0.70
Synechocystis	IntAct	2,625	2,481	5.38
Myxococcus xanthus	IntAct	384	253	0.55
Synechocystis sp.	IntAct	219	220	0.48
Rickettsia sibirica	IntAct	282	24	0.05
Streptococcus pneumoniae	IntAct	193	47	0.10
Drosophila melanogaster	DIP	22,650	1,558	3.38
Saccharomyces cerevisiae	DIP	21,769	2,701	5.86
Caenorhabditis elegans	DIP	3,979	229	0.50
Homo sapiens	DIP	1,485	84	0.18
Mus musculus	DIP	287	36	0.06
Rattus norvegicus	DIP	69	2	0.15
Total: 46,119 interactions in 3,465 proteins (with 530 known PPIs)				

the predicted interacting pairs and that of the same-size random selected protein pairs. Every predicted interaction gets a confidence of evolutionary relationship. Thirdly, we calculated the functional similarities underlying these predicted interactions. We detected the semantic similarity between the GO term pairs of interacting proteins. The boxplots of the three values of GO similarities, i.e., cellular component ('CC'), molecular function ('MF') and biological process ('BP'), in random pairwise proteins and that in predicted pairs are shown in Figure 2 (c), (d) and (e), respectively. The predicted interactions have higher functional similarity than random ones, which further provides evidence for the accuracy of our results.

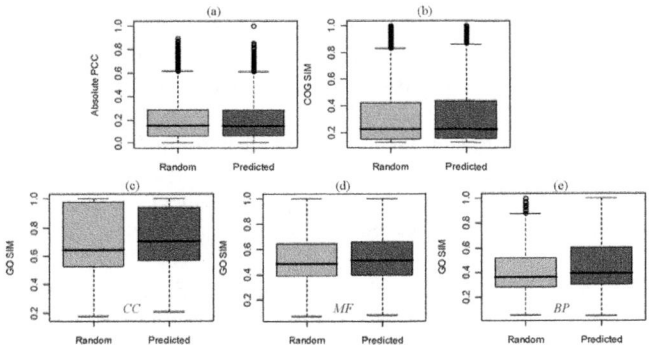

Fig. 2. Boxplot of coexpression (a), coevoluation (b) and cofunction values (c)–(e) of the predicted interactions and that of the same-size random selected protein pairs

4   Conclusion

In conclusion, we established a novel framework to integrate genomic data to infer PPIs in *M. tuberculosis*. We predicted the protein interactions in *M. tuberculosis* by an SVM based classifier by genetic codons. And the documented protein interactions from various species were also mapped to the proteome of *M. tuberculosis* by interolog method. The information from gene expression, evolutionary and functional relationship provided reliable measures of evaluation of our predictions. Our framework can easily be extended to infer the large-scale protein interactions in other species. These predicted interactions provide a valuable reference of interactome for *M. tuberculosis* research. The PPIs are available at: http://www.aporc.org/doc/wiki/MTBPPI.

Acknowledgements. This work was supported by Shanghai NSF under Grants No. 11ZR1443100, and by the Chief Scientist Program of SIBS, CAS under Grant No. 2009CSP002. Part of the authors were also supported by NSFC under Grant No. 61072149 and 91029301.

References

1. Singh, A., Mai, D., Kumar, A., et al.: Dissecting Virulence Pathways of Mycobacterium Tuberculosis Through Protein-Protein Association. Proc. Natl. Acad. Sci. USA 103, 11346–11351 (2006)
2. Andres, L.E., Ezkurdia, I., et al.: EcID. A Database for the Inference of Functional Interactions. E. coli. Nucleic Acids Res. 37, D629–D635 (2009)
3. Shen, J., Zhang, J., et al.: Predicting protein-protein interactions based only on sequences information. Proc. Natl. Acad. Sci. USA 104, 4337–4341 (2007)
4. Yu, H., Luscombe, N.M., et al.: Annotation Transfer Between Genomes: Protein-Protein Interologs and Protein-DNA Regulogs. Genome Res. 14, 1107–1118 (2004)
5. Najafabadi, H.S., Salavati, R.: Sequence-based Prediction of Protein-Protein Interactions by Means of Codon Usage. Genome Biol. 9, R87 (2008)
6. Kerrien, S., Alam-Faruque, et al.: IntAct–open Source Resource for Molecular Interaction Data. Nucleic Acids Res. 35, D561–D565 (2007)
7. Xenarios, I., Salwinski, L., et al.: DIP, the Database of Interacting Proteins: A Research Tool for Studying Cellular Networks of Protein Interactions. Nucleic Acids Res. 30, 303–305 (2002)
8. Altschul, S.F., Madden, T.L., et al.: Gapped BLAST and PSI-BLAST: A New Generation of Protein Database Search Programs. Nucleic Acids Res. 25, 3389–3402 (1997)
9. Lord, P.W., Stevens, R.D., et al.: Investigating Semantic Similarity Measures Across the Gene Ontology: The Relationship Between Sequence and Annotation. Bioinformatics, 1275–1283 (2003)
10. Alfarano, C., Andrade, C.E., et al.: The Biomolecular Interaction Network Database and Related Tools. Nucleic Acids Res. 33, D418–D424 (2005)
11. Vastrik, I., D'Eustachio, et al.: Reactome: A Knowledge Base of Biologic Pathways and Processes. Genome Biol. 8, R39 (2007)

Predicting Long-Term Vaccine Efficacy against Metastases Using Agents

Marzio Pennisi[1], Dario Motta[1],
Alessandro Cincotti[2], and Francesco Pappalardo[1]

[1] University of Catania, Catania, Italy
{mpennisi,francesco}@dmi.unict.it
[2] School of Information Science,
Japan Advanced Institute of Science and Technology, Japan
cincotti@jaist.ac.jp

Abstract. To move faster from preclinical studies (experiments in mice) towards clinical phase I trials (experiments in advanced cancer patients), the chance to predict the outcome of longer experiments represents a key step. We use the MetastaSim model to predict the long-term effects of the Triplex vaccine against metastases. To this end we simulate follow-ups of two and three of three months (equivalent approximately to 5.83 and 8.75 years in humans) to compare the long-term efficacy of the best protocol used "in vivo" against the one found by the MetastaSim model. We also check the efficacy of these two protocols by delaying the time of the first administration, in order to catch up the maximum time delay between the appearing of metastases and the administration of the vaccine needed to guarantee reasonable treatment efficacy.

1   Introduction

In tumor immunology two main approaches can be identified: Immunoprevention and Immunotherapy. The former attempts to train the immune system in recognizing cancer cells before they appear in the host whereas the latter is based on a series of immunologic treatments administered to cancer patients to eradicate existing tumors.

The Triplex vaccine [4,5] represents a clear example of an immunopreventive approach developed to fight against breast cancer. It combines three different signals for the immune system in the same product. The target antigen is administered in combination with two "adjuvants" represented by IL-12 and allogeneic MHC molecules. The IL-12 adjuvant enhances antigen presentation and Th cell activation whereas the allogeneic MHC molecules stimulate different T cell clones and cause a huge production of various cytokines that amplify immune responses.

This cell vaccine has been capable to totally prevent tumor formation in HER-2/neu transgenic mouse models under a Chronic vaccination schedule in a follow-up time of one year. Shorter heuristic protocols failed in fulfilling this job, leaving unanswered the question of whether a protocol capable of guarantee long term survival with a minimal number of administrations exists. SimTriplex [7],

D.-S. Huang et al. (Eds.): ICIC 2011, LNBI 6840, pp. 97–106, 2012.

an Agent Based Model (ABM) developed to model this "in vivo" experiment, has been used in order to to answer at this question. Combining SimTriplex with well known artificial intelligence optimization methods [15,9,6], led to a schedule with fewer injections which has been now tested "in vivo", yielding to encouraging results [14].

In recent studies [8] the same vaccine demonstrated able (in a follow-up of one month) to elicit a considerable therapeutic activity against metastases derived by mammary carcinoma, thus showing immune responses that overlap only partially those at work in long-term immunoprevention of carcinogenesis. To give biologists the chance to better understand the biological behavior and to predict alternative vaccination schedules, a new ABM named MetastaSim has been developed [12,16]. The model has been used to minimize the scheduling and the number and of vaccinations needed to assure almost complete metastases eradication, predicting a protocol with 40% less injections than the best protocol used "in vivo" (hereafter referred as 1-Triplex).

In order to move faster from preclinical studies (experiments in mice) towards clinical phase I trials (experiments in advanced cancer patients), the chance to predict the outcome of longer experiments represents a key step. In this paper we use the MetastaSim model to predict the long-term effects of the vaccine against metastases. To this end we simulate follow-ups of two and three of three months (equivalent approximately to 5.83 and 8.75 years in humans) to compare the long-term efficacy of the best protocol used "in vivo" against the one found by the MetastaSim model.

Moreover we check the efficacy of these two protocols by delaying the time of the first administration, in order to catch up the maximum time delay between the appearing of metastases and the administration of the vaccine needed to guarantee reasonable treatment efficacy.

The paper is structured as follows: Section 1 introduces the problem. In section 2 we will give a brief introduction of the biological background. In section 3 we will recall the MetastaSim model. In Section 4 we will describe the long-term simulations and we will show the obtained results. Finally in section 5 we will draw final conclusions and considerations.

2   Biological Background

Both the immunopreventive and the terapeutic experiments use BALB-neuT female mice models . After birth BALB-neuT mice develop cells hyper-expressing HER-2/neu gene product (p185) in mammary glands. From these cells multiple microscopic lesions arise becoming identifiable as atypical hyperplasia, then progressing to carcinomas in situ, up to macroscopic lesions detectable at around 4-5 months of age.

The therapeutic experiment to test the Triplex efficacy against lung metastases lasts for 32 days. At day 0 all mice receive an intravenous injection of $2.5 \cdot 10^4$ cancer TuBo neu cells (referred to as Neu/H-2), which are used to induce experimental metastases in syngeneic BALB-neuT mice. Then mice are divided in three different sets: an untreated or control set, a first set which is treated with a protocol (vaccination schedule) composed by a twice-weekly vaccination cycle started one day after the injection of the metastatic cells and repeated up to the end of the experiment (1-Triplex protocol), and a set of mice treated with the same cycle started 7 days after the injection of metastases and repeated up to the end of the experiment (7-Triplex protocol).

We note here that instead of waiting later tumor stages where the breast cancer gives rise to the metastatic burden, experimental metastases are induced artificially in healthy mice because mice in late tumor stages present multiple problems, such as an aged immune system. Moreover surgical remotion of primary tumors cannot be achieved easily, and it is also not possible to exactly establish if and when the metastatic process starts.

The Triplex vaccine stimulates immune system responses using the following stimuli:

- The p185neu antigen, product of the rat HER-2/neu gene;
- H-2q MHC molecules (allogeneic for H-2d BALBneuT mice);
- Interleukin-12 (the cells are engineered with the genes coding for murine IL-12).

The p185neu represents the target antigen recognised and by the immune system responses. The H-2q MHC class I molecules favor recognizing by multiple cytotoxic T cell clones and cause a huge production of various cytokines that amplify immune responses. The IL-12 enhances antigen presentation, helper T cell (TH) activation and secretion of interferon-γ (IFN-γ) by natural killer (NK) and TH cells. IFN-γ also has a cytostatic activity on cancer cells (CC) and stimulates granulocytes and macrophages (MP) in infiltrating tumor cell nests in the lungs. Activated TH cells release various cytokines such as interleukin-2 (IL-2) which enhances cytotoxic T cells (TC) activities and releasing of antibodies (Ab) by plasma B cells.

At the end of the experiment (day 32) all mice are killed and lungs are examined to detect the number of formed metastatic lesions. Mice from the untreated set showed ≈ 200 metastatic nodules; mice treated with 1-Triplex and 7-Triplex protocols respectively showed a reduction $> 99\%$ and $\approx 87\%$ in the number of visible lesions.

3   Brief Description of the MetastaSim Dodel

MetastaSim can be defined as "Agent-Based like Model" or as an "extended Cellular Automaton" and uses the same computational framework of the SimTriplex model [7]. An exhaustive description of the model and the modeling framework in general can be found in [16] and in [7], so here we only limit to briefly sketch the model. MetastaSim uses a bi-dimensional 128x128 lattice with hexagonal geometry. The lattice represents a slice of tissue of the frontal ventral surface of mice left lung, and covers a surface of approximately $64mm^2$ and a thickness of $1mm$. Every cell represents an agent with its own life-time, biological behavior, position in the lattice, set of internal states and one or more receptors. Molecules are represented by their concentration per lattice-site, and by their molecular composition in the case of antigens and antibodies. Relevant immune system entities are modeled and randomly distributed on each lattice-site according to their leukocyte formula. Binary strings are used to represent cell receptors and the molecular structures of antibodies and antigens. The interaction probability of two entities is a function of the Hamming distance between their binary strings. This process mimics well real receptor binding and it s able to reproduce relevant biological phenomena, as shown in [10]. At each time-step ($\Delta t = 8$ hours) all entities that lie in the same lattice-site can probabilistically interact with each-others. Obviously only interactions that are immunologically correct and relevant are allowed. As a consequence of an interaction entities can change their internal status, can release other entities (i.e. plasma B cells release antibodies), can duplicate or can be killed. After the interaction phase ends, entities can probabilistically move to a lattice site in their neighborhood.

One biological assumption that has been made is that every nodule has originated from an individual cancer cell. This means that only one cancer cell over 100 is able to pass through lung capillary vessels and settle into the lungs. This settlement process is not modeled and the simulation starts by supposing that cancer cells have already settled in the lungs. In order to model the same nodule kinetics observed in "in vivo" experiments, it is possible to observe that nodule growths are usually proportional to the quantity of nutrients available. However nutrient distribution in lungs is not known and, due to their extremely high vascularization, neither easily predictable. To reproduce the same nodule distribution in sizes observed "in vivo", sizes and number of nodules data from 8 different real mice are then used with the inverse transform sampling method [2] to generate n random nodule measures that are distributed according to the "in vivo" experiment. These sizes are converted into parameters for the Gompertz growth [1] law and used to compute the duplication probabilities that cancer cells belonging to the same nodule have. After a tuning phase where 8 virtual mice were used to determine the optimal value of free parameters, the model has been validated "in silico" by using 100 virtual mice. As result it showed a good agreement with the "in vivo" experiment [16]. Then its first application has been to search for a protocol capable to assure against lung metastases the same protection entitled with the use of the 1-Triplex protocol.

According to biologists' opinion, no more than two vaccinations per week (in pre-established days) can be done. This needed to satisfy some wet biology requirements (i.e. vaccine preparation) and to guarantee a certain level of safeness for the mice (i.e. avoid undesirable effects or exposition to excessive stresses). The 1-Triplex protocol already uses all the available 9 days to vaccinate. Shorter protocols should be therefore obtained by removing some injections from 1-Triplex protocol. MetastaSim has been then used to explore exhaustively the search space of 2^9 of possible protocols, finding a five injections protocol (hereafter referred as Optimal) able to give rise to similar protection entitled with the use of 1-Triplex protocol [16]. The protocol is composed by the three injections in the first 3 available days (days 1,2, and 3), followed by two vaccine recalls (at days 5 and 7). It is worth to note here that all high ranked found protocols share the same structure, i.e. a boost of three injections followed by some (more or less) equally spaced vaccine recalls.

4 Long-Term "in silico" Experiments

In previous simulations both the 1-Triplex and the Optimal protocol were able to elicit almost complete eradication (approx. 99% in the number of prevented nodules), whereas the "in silico" 7-Triplex prevention was estimated to be only around 82-83 %. From a translational point of view the time-length of this first experiment is probably too short to investigate long term experiment results. Life span in BALB - NeuT mice is usually around 2 years whereas medium lifespan in humans is around 70 years. It is therefore possible to consider a scale factor of around 1 to 35. This means that the first experiment would cover a period of approximately 3 years in humans, whereas the critical the time window for the appearing of recidivous cells is usually considered to be 5 years or more.

The first scenario we simulated is therefore composed by an experiment with a time-length of two months. The 1-Triplex and the Optimal protocol are administered only for the first month. This scenario would allow to understand whether metastases are able to start their growth again after the first month or if 1-Triplex and the Optimal protocols are able entitle prevention for longer times, and if they catch the same prevention. At the same time we also checked the 7-Triplex efficacy on the same scenario to show if the protocol is somewhat able to recover the gap between its entitled efficacy and the other two protocols. We will also checked what happens if we administer the 1-Triplex and the optimal protocol for the first two months in a three-months follow up. All the simulations are executed on a randomly selected 100 virtual mice set, and then the median and the mean number of visible metastases over the entire set are taken as outcome. Most important mean entities behaviors are shown as well. Protocols efficacy is measured considering the total number of entitled metastatic nodules a the end of the experiment. Best protocols will have lower medians and means. 1-Triplex and 7-Triplex protocols are extended by repeating the twice-weekly cycle for the required periods. The 2 and 3 months extensions of the Optimal protocol are obtained by repeating the injection schedule of the first month one or two more times respectively.

We remark here that only visible nodules (i.e. nodules that have reached a minimum number of cells and should be visible in "in vivo" lung examinations) are considered. In some cases, even if there is no evidence of visible metastatic burden (such as visible nodules), cancer cells may be still present in a small number but taken under control by the immune system. Moreover tumor multiplicity (presented in tables 1 and 2) is not strictly connected with the total mean number of cancer cells (shown, for example, in figure 2) present in the system. The former represents a measure that can be checked and compared with the "in vivo" results, the latter is only presented to deeply investigate and analyze the system and does not have an equivalent in the "in vivo" experiments. Table 1 reassumes the results coming from the previously described scenarios. The first three rows refer to the one-month experiments already published in [16] and are reported for comparison. In a two months follow-up it's easy to see that the median and mean numbers of metastases entitled with the use of Triplex-1, Triplex-7 and Optimal protocols remain substantially unchanged.

The use of the 1-Triplex and Optimal protocols for two months in a three months follow-up indeed shows how both the two protocols mostly eradicate the metastatic burden. In figure 1 the behavior of some of the involved cells and molecules is showed for this last scenario. Even if the immune response of B and T helper cells is slightly weaker in the Optimal protocol, the mean cancer cells curves of the two protocols are practically indistinguishable. This confirms the redundancy of injections in the 1-Triplex protocol. The lack of some spikes in the Ag graph for the Optimal protocol is justified by the lack of some injections in respect to 1-Triplex protocol. We point out here that these spikes are due to the fact that every vaccine injection introduces new vaccine cells that are easily killed and release antigens as result. Antigens are then rapidly captured by antigen presenting cells and presented to B and T cells to stimulate the immune response. The difference in efficacy between the first two protocols and the 7-Triplex suggests that the latter is probably started too late to entitle total efficacy. The knowledge of the maximum time delay between the injection of metastases and the administration of the vaccine needed to guarantee metastases prevention thus represents a question that should be answered. To this end we simulated a follow-up of three months where the 1-Triplex and Optimal protocols are administered by delaying the start of the treatment of 3, 5, and 7 days. From table 2 it is possible to note that even in this case both protocols allow similar protection rates. A 3 days delay in the start of the treatment does not particularly affect the efficacy of the protocol. Starting from a delay of 5 days it is possible to observe a worsening of the final outcome of the experiment.

To better investigate this scenario it is possible to look at figure 2 where the mean behavior of the total number of cancer cells for the entire time-length of the experiment is displayed. Both the protocols show negligible differences in general and are able to destroy all cancer cells in no more than 65 days when administered with no delay. A delay of 3 days postpones the metastatic cells elimination of approximately 10 days. If a 5-days delay is taken into account, both the treatments are not able to completely eliminate all cancer cells in 3 months even if administered for the entire period. However it is possible to observe that cancer cells curves are decrescent for the 5-days delayed protocol, suggesting a possible complete depletion if longer times are taken under observation. As previously suggested, a protocol delayed by 7 days seems to be not able to contrast the growth of some metastatic nodules. Even if it is possible to observe a decreasing trajectory between days 15 - 30 due to the fact that nodules with low growing rates succumb to immune system responses, nodules with higher growing rates are already too big to be eliminated by the immune system.

5   Conclusions

Cancer represents nowadays one of the most appalling diseases. In particular, metastases represent one of the major concerns in the clinical management of cancer. The majority of cancer mortality is in fact associated with this disseminated disease rather than the primary tumor [3]. Moreover standard protocols for the treatment of cancer usually establish that the risk of recidivous cells last for periods even longer than 5 years. In this optic the use of treatments that can be safely used for long times with minimum risk of collateral effects to substitute, or integrate existing chemotherapy-based treatments can represent a key step in the fight against cancer. However treatment minimization is always advisable. The experimental induction of metastases in tumor-free mice can represent well a typical scenario in human cancer, i.e., the scenario arising after the surgical removal of the primary tumor. Computational modeling of this setup allowed prediction of an Optimal protocol an the "in silico" study of long-term efficacy of the Triplex vaccine.

From results we found that the both 1-Triplex and the Optimal protocols are able to yield to a mostly complete eradication of the metastatic burden if vaccine administrations are continued for two months. From a translational point of view this would suggest that the use of this vaccine in human should be extended to a period of 5 years after surgical intervention. Another major finding comes out from the analysis of the maximum delay needed to avoid the appearing of metastases. MetastaSim suggested a maximum delay of 3 days to achieve complete eradication in a period of three months. This would translate in a maximum delay of slightly more than 3 months in humans. A delay 5 days (which translate to almost 6 months in humans) remains treatable but probably requires longer treatment times. Higher delays may entitle high risk of metastatic occurrence and therefore should be avoided.

We would like to highlight that even if these "in silico" predictions are strictly related to the use of the Triplex vaccine and to the experimental setup it has been tested, such modeling approaches can be applied and integrated to successfully study other diseases and pathologies, such as the ImmunoGrid framework [13,11].

Figures

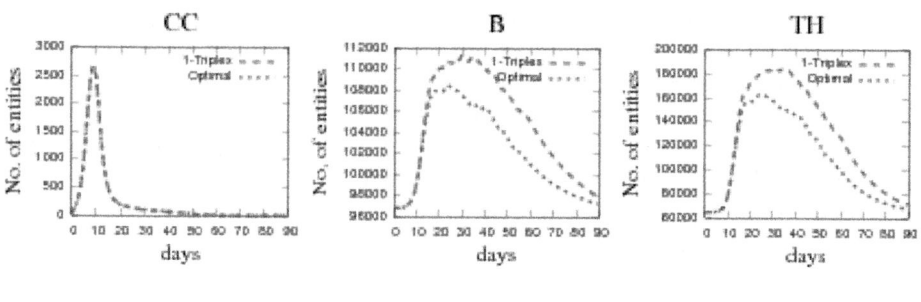

Fig. 1. Cancer Cells (CC), B cells, T helper cells (TH), cytotoxic T cells (TC), Antigens (AG),Interferon-γ (IFN-g) behaviors with 1-Triplex and the Optimal protocols. The time-length of the experiment is 3 months. The two protocols are repeated only the first two months.

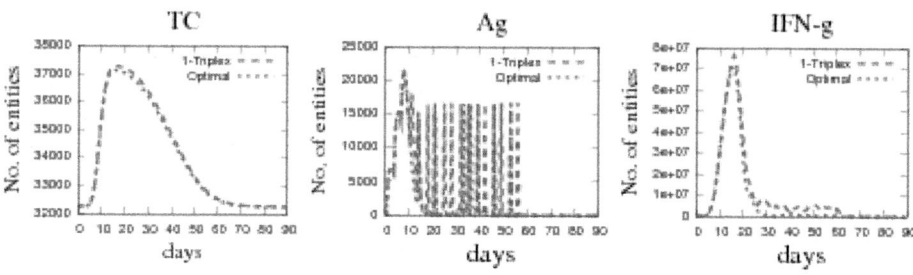

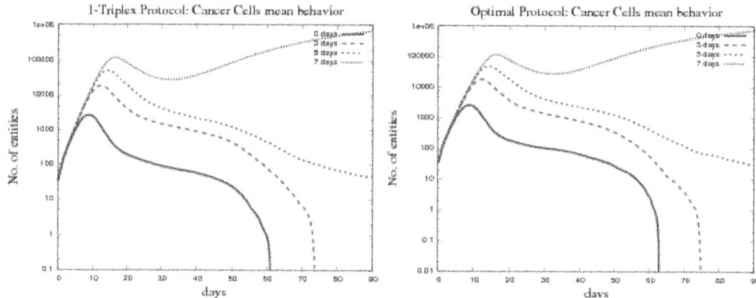

Fig. 2. Cancer Cells (CC), B cells, T helper cells (TH), cytotoxic T cells (TC), Antigens (AG), Interferon-γ (IFN-g) behaviors with 1-Triplex and the Optimal protocols. The time-length of the experiment is 3 months. The two protocols are repeated only the first two months.

Tables

Table 1. 1-Triplex, 7-Triplex and Optimal vaccination protocols predicted efficacy (entitled number of visible metastases at the end of the experiment). The column "Range" reports the minimum and the maximum number of visible metastases observed over the 100-mice random set on which the protocols have been tested. The columns "Median" and "Mean" show the median and the mean number of visible metastases, respectively.

Protocol	Experiment length (Months)	Months of Administration	Range	Median	Mean
1-Triplex	1	1	0 - 2	0	0.33
7-triplex	1	1	0 - 12	5	5.57
Optimal	1	1	0 - 2	0	0.36
1-Triplex	2	1	0 - 2	0	0.35
7-Triplex	2	1	0 - 10	0	5.39
Optimal	2	1	0 - 2	0	0.33
1-Triplex	3	1 - 2	0	0	0
Optimal	3	1 - 2	0	0	0

Table 2. 1-Triplex and Optimal protocols "in silico" efficacy (entitled number of visible metastases at the end of the experiment) with a delay of 3, 5 and 7 days in the time of first injection. The experiment lasts for 3 months. The column "Range" reports the minimum and the maximum number of visible metastases observed over the 100-mice random set on which the protocols have been tested. The columns "Median" and "Mean" show the median and the mean number of visible metastases, respectively.

Protocol	Delay of the first injection (days)	Range	Median	Mean
1-Triplex	0	0 - 0	0	0
	3	0 - 0	0	0
	5	0 - 1	0	0.02
	7	0 - 2	0	0.39
Optimal	0	0	0	
	3	0 - 0	0	0
	5	0 - 1	0	0.02
	7	0 - 4	0	0.46

References

1. Laird, A.K.: Dynamics of tumor growth. Br. J. of Cancer 18, 490–502 (1964)
2. Devroye, L.: Non-Uniform Random Variate Generation. Springer, NY (1986)
3. Liotta, L.A., Stetler-Stevenson, W.G.: Principles of Molecular Cell Biology of Cancer: Cancer Metastasis, 4th edn. JB Lippincott Co., Philadelphia (1993)
4. Nanni, P., Nicoletti, G., De Giovanni, C., Landuzzi, L., et al.: Combined allogeneic tumor cell vaccination and systemic interleukin 12 prevents mammary carcinogenesis in HER-2/neu transgenic mice. J. Exp. Med. 194, 1195–1205 (2001)
5. De Giovanni, C., Nicoletti, G., Landuzzi, L., Astolfi, A., et al.: Immunoprevention of HER-2/neu transgenic mammary carcinoma through an interleukin 12-engineered allogeneic cell vaccine. Cancer Res. 64, 4001–4009 (2004)
6. Pappalardo, F., Mastriani, E., Lollini, P.-L., Motta, S.: Genetic algorithm against cancer. In: Bloch, I., Petrosino, A., Tettamanzi, A.G.B. (eds.) WILF 2005. LNCS (LNAI), vol. 3849, pp. 223–228. Springer, Heidelberg (2006)
7. Pappalardo, F., Lollini, P., Castiglione, F., Motta, S.: Modeling and simulation of cancer immunoprevention vaccine. Bioinformatics 21, 2891–2897 (2005)
8. Nanni, P., Nicoletti, G., Palladini, A., Croci, S., Murgo, A., Antognoli, A., Landuzzi, L., Fabbi, M., Ferrini, S., Musiani, P., Iezzi, M., De Giovanni, C., Lollini, P.-L.: Antimetastatic Activity of a Preventive Cancer Vaccine. Cancer Research 67, 11037, November 15 (2007)
9. Pennisi, M., Catanuto, R., Pappalardo, F., Motta, S.: Optimal vaccination schedules using simulated annealing. Bioinformatics 24(15), 1740–1742 (2008)
10. Forrest, S., Beauchemin, C.: Computer Immunology. Immunol Rev. 216, 176–197 (2007)
11. Pappalardo, F., Halling-Brown, M., Rapin, N., et al.: ImmunoGrid, an integrative environment for large-scale simulation of the immune system for vaccine discovery, design and optimization. Briefings in Bioinformatics 10(3), 330–340 (2009)
12. Pennisi, M., Pappalardo, F., Motta, S.: Agent based modeling of lung metastasis-immune system competition. In: Andrews, P.S., Timmis, J., Owens, N.D.L., Aickelin, U., Hart, E., Hone, A., Tyrrell, A.M. (eds.) ICARIS 2009. LNCS, vol. 5666, pp. 1–3. Springer, Heidelberg (2009)
13. Halling-Brown, M., Pappalardo, F., Rapin, N., et al.: ImmunoGrid: towards agent-based simulations of the human immune system at a natural scale. Philos. T R Soc. A 368(1920), 2799–2815 (2010)
14. Palladini, A., Nicoletti, G., Pappalardo, F., Murgo, A., Grosso, V., et al.: In silico modeling and in vivo efficacy of cancer preventive vaccinations. Cancer Research 70(20), 7755–7763 (2010)
15. Pappalardo, F., Pennisi, M., Castiglione, F., Motta, S.: Vaccine protocols optimization: in silico experiences. Biotechnology Advances 28(1), 82–93 (2010)
16. Pennisi, M., Pappalardo, F., Palladini, A., Nicoletti, G., Nanni, P., Lollini, P.-L., Motta, S.: Modeling the competition between lung metastases and the immune system using agents. BMC Bioinformatics 11(suppl. 7), 13 (2010), doi:10.1186/1471-2105-11-S7-S13

Accurately Predicting Transcription Start Sites Using Logitlinear Model and Local Oligonucleotide Frequencies

Jia Wang[1], Chuang Ma[1,2], Dao Zhou[1], Libin Zhang[1], and Yanhong Zhou[1,*]

[1] Hubei Bioinformatics and Molecular Imaging Key Laboratory, Huazhong University of Science and Technology, Wuhan, Hubei 430074, China
yhzhou@hust.edu.cn
[2] Saban Research Institute of Childrens Hospital Los Angeles, Department of Pediatrics, University of Southern California, Los Angeles, CA 90027, U.S.A

Abstract. In this study, we construct a transcription start site (TSS) prediction model using the logitlinear model and the genomic context features mined in promoter regions. We also develop a computational program named ProKey that is able to accurately predict TSSs in long DNA sequences. Performance evaluation results on the whole human genome show that ProKey could achieve 71.2% sensitivity and 76.3% specificity at the resolution level of 2000bp. Further comparison results exhibit that the correlation coefficient (CC) value of ProKey is higher than that of DragonGSF and Eponine.

Keywords: Transcription start site, Promoter, Oligonucleotide frequency, Context feature, Logitlinear model, Computational biology, Bioinformatics.

1 Introduction

Accurately predicting transcription start sites (TSSs) is important for accelerating the discovery of novel genes and understanding the mechanisms of transcription regulation [1, 2]. Although the prediction accuracy has been improved steadily in recent years, the identification of vertebrate TSSs remains an open and challenging problem in the areas of bioinformatics and computational biology. This is illustrated by the fact that the exact number of human TSSs is still uncertain, even though the drafts of the human genome sequences have become available for many years.

Similarly to the classification of gene prediction methods [3], existing TSS prediction methods can also be broadly grouped into two groups: *ab initio* method and hybrid method. The *ab initio* method relies exclusively on intrinsic features extracted from the genomic DNA sequences. Notable *ab initio* TSS predictors include FirstEF [4], Eponine [5], McPromoter [6], DragonGSF [7], ARTS [8], ProSOM [9], and so on. Hybrid method additionally integrates other evidence resources, such as histone modification data [10], cap analysis of gene expression (CAGE) data [11], and etc. TSS predictors belonging to this method are CoreBoost_HM [10], MetaProm [12], and so on. There is no doubt that integrating evidence resources with the hybrid method is a

---

* Corresponding author.

D.-S. Huang et al. (Eds.): ICIC 2011, LNBI 6840, pp. 107–114, 2012.
© Springer-Verlag Berlin Heidelberg 2012

good way to improve the TSS prediction accuracy. However, further improvement of the *ab initio* method is more significant. On the one hand, the hybrid method can achieve a higher TSS prediction accuracy if more powerful *ab initio* approaches are integrated. More importantly, the improvement of *ab initio* method usually means better understanding of TSSs.

Three classes of features (i.e., signal feature, context feature and structure feature) have been reported in literature for the development of *ab initio* TSS predictors. Signal features describe the functional elements (e.g., CpG-island, TATA box, and etc) located in promoter regions. Context features characterize the nucleotide compositions of promoter sequences. Structure features represent the properties of DNA three-dimensional structures (e.g., DNA flexibility, DNA denaturation values, and etc). Context features are extracted from the genomic context of promoters as a set of k-mers, and structure features are usually calculated by transforming di-, tri- or tetra-nucleotides (i.e., k-mer, k=2,3,4) to physical parameter values. Obviously, context features are the basis of signal and structure features [2]. Therefore, context features are important for developing *ab initio* TSS predictors with high accuracy. However, the prediction accuracy of TSS predictors based on context features is still unsatisfied [13,14]. Two of the most important reasons are: (i) only regions close to the TSSs (e.g., 250bp upstream and 50 bp downstream of the TSS) are usually used to mine context features [15,16], (ii) oligonucleotide positional frequencies are calculated for constructing TSS prediction models [17,18]. This would miss the motifs located in the region farther from the TSS, and may not be efficient to characterize the functional elements which do not occur in specific positions.

In this study, we make an effort to improve the TSS prediction accuracy solely based on the context features. We firstly characterize the local oligonucleotide frequencies in different regions of the promoter, and then construct several models to character the promoter. We subsequently integrate these models with the logitlinear model and finally develop a new *ab initio* TSS predictor named ProKey. We evaluate the prediction performance of ProKey on the whole human genome at different resolution levels, and find that the correlation coefficient (CC) value of ProKey is comparable to that of famous TSS predictors including DragonGSF and Eponine. The source code of ProKey can be requested from the authors.

2   Materials and Methods

2.1   Experimental Data

We construct the training dataset with promoters from eukaryotic promoter database (EPD, http://epd.vital-it.ch/) and non-promoters from two benchmark datasets: Kulp-Reese dataset (http://www.fruitfly.org/seq_tools/datasets/Human) and HMR195 dataset (http://www.cs.ubc.ca/~rogic/evaluation/). EPD provides high-quality promoters which have been experimentally determined. Kulp-Reese dataset and HMR195 dataset are two datasets which have widely used to train and test protein-coding gene predictors. The DNA sequences between the 3'end of initial coding exons and 5'end of terminal coding exons are extracted for non-promoters. Exonic and intronic sequences from Kulp-Reese and HMR195 datasets are also extracted for calculating the background frequencies of oligonucleotides.

To generate TSS data for performance evaluation, we utilize the annotation information from the DBTSS (http://dbtss.hgc.jp/) and RefSeq (http://www.ncbi.nlm.nih.gov/RefSeq/) databases. TSSs provided by DBTSS database were determined with oligo-capped cDNA sequences, and have been widely used to test the performance of TSS predictors [1,8]. 30946 human TSSs and their genomic location information are firstly obtained from DBTSS database. These TSSs are further filtered with the following criteria: (i) the TSS is located in one promoter sequence of the training dataset; (ii) the TSS is not within 50bp from the gene start determined by the alignments between Refseq mRNAs and hg18 genome. This filtration leaves us 7143 genes. The gene sequences and 10k base pairs (bps) of upstream and downstream of sequences are extracted for testing TSS predictors.

2.2   Logitlinear Model

In this study, we divide the promoter sequence (the region [-500, 100] around the TSS, the position of TSS is defined as +1) into six sub-segments, which are [-500, -401], [-400, -301], [-300, -201], [-200, -101], [-100 -1] and [1,100], respectively. For each sub-segment, we construct a context model for the purpose of distinguishing promoter from non-promoter sequences. In brief, the context model of the ith (i=1, 2, ..., 6) sub-segment is defined by the following formula:

$$Model_{(P/B)}(seg_i) = \sum_{j=1}^{j \leq 100-5} Log_2(\frac{f_p(a_j...a_{j+5})}{f_B(a_j...a_{j+5})}),$$ (1)

where $f_p(a_j...a_{j+5})$ and $f_B(a_j...a_{j+5})$ represent the occurring frequencies of the hexamer $a_j...a_{j+5}$ in the ith sub-segments of the promoter sequence (P) and background sequence (B), respectively. The background sequence (B) could be an exonic (E) or intronic (I) sequence. Therefore, the ith sub-segment could be characterized with two scores: $Model_{(P/E)}(seg_i)$ and $Model_{(P/I)}(seg_i)$. We simply select the minimum one as the final score of the ith sub-segment (denoted as $Model(seg_i)$). Comprehensively considering these two scores would be helpful to further improve the accuracy of the TSS predictors.

The extracted promoter sequence is then described with six scores of the context model. The *ab initio* TSS prediction model is constructed by integrating these scores with the logitlinear model [19]:

$$P(TSS) = \frac{e^{\alpha + \sum_{i=1}^{6} \beta_i \times Model(seg_i)}}{1 + e^{\alpha + \sum_{i=1}^{6} \beta_i \times Model(seg_i)}},$$ (2)

where α and β_i (i=1, 2, ..., 6) are the weight coefficients determined with the FITLOGITLINEAR and LISTPARAMETERS functions in statistics package of ISUW (http://ezlearn.cbs.dk/stat/hamat-2/tt).

2.3   TSS Prediction

On the basis of the constructed TSS prediction model, a TSS prediction program named ProKey, is developed in this study. The configuration of ProKey is shown in Figure 1.

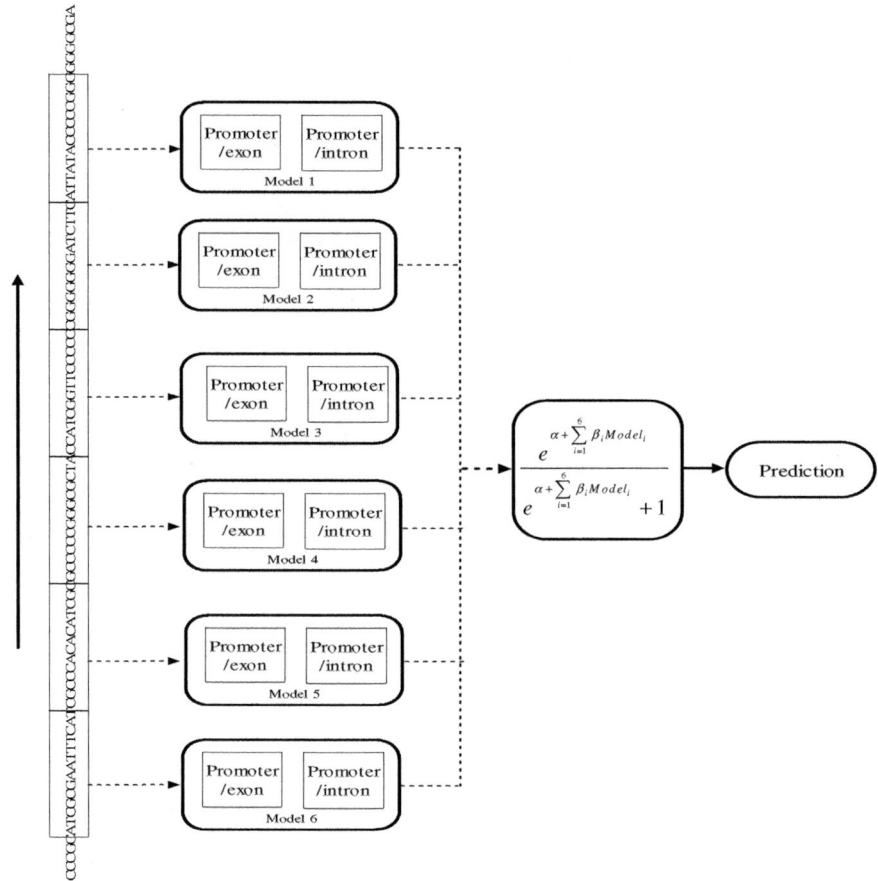

Fig. 1. The configuration of ProKey

For the prediction of TSSs, the given sequence is scanned from the start to the end with a window of 600bp. The segment with 600bp is firstly extracted and divided into six sub-segments. Each of these sub-segments is then characterized by the scores of context model with different background frequencies. The scores of these sub-segments are subsequently integrated into the logitlinear model. Finally, the probability of this segment to be a real TSS is output by ProKey.

ProKey can be easily used to predict TSSs in a long genomic sequence. For any of two prediction results, which are higher than the threshold score (empirically set to 0.94) and closer than 1000bp, are merged to be a predicted TSS.

2.4   Parameter Estimation

To estimate the parameters (α and β_i), we firstly randomly select 1871 exonic and intronic sequence segments from the whole exonic and intronic sequences in the Kulp-Reese and HMR195 dataset, respectively. We then obtain the parameters of

ProKey with the ISUW statistic package. In order to predict TSS more accurately, we subsequently re-train ProKey with the data optimization strategy [20]. In brief, we run ProKey on all non-promoter sequences with the obtained parameters, and then randomly select 1871 high-scoring negative samples (prediction score higher than 0.8). Based on the positive samples and re-selected negative samples, we re-train ProKey and obtain the final parameters of ProKey with the ISUW.

2.5   Evaluation Method

Following the strategy proposed by Bajic et al [1], three measures including the sensitivity (Sn=TP/(TP+FN)), specificity (Sp=TP/(TP+FP)) and correlation coefficient ($CC = (TP \times TN - FN \times FP) / \sqrt{(TP+FN) \times (TN+FP) \times (TP+FP) \times (TN+FN)}$) are applied to evaluate the performance of TSS prediction methods. TP, FP, TN and FN represent the number of true positives, false positives, true negatives and false negatives, respectively. For a known TSS, if there is a predicted TSS falls within the [-L, L] bp region relative to the TSS location, it is considered as a true positive (TP), otherwise it is counted as a false negative (FN). For a predicted TSS, if it locates in the region [L+1, EndofGene], it's considered as a false positive (FP). TN is defined as all the nucleotide tested. The CC measure can be used to indicate the prediction ability of TSS predictor [21]. The closer of CC value to 1.0, the more powerful the TSS predictor is. To comprehensively understand the performance of TSS prediction methods, three levels of resolution, including L=2000, 1000 and 200 bp, are used in this study.

3   Results and Discussions

3.1   Optimal Parameters of ProKey

The parameters of ProKey obtained in the first and second training steps are shown in Table 1. The β_5 and β_6 are much larger than other parameters in both training steps, indicating the context features in the 5th and 6th segments might be important for the TSS prediction. The context features in the region farther from the TSS also make some contribution of distinguishing positive samples from high-scoring negative samples. Take the 1st sub-segment for example, the β_1 is increased from -0.294 to 1.253 (Table 1). This two-step training process is able to improve the prediction accuracy of ProKey. Testing results on the X chromosome show that the sensitivity of ProKey can be improved from 53.9% to 58.7% at the specificity level of 83.0%. Note that performing this training process more times might be helpful to further improve the prediction accuracy of ProKey.

Table 1. The parameters of ProKey

Training	α	β_1	β_2	β_3	β_4	β_5	β_6
1st step	0.395	-0.294	-0.447	0.455	0.015	2.308	2.141
2nd step	0.894	1.253	0.744	1.112	0.380	2.328	1.806

3.2   Performance of ProKey on the Whole Human Genome

With the parameters obtained in the 2nd training step, we evaluate the performance of ProKey on the whole human genome (Table 2). When the resolution level (L) goes from 200bp to 2000bp, the sensitivity of ProKey is improved from 43.1% to 71.2%, and the specificity is increased from 49.5% to 76.3%. For the CC value, ProKey could reach to 0.737 at the resolution level of 2000bp, indicating that ProKey can be used to effectively predict TSSs in the long genomic sequences.

Table 2. Performance of ProKey on the whole human genome at different resolution levels

Resolution (bp)	Sn (%)	Sp (%)	CC
[-2000,2000]	71.2	76.3	0.737
[-1000,1000]	67.6	73.8	0.706
[-200,200]	43.1	49.5	0.461

3.3   Performance Comparison between ProKey and Other Programs

Recently, some TSS predictors have been systematically evaluated on the whole human genome [1,13]. According to these evaluations, DragonGSF (with the highest average accuracy) and Eponine (with high positional accuracy) are selected to compare with ProKey. Since the DragonGSF only provides the web server and limits the length of submitted sequence, we therefore randomly select 358 TSSs in the testing dataset, which also satisfy the length limitation of DragonGSF, for the comparison. The default parameters of these programs are utilized in the testing process, 0.994 for the DragonGSF, 0.999 for Eponion and 0.94 for ProKey. For Eponine, the prediction results are clustered according to the method proposed presented by Bajic et al [1].

From Table 3, it is clear that the CC value of ProKey is much higher than that of DragonGSF at three resolution levels. Compared with Eponine, ProKey achieves a comparable CC value when the resolution level is 200bp, but yields significantly higher CC values at the resolution levels of 1000bp and 2000bp.

Table 3. Performance of ProKey on the whole human genome at different resolution levels

Resolution	Program	Sn (%)	Sp (%)	CC
	ProKey	75.4	78.7	0.771
[-2000,2000]	DragonGSF	65.6	84.2	0.744
	Eponine	43.0	92.8	0.631
	ProKey	71.5	76.6	0.740
[-1000,1000]	DragonGSF	58.7	79.0	0.681
	Eponine	40.2	91.7	0.607
	ProKey	46.4	51.7	0.490
[-200,200]	DragonGSF	27.4	52.4	0.379
	Eponine	31.0	79.3	0.496

4   Conclusion

In this paper, we have proposed an effective *ab initio* TSS prediction method, and developed a computational program called ProKey. In our proposed method, several models are constructed to characterize the genomic context features in different regions relative to the TSS. These models are integrated with the simple logitlinear model. Comparison results demonstrate that the prediction accuracy of ProKey is compared to that of famous TSS predictors including DragonGSF and Eponine.

In the further, we will further improve the TSS prediction accuracy of ProKey with the consideration of modern machine learning methods, and more context features (e.g., the CpG-island information and k-mers (k=1,2,..6)) in promoter regions.

Acknowledgements. This work was supported by the National Natural Science Foundation of China (90608020 and 30971642), the Program for New Century Excellent Talents in University (NCET-060651) and the Natural Science Foundation of Hubei Province of China (2009CDA161).

References

1. Bajic, V.B., Tan, S.L., Suzuki, Y., Sugano, S.: Promoter prediction analysis on the whole human genome. Nat. Biotechnol. 22, 1467–1473 (2004)
2. Zeng, J., Zhu, S., Yan, H.: Towards accurate human promoter recognition: a review of currently used sequence features and classification methods. Brief. Bioinform. 10, 498–508 (2009)
3. Zhou, Y.H., Yang, L., Wang, H., Lu, F., Wan, H.H.: Prediction of eukaryotic gene structures based on multilevel optimization. Chinese Science Bulletin 49, 321–328 (2004)
4. Davuluri, R.V., Grosse, I., Zhang, M.Q.: Computational identification of promoters and first exons in the human genome. Nat. Genet. 29, 412–417 (2001)
5. Down, T.A., Hubbard, T.J.: Computational detection and location of transcription start sites in mammalian genomic DNA. Genome Res. 12, 458–461 (2002)
6. Ohler, U., Niemann, H., Liao, G., Rubin, G.M.: Joint modeling of DNA sequence and physical properties to improve eukaryotic promoter recognition. Bioinformatics 17, S199–S206 (2001)
7. Bajic, V.B., Seah, S.H.: Dragon gene start finder: an advanced system for finding approximate locations of the start of gene transcriptional units. Genome Res. 13, 1923–1929 (2003)
8. Sonnenburg, S., Zien, A., Ratsch, G.: ARTS: accurate recognition of transcription starts in human. Bioinformatics 22, e472–e480 (2006)
9. Abeel, T., Saeys, Y., Rouze, P., Van de Peer, Y.: ProSOM: core promoter prediction based on unsupervised clustering of DNA physical profiles. Bioinformatics 24, i24–i31 (2008)
10. Wang, X., Xuan, Z., Zhao, X., Li, Y., Zhang, M.Q.: High-resolution human core-promoter prediction with CoreBoost_HM. Genome Res. 19, 266–275 (2009)
11. Gupta, R., Wikramasinghe, P., Bhattacharyya, A., Perez, F.A., Pal, S., Davuluri, R.V.: Annotation of gene promoters by integrative data-mining of ChIP-seq Pol-II enrichment data. BMC Bioinformatics 11, S65 (2010)
12. Wang, J., Ungar, L.H., Tseng, H., Hannenhalli, S.: MetaProm: a neural network based meta-predictor for alternative human promoter prediction. BMC Genomics 8, 374 (2007)

13. Abeel, T., Van de Peer, Y., Saeys, Y.: Toward a gold standard for promoter prediction evaluation. Bioinformatics 25, i313–i320 (2009)
14. Bajic, V.B., Brent, M.R., Brown, R.H., Frankish, A., Harrow, J., Ohler, U., Solovyew, W., Tan, S.L.: Performance assessment of promoter predictions on ENCODE regions in the EGASP experiment. Genome Biol. 7, S3.1–S3.13 (2006)
15. Anwar, F., Baker, S.M., Jabid, T., Mehedi Hasan, M., Shoyaib, M., Khan, H., Walshe, R.: Pol II promoter prediction using characteristic 4-mer motifs: a machine learning approach. BMC Bioinformatics 9, 414 (2008)
16. Zeng, J., Zhao, X.Y., Cao, X.Q., Yan, H.: SCS: signal, context and structure features for genome-wide human promoter recognition. IEEE/ACM Trans. Comput. Biol. Bioinform. 7, 550–562 (2010)
17. Narang, V., Sung, W.K., Mittal, A.: Computational modeling of oligonucleotide positional densities for human promoter prediction. Artif. Intell. Med. 35, 107–119 (2005)
18. Xie, X., Wu, S., Lam, K.M., Yan, H.: PromoterExplorer: an effective promoter identification method based on the AdaBoost algorithm. Bioinformatics 22, 2722–2728 (2006)
19. Ma, C., Deng, F.Y., Liu, H., Zhou, Y.H.: Accurate prediction of alternatively spliced cassette exons using evolutionary conservation information and logitlinear model. In: 9th International Joint Conference on Bioinformatics, System Biology and Intelligent Computing, pp. 131–134 (2009)
20. Zhan, Y., Zhou, Y.H., Lu, Z.D.: A new method to improve the sensitivity of support vector machine based on data optimization. In: Proceedings of the 2003 IEEE International Conference on Robotics, Intelligent Systems and Signal Processing, pp. 892–895 (2003)
21. Knudsen, S.: Promoter2.0: for the recognition of PolII promoter sequences. Bioinformatics 15, 356–361 (1999)

A New Method for Identifying Cancer-Related Gene Association Patterns

Hong-Qiang Wang[1,*], Xin-Ping Xie[2], and Ding Li[1]

[1] Intelligent Computing Lab, Hefei Institute of Intelligent Machine,
CAS, P.O. 1130, 230031, Hefei, China
{hqwang126,ld0510104}@126.com
[2] Department of Mathematics and physics, Anhui University
of Architecture, 230022, Hefei, China
xpxie@yahoo.com.cn

Abstract. Gene association plays important roles in complex genetic pathology of cancer. However, development of methods for finding cancer-related gene associations is still in its infancy. Based on a biological concept of gene association module (GAM) comprising a center gene and its expression-related genes, this paper proposes a gene association detection model called kernel GAM (kGAM). In the model, we assume that the expression of the center gene can be predicted by the expression-related genes. Based on defining a cost function, a kernel ridge regression algorithm is developed to solve the kGAM model. Finally, to identify a compact GAM for a given center gene, a heuristic search procedure is designed. Experimental results on three publicly available gene expression data sets show the effectiveness and efficiency of the proposed kGAM model in identifying cancer-related gene association patterns.

Keywords: Microarray data, kernels, ridge regression, gene association.

1 Introduction

Genes in a cell working together and functioning in a coordinated manner plays an important role in the generation of cellular phenotypes and fine coordination between gene activities is essential for the formation of a signaling pathway [12,17]. These coordinated activities are manifested in the form of correlated expression levels of genes [2,3]. Therefore, it is critical and necessary to detect and utilize gene associations to understand complex genetic diseases. Another motivation of this work is that, although the large volume of gene expression data have been accumulated and are available online, it is still difficult and challenging to mine biological knowledge from these data in terms of methodology [10]. Generally, there are two main challenges in analyzing gene expression data: the complexity of invisible biological systems and the non-typical features of gene expression data including high noise, high-dimensionality but small sample size.

---

* Corresponding author.

D.-S. Huang et al. (Eds.): ICIC 2011, LNBI 6840, pp. 115–122, 2012.

Many studies on various model systems have suggested that a gene can be combinatorially regulated by a relatively small number of transcription factors simultaneously or under different conditions, leading to strikingly complex patterns of gene expression. From these findings, we abstract a gene association structure, named gene association module (GAM), which consists of a center gene and its associated (unnecessarily regulating or regulated) elements (genes). In the GAM, the links, only appearing between the center gene and its associated genes, represent the influence of the associated elements on the center gene. As a hub topology, the GAM has been found to be universal in biological systems due to its robustness and sparseness for signal transduction [13,4,6].

In this paper, based on the GAM structure, we develop a kernel GAM model (kGAM) for detecting cancer-related complex gene associations. In the model, the main idea is to use the associated genes to regress the expression of the center gene. To characterize the model, a cost function is defined as the regression error. The cost function allows determining the structural parameters of the kGAM and potentially provides a way to use kGAMs to classify cancer. To find a compact GAM for a given center gene, a heuristic compact-kGAM searching procedure is developed based on the cost function.

In experimental section, we collect three publicly available real-world gene expression data sets, binary or multi-class, to evaluate the performance of the proposed method in detecting gene association patterns. To evaluate the cancer classification performance of the proposed model, we also implement and apply several previous classification methods including Fisher discriminant analysis (FDA), K nearest neighbor(KNN), support vector machines with linear kernel (linear-SVM) and radial basis function kernel (rbf-SVM) to these data sets, and their classification accuracies are compared with those of our model.

2   Methods

2.1   kernel Gene Association Model

Considering a gene association structure composed of a center gene g and p associated elements (genes), we assume that the expression of the center gene can be predicted by the associated genes. Let y denote the expression level of gene g and $\mathbf{x} = [x_1, x_2, \cdots, x_p]$ the expression levels of the p associated genes, such kind of gene association structure can be linearly modeled as:

$$\begin{cases} \hat{y} = f(\mathbf{x}) = A\mathbf{x}^T + b \\ y = \hat{y} + \epsilon \end{cases} \tag{1}$$

where $A = [a_1, a_2, \cdots, a_p]$, b is a constant and $\epsilon \sim N(0, 1)$. The element a_i measures the association of gene i on the center gene g, and its positive value denotes an expression promotion on gene g while its negative value denotes an expression repression.

For such a structure, we define a cost function E as

$$E = \frac{1}{2}(\hat{y} - y)^2 = \frac{1}{2}(\bar{A}\bar{P})^2 \tag{2}$$

where $\bar{A} = [A, -1, 1]$ is referred to as an extended association coefficient vector and $\bar{P} = [x_1, x_2, \cdots, x_p, y, b]$ as an extended expression profile. From Eq.2, the cost function is dependent on the internal relationship of the structure. Given an expression profile sample, only when the association pattern implicity embedded in it, instead of the explicit gene expression values, agrees with the internal relationship will the value of the cost function approach zero. This agrees with the fact that the coordination between the genes, rather than the expression values themselves, plays a crucial role in determining gene activity, and the cost function reflects the level of this activity.

The complexity of biological systems suggests that gene associations may not proceed in a linear manner. We introduce a nonlinear kernel function to approximate the expression value y of the center center in Eq.1, and the resulting gene association structure is referred to as the kernel gene association model (kGAM). The kernel function is constructed as follows. We first consider such a kind of nonlinear transformation

$$\Phi(\mathbf{x}) = [\phi_1(\mathbf{x}^T \mathbf{e}_1), \phi_2(\mathbf{x}^T \mathbf{e}_2), \cdots, \phi_p(\mathbf{x}^T \mathbf{e}_p)]^T, \tag{3}$$

where $\phi_i, i = 1, 2, \cdots, p$ represents a nonlinear function, $\mathbf{e}_i = [e_{ij}; i, j = 1, 2, \cdots, p]^T$ and $e_{ij} = \begin{cases} 1 \ i = j \\ 0 \ i \neq j \end{cases}$. We use the sigmoid function and form ϕ_i as

$$\phi_i(\mathbf{x}) = \left(1 + exp(-\beta(\tfrac{x_i - \mu_i}{\sigma_i})^2)\right)^{-1}, i = 1, 2, \cdots, p \tag{4}$$

where μ_i and σ_i are the location and width parameters, respectively, which can be estimated as the mean and standard deviation of gene expression levels, and $\beta \in (0, 1]$ is a constant. As a result, by combining Eqs.3 and 4, we construct the kernel function as:

$$\kappa(\mathbf{x}, \mathbf{z}) = \sum_{i=1}^{p} (Logsig(\beta, x_i) Logsig(\beta, z_i)) \tag{5}$$

The parameter β is referred to as the kernel parameter, which controls the approximation to gene associations.

To efficiently solve the above gene association model, we introduced a ridge parameter $0 < \lambda < 1$ to impose a sparsity constraint on the values of the association coefficients. The ridge parameter controls the relative trade-off between the sparsity constraint and data approximation fidelity, and a proper value of it will compensate for the information insufficiency so that an effective solution can be found. For the kGAM model, we use the kernel ridge regression technique [14,7] to solve its parameter, A, and the kernel parameter β is optimally chosen by varying its value among (0,1).

2.2 kGAM-Based Cancer Classification and the Searching of a Compact kGAM

As described above, the cost function in Eq.2 reflects the association information encapsulated in a kGAM, and will approach zero when the expression profile of a sample agrees with the internal relationship of the kGAM. This property can be used to design a association-based cancer classification rule as follows. Consider C sample classes, and for each class, with the center gene g and its p associated genes, a kGAM model, $H_i, i = 1, 2, \cdots, C$, has been built, respectively. For a given test sample t, we predict its class to be

$$c = \arg \min_i \{E_i(t)\} \tag{6}$$

where E_i are the cost functions associated to kGAMs H_i.

There is little or no knowledge about how many genes known truly correlated to a given center gene. To find a compact kGAM for a given center gene from a gene pool, we present a heuristic searching procedure. Simply speaking, the procedure begins with, and iteratively searches and attaches the element most associated with the center gene to the list of associated genes. Because gene networks tend to be sparse and only a small group of genes are involved in a particular biological process, the search procedure would converge within a small number of steps, and has a low computational cost.

3   Experimental Results

To evaluate the proposed approach, we collected three publicly available gene expression data sets, two binary datasets, Golub data [9], Singh data [16], and one multi-class dataset, Armstrong data [1]. The three data sets each have a standard training/test split [9,16,1]: For the Golub data, the training and test sets contain 38 and 34 samples, respectively; 102 and 34 samples for the Singh data; and 77 and 15 samples for the Armstrong data.

We analyzed the three data sets based on the standard splits: the training sets are used to detect significant kGAMs and construct classifiers, and the test sets are used for validation. In order to avoid the influence of noisy genes to cancer classification, only 200 genes, with the highest signal-to-noise ratio (SNR) [9] for the binary datasets or the highest variance between samples for the multi-class data set, were used for performance evaluation in our experiments. For each dataset, we tried the 50 genes with the highest SNR/variance values as center genes to search for significant kGAMs for cancer classification.

3.1   Detection of kGAMs for the Three Data Sets

The association coefficients capsulated in a kGAM reflect the gene association patterns in cancer classes. Fig. 1 shows the association coefficients in three kGAMs with "Human common acute lymphoblastic leukemia antigen (CALLA)" being the center gene, for the three classes of the Armstrong data set. Note that, to highlight significant association differences, the three kGAMs are simplified by trimming the association coefficients less than 5% of the maximal values to 0. A number of studies have shown that the CALLA gene plays a potential role as a functional neutral endopeptidase in both normal and malignant lymphoid function. In particular, the gene associates with a number of small secreted peptides whose abnormal misfolding and aggregation may be a cause of a number of diseases [15]. As shown in Fig. 1, the three kGAMs identified suggest that the gene is differently regulated in the three leukemia cancer classes. For the three kGAMs, the TOP2B gene with Accession no. 36571_at is most closely associated with the CALLA gene. The TOP2B gene encodes a DNA topoisomerase, which can control and alter the topological states of DNA during transcription [5]. The three kGAMs disclose that the TOP2B gene represses the expression of the CALLA gene in all the three leukemia classes, as shown in Fig. 1. Some associated genes exhibit remarkably different effects on the CALLA gene in the three leukemia classes. For example, the gene with Accession no. 40797_at, known as ADAM10, promotes the expression of the CALLA gene in Class 2 while represses in Class 3; the gene with Accession no. 1602_at, known as PRKCI (Protein kinase C, iota), promotes the expression of the CALLA gene in Classes 1 and 2 while no significant impact occurs in Class 3. The PRKCI gene has been found to control the dynamics of microtubules within the early secretory pathway [8], and the ADAM10 gene to encode a sheddase, which performs cleaving of the membrane proteins and plays a role in a number of peptide hydrolysis reactions [18].

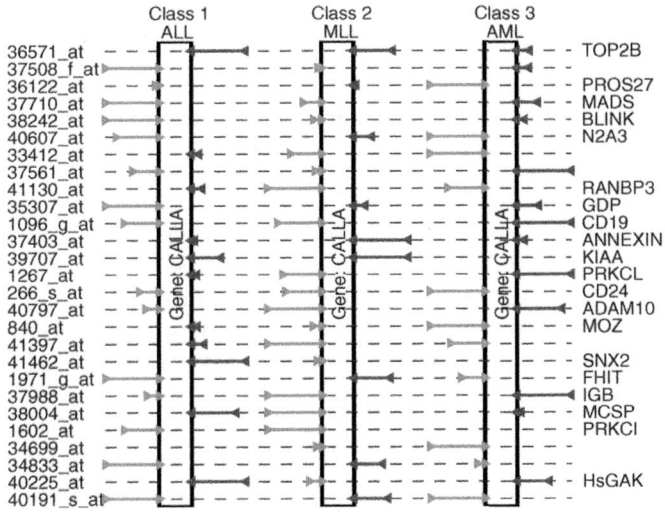

Fig. 1. Association patterns (maximum of the association coefficients is 9) captured in a kGAM model for the three cancer classes of the Armstrong data. The red lines represent negative expression association, the green lines represent positive expression association, and the length of lines represent the association strength. These associated genes have significantly different effects on the hub gene CALLA in the three cancer classes, and these differences in turn determine the characteristics of the three classes.

A kGAM encapsulates a stable association pattern common to a particular cancer class, and its cost function measures how a sample disagree with the class in association patterns. For samples belonging to the class, the values of the cost function will remain low due to the similar association pattern. To illustrate this property, Fig.2 shows the distribution of the cost values of the three classes for the Armstrong data set. From this figure, it can be seen that most of the samples in each class have a low cost value (less than 10^{-4}) according to the corresponding cost functions.

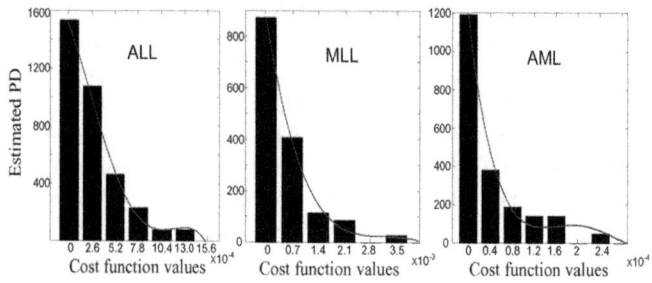

Fig. 2. Distribution of the cost values of the kGAMs found for the three cancer classes of the Armstrong data. The red doted lines are the fitting curves with 4-degree polynomial function. PD is short for probability density.

3.2 Evaluation of the Classification Performance of the kGAM Model

To further show the power of the kGAM model in capturing gene association patterns, we applied the kGAMs identified above to classify cancer according to the kGAM classification rule. Table 1 summarizes the classification accuracies on the test sets by three kGAMs for each of the three data sets. For comparison, based on the same genes as the three kGAMs contained, several conventional methods were implemented to classify the three data sets, which include two support vector machines (SVMs) with linear (linear-SVM) and radial basis function (rbf-SVM) kernels (http://sourceforge.net/projects/svm/), k(k=3))-nearest neighbor (KNN) and Fisher linear discriminant (FLD). The regularization parameter of the linear-SVM was optimally chosen from the range $\{2^{12}, 2^{11}, \cdots, 2^{-1}, 2^{-2}\}$, and the two parameters of the rbf-SVM, regularization factor and kernel width, were optimized based on a two-dimensional grid search technique within the ranges, $\{2^{12}, 2^{11}, \cdots, 2^{-1}, 2^{-2}\}$ and $\{2^4, 2^3, \cdots, 2^{-9}, 2^{-10}\}$. For the multi-class problem, the voting strategy [11] is used along with these previous methods to make optimal classification decision. The results by the conventional methods are compared with ours in Table 1. From Table 1, it can be seen that our kGAM models achieve much better classification accuracies than the other methods, irrespective of the binary problems or the multi-class problem.

Table 1. Comparison of classification accuracies between our kGAM model and several conventional methods for the Golub (binary), Singh (binary) and Armstrong (3-class) data

Datasets	Methods	kGAM I	kGAM II	kGAM III
Golub data	kGAM model	1	1	0.97
	rbf-SVM	0.94	0.97	0.94
	linear-SVM	0.91	0.97	0.88
	KNN	0.94	0.88	0.85
	FLD	0.88	0.88	0.85
Singh data	kGAM model	1	0.97	1
	rbf-SVM	0.94	0.91	0.91
	linear-SVM	0.91	0.76	0.91
	KNN	0.91	0.87	0.97
	FLD	0.38	0.60	0.48
Armstrong data	kGAM model	1	1	0.93
	rbf-SVM	0.93	0.86	0.80
	linear-SVM	0.73	0.60	0.73
	KNN	0.87	0.80	0.80
	FLD	0.67	0.53	0.47

4 Conclusion

We have proposed a model (kGAM) for detecting cancer-related gene associations. The model can flexibly approximate complex association patterns between genes and overcome the problem of small sample in microarray data analysis. The proposed approach was evaluated on three publicly available microarray data sets. The experimental results show the effectiveness and efficiency of the proposed approach in both capturing gene associations. Future work will focus on the optimal construction of the kGAM model and applications on more real-world microarray data sets.

Acknowledgments. This work was supported by the grants of the National Science Foundation of China, Nos. 31071168, 30900321, 60975005, 61005010, 60873012, 60973153 and 60905023.

References

1. Armstrong, S.A., Staunton, J.E., Silverman, L.B., Pieters, R., den Boer, M.L., Minden, M.D., Sallan, S.E., Lander, E.S., Golub, T.R., Korsmeyer, S.J.: Mll Translocations Specify a Distinct Gene Expression Profile That Distinguishes a Unique Leukemia. Nat. Genet. 30(1), 41–47 (2002); 1061-4036 10.1038/ng76510.1038/ng765
2. Basso, K., Margolin, A.A., Stolovitzky, G., Klein, U., Dalla-Favera, R., Califano, A.: Reverse engineering of regulatory networks in human b cells. Nature Genetics 37(4), 382–390 (2005)

3. Calvano, S.E., Xiao, W., Richards, D.R., Felciano, R.M., Baker, H.V., Cho, R.J., Chen, R.O., Brownstein, B.H., Cobb, J.P., Tschoeke, S.K., Miller-Graziano, C., Moldawer, L.L., Mindrinos, M.N., Davis, R.W., Tompkins, R.G., Lowry, S.F.: ProgramInflamm, L.S.C.R., Host Response to, I.: A Network-based Analysis of Systemic Inflammation in Humans. Nature 437(7061), 1032–1037 (2005); 0028-0836 10.1038/nature03985 10.1038/nature03985
4. Carter, S.L., Brechbuhler, C.M., Griffin, M., Bond, A.T.: Gene Co-expression Network Topology Provides a Framework for Molecular Characterization of Cellular State. Bioinformatics 20(14), 2242–2250 (2004)
5. Champoux, J.J.: DNA Topoisomerases: Structure, Function, and Mechanism. Annu. Rev. Biochem. 70(1), 369–413 (2002)
6. Cooper, T., Morby, A., Gunn, A., Schneider, D.: Effect of Random and Hub Gene Disruptions on Environmental and Mutational Robustness in Escherichia Coli. BMC Genomics 7(1), 237 (2006)
7. Du, K.L., Swamy, M.N.S.: Neural Networks in a Soft-computing Framework. Springer-Verlag London Limited, London (2006)
8. Fields, A.P., Regala, R.P.: Protein kinase c: Human Oncogene, Prognostic Marker and Therapeutic Target. Pharmacol. Res. 55(6), 487–497 (2007)
9. Golub, T.R., Slonim, D.K., Tamayo, P., Huard, C., Gaasenbeek, M., Mesirov, J.P., Coller, H., Loh, M.L., Downing, J.R., Caligiuri, M.A., Bloomfield, C.D., Lander, E.S.: Molecular Classification of Cancer: Class Discovery and Class Prediction by Gene Expression Monitoring. Science 286, 531–537 (1999)
10. Lee, I., Lehner, B., Crombie, C., Wong, W., Fraser, A.G., Marcotte, E.M.: A Single Gene Network Accurately Predicts Phenotypic Effects of Gene Perturbation in Caenorhabditis Elegans. Nat. Genet. 40(2), 181–188 (2008); 1061-4036 10.1038/ng.2007.70 10.1038/ng.2007.70
11. Schmann, J.: Pattern Classification: A Unified View of Statistical and Neural Approaches. Wiley Interscience, Hoboken (1996)
12. Segal, E., Friedman, N., Kaminski, N., Regev, A., Koller, D.: From Signatures to Models: Understanding Cancer Using Microarrays. Nat. Genet. 37, S38–S45 (2005)
13. Seo, C.H., Kim, J.R., Kim, M.S., Cho, K.H.: Hub Genes with Positive Feedbacks Function as Master Switches in Developmental Gene Regulatory Networks. Bioinformatics 25(15), 1898–1904 (2009)
14. Shawe-Taylor, J., Cristianini, N.: Kernel Methods for Pattern Analysis. Cambridge University Press, Cambridge (2004)
15. Shipp, M.A., Vijayaraghavan, J., Schmidt, E.V., Masteller, E.L., D'Adamio, L., Hersh, L.B., Reinherz, E.L.: Common Acute Lymphoblastic Leukemia antigen (CALLA) Is Active neutral endopeptidase 24.11 ("enkephalinase"): direct evidence by cDNA transfection analysis. Proceedings of the National Academy of Sciences of the United States of America 86(1), 297–301 (1989), http://www.pnas.org/content/86/1/297abstract
16. Singh, D., Febbo, P.G., Ross, K., Jackson, D.G., Manola, J., Ladd, C., Tamayo, P., Renshaw, A.A., D' Amico, A.V., Richie, J.P., et al.: Gene Expression Correlates of Clinical Prostate Cancer Behavior. Cancer Cell 1, 203–209 (2002)
17. Tlsty, T.: Cancer: Whispering Sweet Somethings. Nature 453(7195), 604–605 (2008); 0028-0836 10.1038/453604a 10.1038/453604a
18. Yang, J., Price, M.A., Neudauer, C.L., Wilson, C., Ferrone, S., Xia, H., Iida, J., Simpson, M.A., McCarthy, J.B.: Melanoma Chondroitin Sulfate Proteoglycan Enhances Fak and Erk Activation by Distinct Mechanisms. J. Cell Biol. 165(6), 881–891 (2004)

A New Mining Algorithm of Association Rules and Applications

Sheng-Li Zhang[*]

School of Computer Science, Wuyi University, Jiangmen, China
shlzhang@126.com

Abstract. It is an important part of research content in data mining to discover association rules from large scale database, the main problem of which is frequent itemsets mining. The classical Apriori algorithm is an efficient one for that. Aimed at the performance bottlenecks of multiply scanning the database and generating a large quantity of candidate itemsets in Apriori algorithm, an improved algorithm of mining association rules is presented for the bottleneck problem. Filtering out the transactions unconcerned with mining targets by a presupposed filter, on the one hand, the improved Apriori algorithm can compresses database and reduces scanning times; on another hand, the number of candidate itemsets also decreases with it, so the improvement strategy can greatly improves the whole performance of the algorithm. The application of improved Apriori algorithm in traffic accident data mining also shows that it is very practical and efficient in data mining.

Keywords: Ailter, Apriori algorithm, Data mining, Application.

1 Introduction

In the recent years, Data mining has become a hotspot for the information fields and gets more and more attention. Association rules analysis is one of the most important tasks and used to find the relationship between itemsets in mass data. Based the analysis of general association rules algorithm, this paper presents a new association rule mining algorithm, which is applied to the data analysis of traffic accidents, experiment results show that this algorithm is effective and efficient.

2 Associated Rules and Algorithm

Association rules analysis has been has been successfully applied in business field which is the most active research of data mining. Association rules analysis usually is used to find the correlation and relationship between different database records,

[*] Supported by Foundation for Distinguished Young Talents in Higher Education of Guangdong, China (LYM10128).

D.-S. Huang et al. (Eds.): ICIC 2011, LNBI 6840, pp. 123–128, 2012.

while rule support and confidence are two measures of rule interestingness; they respectively reflect the usefulness and certainty of discovered rules. Typically, association rules are considered interesting if they satisfy both a minimum support threshold and a minimum confidence threshold [1].The model of association rules are as follows:

Let $I = \{I_1, I_1, I_1 \cdots\}$ be a set of all items, Let D, the task-relevant data, be a set of database transactions where each transaction T is a set of items such that $T \subseteq I$, Each transaction is associated with an identifier, called *TID*. Let A be a set of items. A transaction T is said to contain A if and only if $A\,A \subseteq T$. An association rule is an implication of the form $A \Rightarrow B$, where $A \subset I$, $B \subset I$, and $A \cap B = \varnothing$. The rule $A \Rightarrow B$ holds in the transaction set D with support s, where s is the percentage of transactions in D that contain $A \cup B$, This is taken to be the probability, $P(A \cup B)$ A set of items is referred to as an itemset. An itemset that contains k-items is a k-itemset. If the relative support of an itemset I satisfies a prespecified minimum support threshold, then I is a frequent itemset. *Support* is defined as:

$$Support(A \Rightarrow B) = P(A \cup B) = \left|\{T | A \cup B \subseteq T, T \in D\}\right| / |D| \qquad (1)$$

The rule $A \Rightarrow B$ has confidence c in the transaction set D, where c is the percentage of transactions in D containing A that also contain B, that is $confidence(A \Rightarrow B)$, confidence c is defined as.

$$confidence(A \Rightarrow B) = P(B \cup A) = \left|\{T | A \cup B \subseteq T, T \in D\}\right| / \left|\{T | A \subseteq T, T \in D\}\right| \qquad (2)$$

So, for one transaction set D, mining association rules means to find the association rules whose relative support satisfies a prespecified minimum support threshold and confidence satisfies a prespecified minimum confidence threshold.

In all mining algorithm of association rules, the Apriori algorithm is one of the most influential algorithms proposed by R. Agrawal and R. Srikant in 1994 for mining frequent itemsets for Boolean association rules. The Apriority algorithm is based on the apriority property, which belongs to a special category of properties called antimonotone. By definition, all sub-pattern of frequent pattern must be frequent, all super-pattern, containing infrequent pattern, must be infrequent. The key of this algorithm is to repeatedly scan the transactions in database in order to generate a candidate set of itemsets. As we have seen, in many cases the Apriori candidate generate-and-test method significantly reduces the size of candidate sets, leading to good performance gain. However, usually the volume of database is so great that it may need to generate a huge number of candidate sets when finding frequent itemset, the number of candidate sets will be greater when the support is small, that is also the chief objection of the algorithm.

In order to improve the efficiency of apriori-based mining; many variations of the Apriori algorithm have been proposed that focus on improving the efficiency of the original algorithm. One technology based on transactions reduction is proposed. The basic idea of the technology is that a transaction that does not contain any frequent k-itemsets cannot contain any frequent (k+1)-itemsets, Therefore, such a transaction can be marked or removed from further consideration because subsequent scans of the database for j-itemsets, where j > k, will not require it[2]. In 1995, sampling approach (mining on a subset of the given data) is proposed by Sarasere and Omiecinsky [3], The basic idea of the sampling approach is to pick a random sample S of the given database, and then search for frequent itemsets in sample instead of database. A lower support threshold than minimum support is used to find the frequent itemsets local to sample S (denoted LS), then the rest of the database is then used to compute the actual frequencies of each itemset in LS. A mechanism is also used to determine whether all of the global frequent itemsets are included in LS. Dynamic itemset counting is proposed by Brin in 1997[4]. A dynamic itemset counting technique is proposed in which the database is partitioned into blocks marked by start points. The technique is dynamic in that it estimates the support of all of the itemsets that have been counted so far, adding new candidate itemsets if all of their subsets are estimated to be frequent. Of course, lots of improved algorithms are proposed in the last tens years. Aimed at how to select the minimum support in the association rules mining algorithms with multiple supports, a new mining algorithmwith multiple supports is proposed by Zhao Yuhang and Liu Jianbo. The minimum support is determined by the minimum, maximum support of the itemset and the reference value that is given in the mining process of finding frequent itemset, the problems are avoided that are the high algorithm time complexity and existing invalid rules [5]. He chaobo and Chen qimai used the rough set's feature attributes reduction algorithm to reduce attributes, and then applied the improved Apriori algorithm to mine association rules based on the reduction decision table, and the number of the attributes was reduced by reducing the unimportant attributes[6].

3   Improvement Based on Apriori Algorithm

A huge number of candidate itemset will generate while the Apriori algorithm repeatedly scan the transactions in database, so one of the efficient approaches improving efficiency of algorithms is to reduce the frequency of scanning database and the range of ergodic transaction. If the number of scanned transactions can be reduced during the scanning database repeatedly using the algorithm, the efficiency can be improved in some extent. In this paper, one improved algorithm is proposed based the Apriori algorithm. According to the needs of mining, in order to reduce the number of scanning transactions, the basic idea is to filter out the transactions unconcerned with mining targets during the process of generating frequent itemsets Lk by candidate set of itemsets. As when all frequent itemsets are mined, but we can't find the mining goals in some frequent itemsets, strong association rules will not generate. the improved Apriori algorithm can be described as flowing:

1. In the first iteration of the algorithm, the algorithm simply scans all of the transactions in order to count the number of occurrences of each item and record the TID(identifier of each transaction) while record all frequent itemsets.
2. During the process of generating frequent itemsets Lk by candidate set of itemsets, Let filters filter out the transactions unconcerned with mining targets, then go to the number of transactions need to be scanned will become smaller and the efciency of this algorithm increase.

The flowing shows pseudo-code for the Apriori algorithm and its related procedures.

```
L1= find frequent_1_itemsets( D)
for( k= 2; Lk≠Ø ,k++) {
Ck+1= new_apriori_gen (Lk, filter(S))//S,Set of
attributes value included mining goals
for each transaction)t∈D{ / / scan D for counts
    Ct=subset (Ck, t) // get the subsets of t that are
candidates
    for each candidate transaction c∈Ct
        c.count++}
}
Return L=U k+1L k+1
Gen_Ck+1(Lk; frequent (k)_itemsets; filters(s))
for each itemsetsl1∈Lk
if(l1 [1] = l2 [1]^l1 [2]< l2 [k-2]) then{
c= l1×l2 //generate candidates set.
if(has_infrequent_subset(c,Lk)and (sl∈Ck))then
delete infrequent_subset
else add c to Ck
}
Return Ck
EWT
```

4   Performance Test of the Improved Algorithm

In order to verify the actual effect of the improved Apriori algorithm, this paper did some contrast tests of improved Apriori algorithm and Apriori algorithm using the chess database (chess. Data[7]). The running time was comparative analyzed in different support. In the figure 1, the results show that the execution efficiency of Improved Apriori algorithm is higher than that of Apriori algorithm at low support(as min_sup<32%). The main reason is the Improved algorithm can filter out the transactions unconcerned with mining targets before anew scan candidate sets and the number of candidate sets reduce greatly.

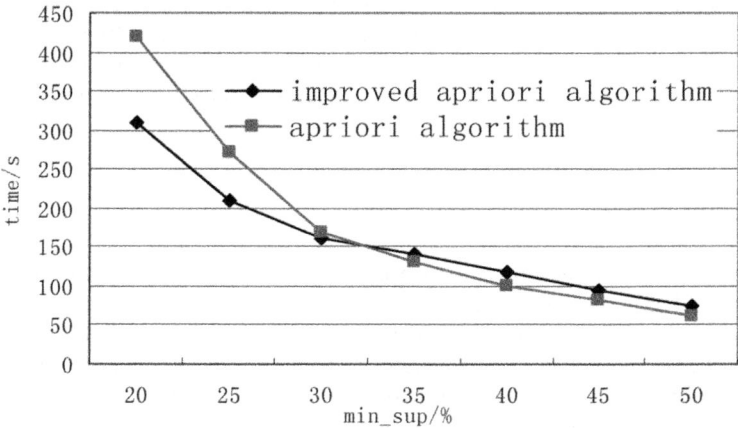

Fig. 1. Time needed for genereating frequent itemsets at different support

5 Application of Improved Algorithm in Traffic Accident Data Mining

In order to authenticate the utility and efficiency, an experiment was conducted using the road traffic accident data of Jiangmen city Guangdong province. The table 1 shows eight large class accident causes by concluding and pooling analysis.

Table 1. Code of accident causes

Code	Accident causes	Code	Accident causes
S1	Speeding driving	S5	Violation of signal
S2	Overloading	S6	Terrible environments
S3	Drunken driving	S7	Unlicensed driving
S4	Fatigue driving	S8	Reverse driving

In addition, according to the situation of casualties and property loss, this paper divided traffic accidents to three grades that are slight accident (A1), common accident (A2) and terrible accident (A3). Let minimum support be 15%, minimum confidence be 70%, data mining results can be shown in table 2.

Table 2. Data mining results

Code of rules	Association rules	Support	Confidence
R1	A1=>S1∧S5	62.6%	75.6%
R2	A2=>S1∧S3	32.3%	81%
R3	A2=>S7	21.4%	78.3%
R4	A3=>S3∧S7	15.8%	80.6%
R5	S3=>S5	35.7%	85%
R6	S5=>S4∧S1	60.5%	82.8%
R7	S2=>S3∧S5	30.6%	73.1%
R8	S3∧S1∧S5=> A3	18%	91.5%

The sixth association rules in table 2 shows that, in all traffic accidents, 60.5% vehicles violated traffic signals adjoint fatigue driving and speeding driving; while the eighth association rules shows that, in all terrible traffic accidents, the probability is over 91.5% that drunken driving, speeding driving and violation of signal occurred at the same time.

6   Conclusion

Filtering out the transactions unconcerned with mining targets by a presupposed filter, On the one hand, the improved Apriori algorithm can compresses database and reduces scanning times, on another hand, the number of candidate itemset also decreases with it, such improvement strategy can greatly improves the whole performance of the algorithm. the application of improved Apriori algorithm in traffic accident data mining shows that it is very practical and efficient in data mining.

References

1. Jiawei, H.: Data Mining Concepts and techniques, 2nd edn., pp. 260–265. China Machine Press, Beijing (2006)
2. Han, J., Pei, J., Yin, Y.: Ming frequent pattern without candidate generation. In: SIGMOD, Dallas, TX, pp. 156–159 (2000)
3. Sarasere, A., Omiecinsky, E., Navathe, S.: An efficient algorithm for mining association rules in large database. In: 21st Int'. Conf. on Very Large Database(VLDB), Zurich, Switzerland, pp. 250–261 (September 1995)
4. Brin, S.: Dynamic itemcount and implication rules for market basket analysis. In: SIGMOD, Tucson, AZ, pp. 255–264 (May 1997)
5. Zhao, Y.H., Liu, J.B.: Association rules mining algorithm with multiple supports based on piecewise function. Computer Engineering and Design 31(21), 4621–4624 (2010) (in Chinese)
6. He, C.B., Chen, Q.M.: Approach for mining association rules based on rough set. Journal of Computer Applications 30(1), 25–28 (2010) (in Chinese)
7. Bart, G.: Frequent itemset mining implementations repository [EB/OL] (2006-04-01)., http://fimi.cs.helsinki.fi/data/

MiRaGE: Inference of Gene Expression Regulation via MicroRNA Transfection II

Y-h. Taguchi[1] and Jun Yasuda[2]

[1] Department of Physics, Chuo University, Tokyo 112-8551, Japan
[2] COE Fellow, Graduate School of Medicine, Tohoku University,
Sendai 980-8575, Japan
jun.yasuda@jfcr.or.jp, tag@granular.com

Abstract. How each microRNA regulates gene expression is unknown problem. Especially, which gene is targeted by each microRNA is mainly depicted via computational method, typically without biological/experimental validations. In this paper, we propose a new computational method, MiRaGE, to detect gene expression regulation via miRNAs by the use of expression profile data and miRNA target prediction. This method is tested to miRNA transfection experiments to tumor cells and succeeded in inference of transfected miRNA as only one miRNA with significant P-values for the first time.

Keywords: MicroRNA, target genes, tumor, computational inference.

1 Introduction

MicroRNAs (miRNAs) are post-transcriptional regulators of gene expression. It binds to target messenger RNAs (mRNAs) through complementary sequences in the three prime untranslated regions (3 UTRs) of the mRNA, and consequently suppresses the expression of the mRNAs. miRNAs are short (19 $\sim$ 22 bases) RNA molecules and abundant in many human cells. The human genome may encode over 1,000 miRNAs, and the coverage by all possible miRNAs may be about 60 % of mammalian genes.

On the other hand, how a miRNA regulates its target genes and which genes are regulated by a miRNA is unclear. Especially, the later is mainly depicted via computational prediction[1], without any biological/experimental validations. There are some direct ways to investigate the bindings of mRNAs to the miRNA-protein complexes, e.g., HITS-CLIP[2] but capability of these methods for identification of miRNA-mRNA relationship is limited since it is unlikely that all the potential target mRNAs for a miRNA simultaneously express in a cell.

Another experimental way to detect miRNA target genes is to analyze the difference of gene expression profiles with or without the transfection of the miRNA to a cell line. However, it is unrealistic to test all the miRNAs with this method because it is time and money consuming.

In this paper, we describe a computational method to detect miRNAs which regulate the cellular transcriptomes in a cell in response to extracellular stimuli by analyzing the difference of gene expression profiles and computational

D.-S. Huang et al. (Eds.): ICIC 2011, LNBI 6840, pp. 129–135, 2012.

miRNA target predictions. By this retrograde method, we can narrow down the miRNAs that actually play some biological roles. Simple expression profiling of miRNAs may not be efficient to find the critical ones. Moreover, we may deduce potentially important miRNAs from the large accumulation of gene expression profile databases such as GEO by our method. It seems to be difficult to go back to the RNA samples for obtaining the miRNA expression profiles corresponds to those old profile data.

We would like to obtain the proof of concept for our methods through the analyses of the gene expression profiles with or without the transfection of single miRNAs, which actually induced cell cycle arrest in the human lung cancer cell lines, and our algorithm can quite frequently predict the transfected miRNA.

2   Materials and Methods

2.1   Gene Expression Data for Transfection Experiment

We have downloaded transfection experiment[3] data set, CBX79, which is deposited and revised at CIBEX data base[4] at Center for Information Biology and DNA Data Bank of Japan (DDBJ), National Institute of Genetics (Mishima, Japan). It includes two biological replicates of negative, mir-107, 185, and let-7a transfection experiments, one day and three days after the transfection. Expression of 45015 genes (probes) are listed.

Since our method is robust for the random noise of gene expression variance and the overall distribution of gene expression between technical replicates should be within the acceptable range, we did not apply any normalization procedure.

2.2   Inference of miRNA which Regulates Target Genes Significantly

The way to detect miRNA whose target genes are significantly differently expressed between negative control and treated one is as follows. First, we have downloaded a list of conserved seed match in 3' UTRs of genes to each miRNA[1][5]. This includes 162 miRNA families. The reason why we do not use major target gene list, e.g., targetScan[6], PITA[7], pictar[8], miranda[9], and others, but use seed match is because Alexiou *et al*[10] recently reported that simple seed match often resuits in higher F-measure than more complicated estimations of target genes (see e. g. Supplementary Data[10]). Then we have picked up genes which has at least one seed match for any miRNAs in those 3' UTRs. Then, 13270 genes remain.

Hereafter, we denote a set of these genes as G. Next, for each miRNA, m, we have listed genes which has at least one seed match in 3' UTRs. We denote this set of genes as G_m, where m denotes one of miRNA families. Also we define a set of genes, $G'_m \equiv G \setminus G_m$, which is a set of genes included into G, but not into

---

[1] http://hollywood.mit.edu/targetrank/hsa_conserved_miR_family_ranked_targets.txt

G_m. After denoting expression of gene g under transfection of miRNA m_0, m_0 is one of mir-107, 185, let-7a, and Negative Control (NC), as $x_g^{m_0}$, we compute gene expression difference between post-miRNA transfection and NC,

$$\Delta x_g^{m_0} \equiv \log x_g^{m_0} - \log x_g^{NC}.$$

Then we apply paired t-test between $\{\Delta x_g^{m_0} \mid g \in G_m\}$ and $\{\Delta x_g^{m_0} \mid g \in G'_m\}$. P-value, P_m, is computed for each miRNA, m. In the previous work[11], we have employed two sided t-test. However, in this study, we employ one-sided t-test, which checks if expression of genes in G_m is significantly *suppressed* than that of genes in G'_m. After applying FDR correction (BH method[12]) to 162 P-values, we have selected ms whose FDR corrected P-value is less than 0.05 as miRNA which regulates target genes significantly. For t-test, we have used t.test module in base package in R[13].

2.3   Coincidence between Biological Replicates

We have also checked if two biological replicates satisfy reproducibility in three ways. Firstly, we employed Pearson correlation coefficients between log transformed P_ms and secondly, Spearman correlation coefficients between them. P-values for these are computed as well as 95 percentile significant interval for the form. Thirdly, we analyzed coincidence between significant miRNAs, ms between two biological replicates. If the first(second) replicates have $m_1(m_2)$ significant miRNAs and m_{12} miRNAs are selected for both replicates, P-value computed by binomial distribution $P(m_1,N,m_2/N)$ or $P(m_2,N,m_1/N)$, where $P(x, N, p)$ is the probability that x among N is selected when the probability of selection is p. N is the number of genes in G.

We have used cor.test module in base package of R for P-values of correlation coefficients and pbinom module for binomial distribution.

2.4   Significant Overlap between Target Genes

To compute P-values of accidental agreement between target genes, $P_{m,m'}^O$ of miRNAs m and m', we have employed binomial distribution $P(n_{mm'}, n_m, n_{m'}/N)$, where $n_{mm'}$ is the number of co-target genes, $n_m(n_{m'})$ is the number of target genes of m(m').

3   Results

Independent of conditions, i.e., date and transfected miRNA, our method almost always gets non-empty set of significant miRNAs, ms (see Table 1). Thus, in principle, our method can detect miRNA regulation of gene expression. Table 1 shows which miRNA significantly regulates target genes. Most remarkably, P_m has the strong tendency to become smallest when $m = m_0$. If we compare the number of the significant miRNAs found in the previous study[11], it is clear that the drastic improvement took place. For example, all analyses of mir-107 transfection experiment gave us several tens of significant miRNAs in the

Table 1. All of significant miRNAs, one day or three days after trasnfection. Bold characters are those transfected. The one in parentheses means not significant but with the smallest P-values.

transfection	day	miRNA	P_m	miRNA	P_m
			replicate 1		replicate 2
mir-107	1	**miR-103/107**	1.32×10^{-8}	**miR-103/107**	4.96×10^{-9}
	3	**miR-103/107**	3.31×10^{-5}	—	—
mir-185	1	**miR-185**	4.32×10^{-12}	**miR-185**	8.74×10^{-18}
		miR-326	1.57×10^{-6}	miR-326	2.42×10^{-6}
		miR-491	1.99×10^{-4}	miR-491	1.05×10^{-4}
		miR-124.2/506	3.78×10^{-4}	miR-34b	1.68×10^{-4}
		miR-7	6.15×10^{-4}	miR-122	2.79×10^{-4}
		miR-485-5p	7.68×10^{-4}	miR-124.2/506	6.77×10^{-4}
		miR-339	1.57×10^{-6}	miR-331	9.34×10^{-4}
		—	—	miR-34/449	1.41×10^{-3}
		—	—	miR-485-5p	1.51×10^{-3}
	3	[**miR-185**]	5.48×10^{-4}	—	—
let-7a	1	**let-7/98**	2.18×10^{-14}	**let-7/98**	4.77×10^{-16}
		miR-196	1.97×10^{-4}	miR-196	1.40×10^{-4}
	3	**let-7/98**	2.66×10^{-8}	**let-7/98**	5.47×10^{-5}

previous study but only the transfected one in the present study, indicating that the improvement in the present study does not deteriorate the sensitivity. Hereafter, we name our present method as MiRaGE, which is, at present, only method to detect transfected miRNA as unique miRNA whose target genes are expressed significantly compared with non-target genes.

MiRaGE has two unique features compared with the previous methods.

1. Comparison between G_m and G'_m. In other method, comparison is done between G_m and all other genes whose expression is measured.
2. Target gene table is obtained based upon only seed match.

The reason why these two are important will be discussed later.

Table 2 shows the results of several statistical tests for the coincidence between biological replicates. For all six cases, at least two out of three tests give the significant P-values < 0.05. For most of cases, P-values is almost 0 within numerical accuracy ($P < 2.2 \times 10^{-16}$). Thus, biological replicates are good enough for inference of miRNA transfection.

4   Discussion

Although P_{m_0} is mostly the smallest, P_m with $m \neq m_0$ also can sometimes take the value as small as P_{m_0}. For example, for two replicates at one day after mir-185 transfection (see Table 1) there are seven and nine significant miRNAs respectively. The materials analyzed in this study are experimental. Hence it is

Table 2. Comparison of two biological replicates. Bold numbers indicate significant P-values (< 0.05). Bold asterisks (*) indicate $P < 2.2 \times 10^{-16}$.

Transfection	mir-107		mir-185		let-7a	
Time	day 1	day 3	day 1	day 3	day 1	day 3
Pearson	0.94	0.25	0.89	0.54	0.95	0.74
			95 % confidence interval			
lower	0.92	0.39	0.92	0.64	0.96	0.81
upper	0.96	0.10	0.86	0.42	0.93	0.67
P-value	*	**0.0016**	*	$\mathbf{1.66 \times 10^{-13}}$	*	*
Spearman	0.63	0.76	0.65	0.52	0.55	0.32
P-value	*	*	*	*	*	$\mathbf{4.73 \times 10^{-5}}$
			# of significant miRNAs			
common	1	0	5	0	2	1
replicate 1	1	1	7	1	2	1
replicate 2	1	0	9	0	2	1
P-value	*	—	$\mathbf{1.96 \times 10^{-7}}$	—	*	*

clearly distinguishable the true and false positives. The seed sequences of false positive miRNAs are different from that of the transfected one. The causes of false positives may be the secondary effect of transfected miRNA. The downregulation of target messengers by miRNA transfection could cause downregulation of other miRNAs targets or upregulation of false positive miRNAs. Alternatively, co-targeting between transfected miRNA and other miRNAs could cause the false positives: the miRNAs with overlapped targets with transfected one would show smaller P-value than other endogenous miRNAs. Theoretically false positives by co-targeting are unavoidable in our method. However, it will be avoidable when the expression profiles of miRNA are available: the unexpressed miRNAs showing small P-value can be filtered out from the list.

One may wonder that co-targeting among miRNAs results in more significant regulation of target genes of non-transfected miRNAs than that of the transfected miRNA. We calculated P-values of significant overlap of target genes by $P(n_{mm_0}, n_m, n_{m_0}/N)$. As a result, even after correction considering multiple comparison, 156, 157 and 156 miRNAs among in total 162 miRNAs have significant overlap ($P < 0.05$) with transfected miRNA of mir-107, mir-185 and let-7a respectively. This means, almost all of non-transfected miRNAs have significant large number of common target genes with those transfected miRNA. However, the correlation coefficient between P_m and P^O_{m,m_0}, do not show significant correlations 11 out of 12 (i.e., two biological replicates $\times$ two time points (day 1 or day 3) $\times$ three transfection) cases (see Table 3). This means, significant regulation of target genes of non-transfected miRNA cannot be explained by the accidental target gene overlap with those of transfected miRNA, m_0. Only one exception among total 12 cases is for let-7 day 3 replicate 2. Since the effects of transfected miRNA become weaker at the day 3, the secondary effect may become apparent and the some fluctuations in experimental conditions might

Table 3. Significance of correlation between P_m and P_{m,m_0}^O. Bold numbers are significant ($P < 0.05$).

Transfection	mir-107				mir-185				let-7a			
Date	day 1		day 3		day 1		day 3		day 1		day 3	
Replicates	1	2	1	2	1	2	1	2	1	2	1	2
correlation	-0.049	-0.086	-0.11	-0.12	0.050	0.098	0.13	-0.11	-0.012	-0.14	-0.092	-0.24
P-value	0.54	0.28	0.15	0.13	0.52	0.22	0.10	0.16	0.88	0.07	0.25	**0.0019**

cause the significance in this case. Actually, P_{m_0} for three days after transfection sometimes does not have small enough P-value (mir-107 replicate 2 and mir-185 replicate 2, see Table 1).

It is also interesting that miRNA target genes are generally more expressed than the other genes in the present datasets (see Fig. 1 in the previous work [11], there are peaks around $\log x_g \simeq 7$, which is far from origin). This tendency cannot be seen in genes not targeted by any miRNA. This fact may also be important to understand how each miRNA regulate genes in cancer formation/suppression.

The reasons why MiRaGE works better than previous methods are worth mentioning. For example, T-REX[14] can detect transfected miRNA as those with the smallest P-value, but cannot avoid having other miRNAs with significantly small P-value. As T-REX, most of such methods compare target genes of a miRNA with all others, while MiRaGE compares target genes with the genes targeted by all the other miRNAs. Since genes targeted by miRNAs are significantly different from genes not targeted by any miRNAs as mentioned above, it is important to employ genes targeted by other miRNAs as negative set. Other critical factor to improve the specificity is to employ simple seed match as miRNA target gene table. Since this table also decides the negative set, highly curated thus those with smaller genes mimic the size of negative set. The success of MiRaGE demonstrates that choice of the negative set is critical to estimate the important miRNAs during the biological processes.

5 Conclusion

In this paper, we have shown that gene expression profile combined with miRNA target genes predicted computationally can often correctly infer transfected miRNA. This suggests that we may be able to infer miRNA regulation of genes solely from gene expressions without considering any other information than computationally predicted target genes.

Acknowledgement. This work was supported by KAKENHI (23300357).

References

1. Brennecke, J., Stark, A., Russell, R.B., Cohen, S.M.: Principles of MicroRNA-targetRecognition. PLoS Biol. 3, 85 (2005)
2. Chi, S.W., Zang, J.B., Mele, A., Darnell, R.B.: Argonaute HITS-CLIP Decodes MicroRNA-mRNA Interaction Maps. Nature 460(7254), 479–486 (2009)

3. Takahashi, Y., Forrest, A.A.R., Maeno, E., Hashimoto, T., Daub, C.O., Yasuda, J.: MiR-107 and MiR-185 Can Induce Cell cycle Arrest in Human Non Small Cell Lung Cancer Cell Lines? PLoS One 4, e6677 (2009)
4. Ikeo, K., Ishii, J., Tamura, T., Gojobori, T., Tateno, Y.: CIBEX: center for information biology gene expression database. C R Biol. 326, 1079–1082 (2003)
5. Nielsen, C.B., Shomron, N., Sandberg, R., Hornstein, E., Kitzman, J., Burge, C.B.: Determinants of Targeting by Endogenous and Exogenous MicroRNAs and siRNAs. RNA 13, 1894–1910 (2007)
6. Friedman, R.C., Farh, K.K., Burge, C.B., Bartel, D.P.: Most Mammalian mRNAs are Conserved Targets of MicroRNAs. Genome Res. 19, 92–105 (2009)
7. Kertesz, M., Iovino, N., Unnerstall, U., Gaul, U., Segal, E.: The Role of Site Accessibility in MicroRNA Target Recognition. Nat. Genet. 39, 1278–1284 (2007)
8. Krek, A., Grün, D., Poy, M.N., Wolf, R., Rosenberg, L., Epstein, E.J., MacMenamin, P., da Piedade, I., Gunsalus, K.C., Stoffel, M., Rajewsky, N.: Combinatorial MicroRNA Target Predictions. Nat. Genet. 37, 495–500 (2005)
9. John, B., Enright, A.J., Aravin, A., Tuschl, T., Sander, C., Marks, D.S.: Human MicroRNA Targets. PLoS Biol. 2, e363 (2004)
10. Alexiou, P., Maragkakis, M., Papadopoulos, G.L., Reczko, M., Hatzigeorgiou, A.G.: Lost in Translation: An Assessment and Perspective for Computational MicroRNA Target Identification. Bioinformatics 25, 3049–3055 (2009)
11. Taguchi, Y.-h., Yasuda, J.: Inference of gene expression regulation via microRNA transfection. In: Huang, D.-S., Zhao, Z., Bevilacqua, V., Figueroa, J.C. (eds.) ICIC 2010. LNCS, vol. 6215, pp. 672–679. Springer, Heidelberg (2010)
12. Benjamini, Y., Hochberg, Y.: Controlling the False Discovery Rate: A Practical and Powerful Approach to Multiple Testing. J. R. Stat. Soc. B 57, 289–300 (1995)
13. R Development Core Team, R: A language and environment for statistical computing. R Foundation for Statistical Computing, Vienna, Austria (2009) ISBN 3-900051-07-0, http://www.R-project.org
14. Volinia, S., Visone, R., Galasso, M., Rossi, E., Croce, C.: Identification of MicroRNA Activity by Targets' Reverse EXpression. Bioinformatics 26, 91–97 (2010)

Syntactic Pattern Recognition from Observations: A Hybrid Technique

Vincenzo Di Lecce and Marco Calabrese[*]

Polytechnic of Bari, DIASS
Taranto, Italy
{v.dilecce,m.calabrese}@aeflab.net

Abstract. This paper presents a novel technique for automated learning from observations. The technique arranges in a row four traditional pattern recognition approaches (numeric, logic, statistical and finally syntactic) within a unifying framework. Each processing step is conceived as a transformation of the input dataset from one state to another. The proposed technique considers measurable observations as inputs and produces a set of formal rules, i.e., a grammar, as final output. To this end, a four-state grammar induction process is described in detail by means of a step-by-step example. As a proof-of-concept for the feasibility of the proposal, references to early experimental validations are given. Finally, possible comparison with other well-known approaches are discussed.

Keywords: Patter recognition, grammar induction, IF THEN rules.

1 Introduction

Automatic learning from observations is a desirable and widely-debated property of intelligent systems. For example, the native ability of infants to recognize complex patterns without prior knowledge (as in early language acquisition skills) is still an open field of research, although promising results are under way [1].

When dealing with unsupervised knowledge acquisition, pattern recognition (PR) techniques start from the assumption that knowledge is present (though not manifest) in the observed data. Consequently, inductive learning procedures are to be found in order to make hidden patterns explicit.

From an artificial intelligence standpoint, two relevant and diverging lines of publications in PR are evident since the late 60s [2]: numerical and logical. Their complementary nature has been seldom emphasized in the literature. More recently, the attention of scholars has moved towards a new dichotomy opposing statistical to syntactic techniques.

This work attempts to fill the existing gap among such different approaches by proposing a novel unifying grammar-induction framework based on a four-state process. The proposed technique is able to find hierarchically nested patterns, hence

---

[*] Corresponding author.

D.-S. Huang et al. (Eds.): ICIC 2011, LNBI 6840, pp. 136–143, 2012.

providing a great flexibility in data classification [3]. Furthermore, since it produces a grammar as output of the learning task, it can be applied straightforwardly in automatic rule-based system modeling.

Paper layout is as follows, Section 2 reports on different approaches to PR; Section 3 presents the proposed PR technique; Section 4 provides a detailed example along with references to first experimental validations and comparisons with other well-known approaches; Section 5 concludes.

2 Related Work

Pattern Recognition (PR) is about classifying measurable observations in either a supervised or an unsupervised manner [4]. According to Fu [5], two main PR approaches can be identified in the literature: decision theoretic and structural. On the one hand, decision theoretic approaches rely on the features extracted from observed data. Patterns are then defined by means of feature vectors considered as points in a multi-dimensional space. On the other hand, structural approaches consider complex patterns as composed of sub-patterns of primitive elements.

Decision theoretic and structural approaches have blossomed a range of PR categories. Jain for example [6] identifies four of them, namely: template matching (the oldest and simplest one), statistical, neural network and syntactic. Neural networks however can be considered as a computational variant of statistical techniques [7], hence, in practice, the actual dichotomy is between statistical and syntactic techniques. Discredit given by Chomsky's assumptions upon the efficacy of statistics in grammar deduction [8][9] probably had a deep influence in fostering this divide.

Another relevant classification, expressed in the form of sharpening divergence between numerical and logical approaches, is referenced in [2].

Shortly after early works of Chomsky, prospective visions attained by Solomonoff [10] and Minsky [11] contributed to shed new light to grammar-induction models, but, at their time, computational models and artificial intelligence techniques were too immature to provide a sound support to their claims.

With the advent of modern computer technologies and the spreading of electronic repositories, statistical techniques have gained much interest among scholars and practitioners. Already in the second half of '80s, there was a widespread consensus about the efficacy of stochastic Markovian models for grammar discovery in practical applications [12]. This trend seems to be confirmed by recent findings in the field of cognitive science, for example in modeling early language acquisition skills [1].

3 Proposed Hybrid Technique

The proposed PR technique can be described as a four-state process. It starts from a numeric dataset (State 1) and ends with a grammar, i.e. a collection of formal rules, extracted from the same dataset (State 4). Intermediate states are logic (State 2) and statistical (State 3). The whole sequence is reported in Table 1.

Table 1. The four states of the proposed PR technique

STATE 1	STATE 2	STATE 3	STATE 4
[numeric]	[logic]	[statistical]	[syntactic]
Series of	Series of	Structured	Formal rules
observations	hypotheses	hypothesis space	

3.1   State 1 - Numeric

This state corresponds to any typical situation where an agent [13] with no *a-priori* knowledge about the external environment collects observations through a set of *n* sensors. State 1 can be formally described as follows.

Consider an array of *n* types of numerical observations:

$$\mathbf{o}_k = \langle \mathbf{o}_k[1], \mathbf{o}_k[2], \dots, \mathbf{o}_k[n] \rangle \tag{1}$$

with $k = 1, 2, 3\dots$ representing the discrete time instants at which the observation samples are taken. The input dataset D is simply defined as the collection of temporal observations:

$$D = \{\mathbf{o}_1, \mathbf{o}_2, \mathbf{o}_3, \dots, \mathbf{o}_k, \dots\} \tag{2}$$

3.2   State 2 - Logic

Transition to this state is fired when a particular "hypothesisation" procedure is applied to the input dataset. The procedure consists in performing numerical hypotheses on data in order to classify them as unary (true) of zero (false) values. Hypotheses can be in fact considered as predicate functions of the type:

$$\mathbf{hyp}_k[i] = \begin{cases} 1 & \text{if hyp}(\mathbf{o}_k[i]) \text{ is TRUE} \\ 0 & \text{if hyp}(\mathbf{o}_k[i]) \text{ is FALSE} \end{cases} \tag{3}$$

It is interesting to notice that a numerical hypothesis can be generated in an unsupervised manner by means of statistical or clustering techniques applied to the input dataset.

From (3), the input dataset is transformed into a set H of hypotheses:

$$H = \{\mathbf{hyp}_1, \mathbf{hyp}_2, \dots, \mathbf{hyp}_k, \dots\} \tag{4}$$

3.3   State 3 - Statistical

State 3 is activated by the application of a classification task to data in H. To this end, classification and regression tree (CART) tools can be applied.

CART is a simple non-linear regression technique [14] that takes a set of data arrays as input and then outputs interclass relationships according to the following form:

$$x_k = CART_k(x_1, x_2, \dots, x_{k-1}, x_{k+1}, \dots) \tag{5}$$

where $CART_k$ defines a decision tree structure. This allows for defining a predicted variable x_k as a function of decision nodes (predictors) x_1, x_2, etc. Each decision node, in our case, is a binary variable corresponding to an IF-THEN-ELSE algorithmic structure. For example, condition $x_1 == 0$ (to be read as 'x_1 is equal to zero') might be described by the following algorithm:

$$CART \text{ output example extract} \tag{6}$$

```
IF x2 == 0 % predictor condition
   IF x3 == 1      % predictor condition
          THEN x1 == 0   % predicted condition
       ELSE …
      END
    ELSE...
  END
```

In the proposed approach, the input for the CART is the set H, while outputs are defined by as many decision trees as there are types of observations. It follows that each decision node corresponds to a hypothesis on a certain type of observed temporal data sequence.

3.4   State 4 - Syntactic

Decision trees in the form described in (6) can be easily transformed into formal syntactic rules. Each branch in fact corresponds to a symbolic-logic pattern as follows:

$$premise \rightarrow conclusion \tag{7}$$

where premise is given by the logic AND of hypotheses (predictors) needed to obtain a certain conclusion (predicted value). With reference to example (6), the rule would be:

$$\overline{x}_2 x_3 \rightarrow \overline{x}_1 \tag{8}$$

The meaning of (7) is that if premises are true then conclusion is true. This premise/conclusion pattern is largely employed in rule-based systems [15]. The four-state process then turns into a completely automated grammar-induction procedure. The human intervention can be limited to deciding which kind of hypotheses to choose in State 2, which is however not necessary if some unsupervised hypothesis generation procedure is implemented. A complete grammar-induction example is detailed in the next Section.

4   Grammar Induction Example

Consider a logic circuit of the type depicted in Figure 1, which implements an OR function with NAND ports. Let A, B, C, D, and E, be five observation points. A and B lines are fed by a random signal generator producing binary values. The truth table is hidden to the observer agent who can only collect 5-length binary words at any time instant. Since the agent has no *a-priori* knowledge about the hidden logic model, the four-state procedure above described is useful for the knowledge extraction task.

In particular, we expect to extract formal rules describing the hidden truth table in a symbolic-logic form. The logic net and the PR technique have been simulated in Matlab® R14 environment.

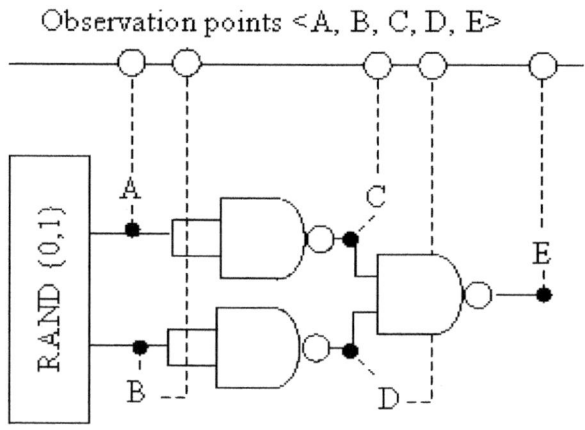

Fig. 1. The NAND logic circuit used as example for the proposed PR technique

4.1   Grammar Induction Procedure

State 1. Looking at the circuit in Figure 1 it is easy to reckon that the input dataset D, after a sufficient number of observations, will be composed of some combination of only four possible patterns: <0,0,1,1,0>, <0,1,1,0,1>, <1,0,0,1,1>, <1,1,0,0,1>.

State 2. The more convenient hypothesisation procedure is to define an hypothesis answering to the binary condition, for any k and any i, of the form:

$$\textbf{hyp}_k[i]: \text{ is it TRUE that } \textbf{o}_k[i] = = 1?$$

Other equivalent binary conditions can be imagined as well. Thanks to the transformation imposed by the chosen hypotheses, dataset is re-mapped into a hypothesis space (in the particular case of this example, due to the binary nature of the observed data, the two sets D and H are coincident if the aforementioned hypothesisation is chosen). Dataset H is then used as input for the CART. A run has been launched for each predicted variable, thus having 5 decision trees.

State 3. Several hundred simulation runs, each one made of one thousand observation samples, have been performed. An excerpt of found data is summarized in Table 2. CART outputs are described through an algorithmic (symbolic) transcription of the logic structures obtained from the decision trees produced by the CART.

Table 2. CARTs (excerpt) for the provided example. $< x_1, x_2, x_3, x_4, x_5 >$ correspond to binary hypotheses on: $< A, B, C, D, E >$

CART for x1 prediction.	CART for x2 prediction	CART for x3 prediction
IF x3==1	IF x4 ==1	IF x1 ==1
THEN x1 == 0	THEN x2 == 0	THEN x3 == 0
ELSE x1 == 1	ELSE x2 == 1	ELSE x3 == 1
END	END	END

CART for x4 prediction	CART for x5 prediction
IF x2 ==1	IF x2 == 1
THEN x4 == 0	THEN x5 == 1
ELSE x4 == 1	ELSE
END	IF x1 == 0
	THEN x5 ==0
	END
	END

State 4. The outputs of state 3 are then transformed according to pattern in (7). Results can be summarized in the form of different premise-conclusion patterns as shown below. It is immediate to notice, for example, the NOT condition among A-C (x_1-x_3) and B-D (x_2-x_4) observation pairs. Rules reported in Table 3 are only those derived from data in Table 2. After simulations, 24 distinct rules were found in total; however, the study of the rule learning rate goes beyond this study.

Table 3. Formal rules obtained from data in Table 2

$\overline{x}_2 \rightarrow x_4$	$\overline{x}_2 \rightarrow x_4$	$\overline{x}_2\overline{x}_1 \rightarrow \overline{x}_5$	$\overline{x}_1 \rightarrow x_3$	$x_2 \rightarrow x_5 \mid \overline{x}_4$
$\overline{x}_2 x_1 \rightarrow x_5$	$x_3 \rightarrow \overline{x}_1$	$\overline{x}_3 \rightarrow x_1$	$x_4 \rightarrow \overline{x}_2$	$\overline{x}_4 \rightarrow x_2$

4.2   Proof-of-Concept Early Implementations

Due to its formulation, the proposed grammar-induction approach is putatively so general to be applied to any setting where measurable observations are available. In very recent times, the IF THEN rule extraction mechanism has been employed with success in discriminating gas sensors behavior in low-cost setups [16] and even in defining smart sensors new design criteria [17]. Early results show that an appropriate mix of low-cost sensors is able to manifest surprising discrimination abilities.

4.3   Comparisons with Other Approaches

The search for recurrent structures, patterns, association rules and, ultimately, knowledge hidden in observations (regardless of the fact they come from a measurement setting or a database) has spawned a great interest in the research community with a special bias on data mining [18] and subsequent evolutions. It is noteworthy that general-purpose approaches to the extraction of useful bits of information from data still represent a challenge since no mature solutions seem to be available at the moment [19]. Although at an early stage, our approach prospectively moves along this track.

A recent research direction dealing with automatic IF THEN rules extraction form data is accounted by linguistic approaches as presented in [20]. On the whole, our work complies with this position. However, we do not make use of fuzzy logic as a theoretical base for our assumptions: we blend instead more basic concepts drawn from the numeric, statistical, logic and syntactic world. This represents, at the best of our knowledge, a novel engagement in the literature.

4.4   Overall Considerations

The proposed four-state grammar induction procedure is appealing since it moves from the numeric world towards symbolic-logic structures without the need of prior knowledge assumptions. It represents a data-driven technique that can be applied in principle to different kinds of measurable observations. This means that, in case of a hidden deterministic model in the observed world (which is a basic assumption for modern science), observed data are able to produce symbolic descriptions of themselves. More complex cases than the one provided by the previous example, such as the inspection of circuits with memory (flip-flop based) or the analysis of analog signals are currently under way and go beyond the scope of this paper. In prospective work, the flexibility and applicability of the presented framework will be specifically assessed.

5   Conclusion

In this work a novel hybrid syntactic PR technique is presented for deriving formal rules (hence a grammar) from a set of measurable observations. The proposed framework underpins on well-grounded PR approaches (numeric, logic, statistical and syntactic) reviewed under a unifying perspective.

This technique has a number of potential applications, like in automatic model discovery problems. In this paper, a detailed example has been provided for automatic induction of formal rules from the inspection of a simple logic circuit. Prospective works will be aimed at handling more complex scenarios.

References

1. Aimetti, G.: Modelling Early Language Acquisition Skills: Towards a General Statistical Learning Mechanism. In: Proc. of the 12th ACM Conference of the European Chapter of the Association for Computational Linguistics: Student Research Workshop, pp. 1–9 (2009)
2. Naylor, W.C.: Some Logical and Numerical Aspects of Pattern Recognition and Rrtificial Intelligence. In: Proc. Of the ACM Spring Joint Computer Conference, pp. 95–101 (1969)
3. Zhang, J., Silvescu, A., Honavar, V.: Ontology-Driven Induction of Decision Trees at Multiple Levels of Abstraction. Abstraction. Reformulation, and Approximation (2002); 316.3. Foster, I., Kesselman, C.: The Grid: Blueprint for a New Computing Infrastructure. Morgan Kaufmann, San Francisco (1999)
4. Jain, A.K., Murty, M.N., Flynn, P.J.: Data Clustering: A Review. ACM Computing Surveys 31(3), 264–323 (1999)
5. Fu, K.S.: Guest Editor's Introduction. Pattern Recognition (1976)
6. Jain, A.K.: Statistical Pattern Recognition: A Review. IEEE Transactions of Pattern Analysis And Machine Intelligence 22(1), 4–37 (2000)
7. Joshi, A.: On Neurobiological, Neuro-Fuzzy, Machine Learning, and Statistical Pattern Recognition Techniques. IEEE Transactions on Neural Networks 8(1), 18–31 (1997)
8. Chomsky, N.: Three Models for the Description of Language. IRE Transactions on Information Theory 2, 113–124 (1956)
9. Chomsky, N.: Reflections on Language. Pantheon Books, New York (1975)
10. Solomonoff, R.: A Progress Report on Machines to Learn to Translate Languages and Retrieve Information. Advances in Documentation and Library Science III (2), 941–953 (1959)
11. Minsky, M.: Steps Toward Artificial Intelligence. Proceedings of the IRE 49(1), 8–30 (1961)
12. Atwell, E., Drakos, N.F.: Pattern Recognition Applied To The Acquisition Of A Grammatical Classification System From Unrestricted English Text. In: Proc. Of the Third Conference of the European Chapter of the ACL, pp. 56–62 (1987)
13. Russell, S., Norvig, P.: Artificial Intelligence: A Modern Approach, 2nd edn. Prentice-Hall, Englewood Cliffs (2002)
14. Riley, M.D.: Some Applications of Tree-based Modelling to Speech and Language. In: Proc. of the Workshop on Speech and Natural Language, pp. 339–352 (1989)
15. Yang, G.: A Syntactic Approach for Building a Knowledge-based Pattern Recognition System. In: Proc. Of the 9th IEEE International Conference on Pattern Recognition, pp. 1236–1238 (1988)
16. Di Lecce, V., Calabrese, M.: Describing Non-selective Gas Sensors Behaviour Via Logical Rules. In: Accepted to the IEEE/ACM 5th International Conference on Sensor Technologies and Applications - Sensorcomm p. 2011(2011)
17. Di Lecce, V., Calabrese, M.: Syntactic Pattern Recognition from Observations: A Hybrid Technique. In: Accepted to the 7th International Conference on Intelligent Computing – ICIC, p. 2011 (2011)
18. Agrawal, R., Imieliński, T., Swami, A.: Mining Association Rules Between Sets of Items in Large Databases. In: Proc. of the ACM SIGMOD International Conference on Management of Data, pp. 207–216 (1993)
19. Cao, L.: Domain-Driven Data Mining: Challenges and Prospects. IEEE Transactions on Knowledge and Data Engineering 22(6), 755–769 (2010)
20. Wu, D., Mendel, J.M.: Linguistic Summarization Using IF–THEN Rules and Interval Type-2 Fuzzy Sets. IEEE Transactions on Fuzzy Systems 19(1), 136–151 (2011)

An Efficient Ensemble Method for Classifying Skewed Data Streams

Juan Zhang, Xuegang Hu, Yuhong Zhang, and Peipei Li

School of Computer and Information, Hefei University of Technology, Hefei 230009, China
zjuan9813@126.com, {sjxhuxg,zhangyh}@hfut.edu.cn,
peipeili.hfut@gmail.com

Abstract. Class distributions of data streams in real application are usually unbalanced, they are hence called Skewed Data Streams (abbreviated as SDS). However, in the classification of SDS, it is a challenge for traditional methods because of the difficulty in the recognition of minority classes. Therefore, many approaches have been proposed to improve the recognition rate of minority classes, while they are time-consuming. Motivated by this, we propose an efficient Ensemble method for Classifying SDS called ECSDS. Our algorithm creates multiple classifiers based on C4.5, and adopts the threshold of F1-value to limit the updating frequency of classifiers. Meanwhile, it adds misclassified positive instances into the training data to guarantee the effectiveness of classifiers when updating. Experimental studies demonstrate that our proposed method enables reducing the time overhead and maintains a good performance on the classification accuracy.

Keywords: Skewed data streams, Ensemble classifiers, F1-value, Misclassified positive instances.

1   Introduction

Data streams emerging from real applications, such as the fields of fraud identification and intrusion detection, present a new characteristic of skewed class distributions. These data streams are hence named as Skewed Data Streams (denoted as SDS), that is, most of instances in SDS belong to negative instances while few ones belong to positive instances. In the classification of SDS, because most of traditional classifiers only address the prediction ability on negative instances, thus, it is challenging for them to classify positive instances accurately. Therefore, how to improve the prediction accuracy on the positive instances is a hot topic in the classification of SDS [1-2].

In fact, researchers have proposed many approaches for handling of SDS, including re-sampling[3-4], cost-sensitive learning[5], feature selection[6] and ensemble learning[7-9]. One of the most popular approaches is the ensemble learning. It is beneficial to improve the predictive accuracy on positive instances, but the time overhead is demanded heavily. To reduce the time overhead, we propose an ensemble method called ECSDS to classify SDS in this paper. In contrast to the algorithm in [8-9], we utilize the value of F_1-value as the threshold to decide the classifier updating. It could

D.-S. Huang et al. (Eds.): ICIC 2011, LNBI 6840, pp. 144–151, 2012.
© Springer-Verlag Berlin Heidelberg 2012

decrease the updating frequency and hence reduce the time overhead. In addition, we add misclassified positive instances in the training data set to improve the classification accuracy when updating the classifiers. Extensive experiments show that our ECSDS method performs better on the time overhead and prediction accuracy on positive instances.

This paper is organized as follows. Section 2 reviews the related work. Section 3 introduces our proposed method and analyzes its time complexity. Parameter settings and experimental results are shown in Section 4, followed by the conclusion and future work in Section 5.

2   Related Work

The issue of SDS always cause the suboptimal classification performance for existing classification algorithms, thus, the handling of SDS has attracted more concerns. For specifically, Gao et al.[8-9] proposed an ensemble framework called SE to classify SDS. It performs better in the classification accuracy on positive instances while its in-time updating mechanism made its time consumption increased. Ouyang et al.[10] introduced a method called IMDWE, which is similar to SE, but its number of base classifiers is much bigger than the former. Song et al.[11] and Wang et al.[12] both utilized the k-means clustering algorithm to deal with imbalanced data sets. However, the k-means clustering algorithm itself demanded heavy time overhead. Ryan et al. [13-14] used simpler method, which deletes negative instances those are classified correctly randomly until meeting the trade-off between negative and positive instances. It will cause the loss of useful information if the data set is extremely unbalanced apparently. Meanwhile, it specifies the weight of basic classifiers using the distance between the ith data chunk ($1 \leq i \leq t$) and the $(t+1)^{th}$ data chunk. It will cause the heavy time consumption.

Regarding the aforementioned approaches, they all utilize the ensemble model to improve the prediction accuracy in the classification of SDS. However, a common weakness relies on the heavy time overhead. Therefore, an efficient Ensemble method for the Classification of SDS, called ECSDS is proposed in this paper. Contrary to the aforementioned algorithms, our algorithm presents the following characteristics. 1) F_1-value is utilized to decide whether updating the classifier or not. It is beneficial to reduce the updating frequency. 2) Misclassified positive instances are added into the training data set to enhance the algorithm's learning ability when updating classifiers. This mechanism makes the algorithm more efficient and guarantees a high accuracy in the classifying positive instances.

3   Our ECSDS Algorithm

3.1   Algorithm Description

Our ECSDS algorithm aims to classify skewed data streams. As shown in Pseudo-code of Algorithm 1, the processing mainly contains three components. First, we generate the ensemble classifier *EC* based on C4.5 using sequential data chunks

$\{D_1, \dots, D_m\}$ with the mechanism in SE [9]. Second, we use it to predict the class labels cl of instances in D_{m+1} and collect misclassified positive instances into the set of misclassified instances MI. Finally, we divide D_{m+1} into the set of positive instances P_{m+1} and the set of negative instances N_{m+1}, then update the training data set TD with them and introduce MI into training data subsets to update classifiers if the updating condition meets the threshold d. The similar processing is repeated till the streaming data is end. In the following subsections, we will give the details of each component.

```
Algorithm1. ECSDS
Input: data chunks D1, D2, … , Dm,… in SDS; the number
of data chunks for initializing EC -m; the threshold of
updating-d
Output: class labels of instances in Di (i>m)
1    The initial F1-value for a data chunk fF = -1;
2    While (Di arrives){
3      if(i==m){
4          Add Pi into set AP, i.e., AP = P1+ P2+…+ Pm;
5        Get the first training set TD, i.e., TD= (AP, Nm);
6        Let Nm=(n1,n2,…,nk), and k = |Nm|/|AP|;
7        EC = ∑_{j=1}^{k} buildClassifer (AP,n_j,C4.5); }
8      if(i>m){
9          for(each instance e in Di){
10             cl = Vote_by_EqualWeihted (EC);
11             if(misclassified(e)&&TrueLabel(e)==positive)
12               Add e into the set MI; }
13           sF=F1-value(EC, Di);
14           if(fF != -1 ){
15           Let delta=fF —sF;
16           if (delta ≥ d) {
17             Divide Di into Pi and Ni;
18             Update TD={AP+Pi, Ni}, repeat step6 once;
19           introduce MI to update
           EC = ∑_{j=1}^{k} RebuildClassifer (MI,AP,n_j,C4.5); }
20           }fF = sF;   } i++;
21    }
```

3.2   Technical Details

Generation of ensemble classifier. Suppose there are m skewed data chunks arrived, and then P_m is too scarce to get a good classifier. Thus, in our algorithm, we use the mechanism in SE [9] to balance the class distribution and build classifier as shown in Steps 4~7. For specifically, we first collect P_i in these $D_1, \dots, D_m$ into the positive instances set AP. Second, we segment N_m into k subsets, and then combine each n_j with AP to form k balanced training data sets. In terms of these balanced data sets, we generate a C4.5-based classifier on each small set (AP, n_j) and then get the ensemble classifiers EC.

Prediction and updating. With the coming streamy data chunk D_i $(i>m)$, we use the current ensemble classifier EC to predict on D_i as shown in Steps10~12. If the instance e in D_i is misclassified and its real class label is positive, then we add e into the set MI. In addition, the F_1-value of current data chunk is calculated, denoted as sF.

After finishing predicting on two data chunks, we estimate the difference *delat* between their F_1-values, if the value of *delta* is no less than the threshold d, the updating is installed as shown in Steps17-19. For precisely, 1) updating the training data TD: add the newest P_i into AP, and replace the old negative instances N_m with N_i. 2)updating EC : add MI into each subset (n_i, AP) to form new subset (n_i, BP, MI) and rebuild EC as described in Step19.

In addition, regarding the setting of the threshold d, if the value of d is too large, the updating frequency of classifier is less. Thus the time overhead could be reduced while the predictive accuracy may be impacted. Otherwise, the updating frequency is still high while the time overhead is increased. Therefore, it is necessary to select an optimal value of d that guarantees the tradeoff between the time complexity and the classification performance of the algorithm (see subsets 4.1 for more details).

3.3   Analysis of Time Complexity

According to the description in Section3.1, the time cost mainly depends on the time consumption of training classifiers and the updating frequency of classifiers. Regarding the former, the time complexity of C4.5 algorithm is $O(x*n*\log n)$[16], where x is the number of attributes in the training data set, and n is the number of instances in that. Suppose the size of each data chunk is same (denoted as n, and n'=|AP|<<n). Hence, the time complexity of getting an ensemble classifier in ECSDS is $k* O(x*n'*\log n')$, where k indicated the number of base classifiers. Meanwhile, if there are N data chunks to be predicted, and the time cost of prediction is linear correlation to the size of instances, denoted as $O(k*N*n)$. Thus, the complexity can be expressed as $O(p *N *k*O(x*n'*log n')+k*N*n)$, where p is the updating frequency of classifiers. If classifiers are updated in real time, the value of p equals to 1. However, in ECSDS, the threshold d is used to decide the updating of classifiers, so $p<<1$. Correspondingly, the time performance of our algorithm could be improved.

4   Experiments

4.1   Experimental Setting

In this section, we select data sets from UCI (http://archive.ics.uci.edu/ml/datasets.html) to validate our EDSDS algorithm. Meanwhile, we also select several synthetic data sets like hyperplane, waveform, LED and SEA, which are generated by the data generators from an open-source experimental tool called MOA [15]. In our experiments, to simulate the skewed class distribution, we resample these synthetic data sets to form SDS only containing $r\%$ positive instances. Details of data sets are shown in Table 1. Regarding the parameter setting d, the optimal value is set to 0.08(see subsection4.1 for details). The number of data chunks m mentioned in Pseudo-code will be set to be half number of data chunks for data sets in UCI, and 10 for synthetic data sets. In the experiments, we use the implementation in the Weka

[16] package, a publicly available data mining software package. All experiments are performed on a P4, 2.93GHz PC with 2G main memory, running Windows XP Professional. Both algorithms are written in Java.

Table 1. Data Sets Descriptions (r=1,3,5; 1k=1000)

Data sets	Classes	Total Number of Instances	Number of Positive Instances	Number Of Data Chunks	Size of Data hunk
thyroid	class1 vs class3	5747	255	6	957
letter	Each class vs rest	20k	734-813	6	3332
covtype	class2 vs class4	28604	274	11	2599
optdigits	each class vs rest	5620	554-572	6	936
krkopt	class4 vs class14	4751	198	9	527
connect-4	class 3 vs class1	50922	6449	10	5009
adult	class 1 vs class2	32561	7841	8	4070
nursery	class3 vs rest	12960	328	6	2160
SEA	class1 vs class2	50k	50k*r%	50	1k
Hyperplane	class1 vs class2	50k	50k*r%	50	1k
LED	class1 vs rest	50k	50k*r%	50	1k
waveform	class1 vs rest	50k	50k*r%	50	1k

4.2   Experimental Analysis

In this subsection, we first consider the optimal value of threshold d by experiments, and then we compare our algorithm with the SE algorithm using the measures of MSE [9] (Mean Squared Error), gmeans[10], F_1-value [13]and time cost respectively. Moreover, F_1-value and gmeans are both the bigger the better, the time overhead and MSE are the smaller the Better.

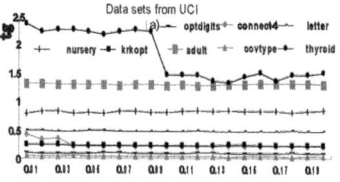 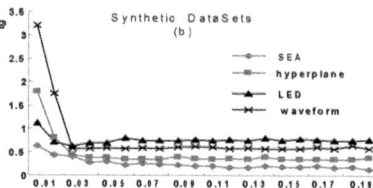

Fig. 1. Relation between tg and d

Analysis of parameter setting −d. To select an optimal value of d, a set of experiments is conducted. In our experiments, we set the value of d varying from 0.01 to 0.2, and add 0.01 each time. Considering the changes of time cost and gmeans varying with the values of d, we use a new measure tg relevant to both of evaluation measures to select the optimal value of d, i.e., $tg=timecost/gmean$. According to this definition, the less the time cost (t) and the larger the value of g-mean (g), the less the value of tg. It indicates the better performance.

Figures 1-a and 1-b report the relationship between the values of tg and the values of d. From the observation, we can see that the impact from different values of d on tg in Figure 1-a is much weaker than that in Figure 1-b. This is because the number of

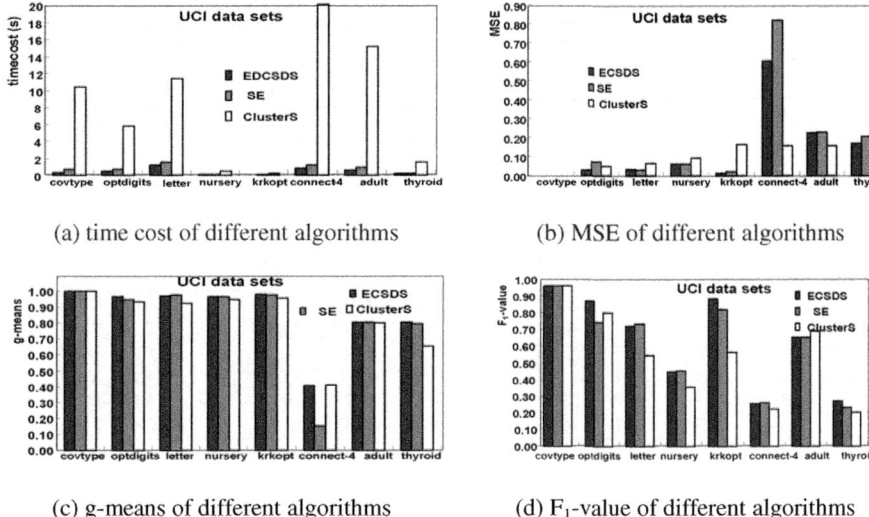

(a) time cost of different algorithms

(b) MSE of different algorithms

(c) g-means of different algorithms

(d) F_1-value of different algorithms

Fig. 2. Results on UCI data sets

data chunks in UCI data sets is much fewer (most of them are no more than 10), which leads to weak relation between the values of *tg* and *d*. However, the size of synthetic data sets is much larger (all of them have 50 data chunks), and the relation between *tg* and *d* are hence more apparent. According to the experimental results, we can obtain that the performance on *tg* is good when the value of *d* exceeds 0.07, namely, the optimal range of *d* [0.07, 0.1]. In our algorithm, we select *d* by 0.08 in the following experiments.

Performance Comparison. In this part, we compare our algorithm with SE [9] and ClusterS[12]. Figure 2 shows the results of comparisons in the time cost, g-means, MSE and F_1-value on UCI data sets. As Figure 2-a and Figure 2-b show, since the value of time cost and MSE are both the smaller the better, ECSDS performances best, and then is SE, and as expected, ClusterS is time-consuming heavily. For example, the time cost in ECSDS is only 39.74% of that in SE and 13.19% of that in ClusterS on data sets of "krkopt", and the MSE on the data sets of "connect-4"("krkopt") is reduced largely about 21% (15.5%) comparing with SE (ClusterS) in our method. On the other hand, our method performs better in the evaluation measures of g-means and F_1-value for most data sets. For example, the values of g-means are improved by 0.69% ~25.46% and 1.17%~15.54% comparing with SE and ClusterS respectively.

Table 2 reports the experimental results conducted on the synthetic data sets. In the observation, we can see that 1) the time overhead (denoted as T, unit: second) of our method is much less than that of SE and ClusterS. For examples, the time overhead of ECSDS is only 7.11% ~ 37.97% of that consumed in SE and 0.64%~8.48% of that consumed in ClusterS. Meanwhile, we observed that the time cost in ClusterS increased when the number of positive instances was become more and more. This is because the value of *k* in k-means clustering algorithm equals to the number of positive instances. However, the change of the time cost in ECSDS and SE are not apparent

Table 2. Results on synthetic data sets

Data Sets	ClusterS				SE				ECSDS			
	T	MSE	g	F_1	T	MSE	g	F_1	T	MSE	g	F_1
SEA$^{1\%}$	4.1	2.7	86.78	45.92	0.9	3.00	96.09	40.31	0.3	2.44	95.96	44.27
SEA$^{3\%}$	8.5	1.27	93.21	81.25	0.9	4.72	96.82	56.27	0.3	3.97	97.29	60.79
SEA$^{5\%}$	12.2	1.89	94.64	83.23	0.9	4.81	96.52	67.57	0.2	3.71	96.76	72.60
Hyper$^{1\%}$	9.4	15.37	64.94	6.52	3.0	23.75	80.12	7.16	0.3	21.61	79.16	6.95
Hyper$^{3\%}$	20.1	29.22	70.74	12.91	3.2	35.44	74.53	13.14	0.3	14.02	86.09	27.05
Hyper$^{5\%}$	29.6	24.99	72.92	22.24	3.1	36.48	74.45	20.06	0.4	29.43	77.10	22.71
LED$^{1\%}$	9.7	6.31	87.14	22.10	1.9	13.28	90.06	12.67	0.7	13.61	89.29	12.06
LED$^{3\%}$	28.2	7.89	90.60	40.91	1.8	16.43	88.33	25.86	0.3	11.23	91.50	33.68
LED$^{5\%}$	44.0	7.64	90.03	53.69	1.7	17.63	87.26	34.98	0.3	16.67	88.07	36.31
wave$^{1\%}$	17.1	17.50	76.78	7.97	6.2	19.69	84.96	8.62	0.5	18.88	83.46	8.39
wave$^{3\%}$	38.4	17.74	80.38	21.32	6.6	29.94	79.74	15.73	0.5	14.84	88.96	27.47
wave$^{5\%}$	55.3	14.30	82.50	36.05	6.3	28.94	80.33	24.72	0.7	23.28	81.30	27.58

with different number of positive instances. Thus, our algorithm is much steadier. 2) Regard the value of MSE, there are 11 records of SE that more than these of ECSDS. These data confirm that our method is superior to SE. Exceptionally, the MSE of SE on data set of "LED" with 1% positive instances has increased by 0.32%. The reason is that classification on this data set is usually not well, and the factor of class distribution makes it more difficult. In addition, there are 9 records of ClusterS are better than these of ECSDS. It may be because the k-means algorithm can choose more representative negative instances and hence improve the effectiveness of classification. 3) As shown in the columns of "g"(gmeans) and "F_1", the values of g-means in ECSDS have improved by 0.9% ~ 15.35% and 0.97%~11.56% contrasted to ClusterS and SE respectively. Besides, the values of F_1-value also improved by 1.33% ~ 13.92% compared to SE, and the results of ECSDS are well-matched. More over, we observed that the performance in ECSDS on these synthetic data sets with 1% positive instances is not better than that of SE. It illustrates that the performance in ECSDS on extremely skewed data sets need to be improved.

In sum, the time performance is improved largely in our algorithm. It states that the threshold of F_1-value do beneficial for reducing time consuming. Though, some results of ClusterS are better than these of ECSDS, the time overhead is considerable heavy and become the most defect of application in streaming data. So consider both time cost and classification effectiveness, the experimental results indicate that ECSDS is an efficient and effective method.

5   Conclusions

In this paper, we proposed the ECSDS method to accelerate classification on SDS while keeping good performance. It utilizes F_1-value to control the updating frequency and adds misclassified positive instances to update classifiers. Experiments show that our method is efficient and effective on the classification of SDS. However, in our method, a single classification model is used as the base classifier. It is disadvantage to get the diversity of decision results. Thus, in the future work, we will utilize different learning methods to generate base classifiers for the handling of SDS.

Acknowledgments. This research is supported by the 973 Program of China under award No. 2009CB326203, the National Natural Science Foundation of China (NSFC) under grants No. 60828005, the Fundamental Research Funds for the Central Universities and Special Funds of Thousand Talents Program under grant 2010HGXJ0715.

References

1. Chawla, N.V., Japkowicz, N., Kotcz, A.: Editorial: Special Issue on Learning from Imbalanced Data Sets. SIGKDD Explorations 6(1), 1–6 (2004)
2. Ye, Z.F., Wen, Y.M., Lu, B.L.: A Survey of Imbalanced Pattern Classification Problems. CAAI Transactions on Intelligent Systems 4(2), 14–156 (2009)
3. Batista, G.E.A.P.A., Prati, R.C., Monard, M.C.: A Study of the Behavior of Several Methods for Balancing Machine Learning Training Data. SIGKDD Explorations 6(1), 20–29 (2004)
4. Laurikkala, J.: Improving Identification of Difficult Small Classes by Balancing Class Distribution. Tech. Rep. A-2001-2,University of Tampere (2001)
5. Fan, W., Stolfo, S., Zhang, J., Chan, P.: AdaCost: MElassification Cost-sensitive Boosting. In: 16th International Conference on Machine Learning, San Mateo, USA, pp. 983–990 (1999)
6. Cardie, C., Howen, H.: Improving Minority Class Predicting Using Case-specific Feature Weights. In: 14th International Conference on Machine Learning, pp. 57–65. Morgan Kaufmann, San Francisco (1997)
7. Hoens, T.R., Chawla, N.V.: Generating Diverse Ensembles to Counter the Problem of Class Imbalance. In: Pacific-Asia Conference on Knowledge Discovery and Data Mining (PAKDD), Hyderabad, India (2010)
8. Gao, J., Fan, W., Han, J.W., Yu, P.S.: A General Framework for Mining Concept-Drifting Streams with Skewed Distribution. Presentation PowerPoint 2007 SIAM International Conference on Data Mining (SDM 2007), Minneapolis. MN (2007)
9. Gao, J., Ding, B., Fan, W., Han, J.W., Yu, P.S.: Classifying Data Stream with Skewed Class Distribution and Concept Drift. IEEE Internet Computing. ieeexplore.ieee.org (2008)
10. Ouyang, Z.Z., Luo, J.S., Hu, D.M., Wu, Q.Y.: An Ensemble Classifier Framework for Mining Imbalanced Data Streams. Acta Electronic Sinica 38(1), 184–189 (2010)
11. Song, Q., Zhang, J., Deng, Z.H.: A Better Intrusion Detection Algorithm Based on Classification of Skewed Data Streams. Journal of Northwestern Polytechnical University 27(6), 859–862 (2009)
12. Wang, Y., Zhang, Y., Wang, Y.: Mining Data Streams with Skewed Distribution by Static Classifier Ensemble. In: 22nd International Conference on Industrial, Engineering and Other Applications of Applied Intelligent Systems, pp. 24–27 (July 2009)
13. Lichtenwalter, R.N., Chawla, N.V.: Adaptive Methods for Classification in Arbitrarily Imbalanced and Drifting Data Streams (2009),
   http://www.nd.edu/~dial/papers/PAKDDICE
14. Lichtenwalter, R.N., Chawla, N.V.: Learning to Classify Data Streams with Imbalanced Class Distributions. In: New Frontiers in Applied Data Mining. LNCS. Springer, Heidelberg (2009)
15. Wang, Y., Li, Z.H., Zhang, Y.: Classifying Noisy Data Streams. In: Wang, L., Jiao, L., Shi, G., Li, X., Liu, J. (eds.) FSKD 2006. LNCS (LNAI), vol. 4223, pp. 549–558. Springer, Heidelberg (2006)
16. Witten, I.H., Frank, E.: Data Mining: Practical Machine Learning Tools and Techniques, 2nd edn. Morgan Kaufmann, San Francisco (2005)

An Association Rules Algorithm Based on Kendall-τ

Anping Zeng[1] and Yongping Huang[2]

[1] School of Computer and Information Engineering,
Yibin University, Yibin, Sichuan 644007, China
[2] Computational Physics Key Laboratory of Sichuan Province,
Yibin University, Yibin, Sichuan 644007, China
Zengap@126.com

Abstract. The disadvantages of apriori algorithm are firstly discussed. Then, a new measure of kendall-τ is proposed and treated as an interest threshold. Furthermore, an improved Apriori algorithm called K-apriori is proposed based on kendall-τ correlation coefficient. It not only can accurately find the relations between different products in transaction databases and reduce the useless rules but also can generate synchronous positive rules, contrary positive rules and negative rules. Experiment has been carried out to verify the effectiveness of the algorithm. The result shows that the algorithm is effective at discovering the association rules in a sales management system.

Keywords: Kendall-τ, association rules, synchronous rules, ontrary rules, apriori, K-apriori.

1 Introduction

Traditional association rules algorithms have only two thresholds, namely, support and confidence. Their limitations have been found since they were proposed. They only generate one type rules and the relationships among the Rule items are very sparse. In order to avoid generating "false impression" rules, some people add new thresholds to strengthen the evaluation of rules. Among them, interest measure is a more classic one point of view. In [1], Srikant et al. gave the interest of the definition of the rules. In [2], Savasere et al. defined the threshold of negative rules. In [3], Azarakhsh et al. presented a geometric measure. In [4], Zeng et al. proposed a Covariance interest. These interest measure can reduce rules, but the rules does not shows the synchronous and contrary trend.

Given a collectivity(X,Y) and sample observations(x_1, y_1), (x_2, y_2), ..., (x_n, y_n), kendall-τ correlation coefficient can calculate the value of correlation between X and Y[5]. The correlation includes synchronous and contrary correlation. According to this property, this paper designs a new interest measure (kendall-τ interest, denoted as I_K) and a new K-apriori algorithm is proposed. According to it, three new types association rules are refined. Relative to other interest algorithms, K-apriori not only has negative rules but also has synchronous

D.-S. Huang et al. (Eds.): ICIC 2011, LNBI 6840, pp. 152–159, 2012.

positive rules and contrary positive rules. This paper is organized as follows: Section 2 provides the basic concepts of Apriori. In Section 3, the kendall-τ interests and their association rule types are defined. In Section 4, K-apriori Algrithm is presented. In Section 5, examples are employed to validate the proposed approach. The paper ends with conclusions and further research topics in Section 6 and 7.

2   An Introduction of Apriori Algorithm

In 1993, Agrawal et al. presented a classical Apriori algorithm, which is an important method for association rules mining [6]. In [6], the problem of association rule mining is decomposed into two subproblems as follows:
(1) Generate all combinations of items that have fractional transaction support above minsupport.
(2) Generate the desired rules on the frequent itemsets.

Algorithm 1. The Procedure of generating Association Rules
---
    (01)   for all $X_k \in L_k, k \geq 2$;
    (02) H_1={the items of the latter part(length=1) of rules generated by X_k };
    (03)   call AP_GenRule(X_k,H_1).

In [6], the procedure of generating the rules is described as Algorithm 1.The Procedure AP_GenRule is described as Algorithm 2.

Algorithm 2. The Procedure AP_GenRule

Input:
    X_k: k-frequent item;
    H_m: the latter part sets, each element's length is 1;
Output: Association rules $X_k - h_{m+1} \Rightarrow h_{m+1}$
Method:
    (01)   if $k > m + 1$ then $H_{m+1} = apriori\_gen(H_m)$;
    (02)   for all $h_{m+1} \in H_{m+1}$;
    (03) $conf = \frac{s(X_k)}{s(X_k - h_{m+1})}$;
    (04)   if $conf \geq MinConf$ then Output Rule $X_k - h_{m+1} \Rightarrow h_{m+1}$;
    (05)   else delete h_{m+1} from H_{m+1};
    (06)   call AP_GenRule(X_k, H_{m+1});
---

3   Definitions

3.1   Synchronous Positive Association Rules and Contrary Positive Association Rules

Table 1 is a invoice bill data set of transaction database and X, Y, Z, W are products. $T_1, T_2, ..., T_n$ are transaction code. $x_1, x_2, ..., x_n$ are the transaction value of product X. $y_1, y_2, ..., y_n$ are the transaction value of product Y. Given one trans-action record T_i, if $x_i \neq null$ and $y_i \neq null$ (i$\in [1, n]$) means that product X and Y occur simultaneously. If $x_i \neq null$ and $y_i = null$ means that product X occurs but Y does not occur. Based on table 1, the following definitions are proposed.

Table 1. The invoice bill data set

TID	X	Y	Z	W
T_1	x_1	y_1	z_1	w_1
T_2	x_2	y_2	z_2	w_2
...	...	...	...	...
T_i	x_i	y_i	z_i	w_i
...	...	...	...	...
T_n	x_n	y_n	z_n	w_n

Definition 1. x_i and y_i $(i=1, 2, ..., n)$ are the transaction value of product X and Y respectively when they occur simultaneously. In [5], Kendall-τ correlation coefficient τ is defined as follows:

$$\tau = \frac{2}{n(n-1)} \sum_{1 \leq i < j \leq n} \mathrm{sgn}((x_j - x_i)(y_j - y_i)) \qquad (1)$$

Where, sgn(.) is a sign function:

$$\mathrm{sgn}((x_j - x_i)(y_j - y_i)) = \begin{cases} 1, & (x_j - x_i)(y_j - y_i) > 0 \\ 0, & (x_j - x_i)(y_j - y_i) = 0 \\ -1, & (x_j - x_i)(y_j - y_i) < 0 \end{cases} \qquad (2)$$

There are three points need to make special note:

(1) When $0 < \tau \leq 1$, the greater the value of τ, the greater the probability of synchronization between X and Y (increase or decrease synchronously).

(2) When $-1 \leq \tau < 0$, the smaller the value of τ, the greater the probability that X increases but Y decreases or X decreases but Y increases.

(3) Since X and Y considered here is happening at the same time, τ can not be equal to 0. The case which X and Y do not occur simultaneously will be dealt with in the definition 4.

Based on the three points, the kendall-τ interest I_K is defined as three types ($I_K^{\uparrow\uparrow}$, $I_K^{\uparrow\downarrow}$ and $\overline{I}_K$). According to the definitions, the synchronous association rules, contrary association rules and negative association rules are generated.

Definition 2. X and Y represent two products. Their transaction values show an synchronous increase or decrease. Then, their kendall-τ interest $I_K^{\uparrow\uparrow}$ is defined as follows:

$$I_K^{\uparrow\uparrow} = \begin{cases} \tau, & 0 < \tau \leq 1 \\ 0, & \tau \leq 0 \end{cases} \qquad (3)$$

Minimum Interest (denoted as the M_i) is the interest threshold. When $I_K^{\uparrow\uparrow} \geq M_i$, some strong association rules are generated, and are named as synchronous positive association rules(denoted by $X \uparrow\uparrow \Rightarrow Y$).

Definition 3. *X and Y represents two products. When the transaction value of X increases but Y's decreases or the transaction value of X decreases but Y's increases, their kendall-τ interest $I_K^{\uparrow\downarrow}$ is defined as follows:*

$$I_K^{\uparrow\downarrow} = \begin{cases} \tau, & -1 \leq \tau < 0 \\ 0, & \tau \geq 0 \end{cases} \tag{4}$$

M_i is the interest threshold. If $I_K^{\uparrow\downarrow} \geq M_i$, some strong association rules are generated, and named as contrary positive rules(denoted by $X \uparrow\downarrow \Rightarrow Y$).

3.2   The Definition of Negative Association Rules

Definition 4. *X and Y represents two products. x_i ($i \in [1, n]$) is the transaction value of X in the invoice bill T_i when X occurs but Y does not occur($x_i \neq null$, $y_i = null$). $\overline{y}_i$ is the average of all the T_i product items' value which does not contain X and Y (for example $\overline{y}_i = \frac{z_i + w_i}{2}$, $z_i \neq null$ and $w_i \neq null$). Kendall-τ correlation coefficient is defined as follows:*

$$\overline{\tau} = \frac{2}{n(n-1)} \sum_{1 \leq i < j \leq n} \text{sgn}((x_j - x_i)(\overline{y}_j - \overline{y}_i)) \tag{5}$$

There are two points need to make special note:

(1) When $0 < \overline{\tau} \leq 1$, the greater the value of $\overline{\tau}$, the greater the probability of synchronization between X and $\overline{Y}$.

(2) When $-1 \leq \overline{\tau} < 0$, the smaller the value of τ, the greater the probability that X increases but $\overline{Y}$ decreases or X decreases but $\overline{Y}$ increases.

Definition 5. *Based on definition 4, kendall-τ interest $\overline{I}_K$ is defined as follows:*

$$\overline{I}_K = \begin{cases} \overline{\tau}, & 0 < \overline{\tau} \leq 1 \\ 0, & -1 \leq \overline{\tau} < 0 \end{cases} \tag{6}$$

M_i is the interest threshold. When $\overline{I}_K \geq M_i$, the negative association rules $X \Rightarrow \overline{Y}$ is generated.

4   *K*-apriori Algrithm

Assume a rule $[Y - I_i] \Rightarrow I_i$ is generated by Apriori algorithm. If the confidence of the rules is greater or equal to the confidence threshold (denoted as *MinConf*), then $[Y - I_i] \Rightarrow I_i$ is a strong association rule.

K-apriori algorithm is an extension of Apriori algorithm. Its idea is as follows:

Step 1: Based on the Apriori algorithm, if the confidence of the rule is greater or equal to *MinConf*, the *K*-apriori algorithm calculates the kendall-τ interest $I_K^{\uparrow\uparrow}$. A synchronous positive rule $[Y - I_i] \uparrow\uparrow \Rightarrow I_i$ is output if $I_K^{\uparrow\uparrow} \geq M_i$. If $I_K^{\uparrow\uparrow} = 0$, *K*-apriori calculates the kendall-τ interest $I_K^{\uparrow\downarrow}$. If $I_K^{\uparrow\downarrow} \leq -M_i$, contrary positive rule $[Y - I_i] \uparrow\downarrow \Rightarrow \overline{I}_i$ is output.

Step 2: Based on the Apriori algorithm, if the confidence of the rule is smaller or equal to the *1- MinConf*, the K-apriori algorithm calculates $\overline{I}_K$. If $\overline{I}_K \geq M_i$, then output negative rule $[Y - I_i] \Rightarrow \overline{I}_i$.

Show as Algorithm 3, to generate positive rules, the K-apriori algorithm replaces the line 04 "Output Rule $X_k - h_{m+1} => h_{m+1}$" of Algorithm 2 as "call $\rho\_rule\ (X_k, h_{m+1}, 1)$". The parameter "1" means to deal with positive rules. And then, line 05 is divided into three lines(05-1, 05-2 and 05-3). In line 05-2, "-1" means to deal with negative rules. The procedure ρ_rule finish the task of generate rules(shown as Algorithm 4).

Algorithm 3. Modify the Procedure AP_GenRule

...

(04)	if $conf \geq MinConf$ then call $\rho\_rule\ (X_k, h_{m+1}, 1)$;
(05-1)	else if $Conf \leq 1 - MinConf$ then
(05-2)	call $\rho\_rule(X_k, h_{m+1}, -1)$;//deal with negative rules
(05-3)	else delete h_{m+1} from H_{m+1};
(06)	call AP_GenRule(X_k, H_{m+1});

---

5   Experiment

To validate the efficiency of the algorithm, the K-apriori DM system selects the transaction data from a sale management system. Showing as Table 2 and Table 3, after noise reduction and product code conversion, the transaction detail D_1 (128, 200 records) and its sum sets D_2 (22,520 records) are created.

Table 2. The transaction detail D_1

TID	Item	Num
C0386	0031	200
C0386	0054	480
C0386	0091	920
C0386	0170	2160
C0386	0207	2440
C0387	0031	380
C0387	0050	530
C0387	0087	350
C0387	0116	420
...	...	...

Table 3. The transaction sets D_2

TID	Items
C0333	0026,0031,0042,0046, 0050,0058,0087,0116,0122
C0334	0042,0046,0087,0097,0116, 0150,0170,0178,0206
C0384	0001,0024,0054
C0386	0031,0054,0091,0170,0207
C0387	0031,0050,0087,0116
C0389	0002,0009,0026,0031,0033, 0037,0050,0116,0170
...	...

5.1   The Example of Synchronous Positive Association Rules

In the system, The Minsupport is 0.2, the *MinConf* is 0.8 and the M_i is 0.9. The frequent items set (0031,0091,0207) is an example to explain the process of generating synchronous positive association rules. Its confidence

$$C_1 = \frac{support(0031, 0091, 0207)}{support(0031, 0207)} = 400/440 = 0.91.$$

$C_1 \geq MinConf$. The kendall-τ interest

$$I_K^{\uparrow\uparrow} = \frac{2}{n(n-1)} \sum_{1 \leq i < j \leq n} \text{sgn}((x_j - x_i)(y_j - y_i)) = 0.94.$$

Because $I_K^{\uparrow\uparrow} > M_i$, (0031,0207) $\uparrow\uparrow \Rightarrow$ 0091 is a synchronization positive rule.

5.2   The Example of Contrary Positive Association Rules

The process of generating contrary positive rules can be illustrated by the frequent items (0026,0042,0046,0087). Its confidence

$$C_2 = \frac{\text{support}(0026, 0042, 0046, 0087)}{\text{support}(0026, 0046, 0087)} = 280/300 = 0.93.$$

$C_2 > 0.8$. The interest measure

$$I_K^{\uparrow\downarrow} = \frac{2}{n(n-1)} \sum_{1 \leq i < j \leq n} \text{sgn}((x_j - x_i)(y_j - y_i)) = -0.98 < -0.9.$$

So, rule (0026,0046,0087) $\uparrow\downarrow \Rightarrow$ 0042 is a contrary positive rule.

5.3   The Example of Negative Association Rules

The process of generating negative rules can be illustrated by the frequent items (0024,0054,0087). Its confidence

$$C_3 = \frac{\text{support}(0024, 0054, 0087)}{\text{support}(0024, 0054)} = 260/1720 = 0.15.$$

$C_3 < 1 - MinConf$. The interest measure

$$\overline{I}_K = \frac{2}{n(n-1)} \sum_{1 \leq i < j \leq n} \text{sgn}((x_j - x_i)(\overline{y}_j - \overline{y}_i)) = 0.92$$

Because $\overline{I}_K > 0.9$, rule (0024,0054) $\Rightarrow$ 0087 is a negative rule.

At last, the K-apriori generates 16 rules including 9 synchronous positive rules, 2 contrary positive rule and 5 negative rules. The rules are applied in a sales management system and show that the result is valid.

6   Analysis and Comparisons

Table 4 is the comparison results of K-apriori algorithm and other algorithms. The advantages of K-apriori algorithm are as follows.

(1) Relative to the Apriori, the K-apriori algorithm has another threshold of kendall-τ interest.

(2) The rules number of K-apriori algorithm is smaller than that of the Apriori. The rules which meet the really interesting of user are output and other rules will be filtered out.

(3) Apriori algorithm only generates positive rules, but the K-apriori algorithm can generate negative rules.

(4) other interest measure algorithms and K-apriori algorithm can generate positive rules and negative rules. But the K-apriori algorithm divides positive rules into synchronous positive rules(denoted as SP in Table 4) and contrary positive rules(denoted as CP in Table 4). In the sale management, this division not only to show that the frequency of the two products at the same time, but also shows that the trend of the two products simultaneously. So, this tight correlation can be better for decision-making.

Table 4. A Comparision Of K-apriori And other Algorithms

Algorithm	Threshold num	Rule num	Rule types	Correlation degree
Apriori	2	84	Positive	sparse
other interest measure	3	≈ 16	SP, Negative	tightness
K-apriori	3	16	SP, CP, Negative	more tightness

7   Conclusions

This paper presented a K-apriori algorithm to generate three types' association rules based on Kendall-τ of view. The proposed method is validated by examples and the superiority can be seen. But how to apply the threshold to other data mining algorithms is one of future works.

Acknowledgments. This work is supported by scientific research fund of sichuan provincial education department Nos.10ZB049; The foundation of yibin technical bureau under grant Nos. 20072036; Scientific research fund of yibin university Nos. 2010Z11.

References

1. Srikant, R., Agrawal, R.: Mining Generalized Association Rules. In: Proc. of the 21th Int'l. Conf. on Very Large Data Bases, pp. 407–419. Morgan Kaufmann, Zurich (1995)
2. Savasere, A., Omiecinski, E., Navathe, S.: Mining for Strong Negative Associations in a Large Database of Customer Transactions. In: Proc. of the 14th Int'l. Conf. on Data Engineering, Orlando, Florida, pp. 494–502 (1998)
3. Jalalvand, A., Minaei, B., Atabaki, G., Jalalvand, S.: A New Interestingness Measure for Associative Rules Based on the Geometric Context. In: Proc of the Third Int'l Conf. on Convergence and Hybrid Information Technology, Busan, Korea, pp. 199–203 (2008)
4. Zeng, A.P., Huang, Y.P., Li, G.j.: FP-Growth Algorithm Based Covariance and its Application in ERP. Mathematics in Practice and Theory 38(12), 11–18 (2008)

5. Li, Y.Q., Zhao, L.W., Wang, Q., Tang, J.Y.: Non-parameter Statistics, pp. 116–119. Southwest Jiaotong University Press (2010)
6. Agrawal, R., Imielinski, T., Swami, A.: Mining Association Rules Between Eet of Items in Large Databases. In: Proceedings of the 1993 ACM SIGMOD Conference on Management of Data, Washington, D. C, pp. 207–216 (1993)
7. Shao, F.J., Yu, Z.Q., Wang, J.L., Sun, R.C.: Principle and Algorithm of Data Mining, pp. 92–99. Science Press, Beijing (2009)
8. Michael, J.A., Gordon, S.L.: Data Mining Technology For Marketing, Sales, and Customer Relationship Managementn. Machinery Industry, 340–350 (2006)

Algorithm 4. The Procedure ρ-rule

Input:

      X_k: k-frequent item; H: the latter part sets;

      $Flag$: if $flag = 1$ it generates positive rules,else it deals with negative rules.

Output:

      $X_k - H \uparrow\uparrow\Rightarrow H$: Synchronous positive rules;

      $X_k - H \uparrow\downarrow\Rightarrow H$: Contrary positive rules;

      $X_k - H \Rightarrow \overline{H}$: Negative rules.

Method:

(01)   int $cov = 0, n = 0, sumsgn = 0, x[n], y[n]$;

(02)   if $flag = 1$ then

(03) D_3=get(X_k);n=count(D_3); // D_3 is the Subsets of transaction sets containing X_k

(04)   for all $t \in D_3$

(05) $x[i]$=get_average($X_k - H, t$); $y[i]$=get_average(H, t);

(06) $i = i + 1$;

(07)   else

(08) D_3=get($X_k - H, \overline{H}$); n=count(D_3);// D_3 is the Subsets of transaction sets containing $X_k - H$, but does not include H

(09)   for all $t \in D_3$

(10) $x[i]$=get_average($X_k - H, t$); $y[i]$=get_average($\overline{H}, t$);

(11) $i = i + 1$;

(12)   for $(j = 2; j <= n; i + +)$

(13)   for$(i = 1; i < j; i + +)$

(14)   sgn=$(x[j] - x[i]) * ((y[j] - y[i])$;

(15)   if $sgn > 0$ then $sumsgn + +$;

(16)   else if $sgn < 0$ then $sumsgn - -$;

(17) $\tau = 2 * sumsgn/(n(n-1))$;

(18)   if $(flag = 1)$ and $(\tau \geq M_i)$ then

(19)   output Rule $X_k - H \uparrow\uparrow\Rightarrow H$; //synchronous positive rule

(20)   else if $(flag = 1)$ and $(\tau \leq -M_i)$ then

(21)   output Rule $X_k - H \uparrow\downarrow\Rightarrow H$; //contrary positive rule

(22)   else if $(flag = -1)$ output Rule $X_k - H \Rightarrow \overline{H}$; //negative rule

---

A Recommendation Algorithm Combining Clustering Method and Slope One Scheme

Zhenzhen Mi and Congfu Xu

Zhejiang University, College of Computer Science and Technology,
Zheda road 38, Hangzhou, Zhejiang, China
mizhenzhen@gmail.com,
xucongfu@zju.edu.cn

Abstract. With the development of electronic commerce, a lot of recommendation techniques has been developed. Collaborative filtering(CF) is one of the most important technologies. However, traditional collaborative filtering suffers sparsity and scalability problems, which results in poor quality of prediction in recommendation systems. To solve these problems, this paper proposed a recommendation algorithm combining clustering method and slope one scheme. This approach uses clustering algorithms to partition the set of items to several clusters based on user rating data, and then we use slope one scheme to predict ratings independently for unknown items based on which cluster the items belong to. We make experiments on the standard benchmark Movielens data sets and compare our approach with the basic slope one scheme. The results show that our algorithm outperforms the slope one scheme.

Keywords: Collaborative filtering, K-means clustering method, Slope one scheme.

1 Introduction

With the explosive growth of electronic commerce sites, information has grown up with exponential rate on the internet, and this leads to the problem of information overload. Facing this situation, users urgently need some mechanism to assist sifting through useless resources and finding the most interested and valuable information to them. Thus, recommendation systems emerge to provide fast and accurate recommendations. However, there are many challenges for traditional recommendation systems as following [13]:

1. Data Sparsity. In real life, only a few users have rated the items which they like or dislike, so the user-item matrix is usually very sparse. Without enough ratings, the prediction quality of CF will be influenced seriously. In addition, data sparsity may lead to the cold start problem.
2. Scalability. Although computer hardware developed quickly recently, it still cannot satisfy the demand of tremendous growing of users and items. In addition, users expect high quality predictions regardless of their purchases or rating history in real time.

D.-S. Huang et al. (Eds.): ICIC 2011, LNBI 6840, pp. 160–167, 2012.

In this paper, we propose a recommendation algorithm combining K-means clustering method and slope one scheme. The method use K-means algorithm to partition the set of items into several small clusters, and then we use slope one scheme to get the prediction rating of target user to items based on which cluster the items belong to. Using clustering algorithm to partition the list of items can increase the scalability of recommendation systems effectively. In addition, it also alleviates the cold start problem by recommending items to users based on other items in the same cluster.

2   Related Work

Collaborative filtering(CF) is the most popular technique in recommendation systems. CF techniques predict the target user's preferences for items by users who have similar interests with him. CF methods generally include model-based CF and memory-based CF. Model-based CF algorithms refer to many math models, such as Bayesian networks models [2], clustering models [14], latent semantic models [3], MDP model [12] and so on. Memory-based CF involves user-based CF and item-based CF. User-based CF [1] firstly identifies the similar users to the target user using some similarity metrics, and then predicts the interest of target user based on the similar users' preferences. While, item-based CF [11] [6] computes the similar items set, and then if the target user likes one item of the set, we will recommend the remaining items in the set to the target user based on some recommendation strategies .

Clustering is used in many recommendation systems. Among them the GroupLens [10] collaborative filtering system for Usenet news was the first system to create a separate item partition for each Usenet discussion group. In most situations, people often use clustering method to pre-process the data and then use the clusters for further analysis or other tasks. In this paper, we also use clustering as an intermediate step.

Slope one scheme [5] works on the intuitive principle of a popularity differential between items rather than similarity between items. Based on the slope one scheme smoothing, a user-based algorithm [15] and an item-based algorithm [16] have already been proposed. The slope one method estimates ratings for new items based on average difference in preference value between a new item and the other items the user co-rated.

3   Algorithms

3.1   Notation

In this paper, we describe algorithms with the following notation. In a typical CF scenario, there is a data structure as user-item matrix R whose rows are a list of m users $\{u_1, u_2, \ldots, u_m\}$ and columns are a list of n items $\{i_1, i_2, \ldots, i_n\}$, with values $r_{u,i}$ being the rating of user u to item i. $U(i)$ is the set of users who have rated the item i, and $I(u)$ is the set of items which have been rated by user

u. $U(i,j)$ indicates the set of users who have co-rated item i and j, while, $I(u,v)$ shows the set of items which have been co-rated by users u and v. The number of elements in a set S is expressed as $|S|$. The average rating of the user u is $\bar{r}_u$, and similarly the average rating of the item i is $\bar{r}_i$. If we partition the items space into a few clusters, and the kth cluster can be presented as C_k. Suppose the ith item belong to C_k, it is expressed as $C_k(i)$. In the process of slope one scheme, $n \times n$ deviation matrix D is necessary, and the value $d_{i,j}$ indicates the average difference between item i and item j. When predicting ratings for unknown item i by user u, we use $p_{u,i}$ to stand for the prediction value.

3.2 Clustering Algorithm

There are many clustering algorithms including hierarchical clustering, K-means clustering and their variants [4]. In this paper, we use the K-means method [7]. The implementation of K-means is shown in the algorithm 1 [8].

Algorithm 1. K-means clustering

Input: The set of n items in the training data set and the number of clusters k.
Output: k clusters
Initialization. We begin with k items which have front kth most rated times as *centroids* rather than k random items. In such a way, we can reduce outliers effectively [9].
Assignment step. Each item i is assigned to the most similar *centroid*. Here, we use Pearson's Correlation Coefficient as following formula 1 to measure the similarity of any two items. The Pearson's Correlation Coefficient between item i and j which have been co-rated by user u will be:

$$w_{i,j} = \frac{\sum_{u \in U(i,j)}(r_{u,i} - \bar{r}_i)(r_{u,j} - \bar{r}_j)}{\sqrt{\sum_{u \in U(i,j)}(r_{u,i} - \bar{r}_i)^2}\sqrt{\sum_{u \in U(i,j)}(r_{u,j} - \bar{r}_j)^2}} \quad (1)$$

The larger the value $w_{i,j}$ is, the stronger the similarity between items i,j is.
Update step. After the assignment, the *centroids* will be removed to the average of all the items assigned to them, and the process is repeated in k clusters.
Repeat the assignment step and update step until the algorithm is deemed to have converged when the assignments stop changing.

3.3 Slope One Scheme

Slope one scheme [5] takes advantage of the linear relationship between items to get the deviation matrix whose values is item-item average difference. It consists of two steps to produce recommendations as following:

1. In the training set, firstly we need to calculate the deviation matrix D whose value $d_{i,j}$ is computed as formula 2.

$$d_{i,j} = \sum_{u \in U(i,j)} \frac{r_{u,i} - r_{u,j}}{|U(i,j)|} \quad (2)$$

where user u rates both item i and item j.

2. After the deviation matrix have been computed, we will use the matrix to produce the prediction. The single prediction rating of user u to item j based on the deviation matrix can be computed as formula 3.

$$p_{u,j} = \frac{\sum_{i \in R_j}(d_{j,i} + r_{u,i})}{|R_j|} \qquad (3)$$

where R_j is the set of all relevant items to item j. The formula 3 is the prediction method of the **basic** slope one scheme. It suppose that the relevant items to items j play the same important role for predicting $p_{u,i}$. However, different relevant items have different influence on producing the predictions. Thus, the **weighted** slope one scheme takes the number of users who co-rated item i and j as the weight, and the prediction formula 4 becomes as following.

$$p_{u,j} = \frac{\sum_{i \in R_j}(d_{j,i} + r_{u,i})|U(i,j)|}{\sum_{i \in R_j}|U(i,j)|} \qquad (4)$$

3.4   Fusion Method

In slope one scheme, utilizing the all items to compute the deviation matrix may take some unrelated information into computation. However, in some situation, computing deviation between unrelated items may finally decrease the the pre-diction accuracy. To filter the noises in computing deviation matrix in slope one scheme, we firstly partition the item-space into smaller clusters, and then for each cluster, we compute the average differences only between the items in this cluster. This method not only has the advantages of clustering algorithm and the slope one scheme, but also makes the time of making on-line recommenda-tions much less than the slope one scheme. Furthermore, in Section 4, we will prove that if the number of clusters are chosen appropriately, the accuracy of prediction will be improved. Our algorithm will be shown in the algorithm 2.

Algorithm 2. The algorithm combing clustering method and slope one scheme

---

Input: The training data set $m \times n$ matrix C and the number of clusters k.
Output: The prediction value.
Clustering the item-space. Apply the K-means clustering algorithm 1 to produce k partitions of items of the training set. Formally, the data set C is partitioned in $C_1, C_2, \ldots, C_k$, where $C_i \bigcap C_j = \emptyset$ $(1 \leq i, j \leq k)$, and $C_1 \bigcup C_2 \bigcup \ldots \bigcup C_k = C$.
Calculating the deviation matrix D. For each cluster $C_i(1 \leq i \leq k)$, we compute the average difference for every pair of items only in C_i cluster according to the formula 2. After the computation for all k clusters, we get a $n \times n$ deviation matrix.
Calculate the prediction rating for unknown items. Once the deviation matrix is obtained, we can use formula 3 or formula 4 to generate predictions for every unknown item i based on the cluster $C_k(i)$.

---

4   Experiments Evaluation and Results

In this section, we describe the data sets, the evaluation metrics and the comparative experiments between our method and the basic slope one scheme and the weighted slope one scheme. In addition, we present the results of the experiments and make a concise analysis of the results.

4.1   Date Set

We used the Movielens data sets[1] including one data set of 100,000 ratings for 1682 movies by 943 users and one data set of 1 million ratings for 3900 movies by 6040 users. The ratings in the data sets are on a numerical scale from 1 to 5 in integers, with the bigger integer representing the stronger preference. We randomly divided the first data set into training set (80% of the data set) and the test set (20% of the data set) five times and the second data set was divided only one time. The two groups of training data sets are respectively named as $G1tr1$, $G1tr2$, $G1tr3$, $G1tr4$,$G1tr5$ and $G2tr1$, similarly, the test data sets as $G1test1$, $G1test2$, $G1test3$, $G1test4$, $G1test5$ and $G2test1$.

4.2   Evaluation Metrics

In this paper, we employ the root mean squared error($RMSE$)which is widely used as the Netflix prize [2] metric for movie recommendation performance. The formula to calculate the $RMSE$ is as following [13]:

$$RMSE = \sqrt{\frac{\sum_{\{u,j\}} (p_{u,j} - r_{u,j})^2}{n}} \qquad (5)$$

where n is the total number of ratings over all users, $\{u, j\}$ is the set of all users and all unknown items. The smaller the $RMSE$ is, the better the prediction accuracy is.

4.3   Experimental Results

To compare the performances of our method, we chose the basic slope one scheme and the weighted slope one scheme as the benchmark systems. All our experiments were coded purely in C++ without any third-party library and we ran our experiments on Windows 7 Professional operating system which was installed on a typical PC with Core 2 Duo E7400 and 4GB of RAM.

Firstly, we need to determine the sensitivity of the parameter of the number of clusters k before running the main experiments on our method and the benchmark systems. We carried out the slope one scheme with clusters and the weighted slope one scheme with clusters on the two different data sets described

---

[1] http://www.grouplens.org/node/73
[2] http://www.netflixprize.com/

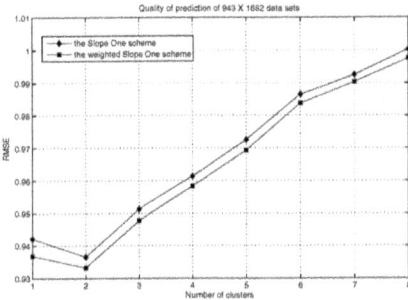
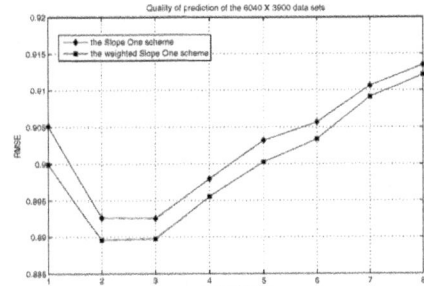

Fig. 1. Impact of the number of clusters on our algorithm combining the slope one or the weighted slope one scheme with clustering method on the 943 × 1682 data sets

Fig. 2. Impact of the number of clusters on our algorithm combining the slope one or the weighted slope one scheme with clustering method on the 6040×3900 data sets

above. Figure 1 and Figure 2 shows the experimental results. In these two figures, we plotted the prediction quality as a function of the number of clusters. We conducted our experiments in Figure 1 by taking the average of the $RMSE$ of 5 training data sets and 5 test data sets. We can observe that the results with the 2 clusters had a clear advantage on the 943 × 1682 data sets and 2 clusters or 3 clusters on the 6040 × 3900 data set. Thus, we selected 2 clusters for the rest experiments.

Secondly, after obtaining the optimal number of clusters, we compared our algorithm with the benchmark systems. The results were presented in table 1. In table 1 the column "average" is the average of $RMSE$ of the front five columns. Table 2 was the improving percent of the prediction accuracy corresponding to table 1. The percent was computed as formula 6:

$$percent = \frac{|r_1 - r_2|}{r_1} \times 100\% \tag{6}$$

where r_1 is the $RMSE$ of the benchmark algorithm, and r_2 is the $RMSE$ of our method. The symbol $|r|$ stands for the absolute value of r. In table 3 we present the prediction time in seconds of our algorithm and the benchmark method.

Table 1. The $RMSE$ of comparative experiments

	G1tr1	G1tr2	G1tr3	G1tr4	G1tr5	average	G2tr1
slope one	0.9527	0.9429	0.9409	0.9382	0.9358	**0.9422**	**0.9051**
slope one with clusters	0.9490	0.9345	0.9326	0.9326	0.9345	**0.9367**	**0.8927**
weighted slope one	0.9474	0.9369	0.9349	0.9337	0.9317	**0.9370**	**0.8999**
weighted slope one with clusters	0.9469	0.9292	0.9288	0.9303	0.9314	**0.9333**	**0.8896**

Table 2. The improving percent of our method compared with benchmark algorithms

	G1tr1	G1tr2	G1tr3	G1tr4	G1tr5	average	G2tr1
improving percent(%) of Table 1	0.3941	0.8972	0.8877	0.6115	0.1481	**0.5878**	**1.3745**
improving percent(%) of Table 1	0.0492	0.8197	0.6495	0.3617	0.03188	**0.3819**	**1.1468**

Table 3. The prediction time in seconds of comparative experiments(s)

	G1tr1	G1tr2	G1tr3	G1tr4	G1tr5	average	G2tr1
slope one	0.665	0.701	0.711	0.72	0.683	**0.696**	**12.47**
slope one with clusters	0.577	0.578	0.579	0.583	0.589	**0.5812**	**8.963**
weighted slope one	0.661	0.705	0.727	0.711	0.686	**0.698**	**12.946**
weighted slope one with clusters	0.568	0.592	0.584	0.578	0.592	**0.5828**	**9.296**

4.4 Discussion

From the experiments mentioned above, we make some important observations. First, the prediction quality of our method is better than the benchmark algorithms when the number of clusters is appropriate. The number of clusters is the most important factor in our method. After the best outcomes are obtained, the more the number of clusters is, the worse the results are. It is because that when the number of clusters becomes greater, the information partitioned into each cluster is less. Without enough information the prediction requires, the accuracy naturally decreases. Second, when the size of data sets is larger, the improvement in prediction accuracy is more obvious. For example, the improvement on 943×1682 data sets is 0.5878% and 0.3819%, while on the 6040×3900 data sets it is the 1.3745% and 1.1468%. Furthermore, in the process of making experiments, we found that the on-line prediction time of our method reduces compared to the benchmark method, although the training time is a little longer. It has practical significance, because the training work can be done concurrently off-line.

5   Conclusion and Future Work

In this paper, we proposed a new approach in improving the sparsity and scalability of recommendation systems by using the clustering algorithms based on the slope one scheme. We evaluated our method by making experiments on different data sets. The results suggest our algorithm produces better recommendations and in the meantime improves the on-line prediction performance.

In the future, we can introduce other better clustering algorithms to partition the item-space to improve the prediction performance further. And we also can take other information of items, such as context information, into consideration to partition the items.

Acknowledgments. This research is supported by the Natural Science Foundations of China (No. 60970081), the 863 Plan project of China (No. 2007AA01Z197) and the National Basic Research Program of China (No. 2010CB327903). The thanks also goes to the anonymous reviewers for their valuable comments.

References

1. Breese, J., Heckerman, D., Kadie, C., et al.: Empirical Analysis of Predictive Algorithms for Collaborative Filtering. In: Proceedings of the 14th Conference on Uncertainty in Artificial Intelligence, pp. 43–52 (1998)
2. Chien, Y., George, E.: A Bayesian Model for Collaborative Filtering. In: Proceedings of the 7th International Workshop on Artificial Intelligence and Statistics (1999)
3. Hofmann, T.: Latent Semantic Models for Collaborative Filtering. ACM Transactions on Information Systems (TOIS) 22(1), 89–115 (2004)
4. Jain, A., Dubes, R.: Algorithms for Clustering Data. Prentice Hall, Englewood Cliffs (1988)
5. Lemire, D., Maclachlan, A.: Slope One Predictors for Online Rating-based Collaborative Filtering. Society for Industrial Mathematics (2005)
6. Linden, G., Smith, B., York, J.: Amazon. com Recommendations: Item-to-item Collaborative Filtering. IEEE Internet Computing 7(1), 76–80 (2003)
7. Lloyd, S.: Least Squares Quantization in PCM. IEEE Transactions on Information Theory 28(2), 129–137 (2002)
8. MacKay, D.: An Example Inference Task: Clustering. Inference and Learning Algorithms on Information Theory, ch. 20, pp. 284–292
9. Quan, T., Fuyuki, I., Shinichi, H.: Improving Accuracy of Recommender System by Clustering Items Based on Stability of User Similarity. In: Proceedings of International Conference on IAWTIC, p. 61. IEEE Computer Society, Washington, DC (2007)
10. Resnick, P., Iacovou, N., Suchak, M., Bergstrom, P., Riedl, J.: GroupLens: An Open Architecture for Collaborative Filtering of Netnews. In: Proceedings of the 1994 ACM Conference on Computer Supported Cooperative Work, pp. 175–186. ACM, New York (1994)
11. Sarwar, B., Karypis, G., Konstan, J., Reidl, J.: Item-based Collaborative Filtering Recommendation Algorithms. In: Proceedings of the 10th International Conference on World Wide Web(WWW), pp. 285–295. ACM, New York (2001)
12. Shani, G., Heckerman, D., Brafman, R.: An MDP-based Recommender System. Journal of Machine Learning Research 6(2), 1265 (2006)
13. Su, X., Khoshgoftaar, T.: A Survey of Collaborative Filtering Techniques. Advances in Artificial Intelligence 2 (2009)
14. Ungar, L., Foster, D.: Clustering Methods for Collaborative Filtering. In: AAAI Workshop on Recommendation Systems, pp. 112–125 (1998)
15. Wang, P., Ye, H.: A Personalized Recommendation Algorithm Combining Slope One Scheme and User Based Collaborative Filtering. In: Proceedings of IEEE International Conference on Industrial and Information Systems(IIS 2009), pp. 152–154. IEEE, Los Alamitos (2009)
16. Zhang, D.: An Item-based Collaborative Filtering Recommendation Algorithm Using Slope One Scheme Smoothing. In: Proceedings of IEEE Second International Symposium on Electronic Commerce and Security(ISECS 2009), vol. 2, pp. 215–217. IEEE, Los Alamitos (2009)

Dimensional Reduction Based on Artificial Bee Colony for Classification Problems

Thananan Prasartvit[1], Boonserm Kaewkamnerdpong[2], and Tiranee Achalakul[1]

[1] Deprtment of Computer Engineering
[2] Biological Engineering Program
King Mongkut's University of Technology Thonburi, Thailand
vc_25_a@hotmail.com

Abstract. High dimensionality of data is a limiting factor to data processing in many fields. It causes ambiguousness in identifying significant factors for data analysis. Dimension reduction is needed to separate irrelevant data from the desired data. This research proposes a novel method for dimension reduction based on artificial bee colony (ABC). The method employs swarm intelligence based on bee foraging model in order to select features that allow us to generate subsets of dimensions from the original high-dimensional data while the resulting subsets satisfy the defined objective. Support vector machine (SVM) is used in this study as fitness evaluation of ABC in classification problems. To evaluate our method, we tested it with five datasets and compared it with other dimension reduction algorithms. The result of this study shows that using ABC and SVM is suitable for reducing the dimension of data. Moreover, this approach provides efficient classification with high accuracy.

1 Introduction

In the past decades, information overload has been an advanced problem of data collection in research works in the fields of science, medicine, engineering, finance, and manufacturing. This brings about large consumption of computation in data analysis process. In many cases, researchers cannot identify the dimensions or variables that are importantly contributed to the phenomena in which they are interested. Thus, the process of reducing data dimensions is needed to accurately and efficiently generate a low-dimensional data from high-dimensional data. Dimension reduction is an important process in data mining for analyzing massively high-dimensional data by eliminating unnecessary properties.

One of dimension reduction processes is called feature selection. This process attempts to maintain the characteristic of the original data by removing non-essential dimensions such as noise or redundancy. There are various feature selection techniques, for example, Branch and Bound (BB) Algorithm. BB was first introduced in [1]. In this work, feature set is represented as a tree structure; each node represents each feature. The process of BB method searches and evaluates all possible feature subsets based on a defined criteria; the results demonstrated that even for the problem with large number of features, BB method can efficiently discover several suboptimal subsets and discard them so that less evaluation time will be required. Another

D.-S. Huang et al. (Eds.): ICIC 2011, LNBI 6840, pp. 168–175, 2012.

technique for feature selection is a stochastic method [2]. This method does not perform a search on all feasible feature sets. It randomly generate subset of features that satisfy some defined parameters, then assign suitable features to the feature subset to achieve the optimal result.

However, both techniques can lead to burdensome and expensive computation because a large amount of data may have to be iteratively analyzed through all feasible cases in the feature selection process. Heuristic techniques are then considered due to their lower computation requirement. Some heuristic techniques in the swarm intelligence algorithm family can even present outstanding results in feature selection. For example, the research of in [3] focused on the feature selection process based on Ant Colony Optimization (ACO). Their work applied ACO in feature selection to improve performance for text categorization. The result shows a lower computation complexity when compared to GA and other algorithms [3]. The study in [4] employed particle swarm optimization (PSO) in conjunction with genetic algorithm (GA) in the feature selection process. Their work used PSO to improve the insufficient performance of populations' GA in each generation. Their proposed method can avoid local optimum and effectively reduce unrelated features.

The dimensional reduction techniques are often used in conjunction with the classification methods. SVM is the recent approach adopted to define categories of data; it provides more robustness and accuracy among all well-known classification algorithms. SVM is adopted in many researches, for example, the work in [5] introduced SVM as a classifier to text classification problem. The work in [6] used SVM as a classifier to evaluate the classification performance for PSO-based feature selection. The results exhibit the capability of PSO-SVM; PSO can reduce the dimension of data via choosing optimal subset of features while SVM decreases classification errors and maintains the accuracy of classification [6].

In this study, we introduced artificial bee colony (ABC) as a feature selection method to improve the performance of dimension reduction processes. SVM is used to evaluate the fitness value of ABC. This rest of the paper is organized as follows: Section 2 describes our proposed method, ABC-SVM. Section 3 shows and discusses the experimental results. Finally, the conclusion of this study is described in section 4.

2   Our Proposed Method: ABC-SVM

Artificial Bee Colony [7] is a meta-heuristic method that imitates an intelligent behavior of honey bees for selecting their food source. Bees in ABC algorithm can be categorized into three groups: employed, onlooker and scout bees. Scout bees are responsible for searching new food sources while the labor of employed bees is updating information of food sources and sharing the information to onlooker bees that are in the hive. Then, the onlooker bees use this information to decide which food source they should exploit when several information are available. In ABC, a food source represents a solution in an optimization problem. The quality of food source can be measured with fitness function which can be related with food source location.

In the feature selection problem, the feature set of data is represented as a food source or solution; each food source is encoded into binary string which is the same as the characteristic of chromosome strings in GA. For example, the food source which represents a feasible set of features can be expressed as $S = F_1F_2F_3F_4...F_d$,; F is the feature or dimension of data and d is the total number of dimensions. The value of each feature F is in binary format [0, 1]. The feature value {1} indicates a selected feature, and the feature value {0} indicates a non-selected feature. In this case, the selected feature is determined as a required feature that cannot be eliminated from the original data while the non-selected feature is the feature that can be removed from the original data.

In our proposed method, the food source (solution) is randomly generated in the initial process. Then, each solution is evaluated through SVM classifier. The solving of SVM is generally a quadratic programming (QP) problem; sequential minimal optimization (SMO) [8] will be applied in this study to optimize the computation time of training period of SVM. SMO divides the large QP problem into series of smallest possible QP problem to avoid the intensive time and memories required [8]. For this study, the radial basis function (RBF) is used as the kernel function of non-linear SVM classifier to learn and recognize pattern of input data from the training set. The equation of RBF function is defined as

$$K(x_i, x_j) = e^{-r(x_i, x_j)^2} \qquad (1)$$

where r is a kernel parameter in RBF function. Then, a testing set is used to determine the classification accuracy for the input dataset. The fitness value is obtained by the classification accuracy of this testing dataset. For a small- and medium-sized data, the accurate value can be estimated by using the 10-fold cross-validation method. The method will randomly separate data into 10 subsets; one subset is used as a testing set while the remaining nine subsets are used as training sets. The process will be performed for 10 times in total. As a result, each subset will be used once as a validation or testing set. The accuracy of classification will be obtained from the average of all correctness values from 10 rounds. For the large-sized data, the holdout method is applied. This method divides the data into two parts for construction training and testing models.

Each food source is modified based on the process of updating feasible solution by employed bees as expressed in the equation (2) where Φ is a random number in the range between negative one and one, [-1,1].

$$v_{ij} = x_{ij} + \Phi(x_{ij} - x_{kj}) \qquad (2)$$

The equation (2) returns a numerical number of a new candidate solution v_{ij} from their current food source x_{ij} and their neighboring food source x_{kj}. However, for dimension reduction these solutions must be binary numbers. Thus, this numeric solution must be converted into either 0 or 1 by using equation (3) and (4) as follows:

$$S(v_{ij}) = \frac{1}{1 + e^{-v_{ij}}} \qquad (3)$$

$$if(rand < S(v_{ij})) \text{ then } v_{ij} = 1; \text{else } v_{ij} = 0 \qquad (4)$$

The equation (3) is a sigmoid limiting function into the interval [0.0, 1.0]. Then, a random number between range [0.0, 1.0], *rand*, is used to determine the binary value of the solution in the equation (4). The new candidate solution must be evaluated with

SVM classifier. If the new fitness value is better than the current one, the employed bees will replace its solution with this new candidate solution; otherwise, the new candidate solution will be ignored.

After employed bees share information of their solutions, onlooker bees will select a food source according to the calculated probability of each food sources based on the equation (5),

$$P_i = \frac{fit_i}{\sum_{i=1}^{N} fit_i} \tag{5}$$

where P_i is the probability value of the solution i, N is the number of all solutions, and fit_i is the fitness value of the solution i. Thus, the solution with higher fitness value will have greater opportunity to be selected by the onlooker bees. After the onlooker bees select their desirable food source, the bees will perform the process of updating feasible solution similar to employed bees.

However, if the fitness value of the current food source has not been improved by a predetermined number of iterations called "limit", the food source will be abandoned. Then, the scout bee will randomly change one dimension of the abandoned solution from one state to another (changing 0 to 1 or changing 1 to 0) in order to avoid sub-optimal solutions similar to mutation in GA. Finally, the whole processes will be repeated until the termination criterion is reached. The dimension reduction process of ABC-SVM is summarized as follows:

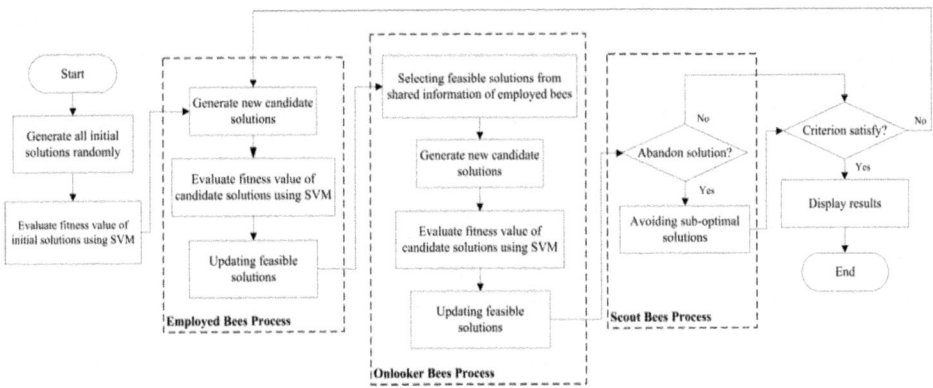

Fig. 1. The flowchart of ABC-SVM method

3   Experimental Results and Discussion

To investigate the performance of ABC-SVM algorithm on classification problems, five datasets from UCI data repository including vowel, wine, Wisconsin diagnostic breast cancer (WDBC), ionosphere, and sonar are used in the experiment. Vowel dataset is a pattern of speech data in eleven vowels of British English. When the speakers speak, each utterance will be recorded in an input array in ten floating-point formats. Wine data are the results of the chemical analysis of wines. 13 chemicals are found in each of the three types of wines. WDBC data are estimated from a digitized image of a fine needle aspirate (FNA) of a breast mass. The characteristic of cell nuclei in the image can be used for diagnostic of the mass into malignant or benign tumors. Ionosphere data includes radar frequencies returned from the ionosphere which can be categorized into two groups: good and bad radar. Good radar will present an evidence of some type of structure in the ionosphere, but bad radar will pass through the ionosphere without any evidence. For sonar dataset, the varieties of signal data are recorded to classify type of the sonar signals. The properties of all datasets used in this study are shown in Table 1.

Table 1. The characteristics of datasets for classification problem used in this study

Datasets	Number of samples	Number of classes	Number of features
Vowel	990	11	10
Wine	178	3	13
WDBC	569	2	30
Ionosphere	201/150	2	34
Sonar	104/104	2	60

Legends: For ionosphere and sonar dataset, x/y format represents the number of samples for train set versus the number of samples for test set.

In order to benchmark our proposed method with the other feature selection methods, we divide all datasets into three categories and use two evaluation methods according to [6]. The input datasets are categorized in terms of problem size which can be considered from the number of features in each datasets. The small size has the number of dimensions between 0 to 19 (vowel and wine datasets). The medium-sized datasets are WDBC and ionosphere datasets with the number of dimensions between 20 and 49. The large-sized dataset includes sonar dataset with the number of dimensions more than 50.

The parameter settings for ABC-SVM algorithm in this study are given as follows: the number of solutions in each iteration is set as 10 which is the same as the number of employed bees and number of onlooker bees. The maximum number of iterations is 1000. The predetermined number of iteration for avoiding the local optima solutions is set to 200.

In ABC-SVM algorithm, bees evaluate the solution based on the fitness value of each solution deriving from accuracy of SVM classifier; the proper settings of SVM parameters including C and r are required for effective and robust classification. r is a kernel parameter in RBF function. The setting of parameter C is a tradeoff between the margin width and training error. Both C and r parameters can affect the computational time and classification accuracy. If both parameters are high, the classification accuracy will be increased but with intensive computation time and the possibility of over-fitting to the training data. If these parameters are low, both the classification accuracy and computational time will be decreased. Hence, the proper parameters will provide an optimized computation time to generate train and test model for SVM with reduced training error and high classification accuracy. From our empirical testing, the values of C and r parameters for SVM classifier are set as presented in Table 2.

Table 2. Value of parameter *r* and *C* for SVM classification

Datasets	Value of *r*	Value of *C*
Vowel	16.0	1024.0
Wine	16.0	1024.0
WDBC	1.0	1024.0
Ionosphere	0.5	1024.0
Sonar	1.0	4096.0

To validate the accuracy of our purpose method with all input datasets, the 10-fold cross-validation method is used for vowel, wine and WDBC datasets. The holdout method is applied to ionosphere and sonar datasets. The experimental results from two groups are shown in Table 3 and Table 4 respectively.

The results of ABC-SVM are compared with those from PSO-SVM method from [6] and other feature selection method from [9]; these methods include sequential forward search (SFS), plus and take away (PTA), sequential forward floating search (SFFS), simple genetic algorithm (SGA), and hybrid genetic algorithm (HGA). HGA is an improved SGA algorithm by embedding the proper local search operation in SGA. The difference between HGA (1), HGA (2), HGA (3) and HGA (4) is the various value of factor called "ripple factor" which is set in the local search operation.

Table 3. The classification accuracy of dataset with 10-fold cross-validation method

Datasets	d*	SFS	PTA	SFFS	SGA	HGA (1)	HGA (2)	HGA (3)	HGA (4)	PSO-SVM d*	PSO-SVM %	ABC-SVM d*	ABC-SVM %
Vowel (D=10)	2	62.02	62.02	62.02	62.02	62.02	62.02	N/A	N/A	7	99.49	8	99.49
	4	92.63	92.83	92.83	92.83	92.83	92.83	92.83	92.83				
	6	98.28	98.79	98.79	98.79	98.79	98.79	98.79	98.79				
	8	99.70	99.70	99.70	99.70	99.70	99.70	99.70	N/A				
Wine (D=13)	3	93.82	93.82	93.82	93.82	93.82	93.82	93.82	N/A	8	100	6	**100**
	5	94.38	94.38	94.94	95.51	95.51	95.51	95.51	95.51				
	8	95.51	95.51	95.51	95.51	95.51	95.51	95.51	95.51				
	10	92.13	92.13	92.70	92.70	92.70	92.70	92.70	92.70				
WDBC (D=30)	6	93.15	93.15	94.20	93.67	94.90	94.90	93.99	93.99	1	95.61	13	**98.07**
	12	92.62	92.97	94.20	94.38	94.38	94.38	94.38	94.38	3			
	18	94.02	94.20	94.20	93.85	94.20	94.20	94.20	94.20				
	24	92.44	93.50	93.85	93.85	93.85	93.85	93.85	93.85				

Legends: d* represent the remaining number of dimensions after dimensional reduction process. The best accuracy values are presented in bold.

Table 4. The classification accuracy of dataset with holdout method

Datasets	d*	SFS	PTA	SFFS	SGA	HGA (1)	HGA (2)	HGA (3)	HGA (4)	PSO-SVM d*	PSO-SVM %	ABC-SVM d*	ABC-SVM %
Ionosphere (D=34)	7	93.45	93.45	93.45	95.44	95.73	95.73	95.73	95.73	15	97.33	13	**99.33**
	14	90.88	92.59	93.79	94.87	95.73	95.73	95.73	95.73				
	20	90.03	92.02	92.88	94.30	94.30	94.30	94.02	94.30				
	27	89.17	91.17	90.88	91.45	91.45	91.45	91.45	91.45				
Sonar (D=60)	12	87.02	89.42	92.31	93.75	94.71	95.67	95.19	95.67	34	96.15	31	**97.12**
	24	89.90	90.87	93.75	95.67	96.63	96.63	97.12	**97.12**				
	36	88.46	91.83	93.27	95.67	96.15	96.15	96.15	96.15				
	48	91.82	92.31	91.35	92.79	92.79	93.27	93.27	93.27				

Legends: d* represent the remaining number of dimensions after dimensional reduction process. The best accuracy values are presented in bold.

The results in Table 3 show that our method can provide the highest accuracy of classification for wine and WDBC datasets. Particularly for wine dataset, ABC-SVM can provide 100% classification accuracy. In WDBC dataset, the accuracy is about 2.5% greater than that of PSO method which is the best method among the aforementioned approaches reported in [6]. The classification accuracy of vowel dataset is 99.49%. However, this result is worse than other methods but it is comparable to that from PSO. In Table 4, the classification accuracy for ionosphere dataset is 99.33% which is about 2% higher than the best result from PSO method, while the classification accuracy for sonar dataset is 97.12% which is equal to the best value from HGA approach.

In this study, ABC-SVM extends equation (2) from the original ABC algorithm to appropriately apply as an effective feature selection method for classification problems. ABC-SVM can select optimal subset as the representative of the original dataset with high accuracy even in case of multi-classification problem like the vowel dataset; the classification accuracy from ABC-SVM is, however, less than that from other methods which may be because of the performance of SVM on multi-classification problems. SVM classifier has traditionally been designed to deal with binary classification problem as in WDBC, ionosphere and sonar dataset. Thus, applying ABC-SVM to multi-classification problems may not gain the full efficiency from SVM classifier. However, our proposed method can be applied to other datasets with various problem sizes. The efficiency of the method is not limited by number of features and samples as shown in the experimental results.

In general, all methods in Table 3 and Table 4 are working in similar processes. They all have the process for searching and selecting optimal subset of features based on the objective function. Nevertheless, the experimental results show that the ABC algorithm gave the better results. Unlike GA and PSO, ABC algorithm has two phases for enhancing the solutions (in the process of updating feasible solution of employed bees and onlooker bees) in one iteration. In PSO, the solution will be modified based on both particle best solution (pbest) and global best solution (gbest) in each iteration. GA uses crossover for improving the chromosome (solution) in each generation. Thus, the process of employed and onlooker phases make the results from ABC method converge to the optimal solution quicker than other methods. Moreover, if some solution traps in local optima, the scout bee phase of ABC will mend it by mutating to a new solution.

4   Conclusion

For data analysis for massively large datasets, dimension reduction is required. This research proposes a method of reducing dimension based on the artificial bee colony algorithm. This method serves as a feature selection method to select the optimal subset of dimensions or features that satisfies the defined objective. In order to evaluate the fitness of a feature set, support vector machine is employed to provide accurate classification performance of the feature set. Similar to other heuristic methods, ABC method does not analyze all possible features but yields highly accurate results with lower operation time. In this study, we demonstrate the performance of our proposed method to reduce the dimension of five datasets from UCI data repository. The results show that the classification accuracy of all tested dataset is over 97%; such high classification performance indicates that ABC-SVM is an appropriate feature selection method for classification problem. The concept of using ABC for dimension reduction is also applicable for other problems beside classification; however, methods that are suitable for the target problem must be used to evaluate the feature subsets.

References

1. Narenadra, P.M., Fukunaga, K.: A Branch and Bound Algorithm for Feature Subset Selection. IEEE Transactions on Computers 26, 917–922 (1977)
2. Dash, M., Liu, H.: Feature Selection for Classification. Intelligent Data Analysis 1, 131–156 (1997)
3. Aghdam, M.D., Ghasem-Aghaee, N.M., Basiri, E.: Text Feature Selection using Ant Colony Optimization. Expert Systems and Applications 36, 6843–6853 (2009)
4. Yang, C., Tu, C., Chang, J., Liu, H., Ko, P.: Dimensionality Reduction Using GA-PSO. In: Proceedings of the 9th Joint Conference on Information Sciences (2006)
5. Joachims, T.: Text categorization with Support Vector Machines: Learning with many relevant features. In: Nédellec, C., Rouveirol, C. (eds.) ECML 1998. LNCS, vol. 1398, pp. 137–142. Springer, Heidelberg (1998)
6. Chung-Ju, T., Li-Yeh, C., Jun-Yang, C., Cheng-Hong, Y.: Feature Selection using PSO-SVM. IAENG International Journal of Computer Science, 138–143 (2006)
7. Karaboga, D., Basturk, B.: A powerful and efficient algorithm for numerical function optimization: Artificial Bee Colony (ABC) algorithm. J. Glob. Optim. 39, 459–471 (2007)
8. Platt, J., Schölkopf, B.C., Burges, J.C., Smola, A.J.: Sequential minimal optimization: A fast algorithm for training support vector machines. In: Advances in Kernel Methods-Support Vector Learning, pp. 185–208 (1999)
9. Oh, I.S., Lee, J.S., Moon, B.R.: Hybrid Genetic Algorithms for Feature Selection. IEEE Transaction on Pattern Analysis and Machine Intelligence 26(11), 1424–1437 (2004)

Robust Controller Design for Main Steam Pressure Based on SPEA2

Shuan Wang[1], Dapeng Hua[1], Zhiguo Zhang[1], Ming Li[1],
Ke Yao[1], and Zhanyou Wen[2]

[1] Henan Nanyang Power Supply Company, Nanyang, Henan, 473000, China
[2] Xuchang Longgang Power Generation co., LTD, Xuchang, Henan, 461000, China
nydlws@163.com

Abstract. Main steam pressure is an important physical quantity that reflects the energy supply-demand relationship between the boiler and turbine. It has a significant role in the unit operation. Because boiler burning behavior varies greatly and the model of main steam pressure is of highly uncertainty, conventional control method can not obtain the expected control effect. In order to improve system control quality, a robust controller for main steam pressure is designed by using the H∞ mixed sensitivity approach in this paper. To better meet the site requirements, adding more restrictions in design, author innovation to put such a complex issue into multi-objective optimization problems. SPEA2 (The Strength Pareto Evolutionary Algorithm 2) are used to optimize the parameters of weighing functions in order to search for the H∞ controller which meets the time-and frequency-domain indexes. Simulation results show that the design of the main steam pressure control system has an excellent robust stability and dynamic quality.

Keywords: main steam pressure, H∞ robust control, SPEA2, weighing function matrix.

1 Introduction

In the fossil-fired power plant, high-pressure and high-temperature steam is used for generation of electric power. Main steam pressure is an important indicator that reflects steam output is in line with the load or not [1], main disturbance comes from fuel (domestic interference) and steam consumption of the turbine (outside disturbance).Its regulating quality and stability can affect the safety and economic operation of the entire unit. Due to the uncertainty of the model, conventional control method can not obtain the expected control effect. The model's uncertainty was considered in the H∞ mixed sensitivity design, the transient performance and anti-jamming capability of the system was also taking into account. In addition, H∞ mixed sensitivity control is designed by choosing weighing function matrix in the frequency domain. It is similar to the classical design method in frequency domain which is familiar with engineers, thus increasing designers' attention.

However, a successful design depends on the appropriate choice of weighing functions, which in turn relies on a designer's experience and familiarity with the

D.-S. Huang et al. (Eds.): ICIC 2011, LNBI 6840, pp. 176–182, 2012.
© Springer-Verlag Berlin Heidelberg 2012

design approach [2]. In [3], the control system design problems were naturally formulated as constrained optimization problems, so that we can use EA to solve. SPEA, an acronym for Strength Pareto Evolutionary Algorithm [4], is an approach that incorporates several of the desirable features of other well-known multi-objective evolutionary algorithms. This algorithm implements elitism through the maintenance of an external set of best solutions found during the whole iteration loop. In SPEA, the fitness of a solution in the population depends on the best solutions in the external set but is independent of the number of solutions this solution dominates, or is dominated by, within the population [5]. To maintain the diversity in the population, a new Pareto-based nicking method is provided.

The paper adopts SPEA2 [6] (an improvement of the original SPEA) to search for the H∞ controller which meets the time-domain and frequency-domain indexes for the main steam pressure.

2   Main Steam Boiler Pressure Control System

Control system as figure 1:

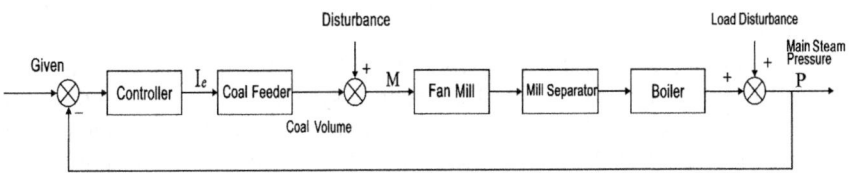

Fig. 1. Schematic diagram of the main steam pressure control system

Steam pressure adjustment process: when the system suffers from disturbance and steam pressure deviates from the given value, then controller adjusts to the current of the coal feeder, and the coal into fan mill correspondingly changes. Through the mill separator, powder is sprayed into furnace, so that the main steam pressure is corrected.

After identification test and optimization fitting, we obtain a generalized object model of the main steam pressure:

$$G(s) = \frac{Ps(s)}{u(s)} = \frac{1.5}{(80s+1)^3} \tag{1}$$

Here: Ps is export steam pressure of the super-heater (main-steam pressure),MPa. u - control signal, also control current of the coal feeder. Generalized objects include: coal feeder, fan mill, mill separator and the heat/steam-pressure conversion process of the boiler. Based on the mathematical model, the control system is designed to the standard robust control problem as follows.

3   H∞ Mixed Sensitivity Design

Consider tracking control of the LTI as figure 2:

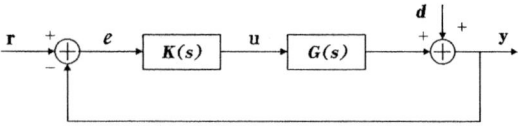

Fig. 2. Tracking Control

Where r is reference input, e is tracking error, u is control input, d is measurement disturbance, y is system output. $G(s)$ represents the physical system to be controlled, $K(s)$ is controller.

Separately weighing to signal e, u and y to compose a generalized system by using $W_S(s)$, $W_R(s)$ and $W_T(s)$, it can be converted into standard H∞ control problems. Once three weighing matrix were determined and generalized object meet certain conditions, we may adopt Riccati equation to solve above H∞ mixed sensitivity problems.

As the weighing matrix reflects performance of both the frequency domain and the time domain, the design procedure of the control system can be expressed as a multi-objective optimization problem when the weighing functions are considered as design parameters and the optimal solution must satisfy a set of inequality constraints.

That is, for a given plant $G(s)$, to find $W_S(s)$, $W_R(s)$ and $W_T(s)$ such that:

$$\min \Psi(W_S, W_R, W_T) \tag{2}$$

And

$$\overline{\sigma}[w_S^{-1}(jw)] + \overline{\sigma}[w_T^{-1}(jw)] \geq 1 \tag{3}$$

$$\gamma_0(W_R, W_S, W_T) \leq \varepsilon_0 \tag{4}$$

$$\Phi_i(W_R, W_S, W_T) \leq \varepsilon_i \quad , i = 1, 2, \cdots, n \tag{5}$$

Where Ψ is the performance indicator that should be minimum, Φ_i are performance indices, which are algebraic or; functional inequalities representing rise time, overshoot, etc., and ε_0 and ε_i are real numbers representing the desired bounds on γ_0 and Φ_i, respectively. This constrained optimization problems are usually non-convex, non-smooth, and multi-objective with several conflicting design aims which need to be simultaneously achieved.

4   Optimization the Weighing Matrix Parameters Using SPEA2

The SPEA2 algorithm is an improvement of the original SPEA. The main differences between them are: an improvement in the fitness assignment scheme, a nearest neighbor density estimation technique and a new archive truncation method that guarantees the preservation of boundary solutions [7]. SPEA2 algorithm using fine-grained assignment policies, which combined into the density information. In this algorithm, each solution, whether it is Pareto greater than or less than other solutions, the individual is always to be taken into account. So SPEA2 may be applied to find solutions to the above optimization problems.

Under the system robustness and performance requirements, determining the structure of the three weighing matrix, the chromosome is a binary string describing parameters of the weighing matrix.

The procedure of objective function Optimization is listed as follows.

(1) Define the plant G(s) and the functions Ψ and Φ_i, decision the values of ε_0 and ε_i.

(2) Define the fundamental form of the weighing functions $W_S(s)$, $W_R(s)$ and $W_T(s)$ and the search domain.

(3) Determine the parameters for SPEA2, such as generate population P of size N, archive set Q of size M, maximum number of generations T.

(4) Generate an initial population P_0 and create the empty archive (external set) $Q_0 = \phi$, set $T = 0$.

(5) Fitness assignment: calculate fitness values of individuals in P_t and Q_t.

(6) Environmental selection: Copy all non-dominated individuals in P_t and Q_t to Q_{t+1}.If size of Q_{t+1} exceeds M, and then reduce Q_{t+1} by means of the truncation operator. If size of Q_{t+1} is less than M, then fill Q_{t+1} with dominated individuals in P_t and Q_t.

(7) If $t \geq T$ or another stopping criterion is satisfied then set *NDSet* to the set of decision vectors represented by the non-dominated individuals in Q_{t+1}, Stop.

(8) Perform binary tournament selection with replacement on Q_{t+1} in order to fill the mating pool.

(9) Apply recombination and mutation operators to the mating pool and set Q_{t+1} to the resulting population. Increment generation counter ($t = t+1$) and go to Step (5).

5   Simulation

Consider the above main steam pressure control, control objectives is to ensure the response speed of the closed-loop system as quickly as possible. At the same time satisfy the following:

(1) Gain margin $G_m \geq 8$;

(2) Phase margin $P_m \geq 75$.

Then may be use the above method to optimize parameters and design a H∞ controller..

5.1   Structure of Weighing Functions and Parameter Coding

According to the system robustness and performance requirements, the structure of the three weighing matrix as follows:

$$W_R(s) = k_2 , \quad W_s(s) = \frac{k_1}{(1/\omega_1)s + 1} , \quad W_T(s) = k_3 s(\frac{1}{\omega_2}s + 1) .$$

When the structure of weighing functions is given, the following parameters set will determine the three weighing matrix:

$$H = \{k_1, \omega_1, k_2, k_3, \omega_2\} .$$

Set search domain:

$$k_1 \in \mathbf{R}_1 = [100,1000] , \; \omega_1 \in \mathbf{R}_5 = [0.00001, 0.01] , \; k_2 \in \mathbf{R}_3 = [0.0001, 10]$$
$$k_3 \in \mathbf{R}_4 = [0.1, 10] , \; \omega_2 \in \mathbf{R}_5 = [100, 1000]$$

5.2   Determination of the Objective Function

i System response speed can be measured by the unit-step response's error function of the closed-loop system. If $y(t)$ is unit-step response, then the minimum function is:

$$\Psi = \int_0^{T_0} [1 - y(t)]^2 \, dt \tag{6}$$

Here, T_0 is a constant time longer than the transition process of system, set $T_0 = 300$.

ii The phase margin requirements: Set

$$\Phi_1 = -G_m \text{ and } \varepsilon_1 = -8 . \tag{7}$$

iii The gain margin requirements: Set

$$\Phi_2 = -P_m \text{ and } \varepsilon_2 = -75 . \tag{8}$$

iv Set the desired border

$$\varepsilon_0 = 1 \tag{9}$$

5.3   Simulation Results

The parameters of the SPEA2 used in the simulation are: $N = 80$, $Q = 10$, $T = 100$.
The weighing functions are:

$$W_s(s) = \frac{951}{10000s + 1} , \quad W_R(s) = 0.0001 , \quad W_T(s) = 7.52s(0.00125s + 1)$$

A four-order controller is also obtained:

$$K(s) = \frac{1.485 \times 10^6 s^3 + 5.567 \times 10^4 s^2 + 695.9s + 2.9}{s^4 + 149.2s^3 + 117.8s^2 + 45.79s + 0.004578}$$

Design results:

$$\gamma = 1.000 , \|S\|_\infty = 1.2473 , \|T\|_\infty = 0.9989 , G_m = 8.2857 , P_m = 75.9431 .$$

Unit-step response, bode diagram, singular values of sensitivity function and singular values of complementary sensitivity function are shown in figure 3~6 :

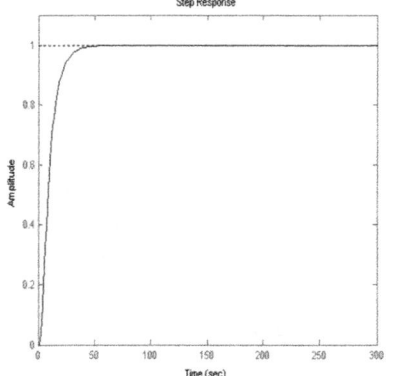

Fig. 3. Unit-step Response of the Closed-loop System

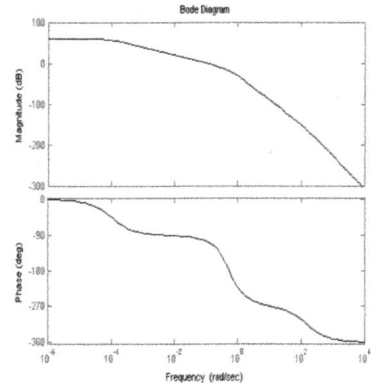

Fig. 4. Bode diagram of open loop system

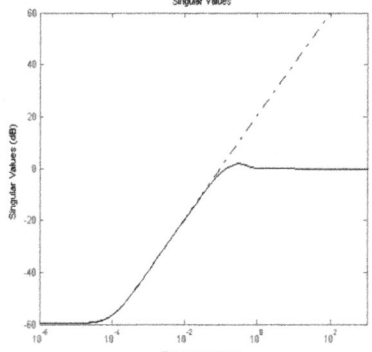

Fig. 5. Singular values curve of sensitivity function

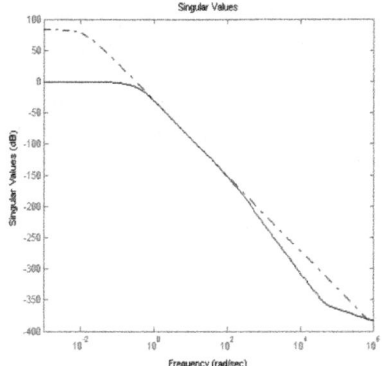

Fig. 6. Singular values curve of complementary sensitivity function

The simulation results showed that: SPEA2 optimization achieved good design effect, not only reach the given design objectives, and avoided the artificial methods of trial and error, which for main steam pressure control of boiler is of great practical significance.

6   Conclusion

In this paper, a robust controller for the main steam pressure system is designed by use of the H∞ mixed sensitivity approach. SPEA2 is used to optimize the parameters of weighing functions in order to search for the H∞ controller which meets the time-domain and frequency-domain indexes. This method is understandable and intuitive, making the selection of weighing matrixes standardized and avoiding the conventional trial and error method. At the same time we can obtain the expected control effect by using the optimizing method.

References

1. Liu, H.B., Li, S.Y., Chai, T.Y.: Intelligent Decoupling Control of Plant Main Steam Pressure and Power Output. Electrical Power and Energy Systems 25(10), 809–819 (2003)
2. Oloomi, H., Shafai, B.: Weight Selection in Mixed Sensitivity Robust Control for Improving the Sinusoidal Tracking Performance. In: Proc IEEE CDC, Hawaii, pp. 300–305 (2003)
3. Tang, K.S., Man, K.F., Gu, D.W.: Structured genetic algorithm for robust H∞ control systems design. IEEE Trans. on Industrial Electronics 43(5), 575–582 (1996)
4. Zitzler, E., Thiele, L.: Multi-objective Evolutionary Algorithms: A Comparative Case Study and the Strength Pareto Approach. IEEE Trans. on Evolutionary Computation 3(4), 257–271 (1999)
5. Claudomiro, S., Roberto, M., Rodrigues, Fredrik, L., Joao, C.: Topology Identification Using Multiobjective Evolutionary Computation. IEEE Transactions on Instrumentation 59(3), 715–729 (2010)
6. Zitzler, E., Lanumanns, M., Thiele, L.: SPEA2: improving the strength Pareto evolutionary algorithm for multiobjective optimization. In: Evolutionary Methods for Design, CIMNE, Barcelona (2002)
7. Kumaja, S., Maheswarapu, S.: Enhanced Genetic Algorithm based Computation Technique for Multi-objective Optimal Power Flow Solution A Parallel Skeleton for the Strength Pareto Evolutionary Algorithm. Electrical Power and Energy Systems 32, 736–742 (2010)

GA-Based Hybrid Algorithm for MBR Problem of FIPP p-Cycles for Node Failure on Survivable WDM Networks*

Der-Rong Din

Department of Computer Science and Information Engineering,
National Changhua University of Education
No. 1, Jin De Road, Changhua 500, Taiwan R.O.C

Abstract. In this paper, the *minimal backup reprovisioning* (MBR) problem is studied, in which, the *failure-independent path protecting p-cycles* (*FIPP p-cycles*) scheme is considered for single node-failure on WDM networks. After recovering the affected lightpaths from a node failure, the goal of the MBR is to re-arrange the protecting and available resources such that working paths can be protected against next node failure if possible. This is a hard problem, a hybrid algorithm which combines heuristic algorithm and genetic algorithm is proposed to solve this problem. The simulation results of the proposed method are also given.

Keywords: WDM, node failure, MBR, FIPP p-cycles, GA.

1   Introduction

In an optical network employing WDM, the failure of a single link (or node) may lead to tremendous data loss since a single link can carry a huge amount of data (on the order of terabits per second). Therefore, several schemes have been proposed to achieve *survivability* by *protection* or *restoration* on the optical layer [1]. In a protection scheme, extra bandwidth is reserved when the connection is provisioned. The primary and backup paths are usually link (or node) disjoint, which guarantees that at least one path is available when any single link (or node) failure occurs in the network [1].

When a node failure occurs, all the traffic flows that go through the node and all the local traffic that is collected and distributed by the node are interrupted. The *path-based protection* techniques can be employed [2]. Setting protection paths for the possible failed nodes may be cheaper and flexible approach because there are huge wavelengths provided by the system and it can be extended more easily than setting up double switches. Note that it is not possible in any scheme to recover lightpaths from a failure at their own source/destination nodes. Node failure protection through network reconfiguration is therefore only concerned with restoring traffic flows that were transiting failed nodes [3].

---

* This work was partly supported by the National Science Council (NSC) of Taiwan, R.O.C. under Grant Number NSC-99-2221-E-018-015.

D.-S. Huang et al. (Eds.): ICIC 2011, LNBI 6840, pp. 183–190, 2012.

1.1 FIPP p-Cycles

The FIPP p-cycles protection method proposed by Kodian and Grover [4] is a pre-connected protection scheme with an end-to-end failure-independent path protection property. In the FIPP p-cycles scheme, several mutually node-disjoint paths can form a *disjoint route set (DRS)* [4]. For a DRS, a single FIPP p-cycle which passes through all end-nodes of the paths in DRS is pre-established. Since any two paths in the same DRS are node-disjoint, only one path in a DRS may be affected by a single node failure. Moreover, the FIPP p-cycles scheme is path-oriented, it can be used to recover traffic transiting through a failed node. More details about the FIPP p-cycles scheme can be found in [4].

1.2 Backup Reprovisioning

Recently, reprovisioning new backups for working paths to combat the effect of multiple concurrent failures (where concurrent means that a new failure occurs before a previous failure is repaired) is the goal of *backup reprovisioning* (BR) [5]. When a failure occurs, the traffic through the affected node is switched to backups. To protect lightpaths against next potential failure, new backups are provided for the paths that become unprotected or vulnerable because of losing their primary or their backup due to the previous failure or due to backup resource sharing. This approach is called *Minimal Backup Reprovisioning* (MBR). The numerical results show that [5], using MBR, the connection vulnerability can be significantly reduced even when the network is heavily loaded.

According to the survey by authors, there is no article studied the backup reprovisioning problem for the FIPP p-cycles on WDM networks against node failure. In this paper, the MBR problem on WDM networks with FIPP p-cycles protection for node failure is studied A hybrid algorithm which combined heuristic algorithm and genetic algorithm [6] is proposed to solve this problem.

2   Problem Definition

In this paper, the physical and virtual topologies of a WDM network, the set of working paths, the disjoint route sets and the set of FIPP p-cycles are fixed and known. For each working path in network, there is a single FIPP p-cycle can be used to protect it in an on-cycle or straddling manner. For each edge (or link) of physical topology, there is a single fiber connects the end-nodes of link such that data can be transmitted bidirectional. All nodes in physical network are with full wavelength converting capabilities. Only the node failure scenario is considered.

2.1 Notations

- $G(V, E)$: the physical topology, where $V = \{v_1, v_2, ..., v_{|V|}\}$ is the set of nodes and $E = \{e_1, e_2, ..., e_{|E|}\}$ is the set of physical links. Each node of the network can also be represented by the number assigned to it.

- $P = \{P_k | k = 1, 2, ..., |P|\}$: the set of working paths, where $P_k \in P$ is the working path. Nodes v_{s_k} and v_{d_k} are the source and destination nodes of P_k, respectively.
- BP_k: the backup path for the working path P_k, $k = 1, 2, ..., |P|$ and path BP_k should be node-disjoint to P_k.
- v_f: the failed node.
- $N(v_f)$: the set of adjacent nodes of node v_f.
- $E(v_f)$: the set of links adjoined to node v_f, $E(v_f) \subseteq E$.
- FC: the set of FIPP p-cycles used to protect all paths in P.
- PoC_j: the set of paths which is protected by the FIPP p-cycle C_j, where $C_j \in FC$, $PoC_j \subseteq P$. Any two paths in PoC_j are mutually node-disjoint.
- P^f: the set of working paths affected by the failed node v_f, $P^f \subseteq P$.

The goal of the problem is to recover the protecting capabilities of the set of FIPP p-cycles to protect working paths such that the total capacities can be minimized. The proposed hybrid method consists of two algorithms, the first one is the *Preprocessing Algorithm* and the second one is the *Genetic Algorithm*.

3   Preprocessing Algorithm

When node v_f fails, the recovering procedure is performed to recover the affected paths. In the recovering procedure, if path P_k is affected by the node failure, first, the primary path is switched to the backup path BP_k which is provided by the protecting FIPP p-cycle. After recovering, the FIPP p-cycle may lose its protecting capability and those paths originally protected by the FIPP p-cycle and the backup path BP_k are vulnerable now. If a node failure occurs again, these paths cannot be protected. In backup reprovisioning, not only the affected working paths are recovered by current FIPP p-cycles, the protecting capabilities of affected FIPP p-cycles are also recovered and working paths can be survived at the sequent node failure such that total capacities can be minimized.

3.1   Cases of Failure

In this subsection, working paths together with the FIPP p-cycles affected by the node failure can be classified into four cases: (1) **End-node failure**: the failed node is the source (or destination) node of the working paths, these working paths cannot be recovered. (2) **Straddling-path failure**: the failed node is an interior node of the working path P_k, but not on the protecting FIPP p-cycle C_i. (3) **On-cycle-path failure**: the failed node is an interior node of the working path P_k and also a node on the protecting cycle. In this case, the working path P_k can still be recovered by using the remaining part of the FIPP p-cycle. (4) **Cycle failure**: the failed node is on the cycle but all paths protected by the cycle are unaffected, then the cycle affected by the failure will lose its protecting capabilities.

3.2   Preprocessing

In this subsection, the details of the proposed algorithm *Preprocessing Algorithm* are described. When node v_f fails, node v_f and the links in $E(v_f)$ are removed from G to form a new network $G'(V \setminus \{v_f\}, E \setminus E(v_f))$. If path P_k is affected by the node failure, the default path BP_k on FIPP p-cycle is used to recover the affected working path. To provide backup reprovisioning, first, an empty set P^f is constructed. Then, for each path (say P_k) together with the protecting cycle (say C_i), the respective case described in previous subsection is checked and applied.

To reduce the total capacities used for protecting all paths in P^f, cycles in the current FC are examined first for whether they can be used to protect paths in P^f. For each working path P_k in P^f, for each existing cycle C_j in FC, if path P_k is node-disjoint to all paths in PoC_j, then cycle C_j can be used to protect path P_k and PoC_j is updated by uniting path P_k. And then path P_k is removed from P^f. This process is repeated performed until no cycle in FC can be found to protected path in P^f. Then, for all unprotected paths in P^f, the Genetic Algorithm is performed to find new FIPP p-cycles to protect them. The details of the Preprocessing Algorithm are shown in Algorithm 1.

Algorithm 1. Preprocessing Algorithm

1: **Input:** G, P, FC, v_f; **Output:** G', FC;
2: Remove failed node v_f and edges in $E(v_f)$ from G to form the graph $G'(V \setminus \{v_f\}, E \setminus E(v_f))$.
3: Construct the empty set P^f.
4: **for all** (working path $P_k \in P$) **do**
5:     Let cycle $C_i \in FC$ be the FIPP p-cycle used to protect path P_k.
6:     According the case of failure, perform the respective case (1)–(4):
       (1) End-node failure
7:         Remove working path P_k. Remove C_i from FC.
8:         Release the network resources of path P_k and cycle C_i. Add all paths in PoC_i to P^f.
9:     **(2) Straddling-path failure:**
10:         Find a new path P_{new} from v_{sk} to v_{dk} on G' which is node-disjoint to all paths in PoC_i.
11:     **if** (P_{new} can be found) **then**
12:             Use P_{new} as the new working paths. Release the network resources of path P_k.
13:             Use BP_k to recover the original FIPP p-cycle C_i. Add path P_{new} to P^f.
14:     **else**
15:             Remove C_i from FC. Release the network resources of path P_k and cycle C_i.
16:             Add all paths in PoC_i and path BP_k to P^f.
17:     **end if**
18:     **(3) On-cycle-path failure:**
19:         Remove C_i from FC. Release the network resources of path P_k and cycle C_i.
20:         Add all paths in PoC_i and path BP_k to P^f.
21:     **(4) Cycle failure:**
22:         Remove C_i from FC. Release the network resources of path P_k and cycle C_i.
23: **end for**
24: **for all** (working path $P_k \in P^f$) **do**
25:     **for all** (working path $C_j \in FC$) **do**
26:         **if** (path P_k is node-disjoint to all paths in PoC_j) **then**
27:             Update $PoC_j = PoC_j \cup P_i$. Remove P_k from P^f.
28:         **end if**
29:     **end for**
30: **end for**

---

4   Genetic Algorithm

4.1   Chromosome Encoding

Because the problem involves representing the protecting cycles for vulnerable paths, a coding scheme that uses positive integer is employed. After performing the Preprocessing Algorithm, the network $G(V, E)$ is updated as $G'(V \setminus \{v_f\}, E \setminus E(v_f))$ and the set P^f of unprotected paths is constructed. For the current physical network $G'(V \setminus \{v_f\}, E \setminus E(v_f))$, the *Cycle Finding Algorithm* proposed in [7] is performed to form the set (NC) of all possible cycles in $G'(V \setminus \{v_f\}, E \setminus E(v_f))$. All cycles in NC are sorted and indexed in non-descending order according to the length of the cycles. For each path P_k in P^f, the set $(NC_k \subseteq NC)$ of cycles which can be used to protect path P_k is formed. A chromosome consisting of an integer array with size $2 \times |P^f|$ is introduced to represent the protecting cycles of vulnerable paths. The chromosome consists of two genes: the *Protecting Cycle Gene (PCG)* and *Copy Gene (CG)*. For each unprotected path P_k in P^f, let v_{sk} and v_{dk} be the source and destination nodes of the path P_k, respectively. The PCG can be represented as $\{PCG_1, PCG_2, ..., PCG_{|P^f|}\}$, where C_{PCG_k} (k=1, 2, ..., $|P^f|$) is a cycle in NC_k and the index of cycle is in $\{1, 2, ..., NC\}$. The CG can be represented as $\{CG_1, CG_2, ..., CG_{|P^f|}\}$, where CG_k (k=1, 2, ..., $|P^f|$) means that the path P_k is protected by the CG_k-th copy of the protected cycle C_{PCG_k}.

4.2   Population Initialization

Let FC^{new} be the set of selected cycles which are used to protected the vulnerable paths in P^f. Initially, FC^{new} is an empty set. Let z_j represent the maximal copy of the cycle $C_j \in NC$, initially $z_j = 0$, for j =1, 2, ..., NC. For each copy of the selected cycle C_j, create a set PoC_{j,z_y} for storing those paths currently protected by the z_y-th (integer $1 \leq z_y \leq z_j$) copy of cycle C_j.

The population of the genetic algorithm is generated by following steps: First, random selected an integer k ($1 \leq k \leq P^f$). For the selected path P_k, a FIPP p-cycle in $NC_k \cap FC^{new}$ is selected randomly to protect it. This means that those cycles currently deployed are selected first if possible. If $FC^{new} \cap NC_k$ is nonempty, then those cycles in $FC^{new} \cap NC_k$ are selected first and tested whether there is a cycle in $FC^{new} \cap NC_k$ can be used to protect the path P_k. It is worth noting that path P_k can be protected by the z_y-th ($1 \leq z_y \leq z_j$) copy of C_j only if path P_k is node-disjoint to all paths in PoC_{j,z_y} of the z_y-th copy of selected cycle $C_j \in FC^{new} \cap NC_k$. If a cycle $C_j \in (FC^{new} \cap NC_k)$ and the z_y copy is found and satisfied the constraint, then add path P_k to the set PoC_{y,z_y}.

If the path cannot be protected by the cycles in $FC^{new} \cap NC_k$, then a new cycle C_j is selected from NC_k to protect path P_k. For the selected cycle C_j, increase z_j by 1, create a set PoC_{j,z_j} to store path P_k, add C_j to the set FC^{new}, then set the value of PCG_k as j and CG_k as z_j. Then, increase k by 1. If $k > |P^f|$, set k to 1. Repeat the above steps to determine the protected cycle of next path until all contents of PCG and CG are set, then the initialization process is terminated.

4.3   Fitness Definition

Generally, GA uses fitness function to map objective to cost and to achieve the goal of a minimal cost. Let $l(C_j)$ be the length of cycle $C_j \in NC$ and z_j be the number of copies of the cycle C_j. An objective function value is associated with chromosome and represented by the following equation: $minimize\ OBJ = \sum_{j=1}^{|NC|} z_j \times l(C_j) = \sum_{\forall C_j \in FC^{new}} z_j \times l(C_j)$. Since the best-fit chromosomes should have a probability of being selected as parents that is proportional to their fitness, they need to be expressed in a maximization form.

4.4   Crossover Operators and Mutation

The *Single Point Crossover* is used to develop th GA. First, randomly select a crossover site (integer) i from interval $[1, |P^f|]$. In SPC, the operator is applied on both the PCG and CG. For the mutation, the *Single Protecting Cycle Mutation* (SPCM) is applied to GA. First randomly select an integer i in the interval $[1, |P^f|]$, assume path P_k be protected by the z_j-th copy of cycle C_j, SPCM changes the protecting cycle of paths in the PoC_{j,z_j} by selecting a cycle C_a with minimal length, if possible. The cycle is selected from the set CS of cycles, where $CS = \bigcap_{P_k \in PoC_{j,z_j}} \{NC_k\}$, with minimal length.

4.5   Chromosome Adjustment Algorithm (CAA)

After performing crossover and/or mutation operations, it may generate infeasible chromosomes. Thus, these chromosomes should be adjusted to a constraint-satisfied solutions by performing the CAA.

To adjust the chromosome to a constraint-satisfied one, first, the chromosome can be considered as two parts separated by the crossover site i. It is worth noting that, before performing crossover, each part is constraint-satisfy. Random select a part (left or right) and a direction (increasing-direction or decreasing-direction) for adjustment. The increasing-direction means that the adjustment is performed from 1 to i and from $i + 1$ to $|P^f|$ for the left part and right part, respectively. And the decreasing-direction is the reverse direction of the increasing-direction. If the left part is selected and the increasing-direction is to be apply, the right part of the chromosome is packed and preserved.

The details of CAA are described as follows. First, construct empty set FC^{new}. For each cycle C_j used in the right part, put it into FC^{new} and construct the sets PoC_{j,z_j}. Second, two paths assigned to same cycle but not to same copies are tried to merge so as the number of copies of cycle can be reduced.

Third, according to the direction of adjustment, the protecting cycle for path P_k $(1 \leq k \leq i)$ is adjusted. For the selected path P_k, a FIPP p-cycle in $NC_k \cap FC^{new}$ is selected randomly to protect it. If $FC^{new} \cap NC_k$ is nonempty, then those cycles in $FC^{new} \cap NC_k$ are selected first and tested whether there is a cycle in $FC^{new} \cap NC_k$ can be used to protect the path P_k. It is worth noting that path P_k can be protected by the z_y-th $(1 \leq z_y \leq z_j)$ copy of C_j only if path P_k is node-disjoint to all paths in PoC_{j,z_y} of the z_y-th copy of selected cycle

$C_j \in FC^{new} \cap NC_k$. If a cycle $C_j \in (FC^{new} \cap NC_k)$ and the z_y copy is found and satisfied the constraint, then add path P_k to the set PoC_{y,z_y}. If the path cannot be protected by the cycles in $FC^{new} \cap NC_k$, then the original protecting cycle (say C_j) is preserved to protect path P_k. For the selected cycle C_j, increase z_j by 1, create a set PoC_{j,z_j} to store path P_k, add C_j to the set FC^{new}, then set PCG_k to j and CG_k to z_j.

5   Experimental Results

The proposed algorithms were conducted by using C++ programming language and Boost Graph Library [8]. All the simulations were run on a PC with Celeron 2 Duo 1.8 GHz CPU, 2.0G RAM, and with Windows Vista operating system. Two physical topologies were used for experiments, they are COST239 and National networks. All nodes in the network may fail and the failed nodes were generated randomly.

The connection requests were generated randomly. A set of FIPP p-cycles is used to protect primary lightpaths. The parameters of GA were determined through simulations, and the crossover probability is set to 1.0, mutation probability is set to 0.1, and population size is set to 500. To know the efficiency of the proposed methods, the *node-disjoint paths* (NDPS) protection method was implemented for comparison.

First, WDM networks with different number of connections were generated. Each case was examined 20 times and the result was evaluated and compared. Let *Backup resources ratio* be the ratio of the total capacities to that of the

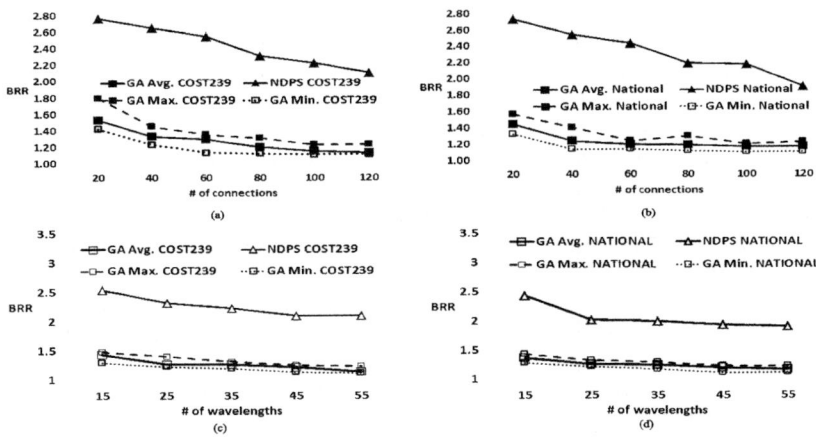

Fig. 1. BRR performance for different number of connections on (a) COST239 with different number of connections, (b) National with different number of connections, (c) COST239 with different number of wavelengths, (d) National with different number of wavelengths

working capacities. Figure 1(a) shows the comparison of the average(Avg.), minimal(Min.) and maximal(Max.) BRR performances of the GA to that of the NDPS methods on the network COST239. The average BRR value of the GA decreases from 1.5 to 1.15 and it is better than that of the NDPS. Even the maximal case of GA can get better performance than NDPS. For the case of the National network, the BRR performance of GA is better than that of the NDPS Figure 1(b).

For the limited number of wavelengths (W in $\{20, 40, 60, 80, 100, 120\}$), WDM networks with 120 connections were generated for these experiments. In Figure 1(c), the BRR value of the GA method decreases about 0.2 on networks and it is better than that of the NDPS. For the case of the National network, the BRR performance of GA is also better than that of the NDPS Figure 1(d).

6   Conclusions

In this article, the minimal backup reprovisioning (MBR) problem for FIPP p-cycles on WDM networks for node failure was studied. After recovering the working paths affected by the failed node, the protecting capabilities of FIPP p-cycles is recovered by performing MBR against the subsequent nodes failure on WDM networks. A hybrid algorithm was proposed to solve this problem. Simulation results show that the performance of GA is better than the traditional node-disjoint paths protecting method.

References

1. Ramamurthy, A., Mukherjee, B.: Survivable WDM Mesh Networks-part II: Restoration. In: Proc. of IEEE ICC, pp. 2023–2030 (1999)
2. Su, H.K., Wu, C.S., Chu, Y.S.: IP Local Node Protection. In: Proc. of Second International Conference on Systems and Networks Communications, ICSNC, vol. 58 (2007)
3. Stamatelakis, D., Grover, W.D.: IP layer Restoration and Network Planning Based on Virtual Protection Cycles. IEEE Journal on Selected Areas in Communications 18(10), 1938–1949 (2000)
4. Kodian, A., Grover, W.D.: Failure-independent Path-protecting P-cycles: Efficient and Simple Fully Pre-connected Optical Path Protection. Journal of Lightwave Technology 23(10), 3241–3259 (2005)
5. Tornatore, M., Lucerna, D., Pattavina, A.: Improving Efficiency of Backup Reprovisioning in WDM Networks. In: Proc. of IEEE Inforcom Phoenix, AZ, USA, pp. 726–734 (15–17) (2008)
6. Holland, J.: Adaptation in Natural and Artificial Systems. Univ. of Michigan Press, Ann Arbor (1975)
7. Liu, C., Ruan, L.: Finding Good Candidate Cycles for Efficient P-cycle Network Design. In: Proc. of Int. Conf. on Computer Communications and Networks, ICCCN 2004, Chicago, USA, pp. 321–326 (11–13) (2004)
8. Boost Graph Library, http://www.boost.org

A New Hybrid Algorithm for the Multidimensional Knapsack Problem

Xiaoxia zhang, Zhe Liu, and Qiuying Bai

College of Software Engineering, University of Science and Technology Liaoning, China
aszhangxx@163.com, liuzhelive@hotmail.com, bqy2004@126.com

Abstract. This paper presents a novel hybrid algorithm to solve the multidimensional knapsack problem. The main feature of this hybrid algorithm is to combine the solution construction mechanism of ant colony optimization (ACO) into scatter search (SS). It considers both solution quality and diversification. A new mechanism of the subset combination method has been applied simultaneity, which hybridizes mechanism of the pheromone trail updating with combination mechanism of scatter search to generate new solutions. Second, an improvement algorithm should be embedded into the scatter search framework to improve solutions. Finally, the experimental results have shown that our proposed method is competitive to solve the multidimensional knapsack problem compared with the other heuristic methods in terms of solution quality.

Keywords: Ant colony optimization, Multidimensional knapsack problem, Scatter search.

1 Introduction

The multidimensional knapsack problem (MKP) has been an important problem in the field in the field of combinatorial optimization. The MKP has many practical applications, such as cutting stock [1], loading problems [2], allocation of databases and processors in a distributed data processing [3]. The objective of the MKP is to find a feasible subset of objects that maximizes the total profit without exceeding the constraint capacities. More formally, a MKP can be defined as follows:

$$\text{maximize} \quad z = \sum_{j=1}^{n} c_j x_j \tag{1}$$

$$\text{s. t.} \quad \sum_{j=1}^{n} a_{ij} x_j \le b_i, i = 1, \cdots, m \tag{2}$$

$$x_j \in \{0,1\}, j = 1, \cdots, n \tag{3}$$

where n is the number of items and m is the number of knapsack's constraints with capacities b_i. a_{ij} is the consumption of resource i for object j, b_i is the available

D.-S. Huang et al. (Eds.): ICIC 2011, LNBI 6840, pp. 191–198, 2012.

quantity of resource i, c_i is the profit associated with object j, and x_j is either 0, implying item j is not selected, or 1 implying item j is selected.

The MKP is clearly NP-hard combinatorial optimization problem and difficult to solve. There have been important advances in the development of exact and approximate algorithms. Exact solution methods can only be used for very small instances, so for real-world problems, researchers have to rely on and resort to approximate or heuristic methods in solving the problem. As for many others combinatorial optimization problems, the MKP has been intensively investigated within metaheuristics such as genetic algorithms (GA), tabu search (TS) and ant colony optimization (ACO) during the last decade. Drexl presented a heuristic based upon simulated annealing [4]. Computational results revealed that the optimal solution was found for most of the test problems. Dammeyer and Voss [5] presented a tabu search heuristic based on reverse elimination and compared static and dynamic strategies for managing the tabu list. In particular, they showed another tabu list dynamic management, called Reactive Tabu Search, has been tested with satisfactory performances by Battiti and Tecchiolli [6]. Thiel and Voss [7] showed that a standard GA using a direct search in the complete search space is not capable of obtaining good solutions for the MKP, except for small problems. Moreover, they suggested a hybrid algorithm based on combination of GA with tabu search and obtained promising results when applied to test problems of moderate size. Chu and Beasley [8] gave the first successful implementation of GAs by restricting the genetic algorithms to search only the feasible search space.

Ant colony optimization (ACO) simulates the behavior of ant colonies in nature as they forage for food and find the most efficient routes from their nests to food sources. The first ACO algorithm was developed by Dorigo, Maniezzo and Colorni [9] and successfully applied to the traveling salesman problem (TSP) based on the path-finding abilities of real ants. Many researchers have proposed new methods to improve the original ACO and applied them successfully to a whole range of different problems. It is a trend to combine ACO with other algorithms to solve combinatorial optimization problems.

To solve the MKP with ACO, the key point is to decide which items should be selected, that is, how to select the best subset of items such that the total profit is maximized and all resource constraints are satisfied. MKP is quite different from ordering problems. In the ordering problems, an ant selects the next item not only depending on the next item pheromone trail, but also consider the current item. On the other hand, in the MKP problem the selection processes of the next item only involves the next item pheromone trail, and we are not interested in solutions giving a particular order. Moreover, in the ordering problem, the pheromone trail is laid on paths while in the MKP no path exists connecting the items, and the pheromone trail is laid on items [10]. The distinctive features of this paper differs from all the papers described above on following points: First, the main idea of this paper is presented a new hybrid algorithm which is to hybridize the solution construction mechanism of the ACO with scatter search (SS) for solving the MKP. Scatter search (SS), first introduced by Glover in 1977 [11], is a population-based meta-heuristic that combines solutions from a reference set to create improved solutions. Within the scatter search framework, after 2-solution combination method for the reference set has been applied, we employ ACO method to generate new solutions through updating the common item pheromone mechanism. Moreover, since some of the parameters in our approach can be adjusted dynamically, the hybrid algorithm will explore different parts of the solution space, and the method search is not trapped at the local optimum. Finally, the experimental results have shown that the hybrid algorithm is to be very efficient and competitive in terms of solution quality.

The paper is organized as follows. In Section 2, we review the basics of ant colony optimization and scatter search algorithms. In Section 3, we describe hybrid algorithm to tackle the multidimensional knapsack problem. The computational results are reported in Section 4. Finally, the conclusion is given in Section 5.

2   Ant Colony Optimization and Scatter Search

The Ant colony optimization (ACO) simulates the behavior of ant colonies in nature as they forage for food and find the most efficient routes from their nests to food sources. As some ants travel, they deposit a constant amount of pheromone trail that other ants are attracted to follow them. The increase in pheromone increases the probability of the next ants selecting the path. Over time, as more ants are able to complete the shorter route, pheromone accumulates faster on shorter paths and longer paths are less reinforced. Ants are capable of not only finding the shortest path from a food source to the nest, but also adapting to changes in the environment once the old one is no longer feasible due to a new obstacle. This natural behavior of ants can be used to explain reason that they can find the shortest path. Bell [12] applies ant colony optimization to an established set of vehicle routing problems. The ACO method includes the two basic steps of construction of solutions and pheromone updating.

Scatter search (SS) is a novel evolutionary method. It is a population-based meta-heuristic that combines solutions from a reference set to create improved solutions. The reference set is a set of feasible solutions and is usually updated by combining these existing solutions to obtain new ones. Scatter search embodies principles and strategies that are still not emulated by other evolutionary methods, and this method has been applied successfully to many combinatorial optimization problems. Russel and Chiang [13] applied scatter search to the standard vehicle routing problem and combined solutions from the reference set through a common arc mechanism. In more recent work, Marti and Laguna [14] provide detailed descriptions and an implementation framework of the scatter search approach.

3   Proposed Hybrid Algorithm

In this section we describe our solution approaches to the multidimensional knapsack problem. Because the multidimensional knapsack problem is extremely complex, this makes the solution methodology more difficult. This stimulates us to develop metaheuristic methods to obtain near optimal solutions to the problem. Ant colony optimization (ACO) and scatter search (SS) are effective to solve the complicated problem. The SS method is not restricted to a single uniform design, and it is very flexible and effective, since each of its elements can be implemented in variety of ways. According to the strength and weakness of ACO and SS, we propose a hybrid algorithm (ACO&SS), which is to incorporate the ant colony optimization (ACO) and threshold accepting into scatter search (SS) framework to solve the multidimensional knapsack problem. The framework of our proposed hybrid algorithm (ACO&SS) consists of the following sections.

3.1 Solution Construction

To solve the MKP with ACO, the key point is to decide which items should be selected, and how to exploit these items when constructing new solutions. At each step of the construction of a solution, an artificial ant k randomly selects the next item j with respect to a probability p_j^k, until a satisfactory solution has been found. The ant k selects the next item j at the current time t based on the following transition rule:

$$j = \begin{cases} \underset{l \in S^k(t)}{\text{argmax}} \{[\tau_l(t)]^\alpha [\eta_l(S^k(t))]^\beta\}, & \text{if } q \le q_0 \\ M, & \text{otherwise} \end{cases} \tag{4}$$

$$P_j^k(t) = \begin{cases} \dfrac{[\tau_j(t)]^\alpha [\eta_j(S^k(t))]^\beta}{\sum_{u \in S^k(t)} [\tau_u(t)]^\alpha [\eta_u(S^k(t))]^\beta}, & j \notin S^k(t) \\ 0, & \text{otherwise} \end{cases} \tag{5}$$

where $\tau_j(t)$ is equal to the amount of pheromone trail on item j, α and β are the weight parameters. The value q is a random uniform variable in $[0,1]$, and the value q_0 ($0 \le q_0 \le 1$) is a parameter which determines the relative importance of exploitation versus exploration. If $q \le q_0$ then the best item is selected (exploitation) depending on Equation (4); otherwise an item is selected according to M (biased exploration). M represents a random variable selected according to the probabilistic rule as in Equation (5). In other words, M gives the probability with which ant k chooses to move to item j. Items already selected by an ant k are stored in the ants working memory $S^k(t)$ and are not considered for selection at time t. Let $\gamma_i^k(t) = b_i - \sum_{l \in S^k(t)} a_{ij}$ be the remaining amount to reach the boundary of the constraint i, where $\sum_{l \in S^k(t)} a_{ij}$ is the amount of resources i consumed at the time t with respect to the solution being built by ant k. The visibility $\eta_j(S_k(t))$ is defined as follows:

$$\eta_j(S^k(t)) = \dfrac{c_j}{\sum_{i=1}^m a_{ij} / \gamma_i^k(t)} \tag{6}$$

The solution construction process in ACO algorithms is equivalent to a greedy heuristic, except that the choice of the next solution component at each step is done probabilistically instead of deterministically.

The pheromone trail is updated both locally and globally. Local updating is performed during solutions construction while global updating is performed at the end of the constructive phase.

3.2   A Reference Set Generation Method

As for MKP, in our representation, let a bit string $S = \{x_1, x_2, \cdots, x_n\}$, $x_j \in \{0,1\}$ represent a potential solution, where n is the number of items in the MKP, $x_j = 0$ means that object j is not selected, while $x_j = 1$ means that the object is selected. The reference set *RefSet* is a set of feasible solutions, in order to achieve sufficient diversification, so we use two different methods to generate the initial solution set. The first one is ACO algorithm, and the second one is a greedy heuristic. The heuristic is similar to the nearest neighbor heuristic for determining the solutions, and the feasible solutions was constructed by the heuristic that randomly selects a variable and sets it to one. These procedures continue until the initial solution set is generated. The reference set consists of a total of b_1 high quality solutions, $RefSet_1 = \{s_1, s_2, \cdots, s_{b_1}\}$, and b_2 diverse solutions, $RefSet_2 = \{s_{b_1+1}, s_{b_1+2}, \cdots, s_b\}$ chosen from the initial solution set. Scatter search does not allow duplications in the reference set. The initialization reference set phase guarantees that various regions of the solution space will be explored.

3.3   A Solution Combination Method

New solutions are generated by a solution combination method from in the reference set. We adopt 2-solution combination method to generated new solutions. Let the reference set $RefSet = \{s_1, s_2, s_3, \ldots, s_{|RefSet|}\}$. 2-solution combination subsets consist of $\{s_1, s_2\}$, $\{s_1, s_3\}$,..., $\{s_2, s_3\}$,..., $\{s_{|RefSet|-1}, s_{|RefSet|}\}$. The total number of subset solutions generated is equal to $\binom{|RefSet|}{2}$. The combination method starts by matching the items from one solution to the items of the other solution. The core behind item matching is to find the common items of two solutions. We start from the principle that, if a solution includes common items that belong to a better solution, then the objective function value is probably better than that of a solution that does not contain such items. Since the best feasible solution can accumulate higher proportion of pheromone, the probability of ants following these more profitable items would be higher than that of those following the less profitable ones.

In the hybrid algorithm procedure, the ACO algorithm is embedded into the scatter search framework to generate new solutions by a solution combination method. The procedure starts with the generation of the initial solution set. Choose the b_1 high quality solutions and b_2 diverse solutions from the initial solution set. We adopt 2-solution combination method to generate the combination subsets, and each combination subset can generate a new solution with ACO according to common item pheromone values. If none of the solutions in the reference set is replaced by the new solutions obtained using 2-solution combination method, the procedure automatically chooses next subset to generate new solutions. The improvement procedure can be executed only if higher quality solutions are obtained. Compute the objective function values and diverse values of new solutions. If there exist high quality solutions or diverse solutions, update the reference set. The procedure is repeated until a terminating criterion is met.

4 Computational Results

In this section, we present computational results of our proposed algorithm, which was coded in the visual C++ and executed on an IBM computer with 512MB RAM and 1600Mhz CPU Speed. To evaluate validity of our proposed algorithm for the multidimensional knapsack problem, the performance of our algorithms was tested on benchmark problems which can be downloaded from the OR Library at the website with http://mscmga.ms.ic.ac.uk/jeb/orlib/mknapinfo.html. For Chu and Beasley instances of the problem, the number of constraints m was set to 5, 10 and 30, and the number of variables n variables to 100 and 250. Each set of benchmark instances contains 30 instances that are divided into three series with the tightness ratio α =0.25, α =0.5 and α =0.75. For these test problems, the best known solutions for benchmark instances 5.100 have been proven optimal and the instances 5.100 contain five constraints and 100 objects. We select one instance from each series as the test instances for parameter setting experiment. The selected instances are 5.100.01, 5.100.11, and 5.100.21. Therefore, most of the parameter values have been determined on these three instances by the numerical experiments. The experiments have been done with the following parameter settings: $q_0 \in$ [0.40,0.80], $\beta \in \{3,4,5,6\}$ and $\rho \in [0.04,0.06]$. Other settings were: $|RefSet|=10$, $\tau_0 = 0.01$ and maximum iterations $N_{max}=2000$.

We compare the performance of hybrid algorithm with a number of the better methods available for the MKP by means of average gaps, and some problems results

Table 1. Performance comparison of the hybrid algorithm with other heuristics

Problem			Average % Gap				
m	n	α	M&O	V&Z	MKHEUR	GA	ACO&SS
5	100	0.25	13.69	10.30	1.59	0.99	0.24
		0.50	6.71	6.90	0.77	0.45	0.21
		0.75	5.11	5.68	0.48	0.32	0.19
5	250	0.25	6.64	5.85	0.53	0.23	0.19
		0.50	5.22	4.4	0.24	0.12	0.11
		0.75	3.56	3.59	0.16	0.08	0.06
10	100	0.25	15.88	15.55	3.43	1.56	0.71
		0.50	10.41	10.72	1.84	0.79	0.68
		0.75	6.07	5.67	1.06	0.48	0.35
30	100	0.25	17.39	17.21	9.02	2.91	1.36
		0.50	11.82	10.19	3.51	1.34	0.46
		0.75	6.58	5.92	2.03	0.83	0.27
	Average		9.09	8.49	2.02	0.84	0.40

are described in Table 1, where M&O refers to the heuristic of Magazine and Oguz [15], V&Z to the heuristic of Volgenant and Zoon [16], MKHEUR to the heuristic by Pirkul [17], GA to genetic algorithm by Chu and Beasley [8], and ACO&SS is the hybrid algorithm we proposed. On these test instances, the ACO&SS algorithm has shown to be more competitive than the other heuristics in terms of solution quality. During this experiment, the ACO&SS algorithm generates solutions that on average have much smaller gaps than the other heuristics in all case, and an average gap is less than 0.40%.

5   Conclusions

In this paper, we propose a novel hybrid algorithm to solve the multidimensional knapsack problem. The main feature of this hybrid algorithm is to hybridize the solution construction mechanism of the ACO with scatter search (SS). The experimental results have shown that the hybrid algorithm produces uniformly higher performance solutions relative to the other competing heuristics on the MKP. The hybrid algorithm is not restricted to a single uniform design, and it is very flexible and effective, since each of its elements can be modified easily according to the practice situation so that the method is convenient enough to be applied to other subset problem.

Acknowledgments. This research is supported by Scientific Research Foundation of Liaoning Educational Committee (Grant No. L2010196).

References

1. Gilmore, P.C., Gomory, R.E.: The theory and computation of knapsack functions. Operations Research 14, 1045–1075 (1966)
2. Beaujon, G.J., Martin, S.P., McDonald, C.C.: Balancing and optimizing a portfolio of R&D projects. Naval Research Logistics 4818–4840 (2001)
3. Gavish, B., Pirkul, H.: Allocation of databases and processors in a distributed data processing. In: Akola, J. (ed.) Management of Distributed Data Processing, pp. 215–231. North-Holland, Amsterdam (1982)
4. Drexl, A.: A simulated annealing approach to the multiconstraint zero–one knapsack problem. Computing 40, 1–8 (1988)
5. Dammeyer, F., Voss, S.: Application of tabu search strategies for solving multiconstraint zero–one knapsack problems. In: Working Paper, Technische Hochschule Darmstadt, Germany (1991)
6. Battiti, R., Tecchiolli, G.: The reactive tabu search. ORSA Journal on Computing 6, 126–140 (1994)
7. Thiel, J., Voss, S.: Some experiences on solving multiconstraint zero–one knapsack problems with genetic algorithms. INFOR 32, 226–242 (1994)
8. Chu, P., Beasley, J.: A genetic algorithm for the multidimensional knapsack problem. Journal of Heuristics 4, 63–86 (1998)
9. Dorigo, M., Gambardella, L.M.: Ant colonies for the traveling salesman problem. BioSystems 43, 73–81 (1997)

10. Leguizamon, G., Michalewicz, Z.: A new version of ant system for subset problem. In: Congress on Evolutionary Computation, pp. 1456–1464 (1999)
11. Glover, F.: Heuristics for integer programming using surrogate constraints. Decision Sciences, 8156–8166 (1977)
12. Bell, J.E., McMullen, P.R.: Ant colony optimization techniques for the vehicle routing problem. Advanced Engineering Informatics 18, 41–48 (2004)
13. Russell, R.A., Chiang, W.C.: Scatter search for the vehicle routing problem with time windows. European Journal of Operational Research 169, 606–622 (2006)
14. Marti, M., Laguna, M., Glover, F.: Principles of scatter search. European Journal of Operational Research 169, 359–372 (2006)
15. Magazine, M.J., Oguz, O.: A heuristic algorithm for the multidimensional zero-one Knapsack Problem. European Journal of Operational Research 16, 319–326 (1984)
16. Volgenant, A., Zoon, J.A.: An improved heuristic for multidimensional 0-1 knapsack problems. Journal of the Operational Research Society 41, 963–970 (1990)
17. Pirkul, H.: A heuristic solution procedure for the multiconstraint zero-one knapsack problem. Naval Research Logistics 34, 161–172 (1987)

Multi-population Cooperative Cultural Algorithms

Yi-nan Guo[1], Dandan Liu[1], and Jian Cheng[1,2]

[1] School of Information and Electrical Engineering, China University of Mining and
Technology, Xuzhou, Jiangsu 221116, China
[2] Department of Automation, Tsinghua University, Beijing 100084, China
nanfly@126.com

Abstract. Based on the dual structure of culture algorithm, a multi-population cooperative cultural algorithm is proposed by embedding the competition cooperative genetic algorithm into the population space of culture algorithm. In each sub-population, genetic algorithm is adopted. And its population size is adjusted in terms of population density so as to enhance the diversity. In belief space, each kind of the knowledge extracted from best individual of all sub-population is utilized to induce each sub-population's mutation operator. Simulation results indicate that this algorithm can effectively speed up the convergence and improve the accuracy and stability of the solutions.

Keywords: Competition co-evolution genetic algorithm, multi-population, cultural algorithm.

1 Introduction

Genetic algorithm is a stochastic global optimization method based on the simulation of nature evolution mechanism. It is easy to fall into premature mature and has slow convergence speed[1]. Aiming at above problems, many improved genetic algorithms are proposed. Although their algorithms' performances are improved, they are not taken the interdependent relationships between the organism and the environment into account[2]. Ehrilich proposed a co-evolution theory [3] based on the phenomena of species' co-evolution mechanisms in nature. Not only the interdependent relationships among the species, but also the relationships between the species and environment are considered in co-evolution theory. According to the adapted biological model, co-evolution algorithms can be classified into competitive co-evolution, predator-prey co-evolution, symbiotic co-evolution, cooperative co-evolution, etc[4]. By combing the co-evolution theory and the biological population density concept, Cao[5] proposed a co-evolutionary algorithm based on the ecological population competition model. In this model, the interaction among species is taken into account. Population size is adjusted according to the ecological population density. Compared with the evolution model without the interaction, this model can maintain the diversity of population better. However, pure co-evolution model without knowledge induction is easy to trend to balance so as to form weakening collaboration problem[6].

By simulating the cultural evolution mechanism in human society, cultural algorithm adapting a dual evolution structure [7] is proposed. In population space, various intelligent optimization algorithms are employed to realize population

D.-S. Huang et al. (Eds.): ICIC 2011, LNBI 6840, pp. 199–206, 2012.

evolution. The individuals with better fitness values are selected as samples by acceptance function. In belief space, implicit valuable information in the evolution is extracted from samples and described as knowledge. Then these knowledge are used to induce the evolution operators in population space. So this dual evolution structure can fully utilize the evolution information to improve the evolution efficiency.

Combined cultural algorithm with the competition co-evolution model, a multi-population cooperative cultural algorithm (MCCA) is proposed in this paper. In population space, the population is divided into several sub-populations. In each sub-population, competition co-evolution genetic algorithm is adopted. Sub-population size is adaptively adjusted in terms of population density so as to improve the diversity. In belief space, two kinds of knowledge are used. The implicit knowledge is extracted from the sample database which is composed of the dominant individuals in each sub-population. Then the knowledge is used to induce the sub-populations' evolution, which fully plays the guidance role of each kind of knowledge.

2   Multi-population Cooperative Cultural Algorithm

The MCCA is composes of belief space, population space and the interface functions. In population space, each sub-population independently evolves by adapting the crossover, mutation and selection operators in genetic algorithm. Sub-population size is adjusted according to population density. In belief space, implicit knowledge is extracted from the better individuals of all sub-populations and then used to induce the sub-populations' evolution.

2.1   Description and Update of Knowledge in Belief Space

Along with the development of cultural algorithms, five kinds of knowledge including normative knowledge, topographic knowledge, domain knowledge and history knowledge[7] have been considered. In this paper, normative knowledge and situational knowledge are adapted.

Normative Knowledge. Normative knowledge memorizes the feasible search space of the problems. It notes a set of variable extent and is used to avoid searching the null space. Suppose u_j and l_j are the upper and lower bounds of j-th variable. f_j^U and f_j^L denote the corresponding fitness value. Here, n is the variable dimension.

$$K1 = \langle L(t), U(t), F^L(t), F^U(t) \rangle \tag{1}$$

$$L(t) = \{l_1(t), l_2(t), \cdots, l_n(t)\}; \quad U(t) = \{u_1(t), u_2(t), \cdots, u_n(t)\} \tag{2}$$

Let $size(I_j(t))$ be the adjustable interval of j-th variable.

$$size(I_j(t)) = u_j(t) - l_j(t) \tag{3}$$

Along with the evolution process, the search space focuses on the advantageous area. So when the dominant individuals are beyond current search space, normative knowledge is updated.

$$l_j(t+1) = \begin{cases} x_{ij}(t) & if \quad (x_{ij}(t) < l_j(t)) or (f(x_i(t)) < F^L(t)) \\ l_j(t) & else \end{cases} \tag{4}$$

$$F^L(t+1) = \begin{cases} f(x_i(t)) & if \quad (x_{ij}(t) < l_j(t)) or (f(x_i(t)) < F^L(t)) \\ F^L(t) & else \end{cases} \tag{5}$$

$$u_j(t+1) = \begin{cases} x_{ij}(t) & if \quad (x_{ij}(t) > u_j(t)) or (f(x_i(t)) < F^U(t)) \\ u_j(t) & else \end{cases} \tag{6}$$

$$F^U(t+1) = \begin{cases} f(x_i(t)) & if \quad (x_{ij}(t) > u_j(t)) or (f(x_i(t)) < F^U(t)) \\ F^U(t) & else \end{cases} \tag{7}$$

Normative knowledge is used to determine the feasibility of offspring individuals and ensures that the evolution process takes place in advantageous area.

Situational Knowledge. Situational knowledge is used to record the dominant individuals in evolution, as shown in formula (8).

$$K_3 = < E_1, E_2, \cdots E_s > \tag{8}$$

Here, s is the capability of situational knowledge. $E_i = \{x_i | f(x_i)\}$ is the i-th dominant individual in sample database. The dominant individuals recorded by situational knowledge are sorted in descending order according to their fitness values. The dominant individuals are selected from the population by acceptance function as the samples. Then situational knowledge is updated based on these samples.

$$< E_1(t+1), E_2(t+1), \ldots\ldots, E_i(t+1) > = \begin{cases} < x^{Best}(t), E_1(t), \ldots\ldots, E_i(t) > & if \ f(x^{Best}(t)) > f(E_i(t)) and(l < s) \\ < x^{Best}(t), E_1(t), \ldots\ldots, E_{i-1}(t) > & if \ f(x^{Best}(t)) > f(E_i(t)) and(l = s) \\ < E_1(t), \ldots\ldots, E_i(t) > & else \end{cases} \tag{9}$$

Here, $x^{Best}(t)$ is the best individual in t-th iteration in population space.

2.2   Knowledge-Inducing Evolution in Population Space

In MCCA, the evolution operation of genetic algorithm is done in the sub-populations. Then the co-evolution process is done among sub-populations. Here, the density of population is changed according to competition equation. And then sub-population size is adjusted by the density so as to change the proportion of each sub-population.

Knowledge-Inducing Genetic Algorithm. Two kinds of knowledge influence the mutation operator. They are respectively used to adjust the change step size and the direction of the variables[9].

$$\overline{x}_{ij}(t) = \begin{cases} x_{ij}(t) + |size(I_j) \cdot N(0,1)| & x_{ij}(t) < E_i(t) \\ x_{ij}(t) - |size(I_j) \cdot N(0,1)| & x_{ij}(t) > E_i(t) \\ x_{ij}(t) + \lambda \cdot size(I_j) \cdot N(0,1) & x_{ij}(t) = E_i(t) \end{cases} \tag{10}$$

Here, $E_i(t)$ is the best individual recorded by situational knowledge. $N(0,1)$ is a random number satisfying uniform distribution. $\lambda = 0.1$ is the constriction factor of step size. In order to limit each variable of individual into the variable extent, the following mutation operator is done. Suppose γ is a random number.

$$\bar{x}_{ij}(t) = \begin{cases} l_j(0) + \gamma \cdot (u_j(0) - l_j(0)) & \text{if } \bar{x}_{ij}(t) < l_j(0) \text{ or } \bar{x}_{ij}(t) > u_j(0) \\ \bar{x}_{ij}(t) & \text{esle} \end{cases} \tag{11}$$

Knowledge-Inducing Competition Co-evolution Model. In biologic evolution process, both the individuals' fitness and the competition among the populations and environment influence the population's evolution. In the paper, Lotka-Volterra[9] competition equation describing the dynamic relationship of competition between environment and populations is adapted. Suppose $\frac{dN^i(t)}{dt}$ is the density increment of sub-population P^i. K^i denotes the environment load of P^i without competition. Let q^i is the maximum increment of individual in P^i. a_j^i is a competitive coefficient, which represents the inhibiting effect from each individual in P^i.

$$N^i(t+1) = N^i(t) + \frac{dN^i(t)}{dt} \tag{12}$$

$$\frac{dN^i(t)}{dt} = q^i N^i(t) \left(\frac{K^i - N^i(t) - \left(\sum_{j=1}^{|P^i|} a_j^i N^j(t) \right)}{K^i} \right) \tag{13}$$

If the density of a sub-population increases, its competition ability is improved. So the dominant individuals shall be added so as to extend population size. It improves the diversity of sub-populations. Moreover, if the density of a population decreases, its competition ability becomes worse. And then the individuals with less competition ability are eliminated so as to reduce sub-population size. This will improve its whole competition ability. There are two instances in the adjustment of sub-population size:

1) When $\frac{dN^i(t)}{dt} > 0$, the competition ability of P^i is good. So there are $\frac{dN^i(t)}{dt}$ individuals added to P^i. Here, the new individuals are generated based on normative knowledge. Let $\sigma \sim U(0,1)$ be the random number meeting the uniform distribution.

$$x_{ij}(t) = l_j(t) + \sigma \cdot (u_j(t) - l_j(t)) \tag{14}$$

2) When $\frac{dN^i(t)}{dt} < 0$, the competition ability of P^i is weak. So sub-population size is decreased so as to improve its competition ability. At this time, $\frac{dN^i(t)}{dt}$ individuals are eliminated from P^i.

2.3   Interface Functions

The interface functions include acceptance function and influence function.

Acceptance Function. The dominant individuals are chosen from sub-populations as samples. They are used to update each kind of knowledge. The number of selected dominant individuals is a key problem. Here, the fixed selection proportion is adapted. That is, the proportion of samples extracted from sub-population is constant. Suppose population size is m, $k_d m$ samples are selected from all sub-populations in decreasing order in terms of their fitness values every time.

Influence Function. Above-mentioned knowledge is used to induce the evolution. Different knowledge records different evolution information and influences the evolution in their own ways. In the competition co-evolution process of population space, normative knowledge and situational knowledge are used to induce the evolution every τ generation. So τ is defined as the knowledge-inducing interval, which directly influences the competition degree among sub-populations.

3   Simulation Results and Analysis

Benchmark functions shown in Tab.1 are taken as examples to validate the rationality of MCCA. The algorithm's performance with different parameters are compared and analyzed. And it is compared with traditional genetic algorithm (GA) and cooperative genetic algorithm (CGA) in section 3.2. Here, main parameters' values are: Population size=60, Pc=0.95, Pm=0.3, Sub-population number=3, Sample-population size=20, Selection proportion=0.2, Termination iteration=1000, Run times=20.

Table 1. Benchmark functions

Function	Variables bounds	Variable dimension	Optimal value
$\min f_1(x) = 100(x_1^2 - x_2) + (x_1 - 1)^2$	[-5.12,5.12]	2	0
$\min f_2(x) = \sum_{i=1}^{n} x_i^2$	[-10,10]	10	0
$\min f_3(x) = 0.5 + \dfrac{(\sin(\sqrt{x_1^2 + x_2^2}))^2 - 0.5}{(1.0 + 0.001(x_1^2 + x_2^2))^2}$	[-100,100]	2	0
$\min f_4(x) = \sum_{i=1}^{n} \left[x_i^2 - 10\cos(2\pi x_i) + 10 \right]$	[-10,10]	10	0
$\min f_5(x) = \dfrac{1}{4000} \sum_{i=1}^{n} x_i^2 - \prod_{i=1}^{n} \cos\left(\dfrac{x_i}{\sqrt{i}}\right)^2 + 1$	[-600,600]	10	0

3.1   Performances Analysis about the MCCA

The parameters about the extraction and utilization of knowledge will influence the algorithm performances.

Sample-Selection Proportion. The sample-selection proportion, expressed by β, denotes how many dominant individuals are selected from the population. It has a direct impact on the update speed of the samples. Here, the algorithm performances with different β are compared, as shown in Tab.2. M1 is average optimal solution. M2 is the mean square error of optimal value. M3 is average convergence generation.

Table 2. Comparison of the performance with different sample-selection proportion

Function		f_1	f_2	f_3	f_4	f_5
	$M1$	1.09×10^{-2}	2.42×10^{-2}	2.80×10^{-3}	3.25	1.46×10^{-2}
$\beta = 0.1$	$M2$	7.57×10^{-3}	4.97×10^{-2}	1.69×10^{-3}	2.60	6.62×10^{-2}
	$M3$	88	230	60	585	236
	$M1$	9.30×10^{-3}	2.29×10^{-2}	2.16×10^{-3}	1.98	1.09×10^{-2}
$\beta = 0.2$	$M2$	6.91×10^{-3}	1.51×10^{-2}	1.53×10^{-3}	1.18	6.20×10^{-3}
	$M3$	72	228	57	503	218
	$M1$	1.06×10^{-2}	3.49×10^{-2}	3.17×10^{-3}	2.18	1.28×10^{-2}
$\beta = 0.4$	$M2$	7.70×10^{-3}	2.15×10^{-2}	2.14×10^{-3}	1.91	8.13×10^{-3}
	$M3$	78	169	50	485	195

Along with the increasing of sample-selection proportion, the convergence speed is faster. However, when β is too large or too small, the accuracy and stability of solutions are worse. The reason for above phenomena is that if β is smaller, the sample database updates slowly. As the knowledge is mainly depended on the former samples, it is easy to induce the evolution to the local convergence and influence the solution accuracy. Moreover, larger β can speed up the update of knowledge. It is good for improving convergence speed. But the algorithm's complexity is increasing and the accuracy of solution is worse. So, $\beta = 0.2$ in the paper.

Knowledge-inducing interval. Knowledge-inducing interval, expressed by τ, is defined as the number of the generation between two knowledge-inducing evolution operations. It directly influences the competition degree among sub-populations and the influence degree of knowledge. Simulation results with different τ are shown in Tab.3. Obviously, suitable knowledge-inducing interval can improve the algorithm performance. By analyzing the results, we know that if τ is smaller, the knowledge can induce the individuals to evolve so as to improve the algorithm performance. However, if knowledge is frequently exchanged, the algorithm's complexity is significantly increased. When τ is larger, the knowledge has little impact on the individuals and can't make the population escape from local solutions in time, which results in slower convergence speed. So, $\tau = 5$ in this paper.

Table 3. Comparison of the performance with different knowledge-inducing iteration

Function		f_1	f_2	f_3	f_4	f_5
	$M1$	4.46×10^{-2}	2.59×10^{-2}	2.62×10^{-3}	1.96	1.09×10^{-2}
$\tau = 2$	$M2$	1.45×10^{-1}	1.35×10^{-2}	2.30×10^{-3}	1.17	4.54×10^{-3}
	$M3$	74	203	56	485	189
	$M1$	9.30×10^{-3}	2.29×10^{-2}	2.16×10^{-3}	1.98	1.09×10^{-2}
$\tau = 5$	$M2$	6.91×10^{-3}	1.51×10^{-2}	1.53×10^{-3}	1.18	6.20×10^{-3}
	$M3$	72	228	58	503	218
	$M1$	8.46×10^{-2}	2.60×10^{-2}	2.68×10^{-3}	2.03	1.11×10^{-2}
$\tau = 8$	$M2$	7.12×10^{-3}	1.52×10^{-2}	2.51×10^{-3}	1.23	5.50×10^{-3}
	$M3$	89	182	59	571	273

3.2 Comparison of the Performances among Different Methods

In order to validate the rationality of MCCA, it is compared with GA and CGA, as shown in Tab.4. In GA, stochastic tournament selection combing with elite strategy is adapted. Single crossover and Gaussian mutation operator are adapted.

It's obvious that MCCA has better solution accuracy, stability and less convergence generation compared with SGA and CGA. The reason is that MCCA adapts the dual structure in cultural algorithm which can make the belief space and the population space influences each other. Moreover, the competition co-evolution model based on population density is introduced into population space. It considers the competition relationship among sub-populations. This improves the population diversity. In a word, MCCA can improve the convergence speed and solution accuracy.

Table 4. Comparison of the performances among different methods

Function		f_1	f_2	f_3	f_4	f_5
	$M1$	5.13×10^{-2}	8.41×10^{-2}	5.62×10^{-3}	7.46	7.04×10^{-1}
GA	$M2$	1.78×10^{-1}	8.36×10^{-2}	8.13×10^{-2}	3.93	3.01×10^{-1}
	$M3$	253	496	225	968	762
	$M1$	1.06×10^{-2}	1.92×10^{-2}	3.00×10^{-3}	1.22×10^{1}	3.28×10^{-2}
CGA	$M2$	7.30×10^{-3}	1.45×10^{-1}	2.07×10^{-3}	4.08	1.86×10^{-2}
	$M3$	99	297	81	629	298
	$M1$	1.04×10^{-2}	2.78×10^{-2}	1.96×10^{-3}	1.72	1.21×10^{-2}
MCCA	$M2$	6.37×10^{-3}	1.13×10^{-2}	1.54×10^{-3}	1.12	6.00×10^{-3}
	$M3$	61	185	64	453	139

4   Conclusion

Inspired by the cultural evolution mechanism and the co-evolution idea, a multi-population cooperative cultural algorithm is proposed. In population space, competition co-evolution genetic algorithm based on population density is introduced. It considers the competition relationship among sub-populations. Sub-population size is adaptively adjusted by their density so as to improve the diversity. In belief space, implicit knowledge is extracted from the sample database which is composed of the dominant individuals in sub-populations. Then the knowledge is utilized to influence the mutation operator in sub-populations' evolution, which fully plays the guidance role of each kind of knowledge. Simulation results show that MCCA can utilize the evolution knowledge to improve the convergence speed and solution accuracy.

Acknowledgment. This work was supported by National Natural Science Foundation of China under Grant 60805025, Natural Science Foundation of Jiangsu under Grant BK2010183, the China Postdoctoral Science Foundation Funded Project under Grant 20090460328 and Qinglan Project of Jiangsu.

References

1. Rudolph, G.: Convergence analysis of canonical genetic algorithms. IEEE Transactions on Neural Networks 5, 96–101 (1994)
2. Gong, D.W., Sun, X.Y.: Co-evolutionary Genetic Algorithms Theory and Applications. Science Press, BeiJing (2009)
3. Ehrlich, P.R., Raven, P.H.: A Study in Coevolution Evolution. Butterflies and Plants 18, 586–608 (1965)
4. Jiao, L.C., Liu, J., Zhong, W.C.: Cooperation Evolution Algorithm and Multi-agent System. Science Press, Beijing (2006)
5. Cao, X.B., Luo, W.J., Wang, X.F.: A Co-Evolution Pattern Based on Ecological Population Competition Mode. Journal of Software 4, 512–556 (2001)
6. Guo, Y.N., Cheng, J., Cao, Y.Y.: A Novel Multi-population Cultural Algorithms Adopting Knowledge Migration. Soft Computing 15, 897–905 (2011)
7. Peng, B.: Knowledge and Population Swarms in Cultural Algorithms for Dynamic Environments. Wayne State University, USA (2005)
8. Shen, W., Zhao, Z.J., Li, X.G.: On the Convergence of Cultural Algorithm with Different Influence Function. Information and Control 37(5), 604–608 (2008)
9. Coelho, L.D.S., Mariani, V.C.: An Efficient Particle Swarm Optimization Approach Based on Cultural Algorithms Applied to Mechanical Design. In: 2006 IEEE Congress on Evolutionary Computation, pp. 1099–1104 (2006)

A Hybrid Quantum-Inspired Particle Swarm Evolution Algorithm and SQP Method for Large-Scale Economic Dispatch Problems

Qun Niu, Zhuo Zhou, and Tingting Zeng

Shanghai Key Laboratory of Power Station Automation Technology,
School of Mechatronic Engineering and Automation
Shanghai University, 200072 Shanghai, China
comelycc@gmail.com

Abstract. This paper proposes a hybrid method QPSO-SQP, which combines a quantum-inspired particle swarm evolution algorithm(QPSO) and the sequential quadratic programming (SQP) method to solve large-scale economic dispatch problems(EDPs). Due to the combination of quantum rotation gates and the updating mechanism of PSO, the QPSO has strong search ability and fast convergence speed, therefore it is employed as a global searcher to obtain good solutions for EDPs. As SQP is a gradient-based nonlinear programming method, it is used as a local optimizer to fine tune the best result of the QPSO. The proposed QPSO-SQP is applied to two large-scale EDPs to validate its effectiveness. The experiment results show that the proposed QPSO-SQP can obtain high-quality solutions and produce a satisfactory performance among most existing techniques.

Keywords: Economic dispatch problem, large scale, quantum-inspired particle swarm algorithm, sequential quadratic programming.

1 Introduction

Economic dispatch problem (EDP) is an important task in the power plant operation, which aims to allocate power generation to match load demand at minimal possible cost while satisfying all power units and system constraints [1]. Given all unit and system equality and inequality constraints, EDPs tend to be a complex and highly constrained non-linear optimization problem, particularly for large power systems.

In the recent years, evolutionary algorithms(EAs), such as evolution programming (EP) [1], particle swarm optimization(PSO)[2], quantum evolution algorithm (QEA)[3], differential evolution(DE)[4] and ant colony optimization(ACO) [5] have became more and more popular in solving EDPs, due to the fact that they do not require the objection function to be differentiable and continuous. Among these methods, QEA and PSO have recently drawn a lot of attention due to their rapid convergence and accuracy compared with other optimization techniques.

D.-S. Huang et al. (Eds.): ICIC 2011, LNBI 6840, pp. 207–214, 2012.

Quantum computing is a novel class of computing based on the concepts of quantum theory, such as quantum states, entanglement and intervention. In 2002, Han et al. [6] proposed the quantum evolutionary algorithm (QEA), which is inspired by the concept of quantum computing. For the standard quantum mutation, the rotation angle is a series of fixed numbers in the whole process of evolution, so the convergence spend is very slow and is undesirable in maintaining current good solution. Particle swarm optimization (PSO) is one of the modern meta-heuristic algorithms, which is based on the analogy of fish schooling and bird flocking. In PSO, each particle makes decision using its own experience together with its neighbor's experiences. Compared with other evolutionary methods, it has advantages of fast convergence and easy implementation. Although the PSO can converge quickly towards the optimal solution, it experiences difficulties in reaching the global optimum and suffer from premature convergence. Considering the advantages and drawbacks of QEA and PSO, Wang et al. [7] proposed a novel quantum swarm evolutionary algorithm (QPSO) in 2007, which uses PSO update strategy to accelerate the convergence speed of QEA. In recent years, QPSO has been used widely and exhibited significant potential in solving various optimization problems, both in mathematics, physics and computer science fields, however only few studies [8] have considered using QPSO to solve economic dispatch problems.

Further, EDPs with more than 40 generators commonly exist in the real world applications and it is difficult to find the optimal solutions when the number of generators arises. Many previous evolutionary algorithms mainly focused on the exploration of the search space, but failed to obtain high-quality solutions for large-scale EDPs.

In this paper, to overcome the drawbacks of these methods, a hybrid method QPSO-SQP is proposed to solve large-scale EDPs. PSO strategy is embedded into quantum mutation to update angle of rotation to improve the convergence speed and avoid premature convergence, thus the information of the currently best individuals can be fully utilized to direct the next searching step and, at the meantime, due to the use of quantum rotation gate, the QPSO can avoid premature convergence and maintain solution diversity. Further, SQP is employed here as a local optimizer to fine tune the best results obtained by the QPSO, as SQP has powerful capabilities in finding local optimistic result, In this way, this kind of combination can take the advantage of both the fast convergence speed of evolutionary algorithm and strong ability of SQP in local search, therefore allowing to find an optimum solution more efficiently and accurately. Two large-scale EDPs are conducted to validate the performance of the proposed QPSO-SQP and the results show that the QPSO-SQP provides satisfactory results for solving large-scale EDPs.

2   Problem Formulation

The economic dispatch problem is one of the most important problems to be solved in the operation and planning of a power system. The main objective of

EDP is to minimize the total generation costs of a power system, while satisfying the operation constraints.

2.1   Objective Function

The traditional EDP minimizes the following incremental fuel cost function associated to dispatch units:

$$min \ F_T = \sum_{i=1}^{n} F_i(P_i) \tag{1}$$

where F_T is the total generation cost; P_i is the power output of the ith generator; F_i is the total fuel cost of the ith generator, which can be defined as:

$$F_i(P_i) = a_i P_i^2 + b_i P_i + c_i + |e_i sin(f_i(P_i^{min} - P_i))| \tag{2}$$

where a_i, b_i and c_i are the cost coefficients. e_i, f_i are the coefficients of generator i, reflecting the valve-point.

Valve-point is the rippling effect added to the generation unit curve when each steam admission value in a turbine starts to open. This curve poses higher non-linearity and discontinuity, so taking the valve-point effects into account will make the problem more difficult for obtaining the optimum and add the number of local minimum points in the fuel cost function.

2.2   System Constraints

(1)Power Balance

$$\sum_{i=1}^{n} P_i - P_{Loss} - P_D = 0, i = 1, 2, ..., n \tag{3}$$

where P_D is the total demand of power system; P_{Loss} is the total system transmission losses, which can be calculated using the B-matrix loss coefficients.

(2)Power Generation Limit

$$P_i^{min} \leq P_i \leq P_i^{max} \tag{4}$$

where P_i^{min} and P_i^{max} are the minimum and maximum output of generator i, respectively.

(3)Ramp Rate Limits

$$P_i - P_i^0 \leq UR_i \ \ and \ \ P_i^0 - P_i \leq DR_i \tag{5}$$

where UR_i and DR_i are the up and down ramp limits of the ith generator, respectively.

3   Implementation of QPSO-SQP for EDPs

3.1   Quantum Evolution Algorithm and Particle Swarm Optimization

The Quantum Evolution Algorithm (QEA) employs quantum computing principles to strengthen conventional evolutionary algorithms (EAs). QEA uses a new representation which is based on the concept of Q-bits, quantum gates and superposition of states. A Q-bit is defined as a pair of complex numbers, which represent the superposition of basic states ("0" state and "1" state) and it is the smallest unit of information for representing individuals, a Q-bit individual with a string of m Q-bits is defined as follows:

$$\begin{bmatrix} \alpha_i \ \alpha_2 \ ... \ \alpha_m \\ \beta_i \ \beta_2 \ ... \ \beta_m \end{bmatrix} \tag{6}$$

where $|\alpha_i|^2+|\beta_i|^2=1$ for $i = 1, 2, ..., m$, $|\alpha|^2$ and $|\beta|^2$ give the probability that the Q-bit will be found in the'0' state and '1' state. In the QEA, the state of Q-bit individual can be updated by a quantum gate. The rotation gate $U(\Delta\theta_i^t)$ and the update operation are presented as:

$$U(\Delta\theta_i^t) = \begin{bmatrix} cos(\Delta\theta_i^t) \ -sin(\Delta\theta_i^t) \\ sin(\Delta\theta_i^t) \ \ cos(\Delta\theta_i^t) \end{bmatrix}, \begin{bmatrix} \alpha_i^{t+1} \\ \beta_i^{t+1} \end{bmatrix} = U(\Delta\theta_i^t)) \begin{bmatrix} \alpha_i^t \\ \beta_i^t \end{bmatrix} \tag{7}$$

where $U(\Delta\theta_i^t)$ is a rotation angle that determines the direction and magnitude of rotation.

As for PSO, each particle is assumed to have two characteristics: a position and a velocity. For example, in a n-dimensional search space, the position and velocity of individual i are represented as the vectors: $X_i = (X_{i1}, X_{i2},, X_{in})$ and $V_i = (V_{i1}, V_{i2},, V_{in})$. It stores the best memory, which records the best fitness value of each individual and the whole group, represented as $pbest_i = (pbest_{1i}, pbest_{2i},, pbest_{ni})$ and $gbest_i$, respectively.

3.2   Sequential Quadratic Programming (SQP)

The SQP method is one of the most effective nonlinear programming methods for constrained optimization problems. It outperforms other nonlinear programming method in terms of efficiency, accuracy and percentage of successful solutions over a large number of test problems [9]. This method closely resembles Newton's method for constrained optimization, just as is done for unconstrained optimization. As the performance of SQP significantly depends on the initial point, hence in this paper, at the beginning, QPSO is used to solve economic dispatch problem as a global search and then the best solution obtained from QPSO is used as initial point for SQP method to fine tune the optimal solution.

3.3   The Proposed QPSO-SQP and Its Implementation for Large-Scale EDPs

The basic QEA uses Q-bit gate rotation in mutation. In a standard quantum mutation operation, the rotation angle is a series of fixed numbers, independent of whether the individual is better or worse, so the converge speed is very slow and is undesirable for maintaining the current good solutions.

The PSO has been proven to be very effective for optimization problems and it has very fast converging characteristics. However in some cases, it converges prematurely without finding local optimum. The inertia weight always drives all velocities to zero before the particle can reach the local optimum.

To overcome those drawbacks, a hybrid algorithm(QPSO-SQP) is proposed in this paper. The PSO strategy is used to update angle of rotation gate in the process of quantum mutation. By combining QEA and PSO, the information of the currently best individual can be fully utilized to guide the next searching step. Then SQP method is employed here to fine tune the best result obtained by the QPSO. Thus, the SQP method can make use of the information from QPSO as its initial point to explorer a global minimum. The procedures of the proposed algorithm QPSO-SQP can be described as follows:

Step 1: Initialize the population $P_Q(t)$. $P_Q(t)$ is a group of Q-bit individuals, $P_Q(t) = [q_1^t, q_2^t, ..., q_n^t]$, where n is the total number of Q-bit, and the *ith* Q-bit individual q_i^t and *ith* quantum angle θ_i^t at generation t are defined as follows:

$$\begin{cases} q_i^t = \begin{bmatrix} \alpha_i^t \\ \beta_i^t \end{bmatrix} = \begin{bmatrix} cos\theta_i^t \\ sin\theta_i^t \end{bmatrix} \\ \theta_i^t = arctan\frac{\beta_i^t}{\alpha_i^t} \end{cases} \tag{8}$$

Step 2: Decode $X_R(t)$ by observing $P_Q(t)$. $X_R(t)$ is the real parameter, which denotes the actual parameter value of the power system that is given as:

$$X_{Ri}^t = (\alpha_i^t)^2 * (P_i^{max} - P_i^{min}) + P_i^{min} \tag{9}$$

where P_i^{min} and P_i^{max} are the minimum and maximum output of generator i, X_{Ri}^t is the real value output of generator i, and in this way, the probability of quantum states can be converted into real parameters.

Step 3: Evaluate $X_R(t)$, calculate quantum angle θ_i^t ,and then store the best individual quantum angles *(θpbest$_i$)* and best group quantum angle *(θgbest)*.

Step 4: Update quantum angles according to the PSO strategy.

$$\Delta\theta_i^{(t+1)} = w*\Delta\theta_i^{(t)} + c_1*rand()*(\theta pbest_i - \theta_i^{(t)}) + c_2*rand()*(\theta gbest_i) - \theta_i^{(t)}) \tag{10}$$

Step 5: Update quantum bits using the Rotation Gate Strategy.

$$q_i^{t+1} = \begin{bmatrix} \alpha_i^{t+1} \\ \beta_i^{t+1} \end{bmatrix} = \begin{bmatrix} cos(\Delta\theta_i^t) & -sin(\Delta\theta_i^t) \\ sin(\Delta\theta_i^t) & cos(\Delta\theta_i^t) \end{bmatrix} \begin{bmatrix} \alpha_i^t \\ \beta_i^t \end{bmatrix} \tag{11}$$

Then we can get the new Q-bit generation $P_Q(t+1) = [q_1^{t+1}, q_2^{t+1}, ...q_i^{t+1} ..., q_n^{t+1}]$

Step 6: Decode $X_R(t+1)$ by observing $P_Q(t+1)$.

Step 7: Evaluate $X_R(t+1)$ and calculate quantum angle θ_i^{t+1}, and then store the new $\theta pbest_i$ and $\theta gbest$.

Step 8: Check if the stop criterion satisfied. If not, then go to step 4. Else, go to step 9.

Step 9: Using the best result of QPSO as the initial point for SQP and fine tune the result and output the SQP solution.

4   Simulation Results

In this section, the proposed QPSO-SQP is applied to two large-scale economic dispatch problems. To compare the solution quality and convergence characteristics, the population size was set to 50. 50 independent runs were executed and the number of iteration was set to 500 for the algorithm, which computational cost was no higher than most of the compared algorithm. Over a large amount of trials, it can be found that the parameters w had an undesirable effect on the solutions of economic dispatch problems. One possible reason is that the inertia weight drives all quantum angle differences too quickly to the local optimal, which suffers from the prematurity, so in this paper, the values of QPSO-SQP parameters were finally set as follows: $w = 0$, $c1 = 1.3$, $c2 = 1.9$. $c2 > c1$ means the population will convergence faster to the global optimal position than the local positions.

Case I: 40-Generating Units. A large-scale power system of 40-generating units with quadratic cost function and valve-point effects is considered. The total load demand of this test system is 10500MW. The system date can be found from [1]. The results obtained from the QPSO-SQP are compared with other algorithm reported in the literature, including: IFEP[2],PSO-GM[10], CBPSO-RVM[10], APSO[11], PSO[12], PSO-SQP[12],EP-SQP[12], DE[4], SQP[4], DEC-SQP[4], GA-PS-SQP[9], IQPSO[8], ST-HDE[13], ACO[5].

Table 1 summarizes the minimum, average and maximum costs of the mentioned methods. The result of the minimum cost 121466.421$/h obtained by QPSO-SQP is comparatively superior to most of the other methods in the literature and the mean cost of 121983.747$/h outperforms other methods as well. Compared with the basic QPSO without SQP fine tuning, SQP have improved the quality of solutions dramatically from 121517.426$/h to 121466.421$/h for minimum cost, which exhibites an excellent local exploration ability. In the reported literature, four evolutionary algorithms have embedded SQP into their methods to solve EDP with the valve-point effect and they are EP [12], PSO [12], DE [4] and GA-PS-SQP[9] . As is shown in Table 1, the proposed QPSO-SQP outperforms the four methods in terms of the best minimum cost, mean cost and maximum cost. It can be concluded that the QPSO provides admirable initial points for the finial search method of SQP to ensure a better fine tuning.

Table 1. Comparison among different methods for case I (40 generating units)

Methods	Min(*$/h*)	Mean(*$/h*)	Max(*$/h*)	St.d
IFEP[2]	122624.35	123382	125740.63	NR
PSO-GM[10]	121845.99	122398.38	123219.220	258.44
CBPSO-RVM[10]	121555.32	122281.14	123094.98	259.99
APSO(2)[11]	121663.52	122153.67	122912.4	NR
PSO[12]	121735.47	122513.92	123467.41	NR
PSO-SQP[12]	122094.67	122245.5	NR	NR
EP-SQP[12]	122323.97	122379.63	NR	NR
SQP[4]	122904.42	124883.77	126585.23	985.54
DE[4]	121900.88	122385.18	122991.25	414.16
DEC(2)-SQP[4]	121741.98	122295.13	122839.29	386.18
GA-PS-SQP[9]	121458.14	122039	NR	NR
IQPSO[8]	121448.21	122225.07	NR	NR
ST-HDE[13]	121698.51	122304.3	NR	NR
ACO[5]	121532.41	121606.45	121679.64	45.58
QPSO	**121517.426**	**122021.264**	**122644.195**	**307.5371**
QPSO-SQP	**121466.421**	**121983.747**	**122634.946**	**291.6111**

Table 2. Comparison among different methods for case II (110 generating units)

Loading	Methods	Min Cost	Mean Cost	Max Cost
	SA[14]	145550.4412	146757.706	147473.4295
Light	SAF[14]	141107.8541	141215.1159	141398.0923
(10000MW)	RQEA[3]	131941.8851	131942.0439	131942.4931
	QPSO	135089.9676	135628.1058	136320.2948
	QPSO-SQP	**131941.8837**	**131941.8838**	**131941.8838**
	SA[14]	314647.0416	315659.1453	317385.2167
Heavy	SAF[14]	314532.8747	314635.3244	314783.5061
(20000MW)	RQEA[3]	313211.5688	313211.5983	313211.8189
	QPSO	313212.0433	313215.872	313217.0138
	QPSO-SQP	**313211.5688**	**313211.5688**	**313211.5688**

Case II: 110-Generating Units. In this case, a 110-generators large-scale system is tested here. There are two kinds of load demands, light (10000MW) and heavy (20000MW). The results are compared with SA [14], SAF [14], RQEA [3]. Table 2 shows the comparison of results of the mentioned methods, it appears that the proposed QPSO-SQP gives the best performance among all the methods and the results of QPSO-SQP is much better than that of QPSO, which once again indicates that the SQP has strong ability of local search.

5   Conclusion

This paper proposes a hybrid methods QPSO-SQP to solve EDPs. In this work, QPSO is used as a global searcher to give a good direction to the global optimal region and then SQP is employed as a local optimizer to fine tune the best solution obtained from the QPSO. The performance has been compared with other optimization methods. From the experimental results, it can be seen that the QPSO-SQP exhibits excellent performance and can obtain high-quality solutions. Thus this hybrid method provides a new effective way to solve economic dispatch problems.

Acknowledgments. This work is supported by the National Natural Science Foundation of China(grant no.60804052), Chen Guang project supported by Shanghai Municipal Education Commission and Shanghai Education Development Foundation, the Projects of Shanghai Science and Technology Community (10ZR1411800 & 08160512100 & 08DZ2272400), Shanghai University "11th Five-Year Plan" 211 Construction Project.

References

1. Sinha, N., Chakrabarti, R.: Evolutionary programming techniques for economic load dispatch. IEEE Transactions on Evolutionary Computation 7(1), 83–94 (2003)
2. Park, J.B., Lee, K.S.: A particle swarm optimization for economic dispatch with nonsmooth cost functions. IEEE Transactions on Power Systems 20(1), 34–42 (2005)
3. Babu, G.S.S., Das, D.B.: Real-parameter quantum evolutionary algorithm for economic load dispatch. IET Generation Transmission Distribution 2(1), 22–31 (2008)
4. Coelho, L.S., Mariani, V.C.: Combining of chaotic differential evolution and quadratic programming for economic dispatch optimization with valve-point effect. IEEE Transactions on Power Systems 21(2), 989–996 (2006)
5. Pothiya, S., Ngamroo, I., Kongprawechnon, W.: Ant colony optimisation for economic dispatch problem with non-smooth cost functions. International Journal of Electrical Power & Energy Systems 32(5), 478–487 (2010)
6. Han, K.H.: Quantum-inspired evolutionary algorithm for a class of combinatorial optimization. IEEE Transactions on Evolutionary Computation 6, 580–593 (2002)
7. Wang, Y., Feng, X.Y., Huang, Y.X., et al.: A novel quantum swarm evolutionary algorithm and its applications. Neuro Computing 70(4-6), 633–640 (2007)
8. Meng, K., Wang, H.G., Dong, Z.Y., et al.: Quantum-Inspired Particle Swarm Optimization for Valve-Point Economic Load Dispatch. IEEE Transactions on Power Systems 25(1), 215–222 (2010)
9. Alsumait, J.S., Sykulski, J.K.: A hybrid GA-PS-SQP method to solve power system valve-point economic dispatch problems. Applied Energy (87), 1773–1781 (2010)
10. Lu, H.Y., Sriyanyong, P., Song, Y.H., et al.: Experimental study of a new hybrid PSO with mutation for economic dispatch with non-smooth cost function. International Journal of Electrical Power & Eenergy Systems 32(9), 921–935 (2010)
11. Selvakumar, A.I.: Thanushkodi:Anti-predatory particle swarm optimization: solution to nonconvex economic dispatch problems. Elect Power Syst. Res. 78(1), 2–10 (2010)
12. Victoire, T.A.A., Jeyakumar, A.E.: Hybrid PSO-SQP for economic dispatch with valve point effect. Electric Power Systems Research 71(1), 51–59 (2004)
13. Wang, S.K., Chiou, J.P., Liu, C.W.: Non-smooth/non-convex economic dispatch by a novel hybrid differential evolution algorithm. IET Generation Transmission Distribution 1(5), 793–803 (2007)
14. Babu, G.S.S., Das, D.B., Patvardhan, C.: Simulated annealing variants for solution of economic load dispatch. Journal of the Institution of Engineers 83, 222–229 (2002)

Recursive and Incremental Learning GA Featuring Problem-Dependent Rule-Set

Haofan Zhang[1], Lei Fang[2], and Sheng-Uei Guan[1]

[1] Department of Computer Science and Software Engineering,
Xi'an Jiaotong-Liverpool University, Suzhou, China
[2] School of Computer Science, University of St Andrews, St Andrews, UK
haofan.zhang08@student.xjtlu.edu.cn

Abstract. Traditional rule-based classifiers training with Genetic Algorithms have their major weaknesses in the classification accuracy and training time. To resolve these drawbacks, this paper reviews Recursive Learning of Genetic Algorithm with Task Decomposition and Varied Rule Set (RLGA) and proposes its variation that features Incremental Attribute Learning (RLGA-I). Experiments show that both the proposed solutions dramatically reduce the training duration with better generalization accuracy.

Keywords: genetic algorithm, task decomposition, domain decomposition, local fitness, incremental attribute learning.

1 Introduction

In the realm of computer science and engineering, pattern classification problem is one of the most essential topics. The task of pattern classification is to organize patterns into group of classes which share the same set of attributes.

Rule-based genetic algorithms (GAs) is one of the most successful approaches for classification, either supervised or unsupervised [1]. There are two main streams of rule-based GAs classifiers: *Pittsburgh* and *Michigan* [2][3]. In the *Pittsburgh (Pitt)* approach, each chromosome is encoded as a complete set of rules. On the other hand, the *Michigan* approach uses instead the whole population as the solution-set. The *Pitt* approach is applied in this paper.

Traditional GAs classifier generally suffers early convergence which indicates its inefficiency in utilising its genetic resources and therefore hinders its further performance improvement. A common approach to overcome this drawback is to decompose the original task into several subtasks and solve them separately, and then combine the sub-solutions into the final solution.

Other problems about GAs classifier are its weakness in efficiency and accuracy, especially when confronting large dimensional problems. Most GA-based solutions in the literature work in the static mode, where all the parameters (e.g. rule number, chromosome length in the Pitt approach, etc.), are determined before the training process starts. However, learning should be more adaptive and the parameters, especially the rule number, should be determined at run-time as the needs of different problems. It is intuitive to see that a larger set of rules may fit for hard problems, on the other hand, a smaller one for simple problems.

D.-S. Huang et al. (Eds.): ICIC 2011, LNBI 6840, pp. 215–222, 2012.
© Springer-Verlag Berlin Heidelberg 2012

2   Traditional GA-Based Classifier

A typical classifier usually contains GA operators, such as selection, crossover, and mutation. A rule is usually represented as an IF-THEN clause. Such a rule is shown as follows:

$$R_i: \text{IF}(V_{1min} \leq x_1 \leq V_{1max}) \wedge (V_{2min} \leq x_2 \leq V_{2max})... \tag{1}$$
$$\wedge (V_{nmin} \leq x_n \leq V_{nmax}) \text{ THEN } y = C$$

where n is the number of attributes, $(x_1, x_2, ...,x_n)$ is the values of the input pattern's attribute set, y is the output class index assigned to the rule, $V_{j\,min}$ and $V_{j\,max}$ are the boundary values for the j th attribute x_j respectively.

Each chromosome CR_j is composed of a number of rules $R_i(i=1,2,...,m)$ by concatenating them together,

$$CR_j = \bigcup_{i=1,m} R_i \quad j = 1,2,...,p \tag{2}$$

where m denotes the number of rules residing in one chromosome CR_j. In most cases, m is set at 30. Moreover, p denotes the population size, i.e. the number of chromosomes processed by one GA. Therefore, one particular chromosome has length of L, if encoded in the way introduced above.

$$L=m*|R| \tag{3}$$

$|R|$ is the length of one rule. A chromosome can be represented by an array with length L.

2.1   Fitness Function (Global Fitness)

The fitness of one chromosome shows its ability in classifying patterns. It reflects the success rate achieved while the corresponding rule set is used for classification and is used to select good candidates for reproduction. The fitness is the percentage of patterns that can be accurately classified by a chromosome.

$$f = \frac{C}{N} = \frac{\text{number of patterns correctly classified}}{\text{total number of patterns}} \tag{4}$$

Since one chromosome represents one entire rule set, the fitness actually measures the collective performance of all the rules in that chromosome. Voting mechanism is usually used here to resolve conflicts among rules in a chromosome.

3   Recursive GA Learning with Task Decomposition and Varied Rule Number

We first give a brief review of our former proposed algorithm which is a recursive learning genetic algorithm (RLGA) that features automatic task and data decomposition with a problem-dependent rule number [9]. We mainly address on its method of elicitation of local fitness.

3.1   Elicitation of Local Fitness and Local Fitness Algorithm

In rule-based GAs, each rule has its own range or scope. The range of a rule is the area enclosed by these boundary values of each dimension (refer to equation (1)). "Each chromosome actually has its own unique sub-domain under which it has domain knowledge; and it is generally unaware of the space outside. Also, good decomposition should be done in connection with the good mapping which means sub-domain knowledge is the main interest here."[9] The solution proposed to overcome this in RLGA is to decompose the search domain and map one chromosome to a specific sub-domain problem until the whole search domain is fully covered. Also, the equation for measuring local performance is designed as shown in equation (5).

$$f = \frac{C}{N} = \frac{\text{number of patterns correctly classified}}{\text{total number of applied patterns}} \tag{5}$$

3.2   The Decomposition Training Algorithm

The RLGA training algorithm learns the problem through recursively decomposing the domain and meanwhile producing a corresponding solution mapped to that decomposed sub-domain, which establishes the linkage between domain decomposition and learner mapping. The algorithm consists of two sections, global control and decomposition training (D-Train). Since our primary interest here is the D-Train, we present the detailed algorithms in Figures 1[9].

1.	Initialize a set of chromosomes with each chromosome encoding m rule. Denote the population as P
2.	Calculate local fitness for each chromosome in P
3.	Select good chromosomes from the population according to their fitness
4.	Perform crossover and mutation operation on the selected chromosomes to reproduce new individuals
5.	Combine the newly reproduced chromosomes with old population, forming a new generation
6.	Repeat Step 2 to Step 5 until termination conditions are met

Fig. 1. Algorithms for the decomposition training part of RLGA

3.3   Integration Phase

According to the algorithm shown in Figure 2, experts in the front will have a higher priority to report answer against the one in the rear, even though both of their ranges cover the input pattern. The reason is elaborated below. In the training phase, the learnt patterns are removed at each round when a D-Train process is finished and the corresponding expert is added to the expert array. Therefore, the later expert(s) is unaware of the removed pattern's belonging class.

> *Pt denotes an input pattern, E denotes the expert array;*
> *for each expert e in E*
>   *decode rule antecedents of e*
>   *if all rule antecedents are valid for the instance then*
>   *return classification result reported by the rule*
> *return don't know*

Fig. 2. Pseudo code of integration algorithm

4   RLGA with Incremental Attribute Learning

To better improve the performance of RLGA, we propose a variation of it which features incremental attribute learning [5] and we name it as RLGA-I. Rather than learning input attributes in batch as with normal GA and original RLGA, RLGA-I learns input attributes one after another. This variation also features dimensional decomposition. Because of RLGA's unique features, i.e. local fitness and domain decomposition, RLGA-I can automatically select the useful features for each decomposed sub-domain. For example, suppose we have a two-dimensional search space as shown in the Figure 3. Obviously, the region covered by Class 1 can be separated by only using one dimension, i.e. $x_0 < x < x_1$. It is not necessary to use both x and y in a rule. After ruling out the region covered by Class 1, the task to locate regions of Class 2 and Class 3 become much simpler. This feature has important meaning in high dimensional problems. Not all of the dimensions are useful when classifying a sub-space. RLGA-I has the potential to filter out useless dimensions so that high-dimensional problem can be solved. Rather than attempting to learn the rules with all the attributes, RLGA-I starts from a subset of attributes. If no satisfactory expert is found, then integrate more attributes until a satisfied expert is found.

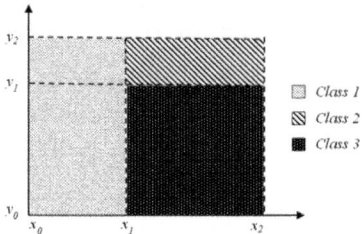

Fig. 3. Class 1 is not necessarily represented by all attributes

4.1   Discriminating Ability of Attributes

To decide the order of the attributes to be added into the chromosomes, we use the concept of Discriminating Ability (DA) [5]. The DA is measured by the training and test classification rates achieved by applying normal GA with only one attribute validated. The single-attribute evolution modules (SEMs) are used to separately evolve a single attribute. The resultant best chromosome in each SEM is used to measure the DA of corresponding attribute. Note that, the resultant groups of chromosomes are saved to be used as new rule segments for the new attribute in the process of formation of initial population. Reuse the resultant populations of SEMs used for estimating DA in the later process can reduce the training time.

4.2   Formation of the Initial Population

The idea of Integration of old chromosomes with new elements is performed to form new population. The process is much the same as IGA [6]. Figure 4 shows how the new element for a new attribute is inserted into an old rule to form a new rule and illustrates the formation of population.

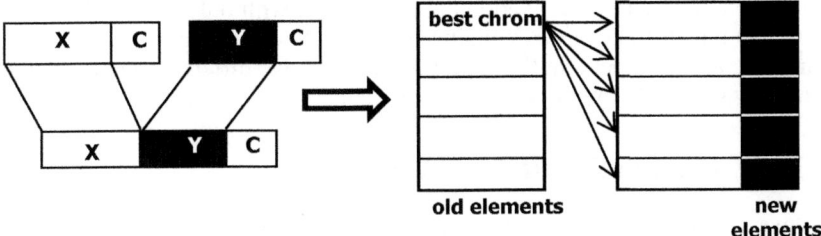

Fig. 4. Formation of new rule in a chromosome and integrating old elements with new elements

Step 1: Evaluate the individual discriminating ability of each attribute

Step 2: Order the attributes in the descending order.

Step 3: Load training patterns from an external file, denote the set of patterns as D

Step 4: Initialization. Set i equal to 1. Initialize parameters for RLGA-I and initialize an expert array, denoted as E.

Step 5: i-th iteration. Introduce the i-th attribute.

Step 6: Evolve the i-th attribute alone in SEM, using the training patterns for the i-th attribute

Step 7: IGA is used to integrate the new elements for the i-th attribute, forming the new population P_i ($i=1$ means the introduction of the first attribute. As there is no old solution, the new elements for the first attribute are simply used as initial population P_1).

Step 8: Invoke the local training method to start weak learner training process. This process evolves P_i with the training patterns including attributes from the 1st to the i-th.

Step 9: Find the best chromosome obtained from local training, denote it as *expert,* identify the learnt patterns from expert and represent them as L
　　　If $i<n$ and $|L|>=1$ then
　　　　　a. Add *expert* into the expert array, E
　　　　　b. Remove the L learnt patterns from D
　　　　　c. Re-evaluate P against the updated D
　　　Else if all patterns are learnt, the training process will be stopped.
　　　Else if $i<n$ and $|L|<1$ then increase i by 1 and go to Step 5

Step 10: If i is equal to n (which is the total number of attributes), apply RLGA until RLGA's termination conditions are met.

Fig. 5. Algorithms for the Global Control of RLGA-I

4.3   The Training Algorithm

Figure 5 gives algorithms of RLGA-I. Note that, comparing with RLGA, RLGA-I has the following major differences:

First of all, different from RLGA, which learns input attributes in their full dimension, RLGA-I learns the attributes one after another under a situation of continuous incremental learning. Also, RLGA-I requires a preprocessing process to sort the attributes to be learnt. The order of the attributes is based on the discriminating ability of each attribute. The discriminating ability is measured by the training and test classification rates achieved by applying normal GA with only one attribute validated. We arrange the attributes in descending order for the future integration process. Most importantly, all the experts in the expert array may have different lengths. Some of the experts may be of full length while others are not.

5   Experimental Results and Analysis

We have conducted a series of experiments on RLGA and RLGA-I via three benchmark datasets. The number of instances / number of attributes / number of classes for each dataset are: Yeast 1484/8/10, Glass 214/9/6 and Wine 178/13/3. All of them are taken from the UCI machine learning repository [7].

All the dataset's data were partitioned into training set and testing set. The training set accounts for 50% of the total instances in each data, and the testing set consist of 50 % of patterns. All the experiments were completed on Intel® Core™2 Duo CPU E7500 PCs with 2.00 GB RAM. The mean values reported were averaged over thirty independent runs. Training time, training classification rate and testing classification rate were measured and compared. The standard deviations (SD) are also reported accordingly. Overall, three different methods were investigated and compared: Normal GA, RLGA and RLGA-I. To set up a comparison model, the traditional normal GA classifier with a fixed number of rules (30) is still investigated. However, since we argue that the number of rules is a problem-dependent parameter, the normal GA with the varied number of rules should be compared as well. The experiment results of NGA-F, NGA-V and RLGA are cited from [9].

Table 1. Experimental Results of Real World Problems

Wine	NGA-F	NGA-V	RLGA	RLGA-I
Training Time	7.50	1.88	**0.41**	3.619 +19.428
Ending CR	0.951 ± 0.022	0.814 ± 0.113	1 ± 0.0	1 ± 0.0
Testing CR	0.875 ±0.052	0.756 ± 0.112	0.928 ± 0.028	**0.944** ± 0.030
No. Rules	30	8	7.23	19.5
Glass	**NGA-F**	**NGA-V**	**RLGA**	**RLGA-I**
Training Time	5.793	8.122	**1.446**	14.484 +13.814
Ending CR	0.529 ± 0.076	0.542 ± 0.078	1 ± 0.0	1 ± 0.0
Testing CR	0.511 ± 0.061	0.509 ± 0.072	0.685 ± 0.057	**0.807** 0.057
No. Rules	30	40	39.93	69.2
Yeast	**NGA-F**	**NGA-V**	**RLGA**	**RLGA-I**
Training Time	30.162	475.836	**78.032**	484.133 +66.953
Ending CR	0.355 ± 0.042	0.4 ± 0.038	1 ± 0.0	0.998 ± 0.002
Testing CR	0.356 ± 0.049	0.387 ± 0.042	0.496 ± 0.02	**0.837** ± 0.025
No. Rules	30	344	343.9	484.13

Experimental results of three benchmark datasets are provided in Table 1. Bold figures indicate the best performance in training time and testing CR among all four algorithms. NGA-F stands for Normal GA with fixed rule number 30, while NGA-V stands for Normal GA with varied rule number. The training time for RLGA-I is represented as *training time + pre-processing time*

As can be seen from the data, for all the problems the testing CR of RLGA-I is dominant among all other approaches. For example, the test CR for Yeast problem is 0.837 while normal GA never exceeds 0.4 and even RLGA can only achieve 0.496. However, we should also see that RLGA-I generally require a considerable longer time for training, because it needs time for pre-processing and incrementally introducing new attributes.

It is also interesting to see that the number of rules used in RLGA and RLGA-I are dependent on the difficulty of the problem. For example, for Wine problem, number of rules used in RLGA-I is only about 20 (19.5). For Yeast problem, it requires about 485 rules to achieve satisfied classification rate.

5   Discussion

According the experiment results, several remarks can be given to RLGA-I. First, RLGA-I show significant capabilities just as RLGA. RLGA-I performs best in learning accuracy. Its performance is even better than other sophisticated approaches, including ECCGA [8] and OIGA [5]. Table 2 lists several successful classifiers in the literature on Yeast and Glass problem respectively.

Table 2. Comparison of various classification methods on Yeast and Glass

Yeast	ECCGA	OIGA	RLGA	RLGA-I
Ending CR	0.477	0.457	1.0	**0.998**
Testing CR	0.435	0.414	0.496	**0.837**
Glass	ECCGA	OIGA	RLGA	RLGA-I
Ending CR	0.737	0.779	1.0	**1.0**
Testing CR	0.47	0.472	0.685	**0.807**

Secondly, by comparing results of RLGA and RLGA-I, we should clearly see there is a trade-off between training time and classification accuracy. In all the experiments RLGA is more cost effective while RLGA-I achieves the best accuracy but requires longer time in training. However, for difficult problems, RLGA-I can be regarded as a more reliable solution than RLGA, like Glass and Yeast problems.

Thirdly, we reckon that RLGA-I has the potential in solving high-dimensional problems that are generally infeasible for traditional GA-based classifiers to solve. For traditional GAs, the enormous size of search space and excessive amount of attributes render the populations almost impossible to evolve. On the other side, automatic task decomposition and incremental attribute learning enable RLGA-I to rapid reduce the search space and adaptively select useful attributes during the training process. However, since the training time for RLGA-I is increasing with the growth of attributes, further investigation on how to improve RLGA-I's efficiency is required to make RLGA-I applicable for high-dimensional problems.

6   Conclusion

This paper first reviews a GA-based classifier, RLGA, with a unique combination in terms of recursive learning, automatic task and data decomposition, and problem-dependent rule number. Then the paper proposed a variation of RLGA denoted as RLGA-I. Besides the merits of RLGA, RLGA-I also features incremental attribute learning. Rather than learning the attributes in batch, it learns the attribute one after the other.

By experimenting on different benchmark tests, RLGA and RLGA-I outperformed normal GA and other approaches in many aspects. RLGA spends much less computation time while achieves significantly better accuracy and RLGA-I's performance achieves the best testing CR among. The improvement shown in these problems are significant, especially for hard problems like yeast, wine, and glass.

References

1. Corcoran, A.L., Sen, S.: Using Real-valued Genetic Algorithms to Evolve Rule Sets for Classification. Evolutionary Computation. In: Proceedings of the First IEEE World Congress on Computational Intelligence (1994)
2. Cordón, O., Herrera, F., et al.: Evolutionary Tuning and Learning of Fuzzy Knowledge Bases, Advances in Fuzzy Systems–Applications and Theory. In: Genetic Fuzzy Systems, vol. 19. World Scientific, Singapore (2001)
3. Smith, S.F.: A Learning System Based on Genetic Adaptive Algorithms. Unpublished doctoral dissertation/thesis University of Pittsburgh Pittsburgh, PA, USA (1980)
4. Michalewicz, Z.: Genetic Algorithms Data Structures. Springer, Heidelberg (1996)
5. Zhu, F., Guan, S.: Ordered Incremental Training for GA-based Classifiers. Pattern Recognition Letters 26(14), 2135–2151 (2005)
6. Guan, S.U., Zhu, F.: An Incremental Approach to Genetic-algorithms-based Classification. IEEE Transactions on Systems, Man, and Cybernetics, Part B: Cybernetics 35(2), 227–239 (2005)
7. Blake, C.L., Merz, C.J.: UCI Repository of Machine Learning Databases (1998), http://archive.ics.uci.edu/ml/ (retrieved March 14, 2010)
8. Zhu, F., Guan, S.U.: Cooperative Co-evolution of GA-based Classifiers Based on Input Decomposition. Engineering Applications of Artificial Intelligence 21(8), 1360–1369 (2008)
9. Fang, L., Guan, S.U., Zhang, H.F.: Recursive Learning of Genetic Algorithms with Task Decomposition and Varied Rule Set. International Journal of Applied Evolutionary Computation, IJAEC (in press, 2011)
10. Tan, C.H., Guan, S.U., et al.: Recursive Hybrid Decomposition with Reduced Pattern Training. International Journal of Hybrid Intelligent Systems 6(3), 135–146 (2009)
11. Ramanathan, K., Guan, S.U.: Recursive Pattern based Hybrid Supervised Training. Engineering Evolutionary Intelligent Systems 82, 129–156 (2008)
12. Yao, W.B., et al.: Self-Evolvable Protocol Design Using Genetic Algorithms. International Journal of Applied Evolutionary Computation (IJAEC) 1(1), 36–56 (2010)
13. Mo, W.T., et al.: Ordered Incremental Multi-Objective Problem Solving Based on Genetic Algorithms. International Journal of Applied Evolutionary Computation (IJAEC) 1(2), 1–27 (2010)

Evaluation of Crossover Operator Performance in Genetic Algorithms with Binary Representation

Stjepan Picek, Marin Golub, and Domagoj Jakobovic

Faculty of Electrical Engineering and Computing, Unska 3, Zagreb, Croatia
stjepan@computer.org, {marin.golub,domagoj.jakobovic}@fer.hr

Abstract. Genetic algorithms (GAs) generate solutions to optimization problems using techniques inspired by natural evolution, like crossover, selection and mutation. In that process, crossover operator plays an important role as an analogue to reproduction in biological sense. During the last decades, a number of different crossover operators have been successfully designed. However, systematic comparison of those operators is difficult to find. This paper presents a comparison of 10 crossover operators that are used in genetic algorithms with binary representation. To achieve this, experiments are conducted on a set of 15 optimization problems. A thorough statistical analysis is performed on the results of those experiments. The results show significant statistical differences between operators and an overall good performance of uniform, single-point and reduced surrogate crossover. Additionally, our experiments have shown that orthogonal crossover operators perform much poorer on the given problem set and constraints.

Keywords: genetic algorithms, crossover, operator comparison, binary representation, nonparametric statistical tests.

1 Introduction

When dealing with optimization problems, one of the challenges is the need to avoid trapping of the algorithm in the local optima. That objective is especially present when there are numerous local optima and the dimensionality of the problem is high. In the last 50 years there have been many metaheuristics coming from the evolutionary computation family that are successful when dealing with such problems. Among those algorithms, genetic algorithms have received considerable attention. During that time, genetic algorithms have been successfully applied to the variety of optimization problems. Since the invention of genetic algorithms in 1960s by J. Holland, crossover operators have played a major role as an exploitation force of genetic algorithms. Holland also used mutation operator in his work, but it was generally treated as subordinate to crossover operator [6]. In the following years, many modifications of crossover operators have appeared [3] [11]. Firstly, it is necessary to answer the question of whether it is possible to find the best search algorithm. The answer is no, since

D.-S. Huang et al. (Eds.): ICIC 2011, LNBI 6840, pp. 223–230, 2012.
© Springer-Verlag Berlin Heidelberg 2012

the "No free lunch" theorem demonstrates that when averaged over all problems, all search algorithms perform equally. However, when working with some knowledge about the problem, it is possible to choose more suitable algorithms.

To evaluate the performance of the algorithms it is not enough to compare mean and standard deviation values [5]. Rather, a proper statistical analysis should be performed. In this paper, we use *nonparametric* statistical tests to evaluate the performance of the crossover operators. For a justification on the selection of the employed statistical methods, and further information about the statistical methods, refer to [2] and [13].

The significance of this comparison lies in the fact that a thorough statistical analysis of these crossover operators has not been done before, to the best of our knowledge. The tests done previously were conducted on smaller sets of crossover operators. The main disadvantage with that approach lies in the fact that it is based on the mean, and standard deviation values, or on the parametric statistical tests. The parametric statistical tests are usually performed without the checking of the necessary conditions for their use, especially in the multi-problem analysis scenario [5]. In this paper we use a mathematically appropriate and thorough approach in conducting statistical analysis.

We begin this paper by giving a short overview of the relevant theory in Section 2. Section 3 defines the parameters used in the experiments and the results obtained from the experiments. Finally, Section 4 draws a conclusion.

2   Preliminary

2.1   Crossover Operator

Crossover, as a process where new individuals are created from the information contained within the parents, is often said to be the distinguishing feature of genetic algorithms. Crossover operators are usually applied probabilistically according to a crossover rate p_c. In this paper crossover will refer to a two-parent case i.e. two individuals are selected as the parents to produce one offspring. Table 1 lists references for all the crossover operators investigated in this paper.

2.2   Test Functions

Test functions from the Table 2 have been selected for the investigation of crossover efficiency. These functions represent well known problems for evaluating the performance of evolutionary algorithms. The table shows the function formulae, dimension of the problem as used in the experiments and the references where additional information on the function can be found.

2.3   Evolutionary Computation Framework - ECF

Evolutionary Computation Framework (ECF), used in this work, is a C++ framework intended for the application of any type of the evolutionary computation. ECF is developed at the Faculty of Electrical Engineering and Computing, Zagreb [8].

Table 1. Crossover operators used in experiments

Crossover operator	Reference
Single-point crossover	[11] [16]
Two-point crossover	[16]
Half-uniform crossover	[4]
Uniform crossover	[11] [16]
Shuffle crossover	[3]
Segmented crossover	[3] [10]
Reduced surrogate crossover	[3]
Non-geometric crossover	[7]
Orthogonal array crossover	[1] [9] [17]
Hybrid Taguchi crossover	[15]

Table 2. Benchmark functions $(D = 30)$

Test function	Domain Range	Reference
$f(x) = \sum_{i=1}^{D} x_i^2$	[-5.12, 5.12]	[12]
$f(x) = \sum_{i=1}^{D} i \cdot x_i^2$	[-5.12, 5.12]	[12]
$f(x) = \sum_{i=1}^{D} 5 \cdot i \cdot x_i^2$	[-5.12, 5.12]	[12]
$f(x) = \sum_{i=1}^{D} \left(\sum_{j=1}^{i} x_j^2 \right)$	[-65.536, 65.536]	[12]
$f(x) = \sum_{i=1}^{D-1} 100 \cdot \left(x_{i+1} - x_i^2 \right)^2 + (1 - x_i)^2$	[-2.048, 2.048]	[1]
$f(x) = 10 \cdot D + \sum_{i=1}^{D} \left(x_i^2 - 10 \cdot \cos\left(2 \cdot \Pi \cdot x_i\right)\right)$	[-5.12, 5.12]	[12]
$f(x) = \sum_{i=1}^{D} -x_i \cdot \sin\left(\sqrt{\lvert x_i \rvert} \right)$	[-500, 500]	[9]
$f(x) = \sum_{i=1}^{D} x_i^2 / 4000 - \prod_{i=1}^{D} \cos\left(x_i / \sqrt{i} \right) + 1$	[-600, 600]	[1]
$f(x) = -20 \cdot e^{-0.2 \sqrt{\sum_{i=1}^{D} x_i^2 / D}} -$ $e^{\sum_{i=1}^{D} \cos(2\Pi x_i)/D} + 20 + e$	[-32.768, 32.768]	[12]
$f(x) = -\sum_{i=1}^{D} \sin\left(x_i\right) \cdot \left(\sin\left(i \cdot x_i^2 / \Pi\right)\right)^{20}$	[0, 3.14]	[12]
$f(x) = \sum_{i=1}^{D} \left(10^6\right)^{(i-1/D-1)} \cdot x_i^2 - 450$	[-100, 100]	[14]
$f(x) = \sum_{i=1}^{D} \lvert x_i \rvert + \prod_{i=1}^{D} \lvert x_i \rvert$	[-10, 10]	[9]
$f(x) = \sum_{i=1}^{D} 2 \cdot D +$ $\sum_{i=1}^{D-1} \left[\sin\left(x_i + x_{i+1}\right) + \sin\left(2 \cdot x_i \cdot x_{i+1}/3\right) \right]$	[3, 13]	[1]
$f(x) = 1/D \cdot \sum_{i=1}^{D} \left(x_i^4 - 16 \cdot x_i^2 + 5 \cdot x_i \right)$	[-5, 5]	[9]
$\sum_{i=1}^{D} \left(\sum_{k=0}^{20} \left[0.5^k \cdot \cos\left(2\Pi \cdot 3^k \left(x_i + 0.5\right)\right) \right] \right) -$ $-D \sum_{k=0}^{20} \left[0.5^k \cdot \cos\left(\Pi \cdot 3^k \right) \right]$	[-0.5, 0.5]	[14]

3   Experimental Results and Comparisons

3.1   Environmental Settings

In all the experiments, binary-coded genetic algorithm with roulette-wheel selection [16] is used. Individuals are binary vectors which represent real values [10]. Parameters of the genetic algorithm that are common to every round of the experiments are as following: simple bit mutation with mutation probability p_m of 0.01 per bit and population size N of 30. Precision is set to 3 digits after the decimal point, which is sufficient to produce large enough number of possible solutions for used test problems. The number of independent runs for each experiment is 30, dimensionality D of all the test problems is set to 30, and the number of fitness evaluation is set to 500000. For all the test functions finding the global minimum is the objective. As a performance measure, we use the error rate obtained for every algorithm.

Factors of an orthogonal array for orthogonal array and hybrid Taguchi crossover operators are dimensions of a problem. For the offspring selection scheme we use the one where the best offspring and the best parent are chosen.

Two rounds of the experiments are conducted in total. Each round is designed to provide the answer to one question. *The first round* is a parameter tuning round where the goal is to find the best value of p_c for every crossover operator and every test function. The goal of *the second round* is to find the best overall algorithm on the set of all the test functions used in this paper.

Additionally, one may want to find the best performing operator for each of the test functions. The answers to these questions are not included in the paper, because knowing the 'best' operator for a single problem may be relevant only for that same problem and the volume of those results exceeds the scope of this work. Naturally, by choosing other selection and mutation operators it is possible to expect different results of analysis. However, in order to adhere to prescribed paper length we decided to use the simplest mutation operator with constant p_m value since the mutation operator is not of the primary interest in this paper. Furthermore, we use only roulette wheel selection as an example of commonly used selection type where the p_c is variable.

3.2   Experiments and Results

In the first round, each operator is run 30 times on each test function with different values of p_c, ranging from 0.1 to 1 in steps of 0.1. The results for every combination of single operator and test function are then processed to find the best crossover probability value. In a large number of combinations there were no significant statistical differences between performances for different p_c values. However, since a single probability value is needed if there is to be an operator comparison, for every combination the p_c value which gives the smallest average error on best individuals in 30 runs was chosen.

In the second round, the operators were compared using the above average error of best individuals from 30 runs with the best probability value. This structure of input data is in accordance with previous analysis performed over multiple algorithms and test problems [5]. A series of non-parametric statistical tests is performed on the data in this round.

The first test is Friedman two-way analysis of variances by ranks, which represents the most well known procedure for testing the differences between more than two related samples [13]. Additionally, we use Iman-Davenport test as a variant of Friedman test that provides better statistics [2]. The objective of the Friedman and Iman-Davenport tests is to show that there is a statistical difference between groups (crossover operators). If there is a statistical difference, then additional *post-hoc* statistical analysis can be performed to discover where those differences are. Table 3 summarizes the rankings obtained by Friedman procedure. The results highlight single-point as the best operator, so the post-hoc analysis is performed with single-point crossover as the control method.

Table 3. Average rankings of crossover operators (Friedman)

Operator	Average ranking
Single-point	3.5667
Two-point	4.3667
Half-uniform	4.7667
Uniform	3.7333
Segmented	4.4333
Shuffle	5.4333
Reduced surrogate	4.3
Non-geometric	5.9333
Orthogonal array	9.4
Hybrid Taguchi	9.0667

With the level of significance α of 0.05 both the Friedman and Iman-Davenport statistic show significant differences in operators with test values of 64.44 and 12.79, respectively, and $p < 0.001$.

In the post-hoc analysis we applied the Bonferroni-Dunn, Hochberg, Finner and Li tests [2] over the results of Friedman procedure. In these tests, a comparison is made between the control operator and the rest of the operators. Adjusted p-values [5] are shown in Table 4. These results indicate whether the control operator is better than each of the remaining operators (i.e. where the null hypothesis is rejected) considering level of significance α of 0.05 and 0.1. For the Bonferroni-Dunn test, a *critical difference* (CD) [5] is calculated which for these data equals 3.0656. The interpretation of this measure is that the performance of two algorithms is significantly different only if the corresponding average ranks differ by at least a critical difference, which is depicted in Fig. 1.

Table 4. Post-hoc comparison (control operator: single-point crossover)

Algorithm	unadjusted p	p_{Bonf}	$p_{Hochberg}$	p_{Finner}	p_{Li}
Orthogonal array	0	0.000001	0.000001	0.000001	0.000001
Hybrid Taguchi	0.000001	0.000006	0.000005	0.000003	0.000005
Non-geometric	0.032296	0.290662	0.22607	0.093792	0.212294
Shuffle	0.091322	0.821901	0.547934	0.193838	0.432492
Half-uniform	0.277726	2.499538	0.880168	0.443246	0.698581
Segmented	0.433081	3.897733	0.880168	0.573144	0.783272
Two-point	0.469295	4.223652	0.880168	0.573144	0.796594
Reduced surrogate	0.507122	4.564102	0.880168	0.573144	0.808867
Uniform	0.880168	7.921516	0.880168	0.880168	0.880168

Table 5. Contrast estimation (pairwise comparison)

	SP	TP	HU	U	Seg	Sh	RS	NG	OA	HT
Single-point (SP)	0	-0.003	-0.004	0.009	-0.021	-0.054	0	-0.094	-3.922	-3.682
Two-point (TP)	0.003	0	-0.001	0.012	-0.018	-0.051	0.004	-0.091	-3.918	-3.679
Half-uniform (HU)	0.004	0.001	0	0.013	-0.017	-0.05	0.005	-0.09	-3.917	-3.678
Uniform (U)	-0.009	-0.012	-0.013	0	-0.03	-0.063	-0.009	-0.103	-3.931	-3.691
Segmented (Seg)	0.021	0.018	0.017	0.03	0	-0.033	0.021	-0.073	-3.901	-3.661
Shuffle (Sh)	0.054	0.051	0.05	0.063	0.033	0	0.055	-0.04	-3.867	-3.628
Reduced surrogate (RS)	-0.001	-0.004	-0.005	0.009	-0.021	-0.055	0	-0.095	-3.922	-3.682
Non-geometric (NG)	0.094	0.091	0.09	0.103	0.073	0.04	0.095	0	-3.827	-3.588
Orthogonal array (OA)	3.922	3.918	3.917	3.931	3.901	3.867	3.922	3.827	0	0.239
Hybrid Taguchi (HT)	3.682	3.679	3.678	3.691	3.661	3.628	3.682	3.588	-0.239	0

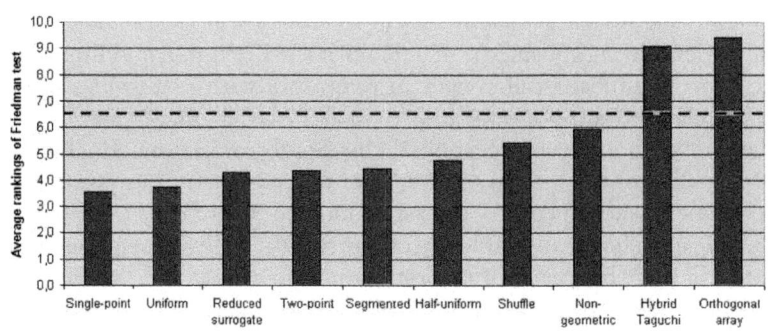

Fig. 1. Bonferroni-Dunn's test (CD = 3.0656, control operator: single-point crossover)

It can be perceived that only orthogonal array and hybrid Taguchi operators may be regarded as significantly worse than the single point crossover with a level of significance α of 0.05. For all the other operators, the null hypothesis cannot be rejected with any of the tests for level of significance $\alpha = 0.05$.

When considering Finner test with level of significance $\alpha = 0.1$, it can be seen that it rejects another hypothesis i.e. single point crossover outperforms the non-geometric crossover.

Since there are no significant differences for the majority of the operators, a procedure *contrast estimation* based on medians [2] can be used to estimate the differences between each two crossover operators. In this test the performance of the algorithms is reflected by the magnitudes of the differences in error rates, and test values are shown in Table 5. A negative value for the operator in a given row indicates that the operator performs better than the operator in a given column. These results highlight the uniform, single point and reduced surrogate operators as the best performing ones.

It is important to state that the chosen benchmark problems have a great influence on the results of the experiments. If the chosen problems are too easy, it can result in too fast convergence to the global optimum for all the crossover operators. On the other hand, choosing problems that are too difficult can result in the trapping of the algorithm in the local optima.

4   Conclusions

In this work we performed an exhaustive search for optimal choice of crossover operators on a well established set of optimization problems. The results clearly show significant statistical differences between certain operators. On the other hand, the conclusions regarding the best overall operators cannot be given with reasonable significance.

A notable difference can be perceived in the case of orthogonal array and hybrid Taguchi crossover, which perform worse than the other operators on the given set of benchmark functions. Nevertheless, this cannot be taken as a general rule, since there certainly exist some problems for which those operators may behave differently. Among other operators, the most successful ones appear to be the uniform, single-point and reduced surrogate crossover.

The presented results may only be considered relevant if the problem at hand bears similarities with the problems addressed in this work. Since in most cases there is not enough time for an exhaustive operator and parameter search, the provided findings may prove useful to researchers in similar optimization environment.

Further work should consider the following issues: additional experiments with different population sizes and mutation probabilities, tournament selection and additional test functions similar to "real-world" problems. When conducting additional, more exhaustive analysis it would be prudent to employ additional statistical tests.

References

1. Chan, K.Y., Kwong, C.K., Jiang, H., Aydin, M.E., Fogarty, T.C.: A new orthogonal array based crossover, with analysis of gene interactions, for evolutionary algorithms and its application to car door design. Expert Systems with Applications (37), 3853–3862 (2010)
2. Derrac, J., Garcia, S., Molina, D., Herrera, F.: A practical tutorial on the use of nonparametric statistical tests as a methodology for comparing evolutionary and swarm intelligence algorithms. Swarm and Evolutionary Computation (1), 3–18 (2011)
3. Dumitrescu, D., Lazzerini, B., Jain, L.C., Dumitrescu, A.: Evolutionary Computation. CRC Press, Florida (2000)
4. Eshelman, L.J.: The chc adaptive search algorithm: How to have safe search when engaging in nontraditional genetic recombination. In: Foundations of Genetic Algorithms, pp. 265–283. Morgan Kaufmann, San Francisco (1991)
5. Garcia, S., Molina, D., Lozano, M., Herrera, F.: A study on the use of nonparametric tests for analyzing the evolutionary algorithms behaviour: a case study on the cec 2005 special session on real parameter optimization. Journal of Heuristics (15), 617–644 (2009)
6. Holland, J.H.: Adaptation in Natural and Artificial Systems: An Introductory Analysis with Applications to Biology, Control, and Artificial Intelligence. The MIT Press, Cambridge (1992)
7. Ishibuchi, H., Tsukamoto, N., Nojima, Y.: Maintaining the diversity of solutions by non-geometric binary crossover: A worst one-max solver competition case study. In: Proceedings of the Genetic and Evolutionary Computation Conference GECCO 2008, pp. 1111–1112 (2008)
8. Jakobovic, D., et al.: Evolutionary computation framework. (March 2011), http://gp.zemris.fer.hr/ecf/, http://gp.zemris.fer.hr/ecf/
9. Leung, Y.W., Wang, Y.: An orthogonal genetic algorithm with quantization for global numerical optimization. IEEE Transactions on Evolutionary Computation 5(1), 41–53 (2001)
10. Michalewitz, Z.: Genetic Algorithms + Data Structures = Evolution Programs, 3rd edn. Springer, New York (1996)
11. Mitchell, M.: An Introduction to Genetic Algorithms. The MIT Press, Cambridge (1999)
12. Pohlheim, H.: Geatbx examples examples of objective functions (2006), http://www.geatbx.com/download/GEATbx_ObjFunExpl_v38.pdf
13. Sheskin, D.: Handbook of Parametric and Nonparametric Statistical Procedures, 4th edn. Chapman and Hall/CRC (2007)
14. Suganthan, P., Hansen, N., Liang, J., Deb, K., Chen, Y., Auger, A., Tiwari, S.: Problem definitions and evaluation criteria for the cec 2005 special session on real parameter optimization. Tech. Report, Nanyang Technological University (2005)
15. Tsai, J.T., Liu, T.K., Chou, J.H.: Hybrid taguchi-genetic algorithm for global numerical optimization. IEEE Transactions on Evolutionary Computation 8(4), 365–377 (2004)
16. Weise, T.: Global Optimization Algorithms Theory and Application (2009), http://www.it-weise.de/
17. Zhang, Q., Leung, Y.W.: An orthogonal genetic algorithm for multimedia multicast routing. IEEE Transactions on Evolutionary Computation 3(1), 53–62 (1999)

Quantum Information Splitting Using GHZ-Type and W-Type States

Lvzhou Li[1] and Daowen Qiu[1,2]

[1] Department of Computer Science, Sun Yat-sen University,
Guangzhou 510006, China
[2] SQIG–Instituto de Telecomunicações, IST, TULisbon,
Av. Rovisco Pais 1049-001, Lisbon, Portugal
lilvzhou@gmail.com, issqdw@mail.sysu.edu.cn

Abstract. Quantum information splitting (QIS) is such a procedure where a sender can distribute a quantum information (state) to several recipients such that all of them can cooperate to recover the quantum state, but any other subset can not. In this paper, we present a uniform method to design QIS schemes using GHZ-type and W-type states. We also give a simple criteria on W states for QIS.

Keywords: Quantum secret sharing, quantum information splitting, W states, GHZ states.

1 Introduction

Secret sharing [1] is a powerful technique in computer science which enables secure and robust communication in information networks, such as the internet, telecommunication systems, and distributed computers. The security of these networks can be enhanced using quantum resources to protect the information. Such schemes have been termed as quantum secret sharing [2]. In 1999, Hillery et al. [2] described a procedure for realizing quantum secret sharing by using multiparticle maximally entangled states, i.e., the Greenberger- Horne-Zeilinger (GHZ) states. For classical information, a shared key can be established between one sender and two or more recipients, all of whom should work together to read the message. In the case of quantum information, Alice can convey a qubit state in such a way that the qubit state can be recovered if and only if all the participants at the receiving end agree to collaborate.

After the seminal work [2], quantum secret sharing has abstracted wide attention from the academic community, and has become a hot issue in the study of quantum information. Roughly speaking, the already known work on this issue can mainly be divided into two lines. The first line is on quantum secret sharing of classical messages, i.e., the secret to be shared is classical informtion, for instance [3-6]. The second line is on quantum secret sharing of quantum messages, i.e., the secret to be shared is a quantum state, and we call this case *quantum information splitting* (QIS). In this paper, we focus on QIS, and thus we do not recall in detail these results on quantum secret sharing of classical information.

D.-S. Huang et al. (Eds.): ICIC 2011, LNBI 6840, pp. 231–238, 2012.

In [2], Hillery et al. first presented a QIS scheme using a tripartite GHZ state where Alice can distribute a qubit state to Bob and Charlie in such a way that by collaborating, Bob and Charlie can restore the qubit state. Later, Cleve et al. [7] made a deeper investigation on QIS, and discussed the general case—(k, n) threshold scheme where a secret quantum state is divided into n shares such that any k of those shares can be used to restore the secret, but any set of $k-1$ or fewer shares contains absolutely no information about the secret. In [8], the authors investigated in detail a $(2, 3)$ threshold scheme.

Generally, in a QIS scheme entangled states are needed. As we know, besides GHZ states, there is another important type of entangled states, that is, W states [9]. In fact, W states have been used for several important quantum information processing tasks; for example, W states have been used for teleportation and superdense coding [10,11], and they have also been used to design QIS schemes. In [12], Zheng proposed a QIS scheme using a tripartite W state. Later, there were a series of papers on QIS scheme using W states [12-17].

In this paper, we revisit QIS schemes based on GHZ states and W states. In particular, we present a uniform method to design QIS schemes using GHZ states or W states, and most of the QIS schemes in the literature based on GHZ states or W states can be rebuilt using the method given by us. Also, we present a criteria for W states to be suitable for QIS.

2   QIS Using GHZ States

2.1   The History QIS Scheme

A three-qubit GHZ state is as follow:

$$|GHZ_2\rangle_{abc} = \frac{1}{\sqrt{2}}(|000\rangle_{abc} + |111\rangle_{abc}). \tag{1}$$

Hillery et al. [2] proposed a QIS scheme using the above GHZ state, and here we recall the main idea. In the QIS scheme, Alice wants to distribute a qubit state

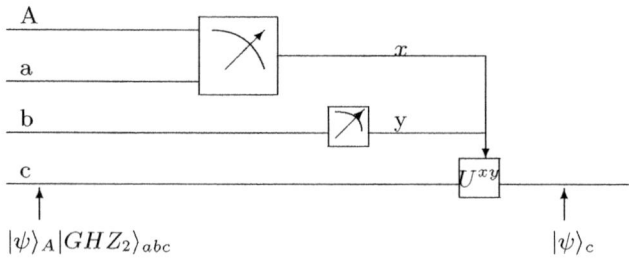

Fig. 1. Hillery et al. [2]'s QIS scheme

to Bob and Charlie in such a way that both of them can collaborate to restore the qubit state, but any one alone can not. Suppose the quantum state to be distributed by Alice is

$$|\psi\rangle_A = \alpha|0\rangle_A + \beta|1\rangle_A. \tag{2}$$

At the beginning, Alice, Bob and Charlie share the GHZ state given in Eq.(1), where particles a, b, c belong to Alice, Bob, and Charlie, respectively. Then the combining state is

$$
\begin{aligned}
|\Omega\rangle &= |\psi\rangle_A |GHZ_2\rangle_{abc} \\
&= \frac{1}{2}[|\Psi_+\rangle_{Aa}(\alpha|00\rangle_{bc} + \beta|11\rangle_{bc}) + |\Psi_-\rangle_{Aa}(\alpha|00\rangle_{bc} - \beta|11\rangle_{bc}) \\
&\quad + |\Phi_+\rangle_{Aa}(\beta|00\rangle_{bc} + \alpha|11\rangle_{bc}) + |\Phi_-\rangle_{Aa}(-\beta|00\rangle_{bc} + \alpha|11\rangle_{bc})] \tag{3}
\end{aligned}
$$

where $|\Psi_\pm\rangle_{Aa} = \frac{1}{\sqrt{2}}(|00\rangle_{Aa} \pm |11\rangle_{Aa})$, $|\Phi_\pm\rangle_{Aa} = \frac{1}{\sqrt{2}}(|01\rangle_{Aa} \pm |10\rangle_{Aa})$.

Now Alice performs a Bell measurement using the base $\{|\Psi_\pm\rangle_{Aa}, |\Psi_\pm\rangle_{Aa}\}$. Then the combining state of Bob and Charlie will reduce to one of four possible encoding states of state $|\psi\rangle$. For example, if Alice's result is $|\Psi_-\rangle_{Aa}$, then the combining state of Bob and Charlie is $\alpha|00\rangle_{bc} - \beta|11\rangle_{bc}$. Alice sends his measurement result to Bob and Charlie using two bits of classical information. Then Bob and Charlie can collaborate to restore $|\psi\rangle_A$ after receiving Alice's measurement result, but if Bob and Charlie do not collaborate, then none alone can restore the state. For example, when Alice's result is $|\Psi_-\rangle_{Aa}$, the combining state of Bob and Charlie is $\alpha|00\rangle_{bc} - \beta|11\rangle_{bc}$. Then by representing $|0\rangle_b, |1\rangle_b$ with $|\pm\rangle = \frac{1}{\sqrt{2}}(|0\rangle_b \pm |1\rangle_b)$, we have $\alpha|00\rangle_{bc} - \beta|11\rangle_{bc} = |+\rangle_b(\alpha|0\rangle_c - \beta|1\rangle_c) + |-\rangle_b(\alpha|0\rangle_c + \beta|1\rangle_c)$. Thus, Bob performs a measurement using the base $\{|+\rangle, |-\rangle\}$, and sends his result to Charlie; then with Bob's result Charlie can restore state $|\psi\rangle_A$ at his end. For example, if Bob's result is $|-\rangle$, then Charlie does nothing; else if Bob's result is $|+\rangle$, then Charlie performs σ_z on his qubit. In summary, the QIS scheme mentioned above can be depicted in Figure 1.

2.2   Our QIS Scheme

Now, we present our QIS scheme for distributing a qubit state to two recipients. The QIS scheme we will give is similar to the one mentioned above, and the difference is the operation performed by Bob and Charlie. In Hillery et al. [2]'s scheme, one of the two recipients does a local measurement and then sends the measurement result to the other one. In our scheme, the two recipients should do a collective unitary operation. At first glance, this difference has no essential effect on QIS, but following our scheme, we can easily generalize it to multiparty case, and based on it we can easily design a QIS scheme using W states. We now depict our scheme in Figure 2.

Similar to Hillery et al.'s scheme, at the beginning, Alice, Bob and Charlie share a GHZ state given in Eq. (1); besides, Alice posses another particle A with state $|\psi\rangle$ given in Eq. (2). Then before Alice's measurement and the CNOT

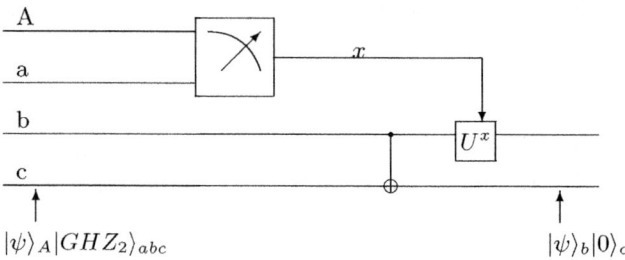

Fig. 2. A QIS scheme for Alice to distribute a qubit state $|\psi\rangle$ to Bob and Charlie in such way that Bob and Charlie should collaborate to restore $|\psi\rangle$. In the scheme, Bob and Charlie collaborate to do a CNOT gate where Bob acts as a control bit, and Charlie acts as a target bit; then $|\psi\rangle$ is restored at Bob 's end.

operation, the combining state is given by Eq. (3). Since Alice's measurement commutes with Bob and Charlie's CNOT gate, we assume that CNOT gate is performed before Alice's measurement. Then after CNOT gate, the combining state changes to

$$|\varXi\rangle_{Aabc} = \frac{1}{2}\Big[|\varPsi_+\rangle_{Aa}(\alpha|0\rangle_b + \beta|1\rangle_b) + |\varPsi_-\rangle_{Aa}(\alpha|0\rangle_b - \beta|1\rangle_b)$$
$$+ |\varPhi_+\rangle_{Aa}(\beta|0\rangle_b + \alpha|1\rangle_b) + |\varPhi_-\rangle_{Aa}(-\beta|0\rangle_b + \alpha|1\rangle_b)\Big]|0\rangle_c. \qquad (4)$$

Now, Alice performs a Bell measurement using the base $\{|\varPsi_\pm\rangle_{Aa}, |\varPsi_\pm\rangle_{Aa}\}$. It is clear that after receiving Alice's measurement result, Bob can perform a suitable unitary operation on his particle to restore $|\psi\rangle$ at his end. The corresponding relation between the possible unitary operation performed by Bob and the measurement result of Alice is given in the following: $|\varPsi_+\rangle \to I, |\varPsi_-\rangle \to \sigma_z, |\varPhi_+\rangle \to \sigma_x, |\varPhi_-\rangle \to \sigma_z\sigma_x$.

In the above procedure, if Bob an Charlie agree to restore the state $|\psi\rangle$ at Charlie's end, then they should perform such a CNOT gate where Charlie's qubit acts as a control bit and Bob's qubit acts as a target bit, and subsequently Charlie performs a suitable unitary operation on his particle according to Alice's measurement result.

Now we can generalize the above scheme to the case of n recipients. Suppose Alice shares the following multiparty GHZ state with n recipients: Bob1, Bob2,$\cdots$, Bob$n$:

$$|GHZ_n\rangle_{a_0b_1\cdots b_n} = \frac{1}{\sqrt{2}}\Big(|00\cdots0\rangle + |11\cdots1\rangle\Big)_{a_0b_1\cdots b_n} \qquad (5)$$

where particle a_0 belongs to Alice and particle b_i belongs to Bob$i$. Alice wants to distribute a qubit state $|\psi\rangle$ given in Eq. (2) to Bob1, Bob2,$\cdots$, Bob$n$ in such a way that all of them can collaborate to restore $|\psi\rangle$, but any other subset of them can not. Following the idea presented in the foregoing scheme, we can now design a scheme for the above task. The main idea is depicted in Figure 3.

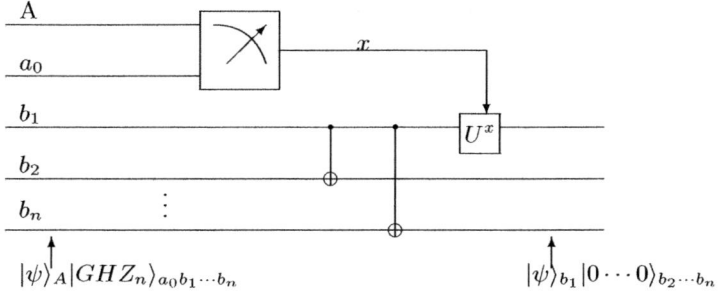

Fig. 3. A QIS scheme using the state $|GHZ_n\rangle$ for Alice distributing a qubit state to n recipients

In the above scheme, if the n recipients agree to restore the state $|\psi\rangle$ at Bob$i$'s end, then they should perform $n-1$ CNOT gates, all of which take their control bits at Bob$i$'s end, and each other Bob$j$ ($j \neq i$) posses a target bit. In Figure 3, we assume the state $|\psi\rangle$ to be restored at Bob1's end. If $n = 2$, then we get the scheme depicted in Figure 2.

3   QIS Using W States

Designing QIS scheme using W states has been discussed in several papers [12-17]. Zheng [12] first proposed a QIS scheme for distributing a quibt state to two recipients using the following W state

$$|W_2\rangle_{abc} = \frac{1}{2}(\sqrt{2}|100\rangle_{abc} + |010\rangle + |001\rangle_{abc}). \qquad (6)$$

Note that the above W state has also been used for perfect teleporation and superdense coding [10,11].

Here we design another different QIS using the above W state, based on the QIS schemes designed in the foregoing section. Firstly we observe the following two facts:

1. The W state can be rewritten as $|W_2\rangle_{abc} = \frac{1}{\sqrt{2}}|1\rangle_a|00\rangle_{bc} + \frac{1}{\sqrt{2}}|0\rangle_a \frac{(|10\rangle_{bc}+|01\rangle_{bc})}{\sqrt{2}}$.
2. The W state can be converted to the GHZ state given in Eq. (1) by such a unitary operation $U = I_a \otimes U_{bc}$ where I_a, an identity, acts on the first qubit, and U_{ab} acts on the other two qubits and has the effect: $|00\rangle \rightarrow |11\rangle, |11\rangle \rightarrow \frac{(|10\rangle-|01\rangle)}{\sqrt{2}}, \frac{(|10\rangle+|01\rangle)}{\sqrt{2}} \rightarrow |00\rangle, \frac{(|10\rangle-|01\rangle)}{\sqrt{2}} \rightarrow \frac{(|10\rangle-|01\rangle)}{\sqrt{2}}$.

With the above observations, we can now design a QIS scheme using state $|W_2\rangle$, and the main idea is depicted in Figure 4.

In the above scheme, Alice, Bob and Charlie share state $|W_2\rangle$ where particles a, b, c belong to Alice, Bob and Charlie, respectively. Besides, Alice also posses particle A whose state is $|\psi\rangle = \alpha|0\rangle + \beta|1\rangle$. Alice's measurement on particles A and a commutes with the unitary operation performed by Bob and Charlie.

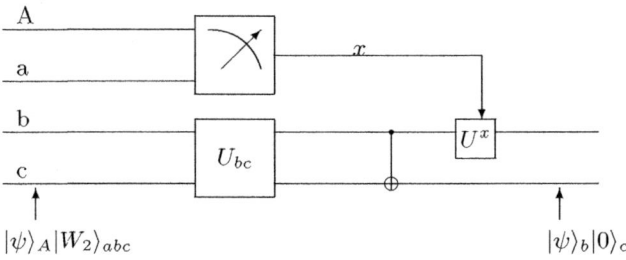

Fig. 4. A QIS scheme using $|W_2\rangle$ for distributing a qubit state $|\psi\rangle$ at Alcie's end to Bob and Charlie in such a way that Bob and Charlie should collaborate to restore $|\psi\rangle$

Thus, without lose of generality we assume that U_{bc} and CNOT are performed before Alice's measurement. Then after the operation of U_{bc} the combining state $|\psi\rangle_A|W_2\rangle_{abc}$ is converted to the state given in Eq. (3). Subsequently, the state is further converted to the state given in Eq. (4) after the performance of CNOT gate. At the end, Bob can restore the state $|\psi\rangle$ after receiving Alice's measurement result. Thus we have presented a QIS scheme using a three-qubit W state, which has some difference from the one given in [12].

In [16], the authors proposed a QIS scheme using the following W state:

$$|W_n^a\rangle_{a_0b_1b_2\cdots b_n} = \frac{1}{\sqrt{2}}|1\rangle_{a_0}\prod_{j=1}^n|0\rangle_{b_j} + |0\rangle_{a_0}\sum_{i=1}^n\frac{1}{\sqrt{2^{i+1-\delta_{in}}}}\prod_{j=1}^n|\delta_{ij}\rangle_{b_j}. \quad (7)$$

Note that if $n = 2$, then the above state reduces to the state $|W_2\rangle$ given in Eq. (6). In the QIS scheme of reference [16], Alice and n recipients first share the above W state where Alice posses the first qubit and each other recipient posses one. Besides, Alice also posses another qubit, say A, who state is $|\psi\rangle$. The aim of Alice is to distribute the qubit state $|\psi\rangle$ to the n recipients such that all the recipients can collaborate to restore $|\psi\rangle$, but any other subset of them can not. Here we do not recall the detail procedure of the QIS scheme presented in [16], but we design a new QIS scheme using the above W state which achieves the same goal as the one in [16]. One will find that our scheme seems not so complicated as the one in [16], and thus it is more understandable.

One first note that the above W state can be converted to a GHZ state as follows: $(I_{a_0} \otimes U_{b_1b_2\cdots b_n})|W_n^a\rangle_{a_0b_1b_2\cdots b_n} = |GHZ_n\rangle_{a_0b_1b_2\cdots b_n}$ where $|GHZ_n\rangle$ is given in Eq.(5), and U, a collective unitary operation acting on n qubits, has the following effect: $|00\cdots 0\rangle \to |11\cdots 1\rangle$, $\sum_{i=1}^n\frac{1}{\sqrt{2^{i+1-\delta_{in}}}}\prod_{j=1}^n|\delta_{ij}\rangle \to |00\cdots 0\rangle$. Thus following the idea inhabiting Figure 4, it is not difficult to design a QIS scheme using this W state, and we depict the scheme in Figure 5.

In [17], the authors proposed a QIS scheme using the following more general W state:

$$|W_n^b\rangle_{a_0b_1b_2\cdots b_n} = \left[\frac{1}{\sqrt{2}}|100\cdots 0\rangle + d_1|010\cdots 0\rangle + d_2|001\cdots 0\rangle + \ldots + d_n|000\cdots 1\rangle\right]_{a_0b_1b_2\cdots b_n}$$

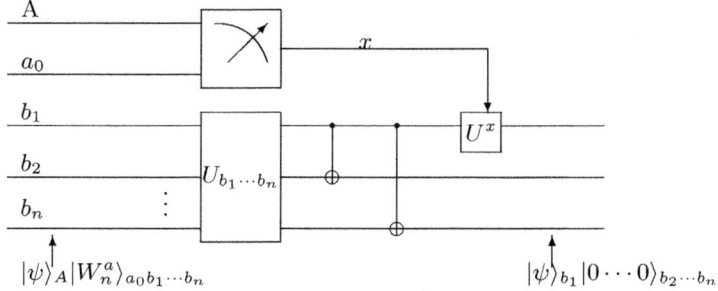

Fig. 5. A QIS scheme using $|W_n^a\rangle$ for Alice distributing a qubit state to n recipients

where $\sum_{i=1}^n d_i^2 = \frac{1}{2}$. In fact, this state can also be converted to a GHZ state as follows: $(I_{a_0} \otimes V_{b_1 b_2 \cdots b_n})|W_n^b\rangle_{a_0 b_1 b_2 \cdots b_n} = |GHZ_n\rangle_{a_0 b_1 b_2 \cdots b_n}$ where $V_{b_1 b_2 \cdots b_n}$, a collective unitary acting on n quibts, has the following effect: $|00 \cdots 0\rangle \rightarrow |11 \cdots 1\rangle$, $\sqrt{2}(d_1|10 \cdots 0\rangle + d_2|01 \cdots 0\rangle + \ldots + d_n|00 \cdots 1\rangle) \rightarrow |00 \cdots 0\rangle$.

By the above observation, we should only in Figure 5 replace $|W_n^a\rangle$ and U by $|W_n^b\rangle$ and V respectively, and then obtain a new QIS scheme for Alice to distribute a qubit state $|\psi\rangle$ to n recipients.

Now we consider the following general W state: $|W_n^g\rangle = a_0|100 \cdots 0\rangle + a_1|010 \cdots 0\rangle + a_2|001 \cdots 0\rangle + \cdots a_n|00 \cdots 1\rangle$. A natural question is: when can the state $|W_n^g\rangle$ be used for designing a QIS scheme where Alice can distribute a qubit state to n recipients? By the foregoing discussions, we can obtain the following criteria:

If the general W state $|W_n^g\rangle$ satisfies $(I_i \otimes U_{\neq i})|W_n^g\rangle = |GHZ_n\rangle$ where I_i is an identity acting on the ith qubit and $U_{\neq i}$ is a unitary operation acting on the left n qubits, then Alice can distribute a qubit state to n recipients such that all the recipients can collaborate to restore the qubit state, but any other subset can not, by sharing $|W_n^g\rangle$ between them where Alice posses the ith qubit and the n recipients poss the left qubits.

4   Conclusions

In this paper, we have presented a uniform way to design QIS schemes using GHZ states or W states, and most of the QIS schemes in the literature based on GHZ states or W states can be rebuilt using the way given by us. Also, we have presented a criteria for W states to be suitable for QIS.

Acknowledgement. This work is supported in part by the National Natural Science Foundation (Nos. 60873055, 61073054), the Natural Science Foundation of Guangdong Province of China (No. 10251027501000004), the Fundamental Research Funds for the Central Universities (Nos. 10lgzd12, 11lgpy36), the Research Foundation for the Doctoral Program of Higher School of Ministry of Education of China (Nos. 20100171110042, 20100171120051), the China Postdoctoral Science Foundation project (Nos. 20090460808, 201003375), and the project of SQIG at IT, funded by FCT and EU FEDER projects Quantlog POCI/MAT/55796/2004 and QSec PTDC/EIA/67661/2006, IT Project Quant-Tel, NoE Euro-NF, and the SQIG LAP initiative.

References

1. Shamir, A.: How to Share a Secret. Commun. ACM 22, 612 (1979)
2. Hillery, M., Buzek, V., Berthiaume, A.: Quantum Secret Sharing. Phys. Rev. A 59, 1829–1834 (1999)
3. Gottesman, D.: Theory of Quantum Secret Sharing. Phys. Rev. A 61, 042311 (2000)
4. Guo, G.P., Guo, G.C.: Quantum Secret Sharing Without Entanglement. Phys. Lett. A 310, 247–251 (2003)
5. Xiao, L., Long, G.L., Deng, F.G., Pan, J.W.: Efficient Multiparty Quantum-Secret-Sharing Schemes. Phys. Rev. A 69, 052307 (2004)
6. Zhang, Z.J., Li, Y., Man, Z.X.: Multiparty Quantum Secret Sharing. Phys. Rev. A 71, 044301 (2005)
7. Cleve, R., Gottesman, D., Lo, H.K.: How to Share a Quantum Secret. Phys. Rev. Lett. 83, 648 (1999)
8. Lance, A.M., Symul, T., Bowen, W.P., Sanders, B.C., Lam, P.K.: Tripartite Quantum State Sharing. Phys. Rev. Lett. 92, 177903 (2004)
9. Dür, W., Vidal, G., Cirac, J.I.: Three Qubits Can Be Entangled in Two Inequivalent Ways. Phys. Rev. A 62, 062314 (2000)
10. Agrawal, P., Pati, A.: Perfect Teleportation and Superdense Coding with W States. Phys. Rev. A. 74, 062320–062324 (2006)
11. Li, L.Z., Qiu, D.W.: The States of W-class as Shared Resources for Perfect Teleportation and Superdense Coding. J. Phys. A. 40, 10871–10885 (2007)
12. Zheng, S.B.: Splitting Quantum Information via W States. Phys. Rev. A 74, 054303 (2006)
13. Zhang, Z.J., Cheung, C.Y.: Minimal Classical Communication and Measurement Complexity for Quantum Information Splitting. J. Phys. B 41, 015503-1-6 (2008)
14. Zuo, X.Q., Liu, Y.M., Zhang, W., et al.: Minimal Classical Communication Cost and Measurement Complexity in Splitting Twoqubit Quantum Information via Asymmetric W States. Int. J. Quantum. Inf. 6, 1245–1253 (2008)
15. Liu, Y.M., Yin, X.F., Zhang, W., et al.: Tripartition of Arbitrary Single-Qubit Quantum Information by Using Asymmetric Fourqubit W State. Int. J. Quantum. Inf. 7, 349–355 (2009)
16. Zhang, W., Liu, Y.M., Yin, X.F., et al.: Partition of Arbitrary Single-Qubit Information Among Recipients via Asymmetric Qubit W State. Sci. China. Ser. G-Phys. Mech. Astron. 52, 1611–1617 (2009)
17. Zuo, X.Q., Liu, Y.M., Zhang, W., Zhang, Z.J.: Simpler Criterion on W State for Perfect Quantum State Splitting and Quantum Teleportation. Sci. China. Ser. G-Phys Mech Astron 52, 1906–1912 (2009)

Attacks and Improvements of QSDC Schemes Based on CSS Codes

Xiangfu Zou[1,2] and Daowen Qiu[1,3]

[1] Department of Computer Science,
Sun Yat-sen University, Guangzhou 510006, China
[2] School of Mathematics and Computational Science,
Wuyi University, Jiangmen 529020, China
[3] SQIG–Instituto de Telecomunicações, IST,
TULisbon, Av. Rovisco Pais 1049-001, Lisbon, Portugal
issqdw@mail.sysu.edu.cn

Abstract. Quantum secure direct communication (QSDC) is a new research direction in quantum cryptography. Recently, a QSDC scheme using quantum Calderbank-Shor-Steane (CSS) error correction codes is proposed [Journal of Software, 17, 509–515 (2006)]. However, in this paper, it is showed that the scheme can not resist the man-in-the-middle attack. To resist man-in-the-middle attacks, an improved QSDC scheme using quantum CSS error correction codes is suggested. Furthermore, we discuss the security of the improved QSDC scheme for an ideal noiseless channel. In particular, it can resist the attack without eavesdropping.

Keywords: Quantum cryptography, quantum secure direct communication, man-in-the-middle attack, CSS error correction code.

1 Introduction

Quantum secure direct communication (QSDC) is a new research direction in quantum cryptography. Different from quantum key distribution [1,2,3,4,5] whose object is to establish a random key between the two parties of communication, QSDC is the direct communication of secret messages without first producing a shared secret key. Note that the secure direct communication is a task that the classical cryptography can not achieve. Because QSDC may be important in some applications, such as urgent circumstances, many scholars have studied and discussed it.

In 2002, Beige *et al.* [6] first proposed the novel notion of QSDC and constructed a QSDC scheme based on sigle-photon. Boström and Felbinger [7] presented a QSDC protocol based on an entangled pair of qubits. However, Wójcik [8] revealed that the QSDC protocol presented by Boström and Felbinger is not secure as far as quantum channel losses are taken into account. Zhang *et al.* [9] improved Wójcik's eavesdropping scheme by constituting a new set of attack operations. Deng and Long *et al.* [10] proposed a Two-Step protocol for quantum secure direct communication using blocks of EPR pairs. The scheme

D.-S. Huang et al. (Eds.): ICIC 2011, LNBI 6840, pp. 239–246, 2012.
© Springer-Verlag Berlin Heidelberg 2012

is secure because an eavesdropper cannot get both sequences simultaneously. Gao *et al.* [11,12] presented a QSDC scheme by EPR pairs and entanglement swapping, and a QSDC scheme by GHZ states and entanglement swapping. Deng and Long [13] proposed a QSDC protocol using quantum one-time-pad and single photons. The QSDC protocol does not need entanglement states as the information carrier. Therefore, quantum entanglement and non-locality are not the necessary requirements for QSDC. Hoffmann *et al.* [14] gave an undetectable attack scheme to show the QSDC protocol using quantum one-time-pad is insecure in a noisy channel. Deng and Long [15] thought that quantum privacy amplification is principally possible. Thereby, their QSDC protocol is secure against the attack strategy in Hoffmann's comment by using quantum privacy amplification directly. Man *et al.* [16] proposed a QSDC protocol by using swapping quantum entanglement and local unitary operations. Lucamarini and Mancini [17] proposed a protocol for deterministic communication that does not make use of quantum entanglement. Nguyen [18] proposed an entanglement-based protocol for two people to simultaneously exchange their messages. Jin *et al.* [19] presented a three-party simultaneous quantum secure direct communication scheme by using GHZ states. Deng *et al.* [20] presented two efficient QSDC network schemes with an N ordered EPR pairs. Any one of the authorized users can communicate with another one on the network securely and directly.

All these existed QSDC schemes need to publish some additional classical messages to check out whether there exist eavesdroppers over the quantum communication channel. Different from these schemes, Lü *et al.* [21] proposed a QSDC scheme using quantum Calderbank-Shor-Steane (CSS) error correction codes which does not use quantum entanglement and needs to release classical information to check the existence of eavesdropping in quantum channel. In this paper, it is showed that the QSDC scheme can not resist the man-in-the-middle (MITM) attack. Furthermore, an improved QSDC scheme which can resist MITM attacks is suggested.

The remainder of this paper is organized as follows. First, in Section 2, we introduce the quantum secure direct communication and the MITM attack. Then, in Section 3, we briefly recall the QSDC scheme using quantum CSS codes [21]. Afterwards, in Section 4, we show that the QSDC scheme [21] can not resist the MITM attack. Subsequently, in Section 5, we improve the QSDC scheme using quantum CSS codes [21] such that it can effectively resist the MITM attack. Furthermore, in Section 6, we prove the security of the improved QSDC scheme and give some of its properties. Finally, in Section 7, we make a conclusion.

2   Preliminaries

For convenience, we briefly introduce the quantum secure direct communication and the MITM attack. Quantum secure direct communication should satisfy the following two requirements [10]:

(1) Communication is direct, i.e., the two communicating parties does not share any (classical or quantum) key, and the secret messages should be read

out directly by the legitimate user and no additional classical information is needed after the transmission of quantum states;

(2) Communication is secure, i.e., the secret messages should not leak even though an eavesdropper controls the quantum channel.

What resources can the two parties of communication use? Generally, they can use the following two resources: (1) a two-way quantum channel between them; (2) a classical public communications channel that can not be blocked. The classical public communications channel are assumed to be susceptible to eavesdropping but not to be the injection or alteration of messages. Note that, the unblocked classical public channel has been widely used in quantum key distribution [1,2,5].

The MITM attack is a very common attack method in modern cryptography. In cryptography, the MITM attack is a form of active eavesdropping in which the attacker makes independent connections with the victims and relays messages between them, making them believe that they are talking directly to each other over a private connection, when in fact the entire conversation is controlled by the attacker. An MITM attack can succeed only when the attacker can impersonate each endpoint to the satisfaction of the other. In quantum cryptography, there also exists the MITM attack. For example, Wang [22] considered the MITM attack on the BB84 protocol.

3   The QSDC Scheme Using Quantum CSS Codes [21]

In the interest of readability, we briefly recall the QSDC scheme using quantum CSS codes [21]. For seeing the QSDC scheme more clearly, we only outline its main steps, and the details can be found in Ref. [21].

Suppose that Alice wants to send a secret message p to Bob by a noiseless quantum channel. Let $C_i = \Gamma(L_i, g_i(z)) = [n, k_i, d_i]$ $(i = 1, 2)$ be both binary Goppa codes such that $C_1^\perp \subseteq C_2$, $d = \min(d_1, d_2)$, the Hamming weight of error vectors $t = \lfloor \frac{d-1}{2} \rfloor$, $k = k_1 + k_2 - n$. $Q : [[n, k, d]]$ denotes the quantum CSS error correcting code constructed by using C_1 and C_2. The protocol can be simply described as follows.

Step 1. **Encoding.** Bob randomly chooses a generator matrix G_i and parity check matrix H_i of C_i $(i = 1, 2)$ such that $G_2 = \begin{pmatrix} H_1 \\ D \end{pmatrix}$, where the rank of D is $k = k_1 + k_2 - n$. Then, he randomly prepares a basis state $|m\rangle$ such that $m \in \boldsymbol{F}_2^k$ and encodes it into $|c\rangle$ using quantum CSS codes Q as

$$|c\rangle = \frac{1}{2^{\frac{n-k_1}{2}}} \sum_{v \in C_1^\perp} |v + m_1 \cdot D^{(1)} + \cdots + m_k \cdot D^{(k)}\rangle, \tag{1}$$

where $|m\rangle = |m_1 m_2 \cdots m_k\rangle$. Bob acts some error $e' = (X'|Z')$ on $|c\rangle$ as

$$|\psi\rangle = \frac{1}{2^{\frac{n-k_1}{2}}} \sum_{v \in C_1^\perp} (-1)^{(v+m \cdot D) \cdot Z'} |v + m \cdot D + X'\rangle, \tag{2}$$

such that the Hamming weight of the error vector e' is $w_q(e') \leq [\frac{t}{2}]$. Bob keeps the matrix G_i, C_i ($i = 1, 2$), D and the bits string e', m as his private keys and sends $|\psi\rangle$ to Alice by the quantum channel.

Step 2. **Encryption.** Suppose that Alice has a privacy message p in hand and wants to transmit it to Bob securely. She firstly applies an algorithm (Algorithm 1 in Ref. [21]) to transform p into a binary error vector $e'' = (X''|Z'')$ such that $t'' = w_q(e'') \leq [\frac{t}{2}]$. Alice receives Bob's qubits $|\psi'\rangle$ and applies the error e'' on them as

$$|\psi'\rangle = \frac{1}{2^{\frac{n-k_1}{2}}} \sum_{v \in C_1^\perp} (-1)^{(v+m \cdot D) \cdot Z' \cdot Z''} |v + m \cdot D + X' + X''\rangle. \tag{3}$$

Alice sends $|\psi'\rangle$ back to Bob.

Step 3. **Decoding.** Bob obtains the error vector $e = (X|Z)$ and recovers $|m'\rangle$ by decoding the quantum codes $|\psi'\rangle$. He measures $|m'\rangle$ using computationally basis $\{|0\rangle, |1\rangle\}$ and compares the measurement result $|m'\rangle$ with his original bits $|m\rangle$. If $m' \neq m$, he believes that eavesdropping happens in the quantum channel. Otherwise, he computes $e'' = (X''|Z'')$, $e'' = e + e'$ and performs an algorithm (Algorithm 2 in Ref. [21]) to recover Alice's secrete bits p.

4   An MITM Attack on the QSDC Scheme Using Quantum CSS Codes

In this section we will consider how to implement MITM attacks on the QSDC scheme using quantum CSS codes [21].

4.1   The Feasibility of MITM Attacks

According to the request of QSDC and the QSDC scheme using quantum CSS codes [21], we can learn that Alice and Bob does not share any secret key or quantum entanglement in the QSDC scheme [21]. Therefore, when Alice receives the quantum information $|\psi\rangle$, she can not confirm that it was sent by Bob. Similarly, Bob can not determine that the received quantum information $|\psi'\rangle$ came from Alice. Furthermore, by the QSDC scheme [21], we know that Alice and Bob do not discuss the measurement results with the classical communication channel. Thereby, at the end of the QSDC scheme, Alice can not be sure that $|\psi\rangle$ is sent by Bob, and Bob can not be sure that $|\psi'\rangle$ came from Alice. Accordingly, what the quantum messages can not be authenticated in the QSDC scheme [21] provides the possibility of MITM attacks.

4.2   An MITM Attack Method

Now, we give an MITM attack on the QSDC scheme [21]. For convenience, we call the malicious active attacker Mallory. Suppose Mallory controls the quantum channel between Alice and Bob and has the abilities of intercepting quantum messages, storing quantum messages, modifying quantum messages, generating new quantum messages, sending quantum messages, and deleting quantum messages. Furthermore, Mallory can imitate Bob when talking to Alice and imitate Alice when talking to Bob.

We embed the middle attack in the QSDC scheme using quantum CSS error correcting codes [21]. The whole attack process includes five steps where Step 2 and Step 4 are the attack operations made by Mallory. The details are specified as follows.

Step 1. **Bob encodes.** It is Step 1 in the QSDC scheme using quantum CSS error correcting codes [21].

Step 2. **Mallory intercepts and encodes.** When Bob sends the quantum message $|\psi\rangle$ to Alice, Mallory intercepts and stores $|\psi\rangle$. Then, he constructs a quantum CSS error correcting code $Q' : [[n, k, d]]$, randomly chooses a basis state $|m'\rangle$ such that $m' \in \boldsymbol{F}_2^k$, and encodes $|m'\rangle$ into $|c'\rangle$ using quantum CSS codes Q' as Bob does in Step 1. Successively, Mallory imitates Bob to send $|c'\rangle$ to Alice (he can also add some error on $|c'\rangle$ before sending it to Alice).

Step 3. **Alice encrypts.** Suppose that Alice wants to transmit privacy message p to Bob securely. When Alice receives $|c'\rangle$, she mistakenly thinks that it is the quantum message $|\psi\rangle$ sent by Bob because she can not distinguish its source. She applies an algorithm (Algorithm 1 in [21]) to transform p into a binary error vector $e'' = (X''|Z'')$ such that $t'' = w_q(e'') \leq [\frac{t}{2}]$. Then, she gets $|c''\rangle$ by applying the error e'' on $|c'\rangle$ as she does in Step 2 of the QSDC scheme [21] and sends $|c''\rangle$ back to Bob.

Step 4. **Mallory decodes and encrypts.** When Alice sends the quantum message $|c''\rangle$ to Bob, Mallory intercepts it and computes the error vector e'' by $|c''\rangle$. Then, he obtains Alice's secrete message p from e'' by Algorithm 2 in [21]. Successively, he gets $|\psi'\rangle$ by applying the error e'' on $|\psi\rangle$ and sends $|\psi'\rangle$ to Bob.

Step 5. **Bob decodes.** When Bob receives $|\psi'\rangle$, he mistakenly thinks that it is sent by Alice because he can not distinguish its source. Bob computes the error vector $e = (X|Z)$ by $|\psi'\rangle$ and recover message m''. If $m'' \neq m$, he believes that eavesdropping happens in the quantum channel; Otherwise, he computes $e'' = e + e'$ and performs Algorithm 2 in [21] to recover Alice's secrete bits p by e''.

4.3   The Effectiveness of the MITM Attack

This MITM attack works because Alice and Bob have no way to verify that they are talking to each other. In Step 3 of the MITM attack, as Alice can not distinguish the source of $|c'\rangle$, she transforms p into a binary error vector e'' and generates $|c''\rangle$ by applying e'' on the quantum state $|c'\rangle$ encoded by Mallory. Therefore, in Step 4 of the MITM attack, Mallory can successfully computes e'' by $|c''\rangle$. Furthermore, Bob can not distinguish the source of $|\psi'\rangle$. Thereby, Mallory's attack can not be found.

The QSDC scheme [21] does not discuss the measurement results with the classical communication channel to detect whether there is eavesdropping. This makes the MITM attacks have a chance. To condone this shortcoming, before

Alice encrypts the message p and sends it to Bob, she must measure partial quantum states and discuss the measurement results with Bob by the unblocked classical public communication channel to determine whether there is eavesdropping. In the next section, we will consider the improvement of the QSDC scheme [21].

5   An Improved QSDC Scheme Using Quantum CSS Codes

In this section, we will improve the QSDC scheme using quantum CSS codes [21] such that it can effectively resist the MITM attack. For the sake of the QSDC scheme being able to resist the MITM attack, Alice must determine the received quantum message was from Bob before she encrypts p and sends it to Bob.

Similar to [21], suppose that Alice wants to transmit a privacy message p to Bob securely by a noiseless quantum channel. The improved QSDC scheme is described as follows.

Step 1. **Encoding**. This step is Step 1 in the QSDC scheme [21].

Step 2. **Adding decoy states**. Bob gets $|\phi\rangle$ by inserting decoy states in sufficient number (such as, inserting k-qubit decoy states) of randomly selecting locations of $|\psi\rangle$. These decoy states are randomly chose from the set $\{|0\rangle, |1\rangle, |+\rangle, |-\rangle\}$, where $|+\rangle = \frac{|0\rangle + |1\rangle}{\sqrt{2}}$, $|-\rangle = \frac{|0\rangle - |1\rangle}{\sqrt{2}}$. Then, Bob sends the quantum message $|\phi\rangle$ to Alice by the quantum channel.

Step 3. **Measuring and comparing**. After Alice receives the quantum message $|\phi\rangle$, Bob announces the locations of decoy states by the classical public channel. Alice randomly selects $\{|0\rangle, |1\rangle\}$ basis or $\{|+\rangle, |-\rangle\}$ basis to measure the decoy states. Then, she tells Bob her choice of measurement basis and the measurement results by the classical public channel. The measurement results must consist with the sending states in where the measurement bases are the sending bases. So, Bob can determine whether the quantum state received by Alice is the quantum state sent by himself. If not, they abort the QSDC scheme; otherwise, they continue to the next steps.

Step 4. **Encryption**. This step is Step 2 in the QSDC scheme [21].

Step 5. **Decoding**. This step is Step 3 in the QSDC scheme [21].

6   Properties of the Improved QSDC Scheme

In this section, we will prove the security of the improved QSDC scheme for an ideal noiseless channel and give some of its properties. The main change is that the improved QSDC scheme adds Step 2 and Step 3 to the QSDC scheme [21]. Therefore, the security discussion in [21] can still be applied to the improved QSDC scheme. Here, we mainly show that the improved protocol can resist the MITM attack.

If Mallory wanted to apply the MITM attack, he must replace the quantum state $|\phi\rangle$ when Bob sent it to Alice. Furthermore, if Mallory desired to escape the detection in Step 3, the replaced quantum states in the positions of the decoy states must consist with the quantum states sent by Bob. Mallory did not know the specific locations of the decoy states when $|\phi\rangle$ is sent to Alice by the quantum channel. In addition, non-orthogonal quantum states can not be accurately measured. Therefore, Mallory can not get all information of the decoy states. Accordingly, Mallory can not replace $|\psi\rangle$ in the quantum states $|\phi\rangle$ and keep the decoy states unchanged. This means that Mallory can not apply an effective MITM attack.

Although, in order to resist MITM attacks, the improved QSDC scheme adds some classical information transfer, it still satisfies two requirements of QSDC [10] mentioned in Preliminaries. In addition, similar to the QSDC scheme in Ref. [21], the improved QSDC scheme does not use entanglement properties and disturbing can be detected by comparing some recovered bits m' and Bob's original bits m. Ref. [23] described an attack without eavesdropping which can not be resisted by many QSDC schemes. It should be noted that the improved scheme can resist the attack without eavesdropping [23] by comparing the recovered bits m' and Bob's original bits m.

7   Conclusion

We constructed an effective MITM attack on the QSDC scheme using quantum CSS error-correcting codes [21]. Furthermore, we improved the QSDC scheme using quantum CSS error-correcting codes [21] such that it can resist the middle attacks. The improved QSDC scheme satisfied the two basic requirements [10] mentioned in Preliminaries. The improved QSDC scheme does not use entanglement properties and the attack without eavesdropping [23] can be detected by comparing the recovered bits m' and Bob's original bits m.

Acknowledgment. This work is supported in part by the National Natural Science Foundation (Nos. 60873055, 61073054), the Natural Science Foundation of Guangdong Province of China (No. 10251027501000004), the Fundamental Research Funds for the Central Universities (No. 10lgzd12), the Research Foundation for the Doctoral Program of Higher School of Ministry of Education of China (No. 20100171110042), and the project of SQIG at IT, funded by FCT and EU FEDER projects Quantlog POCI/MAT/55796/2004 and QSec PTDC/EIA/67661/2006, IT Project QuantTel, NoE Euro-NF, and the SQIG LAP initiative.

References

1. Bennett, C.H., Brassard, G.: Quantum Cryptography: Public Key Distribution and Coin Tossing. In: IEEE International Conference on Computers, Systems and Signal Processing, pp. 175–179 (1984)
2. Ekert, A.K.: Quantum Cryptography Based on Bell's Theorem. Physical Review Letters 67, 661–663 (1991)
3. Lo, H.K., Chau, H.F.: Unconditional Security of Quantumkey Distribution over Arbitrarily Long Distances. Science 283, 2050–2056 (1999)

4. Shor, P.W., Preskill, J.: Simple Proof of Security of The BB84 Quantum Key Distribution Protocol. Physical Review Letters 85, 441–444 (2000)
5. Nielsen, M.A., Chuang, I.L.: Quantum Conputation and Quantum Information. Cambridge University Press, Cambridge (2000)
6. Beige, A., Englert, B.G., Kurtsiefer, C., Weinfurter, H.: Secure Communication With a Publicly Known Key. Acta Physica Polonica A 101, 357–368 (2002)
7. Boström, K., Felbinger, T.: Deterministic Secure Direct Communication Using Entanglement. Physical Review Letters 89, 187902 (2002)
8. Wójcik, A.: Eavesdropping on the "Ping-Pong" Quantum Communication Protocol. Physical Review Letters 90, 157901 (2003)
9. Zhang, Z., Man, Z., Li, Y.: Improving Wójcik's Eavesdropping Attack on the Ping-pong Protocol. Physics Letters A 333, 46–50 (2004)
10. Deng, F.G., Long, G.L., Liu, X.S.: Two-step Quantum Direct Communication Protocol Using the Einstein-Podolsky-Rosen Pair Block. Physical Review A 68, 042317 (2003)
11. Gao, T., Yan, F.L., Wang, Z.X.: Quantum Secure Direct Communication by EPR Pairs and Entanglement Swapping. Nuovo Cimento Della Societa Italiana di Fisica B 119, 313–318 (2004)
12. Gao, T., Yan, F.L., Wang, Z.X.: Deterministic Secure Direct Communication Using GHZ States and Swapping Quantum Entanglement. Journal of Physics A 38, 5761–5770 (2005)
13. Deng, F.G., Long, G.L.: Secure Direct Communication with a Quantum One-time Pad. Physical Review A 69, 052319 (2004)
14. Hoffmann, H., Bostroem, K., Felbinger, T.: Comment on Secure Direct Communication with a Quantum One-time Pad. Physical Review A 72, 016301 (2005)
15. Deng, F.G., Long, G.L.: Reply to Comment on Secure Direct Communication with a Quantum Onetime-pad. Physical Review A 72, 016302 (2005)
16. Man, Z.X., Zhang, Z.J., Li, Y.: Deterministic Secure Direct Communication by Using Swapping Quantum Entanglement and Local Unitary Operations. Chinese Physics Letters 22, 18–21 (2005)
17. Lucamarini, M., Mancini, S.: Secure Deterministic Communication without Entanglement. Physical Review Letters 94, 140501 (2005)
18. Nguyen, B.A.: Quantum Dialogue. Physics Letters A 328, 6–10 (2004)
19. Jin, X.R., Ji, X., Zhang, Y.Q., Zhang, S., Hong, S.K., Yeon, K.H., Um, C.I.: Three-party Quantum Secure Direct Communication Based on GHZ States. Physics Letters A 35, 67–70 (2006)
20. Deng, F.G., Li, X.H., Li, C.Y., Zhou, P., Zhou, H.Y.: Quantum Secure Direct Communication Network with Einstein-Podolsky-Rosen Pairs. Physics Letters A 359, 359–365 (2006)
21. Lü, X., Ma, Z., Feng, D.G.: Quantum Secure Direct Communication Using Quantum Calderbank-Shor-Steane Error Correcting Codes. Journal of Software 17, 509–515 (2006)
22. Wang, Y., Wang, H., Li, Z., Huang, J.: Man-in-the-Middle Attack on BB84 Protocol and its Defence. In: 2nd IEEE International Conference on Computer Science and Information Technology, pp. 438–439 (2009)
23. Cai, Q.Y.: The "Ping-Pong" Protocol can be Attacked Without Eavesdropping. Physical Review Letters 91, 109801 (2003)

Stochastic System Identification by Evolutionary Algorithms

Yi Cao, Yuehui Chen*, and Yaou Zhao

Computational Intelligence Lab, School of Information science and Engineering,
University of Jinan, 106 Jiwei Road, 250022 Jinan, P.R. China
yhchen@ujn.edu.cn

Abstract. For system identification, the ordinary differential equations (ODEs) model is popular for its accuracy and effectiveness. Consequently, the ODEs model is extended to the stochastic differential equations (SDEs) model to tackle the stochastic case intuitively. But the existence of stochastic integral is a rigid barrier. We simply transform the SDEs to their corresponding stochastic difference equations (SDCEs) to eliminate stochastic integrals and propose an easy but effective solution to stochastic system identification. In this solution, the maximum likelihood estimation can be applied and the evolutionary algorithms are used to determine structures and parameters of the unknown system.

Keywords: System identification, stochastic differential equation, evolutionary algorithms, maximum likelihood estimation.

1 Introduction

In many applications, we hope to know the inner laws of a dynamical system through the analysis of the time series data observed. This procedure is called system identification or reverse engineering. A popular solution to this problem is the ordinary differential equations (ODEs) model [2], which employs tree structure evolutionary algorithms such as GP, GEP, MEP .etc to approximate the observed data so as to obtain the structures and parameters of the ODEs equivalently.

An intuitive extension of the ODEs model is to append a stochastic item to each equation, and then this model is converted to a stochastic differential equations (SDEs) model (1).

$$
\begin{cases}
dX_1(t) = f_1(\boldsymbol{x}, t)dt + g_1(\boldsymbol{x}, t)dW_1(t) \\
dX_2(t) = f_2(\boldsymbol{x}, t)dt + g_2(\boldsymbol{x}, t)dW_2(t) \\
\quad\quad\quad \cdots\cdots \\
dX_l(t) = f_l(\boldsymbol{x}, t)dt + g_l(\boldsymbol{x}, t)dW_l(t)
\end{cases}
\tag{1}
$$

where $W_1(t)$, $W_2(t) \cdots W_l(t)$ are independent standard Brown motions. $\boldsymbol{x}$ is a vector containing the system status variables. f_i and g_i are normal functions, called drift coefficient and diffusion coefficient separately.

---

* Corresponding author.

D.-S. Huang et al. (Eds.): ICIC 2011, LNBI 6840, pp. 247–252, 2012.

Unfortunately, this modification introduces stochastic integral and makes the estimation of the parameters in a SDEs model strikingly difficult. One kind of methods are based on mathematical analysis and most of them rely on the solution of the SDEs, e.g. the Vasicek model [6] in finance. There are also some analysis methods for parameter estimation, such as local linearization [4]. But this kind of methods are hard to implement automatically because of the necessity of Itô formula, which can only take effect accompanying a properly constructed function. Another kind of methods are based on stochastic simulation [5], depending on numerous repetitions of numerical solution such as SRK method [7], to estimate the distribution of the sample data. But this kind of methods are extremely time consuming.

Our method is a compromise between analysis methods and stochastic simulation methods, which can be implemented by program with efficiency and can avoid complex analysis theories with less accuracy loss.

2   Methodology

For the SDEs system (1), we define its corresponding SDCEs system as (2),

$$
\begin{cases}
\Delta X_1(t) = f_1(\boldsymbol{x}, t)h + g_1(\boldsymbol{x}, t)J_1 \\
\Delta X_2(t) = f_2(\boldsymbol{x}, t)h + g_2(\boldsymbol{x}, t)J_2 \\
\quad\quad\ldots\ldots \\
\Delta X_l(t) = f_l(\boldsymbol{x}, t)h + g_l(\boldsymbol{x}, t)J_l
\end{cases}
\tag{2}
$$

where h is the time step (or the sampling period for the data drawn from a SDEs system) and $\Delta X(t) = X(t+h) - X(t)$. The differentiations of the standard Brown motion are transformed according to (3),

$$
J_i = W_i(t + h) - W_i(t) \sim \mathrm{N}(0, h) \qquad i = 1, 2 \cdots l
\tag{3}
$$

thus we obtain $J_1,\ J_2 \cdots J_l$, which are independent random variables.

The transformation from a SDEs system to its SDCEs system is approximate, but it is still sensible concerning the following reasons: (1) in SDEs model, the troubles caused by stochastic integral are inevitable, but they have little relation to the systems to recognize; (2) the difference system is widely used; (3) in many cases, the difference system can approximate its corresponding differential system well by shrinking the sampling period (time step); (4) our method can be implemented by program and runs much faster than stochastic simulation methods; (5) our method can reveal both structures and parameters of the unknown system. So the SDCEs system will be our focus.

The sample data are composed of several time series. Let $S = \{(\boldsymbol{x}_n, t_n)\}_{n=0}^{N}$ denote a time series, where t_n is the time and $\boldsymbol{x}_n$ is a l-dimensional vector. $(\boldsymbol{x}_0, t_0)$ is the initial value. While x_{nj} denotes the jth component of $\boldsymbol{x}_n$ and usually stands for a status of a system at the moment t_n. Observations may contain multiple time series, and the whole sample data is noted as $\{S_m\}_{m=1}^{M}$.

For explaining some details about the fitness function, we first introduce our method using the system (4), which is a special non-autonomous system.

$$\begin{cases} \Delta X_1(t) = f_1(t)h + g_1(t)J_1 \\ \Delta X_2(t) = f_2(t)h + g_2(t)J_2 \\ \quad\quad\quad ..... \\ \Delta X_l(t) = f_l(t)h + g_l(t)J_l \end{cases} \tag{4}$$

In (4), function f_i and g_i are normal functions only related to variable t. The fitness function is designed according to the principle of maximum likelihood estimation, so the system that throws these sample data most probably is the one that is closest to the original system and the one we want.

For a time series $S_i = \{(\boldsymbol{x}_n, t_n)\}_{n=0}^{N_i}$ in the sample data, we write the jth equation in the form of (5):

$$X_j(t_{n+1}) - X_j(t_n) = f_j(t_n)h + g_j(t_n)J_j \tag{5}$$

The observation of $X_j(t_n)$ is x_{nj}, so it is obvious that:

$$\frac{[x_{n+1,j} - x_{nj}] - f_j(t_n)h}{g_j(t_n)} \sim N(0, h) \tag{6}$$

then the probability density can be calculated by (7).

$$L_j = -\sum_{n=1}^{N_i} \ln g_j(t_n) - \frac{1}{2h} \sum_{n=1}^{N_i} \frac{[x_{n+1,j} - x_{nj} - hf_j(t_n)]^2}{[g_j(t_n)]^2} \tag{7}$$

The maximum likelihood function in log form is the sum of $L_j, j = 1, 2, \cdots l$, and we obtain the fitness function as (8).

$$L = \sum_{j=1}^{l} L_j \tag{8}$$

One point we want to emphasize is that two or more time series are required. If there is only one time series S and the tree structure evolutionary algorithm can be provided to encode any expression, then the result will be a common difference system, whose trajectory just passes all the sample dots in S. The stochastic terms will be forced to be a constant: 0. Then for each time series $S_m(m = 1, 2 \cdots M)$, $L^{(m)}$ is calculated using (8), and the final maximum likelihood function is the sum of them.

$$L_{final} = \sum_{m=1}^{M} L^{(m)} \tag{9}$$

Another point we want to explain is that the diffusion coefficient should be relevantly small to drift coefficient. If the diffusion coefficient is too large, the algorithm will accustom itself to the stochastic property of the sample data. Thus our results will strongly depend on the sample data.

For the more general case, system (2) at the beginning, the former maximum likelihood function for each dimension (7) is changed to (10),

$$L_j = -\sum_{n=1}^{N_i} \ln g_j(\boldsymbol{x}_n, t_n) - \frac{1}{2h} \sum_{n=1}^{N_i} \frac{[(x_{n+1,j} - x_{nj}) - hf_j(\boldsymbol{x}_n, t_n)]^2}{[g_j(\boldsymbol{x}_n, t_n)]^2} \tag{10}$$

3   Result and Improvement

Given an initial value, the sample data can be constructed by (2) or by SRK method. We choose MEP [3] to encode expressions, and the operator set is $\{+, -, \times, \div, \sqrt{}, \exp, \lg, \sin, \cos\}$. Parameters in the model are optimized by PSO [1] simultaneously. The diffusion coefficient is restricted a small constant and is not encoded by MEP but viewed as a parameter. Table 1 is the descriptions of the result table.

Table 1. Table specifications

Target	the original system, SDEs or SDCEs
Time step	sampling period
Data length	the length of each time series including initial value
Data type	SDEs means we use SRK to generate the sample data, while SDCEs means we use (2) to generate the sample data
Initial value	initial value is written in the form of $(x_1, x_2 \cdots x_l, t)$
Data sequences	the number of time series for each initial value
Systems found	the result given by algorithm in SDCEs form

The first example is shown in Table 2. The algorithm has not found the correct equations. If the algorithm runs thoroughly, we notice that the target system is discarded almost definitely. In stochastic case, it is evident that any finite sample data is insufficient theoretically. This makes it possible that some complex equations that approach to the sample data well have even larger fitness values, which we call them false solutions. And this phenomenon is inevitable if the operators are rich enough. But under the condition that operators and chromosome length are both limited, this reason is secondary.

The corresponding ODEs system of the target in Table 2, shown in (11), is an autonomous system with cyclic sinks, so the range of variable x in the sample data, named effective interval temporarily, is narrow. While only the part of the function on the effective interval will participate in evolution, leading to easy and frequent appearances of false solutions.

$$dX(t) = [-2\sin x + 4\cos x]dt \tag{11}$$

Table 2. Result of a simple system

Target	$dX(t) = [-2\sin x + 4\cos x]dt + 0.1dW(t)$
Time step	0.03
Data length	101
Data type	SDEs
Initial value	0.5
Data sequences 6	
Systems found	$\Delta X(t) = [0.84\cos(1.025x^2 + 8.02) + 3.79]h + 0.089J$
Data sequences 10	
Systems found	$\Delta X(t) = [1.18\sin(5.52\cos x + 2.34) + 1.17]h + 0.088J$

The remedy is to enlarge the effective interval. Rather than sampling more data for a single initial value, it is wiser to prepare different initial values. If the effective interval is wide enough, the original equations can be found. Table 3 gives the results after this improvement.

Table 3. Improved result

Target	$dX(t) = [-2\sin x + 4\cos x]dt + 0.1dW(t)$
Time step	0.03
Data length	101
Data type	SDEs
Initial value	0.5, 3.5, 6
Data sequences 3, 3, 3	
System found	$\Delta X(t) = [4.43\sin(1.0002x + 2.015) - 0.082]h + 0.09J$

However, false solutions cannot be eliminate completely, and additional selection strategy is required to eliminate the meaningless expressions. If there is no priori knowledge about the target system, it is reasonable to presume that the structure of the system is as simple as possible. By making use of this principle, a penalty term is introduced to fitness function.

$$L_{\text{new}} = L_{final} - \alpha C_{\text{nodes}} \tag{12}$$

In (12), L_{final} is the original fitness value computed by (9), and C_{nodes} denotes complexity of the equations, e.g. the quantity of total nodes of the tree structures. α is a small weight. A simple way is to set α to $\beta|L_{final}|$, and β is a small positive constant. So if extra terms cannot bring a significant increase of the fitness value, they will be discarded and the expressions will be simplified during the evolution.

In the following test, as shown in Table 4, we use MEP to encode all the functions in the system and apply (12) to simplify expressions.

Table 4. Result of a 2-dimensional system

Target	$\Delta X_1(t) = [1.5x_1 - 2x_2]h + [0.3x_1/(x_2 + 10)]J_1$
	$\Delta X_2(t) = 3x_1h + 0.1x_1x_2J_2$
Time step	0.03
Data length	101
Data type	SDCEs
Initial value	(1, 1)
Data sequences	4
System found	$\Delta X_1(t) = [1.5040x_1 - 1.98835x_2 - 0.0837]h + [0.0581\sqrt{x1} + 0.0353]J_1$
	$\Delta X_2(t) = [3.00749x_1 + 0.0242]h + [0.107505x_1x_2 + 0.00119579]J_2$

4   Conclusion

Our method is a compromise that can be easily employed in stochastic system identification, accompanying the evolutionary computation. In order to get a satisfying result, it is recommended to apply priori knowledge to exclude the meaningless solutions and to simplify the expressions through appending a penalty term.

Acknowledgments. This research was partially supported by the Natural Science Foundation of China (61070130) and the Shandong Provincial Key Laboratory of Network Based Intelligent Computing.

References

1. Eberhart, R., Kennedy, J.: A new optimizer using particle swarm theory. In: Proceedings of the Sixth International Symposium on Micromachine and Human Science, pp. 87–129 (1995)
2. Iba, H.: Inference of differential equation models by genetic programming. Information Sciences 178(23), 4453–4468 (2008)
3. Oltean, M., Groşan, C.: Evolving evolutionary algorithms using multi expression programming. In: Banzhaf, W., Ziegler, J., Christaller, T., Dittrich, P., Kim, J.T. (eds.) ECAL 2003. LNCS (LNAI), vol. 2801, pp. 651–658. Springer, Heidelberg (2003)
4. Shoji, I., Ozaki, T.: Estimation for nonlinear stochastic differential equations by a local linearization method. Stochastic Analysis and Applications 16(4), 733–752 (1998)
5. Tian, T.: Stochastic models for inferring genetic regulation from microarray gene eexpression data. BioSystems 99(3), 192–200 (2010)
6. Vasicek, O.: An equilibrium characterization of the term structure. Journal of Financial Economics 5(1), 177–188 (1977)
7. Wang, P.: Three-stage stochastic runge ckutta methods for stochastic differential equations. Journal of Computational and Applied Mathematics 222(2), 324–332 (2008)

Do MicroRNAs Preferentially Target the Genes with Low DNA Methylation Level at the Promoter Region?

Zhixi Su[1,*], Junfeng Xia[1,*], and Zhongming Zhao[1,2,3,**]

[1] Department of Biomedical Informatics
[2] Department of Psychiatry
[3] Department of Cancer Biology, Vanderbilt University School of Medicine,
Nashville, TN 37232, USA
zhongming.zhao@vanderbilt.edu

Abstract. DNA methylation in genes' promoter regions and microRNA (miRNA) regulation at the 3' untranslated regions (UTRs) are two major epigenetic regulation mechanisms in majority of eukaryotes. Both DNA methylation of gene's 5'promoter region and miRNA targeting 3' UTR can suppress gene expression and play very important roles in regulating many cellular processes. Although the gene silencing role of both promoter methylation regulation and the miRNA targeting have been well investigated, the relationship between them remains largely unknown. In this study, we used human single base-resolution methylome data of two cell lines to investigate the relationship between them. Our preliminary results suggested that there is a functional complementation between transcriptional promoter methylation and post-transcriptional miRNA regulation, suggesting a possible combined regulation system in the cellular system.

Keywords: DNA methylation, microRNA, gene expression, epigenetic regulation.

1 Introduction

Epigenetics refers to the heritable changes that modify DNA or associated proteins without changing the DNA sequence itself [1]. It has been commonly accepted that epigenetic mechanisms such as DNA methylation and microRNA (miRNA) regulation are important in gene expression regulation. DNA methylation is the best-studied epigenetic modification of the genome that plays important roles in regulating many cellular processes [2, 3, 4, 5, 6, 7]. About 10% of the 28 million methylated CpG dinucleotides are in the promoters of human genes, in which they physically obstruct the binding of transcriptional proteins to the gene or indirectly through the recruitment of methyl-CpG-binding domain proteins. The important role of DNA promoter methylation in gene expression regulation has been reinforced by the current whole genome bisulfite sequencing of more than 20 eukaryotes [8].

---

[*] These authors contributed equally to this study.
[**] Corresponding author.

D.-S. Huang et al. (Eds.): ICIC 2011, LNBI 6840, pp. 253–258, 2012.
© Springer-Verlag Berlin Heidelberg 2012

miRNAs are a class of small noncoding RNAs that act as post-transcriptional regulators of gene expression. They bind mRNAs in their 3' untranslated region (3' UTR) with complementary sequences and lead to translational repression and gene silencing [9, 10]. According to release 17 (April 26, 2011) of the miRNAs database miRBase (http://www.mirbase.org), human genome encodes 1424 miRNA sequences, which may target about 60% of human protein-coding genes. The huge number of miRNAs identified indicated that many biological processes are controlled by miRNAs-medicated gene expression regulation [11].

Although the gene silencing role of both transcriptional promoter methylation and the miRNA regulation has been well established, the relationship between them remains unknown [12, 13, 14]. Since the critical roles of DNA methylation and miRNAs in gene expression regulation, it would be interesting and insightful to investigate the relationship between these two mechanisms. Considering DNA methylation acting on gene's 5' promoter region while miRNA targeting 3' UTR to suppress gene expression, we hypothesize that there is a functional complementation between transcriptional promoter region methylation regulation and post-transcriptional miRNA regulation. Based on this hypothesis, we may predict whether miRNAs preferentially target the genes with low DNA methylation level at the promoter region. We tested our hypothesis by deeply analyzing the human methylome data in two cell lines.

2   Materials and Methods

2.1   Data Set

Human gene structure data was downloaded from Ensembl (version 54; hg18), which include the information about Ensembl Gene ID, Ensembl Transcript ID, Transcript Start (bp), Transcript End (bp), Ensembl Protein ID, 3' UTR Start, 3' UTR End and Chromosome, Strand [15].

The single-base resolution DNA methylation data were derived from [13], including whole genome bisulfite sequencing data for two human cell lines: H1 human embryonic stem cells and IMR90 fetal lung fibroblasts.

miRNA target prediction data was extracted from R package RmiR.hsa (http://www.bioconductor.org/packages/2.8/data/annotation/html/RmiR.hsa.html), including miRNA target site prediction results of 6 resources: miRBase, targetScan, miRanda, tarBase, mirTarget2 and PicTar.

2.2   Measurements of DNA Methylation Level

Based on the single-base resolution bisulfite sequencing data, we used methylation broadness to measure the DNA methylation level in specific genome regions, which represents the fraction of cytosine sites detected as methylated in a given DNA segment. It can be calculated as the proportion of methylated sites over the total sites in a sequence (i.e., #mCG sites/#total CG sites). Since the DNA methylation level of two strands in any given genomics regions are highly correlated, here we used the sense strand to represent the DNA methylation level for a given gene promoter region. In fact, we also used the methylation level of anti-sense and obtained the same conclusion.

2.3   Data Processing

In our work, we extracted the 3'UTR and promoter region information from Gene structure data. If there are multiple 3' UTRs or transcription start site (TSS) for one gene, the most frequently used UTRs and TSS or the longest UTR and the most anterior TSS were chose for further analysis. The methylation data for each promoter were extracted by mapping the promoter region (from -1000 to 200 by upstream from TSS) to the H1/IMR90 cell line DNA methylation data.

3   Results

3.1   Negative Correlation between Gene Promoter Methylation Level and miRNA Target Site Number at Differentially Methylated Genomic Regions

We obtained the differentially methylated regions (DMRs) from Lister *et al.* [13]. In their work, the DMRs were identified as the regions of the genome enriched for sites of higher levels of DNA methylation in IMR90 relative to H1 by Fisher's Exact Test. There are 491 regions were considered as DMRs between H1 and IMR90 cell lines. For the genes located at DMRs and other regions, we calculated the average number of microRNAs and average value of promoter methylation level in H1, respectively. On average, genes located at the DMRs and other regions have 17.2 and 14.3 microRNA targets sites, respectively ($P < 10^{-6}$, Mann-Whitney U test) (Figure 1a). One the other hand, the average value of mCG/CG for genes located at the DMRs and other regions are 0.26 and 0.44 ($P < 10^{-15}$) (Figure 1b). These findings indicate that genes located in DMRs are low methylated while might be regulated by more miRNAs. Thus, there exists a negative correlation between DNA methylation level and miRNA target sites number.

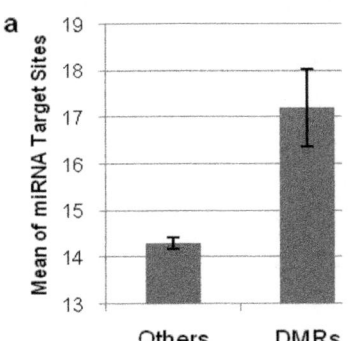 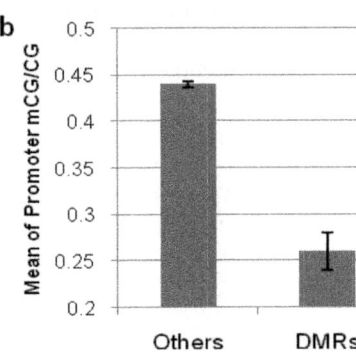

Fig. 1. Differentially methylated regions (DMRs) are low methylated and regulated by more miRNAs. Error bar: standard error.

3.2   Negative Correlation between Gene Promoter Methylation Level and miRNA Target Site Number within Partially Methylated Domains

Since Lister et al. [13] showed a trend of decreased level of methylation level in PMDs (partially methylated domains in IMR90 cell line, contiguous regions with an average methylation level less than 70%), we calculated the average miRNA target site number in PMDs and other regions. As expected, genes located in PMDs were regulated by more miRNAs ($P < 10^{-6}$) and have lower promoter methylation level ($P < 10^{-4}$) (Figure 2). The result still demonstrates that there exists a negative correlation between promoter methylation level and miRNA target site number.

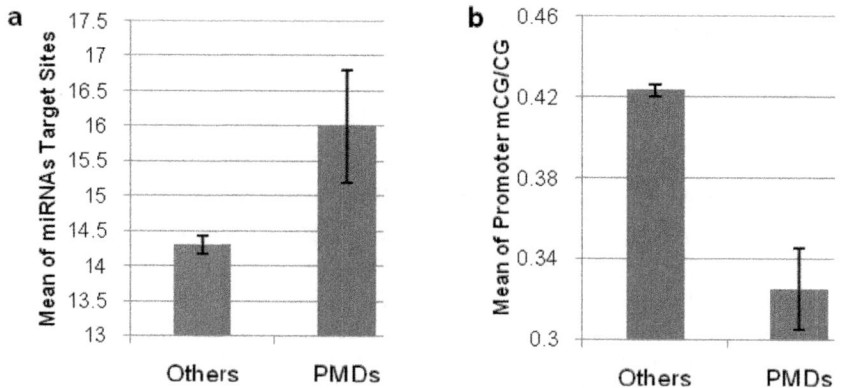

Fig. 2. Genes located in partially methylated domains (PMDs) might be regulated by more miRNAs. Error bar: standard error.

3.3 Correlation between Gene Promoter Methylation Level and miRNA Target Site Number in the Whole Genome

Among 13,529 genes with both promoter methylation level and miRNA target prediction data, we found a weak correlation between gene promoter methylation and miRNA target site (Spearman's $\rho = -0.14$, $P < 10^{-15}$, between mCG/CG in H1 and miRNA target site number; $\rho = -0.14$, $P < 10^{-15}$ in IMR90) (Figure 3).

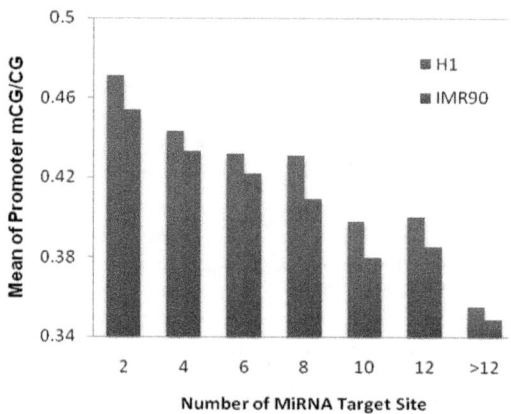

Fig. 3. Correlation between miRNA target site number and gene promoter methylation level

In this study, we also used "normalized" CpG content ($CpG_{O/E}$) to approximately infer the pattern of DNA methylation in human genome. $CpG_{O/E}$ is a robust measure of the level of DNA methylation on an evolutionary time scale due to specific mutational mechanisms of methylated cytosines [16]. We found a highly significant positive correlation between miRNA target site and $CpG_{O/E}$ of gene promoter region ($P < 10^{-15}$).

To sum up, we found the promoter region of protein-coding genes regulated by more miRNAs tend to be highly methylated.

4 Discussion and Future Perspective

To understand how DNA methylation and miRNA regulate the expression of their target genes, many previous exploratory studies have been performed but they all have focused on the effect of each mechanism on the expression of target genes separately. In this study, we investigated the relationship between the promoter methylation and miRNA regulation. Our preliminary results suggest that there is a functional complementation between transcriptional promoter region methylation regulation and post-transcriptional microRNA regulation.

From the evolutionary perspective, both recruitment of DNA methylation in gene promoter region and advent of new miRNA genes in the stage of transition from invertebrate and vertebrate have been suggested to contribute to the high complexity of vertebrate organs and cell types [17, 18, 19]. Although the latest studies have greatly improved our understanding of the evolutionary adaptations and conservation of DNA methylation and miRNA regulation, the relationship between DNA methylation and miRNA regulation and the ways in which these two mechanisms influence each other's evolution and function remain poorly understood. Our results of DNA methylation in relation to miRNA regulation thus provide intriguing angle to better understand the evolution to the organism's complexity.

Acknowledgment. We thank members of the Zhao lab for useful discussion and suggestions. This work was partially supported by NIH grant (LM009598) from the National Library of Medicine. Z. Zhao received additional support from Vanderbilt's Specialized Program of Research Excellence in GI Cancer grant (50CA95103) and the VICC Cancer Center Core grant (P30-CA68485).

References

1. Egger, G., Liang, G., Aparicio, A., Jones, P.A.: Epigenetics in Human Disease and Prospects for Epigenetic Therapy. Nature 429, 457–463 (2004)
2. Bestor, T.H.: The DNA Methyltransferases of Mammals. Hum. Mol. Genet. 9, 2395–2402 (2000)
3. Brown, S.E., Fraga, M.F., Weaver, I.C., Berdasco, M., Szyf, M.: Variations in DNA Methylation Patterns during the Cell Cycle of HeLa Cells. Epigenetics 2, 54–65 (2007)
4. Li, E., Bestor, T.H., Jaenisch, R.: Targeted Mutation of the DNA Methyltransferase Gene Results in Embryonic Lethality. Cell 69, 915–926 (1992)
5. Lippman, Z., Gendre, A.V., Black, M., Vaughn, M.W., Dedhia, N., et al.: Role of Transposable Elements in Heterochromatin and Epigenetic Control. Nature 430, 471–476 (2004)

6. Lister, R., O'Malley, R.C., Tonti-Filippini, J., Gregory, B.D., Berry, C.C., et al.: Highly Integrated Single-base Resolution Maps of the Epigenome in Arabidopsis. Cell 133, 523–536 (2008)
7. Weber, M., Hellmann, I., Stadler, M.B., Ramos, L., Paabo, S., et al.: Distribution, Silencing Potential and Evolutionary Impact of Promoter DNA Methylation in the Human Genome. Nat. Genet. 39, 457–466 (2007)
8. Su, Z., Han, L., Zhao, Z.: Conservation and Divergence of DNA Methylation in Eukaryotes: New Insights from Single Base-resolution DNA Methylomes. Epigenetics 6(2), 134–140 (2011)
9. Filipowicz, W., Bhattacharyya, S.N., Sonenberg, N.: Mechanisms of Post-transcriptional Regulation by MicroRNAs: Are the Answers in Sight? Nat. Rev. Genet. 9, 102–114 (2008)
10. Friedman, L.M., Dror, A.A., Mor, E., Tenne, T., Toren, G., et al.: MicroRNAs Are Essential for Development and Function of Inner Ear Hair Cells in Vertebrates. Proc. Natl. Acad. Sci USA 106, 7915–7920 (2009)
11. Griffiths-Jones, S., Saini, H.K., van Dongen, S., Enright, A.J.: Mirbase: Tools for Microrna Genomics. Nucleic Acids Res. 158, D154–D158 (2008)
12. Li, Y., Zhu, J., Tian, G., Li, N., Li, Q., et al.: The DNA Methylome of Human Peripheral Blood Mononuclear Cells. PLoS Biol. 8, e1000533 (2011)
13. Lister, R., Pelizzola, M., Dowen, R.H., Hawkins, R.D., Hon, G., et al.: Human DNA Methylomes at Base Resolution Show Widespread Epigenomic Differences. Nature 462, 315–322 (2009)
14. Zemach, A., McDaniel, I.E., Silva, P., Zilberman, D.: Genome-Wide Evolutionary Analysis of Eukaryotic DNA Methylation. Science 328, 916–919 (2010)
15. Kasprzyk, A., Keefe, D., Smedley, D., London, D., Spooner, W., et al.: EnsMart: A Generic System for Fast and Flexible Access to Biological Data. Genome Res. 14, 160–169 (2004)
16. Elango, N., Hunt, B.G., Goodisman, M.A., Yi, S.V.: DNA Methylation Is Widespread and Associated with Differential Gene Expression in Castes of the Honeybee, Apis Mellifera. Proc. Natl. Acad. Sci. USA 106, 11206–11211 (2009)
17. Gu, X., Su, Z., Huang, Y.: Simultaneous Expansions of Micrornas and Protein-Coding Genes by Gene/Genome Duplications in Early Vertebrates. J. Exp. Zool. B. Mol. Dev. Evol. 312B, 164–170 (2009)
18. Heimberg, A.M., Sempere, L.F., Moy, V.N., Donoghue, P.C., Peterson, K.J.: Micrornas and the Advent of Vertebrate Morphological Complexity. Proc. Natl. Acad. Sci. USA 105, 2946–2950 (2008)
19. Mandrioli, M.: A New Synthesis in Epigenetics: Towards a Unified Function of DNA Methylation from Invertebrates to Vertebrates. Cell Mol. Life Sci. 64, 2522–2524 (2007)

Ameliorating GM (1, 1) Model Based on the Structure of the Area under Parabola

Cuifeng Li, Jianbo Ye, and Fatai Zheng

Zhejiang Business Technology Institute
cuicui107@hotmail.com

Abstract. According to the research on the structure of background value in the GM(1,1) model, the background value's structure method of GM(1,1) model, a exact formula about the background value of $x^{(1)}(t)$ in the region $[k, k+1]$,which is used when establishing GM(1,1), is established by integrating $x^{(1)}(t)$ from k to $k+1$. The modeling precision and prediction precision of the ameliorating background value can be advanced.

Keywords: grey theory, background value, precisi $[k, k+1]$ on.

1   Introduction

The grey system theory has been caught great attention by researchers since 1982 and has already been widely used in many fields, such as industry, agriculture, zoology, market economy and so on. GM(1,1) has been high improved by many scholars from home and abroad. The grey system theory can effectively deal with incomplete and uncertain information system.

The background value is an important factor in the fitting precision and prediction precision. According to the research on the structure of background value in the GM(1,1) model, the background value's structure method of GM(1,1) model, a exact formula about the background value of $x^{(1)}(t)$ in the region $[k, k+1]$,which is used when establishing GM(1,1), is established by integrating $x^{(1)}(t)$ from k to $k+1$.The modeling precision and prediction precision of the ameliorating background value can be advanced. Moreover, the application area of GM(1,1) model can be enlarged. At last, the model of Chinese per-power is set up. Simulation examples show the effectiveness of the proposed approach.

2   Modeling Mechanism of the Ameliorating GM (1, 1) Model

2.1   GM(1,1) Model

Let the non-negative original data sequence be denoted by:

$$X^{(0)} = \{x^{(0)}(1), x^{(0)}(2),...,x^{(0)}(n)\} \tag{1}$$

D.-S. Huang et al. (Eds.): ICIC 2011, LNBI 6840, pp. 259–266, 2012.
© Springer-Verlag Berlin Heidelberg 2012

Then the 1-AGO (accumulated generation operation) sequence $X^{(1)}$ can be gotten as follow:

$$X^{(1)} = \{x^{(1)}(1), x^{(1)}(2), ..., x^{(1)}(n)\} \tag{2}$$

Where $x^{(1)}(k) = \sum_{i=1}^{k} x^{(0)}(i), k = 1, 2, ..., n$

The grey GM(1,1) model can be constructed by establishing a first-order differential equation for $x^{(1)}(t)$ as:

$$dx^{(1)}(t)/dt + ax^{(1)}(t) = u \tag{3}$$

Where a and u are the parameters to be estimated.

To calculate the integral of (3), we can get the follow equation:

$$\int_k^{k+1} dx^{(1)}(t) + a \int_k^{k+1} x^{(1)}(t)dt = u \int_k^{k+1} dt$$

Then

$$\int_k^{k+1} dx^{(1)}(t) = x^{(1)}(t)\big|_k^{k+1} = x^{(1)}(k+1) - x^{(1)}(k) = x^{(0)}(k+1) \tag{4}$$

Suppose

$$z^{(1)}(k+1) = \int_k^{k+1} x^{(1)}(t)dt \tag{5}$$

is the background value of $x^{(1)}(t)$ in the region $[k, k+1]$.

Thus, (3) can be rewritten into the following form:

$$x^{(0)}(k+1) + az^{(1)}(k+1) = u \tag{6}$$

Form (5), it is observed that the value of $z^{(1)}(k+1)$ can be established by integrating $x^{(1)}(t)$ from k to $k+1$.

Solve a and u by means of LS (least square):

$$\begin{pmatrix} \hat{a} \\ \hat{u} \end{pmatrix} = [B^T B]^{-1} B^T Y \tag{7}$$

Therefore, we can obtain the time response function by solving (3) as follow:

$$\hat{x}^{(1)}(k+1) = [x^{(1)}(1) - \frac{\hat{u}}{\hat{a}}]e^{-\hat{a}k} + \frac{\hat{u}}{\hat{a}} \tag{8}$$

2.2   The Improved Structure of the Background Value

$z^{(1)}(k+1)$ is the average of y and $x^{(1)}(k+1)$ in the traditional GM(1,1) model. We can see from Fig.1 that $z^{(1)}(k+1)$ using the traditional background value can be also regarded as the area of the trapezium $abcd$, but the real background value

$z^{(1)}(k+1)$ is the background value of $x^{(1)}(t)$ in the region $[k,k+1]$.Thus we can know that using the traditional background value to build the model will bring lower precision and higher error.

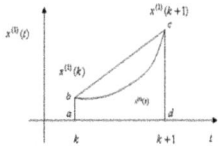

Fig. 1. $Z^{(1)}(k+1)$ using the traditional background value

The background value which is reconstructed by the method of rectangle is proposed by Tan Guanjun. The method has made better precision, but it also has rather bigger error which we can see from fig.2.

A new background value which using high precision interpolation formula and parabola method is proposed. This method can improve the prediction precise of GM(1,1).

The thought of this method is sorted out as follows. The interval of k to k+1 is divided into N space equivalently with the length named Δt is $\dfrac{1}{N}$, and let the values of the function $x^{(1)}(t)$ in the points is $x^{(1)}(k),x_1,x_2,...,x_{N-1,}x^{(1)}(k+1)$ correspondingly as fig.3.

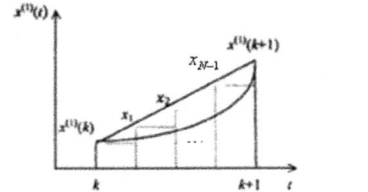
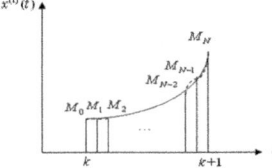

Fig. 2. $z^{(1)}(k+1)$ reconstructed by method of rectangle Fig.3 $z^{(1)}(k+1)$ reconstructed by method of parabo

The total of N areas under parabola is regarded as an approximation of the actual area.

Obviously, the bigger N is, the total of N areas is much closer to the actual area as fig.3.Thus, the background value with this method proposed by this paper is nearer the actual area than the traditional method. Now, the total of N areas named S_N is deduced as follows.

Three points can be determined a parabola as $y = px^2 + qx + r$ .Thus three corresponding point of the curve can be into a curved edge trapezoid.

The area of curved edge trapezium is substituted for rounded line of narrow trapezium in every space.

A parabola $y = px^2 + qx + r$ through three points $A[-h, y_1], B[0, y_2], C[h, y_3]$ in the $[-h, h]$ can be obtained. The area of the parabola named S can be gotten.

Firstly, p, q, r in the parabola equation can be obtained as follows.

$$\begin{cases} y_1 = ph^2 - qh + r \\ y_2 = r \\ y_3 = ph^2 + qh + r \end{cases}$$

$$2ph^2 = y_1 - 2y_2 + y_3$$

The value of S can be gotten as follows.

$$S = \int_{-h}^{h} (px^2 + qx + r)dx = [\frac{1}{3}px^3 + \frac{1}{2}qx^2 + rx]_{-h}^{h}$$

$$= \frac{2}{3}ph^3 + 2rh = \frac{1}{3}h(2ph^2 + 6r) = \frac{1}{3}h(y_1 - 2y_2 + y_3 + 6y_2) = \frac{1}{3}h(y_1 + 4y_2 + y_3)$$

The area of trapezia is relevant to the interval length 2h and the coordinate y_1, y_2, y_3 of the point of A, B and C.

From the above results, the area of parabola though M_0, M_1, M_2 and $M_1, M_2, M_3 \ldots M_{N-2}, M_{N-1}, M_N$ and can be gotten as follows.

$$S_1 = \frac{1}{3}\Delta t[x^{(1)}(k) + 4x_1 + x_2]$$

$$S_2 = \frac{1}{3}\Delta t(x_2 + 4x_3 + x_4) \qquad \text{where,} \Delta t = \frac{1}{N}$$

$$\ldots\ldots$$

$$S_{\frac{N}{2}} = \frac{1}{3}\Delta t[x_{N-2} + 4x_{N-1} + x^{(1)}(k+1)]$$

Adding the above areas, the background of S_N can be obtained as follows.

$$S_N = \frac{1}{3N}\{[x^{(1)}(k) + x^{(1)}(k+1)] + 2(x_2 + x_4 + \cdots + x_{N-2}) + 4(x_1 + x_3 + \cdots + x_{N-1})\}$$

Suppose

$$z_N^{(1)}(k+1) = S_N = \frac{1}{3N}\{[x^{(1)}(k) + x^{(1)}(k+1)] + 2(x_2 + x_4 + \cdots + x_{N-2}) + 4(x_1 + x_3 + \cdots + x_{N-1})\} \quad (9)$$

Where, x_i is the coordinate of the point when $t = k + \frac{i}{N}$.

$$x_i = x^{(1)}(k + \frac{i}{n}), i = 1, 2, \ldots, n-1 \cdot$$

The area of narrow trapezium is substituted for rounded line of narrow trapezium in every space. According to the formula of the area under trapezium, we can obtain as follows.

$$S_N = \int_k^{k+1} x^{(1)}(t)dt \approx \frac{1}{2}[x^{(1)}(k) + x_1]\Delta t + \frac{1}{2}(x_1 + x_2)\Delta t + \frac{1}{2}(x_2 + x_3)\Delta t + \ldots + \frac{1}{2}[x_{N-1} + x^{(1)}(k+1)]\Delta t$$

$$= \frac{1}{2N}[x^{(1)}(k) + 2x_1 + 2x_2 + 2x_3 + \ldots + 2x_{N-1} + x^{(1)}(k+1)] \quad (10)$$

Suppose $\quad z_N^{(1)}(k+1)=S_N=\dfrac{1}{2N}[x^{(1)}(k)+2x_1+2x_2+2x_3+...+2x_{N-1}+x^{(1)}(k+1)]$

Where, x_i is the ordinate value of the corresponding curve when

$i=k+\dfrac{i}{N}$, i=1,2,...,N-1 .

Thus $x_i=x^{(1)}(k+\dfrac{i}{N}),i=1,2...,N-1$

Obviously, the following equality could be gotten when $N=1$.

$z_N^{(1)}(k+1)=S_N=\dfrac{1}{2}[x^{(1)}(k)+x^{(1)}(k+1)]$.

2.3   Calculate the Background Value

From above, if the new background value is to be restructured, we should get the value x_i firstly. But the value x_i is not exit. Now Newton-Cores interpolation formula is introduced to get it.

Suppose $Y(k)=k,k=1,2,...,n$

Let $[Y(k),x^{(1)}(k)],k=1,2,...,n$ be the point of the corresponding curve, then using

Newton-Cores interpolation formula to get the value $x^{(1)}(k+\dfrac{i}{N})$ in light of its

corresponding abscissa $Y(k+\dfrac{i}{N}),i=1,2,...,n-1$.

Definition 3.1. [6] The function $f[x_0,x_k]=\dfrac{f(x_k)-f(x_0)}{x_k-x_0}$ is defined as a first-order

mean-variance of $f(x)$ about x_0,x_k .

The function $f[x_0,x_1,x_k]=\dfrac{f[x_0,x_k]-f[x_0,x_1]}{x_k-x_1}$ is defined as a second-order mean-

variance of $f(x)$ about x_0,x_k .

...

The function $f[x_0,x_1,...,x_{k-1}]=\dfrac{f[x_0,...,x_{k-3},x_{k-1}]-f[x_0,x_1,...,x_{k-2}]}{x_{k-1}-x_{k-2}}$ is defined as

a (k-1) order mean-variance of $f(x)$ about x_0,x_k .

The function $\quad f[x_0,x_1,...,x_k]=\dfrac{f[x_0,...,x_{k-2},x_k]-f[x_0,x_1,...,x_{k-1}]}{x_k-x_{k-1}}$ is defined as

a k order mean-variance of $f(x)$ about x_0,x_k .

Newton-Cores interpolation formula in [6] is as follow:

Suppose x is a point in $[a,b]$, then we can get:

$$f(x) = f(x_0) + f[x, x_0](x - x_0)$$

$$f[x, x_0] = f[x_0, x_1] + f[x, x_0, x_1](x - x_1)$$

...

$$f[x, x_0, x_1, ..., x_{n-1}] = f[x_0, x_1, ..., x_n] + f[x, x_0, ..., x_n](x - x_n)$$

As long as the latter formula has been taken into the former formula, we can get:

$$f(x) = f(x_0) + f[x_0, x_1](x - x_0) + f[x_0, x_1, x_2](x - x_0)(x - x_1) + ... + f[x_0, x_1, ..., x_n](x - x_0)$$

$$(x - x_1)...(x - x_n) + f[x, x_0, ..., x_n]\omega_{n+1}(x) = N_n(x) + R_n(x)$$

$$(16)$$

Where $R_n(x) = f(x) - N_n(x) = f[x, x_0, x_1, ..., x_n]\omega_{n+1}(x)$

$$\omega_{n+1}(x) = (x - x_0)(x - x_1)...(x - x_n)$$

The polynomial of Newton-Cores interpolating formula is as follow:

$$N_n(x) = f(x_0) + f[x_0, x_1](x - x_0) + f[x_0, x_1, x_2](x - x_0)(x - x_1)$$

$$+ ... + f[x_0, x_1, ..., x_n](x - x_0)(x - x_1)...(x - x_n)$$

Then the new background value is held easily as follow

$$z^{(1)}(k+1) = \frac{1}{2N}[x^{(1)}(k) + 2x_1 + 2x_2 + 2x_3 + ... + 2x_{N-1} + x^{(1)}(k+1)] \qquad (11)$$

Generally, the bigger N is, the more accurate GM(1,1) model is.

3 Example

Per-power is the measure of economic development level and people's living standards Thus, it is necessary to build the model of per-power and to predict developmental tendency. Now, using the method proposed by this paper to build China per-power form1980 to 1998 and to predict per-power form 1999 to 2001.

The model by using the method proposed in this paper is as follow:

$$\hat{x}^{(1)}(k) = 4136.36e^{0.070033(k-1)} - 3830.00 , \; k \geq 1$$

$$\hat{x}^{(0)}(k+1) = 279.77e^{0.070033k} , \; k \geq 1$$

$$\hat{x}^{(0)}(1) = 306.35$$

The error inspection of post-sample method can be used to inspect quantified approach .The post-sample error $c = S_1 / S_0$ (where S_1 is variation value of the error and S_0 is variation value of the original sequence) of the model proposed by this paper is $c_1 = 0.0867$, while the post-sample error proposed in [2] is $c_2 = 0.1186$. Then we can come to conclusion that the method proposed by this paper has improved the fitted precision and much better than the method proposed in [2]. The small error probability is $p = P\{|e^{(0)}(i) - e^{-(0)}| < 0.6745S_0\} = 1$ .Thus, the practical application results show the effectiveness of the proposed approach.

Table 1. Comparison of two modeling methods

Year	NO.	Real value	Method proposed in [2]		Method proposed in this paper	
			Model value	Relative error (%)	Model value	Relative error (%)
1980	1	306.35	306.35	0	306.35	0
1981	2	311.2	303.04	2.62	300.07	3.58
1982	3	324.9	325.16	−0.07	321.83	0.94
1983	4	343.4	348.89	−1.60	345.18	−0.52
1984	5	361.61	374.35	−3.52	370.22	−2.38
1985	6	390.76	401.67	−2.79	397.08	−1.62
1986	7	421.36	430.99	−2.29	425.88	−1.07
1987	8	458.75	462.45	−0.81	456.78	0.43
1988	9	494.9	496.2	−0.26	489.91	1.01
1989	10	522.78	532.42	−1.81	525.45	−0.51
1990	11	547.22	571.28	−4.40	563.57	−2.99
1991	12	588.7	612.97	−4.11	604.45	−2.68
1992	13	647.18	657.71	−1.63	648.30	−0.17
1993	14	712.34	705.71	0.93	695.33	2.39
1994	15	778.32	757.22	2.71	745.78	4.18
1995	16	835.31	812.49	2.79	799.87	4.24
1996	17	888.1	871.79	1.84	857.90	3.40
1997	18	923.16	935.42	−1.33	920.14	0.33
1998	19	939.48	1003.7	−6.83	986.89	−5.05
1999*	20	988.60	1076.9	−8.94	1058.49	−7.07
2000*	21	1073.62	1155.5	−7.65	1135.27	−5.74
2001*	22	1164.29	1239.9	−6.49	1217.63	−4.58

(predict value with*)

4   Conclusion

According to the research on the structure of background value in the GM(1,1) model, the structure method of background value, a exact formula about the background value of $x^{(1)}(t)$ in the region $[k,k+1]$, which is used when establishing GM(1,1), is established by integrating $x^{(1)}(t)$ from k to $k+1$. The modeling precision and prediction precision of the ameliorating background value can be advanced. Moreover, the application area of GM(1,1) model can be enlarged. At last, the model of Chinese per-power is set up. Simulation examples show the effectiveness of the proposed approach.

Acknowledgment. This paper is supported by the Natural Science Fundation of Zhejiang Province, P.R. China(NO: 602016).

References

1. Liu, S., Guo, T., Dang, Y.: Grey System Theory and Its Application. Science Press, Beijing (1999)
2. Tan, G.: The Structure Method and Application of Background Value in Grey System GM (1, 1) Model(I). Systems & Engineering-Theory, 98–103 (2004)
3. Chen, T.: A New Development of Grey Forecasting Model. Systems Engineering, 50–52 (1990)
4. Fu, L.: Systematic Theory and Application. Scientific Technical Document Publishing House (1992)
5. Shi, G., Yao, G.: Application of Grey System Theory in Fault Tree Diagnosis Decision. Systems Engineering theory & Practice 144, 120–123 (2001)
6. Gong, W., Shi, G.: Application of Gray Correlation Analysis in the Fe-spectrum Analysis Technique. Journal of Jiangsu University of Science and Technology (Natural Science), 59–61 (2001)

Reliability of Standby Systems

Salvatore Distefano

Dipartimento di Matematica, Università di Messina,
C.da Di Dio, 98166 Messina, Italy
sdistefano@unime.it
Dipartimento di Elettronica e Informazione, Politecnico di Milano,
Via Ponzio 34/5, 20133 Milano, Italy
distefano@elet.polimi.it

Abstract. Reliability theory bases on the concept of boolean components, i.e. of up, operating or down, failed components. But often such assumption is not adequate for modeling specific behaviors of components, units, subsystems and systems. It cannot catch, for example, different operating conditions of components due to dependencies on other components or environment variations.

Aim of this paper is to investigating a specific dynamic behavior, the standby phenomena in reliability contexts, starting from a characterization from both internal and external perspectives. The formal specification of the problem is obtained through the dynamic reliability theory, providing its analytical formulation.

Keywords: Standby, redundancy, standby redundant systems, k-out-of-n standby redundancy.

1 Introduction and Motivations

Standby is a hot topic in reliability as also highlighted in literature. With specific regards to the evaluation of the standby systems reliability several techniques have been used. For example, in [1] and [2] renewal theory and semi-Markov models are exploited to evaluate some specific case studies such as: three-state systems, systems with mixed constant repair time, systems with multi-phase repair, systems with non-regenerative states, two-component systems with cold standby and maintenance, and so on. The method of the supplementary variables and Laplace transform are instead used in [3,4] to evaluate the stationary availability of n-unit parallel redundant systems with correlated failures and single repair facilities.

However, to the best of our knowledge, the specific literature partially faces or lacks of some aspects that can arise by evaluating the problem from different perspectives. In particular, as introduced above, the concept of standby is always mixed to the redundant one. This fact can drive to misunderstanding or even to approximations that can result wrong and dangerous in the system design.

The main aim of this paper is to cover these lacks, focusing on standby and deeply investigating the related phenomena in reliability contexts from both the

D.-S. Huang et al. (Eds.): ICIC 2011, LNBI 6840, pp. 267–275, 2012.

internal and the external viewpoints. In the former case, the unit is observed in isolation, without taking into account the interactions with the external environment, in order to characterize and to evaluate its internal behaviour. Then, the unit behaviour is evaluated also considering such interactions, in a larger system context. With the support of dynamic reliability theory, the characterization thus specified is formalized in terms of specific equations starting from the conservation of reliability principle.

In order to achieve such goals, the remainder of the paper is organized as follows: section 2 provides background on the standby behaviour and related concepts; section 3 characterizes the standby from both the internal and the external points of view also introducing standby redundant systems, while section 4 specifies the standby behaviour in analytical terms. Then section 5 summarizes and closes the paper.

2   Preliminary Concepts

Standby systems are characterized by dual-operating mode: *active* and *sleep*. While in active-mode a standby system is fully operating, able to provide its services. Otherwise, in the sleep-mode no services are provided by the standby system until a specific external call, signal or input switches it from the sleep to the active mode.

Standby systems are widely used in modern technologies due to their capabilities to optimize costs, to reduce the environmental impact, to optimize the system reliability and availability, to adequately manage redundant resources, and so on.

Usually the concept of standby, in technological context, is referred to the energy or to the power consumption of the system. In fact, more and more often standby devices such as standby generators, standby batteries, standby power systems, etc., are used in designs, projects, schemes, data sheets and technical notes. As a confirmation of this, several technical glossaries now include standby and related terms, such as *sleep-mode* or *standby-mode* as in [10]: a mode in which electronic appliances are turned off but under power and ready to activate on command.

The attention attracted by standby and related issues has consolidated a research trend on the topic, especially in recent times in which the sensitiveness on environment, pollution and energy-related problems is strongly grown, giving rise to many government [7,8,9,11] and non-government [5,6,12] initiatives and projects.

This has impacted on the designing approach, preferring low-power devices managed through standby policies with the aim of reducing the power consumption and optimize costs, performance and reliability. In the ICT context, Green computing [16] was born in order to achieve such goals.

A good definition of standby in proposed in [11], in particular since it highlights the *relationship between standby and energy/power*, thus characterizing the active/sleep modes in terms of the load applied to the standby unit: in the former case a full load is applied, while sleep modes are characterized by partial or "phantom" loads. According to such viewpoint, the *hot/warm* standby represents fully or partial powered sleep mode, respectively, while in *cold* sleep mode the system is not powered.

3 Standby Characterization

Standby, in reliability contexts, is usually considered as a specific policy of redundancy. But, as discussed in section 1, it can be interpreted as a more general and complex concept that has to be investigated from a higher level of abstraction, separately from the redundancy.

With this aim, in the following the standby behaviour is studied in deep from two different points of view: *internal*, by observing the effects of the standby from the inside, and *external/system*, taking into account the interactions from a system reliability viewpoint, thus considering the standby unit as a component.

3.1 Inside a Standby Unit

The goal of this section is to observe a standby unit from the inside in order to identify the effects of standby into the unit and to characterize them as a specific state of the standby unit state machine. According to the (static) system reliability theory, two states can be assumed by a component/unit: *up* and *down*. A unit is therefore *Boolean*, i.e. it can only be either operating or failed, respectively. Such classification cannot adequately represent the standby, since the sleep-mode cannot be clearly identified as an up or a down state. It is thus necessary to review such classification by considering the standby behaviour.

A good starting point is the definition provided in [13]:

- **Up** - Pertaining to a system or component that is operable and in service. It can be:
  - *Operating* - Pertaining to a system or component that is operable, in service, and in use.
  - *Idle* - Pertaining to a system or component that is operable and in service, but not in use.
- **Down** - Pertaining to a system or component that is not operable or has been taken out of service.

In this way 3 features characterizing the states of a system can be identified:

- **operable** - if the system is ready for use, for example if it is physically intact;
- **serviceable** - if the system is ready to provide its service to the environment;
- **in use** - if the system is performing its service.

The serviceable property mainly regards the interaction between the unit and the external environment, and therefore it is better considered and evaluated in the next subsection.

Table 1. Standby unit state machine

FEATURE STATE	Operable Yes	Operable Not	Serviceable Yes	Serviceable Not	Use In	Use Not in
operating	X		X		X	
idle	X		X			X
dormant	X			X		X
failed		X		X		X

From such classification it is possible to obtain four meaningful states as reported in Table 1. In this way, the states classification of [13] into operating, idle and failed, is enriched by a new state, the *dormant* one. This latter describes a condition in which an operable unit is not in service due to a particular condition or constraint applied to the unit, for example an external input switching the unit in the sleep-mode as occur in standby unit.

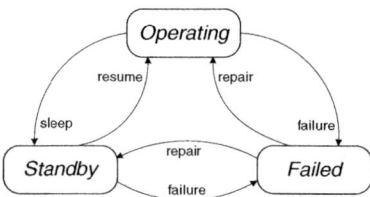

Fig. 1. State machine of a standby unit from inside

From an internal viewpoint, the dormant state cannot be distinguished by the idle one, since the serviceable property, as discussed above, is not taken into account from an internal point of view. Therefore the dormant state has to be considered as an up state. As a consequence, from the internal perspective, idle and dormant states can be merged into a standby state, as shown in Fig. 1. Even though such standby state is an up state as the operating one, it differs by this latter since it is characterized, as introduced in section 2, by a different ("phantom") load that can significantly affect its behaviour. Such distinction is particularly meaningful in case of cold and warm standby but, otherwise, it is meaningless in case of hot standby since the load characterizing such state is the same of the one characterizing the operating state. Therefore the hot standby state, from the internal point of view, is undistinguishable from the operating state and so it can be considered as an operating state.

The standby unit dynamics thus resulting is regulated by four main events: failure, resume, sleep and repair. The *failure* event brings to the failed state. Since in general a unit can fail from both operating and standby states, failures from both such states are possible. The only exception is the cold standby by which it is not possible to fail and therefore no failure events can be specified from it. On the other hand, the *repair* switches from failed to active or standby states, thus representing the standby unit repair. The *resume* represents transitions from the standby to the operating states, while *sleep* the reverse transitions, thus implementing the sleep-active cycles of a standby unit.

3.2 The System Perspective

Once characterized the internal behaviour of a standby unit, the focus is moved towards the system observing the unit from the outside. The standby unit is therefore now considered as a part of the system, thus taking into account the relationships with the other components and with the external environment. In this way the standby can be characterized as a dynamic-dependent behaviour, involving two parts: the driver/trigger side driving the standby, and the standby unit that reacts to the inputs incoming from driver.

From the system reliability point of view, the characterization discussed above and synthesized in Table 1 has to be adequately revised. First of all, it is necessary to take into account the serviceable property above identified and neglected, since it is strictly related to the external viewpoint. This means that the states characterization specified in Fig. 1 has to be modified, and, in particular, that it is no more possible to merge idle and dormant states into a unique standby state. This fact requires further explanations.

Since the dormant state represents the condition in which the standby unit depends upon an external event able to switch it in service, or serviceable, it is no more possible to consider it as an up state as above, but more properly it has to be evaluated as a down state, out of service. In this way, from a system reliability perspective, both the dormant and the failed states are identified as down states.

But there is an important difference between dormant and failed states: in case of failure, an external time-consuming action, such as a replacement or a repair, is required to restore the operating conditions of the unit; while, in the case of the dormant state, the standby unit is not physically failed, it waits for a driver input that can immediately switch it in service.

This further justifies the fact that the one-standby-state characterization of Fig. 1 does not well represent the behaviour of the standby component. It becomes necessary to split the unique standby state into two different states corresponding to the idle and the dormant modes as shown in Fig. 2. In this way,

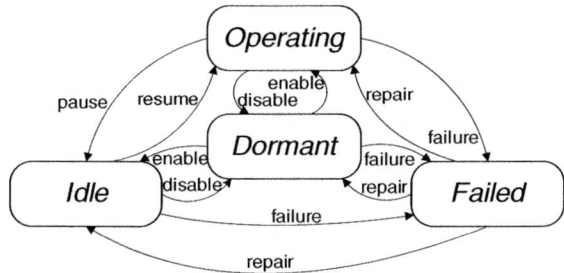

Fig. 2. State machine of a standby unit from outside

a unit can transitions to the dormant state if repaired or from both active and idle states (the *disable* event), vice-versa, a unit can be *enabled* from the dormant state by transitioning to both idle or active states (the *enable* event), or it can fail.

Thus, according to the standby unit characterization of Table 1, from an external point of view it is possible to identify operating and idle as up states, while dormant and failed as down states.

4   Formal Aspects

In section 2, starting from [11], a classification among hot, warm and cold standbys is performed while in section 3, a characterization of the standby behaviour in specific states is proposed. With the aim of quantitatively translating such characterization in terms of reliability, it is possible to base on the heuristic rules that establishes a relationship between the load applied to a generic standby system (subsystem or unit) and the system (un)reliability, as introduced above. According to such rule, to greater load applied to the system corresponds lower reliability or, equivalently, higher unreliability. This is due to the fact that, in case of greater load, the system makes more work and, consequently, its reliability quicker decreases or its failure rate increases. Such heuristic is particularly true when the standby system is subjected to a "phantom" load as in the cold standby case, since it does not work in standby and therefore cannot fail on its own but only if external causes arise.

Following this reasoning, it is possible to provide a reliability characterization of the standby state based on the relationship between reliability and load. From an high level point of view, the standby, the idle and the dormant states (more generally standby) of a standby system, as specified in section 3, can be further characterized as [14,15]:

- *Cold* - if the unit cannot fail autonomously;
- *Warm* - if the unit can fail autonomously, but with a lower failure rate or greater reliability distribution (in the statistical sense) than the operating state one;
- *Hot* - if the unit can fail autonomously as in the operating state, with the same failure rate or reliability distribution.

A standby unit in cold standby is (intrinsically) reliable during its sleep mode. Otherwise (warm and hot standby), the unit can also fail from the sleep mode. From the system point of view, as discussed in section 3.2, a standby can be either an up state in case it is classified as idle, or a down state if identified as dormant.

However, in general a standby state, both idle and dormant, can be characterized by its own reliability function that quantifies the impact of the standby on the system. From the probability theory point of view, a standby system SS can be characterized by, at least, two reliability functions: R_{SS}^O and R_{SS}^S. The former models the behaviour of the unit in the fully operating mode, while the latter characterizes the standby/sleep mode. Assuming the unit is initially, at time $t = 0$, fully operating and at time $t = x > 0$ it switches to the sleep mode, the standby system reliability function $R_{SS}(t, x)$ can be specified as follows:

$$R_{SS}(t, x) = \begin{cases} R^O_{SS}(t) \ t \leq x \\ R^S_{SS}(t) \ t > x \end{cases} \tag{1}$$

where x is associated to the trigger event random variable X driving the standby.

Thus, following the classification of cold, warm and hot standby, cold and warm standby can be more formally specified by eq. (1), in which at change point x there is a change of the reliability CDF from $R^O_{SS}(t)$ to $R^S_{SS}(t)$ with $R^O_{SS}(t) \geq R^S_{SS}(t) \ \forall t \geq 0$, while in the hot standby case $R^O_{SS}(t) = R^S_{SS}(t)$.

As stated above, the standby system active-sleep cycles are triggered by an external event, i.e., in probability terms, the standby system reliability R_{SS} depends on two events as in eq. (1): the standby unit lifetime T and the trigger event X. Assuming to know $R^O_{SS}(t)$ and $R^S_{SS}(t)$ or equivalently the corresponding unreliability CDFs $F^O_{SS}(t)$ and $F^S_{SS}(t)$, and the distribution of the conditioning event X, $F_X(x)$, the aim is to obtain the reliability of the standby system $R_{SS}(t)$ of eq. (1), removing its dependency on x.

Thus, exploiting the *theorem of total probability*, $F_{SS}(t) = 1 - R_{SS}(t)$ can be obtained as follows:

$$\begin{aligned} F_{SS}(t) &= \int_{-\infty}^{+\infty} Pr(T \leq t | X = x) f_X(x) dx = \\ &\int_0^t Pr(T \leq t | X = x) f_X(x) dx + \int_t^{+\infty} Pr(T \leq t | X = x) f_X(x) dx = \\ &= \int_0^t (1 - Pr(T > t | X = x)) f_X(x) dx + F^O_{SS}(t)(F_X(x)|_t^\infty) \end{aligned} \tag{2}$$

where $F^O_{SS}(t) = 1 - R^O_{SS}(t)$ and $F^S_{SS}(t) = 1 - R^S_{SS}(t)$ and $f_X(x) = F'_X(x)$. In order to evaluate $Pr(T > t | X = x)$, it can be observed that the dependent component for $t \leq x$ follows $F^O_{SS}(t)$ and then, at $t = x$, it switches into the sleep mode state characterized by $F^S_{SS}(t)$.

It is therefore necessary to understand what happens at change point x. Starting from the *conservation of reliability principle* [17], also known as the *Markov additive property* [18], the effect of the switching between the two distributions can be quantified in terms of time through τ, thus obtaining the *equivalent time* such that, at change point x:

$$R^O_{SS}(x) = R^S_{SS}(x + \tau) \quad \Rightarrow \quad \tau = R^{S(-1)}_{SS}(R^O_{SS}(x)) - x \tag{3}$$

assuming that $R^O_{SS}(\cdot)$ is strictly decreasing and therefore invertible.

In this way $Pr(T > t | X = x) = R^S_{SS}(t + \tau) = R^S_{SS}\left(t + R^{S(-1)}_{SS}(R^O_{SS}(x)) - x\right)$ since $x \leq t$, and thus substituting it in eq. (2):

$$F_{SS}(t) = F^O_{SS}(t)(1 - F_X(t)) + \int_0^t F^S_{SS}\left(t + R^{S(-1)}_{SS}(R^O_{SS}(x)) - x\right) f_X(x) dx \tag{4}$$

Eq. (4) thus quantifies the unreliability of a standby system switching from operating to sleep modes when triggered by an external event stochastically represented by X.

5 Conclusions

Standby systems are of strategic importance in the actual technologies, being a way for reducing the environmental impact and the costs, by extending the systems' time-to-life. Focusing on reliability, this paper studies in depth the standby behaviour considering different complementary perspectives, the intrinsic one, investigating a standby unit from the inside, and the external/operational viewpoint, considering reliability interactions and dynamics among the standby components of a system.

Starting from such characterization, the behaviour of a generic standby system is analytically investigated, providing the corresponding formal relationships and equations. Moreover, standby redundancy is formally evaluated firstly considering a 2-unit standby redundant system.

References

1. Limnios, N., Oprisan, G.: Semi-Markov Processes and Reliability, ser. Statistics for Industry and Technology. Birkhäuser, Boston (2001)
2. Janssen, J., Manca, R.: Semi-Markov Risk Models for Finance, Insurance and Reliability. Springer, Heidelberg (2007)
3. Itoi, T., Nishida, T., Kodama, M., Ohi, F.: N-unit Parallel Redundant System with Correlated Failure and Single Repair Facility. Microelectronics Reliability 17(2), 279–285 (1978)
4. Nikolov, A.V.: N-unit Parallel Redundant System with Multiple Correlated Failures. Microelectronics and Reliability 26(1), 31–34 (1986)
5. International Energy Agency (IEA). IEA Standby Power Initiative. Task Force 1: Definitions and Terminology of Standby Power, November 17-18, Washington, USA (1999)
6. International Electrotechnical Commission (IEC). IEC 62301 standard: Household electrical appliances - Measurement of standby power. Edition 2.0.
7. Australian Government, Department Of Environment, Water, Heritage and the Arts. Australian standby power program (September 2009),
   http://www.energyrating.gov.au/standby.html
8. U.S. Environmental Protection Agency and U.S. Department of Energy. ENERGY STAR program
9. The European Commission. The Directive 2005/32/EC on the Eco-Design of Energy-using Products (EuP)
10. Meier, A.: Standby Power Use - Definitions and Terminology. In: First Workshop on Reducing Standby Losses, Paris, France (January 1999)
11. Alliance for Telecommunications Industry Solutions (ATIS). American National Standard ATIS Telecom Glossary (2007)
12. Institute of Electrical and Electronics Engineers (IEEE). IEEE Std 446-1995 - IEEE Recommended Practice for Emergency and Standby Power Systems for Industrial and Commercial Applications
13. Institute of Electrical and Electronics Engineers (IEEE). IEEE 610-1991 - IEEE Standard Computer Dictionary. A Compilation of IEEE Standard Computer Glossaries (1991) ISBN:1559370793.

14. Dugan, J.B., Bavuso, S., Boyd, M.: Dynamic Fault Tree Models for Fault-Tolerant Computer Systems. IEEE Trans. Reliability 41(3), 363–377 (1992)
15. Distefano, S., Puliafito, A.: Dependability evaluation with dynamic reliability block diagrams and dynamic fault trees. IEEE Transactions on Dependable and Secure Computing 6(1), 4–17 (2009)
16. Murugesan, S.: Harnessing Green IT: Principles and Practices. IT Professional 10(1), 24–33 (2008), doi:10.1109/MITP.2008.10.
17. Kececioglu, D.: Reliability Engineering Handbook, vol. 1 & 2. DEStech Publications (1991) ISBN Volume 1: 1932078002, ISBN Volume 2: 1932078010
18. Finkelstein, M.S.: Wearing-out of components in a variable environment. Reliability Engineering & System Safety 66(3), 235–242 (1999)

A Kernel Function Based Estimation Algorithm for Multi-layer Soil Structure

Min-Jae Kang[1], Chang-Jin Boo[2], and Ho-Chan Kim[2]

[1] Department of Electronic Engineering
[2] Department of Electrical Engineering, Jeju National University, Korea
{minjk,boo1004,hckim}@jejunu.ac.kr

Abstract. This paper presents an analytic method based estimation scheme to extract soil parameters from the kernel function of integral equation of apparent resistivity. A fast inversion method has been developed for multi-layer earth structure based on the properties of kernel function. The performance of the proposed method has been verified by carrying out a numerical example.

Keywords: Kernel function, Wenner method, Apparent soil resistivity, Multi-layer earth model.

1 Introduction

It is important to know the earth structure in the given area when the grounding system is designed. Because badly designed grounding system cannot ensure the safety of equipment and personnel [1]. For simplifying the problem, in engineering applications, multilayer soils are modeled by N horizontal layers with distinct resistivity and depths [2]. A Wenner configuration method is well known to measure the soil resistivity for this simplified earth model.

The inversion of soil parameter is an unconstrained nonlinear minimization problem [3],[4]. Supposing there are N different layers below the ground surface then $2N-1$ parameters need to be determined, because there exists different $N-1$ thicknesses and N resistivity in the Wenner configuration model. Therefore, many different optimization methods have been carried out to estimate soil parameters in the hope of improving the performance. However, there exist two difficulties in estimating the soil parameters using optimization methods. On one hand, it is hard to obtain the derivatives of the optimized expression. On the other hand, the computing time is hugely consumed. These difficulties can be solved efficiently by using the proposed method in this paper.

This paper presents an analytic method to estimate soil parameters from the kernel function of integral equation of apparent resistivity. The main focus of this paper is to analyze the property of the kernel function for developing a fast inverting algorithm. A three-layer earth structure has been used in a numerical example for simplicity to examine the proposed method.

D.-S. Huang et al. (Eds.): ICIC 2011, LNBI 6840, pp. 276–281, 2012.
© Springer-Verlag Berlin Heidelberg 2012

2   Apparent Soil Resistivity: The Forward Problem

Estimation of soil parameters consists of the forward and inverse problems. The forward problem calculates the apparent resistivity from the assumed soil parameters, $h_i (i = 1, 2, \cdots, N - 1)$ and $\rho_i (i = 1, 2, \cdots, N)$.   In the inverse problem, the soil parameters are recalculated based on the imposed currents and the measured voltages at the electrodes placed cross on the earth surface. A schematic diagram of the apparent resistivity measurement setup has been shown in Fig. 1. h_i and ρ_i are the thickness and the resistivity of the i th layer respectively for an N th layer soil structure.

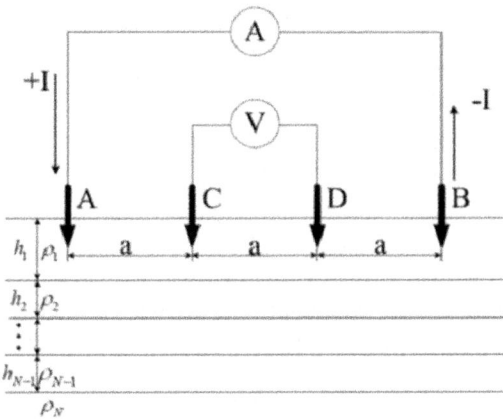

Fig. 1. Wenner configuration for measuring apparent soil resistivity of N-layer earth structure

A current I is injected into the soil by applying the power between electrodes A and B and the potential difference V between electrodes C and D is measured. By changing the test electrode span a, a set of apparent resistivity curves varying with electrode span can be obtained. If a point current source enters at a point A on the surface, the Laplace differential equation can be used to describe the potential distribution by using cylindrical coordinates, r, θ, z . In this case, Laplace differential equation in (1) becomes that there is cylindrical symmetry and so θ is eliminated,

$$\frac{\partial V^2}{\partial r^2} + \frac{1}{r}\frac{\partial V}{\partial r} + \frac{\partial V^2}{\partial z^2} = 0 \tag{1}$$

With appropriate boundary conditions, the surface potential at $z = 0$ becomes

$$\rho(a) = \rho_1 \left| 1 + 2a \int_0^\infty f(\lambda)[J_0(\lambda a) - J_0(2\lambda a)]d\lambda \right| \tag{2}$$

where ρ_1 is the soil resistivity of the first layer, $J_0(\lambda r)$ is the zero order Bessel function and the kernel function $f(\lambda)$ is as follows [1]

$$f(\lambda) = \alpha_1(\lambda) - 1 . \tag{3}$$

$$\alpha_1(\lambda) = 1 + \frac{2K_1 e^{-2\lambda h_1}}{1 - K_1 e^{-2\lambda h_1}} \qquad K_1(\lambda) = \frac{\rho_2 \alpha_2 - \rho_1}{\rho_2 \alpha_2 + \rho_1}$$

$$\alpha_2(\lambda) = 1 + \frac{2K_2 e^{-2\lambda h_2}}{1 - K_2 e^{-2\lambda h_2}} \qquad K_2(\lambda) = \frac{\rho_3 \alpha_3 - \rho_2}{\rho_3 \alpha_3 + \rho_2} \qquad (4)$$

$$\vdots \qquad\qquad\qquad \vdots$$

$$\alpha_{N-1}(\lambda) = 1 + \frac{2K_{N-1} e^{-2\lambda h_{N-1}}}{1 - K_{N-1} e^{-2\lambda h_{N-1}}} \qquad K_{N-1}(\lambda) = \frac{\rho_N - \rho_{N-1}}{\rho_N + \rho_{N-1}}.$$

Since more and more current will flow into the deep layer of the earth with the increasing of i, the corresponding earth apparent resistivity can reflect the property of the deep layer. Then, from a set of measured earth apparent resistivities varying with i, the earth structure can be obtained by solving an inverse problem.

If the potential difference ΔV_{CD} between potential electrodes C and D in Fig. 1 is tested, the apparent resistivity $\rho(a)$ can be expressed by

$$\rho(a) = 2\pi a \frac{\Delta V_{CD}}{I} = 2\pi a \frac{[V(a) - V(2a)]}{I} \qquad (5)$$

where $V(a)$ and $V(2a)$ are the potential of electrodes C and D generated by current test electrodes A and B respectively. By substituting (2) into (5), one can obtain

$$\rho(a) = \rho_1 \left[1 + 2a \int_0^\infty f(\lambda)[J_0(\lambda a) - J_0(2\lambda a)] d\lambda \right] \qquad (6)$$

3 Estimation of Soil Parameters: The Inverse Problem

The analytical formulation is developed by analyzing the kernel function of the integral equation of apparent resistivity. First, it is assumed that the kernel function is obtained by J. Zou's method [2]. Second, from the obtained kernel function the proposed formulation estimates the thickness and resistivity of the earth layers.

3.1 The Proposed Method: Inversion of Soil Parameters from the Kernel Function

The proposed formulation provides the analytic method for estimating thickness and resistivity of the earth layers. The characteristics of the kernel function in the integral equation of apparent resistivity is analyzed to develop the analytic method for estimating soil parameters.

Inspected in (3) and (4), the kernel function has the following characteristics:

(i) $\alpha_i(\lambda)$ $(i = 1, 2, \cdots, N-1)$ converges to 1 as λ increases.

(ii) $K_i(\lambda)$ $(i = 1, 2, \cdots, N-1)$ converges to the constant of $\dfrac{\rho_{i+1} - \rho_i}{\rho_{i+1} + \rho_i}$ as λ increases.

(iii) By rearranging the equation of left hand side in (4), the following equation can be derived

$$K_i(\lambda) = \frac{\alpha_i(\lambda) - 1}{\alpha_i(\lambda) + 1} e^{2\lambda h_i} \tag{7}$$

As seen in (7), if $\alpha_i(\lambda)$ is known, then h_i can be found with a simple method for satisfying the characteristic of the kernel function in (ii). It is noted that $K_i(\lambda)$ changes exponentially increasing or decreasing as λ increases with any other value of h_i except correct one.

(iv) At the same time when h_i is found, also the correct $K_i(\lambda)$ is obtained from (7).

(v) The following equation can be obtained by using the equation of right hand side in (4)

$$\alpha_{i+1}(\lambda) = -\frac{\rho_i}{\rho_{i+1}} \frac{K_i(\lambda) + 1}{K_i(\lambda) - 1} \tag{8}$$

Then $\alpha_{i+1}(\lambda)$ can also be formulated with the $K_i(\lambda)$ obtained in (iv) from (8). Because $\dfrac{\rho_i}{\rho_{i+1}}$ can be found from the $K_i(\lambda)$ which converges to the constant ratio of $\dfrac{\rho_{i+1} - \rho_i}{\rho_{i+1} + \rho_i}$.

Using the above characteristics of the kernel function, soil parameters can be inverted through the following procedures:

1) Obtain $f(\lambda)$, which is the kernel function, from the data of measured apparent resistivity.
2) In general, the resistivity ρ_1 of the first-layer is assumed to be same as the apparent one measured with small electrode span a in Fig. 1. As seen in (6), it is noted that the measured apparent resistivity with small electrode span becomes close to the resistivity ρ_1.
3) Let $i = 1$, $\alpha_1(\lambda) = f(\lambda) + 1$.
4) If $N \geq 2$, go to step 5), otherwise go to step 8).
5) Estimate h_i and $K_i(\lambda)$ using the characteristics of kernel function in (iii) and (iv).
6) Obtain $\alpha_{i+1}(\lambda)$ using the characteristics of kernel function in (v).
7) If $i < N - 1$, let $i = i + 1$, go to step 5), otherwise go to step 8).
8) Obtain ρ_{i+1} using $K_i(\lambda)$ and ρ_1 for $i = 1, \cdots, N$.

Now we are ready to construct the structure of N-layer earth model which is composed of the soil parameters ($h_1, \cdots, h_{n-1}, \rho_1, \cdots, \rho_n$).

3.2   Calculation Results

The performance of the proposed method has been verified by performing a numerical example. For the numerical example a three-layer earth model with the soil parameters(h_1, h_2, ρ_1, ρ_2 and ρ_3) was considered. In order to verify the proposed method, the scenario is started from the kernel function assuming that the kernel function is obtained by J. Zou's method [2].

Table 1. Parameters of a three-layer earth structure

Layer No	Resistivity($\Omega\cdot$m)	Thickness(m)
1	165	3.2
2	2510	17.5
3	336	∞

Fig. 2 shows that K_1, K_2, α_2 and $f(\lambda)$ which are obtained using equation (3) and (4) with the soil parameters in Table 1. Because of three-layer structure K_2 is the constant of ratio of ($\dfrac{\rho_3 - \rho_2}{\rho_3 + \rho_2} = -0.7639$) and K_1 also converges to that of ($\dfrac{\rho_2 - \rho_1}{\rho_2 + \rho_1} = 0.8766$) as seen in (4)

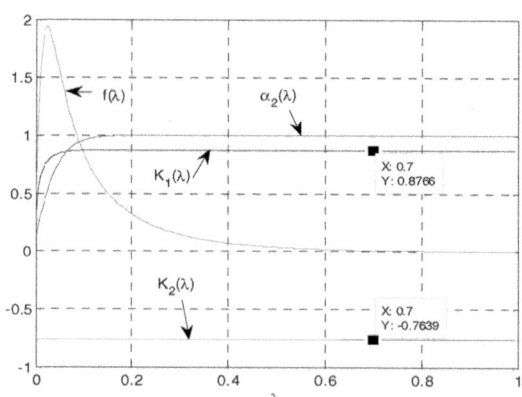

Fig. 2. The kernel and the related functions of 3-layer earth structure

Fig. 3 shows that $K_1(\lambda)$ converges to the constant with the correct value of h_1 which is 3.2 and $K_1(\lambda)$ diverges with any other values of h_1. At the same time when the correct value is found for h_1, also is $K_1(\lambda)$ which converges to 0.8766. Then, $\alpha_2(\lambda)$ is obtained as explained in the proposed procedure step 5). With the same procedures above, h_2 is found as 17.49 where $K_2(\lambda)$ converges to the constant 0.8303. Assuming that ρ_1 is known from the measured apparent resistivity, then all of the soil parameters (h_1 , h_2 , ρ_1 , ρ_2 and ρ_3) are obtained. The calculation result is $h_1 = 3.2$, $h_2 = 17.4$, $\rho_2 = 2511$, $\rho_3 = 335$, and which are obtained with the assumed value $\rho_1 = 165$

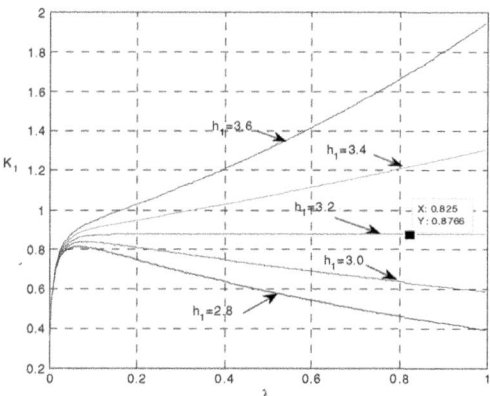

Fig. 3. The change of $K_1(\lambda)$ according to the different values of h_1

4 Conclusion

Analytic method to estimate the soil parameters of multi-layer earth model is presented in this paper. This proposed method is formulated based on the characteristics of the kernel function of integral equation of apparent resistivity. The robustness of the proposed algorithm has been verified by carrying out numerical simulations for different examples even though one numerical example for three-layer model has been shown in this paper. The results show a promising performance of the method.

Acknowledgements. This research was supported by Basic Science Research Program through the National Research Foundation of Korea (NRF) funded by the Ministry of Education, Science and Technology (2010-0025438).

References

1. Zhang, B., Cui, X., Li, L., He, J.L.: Parameter Estimation of Horizontal Multilayer Earth by Complex Image Method. IEEE Trans. On Power Delivery 20, 1394–1401 (2005)
2. Zou, J., He, J.L., Zeng, R., Sun, W.M., Chen, S.M.: Two-Stage Algorithm for Inverting Structure Parameters of the Horizontal Multilayer Soil. IEEE Trans. On Magnetics 40, 1136–1139 (2004)
3. Dawalibi, F.: Earth Resistivity Measurement Interpretation Techniques. IEEE Trans. On Power Apparatus Systems 103, 374–382 (1984)
4. Takahashi, T., Kawase, T.: Analysis of Apparent Resistivity in a Multi-layer Earth Structure. IEEE Trans. On Power Delivery 5, 604–612 (1990)
5. Tagg, G.F.: Earth Resistances. George Newnes Limited (1964)

Computing Technique in Ancient India

Chitralekha Mehera

Institute of Science Education, The University of Burdwan,
Burdwan, West Bengal, India
chitramehera@gmail.com

Abstract. This paper deals with the Sutra "Nikhilam Navatascaramam Dasatah" of Vedic Mathematics in computing mathematical problems viz., 10's complement, multiplication tables, addition, subtraction, multiplication and division quickly considering various examples. The problem solving techniques using the Sutra not only minimize the computational time but also seem to the students as an interesting and effective mathematical learning. The paper also explores the algebraic explanations of these techniques.

Keywords: Vedic mathematics, Bases, Vinculum, Deficiency, 10's complement.

1 Introduction

The advancement of mathematical science has made the computerization imperative, especially, with regard to the number theory rendering thereby tremendous minimization of computational time. The real time computational algorithms are received with appreciations by the stalwarts in computer science. The common domain for the two traditions of computer science and technologies is the mathematical science. The contributions of the Indian ancient mathematicians [6], though of significant importance, have not been well documented in the annals of Mathematical Science at global level. The depth of the mathematical formulation in the form of "Sutras" [2], [6] presented by the Indian ancients during Vedic periods has not been gauged and explored to the fullest extent. The buried treasure in the Indian mythological statements is so wide and great that they contain absolute perfection that can propel the current science to the great height.

This paper provides some methods basing on the ancient Indian mathematics which is both structural and progressive. These processes are actually based on some formulas called "Sutras". The simplicity of this approach is that calculations can be carried out mentally [1]. These processes not only minimize the computational time but also seem to the students as an interesting and effective mathematical learning.

2 Main Features of Vedic System

2.1 Base System

In Vedic Mathematics the base system [2] is very frequently used to reduce the computations involved. The base is considered to be the powers of 10. For example 10, 100, 1000, and so on. That power of 10 should be taken as base which is nearest to the number.

D.-S. Huang et al. (Eds.): ICIC 2011, LNBI 6840, pp. 282–289, 2012.

For the number more than the base, portion exceeding the base is written with a positive sign in front of the number. On the other hand, for the number less than the base, the number is subtracted from the base and the difference is written down in front of the number with a negative sign. This deviation is directly written by using the sutra which means all from 9 and last from 10.

Number	Base	Vedic Style
104	100	104+04
962	1000	962–038

If the difference between the number and the base is very large, then a suitable sub-base or working base can be used. Working base is a convenient multiple of the base.

2.2 Vinculum Operations

In this process the complement of the digits bigger than 5 is taken with a negative sign and a positive carry over to the next higher order digit. For example, $69 = 70 - 1 = 7\bar{1}$, thus 69 can be written as $(7\bar{1})$. But since the digit at the tenth place is also big its compliment $(10 - 7 = 3)$ can be taken and increase the digit previous to this by 1. Then 69 is written as $(1\bar{3}\bar{1})$. If there are few digits larger than 5 appearing together, their collective complement can be written by using the Nikhilum Sutra and increasing the previous digit by 1.

Thus 3789 can be written as $42\bar{1}\bar{1}$, and a vinculum number [2] $3\ \bar{3}\ \bar{4}\ \bar{2}$ can be written as $2\ 6\ 5\ 8$.

3   Idea of 10's Complement

3.1 The Sutra "Nikhilam Navatascaramam Dasatah"

Literally translated, it means *All From 9 and The Last From 10*. The practical application of it is actually quite easy to follow directly from the translation of the sutra. The idea of 10's complement in decimal system and 2's complement in binary system are the straightway application of this Sutra. Essentially, the 10's complement of a number tells how far the number is below the next higher power of 10.

Suppose one has to find the 10's complement of 456. By applying the Sutra *"Nikhilam Navatascaramam Dasatah"* [2], [3], [7] one can get it as 544 without doing any actual subtraction.

But if the number consists of n zeroes at the end, leave them off initially. Find the 10's complement of the remaining number with respect to the power of 10 just above the left-over number. Then add n zeroes back to the right of the answer.

Example: 10's complement of 78,000 is 22,000.

4   Subtraction

4.1 Using the Sutra "Nikhilam Navatascaramam Dasatah"

The most obvious problem that confronts one, when performing subtraction is the concept of borrowing numbers from the left. This can be done very easily using the Sutra *Nikhilam Navatascaramam Dasatah*.

Suppose one has to solve 4000-367. The first digit of 4000 i.e. 4 is simply reduced by 1 and then the Sutra is applied. So, the result is 3633.

For subtraction like 7000 - 4763 the first digit of 4763 i.e. 4 is simply subtracted from the first digit 7 and the result is reduced by 1, then the Sutra is applied. So, the result is 2237.

Problem 1: 18600045-9988989

Problem 2: 14943875-4934893-1938689-6473254

Problem 3: 349849548+56843899-45446842+56367267-28767948-
            37576553+74774798+84765898

Teaching these types of problems to young children, it could become an exercise in frustration both, for them and for the teacher. These types of problems can be easily solved by applying *the Sutra Nikhilam Navatascaramam Dasatah.*

Example: Suppose 87 is to be subtracted from 122 i.e. 122-87.

The problem can be written as 122+ 013 Applying the Sutra *Nikhilam Navatascaramam Dasatah* i.e. 10's complement Now add them up to get 135. From this, subtract the power of 10 with respect to which the 10's complement was taken, that is 100, and 135 - 100 is easy to derive mentally to be 35.

Solution of the aforementioned problems 1, 2 and 3:

Applying *Nikhilam Navatascaramam Dasatah* the problem 1 becomes
            18600045+0011011 = 18611056

From this 10000000 is subtracted and this gives 8611056 which is the required answer.

Applying Nikhilam *Navatascaramam Dasatah* the problem 2 becomes
14943875+5065107+8061311+3526746 = 31597039

From this 30000000 is subtracted and this gives 1597039 which is the required answer.

Following the same procedure the answer of this problem 3 is 510810067.

4.2   The Algebraic Basis of the Method

Suppose b is subtracted from a. Then $(a - b)$ becomes $a - (10^n - d)$, where d is the 10's complement of b with respect to 10^n. This is obviously the same as $a + d - 10^n$. *When the final answer is less than the power of 10 to be subtracted from it:* Consider the simple case of 110 - 988. This can be easily written as 110 + 012 (Applyingthe Sutra *Nikhilam Navatascaramam Dasatah*) which results 122. Now, 1000 is to be subtracted from 122, but 122 is much smaller than 1000. 122 - 1000 is the same as - (1000 - 122). The figure inside the parentheses can immediately be identified as the 10's complement of 122 with respect to 1000. Thus the answer becomes –878, which can be verified to be the correct answer. Thus, the simple rule is that if the final answer results in a number smaller than the power of 10 to be subtracted to give the final answer, then prepend a "-" sign in front and put down the 10's complement of the number as the rest of the answer.

5   Tables, Addition, Multiplication and Division

5.1   Tables

Nikhilam Sutra is used to write the deviations of the numbers from the base for the numbers smaller than the base. This method can be conveniently used to make the process of number tables a very easy task. Using this method there is no need to remember tables beyond 5 and anything beyond 5 can be split down to the level of 5. Note that this method is applicable to any table, even for numbers greater than 100.

Example: *Finding out the table of 1 9.*

Using vinculum method we have, $19 = 2\overline{1}$. Here the digits 2 and $\overline{1}$ are called operators, in writing table of 19 we take the operator $\overline{1}$ i.e. -1 for the unit's place and operator + 2 for the ten's place.

In the present example we apply operator − 1 to the digit 9 at the unit's place and obtain (9 - 1) = 8 as the second level number at unit's place, (8 - 1) = 7 as the third level number at unit's place and so on. Similarly the operator of the ten's place being + 2, we add + 2 at every stage to get the ten's place digit at the next higher level.

Table 1. Table of 19

Level		2	$\overline{1}$ (operators)
1		1	9
2		3	8
3		5	7
4		7	6
5		9	5
6	1	1	4
7	1	3	3
8	1	5	2
9	1	7	1
10	1	9	0

5.2 Addition

The process of addition in Vedic method is converted to a sequence of simple additions, each number being always less than ten. This method is termed as *Shuddhikaran* method in Vedic mathematics. The addition is carried out column by column in the usual manner, moving from bottom to top. Note that the addition process can be carried out in the downward direction with equal case and accuracy.

Example : Add 359, 874, 427.

```
   3   5.  9.
   8.  7   4
   7.  6.  6.
   4   2   7
  ─────────────
 2  4   2   6
```

Step a.   7 + 6 gives 13 which is greater than 9 and therefore *Shuddhikaran* is required. A dot is put beside 6, unity is dropped and only the number at the unit's place 3 is carried forward.

Step b.   3 + 4 = 7, being less than ten, no *Shuddhikaran* is required.

Step c.   Adding 7 to the next number 9, we get 16 which is greater than 9. Put dot beside 9 and 1 at ten's place is dropped and the number in the unit's place 6 is the answer.

Column 2 [ten's place]

Step a.   Before moving to the second column (ten's place) the dots due to *Shuddhikaran* in the first column (unit's place) are to be counted, we get 2 in this example. This 2 is added to the lowest number in the second column and the result is then added to the next number in the second column. Thus, (2+2) + 6 = 10. We have *Shuddhikaran* by putting dot over 6, the unity at ten's place is dropped and the number at unit's place 0 is carried forward.

Step b.   0+7=7, being less than ten, no *Shuddhikaran* is required.

Step c.   Adding 7 to the next number 5 we get 12 which is greater than 9. Put dot above 5 and 1 at ten's place is dropped and the number in the unit's place 2 is the answer.

Column 3 [hundred's place]

Step a.   Following the same procedure 3 is carried forward.

Step b.   Following the same procedure the answer is 4. The number of dots due to *Shuddhikaran* in the third column gives the carry forward number at the thousand's place. Hence, 2426 is the answer.

5.3   Multiplication

For the multiplication [2] by *Nikhilam Navatascaramam Dasatah* that power of 10 is taken as base which is nearest to the number to be multiplied. e.g. for multiplication of 9 by 7

$$
\begin{array}{c c c}
9 & - & 1 \\
7 & - & 3 \\
\hline
6 & / & 3
\end{array}
$$

i)   That power of 10 should be taken as the base which is nearest to the numbers to be multiplied. In this 10 itself is that power.

ii)   The numbers 9 and 7 are to be put vertically above and below on the left-hand side (as shown above)

iii) Each of them is subtracted from the base (10) and the reminders are written down (1 and 3) on the right hand side with a connecting minus sign (-) between them, to show that the both the numbers to be multiplied are less than 10.

iv)   The product will have two parts, one on the left side and one on the right. A dividing line (/) may be drawn for the purpose of demarcation of the two parts.

Now, the left-hand-side digit (of the answer) can be arrived at in one of 4 ways :

a)   The base 10 is subtracted from the sum of the given numbers (i.e. 9+7=16). And put (16-10) i.e. 6 as the left-hand part of the answer ; OR

b)   The sum of the two deficiencies (1+3=4) is subtracted from the base 10. one get the same answer (6) again; i.e. 10-1-3 = 6. OR

c)   Deficiency (3) on the second row is cross subtracted from the original number (9) in the first row. And one has got (9-3) i.e., 6 again. OR

d)   Cross Subtraction is done In the converse way (i.e., I from 7). And one gets 6 again as the left-hand side of the required answer.

v)   Now the two deficit figures (1 and 3) are vertically multiplied. The product is 3 and this is the right-hand-side portion of the answer.

vi)   Thus 9 × 7 = 63

Old historical traditions describe this cross-subtraction [2], [5] process as having been responsible for the acceptance of the × mark as the sign of multiplication.

The rule is for the base 10 the right part of the product is obviously of single digit and that is the unit's digit of the product of the deficiencies. The left surplus part i.e., the tens digit is carried over to the left part of the original product. Similarly, for the base 100 only two digits are entitled for the right part of the product. The left part digit i.e. the hundreds digit is carried over to the left part. Similarly, for the base 1000 and so on.

```
2 5 – 7 5                                    1 1 2 + 1 2
9 9 – 1                                      1 1 2 + 1 2
2 4 / 7 5                                    1 2 4 /  1 4 4 = 1 2 5 / 4 4
1 0 2 6 +    2 6
9 9 7 – 3
1 0 2 3 / 0 7̄ 8̄ = 1 0 2 2 / 9 2 2
```

The algebraic expressions are $(x-a)(x-b)=x(x-a-b)+ab$, $(x+a)(x+b)=x(x+a+b)+ab$ and $(x+a)(x-b)=x(x+a-b)-ab$ respectively with $x = 10, 100, 1000$ and so on.

In all these cases both the multiplicand and the multiplier, or at least one of them is very near to the base taken in each case. When neither of the multiplicand and multiplier is near a convenient base then a *Upasutra Anurupyena* [2], [7] which literally means *Proportionately* is used .

5.4   Upasutra "Anurupyena"

When neither the multiplicand nor the multiplier is sufficiently near a convenient power of 10 which can suitably serve as a base, a convenient multiple or sub-multiple of a suitable base is taken as the "working base" or "Sub-base", the necessary operation is performed with its aid and then the result is multiplied or divided proportionately in the same proportion as the original base may bear to the working base actually used.

Example: Suppose 46 is to be multiplied by 47. Both these numbers are far away form the base 100. The deficiencies are 54 and 53 from the base. 100 is accepted merely as a theoretical base and a sub-multiple 50 is taken as the working base.

```
                           4 6 – 4
1 0 0 / 2 = 5 0            4 7 – 3
                  2 ) 4 3 / 1 2 = 2 1 / ( 5 0 + 1 2 ) = 2 1 6 2
```

The left and the right parts are obtained following the multiplication procedure using the
Sutra *Nikhilam Navatascaramam Dasatah*.

As 50 is a half of 100, 43 is divided by 2 and 21 is written down as the left-hand portion of the answer. Half of the base i.e. 50 is to be added to the right-hand-side portion.

The answer therefore is 2162.

5.5   Division

Suppose 1232 is to be divided by 9.The problem can be written as

```
9              1
1  2  3  /  2
```

(Applying the Sutra *Nikhilam Navatascaramam Dasatah*) [2],

[4] Then put down 0 under the first digit of the dividend. Multiply the sum of 0 and 1 (this is just 1) the 10's complement and put it under the next digit of the dividend. Multiply $2 + 1 = 3$ by the 10's complement of 9 to get 3. Put that under the next digit of the dividend Multiply $3+3= 6$ by the 10's complement of 9. Put this under the next digit of the dividend. Since the next digit of the dividend is to the right of the "|", put a "|" in the third row before putting the 6 to its right. Add up the columns to get 136 |8.

```
9              1
1  2  3  /  2
0  1  3  /  6
─────────────
1  3  6  /  8
```

Examples: 8 9 4 3 7 8 / 7
4 9 8 5 7 / 7 9

```
        7    3
8 9 4 3 7 / 8
0 4 9 9 6 / 2 8 2 9
2 9 0 3
0 3 9
─────────────────────
1 2 7 3 6 3 / 2 8 3 7
1 2 7 7 6 8 / 2
7 9        2 1
4 9 8 / 5 7
0 8 4 / 1 7 0
0 3 4 / 9 2 0
0 0 0 / 4 6
───────────────
6 1 6 / 1 1 9 3
6 3 1 / 8
```

```
1 1 2 3 / 8 8

8 8      1 2
1 1 /    2 3
0 1 /    2 4
0 0 /    2
─────────────
1 2 /    6 7
```

It is in such division that the student finds his chief difficulty, because he/she has to multiply long big numbers by the "trial" digit of the quotient at every step and subtract that result from each dividend at each step; but, in this method by the Nikhilam formula, the bigger the digits, the smaller will be the required complement from 9 or 10 as the case may be; and the multiplication-task is lightened thereby. There is no subtraction to be done at all! And, even as regards the multiplication, no multiplication of numbers by numbers as such but only of a single digit by a single digit, with the pleasant consequence that, at no stage, is a student called upon to multiply more than 9 by more than 9.

A single sample example will suffice to prove this:

```
9 8 1 9 ) 2 0 1    3 7
0 1 8 1       0 2 ₁ 6 2
─────────────────────────
              2 0 4    9 9
```

Note : In this case, the product of 8 and 2 is written down in its proper place, as 16 with "carrying" over to the left and so on.

Thus, in the "division" process by the Nikhilam formula, only small single-digit multiplications are performed; no subtraction and no division is needed at all; and yet the required quotient and the required remainder is readily obtained. In fact, the division-work is accomplished in full, without actually doing any division at all!

5.6 The Algebraic Basis of the Division by Nikhilam Method for Two, Three and Four Digits Respectively

$(bx+c)/(x-a)= b+(c+ab)$, where b is the quotient and $(c+ab)$ is the remainder and $x=10$.

$(bx^2+cx+d)/(x-a)=bx+(c+ab)+\{d+a(c+ab)\}$, where $bx+(c+ab)$ is the quotient and $\{d+(c+ab)\}$ is the remainder and $x=10$.

$(cx^3+dx^2+ex+f)/(x^2-(ax+b))=cx+(d+ac)+[\{(e+bc)+a(d+ac)\}x+\{f+b(d+ac)\}]$,where $cx+(d+ac)$is the quotient and $[\{(e+bc)+a(d+ac)\}x+\{f+b(d+ac)\}]$ is the remainder and $x=10$.

6   Conclusion

The Vedic Sutras can be applied to Algebra, Geometry and other branches of modern Mathematics. Difficult problems or huge sums can often be solved quickly using these methods. These techniques are just a part of a complete system of mathematics which is far more systematic than the 'modern system'. Vedic Mathematics techniques are complementary, direct and easy. The real beauty and effectiveness of Vedic Mathematics cannot be fully appreciated without actually practicing the system. One can then see that it is perhaps the most refined and efficient mathematical system possible. Application of Vedic Sutras can propel the modern mathematical and other sciences to a great height. The Sutras are applicable not only to Decimal system, but also applicable to Binary and other systems.

References

1. Chatterjee, S.: Crunching Numbers the Vedic way. Times of India, Bombay Times (2002)
2. Maharaja, J.S.S.B.T.: Vedic Mathematics. Motilal Banarsidass Publishers, Delhi (1965)
3. Pande, T.G.: Jagatguru Shankaracharya Sri Bharati Krishna Teertha. B.R. Publishing Corporation, Delhi (1997)
4. Thakre, S.G., Karade, M.: Application of Nikhilam Division by 9. Einstein Foundation. International, Nagpur, Vedic Ganit vol. 1, Bull 3 (1985)
5. Williams, K.R.: Vertically and Crosswise, Mathematics in School. Mathematical Association (1999)
6. Vedic.: Mathematics, The History, Discoveries
7. Mehera, C.: Combined Application of Ancient and Modern Methods in Teaching Mathematics
8. Proceedings of the International Conference on Knowledge Globalization, pp. 244–251.EVENTS, Bangladesh (2010)

Smart Sensors: A Holonic Perspective

Vincenzo Di Lecce and Marco Calabrese[*]

Politecnico di Bari, DIASS
Taranto, Italy
{v.dilecce,m.calabrese}@aeflab.net

Abstract. This work introduces a novel perspective in the study of smart sensors technology. The final aim is to develop a new methodology that supports the conception, design and implementation of complex sensor-based systems in a more structured and information-oriented way. A smart sensor can be considered as a hardware/software transducer able to bring the measured physical signal(s) at an application level. However, when viewed through the lens of artificial intelligence, sensor 'smartness' appears to stay in between merely transduction and complex post-processing, with the boundary purposely left blurry and undetermined. Thanks to the recent literature findings on the so-called 'holonic systems', a more precise characterization and modeling of the smart sensor is provided. A 'holon' is a bio-inspired conceptual and computational entity that, as a cell in a living organism, plays the two roles of a part and a whole at the same time. To bring the right evidence of to the advantages of the holonic approach, an example smart application and a related prototype implementation for the disambiguation of low-cost gas sensor responses is shown. The proposed approach unravels the inherent complexity of the disambiguation problem by means of a scalable architecture entirely based on holonic-inspired criteria. Furthermore, the overall setup is economically competitive with other high-selective (hence high-cost) sensor-based solutions.

Keywords: smart sensors, IEEE 1451, holons, holonic systems, artificial intelligence.

1 Introduction

In the latest years, smart sensor technologies have elicited a great interest both on the scientific and OEMs side for their putative benefit of embedding intelligence within sensor circuitry. In response to the original chaos of legacy, incompatible, and often proprietary industrial solutions, the IEEE-1451 family of smart transducer interface standards has offered a common agreement around the definition of "smart sensor", in particular from an architectural viewpoint.

This set of standards is intended to aid transducer manufacturers in developing smart devices and to interface those devices to networks, systems, and instruments by incorporating existing and emerging sensor and networking technologies.

Four main conceptual blocks have been addressed for this aim, namely: Smart Transducer Interface Module (STIM), Network Capable Application Processor

---

[*] Corresponding author.

D.-S. Huang et al. (Eds.): ICIC 2011, LNBI 6840, pp. 290–298, 2012.

(NCAP), Transducer Independent Interface (TII), and Transducer Electronic Data Sheet (TEDS). These modules are arranged as depicted in Figure 1; they account for, at least, two interesting properties that represent a significant standout against traditional sensor technologies:

1. Transparent connection to virtually any networked environment (via NCAP);
2. Self-description ability (via TEDS) in terms of transducer identification, calibration, correction data, measurement range, and manufacture-related information, etc.

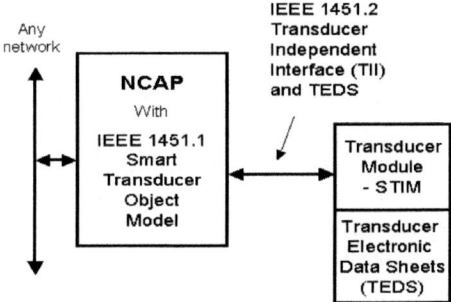

Fig. 1. Principal blocks of the IEEE 1451 set of standards (from [1])

According to modern artificial intelligence stance, it is not a hazard to deem these characteristics as intelligent. Yet, it remains quite undetermined to what extent smart sensors can be considered intelligent entities since they share aspects related to the pure physical world with others related to data acquisition, information processing and communication. Put in other terms, in current smart sensor definition, the boundary lying between physical (and analog) signals and their ontological (and digital) post-elaboration seems to be blurry at the moment.

In this context, our paper proposes a novel viewpoint to the concept of smart sensor by employing some recent findings in the theory of holons and holonic systems. As described later in more detail, holonic systems account for very desirable properties that allow us to provide a coherent framework for dealing with both physical-level and information-level aspects within a single model.

The rest of the paper is organized as follows: Section 2 briefly surveys the basic literature in the field of holonic systems, Section 3 introduces the proposed model, Section 4 presents an example smart application of the holonic model to low-cost gas sensor response disambiguation and a prototype implementation, Section 5 concludes.

2   Related Work

The term 'holon' was first introduced by Arthur Koestler in 1967 [2], meaning an entity capable of information processing and action, which is either atomic or contains parts which are lower-level holons. Hence, a holon plays the role of a whole and a part at the same time.

Koestler's basic idea moved from the concept of organism as a systemic whole to introduce that of multi-level hierarchy conceived as a self-regulating structure of sub-wholes. The latter have to be considered as functional parts of the system they are hosted in; however, at the same time, they also show autonomous behavior which makes them being a whole (intelligent) system as well.

Holons accounts for a recursive interpretation of the concept of system where part and wholes are not considered as separate entities. This part/whole dichotomy reflects on every level of the hierarchy and can be easily observed in the domain of life (e.g., cells in a living organism, human beings in a society, etc…).

2.1 Holons and Artificial Intelligence

Conceptually, holons are rather similar to intelligent agents [3, 4], the basic building blocks in Distributed Artificial Intelligence (DAI) modeling. The same relatedness also holds between holonic systems and multi-agent systems (MAS) [5]. Nevertheless, some important differences are worthy being stressed out.

Following the work of Marik and Pechoucek [6], a comprehensive comparison between holon and agent was presented in [7]. The authors identify three features marking the difference between the two models, namely:

- Information and physical processing: both elements are present in holons while agents are generally considered only as software entities;
- Recursiveness: which is characteristic for holons but not for agents;
- Organization: holons organize themselves according to holarchies, generally represented as dynamic hierarchic structures [8], while agent architectures are fixed and can range from horizontal to vertical organizations [5, 9]

All the three above-mentioned properties are somehow interleaved and account for a very specific characterization of holon as a conceptual and computational entity.

In [10] holon is viewed as an indivisible composition of hardware and software, along with its functional constituent layers. In this sense, it seems that a holon is theoretically closer to the idea of *robots* rather than to that of the so-called *softbots* [4].

As for recursiveness, any (decomposable) holon can be described by a recursive agency according to the model presented in [11]. The authors extend the Unified Modelling Language [12] to support the distinctive requirements of MAS through an object-based description. They state: "[…] *agent systems are a specialization of object-based systems, in which individual objects have their own threads of control and their own goals or sense of purpose*". The holonic (recursive) object-based representation is depicted in Figure 2; with minor adaptations, it is confirmed by more recent works concerning Holonic Manufacturing Systems [13].

A MAS is made of a collection of agents and, according to a holonic perspective, it can be viewed as an agent at the next higher granularity level; the atomic agent corresponds to an agent that cannot be decomposed (hence, it is not another MAS). Since MAS appears to be a significant component in holonic systems implementation, some authors explicitly refer to 'Holonic Multiagent Systems' [14, 15].

The multi-level granularity organization that spans from the previously presented MAS-based architecture is called holarchy. A holarchy is then a hierarchically-nested structure of holons and hence it is a holon itself (when viewed at the highest granularity level).

Some authors [16] also provide a behavioural description of the holarchy depending on the granularity level scoped. They assume that, for an external observer, those simple and reactive acts take place at the base of the holarchy while complex activities and behaviours are observable at the top of the holarchy. In other words, lower levels are more reactive and upper level holons are more proactive. It is useful noticing that this layered viewpoint is the same as described in [5] to MAS.

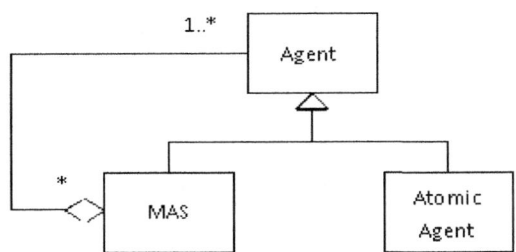

Fig. 2. Agent-based representation of a holonic system (slightly adapted from [11])

2.2   Latest Trends in the Literature of Holonic Systems

Recently, holonic systems as a modeling methodology are witnessing a major change. Albeit first significant use of holonic theories lies in the field of Intelligent Manufacturing [17] and Business Process Management [18], nowadays it seems that the influence of holonic thinking is moving far away its original territory. Probably, on a larger scale, this trend reflects the progressive affirmation of holistic approaches against reductionist ones.

In very recent works, holons have been used also as a computational paradigm [19], thus giving rise to a spectrum of prospective applications in several domains such as complex systems modeling, software engineering, knowledge extraction, etc.. In particular, the authors introduces the concept of holonic granule as a recursive building block to handle information processing at multiple granularity levels.

The novelty of the proposal lies in dealing with data and information processing at any possible level of abstraction with a single computational model. The current work stems from this idea.

3   A Holonic Smart Sensor

The proposed model for smart sensor is based on a single basic building block (the holonic granule) made of two layered parts:

- the Trandsuction Model (Physical Layer) and
- the Ontological Model (Information Layer)

The former accounts for the physical processing and can be identified by all the transduction functions mapping a physical sensed value into a byte-coded datum; the latter regards the information processing, i.e. the interpretation (or equivalently, the semantics) of the datum obtained from the subordinate physical layer. A block view of the proposed model is depicted in Figure 3.

Firstly, it is useful reminding that the holonic granule natively comprises the two parts and none of them makes sense without the other. In other words, the combination of the two models is the primitive representation for any type of more complex holonic granules.

Secondly, it appears evident that the two layers correspond to the hardware and software layers typical of any holonic architecture, as we saw before.

After these preliminary assumptions, a deeper understanding of the two layered models is now needed to better grasp our idea of holonic smart sensor.

The Transduction Model is provided by the sensor manufacturer by means of the datasheet. However, this requires a bit reflection upon. The datasheet is not an exhaustive source of information in the sense that it provides only a general view of the sensor behavior in steady operational condition. To properly account for such uncertainty, tolerance values are reported. In the case of sensor responses affected by other parameters (such as temperature and humidity in low-cost gas sensors), the datasheet provides also additional correction curves. Of course, in order for a datasheet of a really smart sensor to be informative at the maximum extent all possible influences on the measure should be addressed, but this would require a complete modeling of the environment and all the transductions happening from the bare signal to the bit value, thus making the datasheet largely unfeasible.

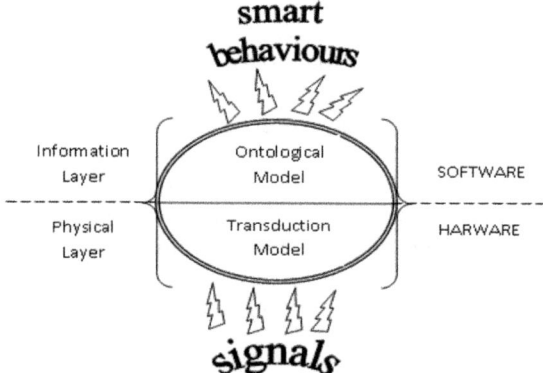

Fig. 3. A holonic conceptual view of a smart sensor. Physical and information processing are tightly intertwined so as to form a unique (holonic) entity: the Ontological Model accounts for a proper description/interpretation of the underlying Transduction Model, which in turn heavily depends on the application domain.

In the Ontological Model, again, transductions do occur but at an ontological level, i.e., between models and meta-models (or, in other words, between description of the target model and description of the source model). According to a holonic perspective, it can be assumed that these ontological descriptions for data interpretation are arranged at multiple nested granularity levels and more precise information can be obtained as long as a more detailed view of the underneath Transduction Model is available. The Ontological Model is in fact based on a representation of the (physical) transduction model. This could sound quite awkward, but, as shown in the subsequent section, it is exactly what is usually done in complex information processing and data fusion as well.

4   Example of a "Smart" Holonic Application

The proposed example is a part of a larger ongoing work aimed at showing how a smart compositions of low-cost sensors is able to manifest surprising discrimination abilities [20]. At the present time, different implementations for e-nose environmental applications have been engineered (see [21] for an example).

Fig. 3. A snapshot of one of the ZigBee-equipped smart sensor node used for our experiments. The whole device is based on low-cost components.

According to the previous holon definition, the general architecture of the solution employed for this paper comprises both hardware and software aspects, namely: sensors, electronic circuits for analog data processing, data storage, "smart" dialogue facilities basing on a multitask processing unit (linux kernel, in our case) and communication subsystems. A prototype is shown in Fig.4. This devices is equipped with two gas sensors, a small switch power supply, a Rabbit CPU card, and a ZigBee (IEEE® 802.15.4) interface. The e-nose can be configured to be a sensing node of a mesh of nodes with local intelligent information processing abilities[22]. Experiments with different type of e-noses (endowed with peer-to-peer communication abilities as in a MAS, or connected to TCP-IP data link) are under way[21].

The experimental setting is built around three extremely low-cost sensors, namely: Hanwei MQ131 and MQ136, Figaro TGS2602. In all these sensors, the sensing material is a metal oxide semiconductor. When the sensing layer is heated at a certain temperature in the air, oxygen is adsorbed on the crystal surface with a negative charge. As quoted in [23], by withdrawing electron density from the semiconductor surface, adsorbed oxygen gives rise to Schottky potential barriers at grain boundaries, and thus increases the resistance of the sensor surface. Reducing gases decrease the surface oxygen concentration and thus decrease the sensor resistance. The overall process causes a decrease in the resistance Rs of the sensing layer which can be measured against a standard value R0 gathered at optimal test condition. Sensor datasheets are given as Rs/R0 values against part-per-million (ppm) concentrations.

Sensor responses are highly ambiguous since a measured Rs/R0 value can be imputed to different concentrations depending on the gas actually being sensed. However, thanks to the proposed model, disambiguation can be viewed as an emerging property, only available at a higher information granularity level.

Suppose to consider five smart sensors, one for each gas sensor (group A) plus two wrapped around respectively a temperature and a humidity sensor (group B). If we considered each smart sensor as working alone, only highly uncertain information, at least from group A, would be obtained. Group A smart sensors would be in fact profoundly affected by Group B values. It can be hypothesized then, at a higher granularity level, temperature and humidity sensors share their ontological model with those in group A thus accounting for temperature/humidity calibration. At this stage however, the problem of ambiguity still remains.

Now, assuming a disambiguation procedure like the one proposed in [23] is employed, i.e., by sharing the ontological models within Group A, it is possible for each sensor to state which gas is actually being measured. The outcome would not be possible unless a knowledge sharing mechanism is employed.

It is noteworthy that the holonic solution is heavily dependent on the particular application domain. In the case of the provided example, the aim was to characterize unambiguously the hazardous emissions in a confined environment with particular reference to SO2, NH3 and CO gases. These all represent, above certain concentrations defined by the American Environment Protection Agency (see http://www.epa.gov/iaq/ for an overview), a potential threat to human health.

Therefore, the employed set of sensors was accurately chosen with reference to the datasheet characteristics in order to achieve two goals at once: 1) minimize the number of employed low-cost sensors and 2) cover the maximum number of measured contaminants of interest.

The analysis led to the so formed sensor triplet:

- MQ131 (supplied by the Hanwei Electronics Co. Ltd company) showing high sensitivity to ozone (O3), but also to NOx, CL2, etc.;
- MQ136 (supplied by the Hanwei Electronics Co. Ltd company) showing high sensitivity to sulfure dioxide (SO2), but also to CO, CH4 and other combustible gases;
- TGS2602 (from the Figaro USA inc. company): showing high sensitivity to volatile organic compounds (VOCs) and odorous gases.

Sensor response patterns provided a complete (albeit ambiguous) coverage of the contaminants for the given problem. Afterwards, the holonic approach made the application requirements achievable thanks to the information processing steps pictorially summarized in Figure 5. A detailed report of our experimental results can be found in [20].

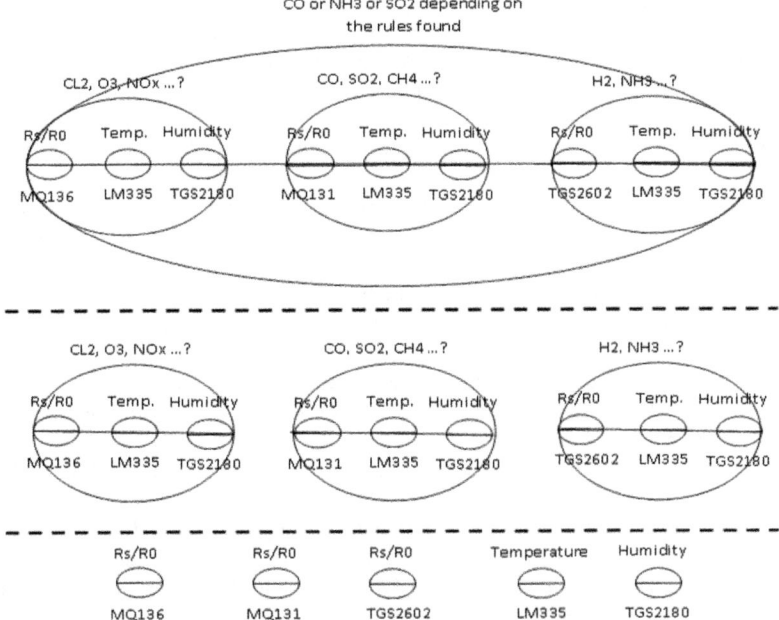

Fig. 5. A holarchy of smart sensors to address the proposed disambiguation problem. Holarchy at the lowermost level accounts for simple transductions. Holarchy at the medium level permits calibration in temperature and humidity but produces ambiguous output. Holarchy at the highest level solves the disambiguation problem by sharing knowledge obtained from the previous level.

5 Conclusion

In this work, a novel perspective in the study of smart sensors has been introduced by employing the concept of holon, a self-similar computational entity able to deliver both physical and information processing at multiple granularity levels. The proposed holonic view complements with existing standards and is highly suitable for modeling complex problems. To this aim, an example smart application and a prototype implementation were devised. In particular, it has been shown that the aggregation of (holonic) smart sensors at multiple granularity levels allows for addressing emergent complex behaviors such as gas discrimination with low-cost setups.

References

1. Lee, K.: A Smart Transducer Interface Standard for Sensors and Actuators. In: Zurawski, R. (ed.) The Industrial Information Technology Handbook. CRC Press, Boca Raton (2004)
2. Koestler, A.: The Ghost in the Machine, 1st edn. Hutchinson, London (1967)
3. Wooldridge, M., Jennings, N.R.: Intelligent Agents: Theory and Practice. Knowledge Engineering Review 10(2), 115–152 (1995)
4. Russell, S., Norvig, P.: Artificial Intelligence: A Modern Approach, 2nd edn. Prentice-Hall, Englewood Cliffs (2003)

5. Sycara, K.: MultiAgent Systems. AI Magazine 19(2), 79–92 (1998)
6. Mařík, V., Fletcher, M., Pěchouček, M.: Holons & agents: Recent developments and mutual impacts. In: Mařík, V., Štěpánková, O., Krautwurmová, H., Luck, M. (eds.) ACAI 2001, EASSS 2001, AEMAS 2001, and HoloMAS 2001. LNCS (LNAI), vol. 2322, pp. 106–233. Springer, Heidelberg (2002)
7. Giret, A., Botti, V.: Holons and agents. Journal of Intelligent Manufacturing 15, 645–659 (2004)
8. Xiaokun, Z., Norrie, D.H.: Dynamic reconfiguration of holonic lower level control. In: Proc. of the 2nd Intern. Conf. on Intelligent Processing and Manufacturing of Materials, vol. 2, pp. 887–893 (1999)
9. Okamoto, S., Scerri, P., Sycara, K.: The Impact of Vertical Specialization on Hierarchical Multi-Agent Systems. In: Proc. of the 23rd AAAI Conference on Artificial Intelligence, pp. 138–143 (2008)
10. Christensen, J.H.: Holonic Manufacturing Systems: Initial Architecture and Standards Directions. In: Proc. of the 1st Euro Workshop on Holonic Manufacturing Systems, HMS Consortium, pp. 1–20 (1994)
11. Parunak, H.V.D., Odell, J.: Representing Social Structures in UML. In: Intern. Workshop on Agent-oriented Software Engineering, vol. 2222, pp. 1–16 (2002)
12. Object Management Group, OMG Unified Modelling Language, Infrastructure, V2.1.2 (2007), http://www.omg.org/spec/UML/2.1.2/Infrastructure/PDF
13. Walker, S.S., Brennan, R.W., Norrie, D.H.: Holonic Job Shop Scheduling Using a Multiagent System. IEEE Intelligent Systems 2, 50–57 (2005)
14. Schillo, M., Fischer, K.: Holonic Multiagent Systems. Zeitschrift für Künstliche Intelligenz (3) (2003)
15. Fischer, K., Schillo, M., Siekmann, J.H.: Holonic Multiagent Systems: A Foundation for the Organisation of Multiagent Systems. In: Mařík, V., McFarlane, D.C., Valckenaers, P. (eds.) HoloMAS 2003. LNCS (LNAI), vol. 2744, pp. 71–80. Springer, Heidelberg (2003)
16. Shafaei, S., Aghaee, N.G.: Biological Network Simulation Using Holonic Multiagent Systems. In: 10th Intern. Conf. on Computer Modelling and Simulation, pp. 617–622 (2008)
17. Leitão, P., Restivo, F.: Implementation of a Holonic Control System in a Flexible Manufacturing System. IEEE Trans. Syst., Man, Cybern., Part C 38(5), 699–709 (2008)
18. Clegg, B.T.: Building a Holarchy Using Business Process-Oriented Holonic (PrOH) Modelling. IEEE Trans. Syst., Man, Cybern.—Part A: Systems And Humans 37(1), 23–40 (2007)
19. Calabrese, M., Piuri, V., Di Lecce, V.: Holonic Systems as Software Paradigms for Industrial Automation and Environmental Monitoring. In: Keynote Speech Paper in the IEEE Symposium Series on Computational Intelligence SSCI 2011 (2011)
20. Di Lecce, V., Calabrese, M.: Describing non-selective gas sensors behaviour via logical rules. To Appear in the Proceedings of the IEEE/ACM International Conference on Sensor Technologies and Applications - Sensorcomm 2011 (August 2011)
21. Di Lecce, V., Dario, R., Amato, A., Uva, J., Galeone, A.: WEGES: A Wireless Environmental Gases Electronic Sensor. To Appear in the Proceedings of the IEEE International Workshop on Advances in Sensors and Interfaces, IWASI 2011 (June 2011)
22. Di Lecce, V., Dario, R., Uva, J.: A Wireless Electronic Nose for Emergency Indoor Monitoring. To Appear in the Proc. Of the International Conference on Sensor Technologies and Applications - Sensorcomm 2011 (August 2011)
23. Lee, A.P., Reedy, B.J.: Temperature modulation in semiconductor gas sensing. Sensors and Actuators B: Chemical 60(1), 35–42 (1999)

Reducing Grammar Errors for Translated English Sentences

Nay Yee Lin[1], Khin Mar Soe[2], and Ni Lar Thein[1]

[1] University of Computer Studies, Yangon, Myanmar
[2] Natural Language Processing Laboratory
University of Computer Studies, Yangon, Myanmar
{nayyeelynn,nilarthein,kmsucsy}@gmail.com

Abstract. One challenge of Myanmar-English statistical machine translation system is that the output (translated English sentence) can often be ungrammatical. To address this issue, this paper presents an ongoing grammar checker as a second language by using trigram language model and rule based model. It is able to solve distortion, deficiency and make smooth the translated English sentences. We identify the sentences with chunk types and generate context free grammar (CFG) rules for recognizing grammatical relations of chunks. There are three main tasks to reduce grammar errors: detecting the sentence patterns in chunk level, analyzing the chunk errors and correcting the errors. Such a three level scheme is a useful framework for a chunk based grammar checker. Experimental results show that the proposed grammar checker can improve the correctness of translated English sentences.

Keywords: Statistical machine translation, grammar checker, context free grammar.

1 Introduction

Grammar checking is one of the most widely used tools within natural language processing applications. Grammar checkers check the grammatical structure of sentences based on morphological processing and syntactic processing. These two steps are parts of natural language processing to understand natural languages. Morphological processing is the step where individual words are analyzed into their components and non-word tokens, such as punctuation. Syntactic processing is the analysis where linear sequences of words are transformed into structures that show grammatical relationships between the words in the sentence [10]. The proposed grammar checker determines the syntactical correctness of a sentence.

There are several approaches for Grammar checking such as syntax-based checking, statistics-based checking and rule-based checking [2]. Among them, we build a chunk based grammar checker by using statistical and rule based approach. In this approach, the translated English sentence is used as an input. Firstly, this input sentence is tokenized and tagged POS to each word. Then these tagged words are

D.-S. Huang et al. (Eds.): ICIC 2011, LNBI 6840, pp. 299–306, 2012.
© Springer-Verlag Berlin Heidelberg 2012

grouped into chunks by parsing the sentence into a form that is a chunk based sentence structure. After making chunks, these chunks relationship for input sentence are detected by using sentence patterns. If the sentence pattern is incorrect, the system analyzes chunk errors and then corrects the grammar errors. The system has currently trained on about 6000 number of sentence patterns for simple, compound and complex sentence types.

This paper is organized as follows. Section 2 presents the overview of Myanmar-English Statistical Machine Translation System. In section 3, the proposed system is described. Section 4 reports the experimental results and finally section 5 concludes the paper and describes future work.

2   Overview of Myanmar-English Statistical Machine Translation System

Myanmar-English statistical machine translation system has developed source language model, alignment model, translation model and target language model to complete translation. Among these models, our proposed system builds target language model to check the grammar errors of translated English sentences.

Input for Myanmar-English machine translation system is Myanmar sentence. After this input sentence has been processed in three models (source language model, alignment model and translation model), translated English sentence is obtained in target language model. However, this sentence might be incomplete in grammar because the syntactic structures of Myanmar and English language are totally different. For example, after translating the Myanmar sentence "စားပွဲပေါ်တွင် စာအုပ်တစ်အုပ် ရှိသည်။", the translated English sentence might be "*is a book on table.*". This sentence has missing words "*There*" and "*the*" for correct English sentence "*There is a book on the table.*". As an another input "သူသည် ရေတစ်ခွက် သောက်နေသည်။", the translated output is "*He is drinking a cup water.*". In this sentence, "of" (preposition) is omitted from "*a cup of water*". These examples are just simple sentence errors. When the sentence types are more complex, reducing grammar errors and correction are more needed. There are many English grammar errors to correct ungrammatical sentences. At present, this grammar checker detects and provides the following errors according to the translated English sentences:

- If the sentence has missing words such as preposition (PPC), conjunction (COC), determiner (DT) and existential (EX) then this system suggests the required words according to the chunk types.
- In Subject-Verb agreement rule, if the subject is plural, verb has to be the plural. We check the verb agreement according to the person and number of the object.
- Sentence can contain inappropriate determiner. Therefore grammatical rules have been identified several kinds of determiner for appropriate noun.
- Translated English sentences can have the incorrect verb form. The system has to memorize all of the commonly used tenses and suggest the possible verb form.

3   Proposed System

There are very few spelling errors in the translation output, because all words are come from the corpus of the SMT system. Therefore, this system proposes a target-dominant grammar checking for Myanmar-English machine translation system as shown in Fig. 1.

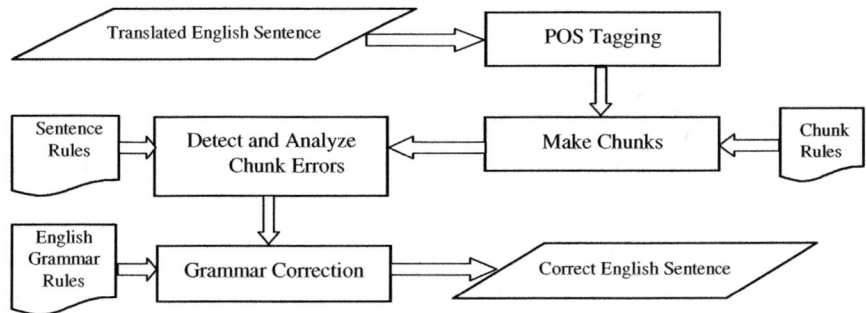

Fig. 1. Overview of Proposed System

3.1   Part-of-Speech (POS) Tagging

POS tagging is the process of assigning a part-of-speech tag such as noun, verb, pronoun, preposition, adverb, adjective or other tags to each word in a sentence. Nouns can be further divided into singular and plural nouns, verbs can be divided into past tense verbs and present tense verbs and so on [5].

POS tagging is the main process of making up the chunks in a sentence. There are many approaches to automate part of speech tagging. This system tags each word by using Tree Tagger which is a Java based open source tagger. However, it often fails to tag correctly some words when one word has more than one POS tags. In this case, refinement of POS tags for these words is made by using the rules according to POS tags of the neighbor words. The example for refinement tags is shown in Table 1.

Table 1. Example of Refinement Tags

Example	Incorrect Tag	POS tags of neighbor words	Refine Tag
He *bit* a rope.	*bit*=NN	Previous tag is PP	*bit*=VBD
He is a *tailor*.	*tailor*=VB	Previous tag is DT	*tailor*=NN

3.2   Making Chunks

A chunk is a textual unit of adjacent POS tags which display the relations between their internal words. Making chunks is a process to parse the sentence into a form that is a chunk based sentence structure. Input English sentence is made in chunk structure by using hand written rules. It represents how these chunks fit together to form the constituents of the sentence.

Context Free Grammar (CFG): Context Free Grammars constitute an important class of grammars, with a broad range of applications including programming languages, natural language processing, bio informatics and so on. CFG's rules present a single symbol on the left-hand-side, are a sufficiently powerful formalism to describe most of the structure in natural language.

A context-free grammar G = (V, T, S, P) is given by

- A finite set V of variables or non terminal symbols.
- A finite set T of symbols or terminal symbols. We assume that the sets V and T are disjoint.
- A start symbol S $\in$ V.
- A finite set P $\subseteq$ V $\times$ (V $\cup$ T)* of productions.

A production (A, α), where A$\in$ V and $\alpha\in$ (V $\cup$ T)* is a sequence of terminals and variables, is written as A$\rightarrow\alpha$. Context Free Grammars are powerful enough to express sophisticated relations among the words in a sentence. It is also tractable enough to be computed using parsing algorithms [9]. NLP applications like Grammar Checker need a parser with an optional parsing model. Parsing is the process of analyzing the text automatically by assigning syntactic structure according to the grammar of language. Parser is used to understand the syntax and semantics of a natural language sentences confined to the grammar. There are two methods for parsing such as Top-down parsing and Bottom-up parsing [3]. In this system, Bottom-up parsing is used to parse the sentences.

Chunking or shallow parsing segments a sentence into a sequence of syntactic constituents or chunks, i.e. sequences of adjacent words grouped on the basis of linguistic properties [7]. The system has used ten general chunk types for parsing. Some chunk types are subdivided into more detail chunk levels such as common chunk types for sentence patterns and specific chunk levels as shown in Table 2.

We make the chunk based sentence structure by assembling POS tags using CFG based chunk rules. For a simple sentence "A young man is reading a book in the library." is chunked as follows:

POS Tagging:
A [DT]   young [JJ]   man [NN]   is [VBZ]   reading [VBG]   a [DT]   book [NN]   in [IN]   the [DT]   library [NN]   . [SENT]

Making Chunks:
[DT][JJ][NN] [VBZ] [VBG] [DT] [NN] [IN] [DT] [NN] [SENT]

NCB1_ [VBZ][VBG] [DT] [NN] [IN] [DT] [NN] [SENT]

NCB1_   PRV1   _ [DT] [NN] [IN] [DT] [NN] [SENT]

NCB1_   PRV1   _   NCB1   _ [IN] [DT] [NN] [SENT]

NCB1_   PRV1   _   NCB1   _   PPC_ [DT] [NN] [SENT]

NCB1_   PRV1   _   NCB1   _   PPC_   NCB1 _ [SENT]

NCB1_   PRV1   _   NCB1   _   PPC_   NCB1 _ END

Chunk Based Sentence Pattern:

S = NCB1_PRV1_NCB1_PPC_NCB1_END

Table 2. Proposed Chunk Types

General Chunk Types	Common Chunk Types	Description	Specific Chunk Types	Example Words
NC	NCS1	Singular Noun Chunk for Subject only	PN1	He, She
	NCS2	Plural Noun Chunk for Subject only	PN2, PNC1	They, We all
	NCS	Singular and Plural Noun Chunk for Subject only	NEC	There
	NCB1	Singular Noun Chunk for both Subject and Object	ANN1,DNN1	A boy, This car
	NCB2	Plural Noun Chunk for both Subject and Object	ANN2,DNN2	The boys, These girls
	NCB	Singular and Plural Noun Chunk for both Subject and Object	NDC	This, These, Those
	NCO	Singular and Plural Noun Chunk for Object only	PN3	him, them, us
VC	PAVC	Past Tense Verb Chunk for both singular and plural noun	DV	wrote, gave
	PAV1	Past Tense Verb Chunk for singular noun	VDZ	was
	PAV2	Past Tense Verb Chunk for plural noun	VDP	were
	PRV1	Present Tense Verb Chunk for singular noun	ZV	is, writes, gives
	PRV2	Present Tense Verb Chunk for plural noun	PV	are, write, give
	FUVC	Future Tense Verb Chunk for singular and plural noun	MVB	will go, will be
TC	TC1	Time Chunk for Present	ADV1	Now, Today
	TC2	Time Chunk for Past	ADV2	Yesterday, last
	TC3	Time Chunk for Future	ADV3	Tomorrow, next
COC	XC	Subordinated Conjunction Chunk	NPR,NDT	Which, who, that
	CC	Coordinated Conjunction Chunk	COC	And, but, or
INFC	INC	Infinitive Chunk with Noun Chunk	IN1	to market, to school
	IVC	Infinitive Chunk with Verb Chunk	IBV	to go, to give
AC	AC	Adjective Chunk	R2A1,A1	more beautiful, old
RC	RC	Adverb Chunk	R1	usually, quickly
PTC	PTC	Particle Chunk	PTC	up, down
PPC	PPC	Prepositional Chunk	PPC	at, on, in
QC	QC	Question Chunk	QDT, QRB	Which, Where

3.3   Detecting Sentence Patterns and Analyzing Chunk Errors

After making chunks, these chunks relationship for input sentence are detected and analyzed chunk errors using trigram language model and rule based model.

Trigram Language Model. The simplest models of natural language are n- gram Markov models. The Markov models for any n-gram are called Markov Chains. A Markov Chain is at most one path through the model for any given input [4]. N-grams are traditionally presented as an approximation to a distribution of strings of fixed length.

According to the n-gram language model, a sentence has a fixed set of chunks, { c_0, c_1, c_2, $\cdots c_n$ }. This is a set of chunks in our training sentences, e.g., {NCB1, PAVC, AC,…, END}. In N-gram language model, each chunk depends probabilistically on the n-1 preceding words. This is expressed as shown in equation (1).

$$p(C_{o,n}) = \prod_{i=0}^{n-1} p(C_i | C_{i-n+1}, \cdots, C_{i-1})\tag{1}$$

Where (c_0) is the current chunk of the input sentence and it depends on the previous chunks. In trigram language model, each chunk (c_i) depends probabilistically on previous two chunks (c_{i-1}, c_{i-2}) and is shown in equation (2) [6].

$$p(C_{o,n}) = \prod_{i=0}^{n-1} p(C_i | C_{i-1}, C_{i-2})\tag{2}$$

Trigram language model is most suitable due to the capacity, coverage and computational power [1]. The trigram model is used in a greater level of some advanced and optimizing techniques such as smoothing, caching, skipping, clustering, sentence mixing, structuring and text normalization. This model makes use of the history events in assigning the current event some probability value and therefore, it suits for our approach.

Rule-Based Model. Rule-based model has successfully used to develop natural language processing tools and applications. English grammatical rules are developed to define precisely how and where to assign the various words in a sentence. Rule-based system is more transparent and errors are easier to diagnose and debug.

Rule-based model relies on hand-constructed rules that are to be acquired from language specialists, requires only small amount of training data and development could be very time consuming. It can be used with both well-formed and ill-formed input. It is extensible and maintainable. Rules play major role in various stages of translation: syntactic processing, semantic interpretation, and contextual processing of language [8]. Therefore, the accuracy of translation system can be increased by the product of the rule based correcting ungrammatical sentences.

3.4   Correcting Grammar Errors

The final step of our proposed system is controlled by English grammar rules. These rules can determine syntactic structure and ensure the agreement relations between various chunks in the sentence. Common chunk types for each general chunk are used to correct grammar errors. When the sentence patterns increased, the grammar rules will be improved. Some correction rules for subject verb agreement are NCS2_PRV2, NCS2_PAVC, NCS2_PAV2, NCS2_FUVC, NCS1_PAVC, NCS1_PAV1, NCS1_FUVC, NCS1_PRV1, NCB1_PAV1, NCB1_PAVC, NCS_FUVC and NCB_PAVC.

4   Experimental Results

For each input sentence, the system has classified the kinds of sentence such as simple, compound and complex. It also describes whether the sentence type is interrogative or declarative. The proposed system is tested on about 1800 number of sentences. The grammar errors mainly found in the tested sentences are subject verb agreement, missing chunks and incorrect verb form. The performance of this approach is measured with precision and recall. Precision is the ratio of the number of correctly reduced errors to the number of reduced errors in equation (3). Recall is the ratio of the number of correctly reduced errors to the number of errors in equation (4). The resulting precision and recall of reducing grammar errors on different sentence types are shown in Table 3.

$$PRECISION= \frac{Number of Correctly \, Reduced Errors}{Number of \, Reduced Errors} \times 100\% \qquad (3)$$

$$RECALL= \frac{Number of Correctly \, Reduced Errors}{Number of Errors} \times 100\% \qquad (4)$$

Table 3. Experimental Results of Reducing Grammar Errors

Sentence Type	Actual	Reduce	Correct	Precision	Recall
Simple	650	570	512	89.83 %	78.77 %
Compound	560	530	440	83.02 %	78.57 %
Complex	530	480	402	83.75 %	75.85 %

5   Conclusion and Future Work

This paper has presented a grammar checker for reducing errors of translated English sentences which makes use of a context free grammar based bottom up parsing, trigram language model and rule based model. It is expected that this ongoing research will yield benefits for Myanmar-English machine translation system. We use our own trained sentence patterns (dataset). Sample sentence patterns are presented as shown in Table 4. The proposed system currently detects the syntactic structure of the sentence and limits the detection of semantic errors.

In the future, we plan to check the semantic grammar errors for translated English sentences. We will expand more trained sentence rules to access all sentence types. If we get more sentence patterns, we can reduce more errors. We also plan to apply more English grammar rules for fully correction. Moreover, we plan to improve the accuracies of detection, analyzing and correction grammar errors.

Table 4. Sample Sentence Patterns

NC_VC_NC (Declarative)	QC_VC_NC_VC (Interrogative)
NCB_PRV1_NCB_END=S	QC_PAV2_NCB2_VC_IEND=S
NCB_PRV1_NCB1_END=S	QC_PAV1_NCB1_VC_IEND=S
NCB_PRV1_NCB2_END=S	QC_PAV1_NCB_VC_IEND=S
NCB_PRV1_NCO_END=S	QC_PAV2_NCB_VC_IEND=S
NCB1_PRV1_NCB_END=S	QC_PAV2_NCS2_VC_IEND=S
NCB1_PRV1_NCB1_END=S	QC_PAV1_NCS1_VC_IEND=S
NCB1_PRV1_NCB2_END=S	QC_PAV1_NCS_VC_IEND=S
NCB1_PAV1_NCB_END=S	QC_PAV2_NCS_VC_IEND=S
NCB1_PAV1_NCB1_END=S	QC_PRV2_NCB2_VC_IEND=S
NCS2_PRV2_NCB2_END=S	QC_PRV1_NCB1_VC_IEND=S
NCS2_PRV2_NCO_END=S	QC_PRV2_NCB_VC_IEND=S
:	:

References

1. Brian, R., Eugene, C.: Measuring Efficiency in High-Accuracy, Broad-Coverage Statistical Parsing. In: Proceedings of the COLING 2000 Workshop on Efficiency in Large-Scale Parsing Systems, pp. 29–36 (2000)
2. Daniel, N.: A Rule-Based Style and Grammar Checker (2003)
3. Keith, D.C., Ken, K., Linda, T.: Bottom-up Parsing (2003)
4. Lawrence, S., Fernando, P.: Aggregate and mixed order Markov models for statistical language processing. In: Proceedings of the Second Conference on Empirical Methods in Natural Language Processing, pp. 81–89. ACM Press, New York (1997)
5. Myat, T.Z.T.: An English Syntax Analyzer for English-to-Myanmar Machine Translation. University of Computer Studies, Yangon (2007)
6. Selvam, M., Natarajan, A.M., Thangarajan, R.: Structural Parsing of Natural Language Text in Tamil Using Phrase Structure Hybrid Language Model. International Journal of Computer, Information and Systems Science, and Engineering, 2–4 (2008)
7. Steven, A.: Tagging and Partial Parsing. In: Ken, C., Steve, Y., Gerrit, B. (eds.) Corpus-Based Methods in Language and Speech. Kluwer Academic Publishers, Dordrecht (1996)
8. Paisarn, C., Virach, S., Thatsanee, C.: Improving Translation Quality of Rule-based Machine Translation. In: 19th International Conference on Computational Linguistics (Coling 2002): Workshop on Machine Translation in Asia, Taipei, Taiwan (2002)
9. Ramki, T.: Context Free Grammars (2005),
   http://web.cs.du.edu/~ramki/courses/3351/2009Fall/notes/cfl.pdf
10. Elaine, R., Kevin, K.: Artificial Intelligent, 2nd edn. McGraw Hill, Inc., New York (1991)

Improve Coreference Resolution with Parameter Tunable Anaphoricity Identification and Global Optimization

Shuhan Qi, Xuan Wang[*], and Xinxin Li

Computer Application Research Center, Shenzhen Graduate School,
Harbin Institute of Technology, 518055 Shenzhen, China
shuhan_qi@qq.com, wangxuan@insun.hit.edu.cn,
lixxin2@gmail.com

Abstract. We build an anaphoric classifier with tunable parameters and realize a global connection between the classifier and coreference resolution. 60 features are used to build the anaphoric classifier. A corpus ratio control method is proposed and a "probability threshold" method is introduced to tune the precision and recall of the anaphoric classifier. The anaphoricity identification joints with the coreference resolution in a way of global optimization, and the parameters of anaphoricity identification are tuned according the result of coreference resolution. Maximum entropy is used for anaphoricity identification and coreference resolution with selected features. The results that combine the coreference resolution with the anaphoric classifier with different recall and precision are analyzed, and a comparison between our system and other coreference resolution systems is taken in the experiments analyze part. Our system improves the baseline coreference resolution system from 50.57 raise up to 53.35 on CoNLL'11 share tasks development data set.

Keywords: coreference resolution, anaphoricity identification, maximum entropy, global optimization.

1   Introduction

Coreference resolution as the core problem of nature language processing (NLP), gains researcher's increasing concern in recent years. Coreference resolution refers to the problem of determining different mentions that represent the same entity in real world.

Coreference resolution is widely used in information extraction, machine translation, automatic abstract system and etc. Coreference resolution has been at the center of several evaluations: MUC-6 (1995), MUC-7 (1998), CoNLL'02, CoNLL'03, CoNLL'11 share tasks. In this paper, we use the data that provided by CoNLL'11 share tasks as our train corpus and experimental data.

Anaphoricity Identification, whether named anaphoric determination or mention detection, is a method that identifies the anaphoricity and non-anaphoricity before the coreference resolution. Although the current coreference resolution especially those

---

[*] Personal Homepage: http://cs.hitsz.edu.cn/teachers/t1/1183037853.html

D.-S. Huang et al. (Eds.): ICIC 2011, LNBI 6840, pp. 307–314, 2012.

statistics based methods have performed reasonably well without mention detection, the resolution system might be improved by anaphoricity identification.

Our goal is to improve the coreference system by identifying the anaphoric with machine learning approach and global optimization. By tuning the parameters of anaphoricity identification according to the result of coreference resolution, we realize a global optimization coreference resolution system. The maximum entropy model is employed as the anaphoric classifier and the coreference classifier. The result shows that the performance of coreference resolution is improved by utilizing the mention detection.

2   Related Work

Since 1990's, the trend of research turn into the classify method with statistics model gradually. Aone et al. and McCarthy et al. treat the coreference problem as classification problem and apply machine learning approach on it [11][12]. Soon proposes a pairwise instance generation and a separate closest-first clustering mechanism, choosing the decision tree C4.5 as statistic model to solve the coreference problem [3]; Ng and Cardie improve the way of instance generation, propose a best-first clustering mechanism and introduce the linguistic features in decision tree[4]. Since dependence exists between mention pairs, the global context model should be more appropriate than the local binary classification model. To build a global model, Luo et al. use the Bell tree[13], Ng propose the global rerankers[14], McCallum and Wellner employ the conditional random fields (CRF) [15].

Ng and Cardie propose a supervised learning based machine learning approach in anaphoricity identification [2]. By taking a comparison between the constraint-based representation and feature-based representation, and combine with the local and global optimization respectively, Ng improve the anaphoricity classifier in a constraint-based representation and global optimization way [1]. Denis and baldridge improve the global optimization; they propose a joint determination of anaphoricity and coreference resolution method by using integer linear programming [16].

3   Anaphoricity Identification

The performance of coreference resolution can be benefited by mention detection since it might eliminate the non-anaphoric phrases. Usually the coreference resolution resolves each noun phrase (NP) in a document, but only a subset of the NP are anaphoric in the document need to be resolved. So the anaphoric detection is quite vital to coreference resolution.

3.1   Learning Model and Feature Selection

A maximum entropy (Maxent) method is carried out in mention detection as the machine learning algorithm. It is not needed to consider about the dependence of features in the feature set. The mention detection is an explicit classifier out of coreference resolution. For each mention N_i in a document, the classifier outcome is the anaphoricity probability $P_A(ANAPH/ N_i)$:

$$P_A(ANAPH \mid N_i) = \frac{\exp(\sum_{k=1}^{n} \lambda_k f_k(N_i, ANAPH))}{Z(i)} \quad . \tag{1}$$

Extending from Ng's 37 features for anaphoric classifier [2], 75 features is used at the beginning.

The features consist of lexical features, grammatical features, morphological features, syntactic features, semantic features etc. We select top 15 features according to the information gain ratio of the features and 5 string features that can't be counted by information gain ratio but helpful to mention detection as well. All these 20 features are listed in Table 1. Assuming the current NP is N_j.

Table 1. The Top 20 Features

Feature Type	Feature	Description
Lexical	STR_MATCH	Y if there exist a NP around N_j that string match with N_j; else N
	HEAD_MATCH	Y if there exist a NP around N_j that have the same head with N_j; else N
	UPPERCASE	Y if N_j is entirely in uppercase; else N
	DT_MATCH	Y if N_j start with "the" and Feature STR_MATCH is "Y" else N
	DT_TABLE	Y if N_j start with "the" and found in DT_table, else N
Grammatical	DEFINITE	Y if N_j start with "the";else N
	PRONOUN	Y if N_j is a pronoun; else N
	PRONOUN_NOT_IT	Y if N_j is a pronoun and it is not "it"; else N
Morphological	NEST_IN	Y if there is a NP nest in N_j; else N
	NEST_OUT	Y if N_j nest in other NP; else N
	HEAD_POS	Head word part of speech(POS) of N_j, String feature
	F_POS	former NP's part of speech(POS), String feature
	B_POS	Succeed NP's POS, String feature
Syntactic	FOLLOW_PREP	Y if N_j is followed by a preposition
	IN_VP	Y if N_j nest in a VP; else N
	IN_PP	Y if N_j nest in a PP; else N
	F_SYN_PARSE	The former syntactic parse, string feature
	B_SYN_PARSE	The succeed syntactic parse, string feature
Semantic	SAME_CLASS	Y if there exist a NP in the same semantic class with N_j; else N
	NUMBER	SINGULAR if N_j is singular in number, PLURAL if N_j is plural, else UNKOWN

Lots of phrases that start with definite article are quite difficult for coreference resolution, such as *the first*, *the people*. Here a DT_table is built in order to solve the problem. The words appear frequently but rarely form coreference in the training corpus are found out, and build the DT_table.

We use a vocabulary tool WordNet to generate semantic features. For the feature SAME_CLASS, WordNet is utilized to search if there is any NP that has the same ancestor semantic class with the current NP.

3.3   Tunable Parameters

A *corpus ratio control* method is proposed in this paper. The ratio of positive examples and negative examples is about 1:3 in the given corpus. The corpus ratio control method controls the ratio of positive and negative samples by setting a ratio threshold before the training file generates. For the positive training example, the features are extracted, and for the negative one, it generates a random number in 0.0 to 1.0, the features extracted when the random number is bigger than threshold. The numbers of negative example as well as precision and recall of mention detection can be controlled by tuning the threshold.

The prediction output of the Maxent model is anaphoricity probability of each NP, we set a probability threshold and we can tune the recall, precision, F-value by changing the threshold of the probability that classifies a entity into anaphoricity. For example, assuming the probability of *family*: P_A (*ANAPH | family*) = 0.45, if the threshold is 0.5, the *family* will be classed into non-anaphoric set, and if the threshold is 0.4, the *family* will be classed into anaphoricity set.

3.4   Global Optimization

The anaphoric classifier cooperate with the coreference resolution in a global optimization way. Global optimization means the anaphoric determination procedure is developed dependently with the coreference resolution; the anaphoric classifier is regulated with respect to coreference performance. By tuning the corpus ratio and the probability threshold we can get anaphoric classifier with different recall, precision and F-value. Combining with different anaphoric classifiers, the result of coreference resolution is different. So the final result of coreference resolution can be improved by changing the parameters of anaphoricity identification.

4   Coreference Resolution

The Maxent model is chosen here as the training and testing model. For each mention pair $< N_i, N_j >$ in a document, the classifier outcome is the anaphoricity probability:

$$P_C(COREF \,|< N_i, N_j >) = \frac{\exp(\sum_{k=1}^{n} \lambda_k f_k(< N_i, N_j >, COREF))}{Z(< N_i, N_j >)} . \qquad (2)$$

We employ Ng and Cardie's method [4] in generating the training instances. We use all the 65 features, the features are selected from all kind of linguistic information, include:

- Distance: Sentence distance, minimum edit distance[5];
- Lexical: String match, Partial String match, Head String match[6];
- Grammar: Gender agreement, Number agreement, Alias agreement [3];

- Syntactic: Head, Path[7];
- Semantic: Semantic class [8], predicate[9].

The mentions will be merged into entities when the classification process is over. There are three strategies for mentions combination: closest first clustering [3][5], best first clustering[4], and aggressive merge clustering[10]. The best-first clustering algorithm means that the current anaphor chooses the candidate antecedent with greatest coreference probability as its antecedent. The best-first clustering is employed in this paper.

5   Result and Analysis

The tool *maximum entropy toolkit* is utilized to realize the maximum entropy model. The experiment data is the development data that provided by CoNLL'11 share tasks. In the anaphoricity identification part, the experiments are built by the anaphoric classifiers with different corpus ratio value and probability threshold. In the coreference resolution part, coreference determination coordinate with various anaphoric classifiers, the results are evaluated on different metrics.

5.1   Anaphoricity Identification Results

Classifiers with different recall, precision and F-value can be constructed by tuning the parameters corpus ratio and probability threshold. The performances of the anaphoric classifiers with different corpus ratio and probability threshold are shown in Table 2. The standard system without tuning is corpus ratio at and probability threshold at 0.5.

Table 2. The Anaphoric Classifiers Performance

corpus ratio	probability threshold	recall	precision	F-value
1.00	0.80	35.71	81.8	49.22
	0.50	66.07	73.54	69.61
	0.30	76.59	65.08	**70.37**
0.85	0.80	44.68	81.6	57.74
	0.75	**50.65**	**80.4**	**62.15**
	0.60	64.24	74.35	68.93
0.65	0.80	53.25	77.69	63.19
	0.60	70.60	70.85	**70.72**
	0.30	82.00	57.68	67.72
0.40	0.80	73.95	53.87	62.33
	0.50	84.03	43.75	57.54
	0.30	86.76	40.21	54.95
Baseline system detection		**75.88**	**49.39**	**59.84**

As the result in Table 2, with the corpus ratio increasing, the recall become smaller and the precision become bigger. Smaller corpus ratio means less negative example will be generated when training. So with less negative example, the classifier will tend to class a mention as positive more easily.

Table 2 shows that the probability threshold also is an impact factor of anaphoricity classification. It is quite reasonable as the probability threshold influences whether a mention is classified as anaphoricity or non-anaphoricity directly and determines the distribution of anaphoric and non-anaphoric. Notice that the performance of anaphoricity identification of the baseline system is measured as well. In this situation, the anaphoric classifier is a part of coreference determination implicitly.

5.2   Coreference Resolution Result

Our system is evaluated on MUC, B-CUBED, CEAF, BLANC metrics. We realize the coreference resolution system of Soon [3] and Ng [4] respectively to make a comparison with our coreference resolution system. For ILP model, experiments on coreference-only baseline system with loose transitivity constraints is performed [17]. *Glpk* package is used to solve the ILP optimization problem. The results of these systems are shown in Table 3.

Table 3. Coreference Resolution performance

metrics	Soon	Ng	Baseline System	ILP
MUC	46.18	45.33	42.93	45.89
B^3	60.03	60.93	59.82	61.85
CEAF	36.37	36.54	36.69	36.85
BLANC	59.90	63.96	62.85	63.92
Average	50.6	51.59	50.57	52.12

The results of baseline coreference system with different anaphoric classifier are listed individually in the Table 4, assuming that the corpus ratio is CR and probability threshold is PT.

We take a metric of average to measure evaluation of the coreference resolution systems. The average can be calculated by equation (3):

$$Average = \frac{(MUC + B^3 + CEAF + BLANC)}{4} \ . \tag{3}$$

We find that our coreference resolution collaborate anaphoric classifier in Table 4 gets a better performance than the systems on average in Table 3, and takes a great improvement on the baseline system. Appling ILP on coreference resolution is a kind of global coreference resolution method. our system still takes a slightly more advance than ILP in average.

A comparison between the local optimization and global optimization is taken here. Local method means that the mention detection is independent from coreference determination. It is a special part of global method at CR=1.0 and PT=0.5. The performance of local optimization is 51.9, while the global one gets the best result up to 53.35.

Table 4 suggest that the F-value of mention detection doesn't have a great impact on the coreference resolution, as CR=1.0, PT=0.3 and CR=0.65, PT=0.6, both of them get the best F-value in anaphoricity identification with 70.37 and 70.72, but the average evaluation is 47.92 and 49.25 respectively. The system gets the best performance

Table 4. The Performance of Baseline System Utilize Different Anaphoric Classifier

systems metrics	CR=1.00 PT=0.80	CR=1.00 PT=0.50	CR=1.00 PT=0.30	CR=0.80 PT=0.80
Mention detection	49.22	69.61	70.37	57.87
MUC	56.96	63.70	59.02	60.23
B^3	61.58	54.40	47.20	61.12
CEAF	26.47	28.22	25.18	27.22
BLANC	61.41	61.28	60.29	62.55
Average	51.60	51.90	47.92	52.78
systems metrics	CR=0.85 PT=0.75	CR=0.85 PT=0.60	CR=0.65 PT=0.60	CR=0.65 PT=0.30
Mention detection	62.15	68.93	70.72	67.72
MUC	62.45	63.51	57.58	53.67
B^3	60.45	55.44	51.53	41.87
CEAF	28.75	28.44	27.19	21.62
BLANC	61.76	61.46	60.69	59.03
Average	53.35	52.21	49.25	44.04

at 53.35 with CR=0.85 and PT=0.75 and the mention detection's F-value is 62.15. So with lower F-value in anaphoric identification, the system may get a better result.

Comparing the Table 2 to the Table 4, we find that the precision of the anaphoricity determination seems more important than the recall. For example, the average metric of CR=0.65, PT=0.3 is 44.04 with mention detection 82 in recall and 57.68 in precision while the average metric of CR=1.00, PT=0.8 is 51.06 with mention detection 35.71 in recall and 81.8 in precision, the result of the latter is much better than the former.

6   Conclusion

In this paper, we presented a global optimization coreference resolution system. With the help of a parameter tunable anaphoric classifier with rich features and global optimization, the performance of the coreference resolution system has been improved from 50.57 to 53.35. We take a comparison between the local optimization and global optimization. We also analyze the condition of the global optimization, and find that the original coreference resolution and the anaphoric classifier should complementary with each other in precision and recall.

In the future, we will do more work on the coreference resolution. We will try to make a combination of anaphoric classifier and coreference resolution in an integer linear programming way.

References

1. Ng, V.: Learning Noun Phrase Anaphoricity to Improve Coreference Resolution: Issues in Representation and Optimization. In: Association for Computational Linguistics Annual Meeting (ACL 2004), 20040721-726. Barcelona(ES) (2004)
2. Ng, V., Cardie, C.: Identifying Anaphoric and Non-Anaphoric Noun Phrases to Improve Coreference Resolution. In: 19th International Conference on Computational Linguistics Coling, vol. 2, Taipei, Taiwan (2002)

3. Soon, W., Ng, H., Lim, D.: A Machine Learning Approach to Coreference Resolution of Noun Phrases. Computational Linguistics 27(4), 521–544 (2001)

4. Ng, V., Cardie, C.: Improving Machine Learning Approaches to Coreference Resolution. In: Isabelle, P. (ed.) Proc. of the 40th Annual Meeting on Association for Computational Linguistics, pp. 104–111. Association for Computational Linguistics, Philadelphia (2001)

5. Strube, M., Rapp, S., Christoph, M.: The Influence of Minimum Edit Distance on Reference Resolution. In: Proceedings of the ACL 2002 Conference on Empirical Methods in Natural Language Processing, EMNLP 2002, vol. 10, pp. 312–319. Association for Computational Linguistics, Stroudsburg (2002)

6. Daum'e III, H., Marcu, D.: A Large-scale Exploration of Effective Global Features for a Joint Entity Detection and Tracking Model. In: Proceedings of Human Language Technology Conference and Conference on Empirical Methods in Natural Language Processing, pp. 97–104. Association for Computational Linguistics, Vancouver (2005)

7. Yang, X.F., Su, J., Tan, C.L.: Kernel-based Pronoun Resolution with Structured Syntactic Knowledge. In: Proceedings of the 21st International Conference on Computational Linguistics and 44th Annual Meeting of the Association for Computational Linguistics, pp. 41–48. Association for Computational Linguistics, Sydney (2006)

8. Ponzetto, S.P., Strube, M.: Exploiting Semantic Role Labeling, Wordnet and Wikipedia for Coreference Resolution. In: Proceedings of the Human Language Technology Conference of the NAACL, Main Conference, pp. 192–199. Association for Computational Linguistics, New York City (2006)

9. Ng, V.: Shallow Semantics for Coreference Resolution. In: Proceedings of IJCAI, pp. 1689–1694 (2007)

10. McCarthy, J., Lehnert, W.: Using Decision Trees for Coreference Resolution. In: Perrault, C.R. (ed.) Proc. of the Fourteenth International Joint Conference on Artificial Intelligence, pp. 1050–1055. Springer, Québec (1995)

11. Aone, C., Bennett, S.W.: Evaluating Automated and Manual Acquisition of Anaphora Resolution Strategies. In: Proceedings of the 33rd Annual Meeting of the Association for Computational Linguistics, pp. 122–129 (1995)

12. McCarthy, J., Lehnert, W.: Using Decision Trees for Coreference Resolution. In: Proceedings of the Fourteenth International Conference on Artificial Intelligence, pp. 1050–1055 (1995)

13. Luo, et al.: A Mention-synchronous Coreference Resolution Algorithm Based on the Bell Tree. In: Scott, D. (ed.) Proc. of the 42th Annual Meeting on Association for Computational Linguistics, pp. 135–142. Association for Computational Linguistics, Barcelona (2004)

14. Ng, V.: Machine Learning for Coreference Resolution: From local classification to global ranking. In: Proceedings of ACL (2005)

15. McCallum, A., Wellner, B.: Conditional Models of Identity Uncertainty with Application to Noun Coreference. In: Proceedings of NIPS (2004)

16. Baldridge, D.J.: Joint Determination of Anaphoricity and Coreference Resolution Using Integer Programming. In: Sidner, C., et al. (eds.) Proc. of Human Language Technologies 2007: The Conference of the North American Chapter of the Association for Computational Linguistics, pp. 236–243. Association for Computational Linguistics, Rochester (2007)

17. Li, X.X., Wang, X., Qi, S.H.: Coreference Resolution with Rich Features and Loose Transitivity Constraints. In: Fifteenth Conference on Computational Natural Language Learning (2011)

A Variable Muitlgranulation Rough Sets Approach

Ming Zhang[1,2,*], Zhenmin Tang[1], Weiyan Xu[3], and Xibei yang[2]

[1] School of Computer Science and Technology, Nanjing University of Science and Technology, Nanjing, Jiangsu, 210094, P.R. China
[2] School of Computer Science and Engineering
[3] School of Mathematics and Physics, Jiangsu University of Science and Technology, Zhenjiang, Jiangsu, 212003, P.R. China
zm_fred@163.com,
tang.zm@mail.njust.edu.cn,
{xwy_yan,yangxibei}@hotmail.com

Abstract. By analyzing the limitations of optimistic multigranulation rough set and pessimistic multigranulation rough set, the concept of the variable multigranulation rough set is proposed. Such multigranulation rough set is a generalization of both optimistic and pessimistic multigranulation rough set. Furthermore, not only the basic properties about the variable multigranulation rough set is discussed, but also the relationships among optimistic, pessimistic and variable multigranulation rough sets are deeply explored. These results are meaningful for the development of multigranulation rough set theory.

Keywords: Optimistic multigranulation rough set, Pessimistic multigranulation rough set, Variable Muitlgranulation Rough Set.

1   Introduction

Rough set [1], proposed by Pawlak, is a powerful tool, which can be used to deal with the inconsistency problems by separation of certain and doubtful knowledge extracted from the exemplary decisions. Though Pawlak's rough set theory has been demonstrated to be useful in the fields such as knowledge discovery, decision analysis, data mining, pattern recognition and so on, it is constructed on the basis of a strict indiscernibility relation . Presently, with respect to different requirements, various extensions of the rough set have been proposed. For example, by generalizing the partition(induced by an indiscernibility relation) to the covering, Zhu et al. provided several models of the covering–based rough sets [2]. To express weaker form of of indiscernibility, Słowiński et al. [4] proposed a generalized definition of rough approximations which is based on the similarity. To deal with the knowledge representation components which may contain incomplete, noisy, and uncertain information, the tolerance–based rough set has been used in feature selection [3], incomplete information system [5], etc.

---

* Corresponding author.

D.-S. Huang et al. (Eds.): ICIC 2011, LNBI 6840, pp. 315–322, 2012.

From discussions above, we can see that most of the expanded rough set models are constructed on the basis of one and only one binary relation, However, it should be noticed that in Ref. [6,7,8,9], Qian et al. argued that we often need to describe concurrently a target concept through multi binary relations on the universe according to a user's requirements or targets of problem solving. Therefore, they proposed the concept of Multigranulation Rough Set (MGRS) model. The first multigranulation rough set model was proposed by Qian et al. in Ref. [6]. Following such work, Qian classified his multigranulation rough set theory into two parts: one is the optimistic multigranulation rough set [7,8] and the other is pessimistic multigranulation rough set [9].

By analyzing Qian's two different multigranulation rough sets, we can see that the optimistic multigranulation rough set is too relax since if only *one* granulation space satisfies with the inclusion condition between the equivalence class and the target concept, then the object should belong to the lower approximation. On the other hand, the pessimistic multigranulation rough set is too strict since if *all* of the granulation spaces satisfy with the inclusion condition between the equivalence classes and the target, then the object belongs to the lower approximation.

The purpose of this paper is to propose a new multigranulation rough set, which is referred to as the variable multigranulation rough set. In our new multigranulation rough set approach, a threshold β is used to control the number of granulation spaces, which satisfy with the inclusion condition between the equivalence class and the target concept.

The paper is organized as following. In Section 2, the rough set and multigranulation rough sets are briefly introduced. In Section 3, the variable multigranulation rough sets models are proposed, the immediate properties about variable multigranulation rough sets are also addressed. Results are summarized in Section 4.

2   Preliminary Knowledge on Rough Sets

In this section, we review some basic concepts on Pawlak's rough set and Qian's multigranulation rough set.

2.1   Pawlak's Rough Set

Formally, an information system can be considered as a 4-tuple $IS =< U, AT, V, f >$, where U is a non–empty finite set of objects(called the universe), AT is a non–empty finite set of attributes,V is regard as the domain of all attributes and $V = V_{AT} = \bigcup_{a \subseteq AT} V_a$, $\forall x \in U$, $f(a, x)$ is the value the x hold on $a(a \in AT)$. In particular, if $AT = C \cup D$ and $C \cap D = \emptyset$, then the C is called condition attributes and the D is called decision attributes, the $< U, C \cup D, V, f >$ is also regard as a target information system.

For an information system I, one then can describe the relationship between objects through their attributes values. With respect to a subset of attributes such that $A \subseteq AT$, an indiscernibility relation $IND(A)$ may be defined as

$$IND(A) = \{(x, y) \in U^2 : f(a, x) = f(a, y), \forall a \in A\}. \tag{1}$$

The relation $IND(A)$ is reflexive, symmetric and transitive, then $IND(A)$ is an equivalence relation. The equivalence relation $IND(A)$ can partitions the set U into disjoint subsets, This partition of the universe induced by $IND(A)$ is denoted by $U/IND(A)$. $\forall x \in U$, We denote the equivalence class including x by $[x]_A$, such $[x]_A = \{y \in U : (x, y) \in IND(A)\}$. so we can derive the lower and upper approximations of an arbitrary subset X of U. They are defined as

$$\underline{A}(X) = \{x \in U : [x]_A \subseteq X\} \text{ and } \overline{A}(X) = \{x \in U : [x]_A \cap X \neq \emptyset\} \tag{2}$$

The pair $[\underline{A}(X), \overline{A}(X)]$ is referred to as the Pawlak's rough set of X with respect to the set of attributes A.

From the viewpoint of the granular computing, the partition $U/IND(A)$ is referred to as a granulation space. Each equivalence class $[x]_A$ may be viewed as a knowledge granule consisting of indistinguishable elements. Obviously, Pawlak's rough set is constructed on the basis of one and only one indiscernibility relation, which can generate a granulation space, and then Pawlak's rough set can be regarded as the single–granulation rough set model.

2.2   Multigranulation Rough Set

In Qian's multigranulation rough set theory, two different models have been defined. The first one is the optimistic multigranulation rough set [7,8], the second one is the pessimistic multigranulation rough set [9].

Optimistic multigranulation rough set. In Qian's optimistic multigranulation rough set, the word "optimistic" is used to express the idea that in multi–independent granulation spaces, we need only at least one granulation space to satisfy with the inclusion condition between equivalence class and target.

Definition 1. *Let I be an information system in which $A_1, A_2, \cdots, A_m \subseteq AT$, then $\forall X \subseteq U$, the optimistic multigranulation lower and upper approximations are denoted by $\sum_{i=1}^{m} A_i^O(X)$ and $\overline{\sum_{i=1}^{m} A_i}^O(X)$, respectively,*

$$\sum_{i=1}^{m} A_i^{O}(X) = \{x \in U : [x]_{A_1} \subseteq X \vee [x]_{A_2} \subseteq X \vee \cdots \vee [x]_{A_m} \subseteq X\}; \tag{3}$$

$$\overline{\sum_{i=1}^{m} A_i}^{O}(X) = \sim \sum_{i=1}^{m} A_i^{O}(\sim X); \tag{4}$$

where $[x]_{A_i}$ $(1 \leq i \leq m)$ is the equivalence class of x in terms of set of attributes A_i, $\sim X$ is the complement of set X.

Theorem 1. *Let I be an information system in which $A_1, A_2, \cdots, A_m \subseteq AT$, then $\forall X \subseteq U$, we have*

$$\overline{\sum_{i=1}^{m} A_i}^{O}(X) = \{x \in U : [x]_{A_1} \cap X \neq \emptyset \wedge [x]_{A_2} \cap X \neq \emptyset \wedge \cdots \wedge [x]_{A_m} \cap X \neq \emptyset\}. \quad (5)$$

Proof. It can be proof easily by Definition 1.

By Theorem 1, we can see that though the optimistic multigranulation upper approximation is defined by the complement of the optimistic multigranulation lower approximation, it can also be considered as a set, in which objects have non–empty intersection with the target in terms of each granulation space.

Theorem 2. *Let I be an information system in which $A_1, A_2, \cdots, A_m \subseteq AT$, suppose that $A = A_1 \cup A_2 \cup \cdots \cup A_m$, then $\forall X \subseteq U$, we have*

$$\sum_{i=1}^{m} A_i^{O}(X) \subseteq \underline{A}(X), \overline{\sum_{i=1}^{m} A_i}^{O}(X) \supseteq \overline{A}(X).$$

Proof. It can be proof easily by theorem 1.

Theorem 2 tells us that the optimistic multigranulation lower approximation is smaller than Pawlak's lower approximation, while the optimistic multigranulation upper approximation is greater than Pawlak's upper approximation.

Pessimistic multigranulation rough set. In Qian's pessimistic multigranulation rough set, the word "pessimistic" is used to express the idea that in multi–independent granulation spaces, we need all the granulation spaces to satisfy with the inclusion condition between the equivalence class and target.

Definition 2. *[9] Let I be an information system in which $A_1, A_2, \cdots, A_m \subseteq AT$, then $\forall X \subseteq U$, the pessimistic multigranulation lower and upper approximations are denoted by $\underline{\sum_{i=1}^{m} A_i}^{P}(X)$ and $\overline{\sum_{i=1}^{m} A_i}^{P}(X)$, respectively,*

$$\underline{\sum_{i=1}^{m} A_i}^{P}(X) = \{x \in U : [x]_{A_1} \subseteq X \wedge [x]_{A_2} \subseteq X \wedge \cdots \wedge [x]_{A_m} \subseteq X\}; \quad (6)$$

$$\overline{\sum_{i=1}^{m} A_i}^{P}(X) = \sim \sum_{i=1}^{m} A_i^{P}(\sim X). \quad (7)$$

Theorem 3. *Let I be an information system in which $A_1, A_2, \cdots, A_m \subseteq AT$, then $\forall X \subseteq U$, we have*

$$\overline{\sum_{i=1}^{m} A_i}^{P}(X) = \{x \in U : [x]_{A_1} \cap X \neq \emptyset \vee [x]_{A_2} \cap X \neq \emptyset \vee \cdots \vee [x]_{A_m} \cap X \neq \emptyset\}. \quad (8)$$

Proof. It can be proved easily by Definition 2.

Different from the upper approximation of optimistic multigranulation rough set, the upper approximation of pessimistic multigranulation rough set is represented as a set, in which objects have non–empty intersection with the target in terms of at least one granulation space.

Theorem 4. *Let I be an information system in which $A_1, A_2, \cdots, A_m \subseteq AT$, suppose that $A = A_1 \cup A_2 \cup \cdots \cup A_m$, then $\forall X \subseteq U$, we have*

$$\underline{\sum_{i=1}^{m} A_i}^{P}(X) \subseteq \underline{\sum_{i=1}^{m} A_i}^{O}(X) \subseteq \underline{A}(X), \overline{\sum_{i=1}^{m} A_i}^{P}(X) \supseteq \overline{\sum_{i=1}^{m} A_i}^{O}(X) \supseteq \overline{A}(X).$$

The above theorem tells us that the pessimistic multigranulation lower approximation is smaller than Pawlak's lower approximation while the pessimistic multigranulation upper approximation is greater than Pawlak's upper approximation. Moreover, the pessimistic multigranulation lower approximation is smaller than optimistic multigranulation lower approximation while the pessimistic multigranulation upper approximation is greater than optimistic multigranulation upper approximation.

3   Variable Multigranulation Rough Set

By Definition 1, we can see that the optimistic multigranulation rough set is too relax since if only *one* granulation space satisfies with the inclusion condition between the equivalence class and the target concept, then the object should belong to the lower approximation. On the other hand, by Definition 2, the pessimistic multigranulation rough set is too strict since if *all* of the granulation spaces satisfy with the inclusion condition between the equivalence classes and the target, then the object belongs to the lower approximation.

To solve such problem, we will propose a new multigranulation rough sets approach, which is referred to as the variable multigranulation rough set. In our multigranulation rough sets approach, a threshold β will be used to control the number of granulation spaces, which satisfy with the inclusion condition between the equivalence classes and the target.

Definition 3. *Let I be an information system in which $A_1, A_2, \cdots, A_m \subseteq AT$, $A = \{A_1, A_2, \cdots, A_m\}$, $0 < \beta \leq 1$, then $\forall X \subseteq U$, the β variable multigranulation lower and upper approximations are denoted by $\underline{\sum_{i=1}^{m} A_i}^{\beta}(X)$ and $\overline{\sum_{i=1}^{m} A_i}^{\beta}(X)$, respectively,*

$$\underline{\sum_{i=1}^{m} A_i}^{\beta}(X) = \{x \in U : \forall A_i \in T, [x]_{A_i} \subseteq X\}; \tag{9}$$

$$\overline{\sum_{i=1}^{m} A_i}^{\beta}(X) = \sim \underline{\sum_{i=1}^{m} A_i}^{P}(\sim X). \tag{10}$$

where $T \subseteq A$ and $\frac{|T|}{m} \geq \beta$.

Theorem 5. *Let I be an information system in which $A_1, A_2, \cdots, A_m \subseteq AT$, suppose that $A = \{A_1, A_2, \cdots, A_m\}$, $0 < \beta \leq 1$, then $\forall X \subseteq U$, we have*

$$\overline{\sum_{i=1}^{m} A_i}^{\beta}(X) = \{x \in U : \forall A_i \in T, [x]_{A_1} \cap X \neq \emptyset\}. \tag{11}$$

where $T \subseteq A$ and $\frac{|T|}{m} \geq \beta$

Proof. It can be proved easily by Definition 3.

Theorem 6. *Let I be an information system in which $A_1, A_2, \cdots, A_m \subseteq AT$, $A = \{A_1, A_2, \cdots, A_m\}$, then $\forall X \subseteq U$, we have*

$$\underline{\sum_{i=1}^{m} A_i}^{\frac{1}{m}}(X) = \underline{\sum_{i=1}^{m} A_i}^{O}(X), \quad \overline{\sum_{i=1}^{m} A_i}^{1}(X) = \overline{\sum_{i=1}^{m} A_i}^{O}(X), \tag{12}$$

$$\underline{\sum_{i=1}^{m} A_i}^{1}(X) = \underline{\sum_{i=1}^{m} A_i}^{P}(X), \quad \overline{\sum_{i=1}^{m} A_i}^{\frac{1}{m}}(X) = \overline{\sum_{i=1}^{m} A_i}^{P}(X). \tag{13}$$

Proof. It can be proof easily by Definition 1 and Definition 3

Theorem 7. *Let I be an information system in which $A_1, A_2, \cdots, A_m \subseteq AT$, suppose that $A = \{A_1, A_2, \cdots, A_m\}$, $0 < \beta \leq 1$, then $\forall X \subseteq U$, we have*

$$\overline{\sum_{i=1}^{m} A_i}^{P}(X) \subseteq \overline{\sum_{i=1}^{m} A_i}^{\beta}(X) \subseteq \overline{\sum_{i=1}^{m} A_i}^{O}(X). \tag{14}$$

$$\underline{\sum_{i=1}^{m} A_i}^{O}(X) \subseteq \underline{\sum_{i=1}^{m} A_i}^{\beta}(X) \subseteq \underline{\sum_{i=1}^{m} A_i}^{P}(X). \tag{15}$$

Proof. It can be derived directly from Theorem 6.

Theorem 8. *Let I be an information system in which $A_1, A_2, \cdots, A_m \subseteq AT$, suppose that $A = \{A_1, A_2, \cdots, A_m\}$, $0 < \beta \leq 1$, then $\forall X \subseteq U$, the following properties hold*

1. $\underline{\sum_{i=1}^{m} A_i}^{\beta}(X) \subseteq X \subseteq \overline{\sum_{i=1}^{m} A_i}^{\beta}(X)$

2. $\underline{\sum_{i=1}^{m} A_i}^{\beta}(\emptyset) = \overline{\sum_{i=1}^{m} A_i}^{\beta}(\emptyset) = \emptyset$, $\underline{\sum_{i=1}^{m} A_i}^{\beta}(U) = \overline{\sum_{i=1}^{m} A_i}^{\beta}(U) = U$

3. $\underline{\sum_{i=1}^{m} A_i}^{\beta}(\sim X) = \sim \overline{\sum_{i=1}^{m} A_i}^{\beta}(X)$, $\overline{\sum_{i=1}^{m} A_i}^{\beta}(\sim X) = \sim \underline{\sum_{i=1}^{m} A_i}^{\beta}(X)$

4. $X \subseteq Y \Rightarrow \underline{\sum_{i=1}^{m} A_i}^{\beta}(X) \subseteq \underline{\sum_{i=1}^{m} A_i}^{\beta}(Y), \overline{\sum_{i=1}^{m} A_i}^{\beta}(X) \subseteq \overline{\sum_{i=1}^{m} A_i}^{\beta}(Y).$

5. $\underline{\sum_{i=1}^{m} A_i}^{\beta}(\underline{\sum_{i=1}^{m} A_i}^{\beta}(X)) = \underline{\sum_{i=1}^{m} A_i}^{\beta}(X)$

   $\overline{\sum_{i=1}^{m} A_i}^{\beta}(\overline{\sum_{i=1}^{m} A_i}^{\beta}(X)) = \overline{\sum_{i=1}^{m} A_i}^{\beta}(X)$

6. $\beta_1 \leq \beta_2 \Rightarrow \underline{\sum_{i=1}^{m} A_i}^{\beta_1}(X) \supseteq \underline{\sum_{i=1}^{m} A_i}^{\beta_2}(X)$, $\overline{\sum_{i=1}^{m} A_i}^{\beta_1}(X) \subseteq \overline{\sum_{i=1}^{m} A_i}^{\beta_2}(X).$

Theorem 8 shows the basic properties about the variable multigranulation rough sets. 1 says that the variable multigranulation lower approximation is included into the target concept and the variable multigranulation upper approximation includes the target concept, 2 shows the normality of the variable multigranulation rough sets, 3 expresses the complement properties of the variable multigranulation rough sets, 4 says the monotonic properties about the variable multigranulation rough sets in terms of the monotonic varieties of the target concepts, 5 says the idempotents of the variable multigranulation rough sets, 6 says the monotonic properties about the variable multigranulation rough sets in terms of the monotonic varieties of the threshold.

Table 1. An example of students' evaluations

U	a_1	a_2	a_3	a_4	d
x_1	Common	God	God	God	Yes
x_2	Bad	Bad	Bad	Bad	No
x_3	Bad	God	God	God	Yes
x_4	Bad	Common	Common	Bad	No
x_5	God	Common	Common	God	Yes
x_6	Bad	Common	Common	Common	No
x_7	Bad	God	Bad	Common	Yes
x_8	Bad	Bad	Common	Common	No
x_9	Bad	God	Bad	Common	Yes

Example 1. We then use an example for evaluation of the β variable multigranulation rough set. Suppose that the director of the school must give a global evaluations to some students, which are showed in Table 1.

In Table 1, $U = \{x_1, x_2, \cdots, x_9\}$ is the nine students described by means of fiver attributes, $AT = \{a_1, a_2, a_3, a_4\}$ is the set of condition attributes, d is the decision attribute.

Let $A = \{a_1, a_2, a_3, a_4\}$, since the decision attribute determines a partition on the universe such that $U/IND(\{d\}) = \{\{x_2, x_4, x_6, x_8\}, \{x_1, x_3, x_5, x_7, x_9\}\}$, then by Definition 1, we have $\underline{\sum_{i=1}^{4} a_i}^{O}(D_1) = \{x_2, x_4, x_8\}$. $\underline{\sum_{i=1}^{4} a_i}^{O}(D_2) = \{x_1, x_3, x_5, x_7, x_9\}$.

$\overline{\sum_{i=1}^{4} a_i}^{O}(D_1) = \{x_2, x_4, x_6, x_8\}$. $\overline{\sum_{i=1}^{4} a_i}^{O}(D_2) = \{x_1, x_3, x_5, x_6, x_7, x_9\}$.

By Definition 2, we have

$\underline{\sum_{i=1}^{4} a_i}^{P}(D_1) = \emptyset$. $\overline{\sum_{i=1}^{4} a_i}^{P}(D_1) = \{x_2, x_3, x_4, x_5, x_6, x_7, x_8, x_9\}$.

$\underline{\sum_{i=1}^{4} a_i}^{P}(D_2) = \{x_1\}$. $\overline{\sum_{i=1}^{4} a_i}^{P}(D_2) = \{x_1, x_2, x_3, x_4, x_5, x_6, x_7, x_8, x_9\}$.

Suppose $\beta = 0.5$, then by Definition 3, we have

$\underline{\sum_{i=1}^{4} a_i}^{\beta}(D_1) = \{x_2\}$. $\overline{\sum_{i=1}^{4} a_i}^{\beta}(D_1) = \{x_2, x_4, x_6, x_7, x_8, x_9\}$.

$\underline{\sum_{i=1}^{4} a_i}^{\beta}(D_2) = \{x_1, x_3, x_5\}$. $\overline{\sum_{i=1}^{4} a_i}^{\beta}(D_2) = \{x_1, x_3, x_4, x_5, x_6, x_7, x_8, x_9\}$.

By the above results, we have $\sum_{i=1}^{m} A_i^{\ P}(D_1) \subseteq \sum_{i=1}^{m} a_i^{\ \beta}(D_1) \subseteq \sum_{i=1}^{m} a_i^{\ O}(D_1)$, $\sum_{i=1}^{m} a_i^{\ P}(D_2) \subseteq \sum_{i=1}^{m} a_i^{\ \beta}(D_2) \subseteq \sum_{i=1}^{m} a_i^{\ O}(D_2)$, and $\overline{\sum_{i=1}^{m} a_i}^{\ P}(D_1) \subseteq \overline{\sum_{i=1}^{m} A_i}^{\ \beta}(D_1) \subseteq \overline{\sum_{i=1}^{m} a_i}^{\ O}(D_1)$, $\overline{\sum_{i=1}^{m} a_i}^{\ P}(D_2) \subseteq \overline{\sum_{i=1}^{m} a_i}^{\ \beta}(D_2) \subseteq \overline{\sum_{i=1}^{m} a_i}^{\ O}(D_2)$, such results demonstrates the correctness of Theorem 7.

4   Conclusions

In this paper, the variable multigranulation rough set approach is proposed. In our approach, a threshold is used to control the number of granulation spaces, which satisfy with the inclusion condition between the equivalence classes and the target. It is shown that our variable multigranulation is between Qian's optimistic and pessimistic multigranulation rough sets. Not only the properties about the variable multigranulation rough sets are explored, but also the relationships among several rough sets are explored.

Acknowledgments. This work is supported by the Natural Science Foundation of China (No.90820306)and Postdoctoral Science Foundation of China (No.20100481149).

References

1. Pawlak, Z.: Rough Sets. International Journal of Computer and Information Sciences 11, 341–356 (1984)
2. Zhu, W.: Topological Approaches to Covering Rough Sets. Information Sciences 177, 1499–1508 (2007)
3. Kryszkiewicz, M.: Rough Set Approach to Incomplete Information Systems. Information Sciences 112, 39–49 (1998)
4. Stefanowski, J., Tsoukias, A.: Incomplete Information Tables and Rough Classification. Computational Intelligence 17, 545–566 (2001)
5. Leung, Y., Li, D.Y.: Maximal Consistent Block Technique for Rule Acquisition in Incomplete Information Systems. Information Sciences 115, 85–106 (2003)
6. Qian, Y.H., Liang, J.Y.: Rough Set Method Based on Multi–granulations. In: 5th IEEE International Conference on Cognitive Informatics, pp. 297–304. IEEE Press, Los Alamitos (2006)
7. Qian, Y.H., Liang, J.Y., Yao, Y.Y., Dang, C.Y.: MGRS: A Multi–granulation Rough Set. Information Sciences 180, 949–970 (2010)
8. Qian, Y.H., Liang, J.Y., Dang, C.Y.: Incomplete Multigranulation Rough Set. IEEE Transactions on Systems, Man and Cybernetics, Part A 20, 420–431 (2010)
9. Qian, Y.H., Liang, J.Y., Wei, W.: Pessimistic Rough Decision. In: Second International Workshop on Rough Sets Theory, Zhoushan, P. R. China, October 19-21, pp. 440–449 (2010)

Incomplete Multigranulation Rough Sets in Incomplete Ordered Decision System

Li-juan Wang[1,2], Xi-bei Yang[1,2], Jing-yu Yang[1], and Chen Wu[2]

[1] School of Computer Science and Technology, NUST, Nanjing 210094, China
[2] Department of Computer Science and Engineering, JUST, Zhenjiang 212003, China
zjwanglijuan@sina.com

Abstract. The tolerance relation based incomplete multigranulation rough set is not able to explore the incomplete ordered decision systems. To solve such problem, similarity dominance relation based rough set approach is introduced into multigranulation environment in this paper. Two different types of models: similarity dominance relation based optimistic incomplete multigranulation rough set model and pessimistic incomplete multigranulation rough set model are constructed respectively. The properties and the relationships of them are discussed. Eight types of decision rules in the two models are proposed. An illustrative example is employed.

Keywords: Similarity dominance relation, incomplete ordered decision system, incomplete optimistic multigranulation rough set, incomplete pessimistic multigranulation rough set.

1 Introduction

Rough set theory (RST) [1-6], proposed by Pawlak[1], is mainly concerned with the approximation of sets described by a single indiscernibility relation on the universe. In the view of granular computing [7-9], the classical RST is based on a single granulation. However, in some circumstances, we often need to describe a target concept through multi binary relations on the universe according to a user's requirements or targets of problem solving. Thus, Qian etc. extend Pawlak's single-granulation rough set model to a multigranulation rough set model (MGRS) [5,6], where the set approximations are defined by using multi equivalence/tolerance relations on the universe.

Knowledge representation in the rough set model is realized via information system (IS). On account of equivalence relation in knowledge base, information system what we got is complete. Nevertheless, since the error of data measuring, or the limitation of comprehension of data, incomplete information systems (IIS) with missing values often occur in knowledge acquisition[2-4,6]. It is worth noting that a particular form of IIS, in which all attributes are considered as criterions. Each criterion indicates an attribute with preference-ordered domain. This form of IIS is called incomplete ordered decision system (IODS). Yang etc. have deeply studied IODS [3-5], and the concept of similarity dominance relation was first proposed in [3].

D.-S. Huang et al. (Eds.): ICIC 2011, LNBI 6840, pp. 323–330, 2012.

The main object of this paper is to study IODS when the decision makers are independent for the same project. Incomplete MGRS (IMGRS) proposed by Qian etc. in [6] is first introduced to study this problem. With the concept of similarity dominance relation [2], similarity dominance relation based IMGRS is presented in this paper. The rest of this paper is organized as follows. In Section 2, we briefly introduce the fundamental concepts and the tolerance relation based pessimistic IMGRS is first proposed. In Section 3, similarity dominance relation based optimistic IMGRS and pessimistic IMGRS are presented. The properties of the two models are deeply studied. In Section 4, rules in similarity dominance relation based IMGRS are explored. An illustrative example is analyzed in Section 5 and results are summarized in Section 6.

2   Preliminary Knowledge on Rough Sets

2.1   Incomplete Ordered Decision System [2]

A decision system is considered as a pair $I = < U, AT \cup D >$, in which AT is the set of the condition attributes, D is the set of the decision attributes and $AT \cap D = \emptyset$. The set of values of all the condition attributes is denoted as V_{AT}, and the set of values of all the decision attributes is denoted as V_D. If all the attributes are criterions, the decision system becomes an ordered decision system (ODS). When the precise values of some of the condition attributes are not known, such ODS is called an incomplete ordered decision system ($IODS$). An $IODS$ is still denoted without confusion by $I = < U, AT \cup D >$. Here, $V = V_{AT} \cup V_D \cup \{*\}$ is the domain of all the attributes, the special symbol "$*$" is used to indicates the unknown value.

2.2   Similarity Dominance Relation Based Rough Set Model [2]

Definition 1. [2] Let I be an $IODS$, $\forall A \subseteq AT$, then the similarity dominance relations of A can be defined as follows:

$$SD_A^{\geq} = \{(x,y) \in U^2 : \forall a \in A, a(x) = * \vee a(x) \geq a(y)\} \tag{1}$$

$$SD_A^{\leq} = \{(x,y) \in U^2 : \forall a \in A, a(x) = * \vee a(x) \leq a(y)\} \tag{2}$$

Obviously, the above two similarity dominance relations $SD_A^{\geq}$ and $SD_A^{\leq}$ are reflexive and transitive, but not necessarily symmetric.

$\forall x \in U$, let us denote by

1. $SD_A^{\geq -}(x) = \{y \in U : (x,y) \in SD_A^{\geq}\}, \forall y \in SD_A^{\geq -}(x)$, then x is similar to y and x dominates y;
2. $SD_A^{\geq +}(x) = \{y \in U : (y,x) \in SD_A^{\geq}\}, \forall y \in SD_A^{\geq +}(x)$, then y is similar to x and y dominates x;
3. $SD_A^{\leq -}(x) = \{y \in U : (x,y) \in SD_A^{\leq}\}, \forall y \in SD_A^{\leq -}(x)$, then x is similar to y and x is dominated by y;

4. $SD_{\bar{A}}^{\leq+}(x) = \{y \in U : (y,x) \in SD_{\bar{A}}^{\leq}\}$, $\forall y \in SD_{\bar{A}}^{\leq+}(x)$, then y is similar to x and y is dominated by x.

Assume that the set of decision attributes D makes a partition of U into a finite number of classes; let $\mathbf{CL} = \{CL_t, t \in T\}, T = \{1, 2, \cdots, l\}$ be a set of these classes which are ordered, that is, $\forall r, s \in T$, if $r > s$ then the objects from CL_r are preferred to the objects from CL_s. The sets to be approximated are an upward union and a downward union of classes, which are defined respectively as $CL_{\bar{r}}^{\geq} = \bigcup_{s \geq r} CL_s$, $CL_{\bar{r}}^{\leq} = \bigcup_{s \leq r} CL_s$, where $r, s \in T$. The statement $x \in CL_{\bar{r}}^{\geq}$ means "x belongs to at least class CL_r", and $x \in CL_{\bar{r}}^{\leq}$ means "x belongs to at most class CL_r" [6].

Definition 2. [2] Let I be an $IODS$, $\forall A \subseteq AT$, $\forall r \in T$.

- The similarity dominance relation based lower and upper approximations of $CL_{\bar{r}}^{\geq}$ are denoted by $\underline{A}_{SD}(CL_{\bar{r}}^{\geq})$ and $\overline{A}_{SD}(CL_{\bar{r}}^{\geq})$ respectively, where

$$\underline{A}_{SD}(CL_{\bar{r}}^{\geq}) = \{x \in U : SD_{\bar{A}}^{\leq-}(x) \subseteq CL_{\bar{r}}^{\geq}\} \tag{3}$$

$$\overline{A}_{SD}(CL_{\bar{r}}^{\geq}) = \{x \in U : SD_{\bar{A}}^{\geq-}(x) \cap CL_{\bar{r}}^{\geq} \neq \emptyset\} \tag{4}$$

- The similarity dominance relation based lower and upper approximations of $CL_{\bar{r}}^{\leq}$ are denoted as $\underline{A}_{SD}(CL_{\bar{r}}^{\leq})$ and $\overline{A}_{SD}(CL_{\bar{r}}^{\leq})$ respectively, where

$$\underline{A}_{SD}(CL_{\bar{r}}^{\leq}) = \{x \in U : SD_{\bar{A}}^{\geq-}(x) \subseteq CL_{\bar{r}}^{\leq}\} \tag{5}$$

$$\overline{A}_{SD}(CL_{\bar{r}}^{\leq}) = \{x \in U : SD_{\bar{A}}^{\leq-}(x) \cap CL_{\bar{r}}^{\leq} \neq \emptyset\} \tag{6}$$

2.3 Tolerance Relations Based Incomplete MGRS [6]

Definition 3. [6] Let I be an IDS in which $A_1, A_2, \cdots, A_m \subseteq AT$, then $\forall X \subseteq U$, the optimistic multigranulation lower and upper approximations in terms of the tolerance relations are denoted by $\underline{\sum_{i=1}^{m} A_i}_T^O(X)$ and $\overline{\sum_{i=1}^{m} A_i}_T^O(X)$ respectively, where

$$\underline{\sum_{i=1}^{m} A_i}_T^O(X) = \{x \in U : T_{A_1}(x) \subseteq X \vee T_{A_2}(x) \subseteq X \vee \cdots \vee T_{A_m}(x) \subseteq X\} \tag{7}$$

$$\overline{\sum_{i=1}^{m} A_i}_T^O(X) = \sim \underline{\sum_{i=1}^{m} A_i}_T^O(\sim X) \tag{8}$$

$T_{A_i}(x)(1 \leq i \leq m)$ is the tolerance class of x in terms of the subset of the attributes A_i, $\sim X$ is the complement of set X.

Theorem 1. Let I be an IDS in which $A_1, A_2, \cdots, A_m \subseteq AT$, then $\forall X \subseteq U$,

$$\overline{\sum_{i=1}^{m} A_i}_T^O(X) = \{x \in U : T_{A_1}(x) \cap X \neq \emptyset \wedge \cdots \wedge T_{A_m}(x) \cap X \neq \emptyset\}. \tag{9}$$

We further introduce the pessimistic MGRS into the incomplete decision system as Definition 4 shows.

Definition 4. Let I be an IDS in which $A_1, A_2, \cdots, A_m \subseteq AT$, then $\forall X \subseteq U$, the pessimistic multigranulation lower and upper approximations in terms of the tolerance relations are denoted by $\underline{\sum_{i=1}^{m} A_i}_T^P(X)$ and $\overline{\sum_{i=1}^{m} A_i}_T^P(X)$ respectively, where

$$\underline{\sum_{i=1}^{m} A_i}_T^P(X) = \{x \in U : T_{A_1}(x) \subseteq X \wedge \cdots \wedge T_{A_m}(x) \subseteq X\} \tag{10}$$

$$\overline{\sum_{i=1}^{m} A_i}_T^P(X) = \sim \underline{\sum_{i=1}^{m} A_i}_T^P(\sim X) \tag{11}$$

Theorem 2. Let I be an IDS in which $A_1, A_2, \cdots, A_m \subseteq AT$, then $\forall X \subseteq U$,

$$\overline{\sum_{i=1}^{m} A_i}_T^P(X) = \{x \in U : T_{A_1}(x) \cap X \neq \emptyset \wedge \cdots \wedge T_{A_m}(x) \cap X \neq \emptyset\} \tag{12}$$

Theorem 3. Let I be an IDS in which $A_1, A_2, \cdots, A_m \subseteq AT$, then $\forall X \subseteq U$,

$$\underline{\sum_{i=1}^{m} A_i}_T^P(X) \subseteq \underline{\sum_{i=1}^{m} A_i}_T^O(X), \overline{\sum_{i=1}^{m} A_i}_T^P(X) \supseteq \overline{\sum_{i=1}^{m} A_i}_T^O(X). \tag{13}$$

3   Similarity Dominance Relation Based IMGRS

3.1   Similarity Dominance Relation Based Optimistic IMGRS

Definition 5. Let I be an $IODS$, $A_1, A_2, \cdots, A_m \subseteq AT$, $\forall r \in T$.

- The similarity dominance relation based optimistic multigranular lower and upper approximations of $CL_{\bar{r}}^{\geq}$ are denoted by $\underline{\sum_{i=1}^{m} A_i}_{SD}^O(CL_{\bar{r}}^{\geq})$ and $\overline{\sum_{i=1}^{m} A_i}_{SD}^O(CL_{\bar{r}}^{\geq})$ respectively, where

$$\underline{\sum_{i=1}^{m} A_i}_{SD}^O(CL_{\bar{r}}^{\geq}) = \{x \in U : SD_{\bar{A}_1}^{\leq -}(x) \subseteq CL_{\bar{r}}^{\geq} \vee \cdots \vee SD_{\bar{A}_m}^{\leq -}(x) \subseteq CL_{\bar{r}}^{\geq}\} \tag{14}$$

$$\overline{\sum_{i=1}^{m} A_i}_{SD}^O(CL_{\bar{r}}^{\geq}) = \{x \in U : SD_{\bar{A}_1}^{\geq -}(x) \cap CL_{\bar{r}}^{\geq} \neq \emptyset \wedge \cdots \wedge SD_{\bar{A}_m}^{\geq -}(x) \cap CL_{\bar{r}}^{\geq} \neq \emptyset\} \tag{15}$$

- The similarity dominance relation based optimistic multigranular lower and upper approximations of $CL_{\bar{r}}^{\leq}$ are denoted as $\underline{\sum_{i=1}^{m} A_i}_{SD}^O(CL_{\bar{r}}^{\leq})$ and $\overline{\sum_{i=1}^{m} A_i}_{SD}^O(CL_{\bar{r}}^{\leq})$ respectively, where

$$\underline{\sum_{i=1}^{m} A_i}_{SD}^O(CL_{\bar{r}}^{\leq}) = \{x \in U : SD_{\bar{A}_1}^{\geq -}(x) \subseteq CL_{\bar{r}}^{\leq} \vee \cdots \vee SD_{\bar{A}_m}^{\geq -}(x) \subseteq CL_{\bar{r}}^{\leq}\} \tag{16}$$

$$\overline{\sum_{i=1}^{m} A_i}_{SD}^O(CL_{\bar{r}}^{\leq}) = \{x \in U : SD_{\bar{A}_1}^{\leq -}(x) \cap CL_{\bar{r}}^{\leq} \neq \emptyset \wedge \cdots \wedge SD_{\bar{A}_m}^{\leq -}(x) \cap CL_{\bar{r}}^{\leq} \neq \emptyset\} \tag{17}$$

- The similarity dominance relation based optimistic boundary regions of $CL_{\bar{r}}^{\geq}$ and $CL_{\bar{r}}^{\leq}$ are denoted by $BN_{\sum_{i=1}^{m} A_i}^O(CL_{\bar{r}}^{\geq})$ and $BN_{\sum_{i=1}^{m} A_i}^O(CL_{\bar{r}}^{\leq})$ respectively, where

$$BN_{\sum_{i=1}^{m} A_i}^O(CL_{\bar{r}}^{\geq}) = \overline{\sum_{i=1}^{m} A_i}_{SD}^O(CL_{\bar{r}}^{\geq}) - \underline{\sum_{i=1}^{m} A_i}_{SD}^O(CL_{\bar{r}}^{\geq}) \tag{18}$$

$$BN^O_{\sum_{i=1}^m A_i}(CL^{\leq}_{\hat{r}}) = \overline{\sum_{i=1}^m A_i}^O_{SD}(CL^{\leq}_{\hat{r}}) - \underline{\sum_{i=1}^m A_i}^O_{SD}(CL^{\leq}_{\hat{r}}) \qquad (19)$$

Theorem 4. Let I be an $IODS$ in which $A_1, A_2, \cdots, A_m \subseteq AT$, then $\forall r \in T$, similarity dominance relation based optimistic IMGRSs have following properties:

- $\underline{\sum_{i=1}^m A_i}^O_{SD}(CL^{\geq}_{\hat{r}}) \subseteq CL^{\geq}_{\hat{r}} \subseteq \overline{\sum_{i=1}^m A_i}^O_{SD}(CL^{\geq}_{\hat{r}})$,

  $\underline{\sum_{i=1}^m A_i}^O_{SD}(CL^{\leq}_{\hat{r}}) \subseteq CL^{\leq}_{\hat{r}} \subseteq \overline{\sum_{i=1}^m A_i}^O_{SD}(CL^{\leq}_{\hat{r}})$;

- $\overline{\sum_{i=1}^m A_i}^O_{SD}(CL^{\geq}_{\hat{r}}) = \bigcap_{i=1}^m (\bigcup_{y \in CL^{\geq}_{\hat{r}}} SD^{\geq+}_{A_i}(y)), \quad \overline{\sum_{i=1}^m A_i}^O_{SD}(CL^{\leq}_{\hat{r}}) = \bigcap_{i=1}^m (\bigcup_{y \in CL^{\leq}_{\hat{r}}} SD^{\leq+}_{A_i}(y))$;

- $\forall r = 2, \cdots, n, \ \underline{\sum_{i=1}^m A_i}^O_{SD}(CL^{\geq}_{\hat{r}}) = U - \overline{\sum_{i=1}^m A_i}^O_{SD}(CL^{\leq}_{\hat{r}-1})$;

- $\forall r = 1, \cdots, n-1, \ \underline{\sum_{i=1}^m A_i}^O_{SD}(CL^{\leq}_{\hat{r}}) = U - \overline{\sum_{i=1}^m A_i}^O_{SD}(CL^{\geq}_{\hat{r}+1})$;

- $\forall r = 2, \cdots, n, \ \overline{\sum_{i=1}^m A_i}^O_{SD}(CL^{\geq}_{\hat{r}}) = U - \underline{\sum_{i=1}^m A_i}^O_{SD}(CL^{\leq}_{\hat{r}-1})$;

- $\forall r = 1, \cdots, n-1, \ \overline{\sum_{i=1}^m A_i}^O_{SD}(CL^{\leq}_{\hat{r}}) = U - \underline{\sum_{i=1}^m A_i}^O_{SD}(CL^{\geq}_{\hat{r}+1})$;

- $\forall r = 2, \cdots, n, \ BN^O_{\sum_{i=1}^m A_i}(CL^{\geq}_{\hat{r}}) = BN^O_{\sum_{i=1}^m A_i}(CL^{\leq}_{\hat{r}-1})$.

3.2 Similarity Dominance Relation Based Pessimistic IMGRS

Definition 6. Let I be an $IODS$ in which $A_1, A_2, \cdots, A_m \subseteq AT$, $\forall r \in T$, then

- The similarity dominance relation based pessimistic multigranular lower and upper approximations of $CL^{\geq}_{\hat{r}}$ are denoted by $\underline{\sum_{i=1}^m A_i}^P_{SD}(CL^{\geq}_{\hat{r}})$ and $\overline{\sum_{i=1}^m A_i}^P_{SD}(CL^{\geq}_{\hat{r}})$ respectively, where

  $$\underline{\sum_{i=1}^m A_i}^P_{SD}(CL^{\geq}_{\hat{r}}) = \{x \in U : SD^{\leq}_{A_1}{}^-(x) \subseteq CL^{\geq}_{\hat{r}} \wedge \cdots \wedge SD^{\leq}_{A_m}{}^-(x) \subseteq CL^{\geq}_{\hat{r}}\} \quad (20)$$

  $$\overline{\sum_{i=1}^m A_i}^P_{SD}(CL^{\geq}_{\hat{r}}) = \{x \in U : SD^{\geq}_{A_1}{}^-(x) \cap CL^{\geq}_{\hat{r}} \neq \emptyset \vee \cdots \vee SD^{\geq}_{A_m}{}^-(x) \cap CL^{\geq}_{\hat{r}} \neq \emptyset\} \quad (21)$$

- The similarity dominance relation based pessimistic multigranular lower and upper approximations of $CL^{\leq}_{\hat{r}}$ are denoted as $\underline{\sum_{i=1}^m A_i}^P_{SD}(CL^{\leq}_{\hat{r}})$ and $\overline{\sum_{i=1}^m A_i}^P_{SD}(CL^{\leq}_{\hat{r}})$ respectively, where

  $$\underline{\sum_{i=1}^m A_i}^P_{SD}(CL^{\leq}_{\hat{r}}) = \{x \in U : SD^{\geq}_{A_1}{}^-(x) \subseteq CL^{\leq}_{\hat{r}} \wedge \cdots \wedge SD^{\geq}_{A_m}{}^-(x) \subseteq CL^{\leq}_{\hat{r}}\} \quad (22)$$

  $$\overline{\sum_{i=1}^m A_i}^P_{SD}(CL^{\leq}_{\hat{r}}) = \{x \in U : SD^{\leq}_{A_1}{}^-(x) \cap CL^{\leq}_{\hat{r}} \neq \emptyset \vee \cdots \vee SD^{\leq}_{A_m}{}^-(x) \cap CL^{\leq}_{\hat{r}} \neq \emptyset\}(23)$$

- Similarity dominance relation based pessimistic boundary regions of $CL^{\geq}_{\hat{r}}$ and $CL^{\leq}_{\hat{r}}$ are denoted by $BN^P_{\sum_{i=1}^m A_i}(CL^{\geq}_{\hat{r}})$ and $BN^P_{\sum_{i=1}^m A_i}(CL^{\leq}_{\hat{r}})$ respectively, where

  $$BN^P_{\sum_{i=1}^m A_i}(CL^{\geq}_{\hat{r}}) = \overline{\sum_{i=1}^m A_i}^P_{SD}(CL^{\geq}_{\hat{r}}) - \underline{\sum_{i=1}^m A_i}^P_{SD}(CL^{\geq}_{\hat{r}}) \qquad (24)$$

  $$BN^P_{\sum_{i=1}^m A_i}(CL^{\leq}_{\hat{r}}) = \overline{\sum_{i=1}^m A_i}^P_{SD}(CL^{\leq}_{\hat{r}}) - \underline{\sum_{i=1}^m A_i}^P_{SD}(CL^{\leq}_{\hat{r}}) \qquad (25)$$

Theorem 5. Let I be an $IODS$ in which $A_1, A_2, \cdots, A_m \subseteq AT$, then $\forall r \in T$, similarity dominance relation based pessimistic IMGRSs have properties:

- $\sum_{i=1}^{m} \underline{A_i}_{SD}^{P}(CL_{\bar{r}}^{\geq}) \subseteq \underline{CL_{\bar{r}}^{\geq}} \subseteq \sum_{i=1}^{m} \underline{A_i}_{SD}^{P}(CL_{\bar{r}}^{\geq})$,

  $\sum_{i=1}^{m} \underline{A_i}_{SD}^{P}(CL_{\bar{r}}^{\leq}) \subseteq \underline{CL_{\bar{r}}^{\leq}} \subseteq \sum_{i=1}^{m} \underline{A_i}_{SD}^{P}(CL_{\bar{r}}^{\leq})$;

- $\overline{\sum_{i=1}^{m} A_i}_{SD}^{P}(CL_{\bar{r}}^{\geq}) = \bigcup_{i=1}^{m} (\bigcup_{y \in CL_{\bar{r}}^{\geq}} SD_{A_i}^{\geq+}(y))$, $\overline{\sum_{i=1}^{m} A_i}_{SD}^{P}(CL_{\bar{r}}^{\leq}) = \bigcup_{i=1}^{m} (\bigcup_{y \in CL_{\bar{r}}^{\leq}} SD_{A_i}^{\leq+}(y))$;

- $\forall r = 2, \cdots, n, \sum_{i=1}^{m} \underline{A_i}_{SD}^{P}(CL_{\bar{r}}^{\geq}) = U - \overline{\sum_{i=1}^{m} A_i}_{SD}^{P}(CL_{\bar{r}-1}^{\leq})$;

- $\forall r = 1, \cdots, n-1, \sum_{i=1}^{m} \underline{A_i}_{SD}^{P}(CL_{\bar{r}}^{\leq}) = U - \overline{\sum_{i=1}^{m} A_i}_{SD}^{P}(CL_{\bar{r}+1}^{\geq})$;

- $\forall r = 2, \cdots, n, \overline{\sum_{i=1}^{m} A_i}_{SD}^{P}(CL_{\bar{r}}^{\geq}) = U - \sum_{i=1}^{m} \underline{A_i}_{SD}^{P}(CL_{\bar{r}-1}^{\leq})$;

- $\forall r = 1, \cdots, n-1, \overline{\sum_{i=1}^{m} A_i}_{SD}^{P}(CL_{\bar{r}}^{\leq}) = U - \sum_{i=1}^{m} \underline{A_i}_{SD}^{P}(CL_{\bar{r}+1}^{\geq})$;

- $\forall r = 2, \cdots, n, BN_{\sum_{i=1}^{m} A_i}^{P}(CL_{\bar{r}}^{\geq}) = BN_{\sum_{i=1}^{m} A_i}^{P}(CL_{\bar{r}-1}^{\leq})$.

Theorem 6. Let I be an $IODS, B = \{b_1, b_2, \cdots, b_t\} \subseteq AT, \forall r \in T$, then similarity dominance relation based IMGRSs have some properties :

- $\sum_{i=1}^{t} \underline{b_i}_{SD}^{P}(CL_{\bar{r}}^{\geq}) \subseteq \sum_{i=1}^{t} \underline{b_i}_{SD}^{O}(CL_{\bar{r}}^{\geq}) \subseteq \underline{B}_{SD}(CL_{\bar{r}}^{\geq})$;

- $\overline{B}_{SD}(CL_{\bar{r}}^{\geq}) \subseteq \overline{\sum_{i=1}^{t} b_i}_{SD}^{O}(CL_{\bar{r}}^{\geq}) \subseteq \overline{\sum_{i=1}^{t} b_i}_{SD}^{P}(CL_{\bar{r}}^{\geq})$;

- $\sum_{i=1}^{t} \underline{b_i}_{SD}^{P}(CL_{\bar{r}}^{\leq}) \subseteq \sum_{i=1}^{t} \underline{b_i}_{SD}^{O}(CL_{\bar{r}}^{\leq}) \subseteq \underline{B}_{SD}(CL_{\bar{r}}^{\leq})$;

- $\overline{B}_{SD}(CL_{\bar{r}}^{\leq}) \subseteq \overline{\sum_{i=1}^{t} b_i}_{SD}^{O}(CL_{\bar{r}}^{\leq}) \subseteq \overline{\sum_{i=1}^{t} b_i}_{SD}^{P}(CL_{\bar{r}}^{\leq})$.

4   Rules in Similarity Dominance Relation Based IMGRS

In similarity dominance relation based optimistic IMGRS models, I is an $IODS$, $A = \{a_1, a_2, \cdots, a_n\} \subseteq AT, \forall r \in T$, then four different types of decision rules should be induced as follows:

- If $x \in \sum_{i=1}^{n} \underline{a_i}_{SD}^{O}(CL_{\bar{r}}^{\geq})$, then $\exists a_i \in A, a_i(y) = * \vee a_i(y) \geq a_i(x) \to y \in CL_{\bar{r}}^{\geq}$ is an ``optimistic at least certain" rule;

- If $x \in \overline{\sum_{i=1}^{n} a_i}_{SD}^{O}(CL_{\bar{r}}^{\geq})$, then $\forall a_i \in A, a_i(y) = * \vee a_i(y) \geq a_i(x) \to y \in CL_{\bar{r}}^{\geq}$ is an ``optimistic at least possible" rule;

- If $x \in \sum_{i=1}^{n} \underline{a_i}_{SD}^{O}(CL_{\bar{r}}^{\leq})$, then $\forall a_i \in A, a_i(y) = * \vee a_i(y) \leq a_i(x) \to y \in CL_{\bar{r}}^{\leq}$ is an ``optimistic at most certain" rule;

- If $x \in \overline{\sum_{i=1}^{n} a_i}_{SD}^{O}(CL_{\bar{r}}^{\leq})$, then $\exists a_i \in A, a_i(y) = * \vee a_i(y) \leq a_i(x) \to y \in CL_{\bar{r}}^{\leq}$ is an ``optimistic at most possible" rule.

Similarly, in similarity dominance relation based pessimistic IMGRS models, four different types of decision rules also be induced as follows:

- If $x \in \sum_{i=1}^{n} a_{i_{SD}}^{P}(CL_{\overline{r}}^{\geq})$, then $\exists a_i \in A, a_i(y) = * \vee a_i(y) \geq a_i(x) \rightarrow y \in CL_{\overline{r}}^{\geq}$
  is an ``pessimistic at least certain" rule;
- if $x \in \overline{\sum_{i=1}^{n} a_{i_{SD}}^{P}}(CL_{\overline{r}}^{\geq})$, then $\forall a_i \in A, a_i(y) = * \vee a_i(y) \geq a_i(x) \rightarrow y \in CL_{\overline{r}}^{\geq}$
  is an ``pessimistic at least possible" rule;
- If $x \in \sum_{i=1}^{n} a_{i_{SD}}^{P}(CL_{\overline{r}}^{\leq})$, then $\forall a_i \in A, a_i(y) = * \vee a_i(y) \leq a_i(x) \rightarrow y \in CL_{\overline{r}}^{\leq}$
  is an ``pessimistic at most certain" rule;
- If $x \in \overline{\sum_{i=1}^{n} a_{i_{SD}}^{P}}(CL_{\overline{r}}^{\leq})$, then $\exists a_i \in A, a_i(y) = * \vee a_i(y) \leq a_i(x) \rightarrow y \in CL_{\overline{r}}^{\leq}$
  is an ``pessimistic at most possible" rule.

5 An Illustrative Example

In this section, we give an example to illustrate the mechanism of similarity dominance relation based optimistic IMGRS. The pessimistic IMGRS can be illustrated analogously.

Example. Let I be an $IODS$, in which $U = \{x_1, x_2, \cdots, x_{12}\}$, $D = \{d\}$, $AT = \{a_1, a_2, a_3\}$. Table.1 is an incomplete ordered decision system, in which the symbol "*" means that an expert cannot decide the level.

Table 1. An incomplete ordered decision system

A/U	x_1	x_2	x_3	x_4	x_5	x_6	x_7	x_8	x_9	x_{10}	x_{11}	x_{12}
a_1	3	2	2	1	1	3	2	3	2	1	2	3
a_2	2	1	1	1	*	1	1	2	1	1	1	*
a_3	2	1	*	1	1	1	1	2	2	2	2	2
d	2	1	1	1	1	2	1	2	1	1	1	2

- As $x_{12} \in \sum_{i=1}^{3} a_{i_{SD}}^{O}(CL_{2}^{\geq})$, then ``optimistic at least certain" rule is acquired:
  $(a_1(y) = * \vee a_1(y) \geq 3) \vee (a_2(y) = *) \vee (a_3(y) = * \vee a_3(y) \geq 2) \rightarrow y \in CL_{2}^{\geq}$;
- As $x_2, x_3, x_7, x_9, x_{11} \in \overline{\sum_{i=1}^{3} a_{i_{SD}}^{O}}(CL_{2}^{\geq})$, then ``optimistic at least possible" rule is acquired:
  $(a_1(y) = * \vee a_1(y) \geq 2) \wedge (a_2(y) = * \vee a_2(y) \geq 1) \wedge (a_3(y) = *) \rightarrow y \in CL_{2}^{\geq}$;
- As $x_5 \in \sum_{i=1}^{3} a_{i_{SD}}^{O}(CL_{1}^{\leq})$, then ``optimistic at most certain" rule is acquired:
  $(a_1(y) = * \vee a_1(y) \leq 1) \vee (a_2(y) = *) \vee (a_3(y) = * \vee a_3(y) \leq 1) \rightarrow y \in CL_{1}^{\leq}$;
- As $x_4, x_{10} \in \overline{\sum_{i=1}^{3} a_{i_{SD}}^{O}}(CL_{1}^{\leq})$, then ``optimistic at most possible" rule is get:
  $(a_1(y) = * \vee a_1(y) \leq 1) \wedge (a_2(y) = * \vee a_2(y) \leq 1) \wedge (a_3(y) = * \vee a_3(y) \leq 1) \rightarrow y \in CL_{1}^{\leq}$.

6   Conclusions

In order to explore IODS, IMGRS has been further generalized in this paper. Two types of similarity dominance relation based IMGRS models are presented. Further, the properties of the two models are deeply studied. Then Eight types of decision rules in two similarity dominance relation based IMGRS models are presented. An example is given to illustrate the mechanism of similarity dominance relation based optimistic IMGRS and the process of rules acquisitions.

In the future, the application of similarity dominance relation based IMGRS to multi–criteria decision analysis is an interesting issue to be addressed.

Acknowledgements. This work is supported by the Natural Science Foundation of China (No. 60632050), China Postdoctoral Science Foundation (No.20100481149).

References

[1] Pawlak, Z.: Rough sets. Int. J. of Computer and Inf. Sci. 11, 341–356 (1982)
[2] Yang, X.B., Yang, J.Y., Wu, C., et al.: Dominance based rough set approach and knowledge reductions in incomplete ordered information system. Inform. Sci. 178(4), 1219–1234 (2008)
[3] Yang, X.B., Xie, J., Song, X.N., et al.: Credible rules in incomplete decision system based on descriptors. Knowledge - Based Syst. 22(1), 8–17 (2009)
[4] Yang, X.B., Yu, D.J., Yang, J.Y., et al.: Dominance based rough set approach to incomplete interval valued information system. Data & Knowledge Engineering 68(11), 1331–1347 (2009)
[5] Qian, Y.H., Liang, J.Y., Yao, Y.Y., et al.: MGRS: a multi-granulation rough set. Inform. Sci. 180, 949–970 (2010)
[6] Qian, Y.H., Liang, J.Y., Dang, C.Y.: Incomplete multi-granulation rough set. IEEE Transactions on Systems, Man and Cybernetics, Part A 40(2), 420–431 (2010)
[7] Zadeh, L.A.: Fuzzy sets and information granularity. In: Gupta, N., Ragade, R., Yager, R. (eds.) Advances in Fuzzy Set Theory and Application, pp. 3–18. North-Holland, Amsterdam (1979)
[8] Lin, T.Y.: Granular computing on binary relations II: rough set representations and belief functions. In: Polkowski, L., Skowron, A. (eds.) Rough Sets and Knowledge Discovery, pp. 122–140. Physica-Verlag, Heidelberg (1998)
[9] Lin, T.Y.: Granular computing on binary relations I: data mining and neighborhood systems. In: Polkowski, L., Skowron, A. (eds.) Rough Sets and Knowledge Discovery, pp. 107–121. Physica-Verlag, Heidelberg (1998)

A Special Class of Fuzzified Normal Forms

Omar Salazar Morales and José Jairo Soriano Méndez

Laboratory for Automation, Microelectronics and Computational Intelligence
Universidad Distrital Francisco José de Caldas
Bogotá D.C., Colombia
osalazarm@correo.udistrital.edu.co,
jairosoriano@udistrital.edu.co
http://www.udistrital.edu.co

Abstract. A fuzzified normal form is obtained when the boolean operators are replaced with a t-norm, t-conorm and fuzzy complement in a boolean normal form. In this paper we present a special class of fuzzified normal forms which has some boolean properties. We can use these properties to design and to simplify the normal forms as in the boolean case. We present some theorems and one application in fuzzy control to justify the selection of fuzzy operators.

Keywords: Boolean normal form, fuzzified normal form, fuzzy complement, triangular conorm, triangular norm.

1 Introduction

Some authors have studied generalizations of boolean concepts to fuzzy theory in order to get new methodologies from classical boolean methodologies [2], [9]. This is the case in the generalization of the boolean normal forms which has been studied as a way to get a mapping $[0, 1]^n \mapsto [0, 1]$ from a n-ary boolean function [7]. In the boolean algebra there are properties that permit to get simplified boolean normal forms to reduce complexity and cost of implementation in the designed systems [4]. In general, all special properties needed to use simplification in a fuzzified normal form, where the boolean operators are replaced by fuzzy operators, are not satisfied. Those include complementation and distributivity properties [5], [6], [8].

This paper presents a special class of fuzzified normal forms where the fuzzy operators, also known as t-norms, t-conorms and fuzzy complements, were chosen carefully in order to get some properties that permit to get simplification as in the boolean case. Some theorems were taken from references and their proofs are omitted. In Sect. 2 we present an introduction to boolean and fuzzified normal forms. In Sect. 3 we give some necessary preliminaries about t-norms, t-conorms and fuzzy complements. In Sect. 4 we present the special class of fuzzified normal forms and some properties. In Sect. 5 we use the theory presented in previous sections to develop a simple application in fuzzy control. Finally, in Sect. 6 we present the conclusion.

D.-S. Huang et al. (Eds.): ICIC 2011, LNBI 6840, pp. 331–344, 2012.

2   Boolean and Fuzzified Normal Forms

In the boolean algebra $\langle \{0,1\} ; \vee, \wedge, ', 0, 1 \rangle$, where the operations disjunction ($\vee$), conjunction ($\wedge$) and complement ($'$) are shown in Table 1, every n-ary function $f : \{0,1\}^n \mapsto \{0,1\}$ of $n \in \mathbb{N}$ finite variables can be represented by the formula

$$f(x_1, \ldots, x_n) = (x_1' \wedge f(0, \ldots, x_n)) \vee (x_1 \wedge f(1, \ldots, x_n)) \tag{1}$$
$$= (x_1' \vee f(1, \ldots, x_n)) \wedge (x_1 \vee f(0, \ldots, x_n)) \tag{2}$$

for all $\langle x_1, \ldots, x_n \rangle \in \{0,1\}^n$. Equations (1) and (2) are known as the *Boole's expansion theorem* [4].

Table 1. Operations in the boolean algebra

$\vee$	0	1		$\wedge$	0	1		$'$	
0	0	1		0	0	0		0	1
1	1	1		1	0	1		1	0

Operations $\vee$, $\wedge$ and $'$ satisfy the following properties for all $a, b, c \in \{0,1\}$. See [3], [4], [6], [8] for more details.

(Idempotency)	$a \vee a = a$	$a \wedge a = a$
(Commutativity)	$a \vee b = b \vee a$	$a \wedge b = b \wedge a$
(Associativity)	$(a \vee b) \vee c = a \vee (b \vee c)$	$(a \wedge b) \wedge c = a \wedge (b \wedge c)$
(Absorption)	$a \vee (a \wedge b) = a$	$a \wedge (a \vee b) = a$
(Distributivity)	$a \vee (b \wedge c) = (a \vee b) \wedge (a \vee c)$	$a \wedge (b \vee c) = (a \wedge b) \vee (a \wedge c)$
(Identity)	$a \vee 0 = a$	$a \wedge 1 = a$
(Absorption by 1 and 0)	$a \vee 1 = 1$	$a \wedge 0 = 0$
(Involution)	$(a')' = a$	
(De Morgan's laws)	$(a \vee b)' = a' \wedge b'$	$(a \wedge b)' = a' \vee b'$
(Kleene's inequality)	$a \wedge a' \leq b \vee b'$	
(Complementation)	$a \vee a' = 1$	$a \wedge a' = 0$

Applying recursively (1) and (2) to a boolean function we get

$$D(f)(x_1, \ldots, x_n) = (f(0, \ldots, 0) \wedge x_1' \wedge \cdots \wedge x_n') \vee (f(0, \ldots, 1) \wedge x_1' \wedge \cdots \wedge x_n) \vee \cdots$$
$$\vee (f(1, \ldots, 0) \wedge x_1 \wedge \cdots \wedge x_n') \vee (f(1, \ldots, 1) \wedge x_1 \wedge \cdots \wedge x_n) \tag{3}$$
$$C(f)(x_1, \ldots, x_n) = (f(0, \ldots, 0) \vee x_1 \vee \cdots \vee x_n) \wedge (f(0, \ldots, 1) \vee x_1 \vee \cdots \vee x_n') \wedge \cdots$$
$$\wedge (f(1, \ldots, 0) \vee x_1' \vee \cdots \vee x_n) \wedge (f(1, \ldots, 1) \vee x_1' \vee \cdots \vee x_n') \tag{4}$$

Equations (3) and (4) are called *disjunctive and conjunctive normal forms* of f. The 2^n values $f(0, \ldots, 0), f(0, \ldots, 1), \ldots, f(1, \ldots, 1)$ are constants in $\{0,1\}$. In the boolean case $D(f) = C(f)$ for all $\langle x_1, \ldots, x_n \rangle \in \{0,1\}^n$. Equations (3) and

(4) can be fuzzified by replacing the triplet $\langle \vee, \wedge,' \rangle$ by a triplet $\langle \triangledown, \triangle, \neg \rangle$, where $\triangledown$ is a t-conorm, $\triangle$ is a t-norm and $\neg$ is a fuzzy complement. The corresponding fuzzified disjunctive and conjunctive normal forms are denoted $D_F(f)$ and $C_F(f)$ and given by

$$D_F(f)(x_1, \ldots, x_n) = (d_1 \triangle \neg x_1 \triangle \cdots \triangle \neg x_n) \triangledown (d_2 \triangle \neg x_1 \triangle \cdots \triangle x_n) \triangledown \cdots$$
$$\triangledown (d_{2^n-1} \triangle x_1 \triangle \cdots \triangle \neg x_n) \triangledown (d_{2^n} \triangle x_1 \triangle \cdots \triangle x_n) \quad (5)$$
$$C_F(f)(x_1, \ldots, x_n) = (c_1 \triangledown x_1 \triangledown \cdots \triangledown x_n) \triangle (c_2 \triangledown x_1 \triangledown \cdots \triangledown \neg x_n) \triangle \cdots$$
$$\triangle (c_{2^n-1} \triangledown \neg x_1 \triangledown \cdots \triangledown x_n) \triangle (c_{2^n} \triangledown \neg x_1 \triangledown \cdots \triangledown \neg x_n) \quad (6)$$

where the variables $x_j \in [0,1]$ $(1 \le j \le n)$ and the constants $d_i, c_i \in [0,1]$ $(1 \le i \le 2^n)$. Then, for each n-ary boolean function f we obtain two $[0,1]^n \mapsto [0,1]$ mappings $D_F(f)$ and $C_F(f)$ because of in general $D_F(f) \ne C_F(f)$ [7]. In order to establish a special class of (5) and (6) and to present some properties, we need some preliminaries about t-norms, t-conorms and fuzzy complements.

3 Preliminaries

1. **T-norms.** A t-norm is defined by a function $\triangle : [0,1]^2 \mapsto [0,1] : \langle a, b \rangle \mapsto a \triangle b$ for all $a, b \in [0,1]$. Then a t-norm is a binary operation on the unit interval that satisfies at least the following axioms for all $a, b, c \in [0,1]$:

   Axiom 1 (Boundary condition). $a \triangle 1 = a$

   Axiom 2 (Monotonicity). $b \le c$ implies $a \triangle b \le a \triangle c$

   Axiom 3 (Commutativity). $a \triangle b = b \triangle a$

   Axiom 4 (Associativity). $a \triangle (b \triangle c) = (a \triangle b) \triangle c$

   Two examples [6] are *standard t-norm* $(\triangle_M)$ and *drastic t-norm* $(\triangle_D)$ given by

   $$a \triangle_M b = \min\{a, b\} \qquad \text{and} \qquad a \triangle_D b = \begin{cases} a & \text{if } b = 1 \\ b & \text{if } a = 1 \\ 0 & \text{otherwise} \end{cases}$$

   for all $a, b \in [0,1]$ (we must use inf if min does not exist). $\triangle_M$ is an upper bound and $\triangle_D$ is a lower bound for any t-norm.

   Theorem 1. [6] *For all* $a, b \in [0,1]$, $a \triangle_D b \le a \triangle b \le a \triangle_M b$

2. **T-conorms.** A t-conorm is defined by a function $\triangledown : [0,1]^2 \mapsto [0,1] : \langle a, b \rangle \mapsto a \triangledown b$ for all $a, b \in [0,1]$. Then a t-conorm is a binary operation on the unit interval that satisfies at least the following axioms for all $a, b, c \in [0,1]$:

   Axiom 5 (Boundary condition). $a \triangledown 0 = a$

   Axiom 6 (Monotonicity). $b \le c$ implies $a \triangledown b \le a \triangledown c$

Axiom 7 (Commutativity). $a \bigtriangledown b = b \bigtriangledown a$

Axiom 8 (Associativity). $a \bigtriangledown (b \bigtriangledown c) = (a \bigtriangledown b) \bigtriangledown c$

Two examples [6] are *standard t-conorm* ($\bigtriangledown_{\mathrm{M}}$) and *drastic t-conorm* ($\bigtriangledown_{\mathrm{D}}$) given by

$$a \bigtriangledown_{\mathrm{M}} b = \max\{a, b\} \quad \text{and} \quad a \bigtriangledown_{\mathrm{D}} b = \begin{cases} a & \text{if } b = 0 \\ b & \text{if } a = 0 \\ 1 & \text{otherwise} \end{cases}$$

for all $a, b \in [0, 1]$ (we must use sup if max does not exist). $\bigtriangledown_{\mathrm{M}}$ is a lower bound and $\bigtriangledown_{\mathrm{D}}$ is an upper bound for any t-conorm.

Theorem 2. [6] *For all* $a, b \in [0, 1]$, $a \bigtriangledown_{\mathrm{M}} b \le a \bigtriangledown b \le a \bigtriangledown_{\mathrm{D}} b$.

3. **Fuzzy Complements.** A fuzzy complement is defined by a function $\neg :$ $[0, 1] \mapsto [0, 1] : a \mapsto \neg a$. This function assigns the value $\neg a \in [0, 1]$ to each $a \in [0, 1]$. All the fuzzy complements satisfy at least the following two axioms.

Axiom 9 (Boundary conditions). $\neg 0 = 1$ *and* $\neg 1 = 0$

Axiom 10 (Monotonicity). *For all* $a, b \in [0, 1]$, *if* $a \le b$, *then* $\neg a \ge \neg b$

Two special cases are the threshold-type complements defined by

$$\neg a = \underline{a} = \begin{cases} 1 & \text{if } a = 0 \\ 0 & \text{if } a > 0 \end{cases} \quad \text{and} \quad \neg a = \overline{a} = \begin{cases} 1 & \text{if } a < 1 \\ 0 & \text{if } a = 1 \end{cases} \quad (7)$$

for all $a \in [0, 1]$. We use the notation $\underline{a}$ and $\overline{a}$ for these two special complements that represent a lower and upper bound for any fuzzy complement.

Theorem 3. *If* $\neg : [0, 1] \mapsto [0, 1] : a \mapsto \neg a$ *is a fuzzy complement then it satisfies* $\underline{a} \le \neg a \le \overline{a}$ *for all* $a \in [0, 1]$

Proof. By definition we have $0 \le \neg a \le 1$. If $a = 0$ then $\underline{0} = \neg 0 = \overline{0} = 1$ and if $a = 1$ then $\underline{1} = \neg 1 = \overline{1} = 0$ (by $\underline{a}$ and $\overline{a}$ definitions and Axiom 9). If $a \in (0, 1)$ then $0 = \underline{a} \le \neg a \le \overline{a} = 1$ □

4   A Special Class of Fuzzified Normal Forms

We restrict our discussion to the following fuzzified normal forms:

$$\widetilde{\mathrm{D}}_{\mathrm{F}}(f)(x_1, \ldots, x_n) = (d_1 \bigtriangleup \overline{x}_1 \bigtriangleup \cdots \bigtriangleup \overline{x}_n) \bigtriangledown_{\mathrm{M}} (d_2 \bigtriangleup \overline{x}_1 \bigtriangleup \cdots \bigtriangleup x_n) \bigtriangledown_{\mathrm{M}} \cdots$$
$$\bigtriangledown_{\mathrm{M}} (d_{2^n - 1} \bigtriangleup x_1 \bigtriangleup \cdots \bigtriangleup \overline{x}_n) \bigtriangledown_{\mathrm{M}} (d_{2^n} \bigtriangleup x_1 \bigtriangleup \cdots \bigtriangleup x_n) \quad (8)$$
$$\widetilde{\mathrm{C}}_{\mathrm{F}}(f)(x_1, \ldots, x_n) = (c_1 \bigtriangledown x_1 \bigtriangledown \cdots \bigtriangledown x_n) \bigtriangleup_{\mathrm{M}} (c_2 \bigtriangledown x_1 \bigtriangledown \cdots \bigtriangledown \underline{x}_n) \bigtriangleup_{\mathrm{M}} \cdots$$
$$\bigtriangleup_{\mathrm{M}} (c_{2^n - 1} \bigtriangledown \underline{x}_1 \bigtriangledown \cdots \bigtriangledown x_n) \bigtriangleup_{\mathrm{M}} (c_{2^n} \bigtriangledown \underline{x}_1 \bigtriangledown \cdots \bigtriangledown \underline{x}_n) \quad (9)$$

where $x_j \in [0, 1]$ $(1 \le j \le n)$, $d_i, c_i \in \{0, 1\}$ $(1 \le i \le 2^n)$, $\bigtriangledown$ and $\bigtriangleup$ are any t-conorm and t-norm, $\bigtriangledown_{\mathrm{M}}$ and $\bigtriangleup_{\mathrm{M}}$ are standard t-conorm and standard t-norm, $\underline{x}_j$ and $\overline{x}_j$ are the complements given in (7). We choose the triplet $\langle \bigtriangledown_{\mathrm{M}}, \bigtriangleup, \overline{a} \rangle$ in (8) and the triplet $\langle \bigtriangledown, \bigtriangleup_{\mathrm{M}}, \underline{a} \rangle$ in (9) according to the properties that we need. The two majors restrictions in the selection of these triplets are complementation and distributivity properties as we will show.

4.1 A Review of Properties

1. **Idempotency.** We have the following theorems.

   Theorem 4. [6] *The standard fuzzy t-norm is the only idempotent t-norm.*

   Theorem 5. [6] *The standard fuzzy t-conorm is the only idempotent t-conorm.*

   Then, we can add or remove repeated terms in the disjunction (8) or the conjunction (9) by using Theorem 4 and Theorem 5.

2. **Commutativity.** All t-norms and t-conorms satisfy commutativity property by Axiom 3 and Axiom 7. The order of terms in (8) and (9) is indifferent.

3. **Associativity.** All t-norms and t-conorms satisfy associativity property by Axiom 4 and Axiom 8. Then we can drop parentheses and just write $a \triangle b \triangle c$ for $(a \triangle b) \triangle c$ and $a \triangle (b \triangle c)$ because of there is no ambiguity. The same comment applies for $\triangledown$. In (8) and (9) we have dropped parentheses by leaving all those that we need.

4. **Absorption.** $\triangle_M$ and $\triangledown_M$ satisfy absorption property. If $a, b \in [0,1]$ and $a \leq b$ then $a \triangle_M (a \triangledown_M b) = a \triangle_M b = a$ and $a \triangledown_M (a \triangle_M b) = a \triangledown_M a = a$. If $b \leq a$ then $a \triangle_M (a \triangledown_M b) = a \triangle_M a = a$ and $a \triangledown_M (a \triangle_M b) = a \triangledown_M b = a$. The inequalities

$$a \triangledown_M (a \triangle b) \geq a \tag{10}$$

$$a \triangle_M (a \triangledown b) \leq a \tag{11}$$

   hold for all $a, b \in [0,1]$ because $a \triangledown_M (a \triangle b)$ is greater than every element in the set $\{a, a \triangle b\}$ and $a \triangle_M (a \triangledown b)$ is lesser than every element in $\{a, a \triangledown b\}$. By Theorem 1 and Theorem 2 we know that $a \triangle b \leq a \triangle_M b$ and $a \triangledown_M b \leq a \triangledown b$ for all $a, b \in [0,1]$. Then by using Axiom 2 and Axiom 6 we have the inequalities

$$a \triangledown_M (a \triangle b) \leq a \triangledown_M (a \triangle_M b) = a \tag{12}$$

$$a = a \triangle_M (a \triangledown_M b) \leq a \triangle_M (a \triangledown b) \tag{13}$$

   By antisymmetry and inequalities (10) with (12), and (11) with (13) we have

$$a \triangledown_M (a \triangle b) = a \tag{14}$$

$$a \triangle_M (a \triangledown b) = a \tag{15}$$

   for all $a, b \in [0,1]$. Equation (14) is the absorption property for (8) and (15) is the absorption property for (9).

5. **Distributivity.** We need the following definition and proposition found in the references.

   Definition 1. *Let $\triangle$ be a t-norm and $\triangledown$ a t-conorm. Then we say that $\triangle$ is distributive over $\triangledown$ if $a \triangle (b \triangledown c) = (a \triangle b) \triangledown (a \triangle c)$ for all $a, b, c \in [0,1]$, and that $\triangledown$ is distributive over $\triangle$ if $a \triangledown (b \triangle c) = (a \triangledown b) \triangle (a \triangledown c)$ for all $a, b, c \in [0,1]$. If $\triangle$ is distributive over $\triangledown$ and $\triangledown$ is distributive over $\triangle$, then $\langle \triangle, \triangledown \rangle$ is called a* distributive pair *(of t-norms and t-conorms).*

Proposition 1. [5] *Let $\triangle$ be a t-norm and $\triangledown$ a t-conorm. Then we have:*
(a) $\triangledown$ is distributive over $\triangle$ if and only if $\triangle = \triangle_{\mathrm{M}}$
(b) $\triangle$ is distributive over $\triangledown$ if and only if $\triangledown = \triangledown_{\mathrm{M}}$
(c) $\langle \triangle, \triangledown \rangle$ is a distributive pair if and only if $\triangle = \triangle_{\mathrm{M}}$ and $\triangledown = \triangledown_{\mathrm{M}}$

We want use distributivity property as in the boolean case. Then by using Proposition 1 we use $\triangledown_{\mathrm{M}}$ in (8) and $\triangle_{\mathrm{M}}$ in (9).

6. **Identity.** All t-norms and t-conorms satisfy identity property by Axiom 1 and Axiom 5. Then we can drop 1 and 0 in (8) and (9) where they are not alone.

7. **Absortion by 1 and 0.** All t-norms and t-conorms satisfy absorption by 1 and 0 property because of the following reasons: $\triangle_{\mathrm{M}}$ and $\triangledown_{\mathrm{M}}$ satisfy the properties $a \triangle_{\mathrm{M}} 0 = 0$ and $a \triangledown_{\mathrm{M}} 1 = 1$ for all $a \in [0,1]$. Then, by using Theorem 1 (with $b = 0$) we have $a \triangle 0 \leq a \triangle_{\mathrm{M}} 0 = 0$, and by using Theorem 2 (with $b = 1$) we have $1 = a \triangledown_{\mathrm{M}} 1 \leq a \triangledown 1$. Then we have $a \triangle 0 = 0$ and $a \triangledown 1 = 1$ for all $a \in [0,1]$.

By using identity and absorption by 1 and 0 properties we can find the constants d_i and c_i ($1 \leq i \leq 2^n$) in (8) and (9):

$$d_1 = \widetilde{\mathrm{D}}_{\mathrm{F}}(f)(0,\ldots,0), d_2 = \widetilde{\mathrm{D}}_{\mathrm{F}}(f)(0,\ldots,1),\ldots,d_{2^n} = \widetilde{\mathrm{D}}_{\mathrm{F}}(f)(1,\ldots,1),$$

$$c_1 = \widetilde{\mathrm{C}}_{\mathrm{F}}(f)(0,\ldots,0), c_2 = \widetilde{\mathrm{C}}_{\mathrm{F}}(f)(0,\ldots,1),\ldots,c_{2^n} = \widetilde{\mathrm{C}}_{\mathrm{F}}(f)(1,\ldots,1) \quad (16)$$

This is not a surprise because all t-norms, t-conorms and fuzzy complements are the same boolean operators when they are restricted to $\{0,1\}$. We must restrict the constants d_i and c_i to values in $\{0,1\}$, i.e., $d_i, c_i \in \{0,1\}$, because we want a n-ary boolean function when $x_1,\ldots,x_n$ are boolean variables in (8) and (9).

8. **Involution.** Complements $\bar{a}$ and $\underline{a}$ in (7) are not involutive. For example $\overline{0.5} = 1$ and $\overline{1} = 0 \neq 0.5$.

9. **De Morgan's Laws.** We need the following definition and theorem found in the references.

Definition 2. *A t-norm ($\triangle$) and a t-conorm ($\triangledown$) are dual with respect to a fuzzy complement ($\neg$) if and only if $\neg(a \triangle b) = \neg a \triangledown \neg b$ and $\neg(a \triangledown b) = \neg a \triangle \neg b$ for all $a,b \in [0,1]$*

Theorem 6. [6] *The triplets $\langle \triangledown_{\mathrm{M}}, \triangle_{\mathrm{M}}, \neg \rangle$ and $\langle \triangledown_{\mathrm{D}}, \triangle_{\mathrm{D}}, \neg \rangle$ are dual with respect to any fuzzy complement $\neg$*

Theorem 7. *The triplet $\langle \triangledown_{\mathrm{M}}, \triangle, \bar{a} \rangle$ is dual with respect to the fuzzy complement $\bar{a}$ and $\langle \triangledown, \triangle_{\mathrm{M}}, \underline{a} \rangle$ is dual with respect to the fuzzy complement $\underline{a}$*

Proof. We prove only the equalities $\overline{(a \triangle b)} = \bar{a} \triangledown_{\mathrm{M}} \bar{b}$ and $\overline{(a \triangledown_{\mathrm{M}} b)} = \bar{a} \triangle \bar{b}$ for the tiplet $\langle \triangledown_{\mathrm{M}}, \triangle, \bar{a} \rangle$. The proof for the triplet $\langle \triangledown, \triangle_{\mathrm{M}}, \underline{a} \rangle$ is analogous. By Theorem 1 we know that $a \triangle b \leq a \triangle_{\mathrm{M}} b$ for all $a,b \in [0,1]$ then by Axiom 10 and Theorem 6 we have

$$\overline{(a \triangle b)} \geq \overline{(a \triangle_{\mathrm{M}} b)} = \bar{a} \triangledown_{\mathrm{M}} \bar{b} \quad (17)$$

If $a < 1$ then $\bar{a} = 1$ and hence by (17) we have $1 = \bar{a} \,\triangledown_M\, \bar{b} = \overline{(a \,\triangle\, b)}$ because by definition $\overline{(a \,\triangle\, b)} \le 1$. If $b < 1$ we have $1 = \bar{a} \,\triangledown_M\, \bar{b} = \overline{(a \,\triangle\, b)}$ for the same reason. If $a = b = 1$ then $\bar{a} = \bar{b} = 0$ and $a \,\triangle\, b = 1$, hence $\overline{(a \,\triangle\, b)} = \bar{1} = 0 = 0 \,\triangledown_M\, 0 = \bar{a} \,\triangledown_M\, \bar{b}$. In conclusion, we have the first equality $\overline{(a \,\triangle\, b)} = \bar{a} \,\triangledown_M\, \bar{b}$ for all $a, b \in [0, 1]$.

If $a = 1$ then $\bar{a} = 0$, hence $\overline{(a \,\triangledown_M\, b)} = \bar{1} = 0 = 0 \,\triangle\, \bar{b} = \bar{a} \,\triangle\, \bar{b}$. If $b = 1$ then $\overline{(a \,\triangledown_M\, b)} = \bar{a} \,\triangle\, \bar{b}$ for the same reason. If $a < 1$ and $b < 1$ then $\bar{a} = \bar{b} = 1$, hence, if $a \le b$ then $\overline{(a \,\triangledown_M\, b)} = \bar{b} = 1 = 1 \,\triangle\, 1 = \bar{a} \,\triangle\, \bar{b}$ and if $b \le a$ then $\overline{(a \,\triangledown_M\, b)} = \bar{a} = 1 = 1 \,\triangle\, 1 = \bar{a} \,\triangle\, \bar{b}$. In conclusion, we have the second equality $\overline{(a \,\triangledown_M\, b)} = \bar{a} \,\triangle\, \bar{b}$ for all $a, b \in [0, 1]$ $\square$

10. **Kleene's Inequality.** We need the following theorem.

Theorem 8. $a \,\triangledown_M\, \bar{a} = 1$ and $a \,\triangle_M\, \underline{a} = 0$ for all $a \in [0, 1]$

Proof. If $a = 0$ then $\bar{a} = \underline{a} = 1$, hence $a \,\triangledown_M\, \bar{a} = 0 \,\triangledown_M\, 1 = 1$ and $a \,\triangle_M\, \underline{a} = 0 \,\triangle_M\, 1 = 0$. If $a = 1$ then $\bar{a} = \underline{a} = 0$, hence $a \,\triangledown_M\, \bar{a} = 1 \,\triangledown_M\, 0 = 1$ and $a \,\triangle_M\, \underline{a} = 1 \,\triangle_M\, 0 = 0$. If $a \in (0, 1)$ then $\bar{a} = 1$ and $\underline{a} = 0$, hence $a \,\triangledown_M\, \bar{a} = a \,\triangledown_M\, 1 = 1$ and $a \,\triangle_M\, \underline{a} = a \,\triangle_M\, 0 = 0$ $\square$

By Theorem 8, then the inequalities $a \,\triangle\, \bar{a} \le b \,\triangledown_M\, \bar{b} = 1$ and $0 = a \,\triangle_M\, \underline{a} \le b \,\triangledown\, \underline{b}$ hold for all $a, b \in [0, 1]$. These two inequalities are the corresponding versions of the Kleene's inequality for (8) and (9) respectively.

11. **Complementation.** In (8) we have expressions as $(b \,\triangle\, a) \,\triangledown_M\, (b \,\triangle\, \neg a)$ and in (9) expressions as $(b \,\triangledown\, a) \,\triangle_M\, (b \,\triangledown\, \neg a)$, with $a, b \in [0, 1]$, where the only difference is one complemented variable. These two expressions can be rewritten as $b \,\triangle\, (a \,\triangledown_M\, \neg a)$ and $b \,\triangledown\, (a \,\triangle_M\, \neg a)$ by using distributivity property. We want the possibility to simplify the fuzzified normal forms as in the boolean case. Then, we choose the complement operation $\bar{a}$ in (8) and the complement operation $\underline{a}$ in (9) by the following theorem.

Theorem 9. Let $\neg : [0, 1] \mapsto [0, 1]$ a fuzzy complement. Then,
(a) $a \,\triangledown_M\, \neg a = 1$ for all $a \in [0, 1]$ if and only if $\neg a = \bar{a}$ for all $a \in [0, 1]$
(b) $a \,\triangle_M\, \neg a = 0$ for all $a \in [0, 1]$ if and only if $\neg a = \underline{a}$ for all $a \in [0, 1]$

Proof. We prove only the first part. The proof of the second part is analogous. The reverse implication follows by Theorem 8. Now suppose that $a \,\triangledown_M\, \neg a = 1$ for all $a \in [0, 1]$. By Theorem 3 we know that $\neg a \le \bar{a}$ for all $a \in [0, 1]$. It is clear that $\neg a = \bar{a}$ when $a \in \{0, 1\}$ by Axiom 9. Now we show that the strict inequality in the open interval $(0, 1)$ is impossible. Suppose that there exists a value $b \in (0, 1)$ such that $\neg b < \bar{b} = 1$.
  - If $b \le \neg b$ then $b \,\triangledown_M\, \neg b = \neg b < 1$, but this is contradictory with the assumption $a \,\triangledown_M\, \neg a = 1$ for all $a \in [0, 1]$.
  - If $\neg b \le b$ then $b \,\triangledown_M\, \neg b = b < 1$, but this is contradictory with the assumption $a \,\triangledown_M\, \neg a = 1$ for all $a \in [0, 1]$.
Then, there is no value $b \in (0, 1)$ such that $\neg b < \bar{b}$, and we have $\neg a = \bar{a}$ for all $a \in [0, 1]$ $\square$

4.2   Additional Properties

If we replace x_1 by 0 and 1 in (8) we have

$$\widetilde{D}_F(f)(0,\ldots,x_n) = (d_1 \triangle \overline{x}_2 \triangle \cdots \triangle \overline{x}_n) \triangledown_M (d_2 \triangle \overline{x}_2 \triangle \cdots \triangle x_n) \triangledown_M \cdots$$
$$\triangledown_M (d_{2^{n-1}-1} \triangle x_2 \triangle \cdots \triangle \overline{x}_n) \triangledown_M (d_{2^{n-1}} \triangle x_2 \triangle \cdots \triangle x_n) \quad (18)$$

$$\widetilde{D}_F(f)(1,\ldots,x_n) = (d_{2^{n-1}+1} \triangle \overline{x}_2 \triangle \cdots \triangle \overline{x}_n) \triangledown_M (d_{2^{n-1}+2} \triangle \overline{x}_2 \triangle \cdots \triangle x_n) \triangledown_M \cdots$$
$$\triangledown_M (d_{2^n-1} \triangle x_2 \triangle \cdots \triangle \overline{x}_n) \triangledown_M (d_{2^n} \triangle x_2 \triangle \cdots \triangle x_n) \quad (19)$$

It is clear that the following equation holds for all $\langle x_1,\ldots,x_n \rangle \in [0,1]^n$ by using distributivity property and (18) and (19)

$$\widetilde{D}_F(f)(x_1,\ldots,x_n) = (\overline{x}_1 \triangle \widetilde{D}_F(f)(0,\ldots,x_n)) \triangledown_M (x_1 \triangle \widetilde{D}_F(f)(1,\ldots,x_n)) \quad (20)$$

In the same way the following equation holds for all $\langle x_1,\ldots,x_n \rangle \in [0,1]^n$ by using a similar argument

$$\widetilde{C}_F(f)(x_1,\ldots,x_n) = (x_1 \triangledown \widetilde{C}_F(f)(0,\ldots,x_n)) \triangle_M (\underline{x}_1 \triangledown \widetilde{C}_F(f)(1,\ldots,x_n)) \quad (21)$$

Equations (20) and (21) are analogous to the Boole's expansion theorem.

4.3   Design of Fuzzified Normal Forms with Boolean Methods

For the design of a fuzzified normal form, all that we need is just specify a t-norm $(\triangle)$ in (8) or a t-conorm $(\triangledown)$ in (9), and the constant values d_i and c_i $(1 \leq i \leq 2^n)$ given in (16). Then, given a n-ary boolean function $f : \{0,1\}^n \mapsto \{0,1\}$ expressed as a table follow the next instructions to get $\widetilde{D}_F(f)$ or $\widetilde{C}_F(f)$.

1. Instructions for $\widetilde{D}_F(f)$:
   (a) For those rows that have value 1 in the column of the expression to be put into fuzzified disjunctive normal form, by using a t-norm $(\triangle)$ form the conjunction of the variables with value equal to 1 with the complements of the variables with value equal to 0 by using the complement $\overline{a}$.
   (b) Form the disjunction of the conjunctions obtained in the last numeral by using the standard t-conorm $(\triangledown_M)$.

2. Instructions for $\widetilde{C}_F(f)$:
   (a) For those rows that have value 0 in the column of the expression to be put into fuzzified conjunctive normal form, by using a t-conorm $(\triangledown)$ form the disjunction of the variables with value equal to 0 with the complements of the variables with value equal to 1 by using the complement $\underline{a}$.
   (b) Form the conjunction of the disjunctions obtained in the last numeral by using the standard t-norm $(\triangle_M)$.

5   An Application in Fuzzy Control

This application was taken from [1]. It uses the Defuzzification Based on Boolean Relations (DBR) model [1], [10]. In Fig. 1 we show a liquid-level system. We want keep the level $h(t)$ constant, where t is the time variable. There are two input valves: V_1 and V_s. First input valve has a constant *large* flow Φ_1 when it is *totally open* and second input valve has a constant *small* flow Φ_s when it is *totally open*, then, the total input flow to tank is the action of these two valves denoted by ϕ_i. There is only one output valve: V_o. Output valve acts as a consequence of gravity action with a coefficient K.

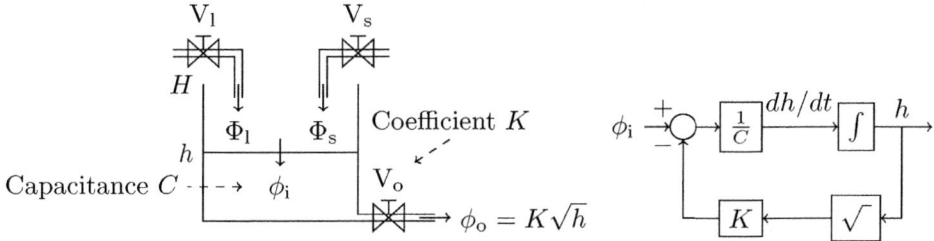

Fig. 1. Liquid-level system and its block diagram

According to the model presented in [1] we have the following differential equation which models the system in Fig. 1:

$$C\frac{dh}{dt} + K\sqrt{h} = \phi_i \quad \text{with} \quad 0 \leq h \leq H \tag{22}$$

where C is the capacitance of tank, K is the coefficient of output valve, ϕ_i is the total input flow, h is the level in the tank and H is the maximum level in the tank. Figure 1 shows a block diagram of (22) and Table 2 shows values for the constants of liquid-level system. The control system in this application is shown in Fig. 2. We must design the static transfer characteristic Ω. This transfer characteristic has two inputs: *error level* and *error level rate*. We design a boolean controller as a first approach and then we fuzzify this controller by using fuzzified normal forms.

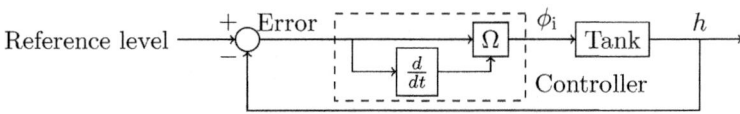

Fig. 2. Block diagram of the control system

Table 2. Constants of the liquid-level system

Symbol	Meaning	Value	Unit
C	Capacitance of tank	126×10^{-3}	m^2
K	Coefficient of output valve	140×10^{-6}	$m^{5/2}/s$
H	Maximum level of tank	1	m
Φ_1	Constant flow in valve V_1 when it is totally open	1.5×10^{-3}	m^3/s
Φ_s	Constant flow in valve V_s when it is totally open	1×10^{-3}	m^3/s

5.1   Boolean Controller

There are four states of the input valves V_1 and V_s if they are totally open or totally closed. We model the state of these valves with boolean variables v_1 and v_s. Then, the total input flow ϕ_i has four values as is shown in the Table 3. By using the Table 3 we can express ϕ_i as a function of v_1 and v_s:

$$\phi_i = v_1 \Phi_1 + v_s \Phi_s \tag{23}$$

Table 3. State of the input valves and values of total input flow

$v_1\ v_s$	ϕ_i	Physical interpretation
0 0	0	If V_1 is totally closed and V_s is totally closed then ϕ_i is 0
0 1	Φ_s	If V_1 is totally closed and V_s is totally open then ϕ_i is Φ_s
1 0	Φ_1	If V_1 is totally open and V_s is totally closed then ϕ_i is Φ_1
1 1	$\Phi_s + \Phi_1$	If V_1 is totally open and V_s is totally open then ϕ_i is $\Phi_s + \Phi_1$

Now we formulate two crisp sets in the universe of error (e) and one in the universe of error rate (de/dt) as is shown in Fig. 3. The sets in the universe of error are: *negative large error* (e_{nl}) and *positive large error* (e_{pl}), and their membership functions are given in (24) and (25). The set in the universe of error rate is: *positive error rate* (de_p), and its membership function is given in (26). We do not need sets called *close to zero* or *negative error rate* because of they can be expressed in terms of the other sets. The set *close to zero* can be expressed as complement of the union of *negative large error* and *positive large error*, and the set *negative error rate* is the complement of *positive error rate*.

$$e_{nl}(e) = \begin{cases} 1 & \text{if } e \leq -0.02 \text{ m} \\ 0 & \text{otherwise} \end{cases} \tag{24}$$

$$e_{pl}(e) = \begin{cases} 1 & \text{if } e \geq 0.02 \text{ m} \\ 0 & \text{otherwise} \end{cases} \tag{25}$$

$$de_p(de/dt) = \begin{cases} 1 & \text{if } de/dt \geq 0 \text{ m/s} \\ 0 & \text{otherwise} \end{cases} \tag{26}$$

There are only six possible combinations of ones and zeros as is shown in the six first rows of Table 4 depending of value in error and error rate. Two last combinations are not possible physically, but they can be used in the design.

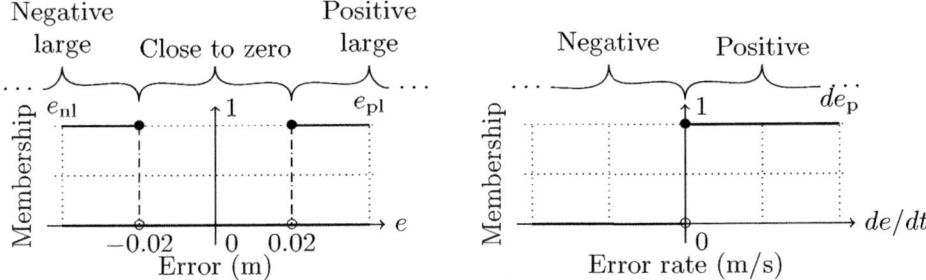

Fig. 3. Membership functions of crisp sets *negative large error* (e_{nl}), *positive large error* (e_{pl}) and *positive error rate* (de_p)

Table 4. Rule base

e_{nl}	e_{pl}	de_p	v_l	v_s	Rule
1	0	0	0	0	If e is negative large and de/dt is negative then ϕ_i is 0
1	0	1	0	0	If e is negative large and de/dt is positive then ϕ_i is 0
0	0	0	0	0	If e is close to zero and de/dt is negative then ϕ_i is 0
0	0	1	0	1	If e is close to zero and de/dt is positive then ϕ_i is Φ_s
0	1	0	1	0	If e is positive large and de/dt is negative then ϕ_i is Φ_l
0	1	1	1	1	If e is positive large and de/dt is positive then ϕ_i is $\Phi_l + \Phi_s$
1	1	0	0	0	
1	1	1	0	0	

We can find the boolean disjunctive normal forms for v_l and v_s from Table 4. We choose disjunctive and not conjunctive normal forms because there are less ones than zeros. The disjunctive normal forms are given in (27) and (28). We can simplify these formulas as is shown in the same equations by using boolean properties.

$$D(v_l) = (e'_{nl} \wedge e_{pl} \wedge de'_p) \vee (e'_{nl} \wedge e_{pl} \wedge de_p) = e'_{nl} \wedge e_{pl} \wedge (de'_p \vee de_p) = e'_{nl} \wedge e_{pl} \quad (27)$$

$$D(v_s) = (e'_{nl} \wedge e'_{pl} \wedge de_p) \vee (e'_{nl} \wedge e_{pl} \wedge de_p) = e'_{nl} \wedge (e'_{pl} \vee e_{pl}) \wedge de_p = e'_{nl} \wedge de_p \quad (28)$$

We obtain the static transfer characteristic Ω of boolean controller given by $\phi_i(e, de/dt) = (e'_{nl}(e) \wedge e_{pl}(e))\Phi_l + (e'_{nl}(e) \wedge de_p(de/dt))\Phi_s$, where e_{nl}, e_{pl} and de_p are the membership functions (24), (25) and (26), by replacing (27) and (28) in (23).

5.2 Fuzzy Controller

The next step is to soft crisp sets *negative large error*, *positive large error* and *positive error rate* by using fuzzy sets as is shown in Fig. 4. Membership functions are given in (29), (30) and (31). We must preserve the combination of ones and zeros of Table 4 with these new membership functions. With that combination of ones and zeros we can fuzzify the original boolean normal forms.

$$e_{\mathrm{nl}}(e) = \begin{cases} 1 & \text{if } e \le -0.03 \text{ m} \\ -(e+0.01)/0.02 & \text{if } -0.03 \text{ m} < e \le -0.01 \text{ m} \\ 0 & \text{otherwise} \end{cases} \quad (29)$$

$$e_{\mathrm{pl}}(e) = \begin{cases} 1 & \text{if } e \ge 0.03 \text{ m} \\ (e-0.01)/0.02 & \text{if } 0.01 \text{ m} \le e < 0.03 \text{ m} \\ 0 & \text{otherwise} \end{cases} \quad (30)$$

$$de_{\mathrm{p}}(de/dt) = \begin{cases} 1 & \text{if } de/dt \ge 0.01 \text{ m/s} \\ (de/dt+0.01)/0.02 & \text{if } -0.01 \text{ m/s} \le de/dt < 0.01 \text{ m/s} \\ 0 & \text{otherwise} \end{cases} \quad (31)$$

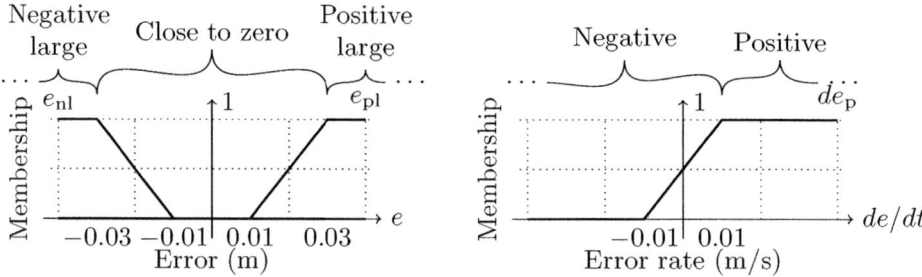

Fig. 4. Membership functions of fuzzy sets *negative large error* (e_{nl}), *positive large error* (e_{pl}) and *positive error rate* (de_{p})

We get the corresponding fuzzified normal forms for v_{l} and v_{s} from Table 4. They are given in (32) and (33). These fuzzified normal forms can be simplified by using their properties in a similar way as the boolean normal forms. The real difference here is that $\widetilde{D}_{\mathrm{F}}(v_{\mathrm{l}})$ and $\widetilde{D}_{\mathrm{F}}(v_{\mathrm{s}})$ are real numbers in $[0,1]$ and not only values in $\{0,1\}$.

$$\widetilde{D}_{\mathrm{F}}(v_{\mathrm{l}}) = (\overline{e_{\mathrm{nl}}} \triangle e_{\mathrm{pl}} \triangle \overline{de_{\mathrm{p}}}) \triangledown_{\mathrm{M}} (\overline{e_{\mathrm{nl}}} \triangle e_{\mathrm{pl}} \triangle de_{\mathrm{p}}) =$$
$$\overline{e_{\mathrm{nl}}} \triangle e_{\mathrm{pl}} \triangle (\overline{de_{\mathrm{p}}} \triangledown_{\mathrm{M}} de_{\mathrm{p}}) = \overline{e_{\mathrm{nl}}} \triangle e_{\mathrm{pl}} \quad (32)$$

$$\widetilde{D}_{\mathrm{F}}(v_{\mathrm{s}}) = (\overline{e_{\mathrm{nl}}} \triangle \overline{e_{\mathrm{pl}}} \triangle de_{\mathrm{p}}) \triangledown_{\mathrm{M}} (\overline{e_{\mathrm{nl}}} \triangle e_{\mathrm{pl}} \triangle de_{\mathrm{p}}) =$$
$$\overline{e_{\mathrm{nl}}} \triangle (\overline{e_{\mathrm{pl}}} \triangledown_{\mathrm{M}} e_{\mathrm{pl}}) \triangle de_{\mathrm{p}} = \overline{e_{\mathrm{nl}}} \triangle de_{\mathrm{p}} \quad (33)$$

We obtain the static transfer characteristic Ω of the fuzzy controller given by $\phi_{\mathrm{i}}(e, de/dt) = (\overline{e_{\mathrm{nl}}}(e) \triangle e_{\mathrm{pl}}(e))\Phi_{\mathrm{l}} + (\overline{e_{\mathrm{nl}}}(e) \triangle de_{\mathrm{p}}(de/dt))\Phi_{\mathrm{s}}$, where e_{nl}, e_{pl} and de_{p} are the membership functions (29), (30) and (31), and $\triangle$ is any t-norm, by replacing (32) and (33) in (23).

5.3 Designed Controllers and Responses on Time

Figure 5 shows the designed controllers. The difference between them is that the output of boolean controller has values in the set $\{0, \Phi_{\mathrm{s}}, \Phi_{\mathrm{l}}, \Phi_{\mathrm{l}} + \Phi_{\mathrm{s}}\}$, but the

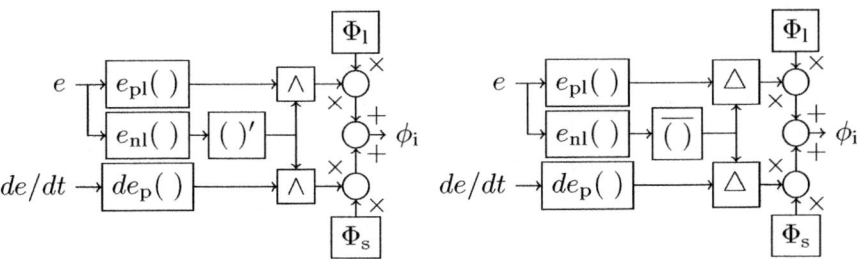

Fig. 5. Boolean controller and fuzzy controller

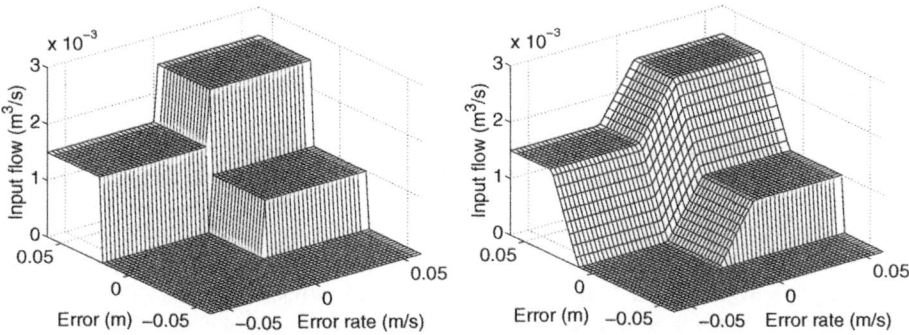

Fig. 6. Boolean transfer characteristic and fuzzy transfer characteristic

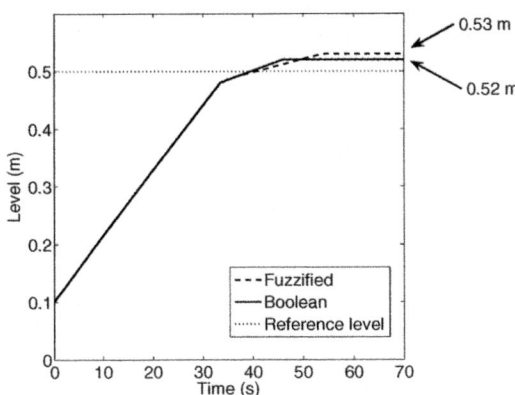

Fig. 7. Level with boolean controller and level with fuzzy controller

output of fuzzy controller has values in the closed interval $[0, \Phi_l + \Phi_s]$. Figure 6 shows the graphic of the boolean transfer characteristic and the fuzzified transfer characteristic by using standard t-norm. Figure 7 shows the responses on time of the control system with boolean and fuzzy controllers. The reference level in that figure is $h_{\text{ref}} = 0.5$ m with a initial condition $h(0) = 0.1$ m.

6   Conclusion

We presented a special class of fuzzified normal forms where the basic methodology in design and some boolean properties are preserved in order to get simplification as in the boolean case. Distributivity and complementation properties are the major restrictions in the selection of fuzzy operators. Distributivity property imposes the selection of standard t-norm and standard t-conorm. Complementation property imposes the selection of a unique threshold-type fuzzy complement which is an upper or lower bound for any fuzzy complement. A possible problem in some applications is the discontinuity in the threshold-type fuzzy complements. In future works we will study possible approximations with different continuous complements and with a permissible error.

References

1. Espitia Cuchango, H.E.: Aplicación Del Concresor Basado En Relaciones Booleanas Para Sistemas De Lógica Difusa Tipo Dos. Master's Thesis in Industrial Engineering, Universidad Distrital Francisco José de Caldas. Engineering Department, Bogotá D.C., Colombia (2009)
2. Gobi, A.F., Pedrycz, W.: Logic Minimization as an Efficient Means of Fuzzy Structure Discovery. IEEE Transactions On Fuzzy Systems 16(3), 553–566 (2008)
3. Grätzer, G.: General Lattice Theory. A Series of Monographs and Textbooks. Academic Press Inc., New York (1978)
4. Hachtel, G.D., Somenzi, F.: Logic Synthesis and Verification Algorithms. Kluwer Academic Publishers, Dordrecht (1996)
5. Klement, E.P., Mesiar, R., Pap, E.: Triangular Norms: Basic Notions and Properties. In: Klement, E.P., Mesiar, R. (eds.) Logical, Algebraic, Analytic, and Probabilistic Aspects of Triangular Norms, 1st edn., vol. ch.1, pp. 17–60. Elsevier B.V., Amsterdam (2005)
6. Klir, G.J., Yuan, B.: Fuzzy Sets and Fuzzy Logic: Theory and Applications. Prentice Hall PTR, New Jersey (1995)
7. Maes, K., De Baets, B.: Facts and Figures on Fuzzified Normal Forms. IEEE Transactions On Fuzzy Systems 13(3), 394–404 (2005)
8. Nguyen, H.T., Walker, E.A.: A First Course in Fuzzy Logic, 3rd edn. Chapman & Hall/CRC, Boca Raton, Florida (2006)
9. Rovatti, R., Guerrieri, R., Baccarani, G.: An Enhanced Two-level Boolean Synthesis Methodology for Fuzzy Rules Minimization. IEEE Transactions On Fuzzy Systems 3(3), 288–299 (1995)
10. Soriano, J., Olarte, A., Melgarejo, M.: Fuzzy Controller for MIMO Systems Using Defuzzification Based on Boolean Relations (DBR). In: Proceedings of The 14th IEEE International Conference on Fuzzy Systems, Reno, Nevada, USA, pp. 271–275 (2005)

Mean-Entropy Model for Portfolio Selection with Type-2 Fuzzy Returns

Ying Liu and Yanju Chen

College of Mathematics & Computer Science, Hebei University
Baoding 071002, Hebei, China
yingliu@hbu.edu.cn

Abstract. Entropy is a measurement of the degree of uncertainty. Mean-entropy method can be used for modeling the choice among uncertain outcomes. In this paper, we consider the portfolio selection problem under the assumption that security returns are characterized by type-2 fuzzy variables. Since the expectation and entropy of type-2 fuzzy variables haven't been well defined, type-2 fuzzy variables need to be reduced firstly. Then we propose a mean-entropy model with reduced variables. To solve the proposed model, we use the entropy formula of reduced fuzzy variable and transform the mean-entropy model to its equivalent parametric form, which can be solved by standard optimization solver.

Keywords: Mean-entropy model, Portfolio selection, Type-2 Fuzzy variable, Mean reduction, Parametric programming.

1 Introduction

Shannon [1] defined entropy as a measure of uncertainty in 1949. Following the Markowitz's pioneer work of portfolio selection [2], Philippatos and Wilson [3] first employed entropy as the risk measure. Compared with the most traditional risk measure: the variance, entropy is more general because it's a nonparametric measure and free from reliance on symmetric probability distribution. Thus, a new model branch of portfolio selection was founded. In 1975, Philippatos and Gressis [4] provided conditions where mean-variance, mean-entropy and second degree stochastic dominance are equivalent. Nawrocki and Harding [5] proposed the state-value weighting of entropy and tested using a portfolio selection heuristic algorithm. These studies have a common character that is all the works were discussed on the basis of probability theory. When the uncertain parameters need to be estimated by experts' experience or decision-makers' subjective judgement, the uncertainty is measured by fuzzy theory [6]. Motivated by the development of fuzzy set theory [7, 8], some researchers have given several definitions of fuzzy entropy. Based on credibility theory [9], Li and Liu [10] proposed a definition of fuzzy entropy in 2008. Then Li and Liu [11] proved some maximum entropy theorems for fuzzy variables. Huang [12] used mean-entropy ideas to fuzzy portfolio selection.

D.-S. Huang et al. (Eds.): ICIC 2011, LNBI 6840, pp. 345–352, 2012.
© Springer-Verlag Berlin Heidelberg 2012

As an extension of an ordinary fuzzy set, Zadeh [13] proposed the concept of type-2 fuzzy set, which is very useful in the circumstance with incorporating uncertainties. Based on fuzzy possibility space, Liu and Liu [14] proposed the concept of type-2 fuzzy variables and introduced the arithmetic about type-2 fuzziness from the analysis viewpoint. After that, some relative researching work [15–17] have been carried out. In 2011, Qin and Liu [18] proposed a mean reduction method, in which a type-2 fuzzy variable can be reduced to a fuzzy variable with a parametric possibility function. Thus, we obtain an effective method to model the problems with incorporating uncertainties.

This paper applies type-2 fuzzy theory to portfolio selection problem, in which the security returns are not known and characterized by type-2 fuzzy variables. Since the expectation and entropy of type-2 fuzzy variable haven't been well defined, type-2 fuzzy variable need to be reduced firstly. Then we deduce the entropy formulas for reduced variables of type-2 triangular fuzzy variable, and apply them to the portfolio selection problem to transform the established mean-entropy model of reduced variables to their equivalent parametric programming ones. For any given parameters, the equivalent programmings are determined linear programmings, which can be solved by standard optimization solvers. Finally, we provide one numerical example to illustrate the proposed model.

The rest of the paper is organized as follows. Section 2 reviews some fundamental concepts. In Section 3, we deduce the entropy formulas for three reductions of type-2 triangular fuzzy variable. In Section 4, we propose a mean-entropy model with the reduced variable. In Section 5, we discuss the equivalent parametric forms for the objective and constraint. Section 6 provides one numerical example to illustrate the modeling idea and efficiency. Section 7 gives our conclusions.

2   Preliminaries

A type-2 fuzzy variable $\tilde{\xi}$ is called triangular if its secondary possibility distribution $\tilde{\mu}_{\tilde{\xi}}(x)$ is

$$\left(\frac{x-r_1}{r_2-r_1} - \theta_l \min\{ \frac{x-r_1}{r_2-r_1}, \frac{r_2-x}{r_2-r_1} \}, \frac{x-r_1}{r_2-r_1}, \frac{x-r_1}{r_2-r_1} + \theta_r \min\{ \frac{x-r_1}{r_2-r_1}, \frac{r_2-x}{r_2-r_1} \} \right)$$

for $x \in [r_1, r_2]$, and

$$\left(\frac{r_3-x}{r_3-r_2} - \theta_l \min\{ \frac{r_3-x}{r_3-r_2}, \frac{x-r_2}{r_3-r_2} \}, \frac{r_3-x}{r_3-r_2}, \frac{r_3-x}{r_3-r_2} + \theta_r \min\{ \frac{r_3-x}{r_3-r_2}, \frac{x-r_2}{r_3-r_2} \} \right)$$

for $x \in [r_2, r_3]$, where $\theta_l, \theta_r \in [0, 1]$ are two parameters characterizing the degree of uncertainty that $\tilde{\xi}$ takes the value x. For simplicity, we denote the type-2 triangular fuzzy variable $\tilde{\xi}$ with the above distribution by $(\tilde{r}_1, \tilde{r}_2, \tilde{r}_3; \theta_l, \theta_r)$.

According to the mean reduction method [18], we have
(i) With E^* reduction method, the reduction ξ_1 of $\tilde{\xi}$ has the following distribution

$$\mu_{\xi_1}(x) = \begin{cases} \frac{(2+\theta_r)(x-r_1)}{2(r_2-r_1)}, & \text{if } x \in [r_1, \frac{r_1+r_2}{2}] \\ \frac{(2-\theta_r)x+\theta_r r_2-2r_1}{2(r_2-r_1)}, & \text{if } x \in (\frac{r_1+r_2}{2}, r_2] \\ \frac{(-2+\theta_r)x-\theta_r r_2+2r_3}{2(r_3-r_2)}, & \text{if } x \in (r_2, \frac{r_2+r_3}{2}] \\ \frac{(2+\theta_r)(r_3-x)}{2(r_3-r_2)}, & \text{if } x \in (\frac{r_2+r_3}{2}, r_3]. \end{cases}$$

(ii) With E_* reduction method, the reduction ξ_2 of $\tilde{\xi}$ has the following distribution

$$\mu_{\xi_2}(x) = \begin{cases} \frac{(2-\theta_l)(x-r_1)}{2(r_2-r_1)}, & \text{if } x \in [r_1, \frac{r_1+r_2}{2}] \\ \frac{(2+\theta_l)x-2r_1-\theta_l r_2}{2(r_2-r_1)}, & \text{if } x \in (\frac{r_1+r_2}{2}, r_2] \\ \frac{(-\theta_l-2)x+2r_3+\theta_l r_2}{2(r_3-r_2)}, & \text{if } x \in (r_2, \frac{r_2+r_3}{2}] \\ \frac{(2-\theta_l)(r_3-x)}{2(r_3-r_2)}, & \text{if } x \in (\frac{r_2+r_3}{2}, r_3]. \end{cases}$$

(iii) With E reduction method, the reduction ξ_3 of $\tilde{\xi}$ has the following distribution

$$\mu_{\xi_3}(x) = \begin{cases} \frac{(4+\theta_r-\theta_l)(x-r_1)}{4(r_2-r_1)}, & \text{if } x \in [r_1, \frac{r_1+r_2}{2}] \\ \frac{(4-\theta_r+\theta_l)x+(\theta_r-\theta_l)r_2-4r_1}{4(r_2-r_1)}, & \text{if } x \in [\frac{r_1+r_2}{2}, r_2] \\ \frac{(-4+\theta_r-\theta_l)x+4r_3-(\theta_r-\theta_l)r_2}{4(r_3-r_2)}, & \text{if } x \in [r_2, \frac{r_2+r_3}{2}] \\ \frac{(4+\theta_r-\theta_l)(r_3-x)}{4(r_3-r_2)}, & \text{if } x \in [\frac{r_2+r_3}{2}, r_3]. \end{cases}$$

The reduced variable of type-2 triangular fuzzy variable is an ordinary fuzzy variable, and obviously it's continuous.

Let ξ be a continuous fuzzy variable. Then, its entropy is defined by[10]

$$H[\xi] = \int_{-\infty}^{\infty} S(\text{Cr}\{\xi = r\})dr$$

where $S(t) = -t\ln t - (1-t)\ln(1-t)$.

For any continuous fuzzy variable ξ with membership function μ, we have $\text{Cr}\{\xi = r\} = \mu(x)/2$ for each $x \in R$. Thus

$$H[\xi] = -\int_{-\infty}^{\infty} \left(\frac{\mu(x)}{2}\ln\frac{\mu(x)}{2} + \left(1 - \frac{\mu(x)}{2}\right)\ln\left(1 - \frac{\mu(x)}{2}\right) \right) dx.$$

3   Entropy for the Reduction of type-2 Triangular Fuzzy Variable

In this section, we deduce the entropy for the reduction of type-2 triangular fuzzy variable, which will be useful in the next section for portfolio selection problem.

Let $\tilde{\xi}$ be a type-2 triangular fuzzy variable defined as $\tilde{\xi} = (\tilde{r}_1, \tilde{r}_2, \tilde{r}_3; \theta_l, \theta_r)$. Then we can obtain the entropy formulas of the reduced fuzzy variables.

Theorem 1. *Let $\tilde{\xi} = (\tilde{r}_1, \tilde{r}_2, \tilde{r}_3; \theta_l, \theta_r)$ be a type-2 triangular fuzzy variable. ξ is the E reduction of $\tilde{\xi}$, and denote $u = \frac{(4+\theta_r-\theta_l)}{16}$. Then $\mathrm{H}[\xi] = \frac{8(r_1-r_3)}{(4+\theta_r-\theta_l)(4-\theta_r+\theta_l)}$ $\{(\theta_r - \theta_l)[(1-u^2)\ln(1-u) - u^2\ln u + \frac{u^2}{2} - \frac{1}{4}] - 1\}$.*

Proof. If ξ is the E reduction of $\tilde{\xi}$, by the distribution of ξ we can obtain

$$\frac{\mu_\xi(x)}{2} = \begin{cases} \frac{(4+\theta_r-\theta_l)(x-r_1)}{8(r_2-r_1)}, & if \ x \in [r_1, \frac{r_1+r_2}{2}] \\ \frac{(4-\theta_r+\theta_l)x+(\theta_r-\theta_l)r_2+4r_1}{8(r_2-r_1)}, & if \ x \in [\frac{r_1+r_2}{2}, r_2] \\ \frac{(-4+\theta_r-\theta_l)x-(\theta_r-\theta_l)r_2+4r_3}{8(r_3-r_2)}, & if \ x \in [r_2, \frac{r_2+r_3}{2}] \\ \frac{(4+\theta_r-\theta_l)(r_3-x)}{8(r_3-r_2)}, & if \ x \in [\frac{r_2+r_3}{2}, r_3]. \end{cases}$$

According to the definition of entropy, we have

$$\mathrm{H}[\xi] = -\int_{-\infty}^{\infty} \left(\frac{\mu(x)}{2}\ln\frac{\mu(x)}{2} + \left(1 - \frac{\mu(x)}{2}\right)\ln\left(1 - \frac{\mu(x)}{2}\right) \right) dx$$
$$= \frac{8(r_1-r_3)}{(4+\theta_r-\theta_l)(4-\theta_r+\theta_l)}\{(\theta_r - \theta_l)[(1-u^2)\ln(1-u) - u^2\ln u + \frac{u^2}{2} - \frac{1}{4}] - 1\},$$

where $u = \frac{(4+\theta_r-\theta_l)}{16}$.

Theorem 2. *Let $\tilde{\xi} = (\tilde{r}_1, \tilde{r}_2, \tilde{r}_3; \theta_l, \theta_r)$ be a type-2 triangular fuzzy variable. ξ is the E^* reduction of $\tilde{\xi}$, and denote $t = \frac{(2+\theta_r)}{8}$. Then $\mathrm{H}[\xi] = \frac{4(r_1-r_3)}{(2+\theta_r)(2-\theta_r)}\{\theta_r[(1-t^2)\ln(1-t) - t^2\ln t + t^2 - \frac{1}{4}] - \frac{1}{2}\}$.*

Proof. The proof is similar to Theorem 1.

Theorem 3. *Let $\tilde{\xi} = (\tilde{r}_1, \tilde{r}_2, \tilde{r}_3; \theta_l, \theta_r)$ be a type-2 triangular fuzzy variable. ξ is the E_* reduction of $\tilde{\xi}$, and denote $s = \frac{(2-\theta_l)}{8}$. Then $\mathrm{H}[\xi] = \frac{4(r_1-r_3)}{(2+\theta_l)(2-\theta_l)}\{\theta_l[(1-s^2)\ln(1-s) - s^2\ln s + s^2 - \frac{1}{4}] - \frac{1}{2}\}$.*

Proof. The proof is similar to Theorem 1.

4   Formulation of Portfolio Selection Problem

Portfolio selection is the problem of how to allocate investor's assets to a many alternative securities so that the investment can bring the maximum return. Entropy was first employed by Philippatos and Wilson [3], in which the expected return of a portfolio was regarded as the investment return and the entropy as the investment risk. The traditional mean-entropy model was built as

$$\begin{cases} \max & \mathrm{E}[x_1\xi_1 + x_2\xi_2 + \cdots + x_n\xi_n] \\ \text{subject to} & \mathrm{H}[x_1\xi_1 + x_2\xi_2 + \cdots + x_n\xi_n] \leq \gamma \\ & x_1 + x_2 + \cdots + x_n = 1 \\ & x_i \geq 0, i = 1, 2, \ldots, n. \end{cases} \tag{1}$$

where x_i is the investment proportions in security i, ξ_i represents the random returns for the ith securities, $i = 1, 2, \ldots, n$, respectively. γ is the predetermined maximum entropy level accepted by the investor.

In this section, we assume that returns of the securities are characterized by type-2 fuzzy variables. Let x_i is the investment proportions in security i, $\tilde{\xi}_i$ represents the type-2 triangular fuzzy returns for the ith security, $i = 1, 2, \ldots, n$, respectively. Then the total return of all securities is $\tilde{\xi} = x_1\tilde{\xi}_1 + x_2\tilde{\xi}_2 + \cdots + x_n\tilde{\xi}_n$. According to Chen and Zhang [19], $\tilde{\xi}$ remains to be a type-2 triangular fuzzy variable. Since the expectation and entropy of type-2 fuzzy variable haven't been well defined, we need reduced these variables before modeling portfolio selection problem. In this paper, we obtain reduced variable ξ of type-2 fuzzy variable $\tilde{\xi}$ by mean reduction method [18], then mean-entropy model with reduced variable is built as

$$\begin{cases} \max & E[\xi] \\ \text{subject to} \, H[\xi] \leq \gamma \\ & x_1 + x_2 + \cdots + x_n = 1 \\ & x_i \geq 0, i = 1, 2, \ldots, n, \end{cases} \tag{2}$$

where γ is the maximum risk level accepted by the investor.

5   Equivalent Parametric Programming of Portfolio Selection Model

To solve model (2), it is required to compute the expectation and entropy of reduced fuzzy variable in the objective and constraint. In the following, we discuss how turn the objective and the constraint to their equivalent parametric forms.

First of all, we discuss the equivalent form of the objective.

Suppose $\tilde{\xi}_i = (\widetilde{r_{1i}}, \widetilde{r_{2i}}, \widetilde{r_{3i}}; \theta_{li}, \theta_{ri})$, $i = 1, \ldots, n$ are mutually independent type-2 triangular fuzzy variables. For simplicity, we assume $\theta_{li} = \theta_l, \theta_{ri} = \theta_r, i = 1, \ldots, n$. Then if $\tilde{\xi} = x_1\tilde{\xi}_1 + x_2\tilde{\xi}_2 + \cdots + x_n\tilde{\xi}_n$, we have

$$\tilde{\xi} = (\widetilde{\sum_{i=1}^{n} x_i r_{1i}}, \widetilde{\sum_{i=1}^{n} x_i r_{2i}}, \widetilde{\sum_{i=1}^{n} x_i r_{3i}}; \theta_l, \theta_r).$$

According to the expectation formula of reduced fuzzy variable [20], the objective of model (2) is equivalent to

$$E[\xi] = \frac{\sum_{i=1}^{n} x_i r_{1i} + 2\sum_{i=1}^{n} x_i r_{2i} + \sum_{i=1}^{n} x_i r_{3i}}{4} + \frac{(\theta_r - \theta_l)(\sum_{i=1}^{n} x_i r_{1i} - 2\sum_{i=1}^{n} x_i r_{2i} + \sum_{i=1}^{n} x_i r_{3i})}{32}. \tag{3}$$

Secondly, we discuss the equivalent form of the constraint. By the Theorems 1, 2 and 3, the equivalent parametric forms of constraint in model (2) are as follow:
(i) If ξ is the E reduction of $\tilde{\xi}$, we have $H[\xi] \leq \alpha \Longleftrightarrow$

$$\frac{8(\sum_{i=1}^{n} x_i r_{1i} - \sum_{i=1}^{n} x_i r_{3i})}{(4 + \theta_r - \theta_l)(4 - \theta_r + \theta_l)}\left\{(\theta_r - \theta_l)[(1 - u^2)\ln(1 - u) - u^2\ln u + \frac{u^2}{2} - \frac{1}{4}] - 1\right\} \leq \alpha, \tag{4}$$

where $u = \frac{(4+\theta_r-\theta_l)}{16}$.

(ii) If ξ is the E^* reduction of $\tilde{\xi}$, we have $H[\xi] \le \alpha \iff$

$$\frac{4(\sum_{i=1}^{n} x_i r_{1i} - \sum_{i=1}^{n} x_i r_{3i})}{(2+\theta_r)(2-\theta_r)}\{\theta_r[(1-t^2)\ln(1-t) - t^2\ln t + t^2 - \frac{1}{2}] - \frac{1}{2}\} \le \alpha, \quad (5)$$

where $t = \frac{(2+\theta_r)}{8}$.

(iii) If ξ is the E_* reduction of $\tilde{\xi}$, we have $H[\xi] \le \alpha \iff$

$$\frac{4(\sum_{i=1}^{n} x_i r_{1i} - \sum_{i=1}^{n} x_i r_{3i})}{(2+\theta_l)(2-\theta_l)}\{\theta_l[(1-s^2)\ln(1-s) - s^2\ln s + s^2 - \frac{1}{2}] - \frac{1}{2}\} \le \alpha, \quad (6)$$

where $s = \frac{(2-\theta_l)}{8}$.

Thus, we obtain the equivalent parametric programming of model (2) in three different cases. For any given parameters, the equivalent programmings are determined linear programmings, which can be solved by standard optimization solvers.

6   Numerical Experiments

In this section, we consider a portfolio selection problem in which the returns of ten securities are assumed to be type-2 triangular fuzzy variables as shown in Table 1. Furthermore, let $\theta_{li} = \theta_l, \theta_{ri} = \theta_r$, $i = 1, \cdots, 10$. Suppose that the

Table 1. The type-2 fuzzy returns of ten securities

security i	return	security i	return
1	(-0.3,2.7,3.4; θ_l, θ_r)	6	(-0.1,2.5,3.6; θ_l, θ_r)
2	(-0.1,1.9,2.6; θ_l, θ_r)	7	(-0.3,2.4,3.6; θ_l, θ_r)
3	(-0.2,3.0,3.9; θ_l, θ_r)	8	(-0.1,3.3,4.5; θ_l, θ_r)
4	(-0.3,2.0,2.9; θ_l, θ_r)	9	(-0.3,1.1,2.7; θ_l, θ_r)
5	(-0.4,2.2,3.3; θ_l, θ_r)	10	(-0.2,2.1,3.8; θ_l, θ_r)

investors accept 1.96 as the maximum tolerable uncertainty level, and require the entropy not great than it when maximizing the expected return. Using E reduction method to reduce type-2 fuzzy returns, then according to (3) and (4), the equivalent parametric programming of model (2) is:

$$\begin{cases} \max & \frac{\sum_{i=1}^{10} x_i r_{1i} + 2\sum_{i=1}^{10} x_i r_{2i} + \sum_{i=1}^{10} x_i r_{3i}}{4} \\ & + \frac{(\theta_r - \theta_l)(\sum_{i=1}^{10} x_i r_{1i} - 2\sum_{i=1}^{10} x_i r_{2i} + \sum_{i=1}^{10} x_i r_{3i})}{32} \\ \text{subject to} & \frac{8(\sum_{i=1}^{10} x_i r_{1i} - \sum_{i=1}^{10} x_i r_{3i})}{(4+\theta_r-\theta_l)(4-\theta_r+\theta_l)}\{(\theta_r - \theta_l)[(1-u^2)\ln(1-u) \\ & -u^2\ln u + \frac{u^2}{2} - \frac{1}{4}] - 1\} \le 1.96 \\ & x_1 + x_2 + \cdots + x_{10} = 1 \\ & x_i \ge 0, i = 1, 2, \ldots, 10, \end{cases} \quad (7)$$

Table 2. Allocation of money to ten securities

$\theta_l = 0.0, \theta_r = 0.0$				$\theta_l = 0.4, \theta_r = 0.8$			
security i	allocation	security i	allocation	security i	allocation	security i	allocation
1	0.00	6	0.00	1	0.00	6	0.00
2	35.79	7	0.00	2	66.74	7	0.00
3	0.00	8	64.21	3	0.00	8	33.26
4	0.00	9	0.00	4	0.00	9	0.00
5	0.00	10	0.00	5	0.00	10	0.00

where $u = (4 + \theta_r - \theta_l)/16$. With the Lingo software, the results about every securities are reported in Table 2, which also shows a comparison under different parameters. Especially, when we set $\theta_l = \theta_r = 0.0$, the type-2 fuzzy variable degenerate to a type-1 fuzzy variable. From these results, we can draw the conclusion that the model (2) is robust with parameters. The corresponding maximum expected returns are 2.329474 ($\theta_l = \theta_r = 0$) and 1.945811 ($\theta_l = 0.4, \theta_r = 0.8$).

7   Conclusions

To model the choice among uncertain outcomes, mean-entropy approach is frequently used in the literature. This paper considered portfolio selection problem with entropy as the measure of risk in type-2 fuzzy decision systems. The major new results include the following three aspects.

(i) Following the mean reduction method, the entropy formulas of reduced fuzzy variable were deduced (Theorems 1, 2 and 3);

(ii) When the returns are characterized by type-2 triangular fuzzy variables, a mean-entropy model with reduced variables was presented (Model (2));

(iii) We transformed the model to its equivalent parametric programming. One numerical example was presented to demonstrated the effectiveness.

Acknowledgements. This work is supported by the National Natural Science Foundation of China (No.60974134), the Natural Science Foundation and Educational Department of Hebei Province (No.A2011201007 and No.2010109), and Science Foundation of Hebei University (No.2010Q28).

References

1. Shannon, C.E.: The Mathematical Theory of Communication. The University of Illinois Press, Urbana (1949)
2. Markowitz, H.: Portfolio Selection. Journal of Finance 7, 77–91 (1952)
3. Philippatos, G., Wilson, C.: Entropy, Market Risk, and The Selection of Efficient Portfolios. Applied Economics 4, 209–220 (1972)
4. Philippatos, G., Gressis, N.: Conditions of Equivalence among E-V, SSD, and E-H Portfolio Selection Criteria: The Case for Uniform, Normal and Lognormal Distributions. Management Science 21, 617–635 (1975)

5. Nawrocki, D., Harding, W.: State-value Weighted Entropy as A Measure of Investment Risk. Applied Economics 18, 411–419 (1986)
6. Zadeh, L.A.: Fuzzy Sets. Information and Control 8, 199–249 (1965)
7. Nahmias, S.: Fuzzy variables. Fuzzy Sets and Systems 1, 97–110 (1978)
8. Wang, P.: Fuzzy Contactability and Fuzzy Variables. Fuzzy Sets and Systems 8(1), 81–92 (1982)
9. Liu, B., Liu, Y.K.: Expected Value of Fuzzy Variable and Fuzzy Expected Value Models. IEEE Transactions on Fuzzy Systems 10(4), 445–450 (2002)
10. Li, P., Liu, B.: Entropy of Credibility Distributions for Fuzzy Variables. IEEE Transactions on Fuzzy Systems 16, 123–129 (2008)
11. Li, X., Liu, B.: Maximum Entropy Principle for Fuzzy Variables. International Journal of Uncertainty, Fuzziness Knowledge-Based Systems 15(2), 43–52 (2007)
12. Huang, X.: Mean-entropy Models for Fuzzy Portfolio Selection. IEEE Transactions on Fuzzy Systems 16(4), 1096–1101 (2008)
13. Zadeh, L.A.: Concept of A Linguistic Variable and Its Application to Approximate Reasoning I. Information Sciences 8, 199–249 (1975)
14. Liu, Z.Q., Liu, Y.-K.: Type-2 Fuzzy Variables and Their Arithmetic. Soft Computing 14(7), 729–747 (2010)
15. Chen, Y., Wang, X.: The Possibilistic Representative Value of Type-2 Fuzzy Variable and Its Properties. Journal of Uncertain Systems 4(3), 229–240 (2010)
16. Qin, R., Liu, Y.K., Liu, Z.Q.: Methods of Critical Value Reduction for Type-2 Fuzzy Variables and Their Applications. Journal of Computational and Applied Mathematics 235(5), 1454–1481 (2011)
17. Wu, X., Liu, Y.K.: Spread of Fuzzy Variable and Expectation-spread Model for Fuzzy Portfolio Optimization Problem. Journal of Applied Mathematics and Computing 36(1-2), 373–400 (2011)
18. Qin, R., Liu, Y.K., Liu, Z.Q.: Modeling Fuzzy Data Envelopment Analysis by Parametric Programming Method. Expert Systems with Applications 38(7), 8648–8663 (2011)
19. Chen, Y., Zhang, L.W.: Some New Results about Arithmetic of Type-2 Fuzzy Variable. Journal of Uncertain Systems 5(3), 227–240 (2011)
20. Liu, Y., Chen, Y.: Expectation Formulas for Reduced Fuzzy Variables. In: Proceedings of The 9th International Conference on Machine Learning and Cybernetics, Baoding, China, vol. 2, pp. 568–572 (2010)

A Linear Regression Model for Nonlinear Fuzzy Data

Juan C. Figueroa-García* and Jesus Rodriguez-Lopez*

Universidad Distrital Francisco José de Caldas, Bogotá - Colombia
jcfigueroag@udistrital.edu.co,
e.jesus.rodriguez.lopez@gmail.com

Abstract. Fuzzy linear regression is an interesting tool for handling uncertain data samples as an alternative to a probabilistic approach. This paper sets forth uses a linear regression model for fuzzy variables; the model is optimized through convex methods. A fuzzy linear programming model has been designed to solve the problem with nonlinear fuzzy data by combining the fuzzy arithmetic theory with convex optimization methods.

Two examples are solved through different approaches followed by a goodness of fit statistical analysis based on the measurement of the residuals of the model.

1 Introduction and Motivation

The linear regression analysis called the *Classical Linear Regression Model (CLRM)* is important statistical tool to establish the relation between a set of independent variables and a dependent one is. A mathematical representation of the CLRM is:

$$y_j = \sum_{i=0}^{m} \beta_i x_{ij} + \xi_j \quad \forall\, j \in \mathbb{N}_m \tag{1}$$

Where y_j is a dependent variable, x_{ij} are the observed variables, β_i is the weight of the i_{th} independent variable and ξ_j is the j_{th} observation. $i \in \mathbb{N}_n$ and $j \in \mathbb{N}_m$.

As Bargiela et al. expressed in [1], the classical linear regression analysis is not able to find the assignment rule between a collection of variables when these are not numerical *(Crisp)* entities i.e. fuzzy numbers (See Zadeh in [9]) To address and solve this problem, Tanaka et al. [8] introduced the fuzzy linear regression (FLR) model.

$$\tilde{y}_j = \sum_{i=0}^{m} \beta_i \tilde{x}_{ij} + \xi_j \quad \forall\, j \in \mathbb{N}_m \tag{2}$$

Where $\tilde{y}_j$ is a fuzzy dependent variable, $\tilde{x}_{ij}$ are fuzzy observations, β_i is the weight of the i_{th} independent variable and ξ_j is the j_{th} observation. $i \in \mathbb{N}_n$ and $j \in \mathbb{N}_m$.

Fuzzy linear regression has the capability of dealing with linguistic variables through different methods such as the least squares method or the gradient-descent algorithm (See Bargiela in [1] and Gladysz in [2]). However, those methods are designed for

---

* Corresponding authors.

D.-S. Huang et al. (Eds.): ICIC 2011, LNBI 6840, pp. 353–360, 2012.

analyzing the most used sets, the symmetrical triangular fuzzy sets. However, the solution routines involve algorithms that require considerable amounts of resources e.g. software, computing machine and time.

The present work presents a Linear Programming (LP) model capable of managing in a simple way, all type-1 fuzzy data. To decompose the information that each fuzzy data contains, we analyze several parameters that represent it. These parameters are called interesting values and are characterized by the following definition.

Definition 1. *Suppose A, B and C be fuzzy sets with membership functions $A(x)$, $B(x)$ and $C(x)$ such that:*

$$A(x) = \sum_{i=0}^{m} \beta_i B_i(x) \tag{3}$$

And suppose $\tau(\cdot)$ a function which output is an interesting value of a given set $(\cdot)$, hence for $A(x)$ we have

$$\tau(A(x)) = \sum_{i=0}^{m} \beta_i \tau(B_i(x)) \tag{4}$$

Where β_i is the weight of the fuzzy set $B_i(x)$, so $\tau(A(x))$ is a linear combination of the weights of $\tau(B_i(x))$. It means that each parameter of $A(X)$ can be expressed as a linear combination of β_i and $\tau(B_i(x))$.

Interesting values have an important attribute: interesting values from a fuzzy set that is a linear combination of a group of fuzzy sets, equal the same linear combination of the interesting values from the second fuzzy set.

An LP model is designed based on the interesting values. Each constraint tries to set an equivalence between the interesting values of the sets Y and X considering slack or surplus; and the objective function is the minimization of their sum.

The LP method is compared to other three proposals. The reason for compairing the four models is to evaluate their efficiency and efficacy. In addition to the comparisons we also discuss a case study, including a statistical analysis of its residuals.

2   A Linear Programming Fuzzy Regression Model

2.1   The Independent Variables

In the classical linear regression model (CLRM) each observation corresponds to a single crisp value which measures a variable; these values, however, cannot encapsulate all information about the variable itself. These variables can bring noise or imprecisions in its measurement, moreover, the measurement process might not be accurate.

These imprecise measures can be represented by fuzzy sets, therefore a regression model should deal with the imprecision involving fuzzy sets. An L-R fuzzy set is composed by the spread, position and shape defined by a central value, by the lower and the upper distance, and by the lower and the upper area.

Let A denotes an L-R fuzzy set, and $A(x)$ its membership function. According to fuzzy number properties (See Klir in [5] and [4]), we have that the *central value* (v_c) of A has 1 as the membership value, this means that $A(v_c) = 1$. In addition, the *lower value* (v_l) and *upper value* (v_u) of A are respectively the left and right boundaries of the support of A.

Let d_l denote the distance from the lower value to the central value of a fuzzy set, i.e. the lower distance of A, hence d_l can be defined as $d_l = v_c - v_l$, and the upper distance can be defined as $d_u = v_u - v_c$. Let a_l denote the area between the lower and the central value of A, i.e. the lower area of A, thus a_l can be numerically expressed as:

$$a_l = \int_{v_l}^{v_c} A(x)dx$$

And analogously a_u i.e. the upper area can be calculated as:

$$a_u = \int_{v_c}^{v_u} A(x)dx$$

The central value of a fuzzy set is the value is the support element which α-cut equals to 1. Figure 1 shows the graphical representation of these values.

2.2   Fuzzy Arithmetic of Interesting Values

The interesting values of a fuzzy number, resulting from operate several fuzzy numbers, can be calculated through the values of the interesting values of each one of the second fuzzy numbers. According to the Definition 1, Klir and Yuan in [5] and Klir and Folger in [4], we derive the following operations on fuzzy sets.

Let B and C denote two L-R fuzzy sets, and $B(x)$ and $C(x)$ their membership functions respectively. Let also v_{cA}, v_{cB} and v_{cC} indicate the central values for the fuzzy sets A, B and C respectively, and let n indicate any real number.

If $C = A + B$ then $v_{cC} = v_{cA} + v_{cB}$. If $C = A - B$ then $v_{cC} = v_{cA} - v_{cB}$. If $C = nA$ then $v_{cC} = n \cdot v_{cA}$.

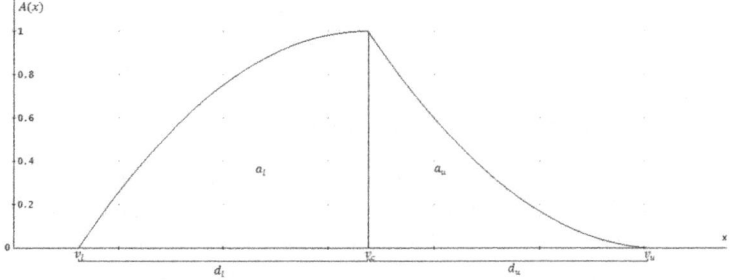

Fig. 1. Solution of the example as a function of α

Then the central value of a fuzzy set is a linear combination of the central values of all fuzzy sets. Let Y and X_i denote fuzzy numbers, and v_Y and v_{cX_i} their respective central values, and let β_i indicate a coefficient that is multiplying each X_i fuzzy number. If $Y = \sum_{i=0}^{n} \beta_i \cdot X_i$ then $v_{cY} = \sum_{i=0}^{n} \beta_i \cdot v_{cX_i}$ On the other hand, conversely let d_{la}, d_{lb} and d_{lc} indicate the lower distances for the fuzzy sets A, B and C respectively, and let d_{ua}, d_{ub} and d_{uc} denote the upper distances for the fuzzy sets A, B and C respectively. If $C = A + B$ then $d_{lC} = d_{lA} + d_{lB}$ & $d_{uC} = d_{uA} + d_{uB}$. If $C = A - B$ then $d_{lC} = d_{lA} + d_{uB}$ & $d_{uC} = d_{uA} + d_{lB}$. If $n \geqslant 0$ & $C = n \cdot A$ then $d_{lC} = n \cdot d_{lA}$ & $d_{uC} = n \cdot d_{uA}$. If $n < 0$ & $C = n \cdot A$ then $d_{lC} = n \cdot d_{uA}$ & $d_{uC} = n \cdot d_{lA}$

On the other hand, let β_i^+ and β_i^- denote the possible values for β_i such that:

$$\beta_i = \begin{cases} \beta_i^+ & if \ \beta_i \geqslant 0 \\ -\beta_i & if \ \beta_i < 0 \end{cases} \tag{5}$$

Let d_{lY} and d_{uY} denote the lower and upper distance for the fuzzy number Y, and d_{lX_i} and d_{uX_i} denote the lower and upper distance for the fuzzy number X_i. Hence

$$d_{lY} = \sum_{i=0}^{n} \beta_i^+ d_{li} - \sum_{i=0}^{n} \beta_i^- d_{uX_i}$$

$$d_{uY} = \sum_{i=0}^{n} \beta_i^+ d_{uX_i} - \sum_{i=0}^{n} \beta_i^- d_{lX_i}$$

Finally, let a_{la}, a_{lb} and a_{lc} be the lower areas for the fuzzy sets A, B and C respectively, and let a_{ua}, a_{ub} and a_{uc} denote the upper areas for the fuzzy sets A, B and C respectively. If $C = A + B$ then $a_{lC} = a_{lA} + a_{lB}$ and $a_{uC} = a_{uA} + a_{uB}$. If $C = A - B$ then $a_{lC} = a_{lA} + a_{uB}$ and $a_{uC} = a_{uA} + a_{lB}$. If $n \geqslant 0$ and $C = n \cdot A$ then $a_{lC} = n \cdot a_{lA}$ and $d_{uC} = n \cdot d_{uA}$. If $n < 0$ and $C = n \cdot A$ then $a_{lC} = |n| \cdot a_{uA}$ and $a_{uC} = |n| \cdot a_{lA}$

Let a_{lY} and a_{uY} denote the lower and the upper distance for the fuzzy number Y, and a_{lX_i} and a_{uX_i} denote the lower and the upper distance for the fuzzy X_i. Thus,

$$a_{lY} = \sum_{i=0}^{n} \beta_i^+ a_{lX_i} - \sum_{i=0}^{n} \beta_i^- a_{uX_i}$$

$$a_{uY} = \sum_{i=0}^{n} \beta_i^+ a_{uX_i} - \sum_{i=0}^{n} \beta_i^- a_{lX_i}$$

2.3   Linear Programming Fuzzy Regression Model

Based on the above results, we need two sets of variables; the first one for slack $s(\cdot)$ and the other one for surplus $f(\cdot)$. These variables are added to each constraint for each j observation. This allows the β_i coefficients to make each equation fits, where Y_j is the dependent variable and X_{ij} are the explanatory variables. Finally, the objective function is the minimization of both the sum of the slack and the surplus variables. Formally,

$$\min z = \sum_{j=1}^{m} s_{vcj} + f_{vcj} + s_{dlj} + f_{dlj} + s_{duj} + f_{duj} + s_{alj} + f_{alj} + s_{alj} + f_{alj}$$

$$s.t.$$

$$v_{cY_j} = \sum_{i=0}^{n} \beta_i \cdot v_{cX_{ij}} + s_{vcj} - f_{vcj} \quad \forall\, j \in \mathbb{N}_m$$

$$d_{lY_j} = \sum_{i=0}^{n} \beta_i^+ d_{lij} - \sum_{i=0}^{n} \beta_i^- d_{uX_{ij}} + s_{dlj} - f_{dlj} \quad \forall\, j \in \mathbb{N}_m$$

$$d_{uY_j} = \sum_{i=0}^{n} \beta_i^+ d_{uXij} - \sum_{i=0}^{n} \beta_i^- d_{lXij} + s_{duj} - f_{duj} \quad \forall\, j \in \mathbb{N}_m \qquad (6)$$

$$a_{lYj} = \sum_{i=0}^{n} \beta_i^+ a_{l_i} - \sum_{i=0}^{n} \beta_i^- a_{uXi} + s_{alj} - f_{alj} \quad \forall\, j \in \mathbb{N}_m$$

$$a_{uYj} = \sum_{i=0}^{n} \beta_i^+ a_{uXi} - \sum_{i=0}^{n} \beta_i^- a_{lXi} + s_{alj} - f_{alj} \quad \forall\, j \in \mathbb{N}_m$$

The first constraint refers to the central value. The second and third constraints are focused in the estimation of the lower and upper distances, and finally the fourth and fifth constraints bound the lower and the upper area values.

The presented model in (6) is defined for Type-1 L-R fuzzy numbers where its main goal is to get a regression model oriented to fit a set of fuzzy dependent variables Y_j through a set of independent fuzzy variables X_{ij}. The model focuses in getting an approximation of the complete membership function of Y_j, $Y_j(x)$ represented by their parameters and its area decomposed into a lower and an upper areas through each constraint of the model presented in (6).

3   Validation of the Model - A Comparison Case

To measure its effectiveness the model is compared to the models proposed by Kao, Tanaka and Bargiela (See[3]). The problem consists of a single variable regression analysis. The input values (vl, vc, vu) characterize symmetrical triangular fuzzy sets, so we need less constraints since the area and distances are linear functions of their shapes.

Table 1. Results for comparison case

Proposal	Bargiela	Tanaka	Kao	Present proposal
β_0	3,4467	3,201	3,565	2,6154
β_1	0,536	0,579	0,522	0,6923
Central value error	0.64627579	0.692704031	0.643133875	1.131409359
Distance error	0.094192	0.077542938	0.09996175	0.041422189
Total error (e)	0.860504381	0.877637151	0.862029944	1.082973475

After computing the interesting values of the variables and applying the LP model (6) for eight observations, the obtained β's, the error values obtained and the RSME-based error $e = \sqrt{1/8 \left(\sum_{j=1}^{8} (v_{cj} - v_{cj}^*)^2 \right) + 1/8 \left(\sum_{j=1}^{8} (d_j - d_j^*)^2 \right)}$ are shown in Table 1 where v_{cj}^* and d_j^* are estimations of v_{cj} and d_j respectively.

Although the error of v_{cj}^* obtained by the LP model is the highest, the error of d_j^* is the lowest, which leads to less area errors. Moreover, its efficiency is improved since the structure of an LP model is even simpler than Tanaka's proposal (See [8]).

3.1   Shipping Company Case Study

In this case, a shipping company wants to identify the role of several factors in the profit incoming. The factors considered are: price of service, shipping time, package weight and the return time of the service vehicle.

The linguistic label for each X_j is its *Expected value*. Each X_j is defined by the average of the observations, and each j constraint uses the i, j observation as v_{cj}, so the membership function for each X_j is defined as follows.

$$\text{Price - } X_1(x) = \begin{cases} 1 - \left(\dfrac{1.05 - x}{0.28} \right)^2 & \text{for } 0.77 < x \leqslant 1.05 \\ 0 & \text{otherwise} \end{cases} \tag{7}$$

$$\text{Shipping time - } X_2(x) = \begin{cases} \left(\dfrac{2.331 - x}{0.669} \right)^2 & \text{for } 2.331 < x \leqslant 3 \\ 1 - \left(\dfrac{3 - x}{0.669} \right)^4 & \text{for } 3 < x \leqslant 3.669 \\ 0 & \text{otherwise} \end{cases} \tag{8}$$

$$\text{Weight - } X_3(x) = \begin{cases} \left(\dfrac{x - 15.07}{9.42} \right)^2 & \text{for } 5.65 < x \leqslant 15.07 \\ 0 & \text{otherwise} \end{cases} \tag{9}$$

$$\text{Return time - } X_4(x) = \begin{cases} \left(\dfrac{x - 6.606}{4.386} \right)^4 & \text{for } 2.22 < x \leqslant 6.066 \\ 0 & \text{otherwise} \end{cases} \tag{10}$$

$$\text{Profit - } Y(x) = \begin{cases} 1 - \left(\dfrac{7.38 - x}{2.265} \right)^2 & \text{for } 5.115 < x \leqslant 7.38 \\ \left(\dfrac{x - 11.793}{4.413} \right)^2 & \text{for } 7.38 < x \leqslant 11.793 \\ 0 & \text{otherwise} \end{cases} \tag{11}$$

The model was applied to 49 observations and 7 X_j, with the following results:

$$Y^* = 0.6216 \cdot X_1 + 7.9743 \cdot X_2 + 0.4685 \cdot X_3 - 0.0073 \cdot X_4 - 5.1762 \tag{12}$$

Figure 2 shows the comparison between the estimated dependent variables (gray area) and expected dependent variables (black line) using the average of Y_j and $\hat{Y}_j$ as central values. At a first glance, Figure 2 shows that the LP model reaches a good approximation of the original $Y_j(x)$, but for decision making, the selected deffuzzification method

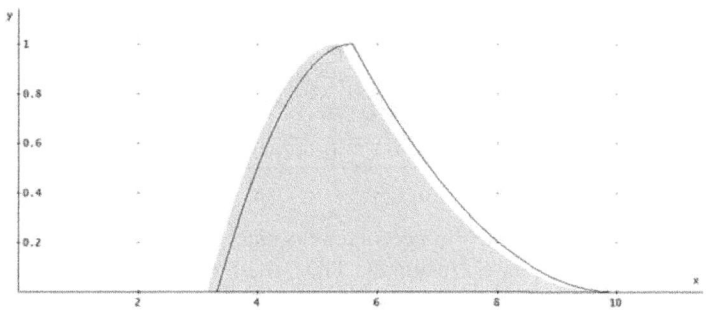

Fig. 2. Graphical comparison of the results of the shipping company case study

is the centroid since it can be obtained by a linear combination of the position, spread and area of the fuzzy sets X_i, viewed as a fuzzy relational equation.

Analysis of the Results. The obtained β's by the regression are used to obtain the estimated centroids, which yields into the following error measures: $RMSE$=2.29 and MSE=5.19 computed through ξ_j (See Equations (1) and (2)).

The determination coefficient obtained is $R^2 = 72.46$, so the 72.476% of the behavior of the dependent variable is explained by the independent variables obtained by the application of the model. In addtion, some desirable properties of the residuals are tested as shown below

Absence of Autocorrelation in the Residuals. Based on a 95% level of significance, the results of the autocorrelation analysis are shown in Table 2. Based on these results, there is no autocorrelation effect, therefore the residuals are randomly distributed.

Table 2. Ljung-Box autocorrelations test on the residuals

Lag	1	2	3	4
Autocorrelation	-0,138	-0,295	-0,12	0,094
Ljung-Box statistic	0,99	5,624	6,409	6,903
Significance	0,32	0,06	0,093	0,141

Normal Distribution in the Residuals. For a 95% confidence level, the Kolmogorov-Smirnov test reaches a p-value of 0.150 and the Shapiro-Wilks test reaches a p-value of 0,121, so we can conclude that the residuals are normally distributed.

Zero Mean Residuals. A One Sample Test was performed to test if $\bar{\xi} = \mu = 0$. A difference test based on a normal distributed asymptotic behavior of the mean of the residuals $\bar{\xi} = 0,320$, with an obtained significance of 0,333. We can conclude that there is no statistical evidence that supports that ξ has no zero mean.

Homoscedasticity of the Residuals. By dividing the residuals in three balanced groups and applying the F-test for variances between each pair, with a 95% confidence level, it is concluded that there is homoscedasticity in the residuals. (See Table 3).

Table 3. Homocedasticity test

Group	1-2	1-3	2-3
F Sample	1,034	1,845	1,783
F Statistic	2,403	2,352	2,352
Significance	0,473	0,117	0,131

Main Conclusions. The LP model had good results since it reached normally independent, zero mean and homocedastic residuals. Thus, the β's and the regression analysis is valid. The β's shows that X_1 and X_3 (See (7) and (9)) have the highest contribution to the profit, so is recommended to review the pricing policy of the company.

4 Concluding Remarks

The LP model presented in this paper focuses on the minimization of the errors between a linear combination of X_j that estimates Y. The method used can deal with nonlinear fuzzy data, therefore our proposal is an alternative formulation for fuzzy regression.

Some real problems involve uncertainty that can be treated as fuzzy sets, therefore the LP model presented in this paper is more efficient than other proposals because it can be handled through mixed fuzzy-convex optimization methods.

Finally, we recommend the use Type-2 fuzzy sets as uncertainty measures to deal with the perception about a linguistic variable of a fuzzy set held by multiple experts. For further information see Melgarejo in [6], and Mendel in [7].

Acknowledgments. The authors would like to thank Jesica Rodriguez-Lopez for her invaluable support.

References

1. Bargiela, A., et al.: Multiple regression with fuzzy data. Fuzzy Sets and Systems 158(4), 2169–2188 (2007)
2. Gladysz, B., Kuchta, D.: Least squares method for L-R fuzzy variable. In: 8th International Workshop on Fuzzy logic and Applications, vol. 8, pp. 36–43. IEEE, Los Alamitos (2009)
3. Kao, C., Chyu, C.: Least-Squares estimates in fuzzy regression analysis. European Journal of Operational Research 148(2), 426–435 (2003)
4. Klir, G.J., Folger, T.A.: Fuzzy Sets, Uncertainty and Information. Prentice Hall, Englewood Cliffs (1992)
5. Klir, G.J., Yuan, B.: Fuzzy Sets and Fuzzy Logic: Theory and Applications. Prentice Hall, Englewood Cliffs (1995)
6. Melgarejo, M.A.: Implementing Interval Type-2 Fuzzy processors. IEEE Computational Intelligence Magazine 2(1), 63–71 (2007)
7. Mendel, J.: Uncertain Rule-Based Fuzzy Logic Systems: Introduction and New Directions. Prentice Hall, Englewood Cliffs (1994)
8. Tanaka, H., et al.: Linear Regression analysis with Fuzzy Model. IEEE Transactions on Systems, Man and Cybernetics 12(4), 903–907 (1982)
9. Zadeh, L.A.: Toward a generalized theory of uncertainty (GTU) an outline. Information Sciences 172(1), 1–40 (2005)

No Reference Image Quality Assessment by Designing Fuzzy Relational Classifier Using MOS Weight Matrix

Indrajit De[1] and Jaya Sil[2]

[1] Department of Information Technology, MCKV Institute of Engineering,
Liluah, Howrah, West Bengal-711204, India
[2] Department of Computer Science and Technology, Bengal Engineering and Science
University, Shibpur Howrah, West Bengal, India
indrajitde@ieee.org, js@cs.becs.ac.in

Abstract. Assessing quality of distorted/decompressed images without reference to the original image is a challenging task because extracted features are often inexact and there exist complex relation between features and visual quality of images. The paper aims at assessing quality of distorted/decompressed images without any reference to the original image by developing a fuzzy relational classifier. Here impreciseness in feature space of training dataset is tackled using fuzzy clustering method. As a next step, logical relation between the structure of data and the soft class labels is established using fuzzy mean opinion score (MOS) weight matrix. Quality of a new image is assessed in terms of degree of membership value of the input pattern corresponding to given classes applying fuzzy relational operator. Finally, a crisp decision is obtained after defuzzification of the membership value.

Keywords: MOS, fuzzy relational classifier, no reference.

1   Introduction

Digital images are subjected to loss of information, various ways of distortions during compression [6] and transmission, which deteriorate visual quality of the images at the receiving end. Quality of an image plays fundamental role to take vital decision and therefore, its assessment is essential prior to application. Modeling physiological and psycho visual features of the human visual system [10, 11, 12] and signal fidelity criteria [9] based quality assessment are reported [10, 11] though each of these approaches has several shortcomings. The most reliable means of measuring the image quality is subjective evaluation based on the opinion of the human observers [7, 14]. However, subjective testing is not automatic and expensive too. On the other hand, most objective image quality assessment methods [5, 8] either require access to the original image as reference [2] or only can evaluate images, degraded with predefined distortions and therefore, lacking generalization approach. Two prominent works have been reported relating to no-reference image quality evaluation, (i) Wang, Bovic and Shiekh's no-reference JPEG image quality index and (ii) H.Shiekh's quality metric based on natural scene statistics (NSS) model applied on JPEG2000 compressed images. Three works are reported very recently to assess quality of an

D.-S. Huang et al. (Eds.): ICIC 2011, LNBI 6840, pp. 361–369, 2012.

image namely, Extreme Learning Machine classifier based mean opinion score (MOS) estimator [16], discrete cosine transform (DCT) domain statistics based metric [17] and blind image quality index [18, 19]. But none of them incorporated computational intelligence approaches, which are human centric computation methods best suited for assessing image quality based on human visual system.

In the paper, human reasoning power is explored to building a fuzzy relational classifier that assesses quality of images by assigning soft class labels using fuzzy MOS based weight matrix. In the proposed method, first impreciseness in feature space is handled by analysing the training dataset using fuzzy clustering method. As a next step logical relation between the feature and the class label is measured by degree of membership value obtained using φ-composition (a fuzzy implication)[1] and conjunctive aggregation methods. Quality of a new image is assessed in terms of degree of membership of the pattern in the given classes by applying fuzzy relational operator. Finally, a crisp decision is obtained after defuzzification of the membership value.

2   Fuzzy Relational Classification

Fuzzy relational classification [20] establishes a correspondence between structures in feature space and the class labels. By using fuzzy logic in classification, one avoids the problem of hard labeling the prototypes and easily captures the partial sharing of structures among several classes. In the training phase of classifier building process, two steps are identified: (a) exploratory data analysis (unsupervised fuzzy clustering), (b) construction of a logical relation between the structures obtained in the previous step and the class labels using fuzzy MOS based weight matrix. In the exploratory step, the available data objects are clustered in groups by the fuzzy c-means (FCM) [13] or a similar algorithm. Clustering results a fuzzy partition matrix, which specifies for each training sample a tuple of membership degrees obtained with respect to each cluster. In the second step, a fuzzy relation is computed, using the memberships obtained in the first step and the target membership of the pattern in the classes. This relation is built by means of the φ-composition (a fuzzy implication) and conjunctive

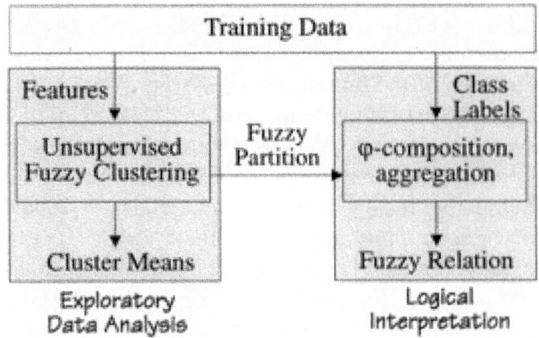

Fig. 1. Fuzzy Relational classifier training phase

aggregation. It specifies the logical relationship between the cluster membership and the class membership. To classify new patterns, the membership of each pattern in the clusters (fuzzy prototypes) is computed from its distance to the cluster centers, giving a fuzzy set of prototype membership. Then, relational composition of this fuzzy set with the fuzzy relation is applied to compute an output fuzzy set. This set provides fuzzy classification in terms of membership degrees of the pattern in the given classes. When a crisp decision is required, defuzzification has to be applied to this fuzzy set. Typically, the maximum defuzzification method is used. The process is given in the diagram (fig-1):

3   Feature Selection

Local features like Scale invariant feature transform (SIFT) features are used to build the classifier. For this purpose David Lowe's [21] algorithm has been used to extract the SIFT features from gray level (PGM format) training images taken from TAMPERE database [23]. Scale Invariant Feature Transform (SIFT) is an approach for detecting and extracting local feature descriptors that are reasonably invariant to changes in illumination, image noise, rotation, scaling, and small changes in viewpoint. The stepwise detection stages for SIFT features are a) scale-space extrema detection. b) keypoint localization. c) orientation assignment and d) generation of keypoint descriptors. These features share similar properties with neurons in inferior temporal cortex that are used for object recognition in primate vision. The first step toward the detection of interest points is the convolution of the image with Gaussian fi lters at different scales, and the generation of difference of Gaussian images from the difference of adjacent blurred images. Interest points (called key points in the SIFT framework) are identified as local maxima or minima of the DoG (Difference of Gaussian) images across scales. Each pixel in the DoG images is compared to its 8 neighbors at the same scale, plus the 9 corresponding neighbors at neighboring scales. If the pixel is a local maximum or minimum, it is selected as a candidate key point. For each candidate key point following steps are performed a) interpolation of nearby data is used to accurately determine its position. b) key points with low contrast are removed. c) responses along edges are eliminated. d) key point is assigned an orientation. To determine the key point orientation, a gradient orientation histogram is computed in the neighborhood of the key point. Once a key point orientation has been selected, the SIFT feature descriptor is computed as a set of orientation histograms on 4×4 pixel neighborhoods.

4   Procedure

The flowchart of the process is:

Step 1: Six training images (see fig.3 and Table 1) from TAMPERE image databases [23] are taken to extract SIFT features from them. Then MOS values are utilized to build the fuzzy classification weight matrix following Keller et al [25] rule given in equation (1).

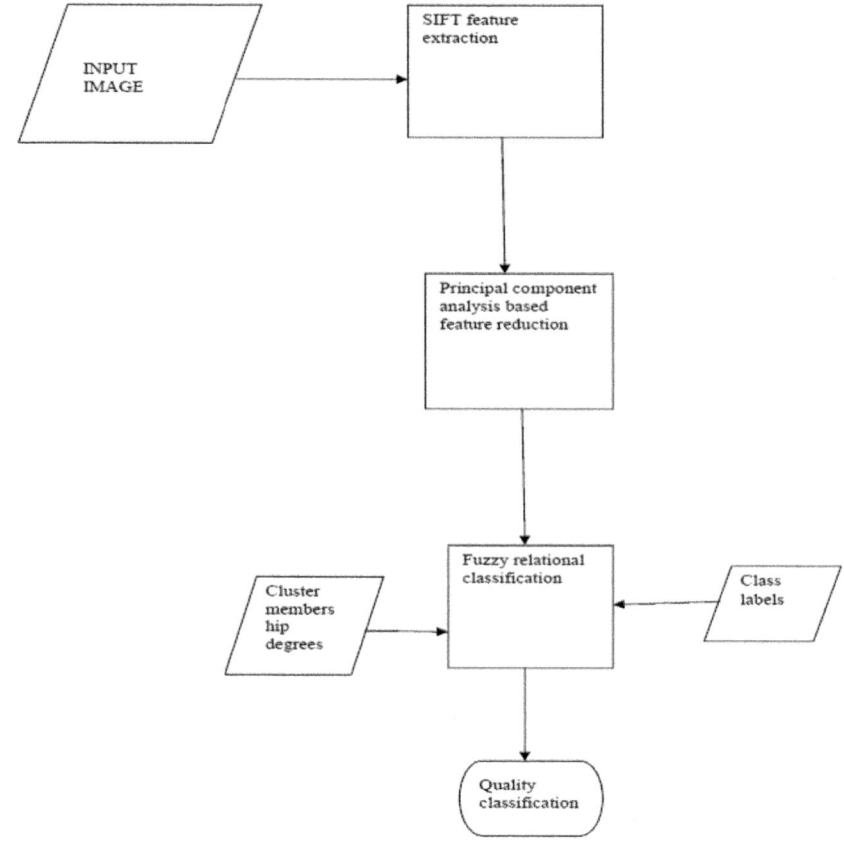

Fig. 2. SIFT based image quality classification

$$y_{li} = \begin{cases} 0.51 + 0.49(n_{li}/k) \\ 0.49(n_{li}/k) \end{cases} \tag{1}$$

where y_{li} represents the class membership of the x_i to the class 1, k is the number of x_i's neighbors and n_{li} stands for x_i's neighbors belonging to class 1.

Step 2: The extracted SIFT features are dimensionally reduced by Principal Component Analysis (PCA). The average reduction is from approximately twenty thousand vectors to one twenty eight eigenvectors. In the process the negative PCA vectors are not taken into consideration.

Step 3: The PCA reduced SIFT feature vectors from the images are combined together in a matrix form where individual image feature vectors are placed along the column of the matrix.

Step 4: Fuzzy C means clustering procedure is operated on the thus prepared data matrix with fuzzyness exponent being 2.5 and number of clusters being 4.

Step 5: The membership degree matrix thus obtained is combined with the classification weight matrix already built up by Lukasiewicz implication method (equation 2) [20] to get the Fuzzy relational matrix for the training images.

$$(r_{ij}) = \min(1, 1 - \mu_{ik} + \omega_{jk})$$
$$k = 1, 2, ...N, \ j = 1, 2...L, i = 1, 2, ...c$$

$$(r_{ij}) = \min_{k=1,2...N} [(r_{ij})_k] \tag{2}$$

where (r_{ij}) is individual relational matrix element, μ_{ik} is membership degree element and ω_{jk} is classification weight element. k is the total number of images (here 6), L is the total number of classes (here 5) and c is the total number of clusters (here 4). N is the total number of patterns (here 6).

Step 6: For nine testing images taken from three image databases namely TAMPERE, LIVE[5] and PROFILE[24] after following Step 1 and Step 2 on it, the membership degrees of the image feature vectors from the four cluster centers already obtained, are computed using equation 3.

$$\mu_i = \frac{1}{\sum_{j=1}^{c} (d(x, v_i) / d(x, v_j))^{2/m-1}} \tag{3}$$

where μ_i is the membership degree of the new image feature vectors with respect to cluster center v_i, d(.) is the Euclidean distance and m is the fuzzyness exponent[22]. The fuzzyness exponent m is changed for each image while computing the membership degrees to observe the effect of fuzzyness on classification accuracy.

Step 7: Given the membership degrees obtained in Step 6 the class membership ω is computed from the relational composition:

$$\omega_j = \max_{1 \leq i \leq c} [\max(\mu_i + r_{ij} - 1, 0)], \ j = 1, 2, ...L \tag{4}$$

where all the variables (c, L) hold there meanings.

5   Results

To build the fuzzy classification weight matrix first crisp MOS based classification weight matrix with binary(0 or 1) values is prepared, then using equation 1 the fuzzy MOS based classification weight matrix is prepared.

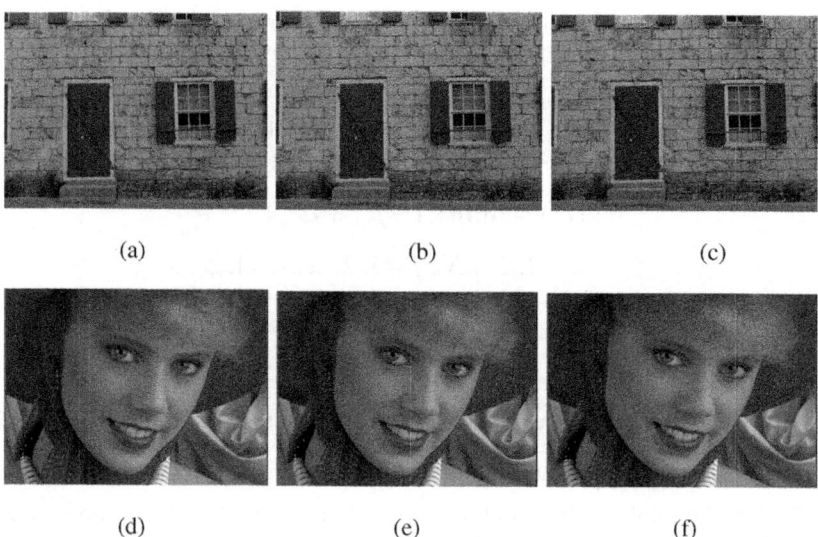

Fig. 3. Training images, (a) Image1, (b) Image2, (c) Image3, (d) Image4, (e) Image5, (f) Image6

Table 1. Training image features

Image Name	Distortion Type	Distortion Level	Mean Opinion Score(MOS)
Image 1	Additive Gaussian Noise	1	5.9706
Image 2	Do	2	5.4167
Image 3	Do	3	4.5556
Image 4	Do	1	4.3793
Image 5	Do	2	3.9655
Image 6	Do	3	3.6

The following table shows the classification weight matrix:

Table 2. MOS related fuzzy classification weight matrix

Images ↓	Class labels →	Excellent	Good	Average	Bad	Poor
Image 1		0.5916	0.0816	0.0816	0.0816	0.0816
Image 2		0.5916	0.0816	0.0816	0.0816	0.0816
Image 3		0.0816	0.5916	0.0816	0.0816	0.0816
Image 4		0.0816	0.5916	0.0816	0.0816	0.0816
Image 5		0.0816	0.0816	0.5916	0.0816	0.0816
Image 6		0.0816	0.0816	0.5916	0.0816	0.0816

After building the classification weight matrix using equation 2 the fuzzy relational matrix is prepared. While preparing the fuzzy relational matrix the number of clusters is taken to be $a \times \sqrt{N}$ where a is a parameter dependent upon training image database (here it is taken as 1.63) and N is the total number of image patterns (here it is 6).

The following table shows the fuzzy relational matrix:

Table 3. Fuzzy relational matrix

Clusters ↓	Class labels →	Excellent	Good	Average	Bad	poor
Cluster1		0.156061	0.156061	0.666061	0.156061	0.156061
Cluster2		0.628505	0.118505	0.118505	0.118505	0.118505
Cluster3		0.119078	0.119078	0.629078	0.119078	0.119078
Cluster4		0.146929	0.656929	0.146929	0.146929	0.146929

The class memberships are computed using equation 4 and the membership is represented as a value present in a particular class label, the others being of zero values.

The following table shows the comparisons:

Table 4. The comparison of fuzzy relational classifier with other quality metrics

Image Name	Fuzzyness Exponent Value For Membership Degree Computation	Fuzzy Relation Based Image Quality In Linguistic Variable Term	Blind Image Quality Index (Linguistic Variable) (krishnamoorthy,bovik,)	Jpegquality Score (Linguistic Variable) (wang et al)	JP2KNR (Linguistic Variable) (Shiekh et al)
Img162 (LIVE database)	2.5	average	good	Excellent	good
Img132 (LIVE)	2.0	average	average	average	good
Chinacongressdistorted (PROFILE DATABASE)	2.5	Average	average	average	good
Chinacongressoriginal (PROFILE)	5.5	Excellent	good	excellent	good
Annanoriginal (PROFILE)	3.0	Excellent	good	Excellent	good
Naipauldistorted (PROFILE)	2.0	Average	poor	average	good
I01 (TAMPERE DATABASE)	5.5	Excellent	good	excellent	good
I04 (TAMPERE)	4.0	Excellent	good	excellent	good
Afghangaussian (PROFILE)	2.0	Average	average	average	good

6   Conclusion

A fuzzy relational classifier is designed to classify distorted images based on SIFT features by incorporating human perception to assess quality of the image. The proposed no reference image quality metric has been compared with the existing quality metric producing methods and the result is satisfactory. In this work Mean Opinion Score Estimation is invoked as a measure of predicted image quality and the effect of fuzzyness exponent on classification process is also studied in assessing quality of image. Future work will include Mean Opinion Score Estimation as a measure of predicted image quality value and the effect of fuzzyness exponent on classification process.

References

1. Lin, C.T., George Lee, C.S.: Neural Fuzzy Systems, pp. 140–174. Prentice Hall Ptr, Englewood Cliffs (1993)
2. Sheikh, H.R., Bovic, A.C.: Image information and visual quality. In: Proc IEEE Int. Conf. Acoust., Speech and Signal Processing (May 2004)
3. Sheikh, H.R., Bovic, A., Cormack, L.: No-Reference Quality Assesment using Natural Scene Statistics: JPEG 2000. IEEE Transactions on Image Processing (2005)
4. Wang, Z., Sheikh, H., Bovic, A.: No Reference Perceptual Quality Assesment of JPEG compressed image. In: Proceedings of IEEE 2002 International Conferencing on Image Processing, pp. 22–25 (2002)
5. Sheikh H. R., Wang Z., Cormack L., Bovic, A.: LIVE image quality assessment database (2003), http://live.ece.utexas.edu/research/quality
6. Sayood, K.: Introduction to data compression, pp. 267–268. Morgan Kauffman Publishers, San Francisco (2000)
7. Wang, Z., Bovik, A.C.: Why is image quality assessment so difficult? In: IEEE Int. Conf. Acoust., Speech, and Signal Processing (May 2002)
8. VQEG, Final report from the video quality experts group on the validation of objective models of video quality assessment (March 2000), http://www.vqeg.org/
9. Sonka, M., Hlavac, V., Boyle, R.: Image processing analysis and machine vision, pp. 254–262. IPT Press (1999)
10. Pappas, T.N., Safranek, R.J.: Perceptual criteria for image quality evaluation. In: Bovik, A. (ed.) Handbook of Image &Video Proc., Academic Press, New York (2000)
11. Watson, B. (ed.): Digital Images and Human Vision. MIT Press, Cambridge (1993)
12. Watson, A.B., et al.: Visibility of wavelet quantization noise. IEEE Transactions on Image Processing 6(8), 1164–1175 (1997)
13. Dunn, J.C.: A Fuzzy Relative of the ISODATA Process and Its Use in Detecting Compact Well-Separated Clusters. Journal of Cybernetics 3, 32–57 (1973)
14. Wang, Z., Bovik, A.C.: Modern Image Quality Assessment, pp. 79–102. Morgan and Claypool Publishers (2006)
15. Wang, Z., Wu, G., Shiekh, H.R., Simoncelli, E.P., Wang, E.Y., Bovik, A.C.: Quality aware images. IEEE Transactions on Image Processing, 1680–1689 (June 2006)
16. Suresh, S., Venkatesh Babu, R., Kim, H.J.: No-Reference image quality assessment using modified extreme learning classifier. Applied Soft Computing, 541–552 (2009)
17. Brandão, T., Queluz, M.P.: No-reference image quality assessment based on DCT domain statistics. Signal Process 88, 822–833

18. Moorthy A. K., Bovik A. C.: BIQI Software Release (2009),
    `http://live.ece.utexas.edu/research/quality/biqi.zip`
19. Moorthy, A.K., Bovik, A.C.: A Modular Framework for Constructing Blind Universal Quality Indices. IEEE Signal Processing Letters (2009)
20. Setnes, M., Babuska, R.: Fuzzy relational classifier trained by fuzzy clustering. IEEE Transactions on Systems, Man, and Cybernetics, Part B 29(5), 619–625 (1999)
21. Lowe, D.G.: Distinctive image features from scale-invariant keypoints. International Journal of Computer Vision 60(2), 91–110 (2004)
22. Yu, J., Cheng, Q., Huang, H.: Analysis of the weighting exponent in the FCM. IEEE Transactions on SMC-Part B 34(1), 634–639 (2004)
23. Ponomarenko, N., Lukin, V., Zelensky, A., Egiazarian, K., Carli, M., Battisti, F.: TID2008 - A Database for Evaluation of Full-Reference Visual Quality Assessment Metrics. Advances of Modern Radioelectronics 10, 30–45 (2009)
24. `http://vasc.ri.cmu.edu/idb/html/face/profile_images/`
25. Keller, J.M., Gray, M.R., Givens, J.A.: A fuzzy k-nearest neighbor algorithm. IEEE Trans. Systems Man Cybernet 15, 580–585 (1985)

Ant Colony Optimization for Nonlinear Inversion of Rayleigh Waves

Jiangqiao Xu[1] and Xianhai Song[2]

[1] School of Energy Resources, China University of Geosciences, Beijing, 100083, China
107300696@qq.com
[2] Changjiang River Scientific Research Institution, Wuhan, Hubei, 430010, China
songxianhaiwcy@sina.com

Abstract. Inversion of Rayleigh wave dispersion curves not only undergoes computational difficulties associated with being trapped by local minima for most local-search methods but also suffers from the high computational time for most global optimization methods due to its multimodality and its high nonlinearity. In order to effectively overcome the above described difficulties, we proposed a new Rayleigh wave inversion scheme based on an ant colony optimization, a commonly used swarm intelligence algorithm. The calculation efficiency and stability of the proposed procedure are tested on a five-layer synthetic model and a real-world example. Results from both synthetic and real field data demonstrate that ant colony optimization applied to nonlinear inversion of Rayleigh waves should be considered good not only in terms of computation time but also in terms of accuracy due to its global and fast convergence in the final stage of exploration.

Keywords: Ant colony optimization, Particle swarm optimization, Rayleigh waves, Shear-wave velocity.

1 Introduction

Rayleigh waves have recently been used increasingly as an appealing tool for near-surface shear(S)-wave velocities [1-3]. Inversion of Rayleigh wave dispersion curves is one of the key steps in Rayleigh wave analysis to obtain a subsurface S-wave velocity profile [4]. However, nonlinear inversion of Rayleigh waves, as with most other geophysical optimization problems, is typically a highly nonlinear, multiparameter, and multimodal inversion problem. Consequently, local optimization methods, e.g. matrix inversion, steepest descent, conjugate gradients, are prone to being trapped by local minima, and their success depends heavily on the choice of the starting model and the accuracy of the partial derivatives [5]. Existing commonly used global optimization techniques, such as Genetic algorithms (GA) [6], Simulated annealing (SA) [7], Artificial neural network (ANN) [8], Wavelet transform (WT) [9], Multi-objective evolutionary algorithms (MOEA) [10] and Monte Carlo (MC) [11], have proven to be quite useful for determining S-wave velocity profiles from dispersion data but are computationally quite expensive due to their slower convergence in the final stage of exploration procedures.

D.-S. Huang et al. (Eds.): ICIC 2011, LNBI 6840, pp. 370–377, 2012.

One of swarm intelligence metaheuristics, namely, ant colony optimization (ACO) is well suitable for the global optimization of highly non-convex objective functions, without the need of calculating any gradient or curvature information, especially for addressing problems for which the objective functions are not differentiable, stochastic, or even discontinuous [12]. ACO are recently been used and tested to optimize complex mathematical problems characterized by the large numbers of local minima and/or maxima [13, 14]. However, few attempts have been made to address nonlinear inversion of Rayleigh wave dispersion data. This paper implemented and tested an improved ant colony optimization hybridized with a particle swarm optimization (PSO) [15] for nonlinear inversion of Rayleigh wave dispersion curves. The proposed scheme is based on the common characteristics of both ACO and PSO algorithms, like, survival as a colony (swarm) by coexistence and cooperation, individual contribution to food searching by an ant (a particle) by sharing information locally and globally in the colony (swarm) between ants (particles), etc. The implementation of the improved ACO algorithm proposed by Shelokar et al. [16], consists of two stages. In the first stage, it applies PSO, while ACO is implemented in the second stage. ACO works as a local search, wherein, ants apply pheromone-guided mechanism to refine the positions found by particles in the PSO earlier stage. The calculation efficiency and stability of the proposed inversion scheme are tested on a five-layer synthetic model and a real example from China. Results from both synthetic and actual field data demonstrate that the ACO algorithm applied to nonlinear inversion of Rayleigh waves should be considered good not only in terms of computation time but also in terms of accuracy.

2   Ant Colony Optimization for Surface Wave Analysis

Ant colony optimization (ACO) is a multiagent approach that imitates foraging behavior of real life ants for solving difficult global optimization problems. Ants are social insects whose behavior is directed more toward the survival of the colony as a whole than that of a single individual of the colony. An important and interesting behavior of an ant colony is its indirect co-operative foraging process. While walking from food sources to the nest and vice versa, ants deposit a substance, called pheromone on the ground and form a pheromone trail. Ants can smell pheromone, when choosing their way; they tend to choose, with high probability, paths marked by strong pheromone concentrations. Also, other ants can use pheromone to find the locations of food sources found by their nest mates.

We implemented a series of MATLAB tools based on MATLAB 7.6 for nonlinear inversion of Rayleigh waves. We have worked on SWIACO, a software package for Surface Wave Inversion via Ant Colony Optimization. In this study, we focus our attention on inversion results of fundamental-mode Rayleigh wave dispersion curves for near-surface S-wave velocities by fixing P-wave velocities (or Poisson's ratio), densities, and thicknesses, to their known values. The improved ACO algorithm parameter settings used in all the simulations is given as: number of particles and ants P=25; cognitive and social scaling parameters, $c1=2$, $c2=2$; maximum and minimum values of inertia weights, wmax=0.7, wmin=0.4; the lower and upper bounds of the search areas are 100 and 600 m/s, respectively, for every layer in all of the latter tests.

The procedure will set out to find the global minimum of rms (root-mean-square) error misfit between the measured and the predicted phase velocities. The algorithm is terminated after 10n iteration in our inversion procedure or when the error misfit reaches a certain previously fixed value (n is the size of solution vector). The inverted models for the 20 calculations are averaged to determine a final solution due to the randomness in the ACO procedure. Forward modeling of Rayleigh wave dispersion curves is based on the fast delta matrix algorithm.

3   Modeling Results

To investigate and evaluate the efficiency and stability of of the ACO inversion procedure, a five-layer synthetic earth model with a soft layer trapped between two stiff layers (Table 1) is used. This model is designed to simulate a real complex pavement structure containing a low velocity layer. In the current analysis, a commonly used frequency range of 5-100 Hz in shallow engineering site investigations was employed.

Table 1. Five-layer subsurface parameters

Layer Number	S-wave velocity (m/s)	P-wave velocity (m/s)	Density (g/cm^3)	Thickness (m)
1	250	713	1.9	3
2	200	735	1.9	3
3	250	917	1.9	4
4	400	1327	1.9	4
5	500	1658	1.9	Infinite

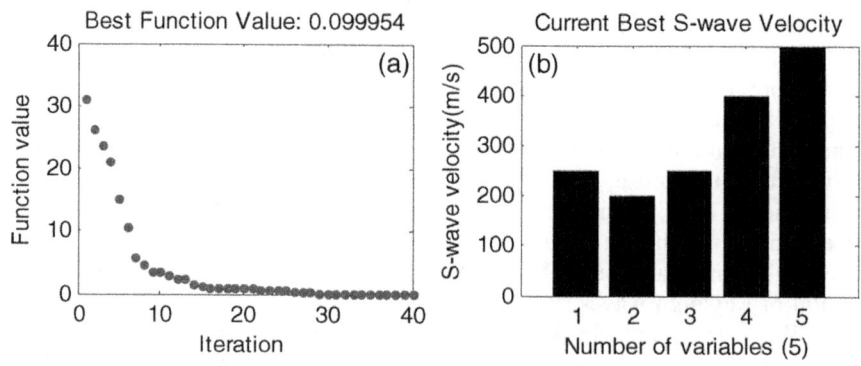

Fig. 1. The performance of the ACO approach for nonlinear inversion of the five-layer model. (a) Objective function value of best point as a function of iteration. (b) Inverted S-wave velocities.

Fig. 1 demonstrates the performance of the ACO algorithm for nonlinear inversion of the five-layer model, which provides valuable insights into the performance of the proposed inversion scheme. The convergence curve in Fig. 1a illustrates a typical characteristic of the ACO algorithm. It shows a very fast initial convergence at the first 25 iterations, followed by rapid improvements as it approaches the optimal solution (Fig. 1b). The inverse process is terminated after 40 iterations because the error misfit converged to approximately zero.

Fig. 2 shows inversion results of the five-layer model using the ACO approach. Clearly, for this multilayer earth model, S-wave velocities are exactly resolved (Fig. 2b). Final phase velocities (solid dots in Fig. 2a) predicted from the inverted S-wave velocity model (dashed line in Fig. 2b) fit measured phase velocities (solid line in Fig. 2a) almost perfectly. In practice, near-surface S-wave velocities inverted by Rayleigh waves may be useful for static correction of oil exploration.

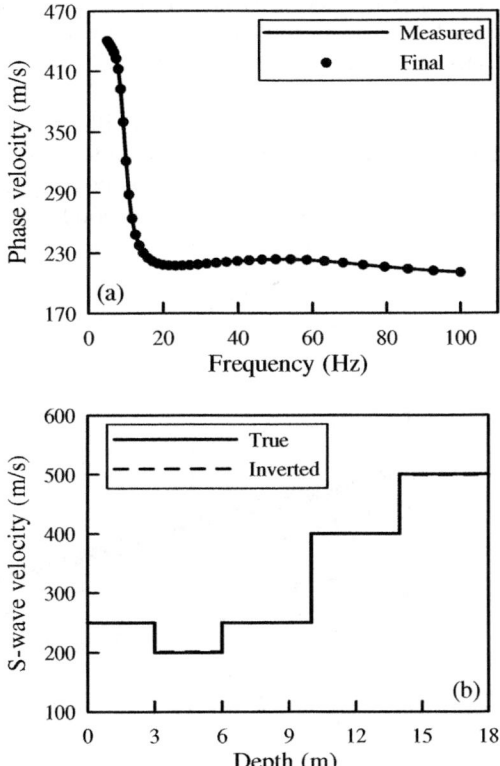

Fig. 2. Inversion of the five-layer model using the ACO algorithm. Dispersion curves (a) labeled as Measured and Final are phase-wave velocities from the True and Inverted models (b), respectively.

4   Field Data Inversion

Modeling results presented in the previous section demonstrated the calculation efficiency and reliability of the ACO inverse procedure. To further explore the performance of the ACO algorithm described above, surface wave data (Fig.3) acquired from a highway roadbed survey in Henan, China have been reanalyzed in the present study using the ACO algorithm. Borehole reveals a 1.4-m-thick soft clay layer between a stiff surficial layer and a gravel soil layer with mixed alluvial deposit.

The real phase velocities (solid dots in Fig. 5a) are extracted by the phase-shift approach within a frequency range of 8-70 Hz. It is worth noticing that the measured dispersion curve is characterized by an inversely dispersive trend (an abrupt variation and discontinuity) within the frequency range of 15-30 Hz, which is likely owing to the inclusion of higher modes in the fundamental mode data when a low velocity layer is present.

Similar to the inverse strategy of the synthetic model, a six-layer subsurface structure was adopted for nonlinear inversion of the observed dispersion curve using the proposed ACO approach. Estimated Poisson's ratio and densities are 0.40 and 1.9 g/cm^3 for each layer of the six-layer model, respectively, and are kept constant in the inverse process because the accuracy of the deduced model is insensitive to these parameters.

The performance of ACO for the real example is illustrated in Fig. 4. Similar to Fig. 1, the objective function values improve rapidly at the first 40 iterations, and then fast converge to a similar constant value (Fig. 4a), which suggests that the improved ACO algorithm has completed the exploration for a good valley (Fig. 4b). The soft

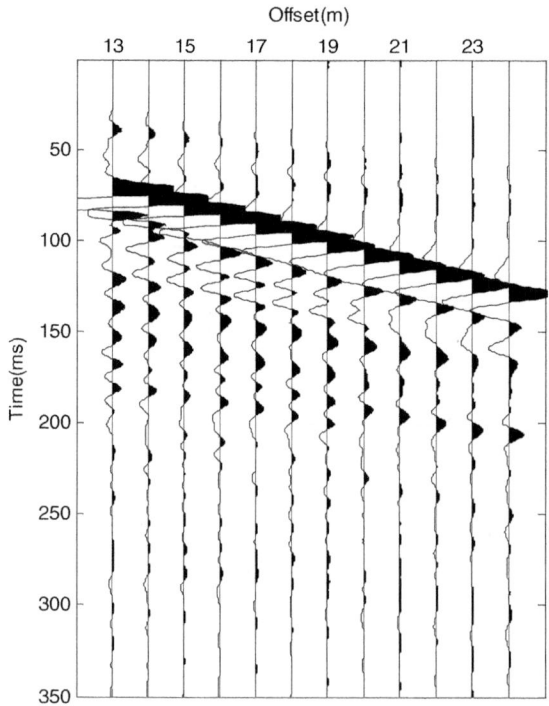

Fig. 3. A 12-channel field raw surface wave shot gather acquired in Central China

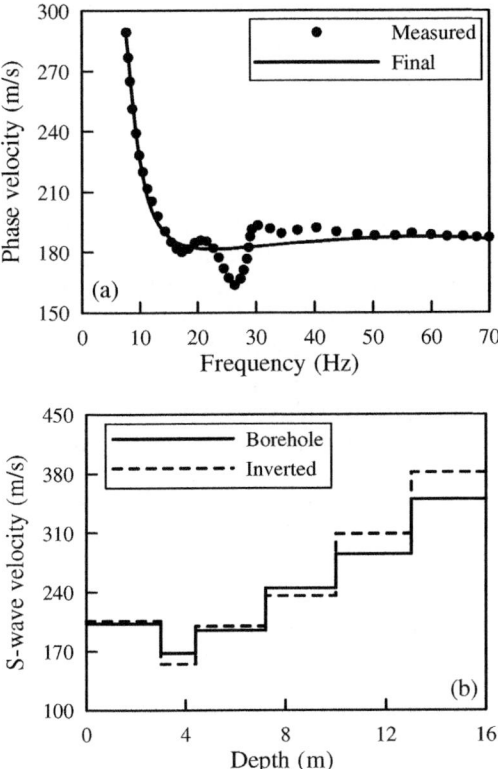

Fig. 5. Inversion of field data using ACO. (a) Measured (solid dots) and Final (solid line) dispersion curves. (b) Borehole measurement (solid line) and inverted (dashed line) S-wave velocity profiles.

clay layer which usually leads to serious subsidence and deformation of a pavement is clearly depicted. We terminate the inversion process after 60 iterations to prevent transferring errors in data into the inverted model.

The inverted model (dashed line in Fig. 5b) is in acceptable agreement with borehole measurements (solid line in Fig. 5b), especially for layer 1, layer 3, and layer 4; misfits are all not more than 4%. However, three greater misfits occur at layer 2, layer 5, and layer 6, which are approximately 8%, 8%, and 11%, respectively, although the calculated dispersion curve (solid line in Fig. 5a) from the best solution (dashed line in Fig. 5b) matches the measured phase velocities (solid dots in Fig. 5a) reasonably well. A couple of factors, including a limited frequency spectrum, estimation errors in P-wave velocities and densities, as well as inclusion of noise in real data, are likely responsible for these discrepancies. Nonetheless, the real example should be considered a successful application to nonlinear inversion of Rayleigh wave dispersion curves using the ACO algorithm. Higher modes are relatively more sensitive to the fine S-wave velocity structure than is the fundamental mode and therefore the accuracy of S-wave velocities can be further improved by incorporating higher-mode data into the inversion process. Efforts should be made to fully exploit higher-mode Rayleigh waves for imaging and characterizing shallow subsurface more accurately, especially for the inversely dispersive site.

5   Conclusions

From our successful inversions of synthetic and observed Rayleigh wave data, we confidently conclude that ant colony optimization (ACO) can be applied to nonlinear inversion of Rayleigh wave dispersion curves. This paper has implemented and tested a Rayleigh wave dispersion curve inversion scheme based on the improved ACO algorithm for solution of highly non-convex Rayleigh wave inversion. A pheromone-guided local search is implemented to improve the performance of ant colony optimization. The results show that PSO helps ACO process not only to efficiently perform global exploration for rapidly attaining the feasible solution space but also to effectively reach optimal or near optimal solution. Results from both synthetic and actual field data show that there is a scope of research in hybridizing swarm intelligence methods to solve difficult Rayleigh wave optimization problems.

Acknowledgements. The authors greatly appreciate Shelokar P.S. for his help on ant colony optimization (ACO) and particle swarm optimization (PSO) programming.

References

1. Park, C.B., Miller, R.D., Xia, J.: Multichannel analysis of surface waves. Geophysics 64, 800–808 (1999)
2. Song, X., Gu, H., Liu, J., Zhang, X.: Estimation of shallow subsurface shear-wave velocity by inverting fundamental and higher-mode Rayleigh waves. Soil Dynamics and Earthquake Engineering 27(7), 599–607 (2007)
3. Song, X., Gu, H.: Utilization of multimode surface wave dispersion for characterizing roadbed structure. Journal of Applied Geophysics 63(2), 59–67 (2007)
4. Song, X., Gu, H., Zhang, X., Liu, J.: Pattern search algorithms for nonlinear inversion of high-frequency Rayleigh wave dispersion curves. Computers & Geosciences 34(6), 611–624 (2008)
5. Song, X., Li, D., Gu, H., Liao, Y., Ren, D.: Insights into performance of pattern search algorithms for high-frequency surface wave analysis. Computers & Geosciences 35(8), 1603–1619 (2009)
6. Yamanaka, H., Ishida, H.: Application of genetic algorithm to an inversion of surface wave dispersion data. Bulletin of the Seismological Society of America 86, 436–444 (1996)
7. Beaty, K.S., Schmitt, D.R., Sacchi, M.: Simulated annealing inversion of multimode Rayleigh-wave dispersion curves for geological structure. Geophysical Journal International 151, 622–631 (2002)
8. Shirazi, H., Abdallah, I., Nazarian, S.: Developing artificial neural network models to automate spectral analysis of surface wave method in pavements. Journal of Computing in Civil Engineering 21(12), 722–729 (2009)

9. Tillmann, A.: An unsupervised wavelet transform method for simultaneous inversion of multimode surface waves. Journal of Environmental & Engineering Geophysics 10(3), 287–294 (2005)
10. Dal Moro, G., Pipan, M.: Joint inversion of surface wave dispersion curves and reflection travel times via multi-objective evolutionary algorithms. Journal of Applied Geophysics 61(1), 56–81 (2007)
11. Maraschini, M., Foti, S.: A Monte Carlo multimodal inversion of surface waves. Geophysical Journal International 182(3), 1557–1566 (2010)
12. Dorigo, M., Blum, C.: Ant colony optimization theory: a survey. Theoretical Computer Science 344(2-3), 243–278 (2005)
13. Socha, K., Dorigo, M.: Ant colony optimization for continuous domains. European Journal of Operational Research 185(3), 1155–1173 (2008)
14. Toksari, M.D.: Ant colony optimization for finding the global minimum. Applied Mathematics and Computation 176, 308–316 (2006)
15. Shaw, R., Srivastava, S.: Particle swarm optimization: a new tool to invert geophysical data. Geophysics 72(2), 75–83 (2007)
16. Shelokar, P.S., Siarry, P., Jayaraman, V.K., Kulkarni, B.D.: Particle swarm and ant colony algorithms hybridized for improved continuous optimization. Applied Mathematics and Computation 188, 129–142 (2007)

Memetic Fitness Euclidean-Distance Particle Swarm Optimization for Multi-modal Optimization

J.J. Liang[1], Bo Yang Qu[1,2], Song Tao Ma[1], and Ponnuthurai Nagaratnam Suganthan[2]

[1] School of Electrical Engineering, Zhengzhou University, Zhengzhou, 450001, China
liangjing@zzu.edu.cn, E070088@e.ntu.edu.sg,
mst19870613@126.com
[2] School of Electrical and Electronic Engineering, Nanyang Technological University,
Singapore 639798, Singapore
epnsugan@ntu.edu.sg

Abstract. In recent decades, solving multi-modal optimization problem has attracted many researchers attention in evolutionary computation community. Multi-modal optimization refers to locating not only one optimum but also the entire set of optima in the search space. To locate multiple optima in parallel, many niching techniques are proposed and incorporated into evolutionary algorithms in literature. In this paper, a local search technique is proposed and integrated with the existing Fitness Euclidean-distance Ratio PSO (FER-PSO) to enhance its fine search ability or the ability to identify multiple optima. The algorithm is tested on 8 commonly used benchmark functions and compared with the original FER-PSO as well as a number of multi-modal optimization algorithms in literature. The experimental results suggest that the proposed technique not only increases the probability of finding both global and local optima but also speeds up the searching process to reduce the average number of function evaluations.

Keywords: evolutionary algorithm, multi-modal optimization, particle swarm optimization, niching.

1 Introduction

Optimization forms an important part of our day-to-day life and evolutionary algorithms (EAs) have been proven to be effective in solving optimization problems. The standard EAs have the tendency of converging into a single solution [1], but are not suitable for the problems which desire multiple solutions. However, in real world optimization, many problem instances fall into the multi-modal category where multiple optima/peaks need to be located in a single run, such as classification problems in machine learning and inversion of teleseismic waves [2].

In literature, various niching methods, such as clearing [3], fitness sharing [4], speciation [5], clustering [6], crowding [7]-[8], and restricted tournament selection [9], are proposed in the past few decades to extend EAs to multimodal optimization. They address the loss of diversity and the takeover effects of traditional EAs by maintaining certain properties within the population of feasible solutions [10]. Compared

D.-S. Huang et al. (Eds.): ICIC 2011, LNBI 6840, pp. 378–385, 2012.
© Springer-Verlag Berlin Heidelberg 2012

to the conventional EAs, the diversity of niching EAs is increased and the probability of getting trapped in a local optimum is lower. Niching EAs are able to search and locate multiple global/local peaks in parallel.   However, more function evaluations are generally needed for niching EAs to reach the single global optima. And most niching techniques are mainly based on grouping of similar populations.

Particle swarm optimization (PSO) algorithm is an effective and robust global optimization technique for solving complex optimization problems. PSO has become one of the ideal optimization algorithms to include niching techniques due to some of its attractive characteristics. Fitness Euclidean-distance ratio PSO (FERPSO)[11] is a commonly used and effective niching PSO algorithm. However, the fine-tuning ability of most PSO algorithms including FERPSO is not good.. Motivated by this observation, a local search method is proposed and combined with FERPSO to improve its fine-searching ability.

The remainder of this paper is organized as follows. A brief overview of PSO and FERPSO is given in Section 2. In Section 3, the proposed local search method is described. The problems definition and the experimental results are presented in Section 4 and 5. Finally, the paper is concluded in Section 6.

2   PSO and Niching

2.1   Particle Swarm Optimization

Particle Swarm Optimization (PSO) was first introduced by Kennedy and Eberhart [13] is 1995. PSO emulates the swarm behavior of insects, birds flocking, and fish schooling where these swarms search for food in a collaborative manner [14]. Similar to other evolutionary algorithms (EAs), PSO is a population-based global search technique. However, the original PSO does not use evolution operators such as crossover and mutation. Each solution in PSO is regarded as a particle that search through the space by learning from its own experience and others' experiences. The position X and velocity V of each particle is updated according to the following equations:

$$V_i^d = \omega * V_i^d + c_1 * rand1_i^d * (pbest_i^d - X_i^d) + c_2 * rand2_i^d * (gbest^d - X_i^d) \tag{1}$$

$$X_i^d = X_i^d + V_i^d \tag{2}$$

where c_1 and c_2 are the acceleration constants, $rand1_i^d$ and $rand2_i^d$ are two random numbers within the range of [0, 1]. ω is the inertia weight to balance the global and local search abilities. $pbest_i$ is the best previous position yielding the best fitness value for the ith particle while $gbest_i$ is the best position found by the whole population.

2.2   Fitness Euclidean-Distance Ratio PSO

The idea of Fitness Euclidean-distance ratio PSO is inspired by FDR-PSO [15] (Fitness-Distance-Ratio based PSO) which is a single global optimization algorithm. In FERPSO, instead of global best the neighborhood best is used to lead the particle. All the particles move towards its personal best as well as their "fittest-and-closest" neighbors *(nbest)*, which are indentified by the FER values.

FER value can be calculated using the following equation:

$$FER_{(j,i)} = \alpha \cdot \frac{f(p_j) - f(p_i)}{\|p_j - p_i\|} .$$

(3)

where p_w is the worst-fit particle in the current population. p_j and p_i are the *personal best* of the *jth* and *ith* particle respectively. $\alpha = \dfrac{\|s\|}{f(p_g) - f(p_w)}$ is a scaling factor. $\|s\|$ is the size of the search space, which is estimated by its diagonal distance $\sqrt{\sum_{k=1}^{d}(x_k^u - x_k^l)^2}$ (where x_k^u and x_k^l are the upper and lower bounds of the *kth* dimension of the search space).

The original velocity update in eqn (1) is modified as:

$$V_i^d = \omega * V_i^d + c_1 * rand1_i^d * (pbest_i^d - X_i^d)$$
$$+ c_2 * rand2_i^d * (nbest^d - x_i^d)$$

(4)

The steps of identifying the *nbest* for particle *i* for FER-PSO is shown in Table 1.

Table 1. The pseudocode of calculating *nbest* for particle *i* for FER-PSO

Input: A list of all particles in the population
Output: Neighborhood best *nbest* based on the *ith* particle's
          FER value.
FER ◀ 0, tmp ◀ 0, nbest_i ◀ pbest_i
For i=1 to NP (Population size)
    For j=1 to NP (Population size)
        Calculate the FER using equation (3).
        If j=1
            tmp=FER
        Endif
        If FER>tmp
            tmp=FER;
            nbest_i =pbest_j
        Endif
    Endfor
Endfor

---

3   Memetic FERPSO

In the original FERPSO, the current particles are responsible for searching new points. However, the current position could be far from the personal best position. Due to this reason, the fine-tuning ability of FERPSO is not good. If the needed accuracy is high, it will be difficult for FERPSO to locate the required number of global/local optima. To overcome this problem, a local search method based on *pbest* member mutation is introduced.

In each iteration of FERPSO searching (after generating new solutions from the current position), the local search method makes use of *pbest* to produce other *NP* (population size) new solutions. The steps of generating new solutions by local search method are shown in Table 2 (assuming maximization problem).

Table 2. Steps of local search

Step 1	Update the current *pbest* by the original FERPSO
	For $i=1$ to *NP* (population size)
Step 2	Find $pbest_j$ (the nearest *pbest* member to $pbest_i$).
Step 3	If $pbest_j$ (objective value) $> = pbest_i$
	$nbest_1 = pbest_j$
	$nbest_2 = pbest_i$
	Else
	$nbest_1 = pbest_i$
	$nbest_2 = pbest_j$
	Endif
Step 4	Temp= $pbest_i + c_1 * rand * (nbest_1 - nbest_2)$
Step 5	Reset Temp within the bounds, if it exceeds the
	Bounds and evaluate Temp.
Step 6	If Temp (objective value) $> pbest_i$
	$pbest_i = $ Temp
	Endif
	Endfor

With the local search technique, more solutions are produced around the personal best which enhance the fine-tuning ability of the original FERPSO.

4   Experiment Preparation

4.1   Test Function

In order to test the proposed algorithm, 8 commonly used multimodal benchmark functions with different properties and complexity are used. The details of the test functions are provided in Table 3.

4.2   Experimental Setup

All the algorithms are implemented using Matlab 7.1 and executed on the computer with Intel Pentium® 4 CPU and 2 Gb of RAM memory. The population size and maximum number of function evaluations are set to be the same for all tested algorithms and the details are listed in Table 4. The parameters setting for PSO are:

$$C_1 = 2.05 \qquad C_2 = 2.05 \qquad \omega = 0.729843788$$

Table 3. Test Functions

Name	Test Function	Range ; Peaks Global/local
F1:Five-Uneven-Peak Trap [8]	$f_3(x) = \begin{cases} 80(2.5-x) & \text{for } 0 \le x < 2.5 \\ 64(x-2.5) & \text{for } 2.5 \le x < 5 \\ 64(7.5-x) & \text{for } 5 \le x < 7.5 \\ 28(x-7.5) & \text{for } 7.5 \le x < 12.5 \\ 28(17.5-x) & \text{for } 12.5 \le x < 17.5 \\ 32(x-17.5) & \text{for } 17.5 \le x < 22.5 \\ 32(27.5-x) & \text{for } 22.5 \le x < 27.5 \\ 80(x-27.5) & \text{for } 27.5 \le x \le 30 \end{cases}$	$0 \le x \le 20$; 2/3
F2:Equal Maxima [17]	$f_4(x) = \sin^6(5\pi x)$	$0 \le x \le 1$; 5/0
F3:Decreasing Maxima [17]	$f_5(x) = \exp[-2\log(2) \cdot (\frac{x-0.1}{0.8})^2] \cdot \sin^6(5\pi x)$	$0 \le x \le 1$; 1/4
F4: Uneven Maxima [17]	$f_6(x) = \sin^6(5\pi(x^{3/4} - 0.05))$	$0 \le x \le 1$; 5/0
F5: Himmelblau's function [16]	$f_8(x,y) = 200 - (x^2 + y - 11)^2 - (x + y^2 - 7)^2$	$-6 \le x, y \le 6$; 4/0
F6: Six-Hump Camel Back [18]	$f_9(x,y) = -4[(4 - 2.1x^2 + \frac{x^4}{3})x^2 + xy + (-4 + 4y^2)y^2]$	$-1.9 \le x \le 1.9$; $-1.1 \le y \le 1.1$ 2/2
F7: Shekel's foxholes [3]	$f_{10}(x,y) = 500 - \dfrac{1}{0.002 + \sum\limits_{i=0}^{24} \dfrac{1}{1+i+(x-a(i))^6 + (y-b(i))}}$ where $a(i) = 16(i \bmod 5) - 2)$, and $b(i) = 16(\lfloor (i/5) \rfloor - 2$	$-65.536 \le x, y \le 65.535$; 1/24
F8: 2D Inverted Vincent [19]	$f(\vec{x}) = \frac{1}{n} \sum\limits_{i=1}^{n} \sin(10.\log(x_i))$ where n is the dimesnion of the problem	$0.25 \le x_i \le 10$; 36

Six different multimodal algorithms are compared in the experiments:

1. FERPSO: Fitness-Euclidean distance ratio PSO.
2. SPSO: Speciation-based PSO.
3. MFERPSO: Memetic Fitness-Euclidean distance ratio PSO.
4. CDE: The crowding DE.
5. r2pso: A *lbest* PSO with a ring topology, each member interacts with only its immediate member to its right.
6. r3pso: A *lbest* PSO with a ring topology, each member interacts with its immediate member on its left and right.

Table 4. Population and no. of function evaluations

Function no.	Population size	No. of function evaluations
F1-F6	50	10000
F7	500	100000
F8	500	200000

25 independent runs are conducted for each of the algorithms. To assess the performance of different algorithms, a level of accuracy which is used to measure how close the obtained solutions to the known global peaks are need to be specified. An optimum is considered to be found if there exists a solution in the population within the tolerated distance to that optimum.

When doing the comparison, following to criteria are used:

1. Success Rate
2. Average number of optima found

5   Experiments and Results

5.1   Success Rate

The success rates of compared algorithms are recorded and presented in Table 5. The boldface indicates the best performance. The ranks of each algorithm are shown in bracket and the last row of the table shows the total rank. As can be seen from the results, the MFERPSO performs better than its original form as well as the other algorithms compared. The better performance is generated by the local search method which is able to increase the accuracy of the solutions found by the original FERPSO.

Table 5. The Success Rate

Test Function	Level of accuracy	MFERPSO	FERPSO	SPSO	R2PSO	R3PSO	CDE
F1	0.05	**0.72(1)**	0 (3)	0(3)	0(3)	0(3)	0.44(2)
F2	0.000001	**1(1)**	0.84(5)	0.88(3)	0.92(2)	0.88(3)	0.28(6)
F3	0.000001	0.24(3)	0(4)	**1 (1)**	0(4)	0(4)	0.48(2)
F4	0.000001	**1(1)**	**1(1)**	0.92(3)	0.88(4)	0.72(5)	0.28(6)
F5	0.0005	**0.76(1)**	0.72(2)	0(5)	0.28(3)	0.24(4)	0(5)
F6	0.000001	**1(1)**	0.96(2)	0(5)	0.56(4)	0.6(3)	0(5)
F7	0.00001	0.84(2)	0(5)	**0.92(1)**	0.6(3)	0.52(4)	0(5)
F8	0.001	0(2)	0(2)	0(2)	0(2)	0(2)	**0.08(1)**
Total Rank	-	**12**	24	23	25	28	32

5.2   Average Number of Optima Found

Beside success rate, the average number of optima found is also one of the most important criteria in comparing different niching algorithms. In order to give a clearer view of the advantage of the local search technique, the number of optima found for each test function is listed in Table 6. The mean value is highlighted in boldface. As revealed by the results, MFERPSO perform either better or comparable to its original form.

Table 6. Locating global and local optima

Test Func.		FERPSOLS	FERPSO	SPSO	R2p so	R3p so	CDE
F1	Min	3	0	2	0	0	4
	Max	5	2	4	2	1	5
	Mean	**4.48**	**0.64**	**3.08**	**0.8**	**0.4**	**4.44**
	Std	0.8718	0.7	0.4933	0.7071	0.5	0.5066
F2	Min	5	4	4	4	4	1
	Max	5	5	5	5	5	5
	Mean	**5**	**4.84**	**4.88**	**4.92**	**4.88**	3.84
	Std	0	0.3742	0.3317	0.2769	0.3317	1.0279
F3	Min	1	1	5	1	1	2
	Max	5	1	5	1	1	5
	Mean	**2.16**	**1**	**5**	**1**	**1**	4.28
	Std	1.675	0	0	0	0	0.8426
F4	Min	5	5	4	4	4	2
	Max	5	5	5	5	5	5
	Mean	**5**	**5**	**4.92**	**4.88**	**4.72**	3.96
	Std	0	0	0.2769	0.3317	0.4583	0.8888
F5	Min	2	3	0	1	1	0
	Max	4	4	2	4	4	1
	Mean	**3.68**	**3.68**	**0.84**	**2.92**	**2.76**	0.32
	Std	0.5568	0.476	0.6245	0.8622	0.8794	0.4761
F6	Min	2	1	0	0	0	0
	Max	2	2	1	2	2	1
	Mean	**2**	**1.96**	**0.08**	**1.44**	**1.56**	0.04
	Std	0	0.2	0.2769	0.7118	0.5831	0.2
F7	Min	20	3	24	23	22	0
	Max	25	7	25	25	25	1
	Mean	**24.44**	**5.16**	**24.92**	**24.44**	**24.32**	24.32
	Std	1.3868	1.28	0.2769	0.7681	0.8524	0.8524
F8	Min	22	18	20	16	17	17
	Max	32	28	30	28	26	26
	Mean	**27**	**23.6**	**25.68**	**21.76**	**22.2**	22.2
	Std	2.48	2.63	2.719	2.9478	2.2546	2.2546

6   Conclusion

In this paper, a memetic FERPSO is introduced to solve multimodal problems. The lack of fine searching ability of the original FERPSO leads to missing of local or global optima. With the proposed local search method, the algorithm possesses enhanced local search ability. The experimental results show, the proposed niching algorithm can outperform its original version as well as a number of other niching algorithms on the benchmark functions.

Acknowledgment. This work is partially supported by National Natural Science Foundation of China (Grant NO. 60905039).

References

1. Mahfoud, S.W.: Niching Methods for Genetic Algorithms, Ph.D. dissertation, Urbana, IL, USA (1995), http://citeseer.ist.psu.edu/mahfoud95niching.html
2. Koper, K., Wysession, M.: Multimodal Function Optimization with a Niching Genetic Algorithm: A Seis-mological Example. Bulletin of Seismological Society of America 89, 978–988 (1999)
3. De Jong, K.A.: An Analysis of the Behavior of a Class of Genetic Adaptive Systems, Ph.D. dissertation, University of Michigan (1975)
4. Mahfoud, S.W.: Crowding and Preselection Revisited. In: Manner, R., Manderick, B. (eds.) Parallel Problem Solving from Nature, vol. 2, pp. 27–36 (1992)
5. Harik, G.R.: Finding Multimodal Solutions Using Restricted Tournament Selection. In: Proc. of the Sixth International Conference on Genetic Algorithms. Morgan Kaufmann, San Francisco
6. Pétrowski, A.: A Clearing Procedure as a Niching Method for Genetic Algorithms. In: Proc. of the IEEE Int. Conf. on Evol. Comp., New York, USA, pp. 798–803 (1996)
7. Goldberg, D.E., et al.: Genetic Algorithms with Sharing for Multimodal Function Optimization. In: Proc. of the Second Int. Conf. on Genetic Algorithms, pp. 41–49 (1987)
8. Li, J.P., et al.: A Species Conserving Genetic Algorithm for Multimodal Function Optimization. Evol. Comput. 10(3), 207–234 (2002)
9. Zaharie, D.: Extensions of Differential Evolution Algorithms for Multimodal Optimization. In: Proc. of 6th Int. Symposium of Symbolic and Numeric Algorithms for Scientific Computing, pp. 523–534 (2004)
10. Cavicchio, D.J.: Adaptive Search Using Simulated Evolution. Ph.D. dissertation, University of Michigan, Ann Arbor (1970)
11. Li, X.D.: A Multimodal Particle Swarm Optimizer Based on Fitness Euclidean-distance Ration. In: Proc. of Genetic and Evolutionary Computation Conference, pp. 78–85 (2007)
12. Qu, B.Y., Suganthan, P.N.: Novel Multimodal Problems and Differential Evolution with Ensemble of Restricted Tournament Selection. In: IEEE Congress on Evolutionary Computation, Barcelona, Spain, pp. 3480–3486 (2010)
13. Thomsen, R.: Multi-modal Optimization Using Crowding-based Differential Evolution. In: Proc. of the 2004 Cong. on Evolutionary Computation, vol. 2, pp. 1382–1389 (2004)
14. Storn, R., Price, K.V.: Differential Evolution-A simple and Efficient Heuristic for Global Optimization over Continuous spaces. J. of Global Optimization 11, 341–359 (1995)
15. Price, K.: An Introduction to Differential Evolution. New Ideas in Optimization, 79–108 (1999)
16. Ackley, D.: An Empirical Study of Bit Vector Function Optimization. In: Genetic Algorithms Simulated Annealing, pp. 170–204. Pitman, London (1987)
17. Deb, K.: Genetic Algorithms in Multimodal Function Optimization, the Clearinghouse for Genetic Algorithms, M.S Thsis and Rep. 89002, Univ. Alabama, Tuscaloosa (1989)
18. Michalewicz, Z.: Genetic Algorithms + Data Structures = Evolution Programs. Springer, Heidelberg (1996)
19. Shir, O.M., Bäck, T.: Niche Radius Adaptation in the CMA-ES Niching Algorithm. In: Runarsson, T.P., Beyer, H.-G., Burke, E.K., Merelo-Guervós, J.J., Whitley, L.D., Yao, X. (eds.) PPSN 2006. LNCS, vol. 4193, pp. 142–151. Springer, Heidelberg (2006)

Clock Drift Management Using Particle Swarm Optimization

Prakash Tekchandani and Aditya Trivedi

ABV-Indian Institute of Information Technology and Management, Gwalior,
474010, Madhya Pradesh, India
prakashtekchandani@gmail.com

Abstract. Time Synchronization is a common requirement for most
network applications. It is particularly essential in Wireless sensor networks (WSN) to allow collective signal processing, proper correlation of
diverse measurements taken from a set of distributed sensor elements
and for an efficient sharing of the communication channel. The Flooding
Time Synchronization Protocol (FTSP) was developed explicitly for time
synchronization of wireless sensor networks. In this paper, FTSP is optimized for clock drift management using Particle Swarm Optimization
(PSO). The paper estimates the clock offset, clock skew and generates
linear line and optimizes the value of average time synchronization error
using PSO. This paper presents implementation and experimental results that produce reduced average time synchronization error optimized
using PSO compared to that of linear regression used in FTSP.

Keywords: Wireless sensor networks, particle swarm optimization, time
synchronization, average time synchronization error, clock drift.

1   Introduction

Rapid advances in information technologies such as very large scale integration
(VLSI), microelectromechanical systems (MEMS) have paved the way for the
proliferation of wireless sensor networks (WSN).The miniaturization of computing and sensing technologies enables the development of tiny, low-power,
and inexpensive sensors, actuators, and controller suitable for many applications. For example, they can be used for structural health monitoring[5], traffic
control, health care[10], precision agriculture [9], underground mining[6]. These
networks require that time be synchronized precisely for consistent distributed
sensing and control. Further more time synchronization is also required for coordination, communication, security, power management, proper correlation of
sensor information from the various sources in WSN. Many synchronization
algorithms[12][1][2][7] have been proposed for WSN[11].

   In this paper PSO is used with FTSP [7] for clock drift management. It
presents detailed analysis of clock offset and clock skew estimation. The proposed
algorithm utilizes the concepts of PSO, it generates linear line and calculates the
value of average time synchronization error. The error is optimized for reduction

D.-S. Huang et al. (Eds.): ICIC 2011, LNBI 6840, pp. 386–393, 2012.

in its value using PSO. While the idea of linear regression has been utilized before, the unique combination with PSO yields significant reduction in average time synchronization error.

We start with a brief discussion covering concepts of particle swarm optimization. Next, methodology for clock drift management using PSO is proposed and discussed. Thirdly, results are discussed for proposed algorithm. Finally comparative study of results generated using PSO and existing approach is presented followed by the conclusion and future work.

2   Particle Swarm Optimization

The particle swarm optimization algorithm, originally introduced in terms of social and cognitive behavior by Kennedy and Eberhart in 1995 [4], solves problems in many fields, especially engineering and computer science. The individuals, called particles henceforth, are flown through the multi-dimensional search space with each particle representing a possible solution to the multi-dimensional optimization problem. Each solution's fitness is based on a performance function related to the optimization problem being solved. After each iteration, the p_{best} and g_{best} are updated for each particle if a better or more dominating solution (in terms of fitness) is found. This process continues, iteratively, until either the desired result is converged upon, or it is determined that an acceptable solution can not be found within computational limits. For an n-dimensional search space, the i_{th} particle of the swarm is represented by an n-dimensional vector $X_i = (x_{i1}, x_{i2}......x_{in})^T$. The velocity of this particle is represented by another n-dimensional vector $V_i = (v_{i1}, v_{i2}......v_{in})^T$. The previously best visited position of the ith particle is denoted as $P_i = (p_{i1}, p_{i2}, ...p_{in})^T$. 'g' is the index of the best particle in the swarm. The velocity of the ith particle is updated using the velocity update equation given by:

$$v_{id} = v_{id} + c_1 r_1 (p_{id} - x_{id}) + c_2 r_2 (p_{gd} - x_{id}) \qquad (1)$$

and the position is updated using

$$x_{id} = x_{id} + v_{id} \qquad (2)$$

where d = 1, 2,...n represents the dimension and i = 1, 2,..S represents the particle index. S is the size of the swarm and c_1 and c_2 are constants, called cognitive and social scaling parameters respectively (usually, $c_1 = c_2$; r_1, r_2 are random numbers drawn from a uniform distribution). The main applications of PSO are training neural networks [3], game learning applications[8], architecture selection[14].

3   Proposed Approach

In this paper clock drift is managed using PSO with FTSP. The time synchronization cycle is initiated by the single, dynamically elected root node. Here we are trying to synchronize the whole network with the time of a single node i.e. the root. The root floods beacons containing global time at fixed intervals of time (say T) to other nodes (local nodes having clock drift) for estimation of root's clock global time. The periodicity of time synchronization beacons is dependent on the amount of time synchronization error acceptable by the application.

For time synchronization the cycle can be broken in to following events:

- The root node starts the process by transmitting the beacons marked with its time stamp (global time). This stamping is done just prior to sending the packet on air (i.e. after its MAC gets access to send packets).
- When any collector node receives a beacon it puts its own local time in the packet just when the radio receives the packet. This pair of global-local time pair acts as a synchronization point for the algorithm and is used to calculate the offset using

$$Offset = Globaltime - Localtime. \tag{3}$$

The time synchronization module creates a table of global-local time pair of each local node for every fixed interval T. Thus we get offset of each node at every fixed interval T. These global-local pair points of a node at each time interval T are used to estimate the drift of that particular local node with respect to root node.

3.1   Clock Drift Management Using Linear Regression

Linear regression line [1] is drawn using synchronization points table. Firstly clock skew(m) and initial offset(c) between a local node and global node is calculated using linear regression[13]. Clock skew is the slope of linear regression line between offset and global time. Initial offset is the intercept of the linear regression line. Using linear regression line local node can estimate the global time, with time synchronization error.

$$m = \frac{N \times \sum globaltime \times offset - (\sum globaltime) \times (\sum offset)}{N \times \sum globaltime^2 - (\sum globaltime)^2} \tag{4}$$

$$c = \overline{offset} - m \times \overline{globaltime} \tag{5}$$

where

$$\overline{offset} = \frac{\sum offset}{N}$$

$$\overline{globaltime} = \frac{\sum globaltime}{N}$$

N=number of times beacons are send.
Hence equation of linear line generated is:

$$y = mx + c \tag{6}$$

3.2 Optimization of Clock Drift Using PSO

In this paper we have used particle swarm optimization algorithm to enhance the FTSP method. In traditional FTSP linear regression method is used to calculate the clock offset or synchronize the nodes. We have used the feature of PSO with FTSP and calculated results show that FTSP with PSO performs better than traditional FTSP which uses linear regression. FTSP uses linear regression to calculate the linear line through which synchronization between the nodes are done. We have proposed a new method using PSO to calculate the linear line for synchronization in place of linear regression. In traditional approach global and local time of nodes are calculated using FTSP and for synchronization of all nodes linear regression method is used. In our proposed approach calculation of local time and global time for nodes are done by FTSP and for synchronization PSO is used instead of linear regression. Through PSO, average time synchronization error is optimized for reduced value. We have assumed short term stability of the clock so the offset varies in linear fashion. The algorithm given in *section* 3.4 is tested for different pairs (root-local) of nodes. The root node is selected randomly. The root node sends beacons at an interval of fixed time (T) to local nodes. The number of times beacon are send is N. PSO runs for each root-local node pair after time synchronization cycle described in *section*3 is completed. To solve the problem using PSO, representation of the individual and fitness value is required. PSO algorithm is based on population (candidate solution) and each population have its own fitness value according to which it is compared from others, so we have to first represent the each individual of PSO according to the problem . In clock drift management using PSO, individual is represented by a linear line because we use PSO in place of linear regression and PSO work same as linear regression by calculating the linear line for synchronization. Lines have two value slopes and intercept so each individual of PSO represents the value of slope and intercept. Dimension of each individual is two and values of dimensions are slope and intercept. The search space consists of individuals (linear lines), parameters to consider are slope and intercept. The particles called individuals (linear lines in our case) henceforth, are flown through the 2-dimensional search space (slope, intercept) with each linear line representing a possible solution. Each solution's fitness is based on average time synchronization error performance function calculating the average time synchronization error.

3.3 Fitness Function

After representation of each individual fitness value of each individual is calculated. On the basis of fitness value optimal solution is determined. Time synchronization error is the absolute difference between actual global time and estimated global time calculated using linear line. To measure the performance of algorithm absolute average time synchronization error is calculated.

$$Fitness = \frac{\sum_{i=0}^{N} mod[(offset_n - y_n)]}{N} \qquad (7)$$

where
$$n = 1, 2, \ldots\ldots N$$

3.4 Algorithm

Algorithm 1 describes the procedure followed by individuals to evaluate minimum average time synchronization error using PSO. First the global and local time for each node is calculated using the basic FTSP. Proposed techniques use the PSO after the processing of basic FTSP. **Algorithm 1** shows how PSO algorithm is used to calculate minimum average time synchronization error.

Algorithm 1. To calculate minimum time synchronization error using PSO

1: **for** $i = 0$ to the number of individual or population **do**
2:     **for** $j = 0$ to the dimension of individual **do**
3:         Randomly initialize the individuals within the search space
4:     **end for** j
5:     Evaluate average time synchronization error for each individual
6:     Compute the global best slope, intercept and local best slope, intercept
7: **end for** i
8: **repeat**
9:     **for** $i = 0$ to the number of individual or population **do**
10:        **for** $j = 0$ to the dimension of individual **do**
11:            Update velocity according to equation (1)
12:            Update position according to equation (2)
13:        **end for** j
14:        Calculate average time synchronization error of updated position
15:        If required, update previous information for global best slope, intercept and local best slope, intercept
16:    **end for** i
17: **until** (maximum number of iteration)

The size of the swarm S is 200, dimension of individual is 2, control parameter c1=c2=c is 1.7. Maximum iteration is 50, maximum evaluation for each iteration is 50000. First initialization process is completed by randomly initializing each individual's velocity and position from the search space. Fitness value of each individual is calculated according the average time synchronization error performance function according to equation (7). According to fitness value of each individual global best slope, intercept and local best slope, intercept is also initialized. After completion of initialization process for each iteration velocity and position of each individual is computed using equation (1), (2) respectively i.e. next slope and intercept of each line is calculated. Now each individual fitness value for next position is calculated according the average time synchronization error performance function. Based on fitness value local best slope, intercept and global best slope, intercept is updated if they are giving less average time synchronization error. This updation is done until the maximum number of iterations is completed. After all the iterations global best slope, intercept is obtained for which average time synchronization error is minimum.

---

4   Results

Algorithm 1 is implemented on Visual C++. It is tested for different pairs of (root-local) nodes. The root node is selected randomly. We have assumed node number 2 is root node, local node 3 is estimating global time. The root node sends beacons at an interval of 100 units (T). The number of times (N) the beacon message is sent is taken to 5. Suppose 5 nodes are in network and 2nd node is selected as root node then result pair comprises of 2-1, 2-3, 2-4, 2-5. Results are shown in next section for one pair i.e. 2-3 as one pair only is sufficient to proof that using PSO average time synchronization error is optimized for reduced value. The root node transmits the beacons marked with global timestamp at an interval of 100 units and local node having clock drift receives it and puts its own local time stamp shown in Table1, also offset calculated by equation (3) at local node for each send of beacon is also shown.

4.1   Result of Clock Drift Management Using Linear Regression and PSO

Figure 1(a) shows a Linear Regression line is generated using the synchronization points of Table 1. The slope calculated using equation (4) is 0.045 and intercept calculated using equation (5) is 161.365. The average time synchronization error calculated using fitness function equation (7) is 0.92.

Figure 1(b) shows the linear line generated using PSO. PSO is run for each pair after time synchronization cycle get completed described in section 3. The parameters for the algorithms are standard. Some of the parameters which have been varied are the population size is taken as 200, number of iterations is 50, number of times (N) the beacon message is sent by root node is 5. The slope is 0.040 and intercept is 163.294. The average time synchronization error using PSO is 0.82. The parameters of linear line (i.e., slope, intercept) for different number of nodes and average time synchronization error using Linear Regression and PSO is shown in Table 2.

4.2   Comparative Result of Particle Swarm Optimization with Linear Regression

Table 2 last coloumn depicts the (%) reduction in average time synchronization error using PSO compared to Linear Regression. Using PSO average time synchronization error is reduced in the range between 7.89% to 13.46% compared to Linear Regression.

Table 1. Time stamping for global and local node at intervals of 100 units of time

Number of Nodes	Global node Timestamp	Local node Timestamp	Offset at Local Node
1	183	15	168
2	283	107	176
3	383	204	179
4	483	300	183
5	583	396	187

Table 2. Parameters of linear line calculated using Linear Regression(LR) and PSO

Number of Nodes	m(LR)	c(LR)	Error(LR)	m(PSO)	c(PSO)	Error(PSO)	Reduction(%)
5	0.045	161.365	0.92	0.040	163.294	0.82	10.10
10	0.045	161.965	1.96	0.048	160.893	1.80	8.16
20	0.062	154.854	1.52	0.055	156.849	1.40	7.89
50	0.080	154.760	2.08	0.091	151.182	1.80	13.46

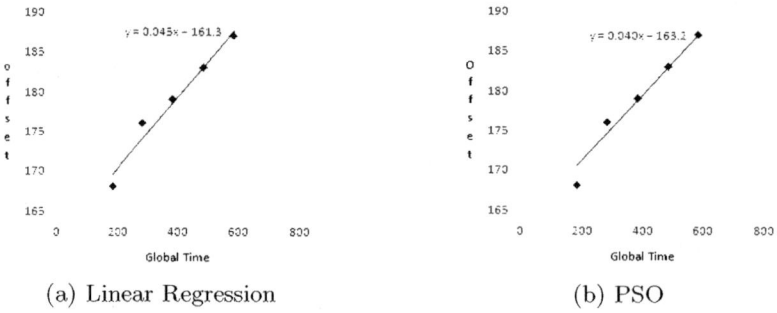

(a) Linear Regression   (b) PSO

Fig. 1. Linear line generated using Linear Regression and PSO

5   Conclusion

Time Synchronization is critical in Wireless Sensor Networks (WSN) to allow collective signal processing, proper correlation of diverse measurements taken from a set of distributed sensor elements and for an efficient sharing of the communication channel. Flooding Time Synchronization Protocol was developed explicitly for time synchronization of wireless sensor networks. This work estimates the drift of the receiver clock with respect to the sender clock, optimizes the value of average time synchronization error using Particle Swarm Optimization (PSO). The algorithm is implemented on Visual C++ platform. The average time synchronization error calculated by PSO is compared with the value of average time synchronization error calculated by Linear Regression used in Flooding Time Synchronization Protocol. Using PSO average time synchronization error is reduced in the range between 7.89% to 13.46% compared to Linear Regression.

References

1. Elson, J., Girod, L., Estrin, D.: Fine-grained Network Time Synchronization Using Reference Broadcasts. ACM SIGOPS Operating Systems Review 36(SI), 147–163 (2002)
2. Ganeriwal, S., Kumar, R., Srivastava, M.: Timing-sync Protocol for Sensor Networks. In: Proceedings of the 1st International Conference on Embedded Networked Sensor Systems, pp. 138–149. ACM, New York (2003)

3. Kennedy, J.: The Particle Swarm: Social Adaptation of Knowledge. In: Proceedings of International Conference on Evolutionary Computation, pp. 303–308. IEEE, Los Alamitos (1997)
4. Kennedy, J., Eberhart, R.: Particle Swarm Optimization. In: Proceedings of International Conference on Neural Networks, vol. 4, pp. 1942–1948. IEEE, Los Alamitos (1995)
5. Kim, S., Pakzad, S., Culler, D., Demmel, J., Fenves, G., Glaser, S., Turon, M.: Health Monitoring of Civil Infrastructures Using Wireless Sensor Networks. In: Proceedings of the 6th International Conference on Information Processing in Sensor Networks, pp. 254–263. ACM, New York (2007)
6. Li, M., Liu, Y.: Underground Coal Mine Monitoring with Wireless Sensor Networks. ACM Transactions on Sensor Networks (TOSN) 5(2), 1–29 (2009)
7. Maróti, M., Kusy, B., Simon, G., Lédeczi, Á.: The Flooding Time Synchronization Protocol. In: Proceedings of the 2nd International Conference on Embedded Networked Sensor Systems, pp. 39–49. ACM, New York (2004)
8. Messerschmidt, L., Engelbrecht, A.: Learning to Play Games Using a PSO-based Competitive Learning Approach. IEEE Transactions on Evolutionary Computation 8(3), 280–288 (2004)
9. Panchard, J., Rao, S., TV, P., Hubaux, J., Jamadagni, H.: Commonsense net: A Wireless Sensor Network for Resource-poor Agriculture in the Semiarid Areas of Developing Countries. Information Technologies and International Development 4(1), 51–67 (2007)
10. Schwiebert, L., Gupta, S., Weinmann, J.: Research Challenges in Wireless Networks of Biomedical Sensors. In: Proceedings of the 7th Annual International Conference on Mobile Computing and Networking, pp. 151–165. ACM, New York (2001)
11. Sohraby, K., Minoli, D., Znati, T.: Wireless Sensor Networks: Technology, Protocols, and Applications. Wiley-Blackwell (2007)
12. Stojmenoviâc, I.: Handbook of Sensor Networks: Algorithms and Architectures. Wiley, Chichester (2005)
13. Woodbury, G.: An Introduction to Statistics. Duxbury Pr., Boston (2001)
14. Zhang, C., Shao, H.: An ANN's Evolved by a New Evolutionary System and Its Application. In: Proceedings of the 39th IEEE Conference on Decision and Control, vol. 4, pp. 3562–3563. IEEE, Los Alamitos (2000)

Stem Cells Optimization Algorithm

Mohammad Taherdangkoo[1,*], Mehran Yazdi[1], and Mohammad Hadi Bagheri[2]

[1] Department of Communications and Electronics,
Faculty of Electrical and Computer Engineering,
Shiraz University, Shiraz, Iran
[2] Radiology Medical Imaging Research Center, Shiraz University
of Medical Sciences, Shiraz, Iran
mtaherdangkoo@yahoo.com, yazdi@shirazu.ac.ir
bagherih@sums.ac.ir

Abstract. Optimization algorithms have been proved to be good solutions for many practical applications. They were mainly inspired by natural evolutions. However, they are still faced to some problems such as trapping in local minimums, having low speed of convergence, and also having high order of complexity for implementation. In this paper, we introduce a new optimization algorithm, we called it Stem Cells Algorithm (SCA), which is based on behavior of stem cells in reproducing themselves. SCA has high speed of convergence, low level of complexity with easy implementation process. It also avoid the local minimums in an intelligent manner. The comparative results on a series of benchmark functions using the proposed algorithm related to other well-known optimization algorithms such as genetic algorithm (GA), particle swarm optimization (PSO) algorithm, ant colony optimization (ACO) algorithm and artificial bee colony (ABC) algorithm demonstrate the superior performance of the new optimization algorithm.

Keywords: Optimization Algorithm, Genetic Algorithm (GA), Particle Swarm Optimization (PSO), Ant Colony optimization (ACO), Artificial Bee Colony (ABC) Algorithm, Stem Cells Algorithm (SCA).

1 Introduction

Optimization Algorithms have a special place in solving very complex mathematical problems. The introduction of optimization algorithms into the engineering world has also greatly contributed to the researchers. Most optimization algorithms are considered as general solution to many problems and perhaps the main reason for this is in simple formulating and easy understanding the evolution of these algorithms.

Optimization algorithms were mainly inspired by natural evolution. The most famous one is genetic algorithm, firstly proposed by Holland [1]. This algorithm was inspired by the natural behavior of human reproduction and goes to reach to an expert series. This algorithm does not need neither a memory which typically used in other

---

* Corresponding author.

D.-S. Huang et al. (Eds.): ICIC 2011, LNBI 6840, pp. 394–403, 2012.
© Springer-Verlag Berlin Heidelberg 2012

optimization algorithms, nor the existence of performances such as mutation, crossover etc. It caused that the algorithm is capable to be applied on a large number of issues. It should be noted that the performance of this algorithm when applying to the binary issues is excellent and is more preferred than other algorithms, and this is actually the reason why until now the generic based algorithms have found many applications in optimizing functions and identifying systems. In recent years, many researchers have enhanced this algorithm to solve problems raised by it such that its performance used today is much better than the first proposed idea.

Another optimization algorithm which has been used by many researchers is Particle swarm optimization (PSO) algorithm. This algorithm was proposed for the first time by Eberhart & Kennedy [2] in 1995.This algorithm that belongs to population based algorithms was inspired by the behavior of animals' social life rather than evolutionary mechanisms. The behavior of particles forming the group is based on a series of constraints such as the speed coordination with near neighborhoods and also acceleration according to the distance. The implementation of particles swarm optimization algorithm is done with a criterion function and each particle in the algorithm represents a unique answer for the investigated problem. The advantages of this algorithm with respect to genetic algorithm are the limited computation, a simple implementation, having memory for particles and easy encoding. But this algorithm comprises the problem of premature convergence. This problem is due to the fact that the information exchange very quickly among the particles and thus the particles form groups very rapidly, and consequently the diversity of particles drops quickly in search space and make it difficult to escape from local minimums. However, the problem of premature convergence can be solved in certain applications [3].

The third optimization algorithm that has been considered by many researchers is ant colony optimization (ACO). This algorithm was proposed by M. Dorigo *et al.* [4] to solve the salesman problem. This algorithm was inspired by studies and observations on the behavior of ants in the colony and its implementation is based on the behavior of ants when finding food for survival of their colony. Considering that ants scatter a chemical material called pheromone as they move, this is the pheromone that evaporates and impresses their move and eventually ants reach the food source (optimal response). This algorithm has a good performance in solving the issues related to routing [5]. Disadvantages of this algorithm can be mentioned as poor criterion for its iteration ending, because the process will continue until reaching the maximum number of ants to search or lake of improvement.

With this problem, applying the algorithm to many issues is faced with difficulty and in addition it is very dependent on the move transfer function and pheromone evaporation rate. Although, this problem has been solved for many studies, it has been the reason of not introducing this algorithm widely into the medical engineering world especially in medical image segmentation where generally a unique measure function that is consistent with the terms of this algorithm cannot be designed.

The next optimization algorithm that is considered as one of the newest optimization algorithms is artificial colony of honey bees' algorithm that was proposed for the first time by B. Basturk & D. karaboga [6]. This algorithm was inspired by the social behavior of bees to find food. The algorithm is somehow faced with the problem of premature convergence less than other algorithms, however its implementation is harder than other optimization algorithms. This algorithm has been focused less than other algorithms. In this algorithm, three types of bees (employer bees, onlooker bees and scout bees) are used, but it is the employer bees that are capable of becoming scouts. Employer bees are responsible for searching a considered space and they come back to the hive and share the resulted information with onlooker bees after storing the location of food sources and its nectar content in a memory. Onlooker bees select an employer bee that contains the highest amount of nectar according to classification of food sources and then this employer bee becomes a scout bee and eventually begins to recruit other bees and then moves with recruited bees toward food source in order to drain the nectar. This process will continue until the entire space is covered and the optimal response is achieved. Having low level of computation and fairly good flexibility are considered as the advantages of this algorithm. However, the complexity of its implementation is still a drawback of this algorithm.

In this paper, we propose a new optimization algorithm based on population, we called it stem cell algorithm (SCA), which does not have the existing problems in the previous optimization algorithms. The very easy implementation of this algorithm leads to use it very easily in many applications.

This algorithm was inspired by the behavior of stem cell function in the body. Having high speed, low level of computation and simple implementation are considered as the main advantages of the proposed algorithm.

2   Stem Cells

Stem cells are found in all multi cell organs of the body. These cells are identified according to their ability to self-renewal through indirect cell division and with respect to other cell types.

Research on stem cells was proposed for the first time by James E. Till and Ernest A. McCulloch [7,8] at the University of Toronto in 1960. The very important researched cells were embryonic tissues, which are considered as the blastocysts inner cell group and adult stem cells that are found in adults. It should be noted that in a developing embryo, stem cells are found in all fetal tissues but in an adult body, stem cells act as a repairing system for the body and equip some cells until they do not have any problem in their survival. Stem cells may also result in changes in organs such as blood, skin, etc.

In addition, stem cells have the ability to grow and can be transferred into a particular cell.

Two features are considered as the main characteristics in the definition of stem cells that are as follows:

1. Self-renewal: They have the ability of producing from the various cycles of cell division with maintaining the characteristics of that cell.
2. Power: They have the capacity of resolution from different types of specific cells, but it is possible that a cell has also the ability to be separated into several cells.

The ability of self-renewal can be expressed in two ways: In the first case proliferation is done asymmetrically so that a stem cell is divided into two cells; a daughter cell which is similar to original stem cell and another cell with a fundamental difference with original stem cell. Another form of renewal will be symmetrical. This renewal is such that the original stem cell is divided into a daughter cell that is similar to the original stem cell and other cell is converted to mitosis and generates two cells similar to the original stem cell.

Power, another feature of stem cells, is related to the capacity to detect and separate stem cells so that they have the capacity of separation to different types of cells. Stem cells have the ability of producing an organ from a component of that organ. These cells can also be converted to a large number of cells with similar properties of the original cells.

3   Proposed Stem Cells Optimization Algorithm

This algorithm is a population-based evolutionary algorithm. Of course stem cells algorithms are not a completely new idea in comparison with other optimization algorithms, but we have tried to improve the speed of convergence and escape local minimums to a very acceptable level with a nearly new implementation method. This algorithm has the ability of high flexibility and has an acceptable speed compared with other optimization algorithms in a very large extent.

It should be stated that similar and dissimilar proliferation of selecting the best stem cell in each iteration and use it in an iteration way are the main core of this algorithm. The ability of stem cells to become desired aim forms also another important part of this algorithm. Unlike other algorithms in which the initial population is fixed, the population of this algorithm is defined in a range such that as the algorithm progresses the population gradually increases at each iteration until it reaches to the best answer. Considering the fact that each member of the population is an optimized response for the problem to confound with, the defined population and its increase in each iteration depends on the space of the given problem, in such a way that the space of the problem is divided into several regions (Each region specifies the variables of investigated problem) and a stem cell is sent for each region to search it. Considering that the aim of any optimization algorithm is obtaining an optimal answer relative to the variables of the problem, a matrix of variables must first be created. The initial matrix of variables in Stem Cells Algorithm (SCA) consists of the features of stem cell that transfer to organ or tissue of adult person.

Stem feature matrix is defined as follows.

$$Stem\ Cells = [SC_1, SC_2, ..., SC_N]. \tag{1}$$

Where N is the number of features related to the number of discussed problem variables. So the number of stem cell features is proportional to the existing variables in the problem. The cost of each of the existing stem cells is then calculated by a measure function. Two memories are considered for each stem cell: local memory and global memory. The cost of each cell is calculated by the criterion function. The cost value of each stem cell is stored in the local memory. Now the cell having the lowest cost is considered as the best cell in the first iteration. However, this choice is valid only in one region of cells. In each region, the best cell is then selected. The location and cost of the best cells in each region are stored in the global memory and then data from each region will be shared with the best cells in other areas and will be selected between them and used in subsequent iteration. The best cell in previous iteration renewals its own similar and opposite based on Eq. (2). It should be noted that, the number of self-renewals from the best cell depends on the type of the problem and each selected cell in previous step can produce only its own similar or just produces its opposite point or both of them.

$$SC_{Optimum}(t+1) = \zeta SC_{Optimum}(t) \tag{2}$$

Where t denotes the iteration number, $SC_{Optimum}$ is the best stem cell in each iteration and ζ is a random number where $\zeta \in [0,1]$, however, it can be considered accidental such as $\zeta = 0.96$. If selected cell is to be exactly renewal of the same cell as itself, we consider $\zeta = 1$ and if it has its opposite point, the value $\zeta = 0.2$ can be replaced. The self-renewal process is defined as:

$$x_{ij}(t+1) = \mu_{ij} + \varphi(\mu_{ij}(t) - \mu_{kj}(t)) \tag{3}$$

Where x_i denotes the ith stem cell position, and t is the iteration number, μ_k is the random selected stem cell, j is the solution dimension and if $\mu_{ij}(t) - \mu_{kj}(t) = \tau$ then $\varphi(\tau)$ produces a random variable in the interval of $[-1, 1]$.

It should be noted that choosing the best cell and using it in the next iteration and its self-renewal are just a part of that iteration population that should be determined at the beginning of algorithm process. Another part of the population is considered completely random. This action is likely to have an advantage that minimizes the probability of trapping in a local minimum, because the orientation of algorithm into obtained point of convergence is always tested. This process will continue until its objective to find the best stem cell is achieved and this problem is equivalent to find the best parameters in the case of the lowest value of cost function. Finally, the best stem cell is selected when it has the most power relative to the other cells. The comparative power relation can be obtained by:

$$\varsigma = \frac{SC_{Optimum}}{\sum_{i=1}^{N} SC_i} \tag{4}$$

Where ς denotes the comparative power of stem cells, $SC_{Optimum}$ is stem cells selected in terms of cost, and N is the final number of population regarding to which the best answer is obtained and the algorithm stops. Finally, the algorithm continues until the convergence criterion will be true or the specified number of iterations will be reached.

Pseudo code of the stem cells optimization algorithm is defined as follows:

1 . *Initialize   parameters*

{ M = *Maximum   number   of   stem   cells* ,

P = *number   of   population*   ,{ $10 < P \le M$ }

$C_{Optimum}$ = *best   of   stem   cell   in   each   iteration*   ,

$SC_{Optimum}(t + 1) = \zeta * SC_{Optimum}(t)$, *t represent   of   iteration* .

χ = *penalty   parameter   for   prevent   grow   stem   cell* ,

sc^{i} = i^{th} *stem   cell   in   the   population*   ,

$x_{ij}(t + 1) = \mu_{ij} + \varphi(\mu_{ij}(t) - \mu_{kj}(t))$

x_{i} = i^{th} *stem   cell   position*   , $\varphi(\tau)$ = { *random   var iable* }$\in [-1,1]$

$\varsigma = \dfrac{SC_{Optimum}}{\sum_{i=1}^{N} SC_{i}}$:*Comparativ   e Power*

T_{j} = *is   the   set   of   task   assigned   to   agent   j*

Let $T_{j} = \phi \forall j = 1,..., m$

Construct   a list   of   agents   for   each   task

while (*not   all   task   have   been   assigned*)

repeat

Fitness   Functions   for   Minimizati   on { $f_{1}(\vec{x}), f_{2}(\vec{x}), f_{3}(\vec{x}), D = 10, 20, 30, 50$ }

2 . *Uniform   distributi   on of   stem   cells   in   the   search   space*

3 . *Construct   initial   stem   cell   a* lg *orithm   by u* sin *g randomized   search   heuristic*

4 . *Maximum   iteration* = 0

6 . $P = 10$,

repeat

6 . *Evaluate   fitness   value   for   each   stem   cell   in   the   population*

7 . *Find   best   stem   cell   according   to fitness   value*

(*for   example*   , *Minimizati   on   problems*)

8 . *In   each   iteration*   : *Find   best   stem   cell* , *replace   with   respective   stem   cell*

and   campare   with   stem   cells   in   previous   iteration .

{ *If   the   best   fitness   value   of   stem   cell   is better   than   the   stem   cells*

in   previous   iteration   , *the   best   stem   cell   do   Self* − *renewal   by*

$SC_{Optimum}(t + 1) = \zeta * SC_{Optimum}(t)$

and   , *So compute   Comparativ   e Power   for   stem   cells   and   choose   best   stem*

cell   according   to Comparativ   e Power   and   fitness   value ,

9 . *Best   Solution*

Until ($i = M$)

10 . *Cycle* = *Cycle* + 1

Until (*Cycle* = *Max* _ *Iteration*)

4   Experimental Results

In order to prove our claim related to the better performance of our proposed algorithm than other introduced optimization algorithms, we have compared the efficiency and the accuracy of the SCA using the Benchmark functions. For all Benchmark functions, 30 runs were applied with different random seeds for generating the random variables, each run contains 5000 iterations, and the population size was set to 100 for all optimization algorithms mentioned before except the SCA. We have extracted these functions from well-known papers [9] that are of more attention and many introduced algorithms are faced to specific problems to solve. Note that when the space size goes up (for example *Dimension*> 30) the performance of optimization algorithms drops dramatically because the algorithm will face with several optimal answers. In this case of distribution, selected answer in considered space also adds complexity to the subject randomly. Initialization range for the first function is $[-600, 600]$, initialization range for the second function is $[-15, 15]$ and initialization range for the third function is $[-32.76, 32.76]$, initialization range for fourth function is $[-0.52, 0.52]$. The global minimum is set to zero for all functions.

The four Benchmark functions are listed in the equations (5) through (8).

$$f_1(\vec{x}) = \frac{1}{4000}\left(\sum_{i=1}^{D}(x_i - 100)^2\right) - \left(\prod_{i=1}^{D}\cos\left(\frac{x_i - 100}{\sqrt{i}}\right)\right) + 1 \tag{5}$$

$$f_2(\vec{x}) = \sum_{i=1}^{D}(x_i^2 - 10\cos(2\pi x_i) + 10) \tag{6}$$

$$f_3(\vec{x}) = 20 + e - 20e^{\left(-0.2\sqrt{\frac{1}{D}\sum_{i=1}^{D}x_i^2}\right)} - e^{\frac{1}{D}\sum_{i=1}^{D}\cos(2\pi x_i)} \tag{7}$$

$$f_4(\vec{x}) = \sum_{i=1}^{D}\left|\frac{\sin(10 x_i \pi)}{(10 x_i \pi)}\right| \tag{8}$$

Figures 1, 2, 3 and 4 show the experimental results of applying all the algorithms to three Benchmark functions with different values of parameters. As can be seen, the SCA for less parameter has better convergence to minimum of Benchmark function than the other optimization algorithms. In the other words, this proves that SCA has the ability of getting out of a local minimum in the search space and achieving the global minimum. In the stem cells algorithm, the cells having higher comparative power than the other cells are very good for global optimization and normal self-renewal process for best selected cells is very efficient for local optimization. So, we get better performance using SCA in optimizing multimodal and multivariable functions.

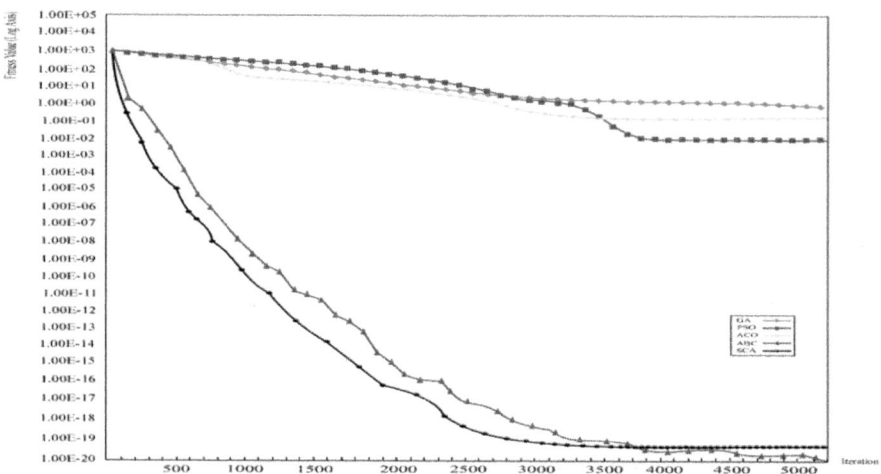

Fig. 1. The results of applying different algorithms on Benchmark function of Eq (5)

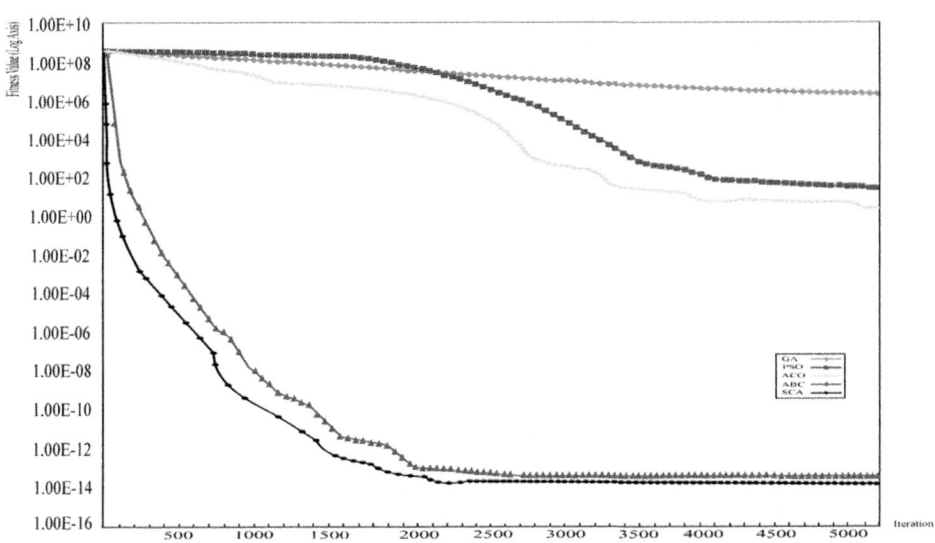

Fig. 2. The results of applying different algorithms on Benchmark function of Eq (6)

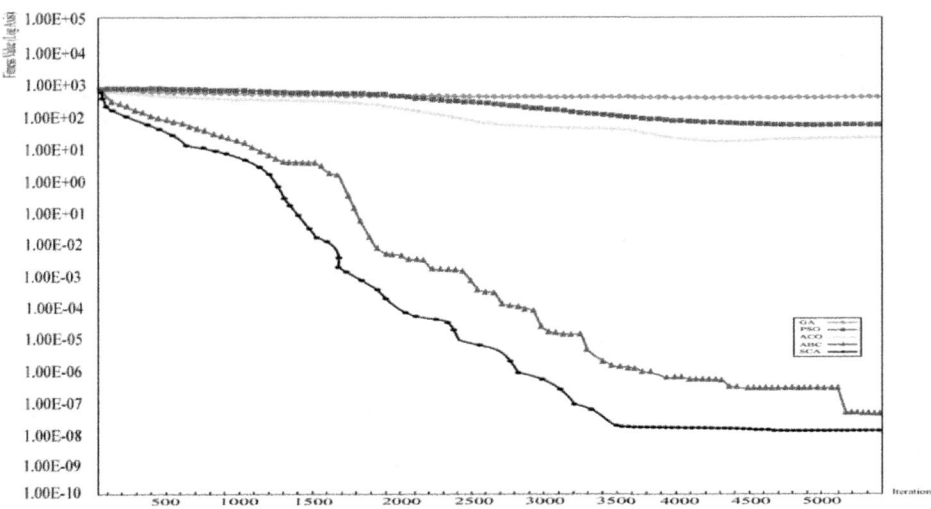

Fig. 3. The results of applying different algorithms on Benchmark function of Eq (7)

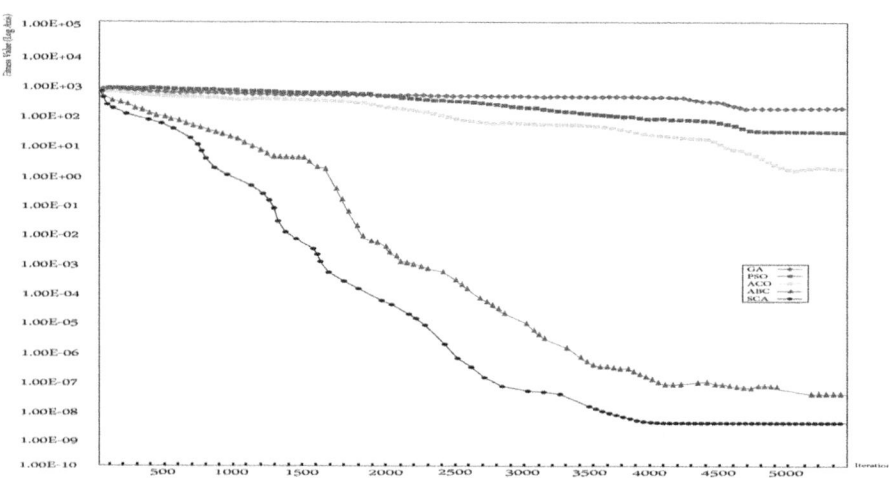

Fig. 4. The results of applying different algorithms on Benchmark function of Eq (8)

5   Conclusion

We have introduced a new optimization algorithm that has overcome the existing
problems in the previously introduced optimization algorithms such as genetic
algorithm, ant colony algorithm, particle swarm optimization algorithm and artificial
bee colony algorithm. The new algorithm is based on natural behavior of stem cells in
reproducing themselves. We have tested the proposed algorithm on several
Benchmark functions and the obtained results compared with those of applying other
optimization algorithms have demonstrated the superior performance of the proposed
algorithm.

References

1. Holland, J.H.: Adaptive in Natural and Artificial Systems. University of Michigan Press,
   AnnArbor (1975)
2. Kennedy, J., Eberhart, R.C.: Particle Swarm Optimization. In: Proc. IEEE International
   Conference on Neural Network, Australia, pp. 1942–1948 (1995)
3. Zheng, Y.L., Ma, L.H., Zhang, L.Y., Qian, J.X.: On the Convergence Analysis and
   Parameter Selection in Particle Swarm Optimization. In: International Conference on
   Machine Learning and Cybernetics, pp. 1802–1807 (2003)
4. Dorigo, M.: Optimization, Learning and Natural Algorithms. PhD thesis. Politecnico di
   Milano (1992)
5. Di, C.G., Dorigo, M.: Ant Colonies for Adaptive Routing in Packet-switched.
   Communications Networks 1498, 673–682 (1998)
6. Karaboga, D., Basturk, B.: A Powerful and Efficient Algorithm for Numerical Function
   Optimization. Journal of Global Optimization 39(3), 459–471 (2007)

7. Becker, A.J., McCulloch, E.A., Till, J.E.: Cytological Demonstration of the Clonal Nature of Spleen Colonies Derived from Transplanted Mouse Marrow Cells. Nature 197, 452–454 (1963)
8. Siminovitch, L., McCulloch, E.A., Till, J.E.: The Distribution of Colony-forming Cells Among Spleen Colonies. Journal of Cellular and Comparative Physiology 62, 327–336 (1963)
9. Srinivasan, D., Seow, T.H.: Evolutionary Computation. In: CEC 2003, Canberra, Australia, pp. 2292–2297 (2003)

Constrained Clustering via Swarm Intelligence

Xiaohua Xu[1], Zhoujin Pan[1], Ping He[2], and Ling Chen[1]

[1] Department of Computer Science and Engineering, Yangzhou University,
Yangzhou 225009, China
[2] Department of Computer Science and Engineering,
Nanjing University of Aeronautics and Astronautics, Nanjing 210016, China
{arterx,pprivulet,angeletx,yzlchen}@gmail.com

Abstract. This paper investigates the constrained clustering problem through swarm intelligence. We present an ant clustering algorithm based on random walk to deal with the pairwise constrained clustering problems. Our algorithm mimics the behaviors of the real-world ant colonies and produces better clustering result on both synthetic and UCI datasets compared with the unsupervised ant-based clustering algorithm and the cop-kmeans algorithm.

Keywords: swarm intelligence, constrained clustering, ant clustering, random walk.

1 Introduction

Clustering is one of the most important data analysis methods in machine learning and data mining. Clustering is often regarded as unsupervised learning. In the real-world applications, we can obtain some prior knowledge for our problems at hand in the form of pairwise constraints. How to utilize these constraints has become more important in recent days [3].

Swarm intelligence [7] is a very popular research area in artificial intelligence. During the development of applying swarm intelligence to clustering, the earliest model proposed by Deneubourg [2], often called the basic model (BM, Basic Model), is used to explain the ants' behavior of piling bodies together to form an ants' grave. By adding real data vectors that contain the similarity of data objects measurement, Lumer and Faieta modified the BM model, often called LF [2], to form the de facto clustering model. Recently Xu et al. presented an ant sleeping model [6] to improve performances of ant-based clustering. He and Hui used ant-based clustering (Ant-C) [9] algorithms to analyze gene expression data; El-Feghi et al. presented AACA [10] algorithm to improve its convergence. Mohamed Jafar Abul Hasan gave out a survey about the huge evolution of clustering based on swarm intelligence [11]. Similar to the ant clustering algorithm, Lutz Herrmann and Alfred Ultsch also gave out an artificial life system (Artificial Life System) [4] based on ESM [5][8] to deal with clustering problem.

In our previous work, we have invented a new ant clustering framework RWAC [6]. In this paper, based on RWAC, we propose a constrained ant clustering framework by embedding the constrained migration mechanism to deal with the situation

D.-S. Huang et al. (Eds.): ICIC 2011, LNBI 6840, pp. 404–409, 2012.

when the domain knowledge is provided in the form of pairwise constraints. CAC is a simple and effective ant-based semi-supervised clustering algorithm.

2   Constrained Clustering

In the real-world applications, we can sometimes obtain a small amount of domain information, such as labels or constraints. Constraints, in another word, two data points that must be assigned into the same cluster or different clusters, can be easily accessible. Generally, we consider two types of constraints.

- Must-link constraints $c_=(i,j)$: data points i and j must be assigned into the same cluster after clustering.

- Cannot-link constraints $c_{\neq}(i,j)$: data points i and j can not be placed in the same cluster.

The task of constrained clustering is to partition the dataset with the help of pairwise constraints to achieve a higher clustering performance and satisfying as many constraints as possible.

3   Constrained Ant Clustering Algorithm

3.1   The Framework of Ant Constrained Clustering

CAC we proposed in our paper is based on the random walk ant clustering algorithm (RWAC). RWAC simulated the behavior that ants in their environment finding a safe place to sleep.

- Definition 1: no-force distance -τ_1, τ_2

If two ants are constrained by cannot-link and their grid distance is larger than τ_1, there is no force between them. If two ants are constrained by must-link and their grid distance is short than τ_2, there is no force between them, either.

- Definition 2: exclusive factor- a

Constrained by the cannot-link, the closer two ants are to each other, the more likely they will move toward opposite direction.

$$a = \frac{\tau_1 - g(i, j)}{\tau_1} \tag{1}$$

Where $g(i, j)$ is the grid distance (the least steps of walk).

- Definition 3: attractor – r

Constrained by must-link the greater distance between two ants, the easier it will move towards each other.

$$r = \frac{g(i, j)}{\tau_2 / 2 + g(i, j)} \tag{2}$$

3.2   Algorithm Description

Based on the above definition and description, we gave out the CAC algorithm:

Table 1. Algorithm CAC

Algorithm CAC
Input: ants with adjusted affinity matrix
Output: clustering information of all ants
Initialize parameters.
for each ant *do*
Randomly place the ant in grid of G and initialize its class label
endfor
while not reach the termination *do*
for each ant *do*
if ant is constrained by must-link and grid-distance $>\tau_2$
Moving towards each other with attraction probability
endif
if ant is constrained by cannot-link and grid-distance $<\tau_1$
Moving backwards with repulsion probability
endif
Calculate ant's fitness f(ant)
Calculate the probability of awaken p_a(ant)
Generated random number r=random([0,1))
if r <= p_a(ant)
Active the ant and let it walk into a free location
else
Stay in sleep in the original grid
endif
Update the label vector according local random walk
endfor
Update the parameters
endwhile

CAC is much like RWAC. Ants are randomly place on the grid during the initializations. In the main loop we first deal with the case of constraints to performing constrained migration which we will explain in next subsection.

3.3   Constrained Migration Mechanism

Constrained by must-link two ants on the grid with a distance greater than τ_2, they will be forced by attractor and walk toward each other. Relatively constrained by cannot-link two ants on the grid with a distance less than the τ_1 , they will be forced by repulsive factor and walk away from each other. Constrained migration strategy in detail is as follows:

- Calculate the active probability of two ants. Choose the ant with relatively higher active probability as a candidate for migration.

- Generate a random value between 0 and 1. If the random value is smaller than attractor of the candidate that constrained by must-link, the candidate will move closer to the other one. If the random value is smaller than exclusive factor of the candidate that constrained by cannot-link, the candidate will move away from the other one.

- With the increase or decrease of grid distance between two ants, the force is also decreasing. If the grid distance is equal to the no-force distance, the ant constrained by must-link would sleep while the ant constrained by cannot-link would random walk as non-constrained ants.

4   Experiments

4.1   Illustration on Artificial Dataset

We use two circles to test CAC algorithm. As we can see in Figure 1, six constraints containing 2 cannot-link constraints and 4 must-link constraints are added to the data sets. Figure 2(left) shows the results of CAC. Figure 2(right) shows the results of RWAC algorithm with same parameters. The clustering performance of CAC algorithm greatly outperforms that of   RWAC algorithm on this dataset.

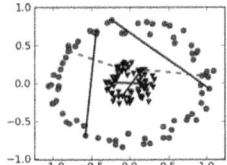
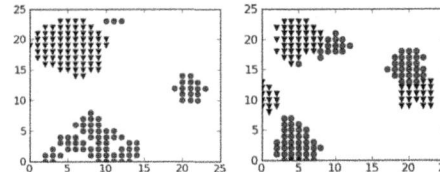

Fig. 1. Two circles   **Fig. 2.** Clustering result by CAC (left) and by RWAC (right)

4.2   The Evaluation Methods

In this paper, we use purity and F-score to evaluate the clustering quality of the three algorithms. Purity is a relatively straightforward evaluation criterion. While calculate of purity, we have

$$purity\ (\Omega, C) = \frac{1}{N} \sum_k \max_j \mid \omega_k \cap c_j \mid \tag{3}$$

Where $\Omega = \{\omega_1, \omega_2, \ldots, \omega_k\}$ is the clusters gained by the clustering algorithms and $C = \{c_1, c_2, \ldots, c_k\}$ represents the classes which data points originally belonging.

As rand index punishes both false positive and false negative, F measure (F-score) add a different weight for these two types of errors.

$$F = \frac{2PR}{P + R} \tag{4}$$

where $P = TP/(TP + FP), R = TP/(TP + FN)$.

4.3 Experiments on Artificial Datasets and UCI Datasets

We randomly generated 1%, 2%, 4%, 7%, 11%, 16%, 20% constrained data points for each dataset. Table 2 summarized the four test datasets from UCI for the comparison of CAC, RWAC [6] and cop-kmeans [1]. Figure 3-7 shows that CAC algorithm gives better clustering performance than the other two algorithms with both purity and F-score criteria.

Table 2. UCI datasets for testing

Datasets	#Objects	#Attributes	#Classes
Iris	150	4	3
Wine	178	13	3
Breast	683	9	2
Balance	625	4	3

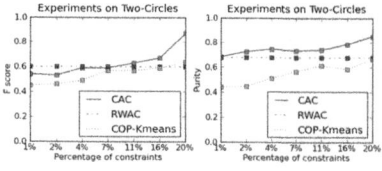

Fig. 3. Two-circles   **Fig. 4.** Iris

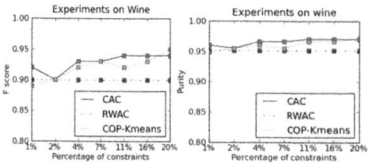   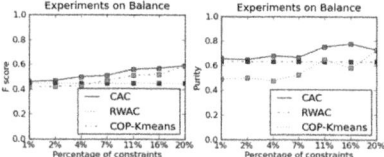

Fig. 6. Wine   **Fig. 7.** Balance

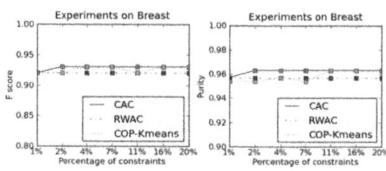

Fig. 5. Breast

5 Conclusions

In this paper, we proposed a constrained ant clustering framework by embedding the migration mechanism to deal with the constrained clustering problem on the basis of RWAC algorithm. The experimental results illustrate that our CAC algorithm outperforms RWAC and cop-kmeans algorithms on both artificial dataset and UCI datasets.

Acknowledgements. This work is supported by the National Natural Science Foundation of China under grant No.61003180 and No.61070047, Natural Science Foundation of Education Department of Jiangsu Province under contract 09KJB20013 and Natural Science Foundation of Jiangsu Province under contract BK2010318.

References

1. Wagstaff, K., Cardie, C., Rogers, S., Schroedl, S.: Constrained k-means clustering with background knowledge. In: Proceedings of the Eighteenth International Conference on Machine Learning, pp. 577–584 (2001)
2. Dorigo, M., Bonabeau, E., Théraulaz, G.: Ant algorithms and stigmergy. Future Generation Computer Systems 16(8), 851–871 (2000)
3. Zhu, X.: Semi-supervised learning with graphs. Doctoral dissertation, Carnegie Mellon University. CMU-LTI-05-192
4. Herrmann, L., Ultsch, A.: An Artificial Life Approach for Semi-supervised Learning. In: Data Analysis, Machine Learning and Applications Studies in Classification, Data Analysis, and Knowledge Organization, vol. II, pp. 139–146 (2008)
5. Ultsch, A., Herrmann, L.: Automatic Clustering with U*C. Technical Report, Dept. of Mathematics and Computer Science, University of Marburg (2006)
6. Xu, X., Chen, L., He, P.: A novel ant clustering algorithm based on cellular automata. Web Intelligence and Agent Systems 5(1), 1–14 (2007)
7. Bonabeau, E., Dorigo, M., Theraulaz, G.: Swarm Intelligence: From Natural to Artificial Systems, 1st edn. Oxford University Press, USA (1999)
8. Ultsch, A.: Emergence in Self-Organizing Feature Maps. In: Proc. Workshop on Self-Organizing Maps (WSOM 2007), Bielefeld, Germany (2007)
9. He, Y., Hui, S.C.: Exploring ant-based algorithms for gene expression data analysis. Artifl. Intell. Med., 105–119 (2009)
10. El-Feghi, I., Errateeb, M., Ahmadi, M., Sid-Ahmed, M.A.: An adaptive ant-based clustering algorithm with improved environment perception. In: Proceedings of the 2009 IEEE International Conference on Systems, Man, and Cybernetics, San Antonio (2009)
11. Abul Hasan, M., Ramakrishnan, S.: A survey: hybrid evolutionary algorithms for cluster analysis. Artificial Intelligence Review 1–26 (2011) Issn: 0269-2821

A Population-Based Hybrid Extremal Optimization Algorithm

Yu Chen, Kai Zhang, and Xiufen Zou*

School of Mathematics and Statistics, Wuhan University, Wuhan, 430072, China
xfzou@whu.edu.cn

Abstract. The extremal optimization (EO) algorithm is a kind of evolutionary algorithm which has been applied successfully in combinatorial optimization, while its application on continuous optimization encounters the problems of heavy complexity and weak exploration ability. This paper proposes a new hybrid population-based EO algorithm, named as the adaptive co-evolution population-based extremal optimization (ACPEO) algorithm, in which all individuals co-evolve adaptively with each other and the differential evolution (DE) operator is incorporated to improve the global search ability. By employing a novel evaluation method of variables, the ACPEO algorithm performs well on several kind of benchmark problems. Experimental results show that the ACPEO algorithm is robust due to the capability for solving different problems with the same parameter setting, and it is also stable because changes in the parameters' values do not influence its performances seriously.

Keywords: Evolutionary algorithm, extremal optimization, memetic algorithm.

1 Introduction

Recently, the extremal optimization (EO) algorithm, which is a new kind of evolutionary algorithm (EA) based on the celebrated Bak-Sneppen (BS) model [1] showing the emergence of self-organized criticality (SOC) [2] in ecosystem, was proposed by Boettcher and Percus [3]. Different from some classical EAs such as genetic algorithms (GAs), the EO algorithms employ mechanisms that mutate genes (or variables) with relatively poor fitness, which leads to the chain reactions called as "avalanches". Thus, an EO algorithm converges to not a equilibrium state, but a critical state that can be break away from easily. Consequently, the EO algorithms have good exploitation ability and theoretical ability of global exploration, leading to their successful utilizations in solving combinational optimization problems [3–5]. Recently, they are also extended to solve continuous optimization problems [6, 7], and the multi-objective models are competitive in multi-objective optimization (MO) [8–11], too.

---

* Corresponding author.

D.-S. Huang et al. (Eds.): ICIC 2011, LNBI 6840, pp. 410–417, 2012.
© Springer-Verlag Berlin Heidelberg 2012

However, the original EO algorithm comes to its dead ends in some implementations, because only variables (genes) with the worst fitness are mutated at each generation [12]. The individuals in the τ-EO algorithm [12, 13] can escape from the local optimal solutions, but it is unsuitable for continuous optimization problems because the evaluation method of variables cannot be extended to continuous models. Although the real-coded continuous extremal optimization (CEO) [7] algorithm can be applied to continuous optimization problems, the computational expense increases greatly with the dimension of the decision space because every variable (gene) has to be scored by one independent individual evaluation. Moreover, some global search algorithms [14] and the population scheme [11] are also proposed to obtain better global exploration ability.

In this paper, the adaptive co-evolution population-based extremal optimization (ACPEO) algorithm is proposed to solve the continuous optimization problems efficiently. All individuals in the ACPEO algorithm co-evolve with each other adaptively, and no additional function evaluations are needed in the process of new individual generation. Meanwhile, there are also no sorting procedures in the procedure of mutation. Then, by employing the differential evolution (DE) operator to improve its global search ability, two different schemes of the ACPEO algorithm are devised for different problems: one can converge to the optimal solutions quickly, while the other can keep good diversity of the population during the evolving process. The rest of the paper is structured as follows: Section 2 introduces the EO algorithm and describes the proposed ACPEO algorithm in detail. Comparisons between the ACPEO algorithm and PSO-EO are made and the influences of parameters are investigated in Section 3. Finally, the conclusions and the future work are addressed in Section 4.

2   The Adaptive Co-evolution Population-Based Extremal Optimization (ACPEO) Algorithm

The most apparent distinction between the EO algorithms and other EAs is the definition of local cost contributions λ_i for each variable, instead of merely a global fitness of the whole individual [12]. According to the fitness values of variables[1] in an individual, mutations can be performed on different variables respectively. However, there also exist some common deficiencies in the existing real-coded EO algorithms:

1. The EO algorithms have to employ $\Theta(n)$ function evaluations to evaluate the variables of each individual [7, 10, 11, 14].
2. The evaluation processes of the EO algorithms contain sorting processes whose computational complexities are at least $\Theta(n \log n)$.
3. The existing population-based EO algorithms sometimes converge to the local optimal solutions because of their weak exploration abilities.

---

[1] Because each variable (gene) of an individual must be evaluated in EO algorithms, the function value of a feasible solution **x** is called the "cost value", and the "fitness value" in EA is used to represent the goodness of variables.

Thus, we are devoted to propose an improved population-based hybrid algorithm named as the adaptive co-evolution population-based extremal optimization (ACPEO) algorithm. To eliminate the redundant function computations in variable evaluations, the ACPEO algorithm evaluates the variables of individual $\mathbf{x} = (x_1, \ldots, x_n)$ by randomly selected $\mathbf{y} = (y_1, \ldots, y_n) \in X$ via

$$\lambda_i = \left| \frac{f(\mathbf{x+h}) - f(\mathbf{x})}{h_i} \right|, \quad i = 1, \ldots, n, \tag{1}$$

where $\mathbf{h} = (h_1, \ldots, h_n)$ is a vector generated by

$$h_i = \begin{cases} x_i - y_i & \text{with probability } \dfrac{1}{2}, \\ y_i - x_i & \text{with probability } \dfrac{1}{2}, \end{cases} \tag{2}$$

$i = 1, \ldots, n$. After the traversal of all variables in $\mathbf{x}$, the new candidate individual $\mathbf{x}' = (x'_1, \ldots, x'_n)$ can be generated by mutation

$$x'_i = x_i + \sigma \tag{3}$$

with probability

$$P_i = \exp\{K \times \frac{\lambda_i - \lambda_w}{\lambda_b - \lambda_w}\} \tag{4}$$

$i = 1, \ldots, n$, where σ is a random number obeying some kind of probability distribution, λ_b and λ_w are the respective fitness values of the best and the worst variables, and K is the impact factor of $\frac{\lambda_i - \lambda_w}{\lambda_b - \lambda_w}$.

When the EO search is performed on $\mathbf{x}$, μ individuals $\mathbf{y}^{(1)}, \ldots, \mathbf{y}^{(\mu)}$ are selected randomly from the population. Then, several different probability distributions that σ obeys are employed to achieve both good exploration and strong exploitation. Here μ is set to be times of five, and the distribution of σ is

$$\sigma \sim \begin{cases} Cauchy(0,1) & if \ j \equiv (0 \mod 5), \\ Guass(0,0.1) & if \ j \equiv (1 \mod 5), \\ Guass(0,1) & else, \end{cases} \tag{5}$$

where $j = 1, \ldots, \mu$. The best one of the μ candidates $\mathbf{x}'^{(1)}, \mathbf{x}'^{(2)}, \ldots, \mathbf{x}'^{(\mu)}$ generated according to (1), (2), (3), (4) and (5) is selected to update $\mathbf{x}$. Repeat the aforementioned process for T generations. Then, $\mathbf{x}_{best}$ is returned by the function $IEO(\mathbf{x}, X)$ described in Algorithm 1. Afterwards, the DE operator $DE/rand/1$ [16] is employed to constitute two different schemes of ACPEO, which can be depicted consistently as Algorithm 2. Here $rand()$ generates a random number distributed uniformly in (0,1).

1. *Scheme 1*: When applied to solve some problems with enormous local optima, the improved EO search $IEO(\mathbf{x}, X)$ is performed as a local search strategy. For this case, the ACPEO algorithm performs the DE operator with $P_{DE} = 1$, and $IEO(\mathbf{x}, X)$ is only implemented on $\mathbf{x}$ with a small probability;

2. *Scheme 2*: When the ACPEO algorithm is utilized to solve optimization problems with large absorbing regions, the EO search $IEO(\mathbf{x}, X)$ is performed on every individual in the population, and the DE operation is only carried out with probability $P_{DE} < 1$.

Algorithm 1. $IEO(\mathbf{x}, X)$

1: Input $\mathbf{x}$; Let $\mathbf{x}_{best} = \mathbf{x}$;
2: **for** $t = 1, \ldots, T$ **do**
3:     **for** $i = 1, \ldots, \lambda$ **do**
4:         Randomly select an individual $\mathbf{y}^{(i)}$ from X;
5:         Generate a candidate $\mathbf{x}'^{(i)}$ according to (1), (2), (3), (4) and (5).
6:     **end for**
7:     Set $\mathbf{z} = \arg \min_{i=1,\ldots,\lambda} \{f(\mathbf{x}'^{(i)})\}$;
8:     **if** $f(\mathbf{z}) < f(\mathbf{x}_{best})$ **then**
9:         $\mathbf{x}_{best} = \mathbf{z}$;
10:     **end if**
11: **end for**
12: Output $\mathbf{x}_{best}$.

---

3   Numerical Experiments

To highlight the efficiency of the ACPEO algorithm, we first compare the ACPEO algorithm with PSO-EO by some celebrated unconstrained optimization problems. Then, the Bump problem, a constrained optimization problem which is hard to solve and sensitive to the parameter settings, is used to investigate the impacts of parameters in the ACPEO algorithm.

Algorithm 2. ACPEO

1: Generate the population X randomly, and denote the individuals with greatest and least cost values by $\mathbf{x_w}$ and $\mathbf{x_b}$ respectively;
2: **if** $rand() < P_{DE}$ **then**
3:     Generate a new candidate $\mathbf{x}'$ by $DE/rand/1$;
4:     Replace $\mathbf{x_w}$ with $\mathbf{x}'$ if $\mathbf{x}'$ is better than $\mathbf{x_w}$;
5: **end if**
6: **for all** $\mathbf{x} \in X$ **do**
7:     **if** $rand() < P_{EO}$ **then**
8:         $\mathbf{x} = IEO(\mathbf{x},X)$;
9:     **end if**
10: **end for**
11: Go back to step 2 until the stopping criterion is satisfied.

3.1   Comparison with the PSO-EO Algorithm

To compare ACPEO with PSO-EO, we perform numerical experiments on a series of function optimization problems [14]. The population size of ACPEO is set to be 50, and both Cr and F are set to be 0.5 in the DE operator. For problems f_2 and f_3, *Scheme 2* is employed, where the values of P_{DE}, P_{EO}, μ, K and T are set to be 1, 1, 10, 1 and 10, respectively. *Scheme 1* of the ACPEO algorithm is employed to solve problems f_1, f_4, f_5 and f_6, and the respective values of P_{DE}, P_{EO}, μ, K and T are set to be 1, 0.05, 5, 1 and 10. Due to the different frameworks of PSO-EO and ACPEO, the comparisons are based on the same number of function evaluations (FEs).

After 20 independent runs of ACPEO, the experimental results are listed in Table 1, where the results of PSO-EO are cited from [14]. Although there are several parameters in the ACPEO algorithm, only two different sets in parameter settings are needed for the six benchmark functions. The benchmark functions f_1, f_4, f_5 and f_6 belong to the same class, even if f_6 is just a unimodal problem while the other three are problems with large amounts of local optima. This is because only optimizers with strong global exploratory abilities can be efficient for these four problems. Thus, in *Scheme 1* the EO search acts as a local search strategy when solving these four problems, and it is performed with a small probability 0.05. However, f_2 and f_3 are defined in a relatively large region, and some of the absorbing regions of local optima are extraordinarily easy to be tapped into. Then, keeping the diversity of the population is especially important. The ACPEO algorithm of *Scheme 2* can search the local neighborhoods of the individuals efficiently, and the diversity of the population is always preserved well. Thus, for f_2 and f_3, the probability of performing the EO search is set to 1 and the DE operator functions only once every 10 generations. Although the results of ACPEO are a little worse than those of PSO-EO for f_3, f_4 and f_5, it is still competitive to PSO-EO because no extra function evaluations are performed in the process of variable evaluation. Consequently, the ACPEO algorithm has much less complexity than PSO-EO and performs better than PSO-EO if the extra function evaluations in the process of variable evaluation are also counted in the total amount of FEs.

3.2   Parameter Study

The previous study shows that the ACPEO algorithm is competitive to PSO-EO algorithm when it is utilized to solve function optimization problems, where only two different kinds of parameter settings are needed to solve six different unconstrained problems. However, for some constrained optimization problems, the parameter settings are much more complicated. Thus, the ACPEO algorithm is tested for different parameter settings by the Bump problem [17], a constrained optimization problem with a great number of local optimal solutions. The original Bump problem is a nonlinear maximum problem with an unknown global optimal solution, and by multiplying the objective function by -1, it is converted to a minimum problem in this paper. Because the Bump problem is a constrained optimization with a great deal of local optimal solutions distributed in a relative narrow region, *Scheme 1* of the ACPEO algorithm is implemented here.

Table 1. Comparisons between ACPEO and PSO-EO for unconstrained OPs

Function	Algorithm	Worst	Mean	Best	St.dev.
f_1	PSO-EO	-9.66	-9.66	-9.66	2.15E-3
	ACPEO	-9.66	-9.66	-9.66	3.2E-15
f_2	PSO-EO	-12562.6	-12568.0	-12569.5	2.01
	ACPEO	-12569.5	-12569.5	-12569.5	6.8E-5
f_3	PSO-EO	0	0	0	0
	ACPEO	3.5E-15	3.8E-16	0	8.4E-16
f_4	PSO-EO	0	0	0	0
	ACPEO	3.5E-015	1.2E-015	0	1.4E-015
f_5	PSO-EO	-8.88E-16	-8.88E-16	-8.88E-16	0
	ACPEO	6.3E-014	1.7E-015	6.6E-015	1.3E-014
f_6	PSO-EO	9.99E-4	9.88E-4	9.54E-4	2.39E-5
	ACPEO	4.0E-26	6.2E-27	3.2E-28	9.7E-27

Table 2. Parameters Settings for the Bump Problem

Parameter Setting	P_{DE}	P_{EO}	μ	K	T	Parameter Setting	P_{DE}	P_{EO}	μ	K	T
I	1	0.2	10	1	25	II	1	0.1	10	1	25
III	1	0.3	10	1	25	IV	1	0.2	5	1	25
V	1	0.2	15	1	25	VI	1	0.2	10	0	25
VII	1	0.2	10	2	25	VIII	1	0.2	10	1	20
IX	1	0.2	10	1	30						

The parameters C_r and F are also set to be 0.5, and the influence of the other five parameters, P_{DE}, P_{EO}, μ, K and T, are investigated by numerical experiments with different parameter settings listed in Table 2.

After 50 independent runs of ACPEO for each parameters' setting listed in Table 2, the best obtained approximate objective value is -0.8352622601, which is better than the result -0.835262013 in [17]. In Figs. 3.2, 3.2, 3.2 and 3.2, four illustrations indicate the changing tracks of the statistical results of best function values, mean function values, standard deviations of function values and worst function values in the population, respectively. The horizon axis is the number of FEs, and the vertical axis is the average value for 50 independent runs.

Fig. 1 shows that the best value decrease monotonously as the population evolves, for the reason that the algorithm is a kind of elitist algorithm. However, the mean values of the population decrease wavily, where the weak fluctuations are created by the mutation on the worst individual $\mathbf{x_w}$. Meanwhile, the standard deviation and worst value in the population also fall wavily and then oscillate around some exact value, which means that the diversity of the

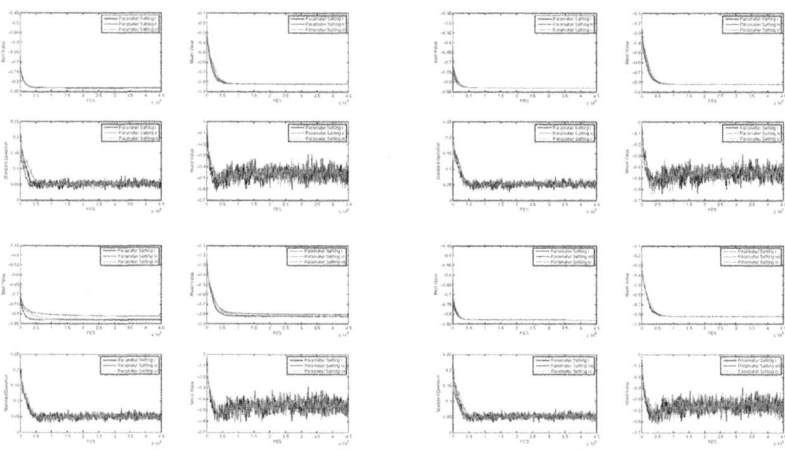

Fig. 1. The comparison diagrams for different values of parameters

population can keep around some exact value after some generations. Consequently, the premature does not occur, and the good exploitation is also achieved. In this way, the mean function value of the population also decreases wavily by the alternate functions of the mutation and co-evolution between individuals. The numerical results also demonstrate that the ACPEO algorithm is not sensitive to the regulations of parameters except for K. When $K = 0$, all variables in the individual mutate, and the EO optimizer degenerates into the ES search. So the convergence rate and convergence quality are relatively worse than those in the cases $K = 1$ and $K = 2$ (illustrated by red and green curves respectively in Fig. 3.2); When $K = 2$, the great value of K leads to the relatively weak ability of breaking away from the local absorbing region. Thus, both convergence rate and quality are worse than those in the case $K = 1$.

4   Conclusions and Future Works

EO is a new proposed evolutionary algorithm good at local exploitation, but the heavy complexity of variable evaluation restricts severely its applications on continuous optimization problems. These motivate the present work on population-based EO algorithms. The proposed ACPEO algorithm incorporates a co-evolution population-based framework with the DE operator, and the complexity of ACPEO is much less than that of the existing EO algorithms.

Numerical experiments demonstrate that the ACPEO algorithm performs well in some celebrated benchmark problems. Although there are several parameters in the ACPEO algorithm, only two different sets of parameters' settings are needed to solve these problems effectively, which shows that the ACPEO algorithm is robust. Moreover, parameter studies also show that the ACPEO algorithm is insensitive to the parameters except for the value of K. The future work lies in the theoretical analysis of EO algorithm, and we will extend the ACPEO algorithm to solve multi-objective optimization problems as well.

Acknowledgements. This work was supported by the Chinese National Natural Science Foundation under Grant 61070007 and the Key Program of National Natural Science Foundation of China under Grant 51039005.

References

1. Bak, P., Sneppen, K.: Punctrated Equilibrium and Criticality in a Simple Model of Evolution. Phys. Rev. Lett. 71(24), 4083–4086 (1993)
2. Bak, P., Tang, C., Wiesenfeld, K.: Self-organized Criticality. Phys. Rev. A. 38(1), 364–374 (1988)
3. Boettcher, S., Percus, A.G.: Nature's Way of Optimizing. Artificial Intelligence 199, 275–286 (2000)
4. Boettcher, S., Percus, A.G.: Extremal Optimization at the Phase Transition of the Three-coloring Problem. Phys. Rev. E. 69(6), 1–8 (2004)
5. Chen, Y., Lu, Y., Chen, P.: Optimization with Extremal Dynamics for the Traveling Salesman Problem. Physica A 385(1), 115–123 (2007)
6. Sousa, F.L., Ramos, F.M., Paglione, P., et al.: New Stochastic Algorithm for Design Optimization. AIAA J. 41(9), 1808–1818 (2003)
7. Zhou, T., Bai, W.J., Chen, L.J., et al.: Continuous Extremal Optimization for Lennrd-Jones clusters. Phys. Rev. E. 72(1), 1–5 (2005)
8. Ahmed, E., Elettreby, M.F.: On Multiobjective Evolution Model. Int. J. Mod. Phys. C. 15(9), 1189–1195 (2004)
9. Galski, R.L., De Sousa, F.L., Ramos, F.M., Muraoka, I.: Spacecraft Thermal Design with the Generalized Extremal Optimization Algorithm. Inverse Probl. Sci. Eng. 15(1), 61–75 (2004)
10. Chen, M.R., Lu, Y.Z.: A Novel Elitist Multiobjective Optimization Algorithm: Multiobjective Extremal Optimization. Eur. J. Oper. Res. 188, 637–651 (2008)
11. Chen, M.R., Lu, Y.Z., Yang, G.: Multiobjective Optimization Using Population-based Extremal Optimization. Neural Comput. & Appl. 17, 101–109 (2008)
12. Boettcher, S.: Extremal Optimiziton: Heuristics via Coevolutionary Avalanches. Comput. Sci. & Eng. 6(2), 75–82 (2000)
13. Boettcher, S., Percus, A.G.: Optimization with Extremal Dynamics. Complexity 8(2), 57–62 (2003)
14. Chen, M.R., Li, X., Zhang, X., Liu, Y.Z.: A Novel Particle Swarm OptimizerHybridized with Extremal Optimization. Appl. Soft Comput. 10, 367–373 (2010)
15. Zhao, R.Q., Tang, W.S.: Monkey Algorithm for Global Numerical Optimization. J. Uncertain Sys. 2(3), 165–176 (2008)
16. Stron, R., Price, K.: Differential Evolution - a Simple and Efficient Heuristic for Global Optimization over Continuous Spaces. J. Global Optim. 11, 341–359 (1997)
17. Liu, P., Lau, F., Lewis, M.J., Wang, C.-l.: A New Asynchronous Parallel Evolutionary Algorithm for Function Optimization. In: Guervós, J.J.M., Adamidis, P.A., Beyer, H.-G., Fernández-Villacañas, J.-L., Schwefel, H.-P. (eds.) PPSN 2002. LNCS, vol. 2439, pp. 401–410. Springer, Heidelberg (2002)

An Effective Ant Colony Algorithm for Graph Planarization Problem

Li-Qing Zhao[1], Cui Zhang[2], and Rong-Long Wang[1]

[1] Graduate School of Engineering, University of Fukui, Bunkyo 3-9-1,
Fukui-shi, Japan
[2] Department of Autocontrol, Liaoning Institute of Science and Technology,
Benxi, China
nkzlq@hotmail.com, bxlkyzhangcui@163.com, wang@u-fukui.ac.jp

Abstract. In this paper, an effective ant colony algorithm is proposed for solving the graph planarization problem. In the proposed algorithm, two kinds of pheromone are adopted to reinforce the search ability, and each kind of pheromone consists of two elements. The proposed algorithm is verified by a large number of simulation runs and compared with other algorithms. The experiment results show that the proposed algorithm performs remarkably well and outperforms its competitors.

Keywords: Graph planarization problem, NP-complete problem, Ant colony algorithm, Pheromone.

1 Introduction

A graph is said to be planar, or embeddable in the plane, if it can be drawn on a plane, such that no two edges intersect except at their end vertices. Given a non-planar graph $G = (V, E)$,($|V| = n$, $|E| = m$), the graph planarization problem is to find a spanning planar subgraph G'=(V, F) with a maximum number of edges, which is known as the maximum planar subgraph problem(MPSP). The graph planarization problem has important applications in many areas, for example circuit and VLSI design, networks design and analysis, computational geometry and it is one of the most intensively studied classed of graphs. However, the graph planarization problem is NP-complete for general graphs. Therefore, no tractable algorithm is known for solving it, which is the motivation for finding fast algorithms that yield approximate solutions [1].

Since finding a maximum planar subgraph is NP-hard [2], the planarization problem has been widely studied. Several graph planarization heuristics, neural network learning methods and genetic algorithms have been proposed in the literature. Based on the PQ-tree technique, Jayakumar et al. [3] proposed a $O(n^2)$ near-maximal planarity testing algorithm. Kant [4] presented a corrected and more generalized version of Jayakumerfs algorithm later. Cai et al. [5] developed an $O(mlog^n)$ algorithm for the problem based on the Hopcroft-Tarjan planarity testing algorithm [6]. Another algorithm with the same complexity

D.-S. Huang et al. (Eds.): ICIC 2011, LNBI 6840, pp. 418–425, 2012.

can also be derived from the incremental planarity testing algorithm of Di Battista and Tamassia [7]. Linear-time $O(m + n)$ algorithms can be considered in reference [8]. Goldschmidt and Takvorian [9] presented a two-phase graph planarization heuristic. Further Junger and Mutzel [10] reported a branch and cut algorithm for finding maximum planar subgraph. Resende and Ribeiro [11] gave a greedy randomized adaptive search procedure (GRASP), a meta-heuristic for graph planarization problem. Using the neural network techniques [12], Takefuji and Lee [13] [14] presented a parallel planarization algorithm for generating a near-maximal planar subgraph within $O(1)$ time. Wang et al. proposed a Hopfield network learning algorithm for the problem [15]. Besides, genetic algorithm (GA) based algorithm [16] was also proposed to solve the problem and very good results were reported.

For solving such discrete combinatorial problems, swarm intelligence algorithm have been shown to be good problem solvers with various application domains[17]. The ACO algorithm uses pheromone as an indirect communication medium among the individuals of a colony of ants, and the procedure of converging to the global optimum is a dynamic positive feedback of pheromone. In the existing ACO algorithms, only one kind of pheromone is used, or a two-state ant colony was adopted with one kind of pheromone [18]. However for some combinatorial optimization problems, for example, the graph planarization problem, it is not efficient to show the quality of solution using only one kind of pheromone. In this paper we propose an effective ant colony algorithm with two kinds of pheromone for the graph planarization problem. Besides, we also define two elements for each kind of pheromone to reinforce the search ability.

2   Problem Formulation

A graph is planar if it can be drawn in the plane so that no two edges intersect except at a common endpoint. For a given n-vertex m-edge graph $G = (V, E)$, a planar sub-graph G' of G such that adding to G' any edge of $E(G) - E(G')$ results in a non-planar graph is called a maximal planar sub-graph of G. the graph planarization problem is to find a maximum planar sub-graph from a general non-planar graph [1]. For solving this problem with ant colony algorithm, we need encode the problem. For an m-edge graph, we can use a list $(x_1, x_2, \ldots, x_m)$ to represent the solution of graph planarization problem. Each element (x_i) in the list corresponds to an edge of the graph and has the value $-1, 0$ or 1 according to whether, in this solution the edge is a lower edge, not considered or an upper edge. Thus the problem can be mathematically transformed into the following optimization problem:

Optimization Description:

$$\text{Maximize} : \sum_{i=1}^{m} |x_i| \tag{1}$$

$$\text{Constraint condition}: \frac{1}{2} \sum_{i=1}^{m} \sum_{j=i+1}^{m} |x_i + x_j| \cdot x_i \cdot x_j \cdot d_{ij} = 0 \qquad (2)$$

where m is the number of edges, $d_{ij} = 1$ if two edges i and j intersect; otherwise, it equals 0. The existence of a crossing between two upper edges (or two lower edges) is easy to determine by using the determination conditions mentioned before.

According to the problem description in Eq. (1) and Eq. (2), in this algorithm we can evaluate the quality of the solution by the following equation:

$$f(\overline{x}) = \frac{A \cdot \sum_{i=1}^{m} |x_i|}{1 + \frac{B}{2} \sum_{j=i+1}^{m} |x_i + x_j| \cdot x_i \cdot x_j \cdot d_{ij}} \qquad (3)$$

where A and B are parameters. We can adjust A, B to get the appropriate evaluation function for the ant colony algorithm.

3 An Effective Ant Colony Algorithm for Graph Planarization Problem

Artificial ants in ACO algorithms can be seen as probabilistic construction heuristics that generate solutions iteratively by taking into account accumulated past search experience: pheromone trails and heuristic information on the instance under solution. In the existing ACO algorithms, only one kind of pheromone is used, or a two-state ant colony was adopted with one kind of pheromone [18]. However for some combinatorial optimization problems, for example, the graph planarization problem, it is not efficient to show the quality of solution using only one kind of pheromone. In view of the speciality of the graph planarization problem which is addressed in section 2, it is very difficult to solve it with traditional ACO or other PSO algorithms. As a result, ant colony algorithm has not been used for solving graph planarization problem before, so an effective ant colony algorithm with two kinds of pheromone is proposed in this paper for graph planarization problem.

In this section we will introduce the proposed algorithm in detail, in which two kinds of pheromone has been defined for graph planarization problem. In the proposed algorithm, each ant deposits two kinds of pheromone (τ_1, τ_2) on every edge, and each kind pheromone has two elements $(\tau_1^1, \tau_1^{-1}$ and $\tau_2^1, \tau_2^{-1})$. For one kind of pheromone, for example τ_1, pheromone elements τ_1^1 and τ_1^{-1} indicate the learned desirability of setting the edges into upper group and under group respectively. Actually the proposed algorithm can be developed to n kinds of pheromone with n elements for other problems $((\tau_1^1, \tau_1^2, \dots, \tau_1^n), (\tau_2^1, \tau_2^2, \dots \tau_2^n),$ $\dots, (\tau_n^1, \tau_n^2, \dots, \tau_n^n))$. What the definitions of the two kinds of pheromone are and how to calculate it for the graph planarization problem will be addressed later. Each ant starts with an empty solution and construct a complete solution by iteratively setting the edges to upper one $(x_i = 1)$, under one $(x_i = -1)$ or unconsidered one $(x_i = 0)$ until all edges are set. The edges is set according to certain probability by taking into account accumulate experience including pheromone and heuristic. In iteration t an ant k selects $\sharp i$ edge for upper group and under group with probability $P_k^1(i)$ and $P_k^{-1}(i)$, respectively.

$$
P_k^1(i) = \begin{cases} \dfrac{[\tau_{1i}^1(t)]^{r_1}[\tau_{2i}^1(t)]^{r_2}[\eta_i(t)]^{r_3}}{\sum_{j \in J^k}[\tau_{1j}^1(t)]^{r_1}[\tau_{2j}^1(t)]^{r_2}[\eta_j(t)]^{r_3}}, & \text{if } i \in J^k \\ 0, & \text{otherwise} \end{cases} \tag{4}
$$

$$
P_k^{-1}(i) = \begin{cases} \dfrac{[\tau_{1i}^{-1}(t)]^{r_1}[\tau_{2i}^{-1}(t)]^{r_2}[\eta_i(t)]^{r_3}}{\sum_{j \in J^k}[\tau_{1j}^{-1}(t)]^{r_1}[\tau_{2j}^{-1}(t)]^{r_2}[\eta_j(t)]^{r_3}}, & \text{if } i \in J^k \\ 0, & \text{otherwise} \end{cases} \tag{5}
$$

Where τ_{1i}^1, τ_{1i}^{-1}, τ_{2i}^1, τ_{2i}^{-1} are pheromone trails and η_j is heuristic information in edge $\sharp i$. J^k is the set of edges that remain to be set by ant k, the parameters r_1, r_2 and r_3 determine the relative influence of two kinds of pheromone and heuristic information.

After the characteristics of a given n-vertex m-edge graph $G = (V, E)$ are considered, the heuristic for the novel ACO algorithm is proposed. Assuming all the edges are on upper side, and no one edge is unconsidered, in other words $x_i = 1, (i = 1, 2, \ldots, m)$, so the heuristic is related to the number of edges which intersect. For edge $\sharp i$, the larger the intersecting number the smaller the heuristic is. As a result the heuristic for the proposed algorithm are as following:

$$
\eta_i = \frac{n}{\delta_1 + \sum_{j=1}^{m} x_i \cdot x_j \cdot d_{ij}} \tag{6}
$$

Where n is the number vertex in graph G, δ_1 is a parameter which avoid denominator becoming 0, here δ_1 is set to 1, and the definition of d_{ij} has been introduced in section 2.

In the proposed algorithm, each ant deposits two kinds of pheromone (τ_1, τ_2) on every edge. Firstly, the first kind of pheromone τ_1 is introduced. The difference between two kinds pheromone is how to update it. For the first kind pheromone τ_1, the update method is based on the total quality of the solution which induced from traditional ACO algorithm. Because the ant colony has two states, for the first kind of pheromone, there are two elements which will be deposited on edges. The ant deposits pheromone elements τ_1^1 and τ_1^{-1} on every edge. The amount of the first kind of pheromone an ant k deposits on edge $\sharp i$ is defined as following:

$$
\Delta \tau_{1i}^{1,k} = \frac{1}{m} \cdot f(\overline{x})_k \cdot G_k^{i,1} \tag{7}
$$

$$
\Delta \tau_{1i}^{-1,k} = \frac{1}{m} \cdot f(\overline{x})_k \cdot G_k^{i,-1} \tag{8}
$$

where m is the number of edges in graph G, $f(\overline{x})_k$ (defined in Eq. (3))is used to evaluate the total quality of the solution which have been found by ant k, and for the solution of ant k, if $x_i = 1$, $G_k^{i,1} = 1$, if $x_i = -1$, $G_k^{i,-1} = 1$. In this way, the increase of the first kind of pheromone update on edge $\sharp i$ by ant k

depends on the total quality of the solution found by the ant, so the pheromone update rule in AS_{rank} is used here. But, some improvements have been done. To avoid search stagnation, after selected the global best one and $(\omega - 1)$ best ones, and a number of random ants were select to deposit pheromone. As a result the balance of diversity and intensity is achieved.

$$\tau_{1i}^1(t+1) = \rho_1 \cdot \tau_{1i}^1(t) + \sum_{k=1}^{\omega-1}(\omega - k) \cdot \Delta\tau_{1i}^{1,k}(t)$$

$$+\omega \cdot \Delta\tau_{1i}^{1,gb}(t) + \sum_{k=Rd} \Delta\tau_{1i}^{1,k}(t) \tag{9}$$

$$\tau_{1i}^{-1}(t+1) = \rho_1 \cdot \tau_{1i}^{-1}(t) + \sum_{k=1}^{\omega-1}(\omega - k) \cdot \Delta\tau_{1i}^{-1,k}(t)$$

$$+\omega \cdot \Delta\tau_{1i}^{-1,gb}(t) + \sum_{k=Rd} \Delta\tau_{1i}^{-1,k}(t) \tag{10}$$

Where ρ_1 is the pheromone decay coefficient from t to $t + 1$, and not only the global best solution and the $(\omega-1)$ best ants are allowed to deposit pheromone for updating, some of ants are picked randomly are allowed to deposit pheromone, and Rd is the ant randomly picked.

In this paper, another kind of pheromone τ_2 is proposed. From above we can see that the first kind of pheromone is related to the quality of the solution, it can only reflect the integrated quality of the solution, but the local quality of the solution can't be reflected. As a result, a second kind of pheromone τ_2 is designed to offset that shortage. The second kind of pheromone also has two different elements τ_2^1 and τ_2^{-1}. The amount of the second kind of pheromone τ_2 an ant k deposits on edge $\sharp i$ is defined as following:

$$\Delta\tau_{2i}^{1,k} = \frac{\sum_{x_i=-1} x_i \cdot x_j \cdot d_{ij}}{\delta_2 + \sum_{x_i=1} x_i \cdot x_j \cdot d_{ij}} \tag{11}$$

$$\Delta\tau_{2i}^{-1,k} = \frac{\sum_{x_i=1} x_i \cdot x_j \cdot d_{ij}}{\delta_2 + \sum_{x_i=-1} x_i \cdot x_j \cdot d_{ij}} \tag{12}$$

Where δ_2 is a parameter which avoid denominator becoming 0, here we set δ_2 to 1, and the definition of d_{ij} is introduced in section 2. In this way, the increase of the second kind of pheromone update on edge $\sharp i$ by ant k depends on the solution found by ant k, and the update rule is the same as the first kind of pheromone.

$$\tau_{2i}^1(t+1) = \rho_2 \cdot \tau_{2i}^1(t) \cdot \prod_{k=1}^{\omega-1}(\omega - k) \cdot \Delta\tau_{2i}^{1,k}(t) \cdot \omega \cdot \Delta\tau_{2i}^{1,gb}(t) \cdot \sum_{k=Rd} \Delta\tau_{2i}^{1,k}(t) \tag{13}$$

$$\tau_{2i}^{-1}(t+1) = \rho_2 \cdot \tau_{2i}^{-1}(t) \cdot \prod_{k=1}^{\omega-1}(\omega - k) \cdot \Delta\tau_{2i}^{-1,k}(t) \cdot \omega \cdot \Delta\tau_{2i}^{-1,gb}(t) \cdot \sum_{k=Rd} \Delta\tau_{2i}^{-1,k}(t) \tag{14}$$

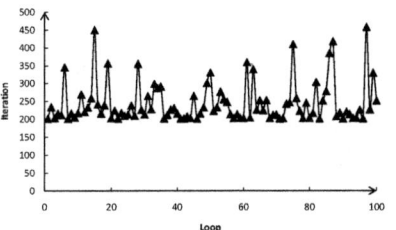
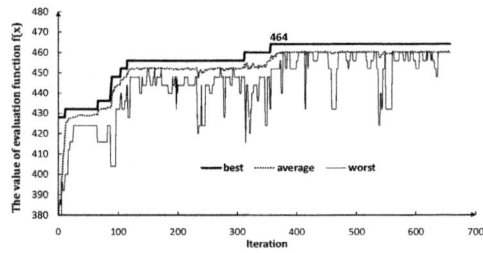

(a) The convergence of graph 12 in 100 loops

(b) The convergence property of graph 14

Fig. 1. The convergence property

4   Simulate Result

In order to widely verify the proposed method, we tested the method on a total of 19 benchmark graphs. In the simulation, the size of ant colony was set to 100. In the preliminary experiments we tried to find reasonable parameter settings for the proposed ant colony algorithm. In the evaluation function, the parameter A, B will influence the quality of the solution seriously, from the preliminary experiments we set $A = B = 4$, which means both reserving number and intersecting number will influence the evaluation function, even when the intersecting number is very small. The values of parameters r_1, r_2, r_3 decide which is important among the first kind of pheromone, the second kind of pheromone and the heuristic, from the preliminary experiments, we set $(r_1, r_2, r_3) = (1, 8, 1)$. The pheromone decay coefficients are set as following: $\rho_1 = \rho_2 = 0.1$.

For every graph, we simulated for 100 loops, the average numbers of the convergence iterations for graph 1 to graph 19 are shown in the Table 1. From Table 1., we can see that the best solution can be found within hundreds of iterations using the proposed ant colony algorithm, but it needs at least ten thousand iterations to find the best solution using other algorithms, for example HNL and GA. To express the convergence property in detail, the convergence iteration of every loop is shown in Fig.1(a), which induced from graph 12 (20 vertices, 90 edges benchmark graph). We can see that the algorithm can converge within hundreds of iterations. And although it converges quickly, the solution which has been found is also very good. And to see the process of the convergence in detail, we tested the convergence of the graph 14 (50 vertices, 491edges benchmark graph), which is shown in Fig.1(b). We can see that the best solution was found within hundreds of iteration.

To evaluate our results, we compare the results of the method with the Takefuji-Lee's [13] neural network (T-L), Hopfield network learning method (HNL) [15] ant the improved genetic algorithm proposed by Wang et al. [16]. Since all these methods used the same vertex ordering, we can say that this

Table 1. The simulation results

Graph	Vertices	Edges	T-L [13]	HNL [15]	IGA [16]	proposed algorithm	convergence iterations
G_1	10	22	20	20	20	20	212.61
G_2	45	85	80	80	80	80	203.53
G_3	10	24	21	22	22	22	204.89
G_4	10	25	22	22	22	22	203.42
G_5	10	26	22	22	22	22	204.17
G_6	10	27	22	22	22	22	204.06
G_7	10	34	23	23	23	23	207.68
G_8	25	69	58	61	61	61	251.30
G_9	25	70	59	61	61	61	246.60
G_{10}	25	71	58	61	61	61	234.58
G_{11}	25	72	60	61	61	61	244.93
G_{12}	25	90	61	63	63	63	244.92
G_{13}	50	367	70	82	84	83	356.39
G_{14}	50	491	100	109	114	116	357.77
G_{15}	50	582	101	115	119	118	343.87
G_{16}	100	451	92	100	101	102	364.52
G_{17}	100	742	116	126	127	132	437.85
G_{18}	100	922	115	135	138	141	442.70
G_{19}	150	1064	127	138	145	145	447.89

comparison is a suitable. Information on the test graph as well as all simulation results are shown in Table 1. The results that we recorded for each graph the number of nodes, the number of edges, and the size of the planar subgraph produced by each algorithm. We can see in Table 1. that for graphs $G_1 \sim G_{12}$, both the proposed method and HNL obtained the same solutions. On the other hand, for most of the other graphs, the proposed method performs much better than other algorithm.

5   Conclusions

We have proposed an effective ant colony algorithm for graph planar. In the proposed algorithm two kinds of pheromone have been defined, and each kind of pheromone consists of two elements, it makes the ACO algorithm more complete. To evaluate the proposed ant colony algorithm, a larger number of simulations have been executed, and the simulation results show that the proposed algorithm works remarkably well and superiorly to its competitors to solve the graph planarization problem. Besides, it is worth noting that the proposed algorithm can be expanded to ant colony algorithm with n kinds and n-elements pheromone according to a problem to be solved.

References

1. Nishizeki, T., Chiba, N.: Planar Graphs: Theory and Algorithm. North Holland, Amsterdam (1988)
2. Liu, P.C., Geldmacher, R.C.: On the deletion of non-planar edges of a graph. In: Proc. 10th South-East Conference on Combinatorics, Graph Theory, and Computing, Boca Raton, FL, USA, pp. 727–738 (1977)
3. Jayakumar, R., Thulasiraman, K., Swamy, M.N.S.: $O(n^2)$ algorithms for graph planarization. IEEE Trans.Comput.-Aided Des.Integr.Circuits Syst. 8(3), 257–267 (1989)
4. Kant, G.: An $O(n_2)$ maximal planarization algorithm based on PQ-tree, Technical Report RUU-CS-92-03. Dept. of Computer Science, Utrecht University, Utrecht, the Netherlands (1992)
5. Cai, J., Han, X., Tarjan, R.E.: An $O(mlog^n)$ time algorithm for the maximal planar subgraph. SIAM J.Comput. 22, 1142–1162 (1993)
6. Hopcroft, J., Tarjan, R.E.: Efficient planarity testing. J. ACM 21(4), 549–568 (1974)
7. Battista, G.D., Tamassia, R.: Incremental planarity testing. In: Proc. IEEE Symp. on Found. of Comp. Sci., pp. 436–441 (1989)
8. Hsu, W.L.: A linear time algorithm for finding maximal planar subgraphs. In: Staples, J., Katoh, N., Eades, P., Moffat, A. (eds.) ISAAC 1995. LNCS, vol. 1004, pp. 352–362. Springer, Heidelberg (1995)
9. Goldschmidt, O., Takvorian, A.: An efficient graph planarization two-phase heuristic. Networks 24(2), 69–73 (1994)
10. Junger, M., Mutzel, P.: Maximum planar subgraphs and nice embeddings: Practical layout tools. Algorithmica 16, 33–59 (1996)
11. Resende, M.G.C., Ribeiro, C.C.: A GRASP for graph planarization. Networks 29, 173–189 (1997)
12. Hopfield, J.J., Tank, D.W.: Neural computation of decisions in optimization problems. Biol.Cybern. (52), 141–152 (1985)
13. Takefuji, Y., Lee, K.C.: A near-optimum parallel planarization algorithmh. Science 245(4922), 1221–1223 (1989)
14. Takefuji, Y., Lee, K.C., Cho, Y.B.: Comments on $O(n^2)$ algorithm for graph planarization. IEEE Trans. Comput.-Aided Des. Integer. Circuits Syst. 10(12), 1582–1583 (1991)
15. Wang, R.L., Tang, Z., Cao, Q.P.: An efficient parallel algorithm for planarization problem. IEEE Trans. Circuit Syst. I, Fundam., Theory Appl. 49(3), 101–397 (2002)
16. Wang, R.L., Okazaki, K.: Solving the Graph Planarization problem Using an Improved Genetic Algorithm. IEICE Trans. Fundamentals E89-A(5) (May 2006)
17. Poli, R., Kennedy, J., Blackwell, T.: Particle swarm optimization—An overview. Swarm Intell. 1, 33–57 (2007)
18. Wang, R.-L., Okazaki, K.: A two-state ant colony algorithm for solving the minimum graph bisection problem. International Journal of Computational Intelligence and Applications 8(4), 487–498 (2009)

A Novel Chaos Glowworm Swarm Optimization Algorithm for Optimization Functions

Kai Huang[1] and Yong quan Zhou[1,2]

[1] College of Mathematics and Computer Science Guangxi University for Nationalities
Nanning, Guangxi 530006, China
[2] Guangxi Key Laboratory of Hybrid Computation and IC Design Analysis,
Nanning Guangxi 530006, China
yongquanzhou@126.com

Abstract. This paper a novel chaotic glowworm swarm optimization algorithm (CGSO) is proposed. In CGSO algorithm, the chaotic search strategies are incorporated in GSO to initialize the first iteration solutions, so that it can obtain high-quality and evenly distributed initial solutions, and avoids GSO being trapped in local optima, each glowworm disturbs by chaos in a disturbance range can get more precise global solution. Compared with GSO algorithm, experiments with six test functions shows that convergence quality and precision are improved, which testify that CGSO are valid and feasible.

Keywords: Glowworm swarm optimization, chaotic search strategy, chaotic glowworm swarm optimization, test functions.

1 Introduction

Many problems of the natural sciences and the engineering technology can be summarized as global optimization problems. About global optimization problem, so far, some researchers have already proposed a variety of optimization algorithms, including traditional algorithms based on gradient and various heuristic algorithms. Traditional algorithms such as DFP variable-dimension algorithm, gold segmentation method, the conjugate direction method, Powell accelerated method, interval method and so on [1]. The optimization results of these algorithms depend highly on the initial conditions and the modality of functions. The modality of many objective functions of engineering problems is complicated, so it is very hard to solve this problem with traditional algorithms based on gradient. In recent years, with the development of computational intelligence technology, various new bionic intelligent algorithms have been proposed, such as particle swarm algorithm, simulated annealing algorithm, genetic algorithm and so on.

Glowworm swarm optimization (GSO) [2,3] is proposed by Krishnanad K.N. and Ghose.D in 2005. The algorithm has been successfully applied to the sensor's noise test [4] and simulation robot [5], etc. The algorithm is swift and efficient in capturing optimal regions and has high commonality, etc. But it also has some problems, such as the highly dependence on initial solutions, slow convergence, easy to be trapped in local optima and the imprecise solutions. So this paper proposed a chaotic glowworm

D.-S. Huang et al. (Eds.): ICIC 2011, LNBI 6840, pp. 426–434, 2012.

swarm optimization (CGSO) to conquer the defects of basic GSO. Simulation results demonstrate that the improved algorithm can get high-quality and evenly distributed initial solutions, which is effective to avoid being trapped in individual local optima.

2   Basic Glowworm Swarm Optimization Algorithm (GSO)

In the GSO, a swarm of glowworms are initially randomly distributed in the solution space. They carry their own luciferin respectively which has equal initial value. The glowworms emit a light whose intensity is proportional to the associated luciferin. The luciferin quantity is tightly associated with the position the glowworms locate in their movement. The glowworm whose luciferin quantity is higher has stronger attraction to the other glowworms in its neighborhood. The neighborhood whose size is decided by radius (r_d) is called local-decision range in GSO. The size of r_d dynamically changes between 0 and r_s. r_s is called radial sensor range in GSO. In the movement, each glowworm moves to another glowworm which is in its neighborhood at a certain probability. Glowworm j who wants to become the neighbor of glowworm i must be located in the neighborhood of i and has higher luciferin quantity than i. Through the movement of glowworms, most glowworms will converge to the glowworms that have higher luciferin quantity. Each iteration of GSO is constituted by two stages. The first one is update stage, the other one is movement stage.

Luciferin update stage: In this stage, every glowworm updates their luciferin by formula (1).

$$l_i(t) = (1-\rho)l_i(t-1) + \gamma J(x_i(t)) \qquad (1)$$

Where $l_i(t)$ is the luciferin quantity of i at iteration t, $\rho \in (0,1)$ is the modulus to control luciferin quantity, γ is the modulus to evaluate objective value of function, $J(x_i(t))$ and is the objective value of function。

Movement of Glowworm: In this stage, glowworm i selects another glowworm j which is located in its neighborhood and move to it. The probability formula is given by (2), the next position of i is decided by (3), the update of r_d, which is in the end of movement stage, is given by (4).

Probability formula used to select a neighbor is as follow:

$$p_{ij}(t) = \frac{l_j(t) - l_i(t)}{\sum_{k \in N_i(t)} l_k(t) - l_i(t)} \qquad (2)$$

Update formula of position:

$$x_i(t+1) = x_i(t) + s * \left(\frac{x_j(t) - x_i(t)}{\|x_j(t) - x_i(t)\|} \right) \qquad (3)$$

Where $x_i(t) \in R^m$ is the position of i, at time t, in the m-dimensional real space, $\| \bullet \|$ represents the Euclidean norm operator, and s (>0) represents moving step of glowworm.

Update formula of local-decision range:

$$r_d^i(t+1) = \min\{\ r_s, \max\{\ 0, r_d^i(t) + \beta(n_t - |N_i(t)|)\}\} \tag{4}$$

Where β is the proportion modulus, n_t is the modulus used to control the number of neighbors, $|N_i(t)|$ is the number of neighbors of i.

3   Chaotic Glowworm Swarm Optimization Algorithm (CGSO)

3.1   Defects of GSO

Analyzing defects of GSO in its search process: (1) Initialization process of GSO is random. Although random initialization can guarantee the initial solutions distributed evenly in the solution space, the quality of solutions are unreliable, because a part of solutions apart from the global optimum. If the initial solutions are not only distributed evenly but also high-quality, it will contribute to the quality and efficiency of solutions, and prevent algorithm to be prematurely trapped in local optima in a certain extent. (2) The search process of GSO is by constant step. When a glowworm who locates at the local optima has no neighbors, it will become stationary. All these reasons lead to the inaccuracy of GSO.

3.2   Chaotic Search Strategy

Chaos is widespread in nature [6], a nonlinear phenomenon presents in the vast majority of nonlinear systems. Chaos is a random motion mapped by deterministic equation, but it is different from the phenomenon of disorder and irregularity. Chaos looks like a random phenomenon, but it has a fine internal structure. Chaos has such properties: (1) random; (2) Ergodic: it can search all of the states unrepeated by its own rules within certain range; (3) regularity. So it is advantageous for the optimization with chaos undoubtedly. Because of all these advantages, can chaos became a good strategy for search process to avoid being trapped in local optima and increase the ability of searching global optimum.

Chaos search is usually by Logistic function, the formula is as follow:

$$z_{i+1,d} = \mu z_{i,d}(1 - z_{i,d}) \tag{5}$$

Where μ is a control parameter. When $\mu = 4, 0 \le z_0 \le 1$, Logistic is totally in chaotic state. In this text, μ assign 4, formula (5) is used as the chaotic signal generator.

3.3   CGSO Algorithm

Detail steps of CGSO are as follows:

Step 1. Initialize $\rho, \gamma, \beta, s, l_0, m, N, D, \ p, n_t, r_s$, and initialize the maximum iteration number T_{max} .

Step 2. Initialize glowworms by chaos.

Step 2.1. Generate a vector of D -dimension $z_1 = [z_{1,1}, z_{1,2}, z_{1,d} \cdots, z_{1,D}]$, $z_{1,d} \in [0,1]$, each dimension has tiny difference.

Step 2.2. Use vector z_1 as initial iteration vector of chaos, according to formula (5), $z_{i+1,d} = \mu z_{i,d}(1-z_{i,d})$, $(d = 1,2,...D; i = 1,2,...N-1)$ can get N number of $z_1, z_2, \cdots, z_N$.

Step 2.3. Map each dimension of z_i to the range of the objective variable by $x_{id} = a_d + (b_d - a_d)z_{id}, (d = 1,2, \cdots D; i = 1,2, \cdots N)$, and then calculate the fitness of $x_i, i = 1,2, \cdots N$, select m individuals which are better from the N initial individuals.

Step 3. generate a vector of D -dimension, each dimension is between 0 and 1, such as $u_0 = (u_{0,1}, u_{0,2}, \cdots, u_{0,n})$.

While (iteration $t \le T_{max}$) do

For $i = 1 : m$

Step 4. Update luciferin of all the glowworm according to formula (1).

Step 5. Calculate the neighbors of each glowworm.

Step 6. Select $j(j \in N_i(t))$ as the movement direction of i by roulette, and update the position of i by formula (3).

Step 7. Update r_d by formula (4) .

Step 8. Disturb to the position each glowworm locates.

Step 8.1. Generate

$$u_1 = (u_{1,1}, u_{1,2}, ..., u_{1,n}), \ u_{1,d} = 4u_0(1 - u_{0,d}), d = 1,2,...,D.$$ and map each dimension of u_1 to the range of disturbance $[-r_d^i, r_d^i]$.

Step 8.2. Get the new position x_i' of x_i according to step 8.2 by

$$x_i' = x_i - r_d^i + 2r_d^i u_1, \ u_0 = u_1 .$$

Step 8.3. Calculate the fitness of x_i' . If x_i' is better than x_i , then replace x_i , else not.

End While

Step 9. Let $t = t + 1$, complete an iteration, and judge whether the end condition is satisfied,. If satisfied, record the result and exit iteration, or return to step 4 for the next iteration.

4   Experimental Results and Analysis

4.1   Test Functions

In order to demonstrate that CGSO has better convergence speed and can get more precise solution than GSO. The six functions are as follows:

$$F_1(x) = -13 + x_1 + ((5 - x_2) \times x_2 - 2) \times x_2)^2 + -29 + x_1 + ((x_2 + 1) \times x_2 - 14) \times x_2)^2$$

$$F_2(x) = (x_1^2 + x_2^2)^{\frac{1}{4}} (\sin^2 (50 (x_1^2 + x_2^2)^{\frac{1}{10}} + 1 \cdot 0)$$

$$F_3(x) = \sum_i^{10} x_i^2$$

$$F_4(x) = 100 (x_1^2 - x_2^2)^2 + (1 - x_1)^2$$

$$F_5(x) = g(x)h(x)$$

$$g(x) = 1 + (x_1 + x_2 + 1)^2 (19 - 14 x_1 + 3 x_1^2 - 14 x_2 + 6 x_1 x_2 + 3 x_2^2)$$

$$h(x) = 30 + (2 x_1 - 3 x_2)^2 (18 - 32 x_1 + 12 x^2 + 48 x_2 - 36 x_1 x_2 + 27 x_2^2)$$

$$F_6(x) = 4 + 4 \cdot 5 x_1 - 4 x_2 + x_1^2 + 2 x_2^2 - 2 x_1 x_2 + x_1^4 - 2 x_1^2 x_2$$

Table 1. Standard test functions

Function	Search range	Optimum
$F_1(x)$	[-10, 10]	0
$F_2(x)$	[-100, 100]	0
$F_3(x)$	[-100, 100]	0
$F_4(x)$	[-2.048, 2.048]	0
$F_5(x)$	[-2, 2]	3
$F_6(x)$	[-100, 100]	-0.5134

4.2   Experimental Parameters

Experimental identical parameters are set as table 2:

Table 2. Parameters of CGSO

ρ	γ	β	n_t	l_0
0.4	0.6	0.08	5	5

For the functions of F_1, F_2, and $F_4 - F_6$, the glowworm number is 50 and maximum iteration is 400 and step is 0.03 and r_s is 3. For the functions of

F_3, the glowworm number is 100 and maximum iteration is 200 and step is 0.3 and r_s is 10.

4.3 Test Environment

The GSO and CGSO are coded in MATLAB2008a and implemented on Intel® Core™2 Duo CPU E4500 2.20GHz PC with 2G RAM under windows XP operation system.

4.4 Simulation Results and Analysis

We test CGSO and GSO by two methods: the first one is that we consider it convergent while $\left| fun_{bes} - fun^* \right| < \varepsilon$ is satisfied under the given precise ε, or we consider it not convergent. For the low-dimension functions of F_1, F_2 , and $F_4 - F_6$, the ε is 1×10^{-5}. For the high-dimension functions of F_3, the ε is 1×10^{-1}. fun_{best} is the optimum we get, while fun^* is the theory optimum; the second one is that CGSO and GSO are implemented respectively to obtain the best, worst and average objective function values, so we can compare the two algorithm through those data and following convergent curve graph.

Table 3. The comparison of the required iterations under setting precise in 20 times experiments

Function	Algorithm	The lest iterations are needed for 20 experiments	The most iterations are needed for 20 experiments	The average iterations are needed for 20 experiments	The amount of time of global optima are obtain	Convergence rate
$F_1(x)$	GSO	289	379	3.3871 e+002	0	7/20
	CGSO	89	263	1.8065e+002	0	20/20
$F_2(x)$	GSO	61	287	1.9720e+002	0	5/20
	CGSO	1	35	11.8000	2	20/20
$F_3(x)$	GSO	A/N	A/N	A/N	0	0/20
	CGSO	43	195	1.2666e+002	0	12/20
$F_4(x)$	GSO	51	149	92.8000	0	20/20
	CGSO	18	92	51.3000	0	20/20
$F_5(x)$	GSO	122	353	213	0	5/20
	CGSO	13	342	1.6160e+002	0	10/20
$F_6(x)$	GSO	A/N	A/N	A/N	0	0/20
	CGSO	76	147	1.0990e+002	20	20/20

From table 3 and table 4, we can see that GSO is trapped in local optima for F_6, while CGSO is well convergent. For F_2 and F_6, CGSO needs less iteration to converge and has more precise solution than GSO.

Table 4. The test results compared between GSO and CGSO

Function	Algorithm	Best objective value	Worst objective value	Average objective value
$F_1(x)$	CGSO	1.263369205817174e-008	1.170754437513471e-006	3.346502236808668e-007
	GSO	2.103691420791615e-007	36.208308877831712	4.403960750989402
$F_2(x)$	CGSO	0	4.073643334612667e-009	3.958345307383755e-010
	GSO	1.243690386468097e-007	0.046894577920103	0.014444826943134
$F_3(x)$	CGSO	0.023628956053632	0.263412643469820	0.113698630782524
	GSO	0.122486743156930	0.494959851739065	0.320615232721686
$F_4(x)$	CGSO	3.345849580315825e-009	2.887969696359693e-007	7.229059331287383e-008
	GSO	2.592568694356495e-009	6.813296340394069e-007	9.533037909065978e-008
$F_5(x)$	CGSO	3.000000050935355	3.000059917677942	3.000012844101273
	GSO	3.000000473184144	3.000082722650252	3.000029193257929
$F_6(x)$	CGSO	-0.513409256498349	-0.513407051098007	-0.513408950527014
	GSO	57.391904470722721	8.821796312295113e+003	1.835902344154499e+003

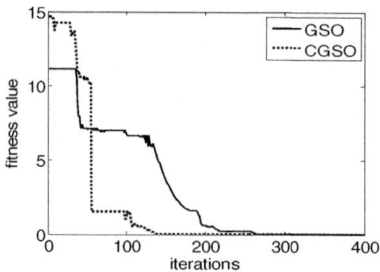

Fig. 2. Comparison of curve graph for F_1

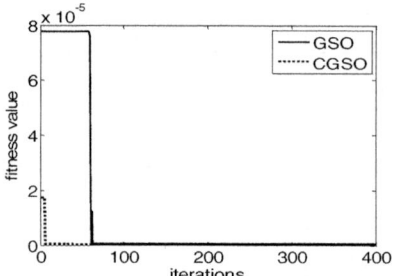

Fig. 3. Comparison of curve graph for F_2

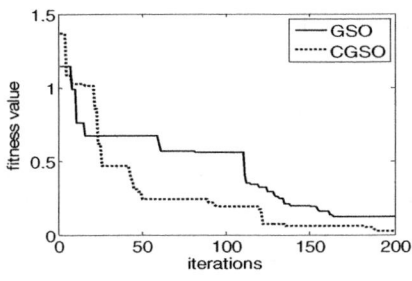

Fig. 4. Comparison of curve graph for F_3 **Fig. 5.** Comparison of curve graph for F_4

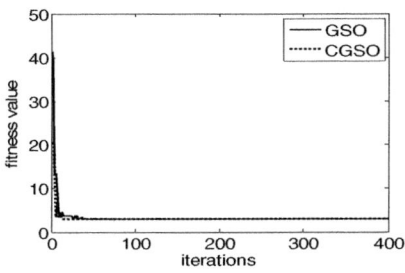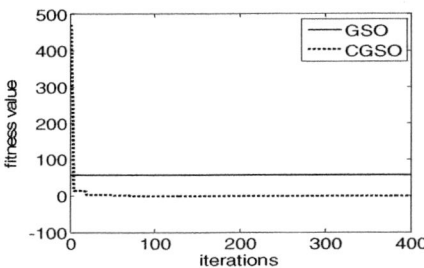

Fig. 6. Comparison of curve graph for F_5 **Fig. 7.** Comparison of curve graph for F_6

From above convergence graph, we can see that CGSO has better convergence speed and higher precise than GSO.

5   Conclusions

This paper proposed a chaotic glowworm swarm optimization algorithm. Strategy of chaotic initialization incorporated to GSO can increase the quality of initial solutions and avoid being in local optima in a certain extent and enhance the ability of capturing the global optimum. Moreover, chaotic disturbance added to GSO can enhance the convergence speed and the precise of solution.

Acknowledgement. This work is supported by Grants 0991086 from Guangxi Science Foundation.

References

1. Csendest: Numerical experiences with a new generalized subinterval selection criterion for interval global optimization. Reliable Computing 9(2), 109–125 (2003)
2. Krishnanand, K.N., Ghose, D.: Glowworm swarm optimization: a new method for optimizing multi-modal functions. Computational Intelligence Studies 1(1), 93–119 (2009)

3. Krishnanand, K.N.: Glowworm swarm optimization: a multimodal function optimization paradigm with applications to multiple signal source localization tasks. Indian: Department of Aerospace Engineering, Indian Institute of Science (2007)
4. Krishnanand, K.N., Ghose, D.: A glowworm swarm optimization based multi-robot system for signal source localization. Design and Control of Intelligent Robotic Systems, 53–74 (2009)
5. Krishnanand, K.N., Ghose, D.: Chasing multiple mobile signal sources: a glowworm swarm optimization approach. In: Third Indian International Conference on Artificial Intelligence (IICAI 2007), Indian (2007)
6. Liu, B., Wan, L., Jin, Y.H., et al.: Improved particle swarm optimization combined with chaos. Chaos, Solitons and Fractals 25, 1261–1271 (2005)

Semi-supervised Protein Function Prediction via Sequential Linear Neighborhood Propagation

Jingyan Wang, Yongping Li*, Ying Zhang, and Jianhua He

Shanghai Institute of Applied Physics, Chinese Academy of Sciences,
2019 Jialuo Road, Jiading District, Shanghai 201800, P.R. China
{wangjingyan,ypli,zhangying1,hejianhua}@sinap.ac.cn

Abstract. Predicting protein function is one of the most challenging problems of the post-genomic era. The development of experimental methods for genome scale analysis of molecular interaction networks has provided new approaches to inferring protein function. In this paper we introduce a new graph-based semi-supervised classification algorithm Sequential Linear Neighborhood Propagation (SLNP), which addresses the problem of the classification of partially labeled protein interaction networks. The proposed SLNP firstly constructs a sequence of node sets according to their shortest distance to the labeled nodes, and then predicts the function of the unlabel proteins from the set closer to labeled one, using Linear Neighborhood Propagation. Its performance is assessed on the *Saccharomyces cerevisiae* PPI network data sets with good results compared with three current state-of-the-art algorithms.

Keywords: Protein Function Prediction, Graph-base Semi-supervised Classification, Sequential Linear Neighborhood Propagation.

1 Introduction

A major challenge in the post-genomic era is to determine protein function at the proteomic scale. Traditionally, computational methods to assign protein function have relied largely on sequence homology. In recent years, the problem of predicting protein functional categories given the set of interactions and the knowledge of the function for a subset of interacting proteins has received increasing attention. In this paper we address these issues by proposing a new approach to function prediction from protein-protein interaction (PPI) network. Many works have been done to learn the protein function from the PPI network [2,3,4,10]. However, in general, the above protein function prediction methods are supervised methods, which require many function labeled proteins to establish a learner of the satisfactory generalization capability. In many practical applications, the annotation protein is often difficult, expensive and time consuming, especially when it has to be done manually by experts. On the other hand, there is often a massive amount of unannotated proteins available. In order to exploit

---

* Corresponding author.

D.-S. Huang et al. (Eds.): ICIC 2011, LNBI 6840, pp. 435–441, 2012.

unannotated proteins in PPI, semi-supervised protein function prediction has become a novel paradigm by using a large number of unannotated points together with a small number of annotated points to build a better learner. Freschi [2] proposed the semi-supervised protein function prediction in a view of machine learning. In this framework, algorithms exploit both labeled and unlabeled data by leveraging the relationships provided by edges of the graph.

Graph-based semi-supervised learning is first introduced by [5], which is based on kernel smoothing. Linear Neighborhood Propagation (LNP) [6] is proposed to address two little-studied issues: graph structure construction and edge weight estimation. However, for the original LNP, the certainty about predictions of LNP is equal for all the unlabeled proteins in the PPI network. This method neglected the fact that when the LNP protein function predictor moves away from the labeled proteins, uncertainty about predictions increases and the class prior becomes more informative. We address this problem by defining a sequence of node sets of PPI according to their shortest distance to the function labeled node in PPI. Then from the labeled proteins, the function labels are propagated using LNP from the inner sets to the outside sets step by step, using LNP. Thus the proposed semi-supervised protein function method is called Sequential Linear Neighborhood Propagation (SLNP).

The paper is organized as follows: Section 2 introduces Sequential LNP function prediction algorithm. We show the performance and consequences of our method on Saccharomyces Cerevisiae PPI network in Section 3, whereas Section 4 concludes the paper.

2   Sequential Linear Neighborhood Propagation

The aim of our approach is the derivation of an algorithm for function prediction (label classification) in PPI networks. A first input of this algorithm is a network of physically interacting proteins that we represent as a graph whose nodes are the proteins and whose (possibly weighted) edges represent the strength of such interactions. A second input is the set of label annotations associated to each protein of a given subset of the nodes. The output of the algorithm is a prediction of the function(s) for each of the proteins whose label is unknown [2].

Suppose there are n proteins $X = \{x_i\}_{i \in N}$ and l proteins $\{x_i\}_{i \in L}$ are labeled as $\{\overline{y}_i\}_{i \in L}, L = \{1, \cdots, l\}$. Consider a two-class classification problem, we task is to predict whether one protein x_i have a special function or not. Denote $\overline{\mathbf{y}}_l = [y_1, y_2, \cdots, y_l]^T$, then

$$y_i = \begin{cases} 1, & if \ x_i \ have \ this \ function \\ -1, & otherwise \end{cases} \tag{1}$$

The task is to assign the labels $\{y_i\}_{i \in U}$ to the remaining points $\{x_i\}_{i \in U}$ with $U = \{l+1, \cdots, n\}$. Let $u = n - l$ be the number of unlabeled proteins, $\mathbf{y}_u = [y_{l+1}, y_{l+2}, \cdots, y_n]^T$, and $\mathbf{y} = [y_1, y_2, \cdots, y_l, y_{l+1}, y_{l+2}, \cdots, y_n]^T = \begin{bmatrix} \overline{\mathbf{y}}_l \\ \mathbf{y}_u \end{bmatrix}$.

2.1   Sequential LNP

PPI network is organized in the form of a graph $G = \{V, E, W\}$ [6]. The node set V corresponds to n proteins in PPI , $V = V_L \cup V_U$, where V_L is the set of function labeled proteins, while V_U is the set of unlabeled ones. E is the edge set, and $E = \{e_1, \cdots, e_k\}$ with edge e_i corresponding to hyperedge in PPI, which is constructed by connecting a vertex x_i and its neighboring vertices $N(i)$, as $e_i = \{x_i\} \cup N(i)$ [6]. $W = \{w_1, \cdots, w_k\}$ with w_i equal to the weights of pairs of proteins in edge e_i, weights should satisfy $\sum_{j \in N(i)} w(i, j) = 1$.

Let $d(i, j)$ denote the shortest path distance between two nodes x_i and x_j in the PPI networks. Further, let $d(i, S)$ denote the shortest path distance between node x_i and a node set S, formally, $d(i, S) = min_{j \in S}\{d(i, j)\}$. Consider next a sequence of sets $\{V^r\}_{r=0}^k$, such that $V_L = V^0 \subseteq V^1 \subseteq V^2 \subseteq \cdots V^k = V$. The sets V^r contain all the unlabeled nodes that are within a certain distance of labeled ones and are formally defined as:

$$V^r = \{x_i \in V | d(i, V_L) \leq q^r\} \tag{2}$$

where q^r a prespecified nonnegative threshold with $q^0 = 0$ and $q^k = max_{x_i \in V_U} d(i, V_L)$. We set $q^0 = 0$ so that only the labeled nodes are included in $V_0 = V_L$, and $q^k = max_{x_i \in V_U} d(i, V_L)$ so that all the nodes can be included in the last set $V^k = V$. In the case of an unweighted graph, the sets V^r contain all the unlabeled nodes that are at most q^r hops away from labeled ones.

Finally, define the sequence of symmetric adjacency matrix A^r of size $|V^r| \times |V^r|$, whose (i, j) element $A^r(i, j)$ contains the similarity measure between nodes x_i and $x_j \in V^r$. This matrix A^r is a submatrix of the weighted similarity matrix A, whose values contain similarities between nodes in V. Given a adjacency matrix A^r, we construct the smoother weight matrix $W^r = [w^r(i, j)]$ for V^r,

$$w^r(i, j) = \frac{A^r(i, j)}{\sum_{k \in N^r(i)} A^r(i, k)} \tag{3}$$

where $N_r(i) = N(i) \cap V^r$, is the neighbors of x_i within V^r. Notice that W^r is also a stochastic matrix as W.

A biharmonic matrix Q^r [6] of size $|V^r| \times |V^r|$ that is designed to reflect the local graph structure, is then defined for any $x_i \in V^r$ as

$$Q^r = (I - W^r)^T (I - W^r) \tag{4}$$

The Sequential Linear Neighborhood Propagation (SLNP) proceeds in an iterative fashion and at step r uses the labeled proteins $\mathbf{y}^{r-1} = \begin{bmatrix} \mathbf{y}_l^{r-1} \\ \mathbf{y}_u^{r-1} \end{bmatrix}$ (both known and estimated ones) in set V^{r-1} to predict those unlabeled proteins $\mathbf{y}_u^r$ in set V^r. The final output of the algorithm is an estimate of the entire vector $\mathbf{y}$. In many cases, it is common to clamp the response at each iteration, by setting $\mathbf{y}_l^r = \mathbf{y}^{r-1}$ for all nodes $v \in V^{r-1}$. Then, in step r, we can estimate y_u^r by

$$E(\mathbf{y}_u^r | \mathbf{y}_l^r = \mathbf{y}^{r-1}) = -Q_{uu}^r {}^{-1} Q_{ul}^r \mathbf{y}^{r-1} \tag{5}$$

where $Q^r_{uu} = [q_{ij}], i, j \in V^r_u$ is a submatrix of Q^r and $Q^r_{ul} = [q_{ij}], i \in V^r_u, j \in V^r_l = V^{r-1}$.

2.2 Predicting the Protein Functions in PPI Using SLNP

We try to develop a new protein function predicting method using SLNP among the PPI in this section. One problem is that the protein function prediction is a multi-class problem. We adopt the one-to-the-rest methodology to calculate y_u for each protein class. The algorithm of the predicting the proteins' functions using SLNP is given Algorithm 1.

Algorithm 1. SLNP Protein Function Prediction Algorithm

Require: $G = \{V, E\}$: the graph representing the PPI network;
Require: $A = [A(i, j)]$: the adjacency matrix of the graph;
Require: $\overline{F}_l$: the matrix of function label annotations associated to each protein of a given subset of the nodes in PPI.
  Initialize $F^0 = \overline{F}_l$;
  Generate the sequence of sets $\{V^r\}^k_{r=0}$ as (2);
  for $r = 1, 2, \cdots, k$ **do**
    $F^r_l = F^{r-1}$;
    Construct the adjacency matrix A^r, whose (i, j) element $A^r(i, j)$ contains the similarity measure $A(i, j)$ between nodes x_i and $x_j \in V^r$;
    Construct the smoother weight matrix $W^r = [w^r(i, j)]$ for V^r as (3);
    Compute the biharmonic matrix Q^r for V^r using (4): $Q^r = (I - W^r)^T(I - W^r)$;
    Compute the transform matrix $P^r = -Q^r_{uu}Q^r_{ul}$;
    Predict the function of proteins $x_i \in V^r_u$: $F^r_u = P^r F^r_l$,
    $$y_u(i, k) = \begin{cases} 1, & if \ f^r_u(i, k) \geq th_k \\ 0, & otherwise. \end{cases}$$
  end for
  Output $y_i, i \in U$: the function annotations of unlabeled proteins in PPI.

---

3   Experiments

We experimentally compare SLNP against the traditional semi-supervised function prediction approaches [1,2] for the Saccharomyces Cerevisiae PPI network.

3.1 Data Set and Experimental Setup

We tested the proposed approach on the Saccharomyces Cerevisiae PPI network following [2] and [1]. It consists of a network of 12,531 interactions among 4,495 proteins that are known to physically interact. The weights that are assigned to the edges between pair of nodes are computed by evaluating the probability (i.e. the reliability) of the interactions after separation of functional linkage by experimental source of evidence[1].

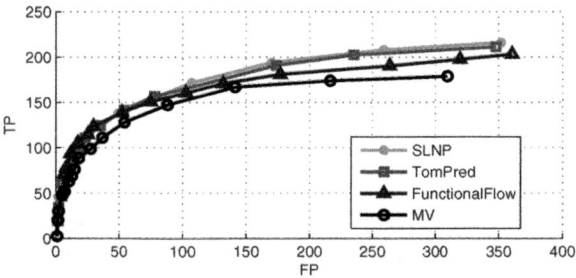

Fig. 1. ROC curve analysis for SLNP, TomPred, FunctionalFlow and MV (SC_w dataset, unknown proteins: 20%)

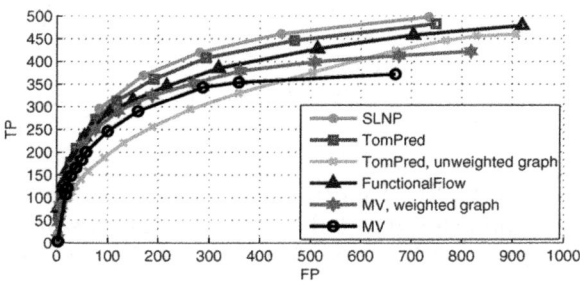

Fig. 2. ROC curve analysis for SLNP, TomPred, FunctionalFlow and MV (SC_w dataset, unknown proteins: 50%)

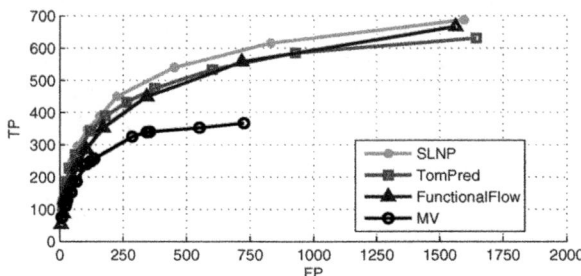

Fig. 3. ROC curve analysis for SLNP, TomPred, FunctionalFlow and MV (SC_w dataset, unknown proteins: 80%)

We followed the experimental setup defined in [1] and [2] , also for what concerns the reference set of labels to be used for annotating the PPI network: in particular we used the MIPS controlled vocabulary for biological processes (second hierarchy level) which consists of 72 labels [1,8]. Cross-validating semi-supervised protein labeling algorithms is an issue that entails the choice of proper benchmarking evaluations. To obtain reliable results, we repeat the experimental process 5 times. We select 5 random subsets from the data to create 5 pairs of annotated and cleared protein subsets. For each of these pairs, algorithms are tested on their capability of recovering correct labels given the remaining subset of annotated nodes.

3.2   Results

We implemented our SLNP protein function method and compare the MV, FunctionalFlow, TomPred, original LNP algorithms on the same benchmark (SC_w dataset). We tested three different levels of sparsity in the annotation of the network to be labeled by clearing, respectively, 10%, 20%, 50%, and 80% of the known annotations. Note that our SLNP construct the sequence of data set using the binary unweighted networks G, while predicting the function categories using the weighted networks with affinity matrix A. We decided to evaluate our approach according to a leave-a-percentage-out cross-validation method. A first degree of freedom is the number of cleared annotated proteins. In the leave-a-percentage-out cross-validation a given fraction of known annotations is cleared and used for testing [1,2,3].

Figs. 1, 2 and 3 show the ROC curves of SC_w protein function prediction performance using the SLNP, TomPred, FunctionalFlow and MV respectively. (Throughout, our approach is denoted by "SLNP.) The results support that our method produces significant better performance for the 2 two experiments (with 50% and 80% proteins are cleared) when compared with the other works. In the experiment with 20% proteins cleared, our method achieved a score very close to the TomPred presented in [2]. By predicting the function label of a protein in according to its PPI neighbors in a linear way sequentially , the learned function shows clear accuracy gains over the traditional function prediction methods. With the SCw dataset when 50% and 80% of proteins are cleared, SLNP yields the most accurate predictions of all methods in almost all the range of the stringency threshold, and with 20% cleared it outperforms the TomPred and FunctionalFlow techniques, except for the leftmost part of the ROC curve.

4   Conclusion

This paper presented a improvement on the LNP semi-supervised classification model for protein function prediction using PPI network. We improved the LNP approach by introducing different uncertainty sequence set, which is appropriated according to their shortest distance to the labeled proteins in PPI. Experiments Comparative protein function prediction experiments on public Cerevisiae PPI network sets show the validity of our algorithm.

Acknowledgment. This work was supported by the National Grand Fundamental Research (973) Program of China under Grant No. 2011CB911102 and Natural Science Foundation of Shanghai under Grant No. 08JC1422500.

References

1. Nabieva, E., Jim, K., Agarwal, A., Chazelle, B., Singh, M.: Whole-proteome Prediction of Protein Function via Graph-theoretic Analysis of Interaction Maps. Bioinformatics 21, i302–i310 (2005)
2. Freschi, V.: A Graph-Based Semi-supervised Algorithm for Protein Function Prediction from Interaction Maps. In: Stützle, T. (ed.) LION 3. LNCS, vol. 5851, pp. 249–258. Springer, Heidelberg (2009)
3. Murali, T.M., Wu, C.J., Kasif, S.: The Art of Gene Function Prediction. Nature Biotechnology 24(12), 1474–1476 (2006)
4. Sharan, R., Ulitsky, I., Shamir, R.: Network-based Prediction of Protein Function. Molecular System Biology 3(88), 1–13 (2007)
5. Culp, M., Michailidis, G.: Graph-Based Semisupervised Learning. IEEE Transactions on Pattern Analysis and Machine Intelligence 30(1), 174–179 (2008)
6. Wang, J.D., Wang, F., Zhang, C.S., Shen, H.C., Long, Q.: Linear Neighborhood Propagation and Its Applications. IEEE Transactions on Pattern Analysis and Machine Intelligence 31(9), 1600–1615 (2009)
7. Srinivasan, B.S., Novak, A.F., Flannick, J.A., Batzoglou, S., McAdams, H.H.: Integrated Protein Interaction Networks for 11 Microbes. In: Apostolico, A., Guerra, C., Istrail, S., Pevzner, P.A., Waterman, M. (eds.) RECOMB 2006. LNCS (LNBI), vol. 3909, pp. 1–14. Springer, Heidelberg (2006)
8. Mewes, H.W., Frishman, D., Guldener, U., Mannhaupt, G., Mayer, K., Mokrejs, M., Morgenstern, B., Munsterkotter, M., Rudd, S., Weil, B.: Mips: A Database for Genomes and Protein Sequences. Nucleic Acid Research 30, 31–34 (2002)
9. Sharan, R., Ulitsky, I., Shamir, R.: Network-based Prediction of Protein Function Author(s): Sharan, R., Ulitsky, I., Shamir, R Source. Molecular Systems Biology 3, 88 (2007)
10. Chua, H.N., Sung, W.K., Wong, L.: Exploiting Indirect Neighbours and Topological Weight to Predict Protein Function from Protein-Protein Interactions. Bioinformatics 22(13), 1623–1630 (2006)

An Improved Newman Algorithm for Mining Overlapping Modules from Protein-Protein Interaction Networks

Xuesong Wang, Lijing Li, and Yuhu Cheng

School of Information and Electrical Engineering
China University of Mining and Technology, Xuzhou, Jiangsu 221116, P.R. China
{wangxuesongcumt,lilijing_29,chengyuhu}@163.com

Abstract. With the development of high-throughput technologies in recent years, more and more scientists focus on protein-protein interaction (PPI) networks. Previous studies showed that there are modular structures in PPI networks. It is well known that Newman algorithm is a classical method for mining associations existed in complex networks, which has advantages of high accuracy and low complexity. Based on the Newman algorithm, we proposed an improved Newman algorithm to mine overlapping modules from PPI networks. Our method mainly consists of two steps. Firstly, we try to discover all candidate nodes whose neighbors belong to more than one module. Secondly, we determine candidate nodes that have positive effects on modularity as overlapping nodes and copy these nodes into their corresponding modules. In addition, owing to the features of existing system noise in PPI networks, we designed corresponding methods for de-noising. Experimental results concerning MIPS dataset show that, the proposed improved Newman algorithm not only has the ability of finding overlapping modular structure but also has low computational complexity.

Keywords: Protein-protein interaction network, Overlapping module, Newman algorithm, Noise, Hub protein.

1 Introduction

As we all know, protein-protein interaction (PPI) network is comprised of pairwise interaction proteins, in which a node represents a protein and an edge represents the pairwise interaction between two proteins. Previous studies showed that there are some modular structures that are densely connected internally but sparsely interacting with the rest of the network in PPI networks [1-2]. In the past decade, many scientists paid attention to the study of module mining methods for protein function prediction. For example, Bader and Hogue proposed a molecular complex detection method [3]; Zhang et al. used the clique percolation method (CPM) to PPI networks [4]. However, most of the methods either can not find overlapping modules in PPI network or have too large computation complexity.

Based on above considerations, we proposed an improved Newman algorithm. We made some local adjustments on the modular structure obtained from traditional

D.-S. Huang et al. (Eds.): ICIC 2011, LNBI 6840, pp. 442–447, 2012.

Newman algorithm [5]. Nodes that have positive effects on modularity of different modules were used to achieve the goal of discovering overlapping modules [6]. The improved Newman algorithm not only inherits the high accuracy and low complexity of Newman algorithm, but also has the ability of mining overlapping modules. Therefore, it is very suitable for large scale network data mining. In addition, we designed de-noising steps according to the concept of clustering coefficient to remove false-positive data in PPI networks [7].

2   Methods

2.1   De-noising

Generally speaking, there are large amount of noise data in PPI networks [8]. Here, we introduce a novel de-noising method, which assigns a weight to each edge of a PPI network to reflect the reliability of the corresponding interactions [9]. The weight $SCC(i, j)$ between node i and j can be defined as

$$SCC(i, j) = CC_i + CC_j - CC_i' - CC_j' \tag{1}$$

where CC_i and CC_j are clustering coefficients of node i and j. CC_i' and CC_j' represent the clustering coefficients of i and j excluding the edge between them. According to the viewpoint of Asur *et al.* [9], if two nodes are not actually connected in the original network, then, the $SCC(i, j)$ value should be smaller or equal to zero. Consequently, we can infer that, if the value of $SCC(i, j)$ is lower than a pre-defined threshold value α, the edge can be removed as noise.

2.2   Improved Newman Algorithm

Newman algorithm is a hierarchical agglomerative method which is based on the idea of modularity [5]. As we know, modularity is a measure of the quality of a particular division of a network, and large value of modularity always corresponds to good network division [6]. If we let e_{rk} to be the fraction of edges in the network that connect vertices in group r to those in group k, and let $a_r = \sum_k e_{rk}$. Then the modularity Q can be described as

$$Q = \sum_r e_{rr} - a_r^2 \tag{2}$$

The physical meaning of equation (2) is that, modularity is equal to the fraction of edges that fall within modules, minus the expected value of the same quantity if edges fall at random with no regard to the modular structure [6]. Through optimizing Q, the best modular structure can be discovered.

In order to speed up the optimization procedure, Newman showed that the application of the change of modularity, ΔQ, can reduce the computation complexity greatly. If there are m edges in the initial network, and k represents the degree, then ΔQ can be calculated.

$$e_{ij} = \begin{cases} 1/2m, & \text{nodes } i \text{ and } j \text{ are connected} \\ 0, & \text{else} \end{cases} \tag{3}$$

$$a_i = k_i / 2m \tag{4}$$

$$\Delta Q = 2(e_{ij} - a_i a_j) \tag{5}$$

By maximize the value ΔQ, a best modular structure could be got.

Generally speaking, the existing algorithms for discovering modular structure in PPI networks are unable to simultaneously satisfy the requirements of low complexity and discover overlapping modules. In this paper, we improve the traditional Newman algorithm to satisfy the above two conditions. The detailed steps of the improved Newman algorithm are listed as follows.

Step 1. Perform Newman algorithm.

Step 2. Discover the nodes that whose neighbors in the initial network are not all included in its module, and collect them to be candidate nodes set.

Step 3. Randomly select a node i from the candidate nodes set. Suppose that one of its neighbors, j, locates in module B. Copy i to B and then we get a new module B'. If the criterion in equation (6) is satisfied, then i is an overlapping node.

$$Q_{B'} > Q_B \tag{6}$$

Q_B and $Q_{B'}$ is the modularity of B and B' respectively.

Step 4. Repeat steps 2 ~ 3 till all overlapping nodes are discovered.

2.3   Algorithm Steps

The steps of using the improved Newman algorithm to mine overlapping modules from PPI networks can be summarized as follows.

(a) Pre-process

Step 1. Import the input data.

Step 2. Remove the edges that whose SCC values are lower than our pre-defined value α.

(b) Mine Overlapping Modular Structure from PPI Networks

Step 3. Perform the traditional Newman algorithm, and get a modular structure C.

Step 4. Discover the candidate nodes from C.

Step 5. Define nodes that satisfy equation (6) to be the overlapping nodes.

Step 6. Copy overlapping nodes and put them into their corresponding modules so as to obtain the overlapping structure C'.

Step 7. Export the output data.

According to the above algorithm steps, the computational complexity of our algorithm is $O(n^2 + mn + m + n)$ with n nodes and m edges in the initial PPI network. Compared with the current existing algorithm CPM ($O(\exp(n))$) and MCL ($O(n^3)$) [10], both of which can detect overlapping modular structure, our algorithm is much faster.

3   Experiments and Analysis

In order to study the performance of our algorithm on large scale PPI networks, we did experiments on a real PPI dataset from Munich Information Center for Protein Sequences (MIPS) [11].The initial dataset contains 5090 proteins and 15662 edges. After we set α to be zero and the minimum module size to be 3, we got 210 modules including 8 overlapping modules.

3.1   Distribution of Modular Structure

Fig. 1 is the module distribution map of the network obtained by the improved Newman algorithm. Fig. 1 shows that most of the modules have smaller size, whereas a very few modules have extremely large one. This coincides with the scale-free property of PPI networks which can be explained by Fig. 2.

Fig. 2 is the degree distribution curve of the MIPS dataset used in our study. Here, the abscissa k represents degree of protein, and the ordinate $P(k)$ represents the fraction of proteins in the network with degree k. Here, we can see that, MIPS networks follows the power law relationship: $P(k) \sim k^{-r}$ with $r \approx 2.5$.

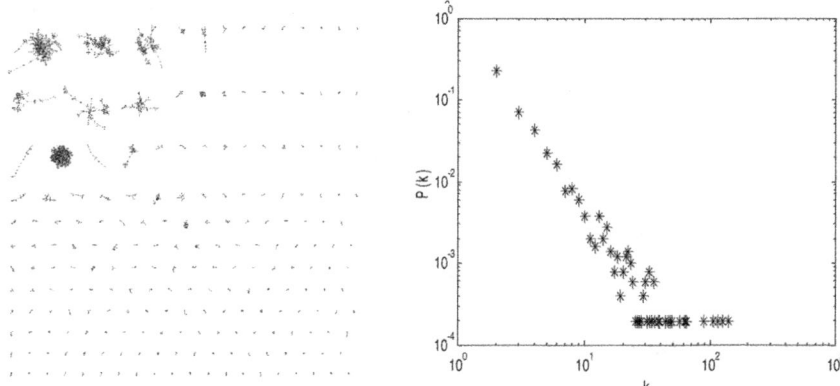

Fig. 1. Module distribution map   **Fig. 2.** Internal structure of module 44

3.2   Comparative Analysis of CPM and the Improved Algorithm

To compare with the CPM algorithm, we firstly defined three evaluation functions. They are data recognition rate (DRR, the proportion of nodes in input data to that of output data), best matching module rate (BMMR, the proportion of modules whose $\log(P_{ol})$ is lower than -3 to all identified modules), and module annotation rate (MAR, the proportion of modules whose match rate is 1 to all identified modules) respectively. Here, P_{ol} is the probability of a random overlap between the experimental complex and the computational complex [2]. By minimizing P_{ol}, we can determine the best-matching function for a module.

$$P_{ol} = \frac{\binom{n_2}{ol}\binom{n - n_2}{n_1 - ol}}{\binom{n}{n_1}} \tag{7}$$

where n_1 and n_2 represent the sizes of experimental complex and computational complex respectively, ol is the number of proteins having similar functions. n is the total number of proteins in a network.

Table 1. Comparison results on the MIPS dataset

Algorithm	Num_M	DRR	MAR	BMMR	Min $\log(P_{ol})$
CPM	125 (k=4)	18.39%	88.80%	45.60%	-77.77
Improved Newman	210	72.80%	67.62%	45.77%	-84.07

(Note: Num_M and Min $\log(P_{ol})$ are the mean number of modules and the minimum value of $\log(P_{ol})$ respectively)

Table 1 shows that, CPM and improved Newman algorithms have nearly the same ability to discover the best matching modules with BMMR being 45.60% and 45.77% respectively. The module annotation rate of CPM is higher than that of the improved Newman algorithm, while the minimum value of $\log(P_{ol})$ of the improved Newman algorithm is better than that of CPM. An overall view of table 3 shows that there exists a huge disparity in data recognition rate. The reason for this is that, CPM is only able to discover fully connected networks, which leads it to neglect many functional divergence networks.

In summary, the above study verifies that the Newman algorithm is a very suitable method for mining modules in large scale PPI networks.

4   Conclusions

With the development of biological systems in recent years, PPI networks are given more and more attention. Using biological or complex network methods for module mining from PPI networks is one of the most popular fields in bioinformatics. In this paper, we proposed an improved Newman algorithm with de-noising steps to mine overlapping modular structure from PPI networks. The experiments on MIPS dataset proves its effectiveness for module mining in large scale PPI networks. However, it must be noted that, the sizes of modules obtained by our method are very different. This is because of the scale-free property of PPI networks, which makes it is difficult for functional annotation. In order to further perfect our algorithm, we will try to settle the problem of unbalanced clustering in our next work.

Acknowledgements. This work was supported by National Nature Science Foundation of China (60804022, 60974050, 61072094), Program for New Century Excellent Talents in University (NCET-08-0836, NCET-10-0765), Fok Ying-Tung Education Foundation for Young Teachers (121066), Nature Science Foundation of Jiangsu Province (BK2008126).

References

1. Schwikowski, B., Uetz, P., Fields, S.: A network of interacting proteins in yeast. Nature Biotechnology 18(12), 1257–1261 (2000)
2. Spirin, V., Mirny, L.A.: Protein complexes and functional modules in molecular networks. Proceedings of the National Academy of Sciences 100(21), 12123–12128 (2003)
3. Bader, G.D., Hogue, C.W.: An automated method for finding molecular complexes in large protein interaction networks. BMC Bioinformatics 4(1), 2 (2003)
4. Zhang, S.H., Ning, X.M., Zhang, X.S.: Identification of functional modules in a PPI network by clique percolation clustering. Computational Biology and Chemistry 30(6), 445–451 (2006)
5. Newman, M.E.J.: Fast algorithm for detecting community structure in networks. Physical Review E 69(62), 066133-1–066133-5(2004)
6. Newman, M.E.J., Girvan, M.: Finding and evaluating community structure in networks. Physical Review E 69(22), 026113-1–026113-15 (2004)
7. Taylor, I.W., Linding, R., Warde, F.D.: Dynamic modularity in protein interaction networks predicts breast cancer outcome. Nature Biotechnology 27(2), 199–204 (2009)
8. Kuchaiev, O., Rašajski, M., Higham, D.J., Pržulj, N.: Geometric de-noising of protein-protein interaction networks. PLoS Computational Biology 5(8), e1000454 (2009)
9. Asur, S., Ucar, D., Parthasarathy, S.: An ensemble framework for clustering protein-protein interaction networks. Bioinformatics 23(13), 29–40 (2007)
10. Danon, L., Diaz-Guilera, A., Duch, J., Arenas, A.: Comparing community structure identification. Journal of Statistical Mechanics: Theory and Experiment, P09008 (2005)
11. Mewes, H.W., Frishman, D., Mayer, K.F.: MIPS: analysis and annotation of proteins from whole genomes in 2005. Nucleic Acid Research 172, D169–D172 (2006)

Protein Interface Residues Prediction Based on Amino Acid Properties Only

Bing Wang[1,2], Peng Chen[3], and Jun Zhang[2]

[1] School of Electrical Engineering and Information, Anhui University of Technology,
Maanshan, Anhui 243002, China
[2] Department of Chemistry, University of Louisville, Louisville, KY 40292, USA
[3] Hefei Institute of Intelligent Machines, Chinese Academy of Sciences,
Hefei, Anhui 230031, China
wangbing@ustc.edu

Abstract. Protein-protein interactions play essential roles in protein function implementation. A computational model is introduced in this work for predicting protein interface residues based on amino acid chemicophysical properties only. 17 amino acid properties are selected from AAindex database and used as input features of a prediction model which is constructed by support vector machines method to infer protein interface residues in protein hetero-complexes. The results achieved in this work demonstrated the properties used in this work can actually capture up the difference between interface and noninterface residues.

Keywords: Amino Acid Property, Protein Interface Residues, Support Vector Machines, Hetero-complexes.

1 Introduction

The interactions between proteins play a very important role for the majority of biological functions, such as DNA replication, signal transduction, immunological recognition and protein synthesis [1]. In recent years, many methods have been developed to predict protein interaction sites or location of interface residues. Those approaches have addressed various aspects of protein structure and properties, such as amino acid composition [2-7], solvent accessibility [8], sequence entropy and secondary structure [9, 10], evolutionary profiles and conservation score [4].

In this paper, a method based on the amino acid chemical and physical properties is proposed for prediction of protein-protein interface residues using a a support vector machines (SVMs) predictor. The amino acid properties were extracted from the amino acid index (AAindex) database. Then has been constructed for differ protein interface residues from non-interface residues in protein chains. The results based on a non-redundant protein chains set show that the amino acid properties what we used in this work are effective to capture the difference between interface and non-interface residues.

D.-S. Huang et al. (Eds.): ICIC 2011, LNBI 6840, pp. 448–452, 2012.

2   Methods

2.1   Amino Acid Properties

The complexes used in this work were same as our previous study [4]. The amino acid properties are extracted from AAindex database, which is a database of numerical indices representing various physiochemical and biochemical properties of amino acids. Only the AAindex1 database was used in this work which comprises 544 sets of numerical indices for the 20 amino acids, and all of them are derived from published literature. Firstly we remove the 13 property items for there are some N/A values within it. Then all similar properties are eliminated for information-redundancy.   The Pearson's correlation coefficient was used here to calculate similarity values among the properties and 0.3 is set up as the similarity threshold. As a result, 17 amino acid properties are used in this work for identifying protein interface reside

$$S(p_i, p_j) = \frac{\text{cov}(p_i, p_j)}{\sigma_{p_i} \sigma_{p_j}} \quad (1)$$

where p_i and p_j denotes amino acid property which is a 20-dimensional vector where the value in each dimension is the index of one of 20 amino acids, and cov means covariance, σ means standard deviations.

Table 1. The selected amino acid properties

Property id	Property description	Property id	Property description
ANDN920101	alpha-CH chemical shifts	JOND920102	Relative mutability
ARGP820101	Hydrophobicity index	KHAG800101	The Kerr-constant increments
BEGF750101	Conformational parameter of inner helix	FAUJ880104	STERIMOL length of the side chain
BUNA790103	Spin-spin coupling constants 3JHalpha-NH	PALJ810107	Normalized frequency of alpha-helix in all-alpha class
BHAR880101	Average flexibility indices	RACS820114	Value of theta(i-1)
BURA740102	Normalized frequency of extended structure	WERD780103	Free energy change of alpha(Ri) to alpha(Rh)
GEOR030101	Linker propensity from all dataset	YUTK870102	Unfolding Gibbs energy in water, pH9.0
CHOP780204	Normalized frequency of N-terminal helix	CHAM830102	A parameter defined from the residuals obtained from the best correlation of the Chou-Fasman parameter of beta-sheet
CHOP780215	Frequency of the 4th residue in turn		

2.2   Predicting Model Construction

Similar to previous works, a sliding window technique is used in this study in order to involve the association among neighboring residues because protein interface is formed by some residues which closed to each other in spatial position. Therefore, the input vector of predicting model is fed with a window of 11 residues, centered on the target residue and including the five spatially neighboring residues on each side. As a result, each residues is represented by a $11 \times 17 = 187$ components vector.

A ten-fold cross-validation strategy was employed to conduct the subsequent experiments. In this strategy, proteins in the dataset are divided into 10 subsets which consist of roughly the same number of proteins, one of them is for the test process and the other ones are for the model training process. The number of positive samples or so-called interface residues is much smaller than that of negative samples or non-interface residues. Only 34.3% of the samples are interface residues in this work, which leads to a rather imbalanced data distribution. To overcome this problem, we will randomly select negative samples in the model training process to make sure the number negative and positive samples is same.

3   Results

3.1   Correlation of the Selected Properties

Obviously, More discriminative power what the property can differentiate interface and non-interface residues is, more successful the prediction model will be. The similarities among 17 amino acid properties can be seen in Table 2.

It can be seen that the maximum similarity among the selected amino acid properties is 0.28, the minimum value is -0.62, and the mean value is -0.05. The very low correlation among the selected amino acid properties means that there is no information-redundancy in the feature set what we used in this work. Reducing the number of amino acid properties from the original 544 to present 17 decreases the computational complexity drastically and speeds up the model learning process. Meanwhile, removing most irrelevant and redundant features from the data can enhance generalization capability of our proposed model.

3.2   Prediction Performance

Among 10329 protein surface residues, our prediction shows 4484 of them are assigned to +1 (interface residue), and 5845 of them are assigned to -1 (non-interface residue) by our proposed prediction model, respectively. Based on the definitions of performance measures, our proposed model can obtain a Sen of 57.9%, a Spec of 65.0%, a Prec of 46.9%. Furthermore, the value of MCC our model achieved in this work, 0.22, denotes that the selected properties can actually captures up the difference between the interface and non-interface residues.

In order to further evaluate our presented model, we compared it with two previous models: one is backpropagation (BP) neural network model, another is radial basis function (RBF) neural network model which is optimized by expectation maximum algorithm (EM), and these two models use amino acid residue sequence profile as input features for predicting protein interface residues. The comparison results can be found in Table 2. It can be seen that our proposed model can obtain best performance among this three computational models, especially for sensitivity measure. Furthermore, the MCC of 0.22 shows the importance of our selected amino acid properties in prediction of protein interface residues.

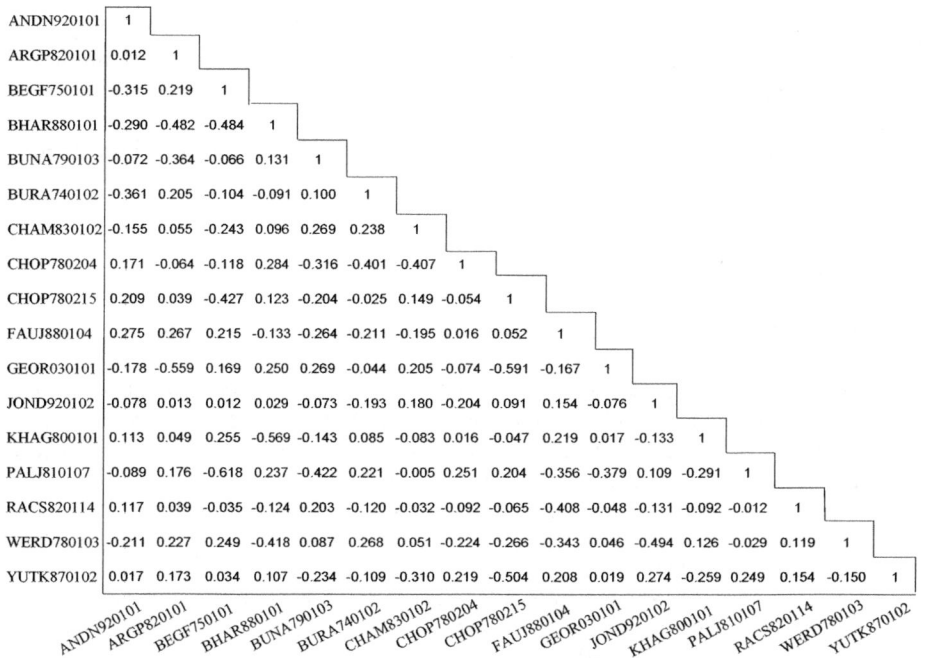

Fig. 1. The pairwise correlation values among the 17 selected amino acid properties

Table 2. Comparisons of prediction performance among three models

Model[a]	Sen	Spec	Prec	Acc	F1	MCC
BP NN	44.9%	62.9%	49.2%	0.61	0.47	0.12
RBF_EM	42.5%	74.8%	57.5%	0.60	0.49	0.18
SVM_AA	57.9%	65.0%	46.9%	0.625	0.52	0.22

[a]BP NN denotes backpropagation neural network; RBF_EM is RBF neural network optimized by EM algorithm; SVM_AA is present model in this work.

4   Conclusions

This work proposed a promising method which can infer protein interface residues from protein surface in hetero-complexes. The results achieved here demonstrated the effectiveness of our proposed method by comparison of two computational models which address the same job. Pearson correlation approach has been employed to the selection of amino acid properties whose original number is 544 and current number is 17, which enhance the model generalization capability and in the same time deduce the computational complexity obviously. Furthermore, the MCC value we achieved in this work shows that the selected amino acid properties we used actually differentiate the interface and non-interface residues.

Acknowledgement. This work was supported by the National Science Foundation of China (Nos. 60803107).

References

1. Alberts, B.D., Lewis, J., Raff, M., Roberts, K., Watson, J.D.: Molecular Biology of the Cell, 2nd edn. Garland, New York (1989)
2. Fariselli, P., Pazos, F., Valencia, A., Casadio, R.: Prediction of protein–protein interaction sites in heterocomplexes with neural networks. Eur. J. Biochem. 269(5), 1356–1361 (2006)
3. Ofran, Y., Rost, B.: Predicted protein-protein interaction sites from local sequence information. Febs Letters 544(1-3), 236–239 (2003)
4. Wang, B., Chen, P., Huang, D.S., Li, J.J., Lok, T.M., Lyu, M.R.: Predicting protein interaction sites from residue spatial sequence profile and evolution rate. FEBS Lett. 580(2), 380–384 (2006)
5. Wang, B., Chen, P., Wang, P., Zhao, G., Zhang, X.: Radial basis function neural network ensemble for predicting protein-protein interaction sites in heterocomplexes. Protein Pept. Lett. 17(9), 1111–1116 (2010)
6. Wang, B., Wong, H.S., Chen, P., Wang, H.Q., Huang, D.S.: Predicting Protein-Protein Interaction Sites Using Radial Basis Function Neural Networks. In: International Joint Conference on Neural Networks, pp. 2325–2330 (2010)
7. Yan, C., Dobbs, D., Honavar, V.: A two-stage classifier for identification of protein-protein interface residues. Bioinformatics 20(suppl. 1), 371–378 (2004)
8. Porollo, A., Meller, J.: Prediction-based fingerprints of protein-protein interactions. Proteins 66(3), 630–645 (2007)
9. Yan, C., Wu, F., Jernigan, R.L., Dobbs, D., Honavar, V.: Characterization of protein-protein interfaces. Protein J. 27(1), 59–70 (2008)
10. Neuvirth, H., Raz, R., Schreiber, G.: ProMate: a structure based prediction program to identify the location of protein-protein binding sites. J. Mol. Biol. 338(1), 181–199 (2004)

Hybrid Filter-Wrapper with a Specialized Random Multi-Parent Crossover Operator for Gene Selection and Classification Problems

Edmundo Bonilla-Huerta[1], Béatrice Duval[2], José C. Hernández Hernández[1], Jin-Kao Hao[2], and Roberto Morales-Caporal[1]

[1] LITI, Instituto Tecnológico de Apizaco,
Av. Instituto Tecnológico s/n, 93000 Apizaco, Tlaxcala, México
{edbonn,josechh,robeerto-morales}@itapizaco.edu.mx
[2] LERIA, Université d'Angers,
2 Boulevard Lavoisier, 49045 Angers, France
{bd,hao}@info.univ-angers.fr

Abstract. The microarray data classification problem is a recent complex pattern recognition problem. The most important goal in supervised classification of microarray data, is to select a small number of relevant genes from the initial data in order to obtain high predictive classification accuracy. With the framework of a hybrid filter-wrapper, we study in this paper the role of the multi-parent recombination operator. For this purpose, we introduce a Random Multi Parent crossover (RMPX) and we analyze their effects in a genetic algorithm (GA) which is combined with Fisher's Linear Discriminant Analysis (LDA). This hybrid algorithm has the major characteristic that the GA uses not only a LDA classifier in its fitness function, but also LDA's discriminant coefficients to integrate a multi-parent specialized crossover and mutation operation to improve the performance of gene selection. In the experimental results it is observed that RPMX operator work very well by achieving lower classification error rates.

Keywords: RPMX, multi-parent recombination, hybrid, filter-wrapper, linear discriminant analyze, genetic algorithm, gene selection, microarray.

1 Introduction

Recently in many evolutionary algorithms (EAs), different multi-parent recombinations have been proposed to create offspring. Scanning crossover and diagonal crossover are proposed in [6,7]. These two methods allows to adopt more than two parents in the process of recombination. In [12], a gene pool recombination operator is proposed, where the gene pool consists of several pre-selected parents. In [19], a real-coded center of mass crossover (CMX) and the multi-parent feature-wise crossover (MFX) as two multi-parents recombination operators that can lead to obtain a better performance than the crossover operators used in the genetic algorithms. In [9], a fitness-weighted crossover (FWX) is proposed with

D.-S. Huang et al. (Eds.): ICIC 2011, LNBI 6840, pp. 453–461, 2012.

an original random threshold mechanism which is used to determine the parent-numbers to reproduce offspring. Patel et al, [13] suggest two multi-parent operators: 1) MPX (multi-parent crossover with polynomial distribution) and 2) MLX (multi-parent cross-over with lognormal distribution) to solve multi-objective optimization problems by using a genetic algorithm NGSC-II. Multi-parent partially mapped crossover operator MPPMX is proposed in [17], which generalizes the partially mapped crossover operator (MPX) to a multi-parent crossover. A Multi-parent uniform crossover operator with short term memory is reported in [10]. More recently, in [3], Bongirwar et al presents two multi-parent recombination operators (MPX and MLX) in order to solve multimodal problems. In all these methods the number of parents plays a key role to keep the population diversity and to avoid the premature convergence, although the optimal parents number is actually a problem still open.

The DNA microarray technique has made possible to monitor and to measure simultaneously thousands of gene expressions in a cell mixture. This technology enables to consider cancer diagnosis based on gene expressions [2,4,1,8]. Given the very high number of genes, it is useful to select a limited number of relevant genes for classifying tissue samples.

This paper presents two multi-parent operators LDA-GA based for gene selection. In this paper, we analyze the effect of the number of parents for our Random Multi Parent crossover (RMPX), this number varies from 2 to 12 parents. We argue that more parents are used, more good solutions (gene subset) are obtained, because more parents explore and exploit into space search. For this study a gene selection filter is used on the hybrid approach and the Fisher's Linear Discriminant Analysis (LDA) is used to provide useful information to a Genetic Algorithm (GA) for an efficient exploration of gene subsets space. It has been used for several classification problems and recently for microarray data [5].

The organization of the rest of this paper is as follows: Section 2 presents our wrapper LDA-based GA for gene selection. Section 3 shows the experimental results on seven microarray datasets and presents a table of comparisons with other well-known approaches. Finally, conclusions are presented in Section 4.

2   Hybrid Filter-Wrapper Method

In this section we describe our hybrid filter-wrapper method LDA-based Genetic Algorithm (LDA-GA) for gene subset selection. First, we apply a filter BSS/WSS (B/W)[5] to retain a group G_p of different p top ranking genes. Then, the LDA-based GA is used to reduce the search space of 2^p. The purpose of this search is to find good solutions (gene subsets) with high classification performance. In what follows, we present the general procedure and then show the components of the LDA-based Genetic Algorithm. In particular, we explain how LDA is combined with the Genetic Algorithm.

2.1   General GA Procedure

2 Our LDA-based Genetic Algorithm is defined as follows:

- Initial population: The initial population is generated randomly in such a way that each chromosome contains a number of genes ranging from $p \times 0.6$ to $p \times 0.75$. The population size is fixed at 100 in this work.
- Evolution: The chromosomes of the current population P are sorted according to the fitness function (see Section 2.3). The "best" 10% chromosomes of P are directly copied to the next population P' and removed from P. The remaining 90% chromosomes of P' are then generated by using crossover and mutation.
- Crossover and mutation: Mating chromosomes are determined from the remaining chromosomes of P by considering each pair of adjacent chromosomes. By applying our multi-parent recombination operator (see Section 2.4), one child is created each time. This child undergoes then a mutation operation (see Section 2.5) before joining the next population P'.
- Stop condition: The evolution process ends when a pre-defined number of generations is reached.

2.2 Chromosome Encoding

In our model, a chromosome encodes: 1) a gene subset (τ) and 2) Their LDA coefficients(ϕ). Both are defined as:

$$I = (\tau; \phi)$$

where τ and ϕ have the following meaning. The first part (τ) is a *binary vector* and represents effectively a *candidate gene subset*. Each allele τ_i indicates whether the corresponding gene g_i is selected ($\tau_i{=}1$) or not selected ($\tau_i{=}0$). The second part of the chromosome (ϕ) is a real-valued vector where each ϕ_i corresponds to the *discriminant coefficient* of the eigen vector for gene g_i. We use the LDA discriminant coefficient to define the contribution of gene g_i to the projection axis w_{opt}. A chromosome can be thus represented as follows:

$$I = (\tau_1, \tau_2, \ldots, \tau_p; \phi_1, \phi_2, \ldots, \phi_p)$$

The length of τ and ϕ is defined by the number of the pre-selected genes (p) obtained with the filter BSS/WSS.

2.3 Fitness Evaluation

The purpose of the genetic search in our LDA-GA approach is to seek gene subsets having the minimal size and the highest prediction accuracy. To achieve this double objective, we devise a fitness function taking into account these (somewhat conflicting) criteria.

To evaluate a chromosome $I{=}(\tau; \phi)$, the fitness function considers the classification accuracy of the chromosome (f_1) and the number of selected genes in the chromosome (f_2). More precisely, f_1 is obtained by evaluating the gene subset τ using the LDA classifier on the training dataset with replacement from the original dataset through the 10-fold cross validation method. The second part of the fitness function f_2 is calculated by the formula:

$$f_2(I) = \left(1 - \frac{m_\tau}{p}\right) \qquad (1)$$

where m_τ is the number of bits having the value "1" in the candidate gene subset τ, i.e. the number of selected genes; p is the length of the chromosome corresponding to the number of the pre-selected genes from the filter ranking.

Then the fitness function f is defined as the following weighted aggregation:

$$f(I) = \alpha f_1(I) + (1 - \alpha)f_2(I)$$
$$\text{subject to } 0 < \alpha < 1$$

where α is a parameter that allows us to allocate a relative importance factor to f_1 or f_2. Assigning to α a value greater than 0.5 will push the genetic search toward solutions of high classification accuracy (probably at the expense of having more selected genes). Inversely, using small values of α helps the search toward small sized gene subsets. So varying α will change the search direction of the genetic algorithm.

2.4   Multi-Parent Recombination

We use the discriminant coefficients obtained from LDA classifier to design our crossover and mutation operators. Here, we explain how our LDA-based specialized genetic operators work. Our method is based on a modified random threshold (θ) proposed in [9] to select the parents of crossover operation (denoted by RMPX hereafter). The number of parents np involved in the recombination. The number of parents is determined by the random threshold as shown below:

$$np = \begin{cases} 2 & \text{if } \theta \leq 0.166 \\ 4 & \text{if } 0.166 \geq \theta \geq 0.333 \\ 6 & \text{if } 0.333 \geq \theta \geq 0.50 \\ 8 & \text{if } 0.50 \geq \theta \geq 0.666 \\ 10 & \text{if } 0.666 \geq \theta \geq 0.833 \\ 12 & \text{if } \theta \geq 0.833 \end{cases}$$

RMPX combines randomly different parent chromosomes $I^1 \ldots I^{np}$, we take the majority of parental genes following the definition of OB-SCAN [18] to create a new chromosome I^c. Given np parents based in the random threshold, OB-SCAN reproduce the child I^c as follows:

$$I_i^{c'} = \begin{cases} 0 & \text{if } \sum_{j=1}^{np} (I_j^c)_i < \frac{n}{2} \\ 1 & \text{if } \sum_{j=1}^{hp} (I_j^c)_i > \frac{n}{2} \\ rand(0,1) & otherwise \end{cases}$$

where $rand(0, 1)$ denotes a binary random function and $(I_j^c)_i$ the ith gene of the chromosome (I_j^c). In order to obtain a subset of informative genes, we propose to create a specialized recombination operator AND (denoted as $\otimes$). This genetic operator preserves the genes obtained by the multi-parent crossover (I_j^c) and the genes that have the most frequently appearing genes by the LDA coefficients (J_j^c). That is denoted as $K^c = I^c \otimes J^c$, where K^c is the child that contains the best information of the multi-parent recombination using LDA coefficients. Before inserting the child into the next population, K^c undergoes a mutation operation based in the LDA-coefficients to remove the gene having the lowest discriminant coefficients.

2.5   LDA-Based Mutation

In a conventional GA, the purpose of mutation is to introduce new genetic materials for diversifying the population by making local changes in a given chromosome. For binary coded GAs, this is typically realized by flipping the value of some bits ($1 \rightarrow 0$, or $0 \rightarrow 1$). In our case, mutation is used for dimension reduction; each application of mutation eliminates a single gene ($1 \rightarrow 0$). To determine which gene is discarded, one criterion is used, leading to the next mutation operator.

– *Mutation using discriminant coefficient (M1)*: Given a chromosome $K=(\tau; \phi)$, we identify the smallest LDA discriminant coefficient in ϕ and remove the corresponding gene, that is, the least informative gene among the current candidate gene subset τ). That leads to minimize the running time of our algorithm.

3   Experiments on Microarray Datasets

3.1   Microarray Gene Expression Datasets

Table 1. Summary of datasets used for experimentation

Dataset	Genes	Samples	References
Leukemia	7129	72	Golub et al [8]
Colon	2000	62	Alon et al [2]
Lung	12533	181	Gordon et al [11]
Prostate	12600	109	Singh et al [16]
CNS	7129	60	Pomeroy et al [15]
Ovarian	15154	253	Petricoin et al [14]
DLBCL	4026	47	Alizadeh et al [1]

3.2   Experimental Results

All experiments were made on a DELL precision M4500 laptop with Intel Core i7, 1.87 Ghz processor and 4 GB of RAM. Our model was implemented in MAT-LAB. The following parameters were used in the experiments: a) population size $|P| = 50$, b) maximal number of generations is fixed at 250, c) individual

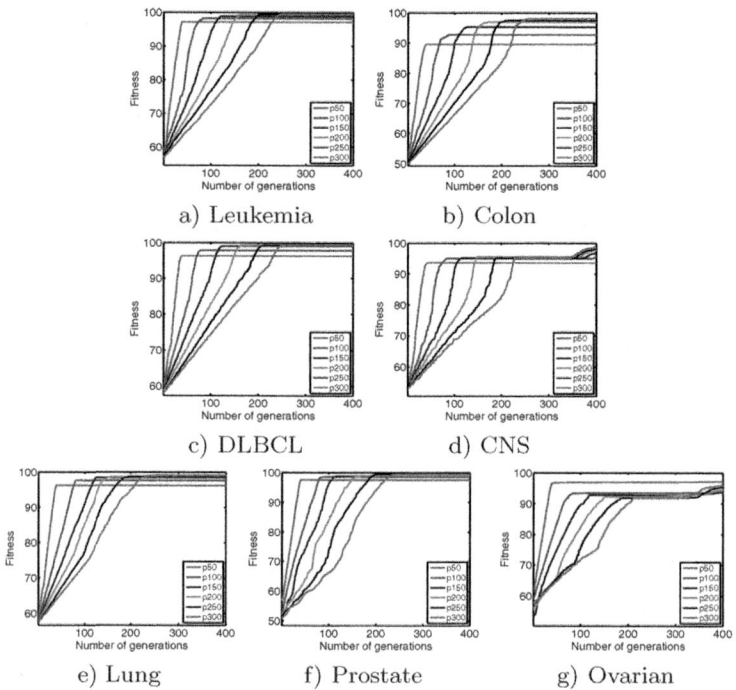

Fig. 1. A comparison between different values of selected genes with $\alpha = 0.50$

length (number of pre-selected genes) $p = 50, 100, 150, 200, 250$ and $p = 300$ are evaluated in this experimental protocol. We study the effects of multi-parent LDA-based operators (crossover and mutation). These operators were tested under the same conditions on seven microarray datasets (Leukemia, Colon cancer, Lung cancer, Prostate cancer, Ovarian, CNS and DLBCL). Figure 1, list the clear influence of the number of genes used in each dataset. In order to avoid the overfitting, we use a value of $\alpha = 0.50$ and 10-fold cross validation method.

We observe that our model achieve the highest accuracy with $p = 300$ in all datasets. For leukemia we obtain 99.50% of accuracy with a gene subset of 3 genes. Colon tumor offers a very good performance (98.83%) with 7 genes. DLBCL provides a reduced gene subset (3 genes) with a classification of 99.5%. The number of genes for CNS dataset is 4 with a high recognition rate. Lung and Prostate cancer with 4 genes gives a classification performance greater equal to 90%. For the Ovarian cancer we have a good performance with a small number of genes.

Table 2 summarizes the best accuracies (in bold) obtained by other methods and by our filter-wrapper approach on the seven datasets presented previously. An entry with the symbol (–) in this table means that the paper does not treat the corresponding dataset. All the methods reported in this table use a process of

Table 2. Comparison of our results against methods on cancer classification

Author	Leukemia	Colon	Lung	Prostate	CNS	Ovarian	DLBCL
[29]	97.5	85.0	–	92.5	–	–	–
[24]	**100**(30)	91.9(30)	**100**(30)	97.0(30)	–	99.2(75)	98(30)
[27]	91.1	95.1	93.2	73.5	88.3	–	–
[21]	**100**	93.5	97.2	–	–	–	–
[20]	95.9(25)	87.7(25)	–	–	–	–	93.0(25)
[26]	98.6(5)	87.0(4)	**100**(3)	–	–	–	–
[28]	95.8(20)	**100**(20)	–	–	–	–	95.6(20)
[25]	94.1(35)	83.8(23)	91.2(34)	–	65.0(46)	98.8(26)	–
[22]	97.1(20)	83.5(20)	–	91.7(20)	68.5(20)	**99.9**(20)	93.0(20)
[31]	**100**(30)	90.3(30)	**100**(30)	95.2(30)	80(30)	–	92.2(30)
[30]	83.8(100)	85.4(100)	–	–	–	–	–
[23]	**100**(4)	93.6(15)	–	–	–	–	–
Our model	99.5(3)	98.83(7)	99.1(**3**)	**99.5(3)**	**99.3(4)**	97.4(6)	**99.5(3)**

cross validation. Each cell contains the classification accuracy and the number of genes when this is available. We remark that each cell contains the classification accuracy and the number of genes when this is available.

We observe that our model achieves the highest accuracy with $p = 300$ in all datasets by using a number of parents defined on the interval $\{2, \ldots, 12\}$. For leukemia we obtain 99.50% of accuracy with a gene subset of 3 genes. Colon tumor offers a very good performance 98.83% with 7 genes. DLBCL provides a reduced gene subset with only three genes and a classification of 99.50%. The number of genes for the CNS dataset is 4 with a high recognition rate (99.33%). Lung and Prostate cancer with three genes give a classification performance greater to 99%. For the Ovarian cancer we obtain with 5 genes having a performance of 97.43% according to table 2.

4   Conclusions and Future Work

In this paper we proposed a study of multi-parent using a filter-wrapper framework with specialized genetic operators RMXP for the gene selection and classification of microarray gene expression. Our model work very well when the number of parents is randomly selected from the set $\{2, \ldots, 12\}$. The average runtime of our model (GA/SVM) lies between 30 and 35 minutes for the datasets: Colon, Leukemia, DLBCL and CNS. In contrast for the datasets: Lung, Prostate and Ovarian our model needs less than 70 minutes. We have extensively evaluated our model on seven public datasets using a 10-fold cross-validation process. We confirm that our model can be used to obtain very small gene subsets with high performances. The future work, consist on testing our model with the .632 bootstrap method to assess the performance of our model. In the stop condition of the GA, we have fixed at 250 generations in this work, in future we can stops our algorithm if the value of the fitness function for the best gene subset in the current population is less than or equal to fitness limit.

Acknowledgments. This work is partially supported by the PROMEP project ITAPI-EXB-000.

References

1. Alizadeh, A., Eisen, M.B., et al.: Distinct Types of Diffuse Large (b)–Cell Lymphoma Identified by Gene Expression Profiling. J. Nature 403, 503–511 (2000)
2. Alon, U., Barkai, N., et al.: Broad Patterns of Gene Expression Revealed by Clustering Analysis of Tumor and Normal Colon Tissues Probed by Oligonucleotide Arrays. Proc. Nat. Acad. Sci. USA 96, 6745–6750 (1999)
3. Bongirwar, V.K., Agarwal, V.H., Raghuwanshi, M.M.: Multimodal Optimization Using Real Coded Self-Adaptive Genetic Algorithm. J. International Journal of Engineering Science and Technology 1, 61–66 (2011)
4. Ben-Dor, A., Bruhn, L., et al.: Tissue Classification with Gene Expression Profiles. J. Computational Biology 7(3-4), 559–583 (2000)
5. Dudoit, S., Fridlyand, J., Speed, T.P.: Comparison of Discrimination Methods for the Classification of Tumors Using Gene Expression Data. J. The American Statistical Association 97(457), 77–87 (2002)
6. Eiben, A.E., Raue, P.E., Ruttkay, Z.: Genetic algorithms with multi-parent Recombination. In: Davidor, Y., Männer, R., Schwefel, H.-P. (eds.) PPSN 1994. LNCS, vol. 866, pp. 78–87. Springer, Heidelberg (1994)
7. Eiben, A.E.: Multiparent Recombination in Evolutionary Computing. In: Ghosh, A., Tsutsui, S. (eds.) Advances in Evolutionary Computing: theory and applications 2003. Natural Computing IX, pp. 175–192. Springer, Heidelberg (2003)
8. Golub, T., Slonim, D., et al.: MolecularClassification of Cancer: Class Discovery and Class Prediction by Gene Expression Monitoring. J. Science 286(5439), 531–537 (1999)
9. Gong, D., Ruan, X.: A New Multi-Parent Recombination Genetic Algorithm. In: Fifth World Congress on Intelligent Control and Automation, pp. 531–537. IEEE Press, New York (2004)
10. Garcia-Martinez, C., Lozano, M.: Evaluating a Local Genetic Algorithm as Context-Independent Local Search Operator for Metaheuristics. J. Soft Computing 14(10), 1117–1139 (2010)
11. Gordon, G.J., Jensen, R.V., et al.: Translation of Microarray Data into Clinically Relevant Cancer Diagnostic Tests Using Gene Expression Ratios in Lung Cancer and Mesothelioma. J. Cancer Research 17(62), 4963–4967 (2002)
12. Muhlenbein, H., Voigt, H.M.: Gene Pool Recombination for the Breeder Genetic Algorithm. In: The Metaheuristics International Conference, pp. 19–25. Kluwer Academic Publishers, Norwell (1995)
13. Patel, R., Raghuwanshi, M.M.: Multi-objective Optimization Using Multi Parent Crossover Operators. J. Journal of Emerging Trends in Computing and Information Sciences 2(2), 33–39 (2010)
14. Petricoin, E.F., Ardekani, A.M., Hitt, B.A., Levine, P.J., Fusaro, V.A., Mills, G.B., Simone, C., Fishman, D.A., Kohn, E.C., Liotta, L.A.: Use of Proteomic Patterns in Serum to Identify Ovarian Cancer. J. Lancet. 359(9306), 572–577 (2002)
15. Pomeroy, S.L., Tamayo, P., et al.: Prediction of Central Nervous System Embryonal Tumour Outcome Based on Gene Expression. J. Nature 415, 436–442 (2002)
16. Singh, D., Febbo, P., Ross, K., Jackson, D., Manola, J., Ladd, C., Tamayo, P., Renshaw, A., D'Amico, A., Richie, J.: Gene Expression Correlates of Clinical Prostate Cancer Behavior. J. Cancer Cell 1, 203–209 (2002)

17. Ting, C.K., Su, C.H., Lee, C.N.: Multi-parent Extension of Partially Mapped Crossover for Combinatorial Optimization Problems. J. Expert Systems with Applications 37(3), 1879–1886 (2010)
18. Ting, C.K.: On the Convergence of Multi-parent Genetic Algorithms. In: The IEEE Congress on Evolutionary Computation, pp. 396–403. IEEE Press, Edinburgh (2005)
19. Tsutsui, S., Ghosh, A.: A Study of the Effect of Multi-parent Recombination with Simplex Crossover in Real Coded Genetic Algorithms. In: The IEEE World Congress on Computational Intelligence, pp. 828–833. IEEE Press, Anchorage (1998)
20. Cho, S.B., Won, H.H.: Cancer Classification Using Ensemble of Neural Networks with Multiple Significant Gene Subsets. Applied Intelligence 26(3), 243–250 (2007)
21. Ding, C., Peng, H.: Minimum Redundancy Feature Selection From Microarray Gene Expression Data. Bioinformatics and Computational Biology 3(2), 185–206 (2005)
22. Li, X.Q., Zeng, J.Y., Yang, M.Q.: Partial Least Squares Based Dimension Reduction with Gene Selection for Tumor Classification. In: Proceedings of IEEE 7th International Symposium on Bioinformatics and Bioengineering, pp. 1439–1444 (2007)
23. Li, S., Wu, X., Hu, X.: Gene Selection Using Genetic Algorithm and Support Vectors Machines. Soft Comput. 12(7), 693–698 (2008)
24. Liu, B., Cui, Q., Jiang, T., Ma, S.: A Combinational Feature Selection and Ensemble Neural Network Method for Classification of Gene Expression Data. BMC Bioinformatics 5:136(138), 1–12 (2004)
25. Pang, S., Havukkala, I., Hu, Y., Kasabov, N.: Classification Consistency Analysis for Bootstrapping Gene Selection. Neural Computing and Applications 16:527, 539 (2007)
26. Peng, Y., Li, W., Liu, Y.: A Hybrid Approach for Biomarker Discovery from Microarray Gene Expression Data. Cancer Informatics 2, 301–311 (2006)
27. Tan, A.C., Gilbert, D.: Ensemble Machine Learning on Gene Expression Data for Cancer Classification. Applied Bioinformatics 2(2), 75–83 (2003)
28. Wang, Z., Palade, V., Xu, Y.: Neuro-fuzzy Ensemble Approach for Microarray Cancer Gene Expression Data Analysis. In: Proc. Evolving Fuzzy Systems, pp. 241–246 (2006)
29. Ye, J., Li, T., Xiong, T., Janardan, R.: Using Uncorrelated Discriminant Analysis for Tissue Classification with Gene Expression Data. IEEE/ACM Trans. Comput. Biology Bioinform. 1(4), 181–190 (2004)
30. Yue, F., Wang, K., Zuo, W.: Informative Gene Selection and Tumor Classification by Null Space Lda for Microarray Data. In: Chen, B., Paterson, M., Zhang, G. (eds.) ESCAPE 2007. LNCS, vol. 4614, pp. 435–446. Springer, Heidelberg (2007)
31. Zhang, L., Li, Z., Chen, H.: An Effective Gene Selection Method Based on Relevance Analysis and Discernibility Matrix. In: Zhou, Z.-H., Li, H., Yang, Q. (eds.) PAKDD 2007. LNCS (LNAI), vol. 4426, pp. 1088–1095. Springer, Heidelberg (2007)

Similarity Analysis of DNA Sequences Based on Three 2-D Cumulative Ratio Curves

Hong-Jie Yu

[1] Intelligent Computing Laboratory, Hefei Institute of Intelligent Machines,
Chinese Academy of Sciences, P.O. Box 1130, Hefei Anhui 230031, China
School of Science, Anhui Science and Technology University, Fengyang, Anhui 233100, China
and
[2] Department of Automation, University of Science and Technology of China, Hefei, China
yhj70@mail.ustc.edu.cn

Abstract. Based on three classifications of the DNA bases, three 2-D graphical representations of DNA primary sequences were proposed. These are called R/Y-ratio curve, M/K-ratio curve and W/S-ratio curve, respectively. The presented method has the advantages that (a) there is no loss of information when transferring a DNA sequence to its mathematical representation; and (b) the coordinates of every node on the three 2-D curves have clear biological implication. As an example, the similarities among the coding sequences of the first exon of beta-globin gene from eight species were examined.

Keywords: Cumulative ratio, Curve, Comparison, Analysis, Similarity.

1 Introduction

Advances in sequencing technology have greatly facilitated biological research involving DNA sequences. It is one of the challenges for bio-scientists to mathematically analyze the large volumes of biological data to find biological interests [1]. Graphical representations of DNA sequences provide a simple way of viewing, sorting and comparing various gene structures with their intuitive feel [2].

The DNA sequences are usually expressed in terms of a series of four letters which may be called the Letter Sequence Representation (LSR) of the DNA sequences. The LSR of the DNA sequences is necessary for the information storage and some statistical analysis. However, it is extremely difficult to recognize, to remember and to compare the sequences intuitively, especially for the long sequences, based merely on the LSR of the sequences [3]. However, it has been acknowledged that information contained in DNA sequences is difficult for humans to comprehend without careful extraction and processing. Many methods have been proposed to characterize DNA sequences, with special efforts given to representing the sequence graphically [4]. Meanwhile, to overcome the disadvantages of LSR, an idea of mapping the DNA sequence by a geometrical curve in 3-dimensional space. Hamori and Ruskin [5] first proposed a 3D graphical representation for DNA sequences, some different graphical approaches representing DNA sequences have been reported by several authors [4, 6-9].

D.-S. Huang et al. (Eds.): ICIC 2011, LNBI 6840, pp. 462–469, 2012.

Using graphical approaches to study biological problems can provide an intuitive picture or useful insights for helping analyzing complicated relations in these systems, as demonstrated by many previous studies on a series of important biological topics, such as analysis of DNA and protein sequence [8, 10, 11]. Moreover, Douglas J. Cork [12] presented a description of the W-curve generation process, including a comparison technique of aligning extremes of the curves to effectively phase-shift them past the HIV-1 gap problem. With W-curve heuristic alignment, it obtain clinically useful results in a short time—short enough to affect clinical choices for acute treatment. So this graphical methods can also deal with some biological and medical related problems. Recently, the images of cellular automata were also used to represent biological sequences [13].

In addition, several 2D graphical representations of DNA sequences were proposed [14-17]. These methods are straight forward but are accompanied with some loss of information due to overlapping and crossing of the curve representing DNA with itself and degeneracy generated by the circuit. Randic et al. [15] developed a novel 2D representation method in which there is no loss of information in transferring a DNA sequence to its mathematical representation. Subsequently, Randic et al. put forward a novel 2-D graphical representation of DNA sequences based on a four-color map. The novel representation is rather compact and it not only enables one to carry out a visual inspection of similarity between DNA sequences, but also leads to their numerical characterization [18]. Moreover, several other 2D representations have been proposed [19-22]. Recently, more other 3D representations were developed by several authors [8, 16] to overcome the problem of degeneration in graphical representation. These methods, however, do not seem to possess apparent biological meanings.

In this study, we proposed a novel approach to represent DNA primary sequences graphically. Based on three classifications of the four DNA bases A, G, T and C, respectively, three 2-D cumulative ratio curves, namely, the RY-ratio curve, the K/M-ratio curve and the W/S-ratio curve were constructed. The coordinates of every node on these 2-D cumulative ratio curves have clear biological implication. In Section 3, we will present three applications developed based on the proposed representations.

2   Materials and Methods

2.1   Dataset Preparation

The experimental data in this study were derived from the dataset used by Ashesh Nandy, M.H. et al. [2]. The dataset comprises the bases of the first exon in the beta globin gene for the eight species. (Note: All the papers have used 90 bases for the rabbit exon 1 but it should be 92 bases. Here we report the corrected sequence.).

2.2   Construction of Three Ratio Curves

The four DNA bases(A,G,T and C) can be classified by the following three ways according to their chemical properties [4]:

(a) Chemical structures of the bases: puRine(A,G)/pYrimidines(T,C);

(b) Functional groups of the bases: aMino(A,C)/Keto(G,T);
(c) The strength of the H-bonds between paired bases: Weak(A,T)/Strong(G,C).

Table 1. Sequences of the first exon in the beta globin gene for the eight species

Species	Sequences
Human (92 bases)	ATGGTGCACCTGACTCCTGAGGAGAAGTCTGCCGTTACTGCCCTGTG GGGCAAGGTGAACGTGGATGAAGTTGGTGGTGAGGCCCTGGGCAG
Goat (86 bases)	ATGCTGACTGCTGAGGAGAAGGCTGCCGTCACCGGCTTCTGGGGCA AGGTGAAAGTGGATGAAGTTGGTGCTGAGGCCCTGGGCAG
Opossum (92 bases)	ATGGTGCACTTGACTTCTGAGGAGAAGAACTGCATCACTACCATCT GGTCTAAGGTGCAGGTTGACCAGACTGGTGGTGAGGCCCTTGCCAG
Gallus (92 bases)	ATGGTGCACTGGACTGCTGAGGAGAGGCAGCTCATCACCGGCCTCT GGGGCAAGGTCAATGTGGCCGAATGTGGGGCCGAAGCCCTGGCCAG
Lemur (92 bases)	ATGACTTTGCTGAGTGCTGAGGAGAATGCTCATGTCACCTCTCTGTG GGGCAAGGTGGATGTAGAGAAAGTTGGTGGCGAGGCCTTGGGCAG
Mouse (92 bases)	ATGGTGCACCTGACTGATGCTGAGAAGGCTGCTGTCTCTTGCCTGTG GGGAAAGGTGAACTCCGATGAAGTTGGTGGTGAGGCCCTGGGCAG
Rabbit (92 bases)	ATGGTGCATCTGTCCAGTGAGGAGAAGTCTGCGGTCACTGCCCTGT GGGGCAAGGTGATTGTGGAAGAAGTTGGTGGTGAGGCCCTGGGCAG
Rat (92 bases)	ATGGTGCACCTAACTGATGCTGAGAAGGCTACTGTTAGTGGCCTGT GGGGAAAGGTGAACCCTGATAATGTTGGCGCTGAGGCCCTGGGCAG

2.2.1   R/Y-ratio Curve. First considering the R/Y classification, we represent the ratio of puRine bases R(A,G) and pYrimidine bases Y(T,C) as RY(i, x_i). Given a DNA sequence with n bases, we look at one base at a time. For the i-th position of sequence, a corresponding point RY(i, x_i), i=1,2,…,n, can be determined in the 2-D plane as follows:

$$x_i = \frac{A+G}{C+T} \tag{1}$$

where the A,C,G,T, means the cumulative occurrence number of those four type nucleotides from the first position to the i th position of sequence, respectively. All n bases on the DNA sequence are examined consecutively, and in the end we will obtain n points: RY_1, RY_2, … , RY_n in the 2-D plane. Then, starting from the starting point, connecting adjacent points, we will obtain a 2-D curve, called as the R/Y-ratio curve.

2.2.2   M/K-ratio Curve. Likewise, for the M/K classification, we represent the ratio of aMino bases M(A,C) and Keto bases K(G,T) as MK(i, y_i), where y_i is determined by

$$y_i = \frac{A+C}{G+T} \tag{2}$$

A different way of representing the DNA sequence graphically is thus established. We call the 2-D curve generated under this definition the M/K-ratio curve. Thus we get the M/K-ratio curve.

2.2.3 W/S-ratio curve. Similarly, for the W/S classification, we calculate the ratio of weak hydrogen bases W(A,T) and the strong hydrogen bases S(G,C) as WS(i, z_i), where z_i is given by

$$z_i = \frac{A+T}{G+C} \tag{3}$$

We then obtain the third 2-D graphical representation of the DNA sequence. 2-D curve generated under this definition is called the W/S-ratio curve.

3   Results and Discussion

3.1   Calculation of the Cumulative Ratio of a DNA Sequence

Based on formulas (2), (3), and (4), we can obtain three coordinates for each curve of eight species' exon1 sequence. In order to view the difference among eight curves clearly, we consider three projections on three 2-D planes, where the x_i, y_i, z_i are projected onto y-axis in three subplots (Fig. 1A, Fig. 1B, Fig. 1C), respectively. However, each x-axis in three subplots of Fig. 1 means the i-th site, i=1, 2, ... , . Comparing the length of eight sequences, we let the minimal length be n. Here, n is eighty-six.

Fig. 1 A shows, from the 30th site to the last site, the trend of the R/Y-ratio fall into the interval [1, 1.5], except for goat and mouse. With the exception of mouse, at the median site (around the 44th site), the other seven species' R/Y-ratio arrive at the lowest value, which approximate to one. At this time, the content of puRine is equivalent to the content of pYrimidine. Lately, at the 81st site, we can see the all R/Y-ratio values of eight sequences arrive at another new high, which is greater than one. As for the whole sequences, the contents of puRine are greater than the contents of pYrimidine.

By comparison, the variation range of eight M/K-ratio curves is small, for the M/K-ratio value interval is [0.5, 1], as shown in Fig. 1B. Except for the opossum and gallus, the rest six curves are close to each other. In particular, the corresponding M/K-ratio curve of opossum is far away from the others. With the exception of goat, at the median site (around the 42nd site), the other seven species' M/K-ratio arrive at the ceiling value. Lately, at the 81st site, we can see the all M/K-ratio values of eight sequences reach a new low, where the fluctuation is opposite to that of R/Y-ratio. As for the whole sequences, the contents of aMino bases are less than the contents of Keto bases.

For eight W/S-ratio curves, the coordinates of eight W/S-ratio curves range from 0.5 to 1. A close look at Fig. 1C, we can find that the GC-content of opossum is the lowest among the eight species and the GC-content is ascending, whereas the sequence of gallus is GC-rich. Table 2 illustrates the GC-content of gallus is the largest (62.79%) among the eight sequences, while the GC-contents of opossum and lemur are the lowest (52.33%) among these eight sequences. Meanwhile, with the exception of opossum, around the 71st site, the other seven species' W/S-ratio amount to the peak value. Lately, we can see the all W/S-ratio values of eight sequences are descending. Particularly, the corresponding W/S-ratio curves of opossum and gallus are both far away from the others.

A

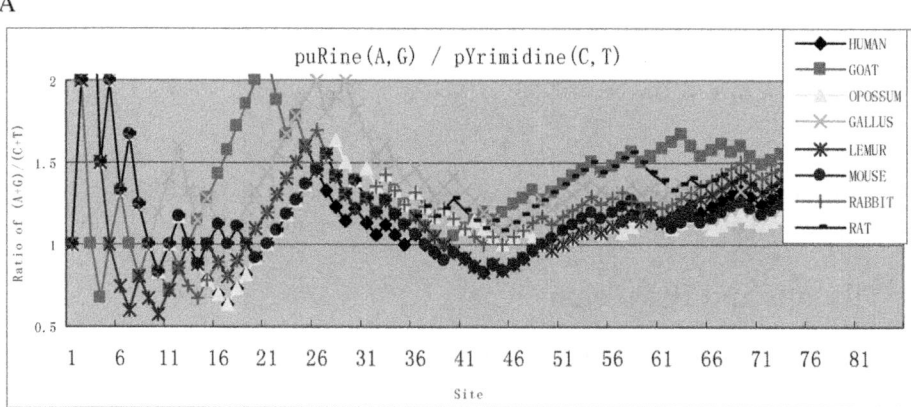

B

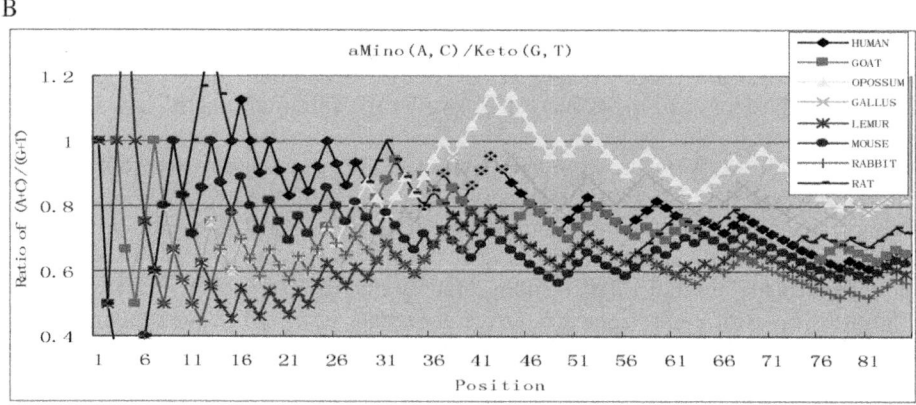

C

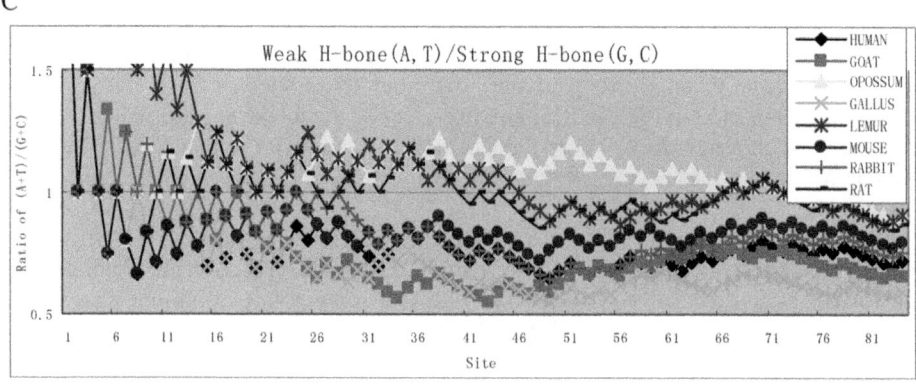

Fig. 1. The three 2-D ratio curves of the sequences of eight species' exon1 of beta-globin gene. (A) eight cumulative Purine / Pyrimidine -ratio curves; (B) eight cumulative aMino/Keto-ratio curves; (C) eight cumulative Weak H-bond/Strong H-bond ratio curves.

Table 2. GC-content of sequences listed in table 1

Species	Mean	Ensemble
Human	56.01%	58.14%
Goat	56.64%	60.47%
Opossum	47.72%	**52.33%**
Gallus	**58.61%**	**62.79%**
Lemur	46.18%	**52.33%**
Mouse	53.84%	55.81%
Rabbit	53.86%	56.98%
Rat	49.99%	53.49%

Table 3. The pair distance between the beta globin exon sequences of eight species

	Human	Goat	Opossum	Gallus	Lemur	Mouse	Rabbit	Rat
Human	0	2.3588	3.1986	2.4677	2.2987	1.3169	1.7327	2.4947
Goat		0	4.3554	2.0910	3.3772	2.6331	1.8683	2.6570
Opossum			0	3.9138	2.6201	3.1471	3.6404	2.6379
Gallus				0	3.6559	2.8358	2.2854	3.1000
Lemur					0	1.7153	2.0306	1.9053
Mouse						0	1.5329	2.1036
Rabbit							0	2.1309
Rat								0

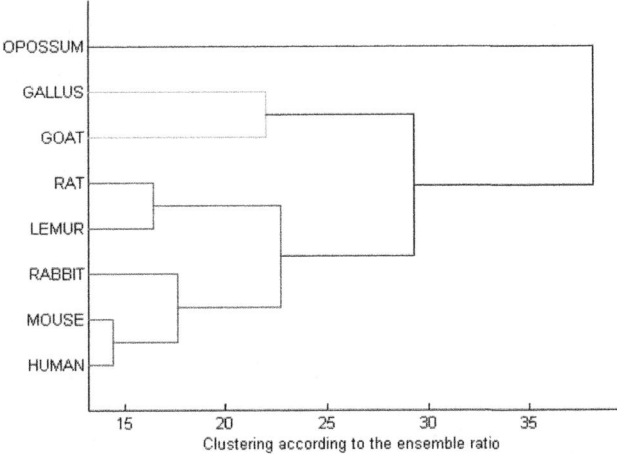

Fig. 2. The dendrogram of eight species based on their three type cumulative ratio value from the thirtieth position to the last position

On the other hand, in order to compare the relationship among the eight species, we calculated the pair distances between each other according to the ratio values from the 30th site to the last site of each sequence. Thus, the pair distances are given in Table 3.

Compared to the related study [2], similar results are obtained. From the Fig. 2, we can find that opossum, gallus and goat are far from the rest five species, while mouse and rabbit are close to human. The dendrogram reflects the relationship between the species each other.

4   Conclusion

The results presented here illustrate a fact that the base's three chemical properties of DNA sequence are helpful for similarity analysis based on three cumulative ratio. The three cumulative ratio involve the following respects: (a) puRine / pYrimidines: $(A+G)/(T+C)$; (b) aMino / Keto: $(A+C)/(G+T)$; (c) Weak / Strong: $(A+T)/(G+C)$. The proposed method is very simple, efficient and effective. Demonstrated results show that the method has distinguishable ability to compare similarity of DNA sequences. Additionally, the proposed method has a wide range of applicability for analysis of biological sequence.

Acknowledgement. This work was supported by the grants of the National Science Foundation of China, Nos. 31071168, 30900321, 60975005, 61005010, 60873012 & 60905023.

References

1. Huang, G., et al.: H–L curve: A novel 2D graphical representation for DNA sequences. Chemical Physics Letters 462(1-3), 129–132 (2008)
2. Nandy, A., Harle, M., Basak, S.C.: Mathematical descriptors of DNA sequences development and applications. ARKIVOC ix, 211–238 (2006)
3. Zhang, R., Zhang, C.-T.: Z Curves, An Intuitive Tool for Visualizing and Analyzing the DNA Sequences. Journal of Biomolecular Structure & Dynamics 11(4), 767–782 (1994)
4. Xie, G., Mo, Z.: Three 3D graphical representations of DNA primary sequences based on the classifications of DNA bases and their applications. Journal of Theoretical Biology 269(1), 123–130 (2011)
5. Hamori, E., Ruskin, J.: H Curves, A Novel Method of Representation of Nucleotide Series Especially Suited for Long DNA Sequences. Journal of Biological Chemistry 258(2), 1318–1327 (1983)
6. Qi, Z.-H., Fan, T.-R.: PN-curve: A 3D graphical representation of DNA sequences and their numerical characterization. Chemical Physics Letters 442(4-6), 434–440 (2007)
7. Cao, Z., Li, R., Chen, W.: A 3D graphical representation of DNA sequence based on numerical coding method. International Journal of Quantum Chemistry 110(5), 975–980 (2009)
8. Yu, J.F., Sun, X., Wang, J.H.: TN curve: A novel 3D graphical representation of DNA sequence based on trinucleotides and its applications. Journal of Theoretical Biology 261(3), 459–468 (2009)

9. Zhang, Z.J.: DV-Curve: a novel intuitive tool for visualizing and analyzing DNA sequences. Bioinformatics 25(9), 1112–1117 (2009)
10. Qi, X.-Q., Wen, J., Qi, Z.H.: New 3D graphical representation of DNA sequence based on dual nucleotides. Journal of Theoretical Biology 249(4), 681–690 (2007)
11. Wu, Z.C., Xiao, X., Chou, K.C.: 2D-MH: A web-server for generating graphic representation of protein sequences based on the physicochemical properties of their constituent amino acids. Journal of Theoretical Biology 267(1), 29–34 (2010)
12. Cork, D.J., et al.: W-Curve Alignments for HIV-1 Genomic Comparisons. PLoS ONE 5(6), e10829 (2010)
13. Xiao, X., et al.: Using cellular automata to generate image representation for biological sequences. Amino Acids 28(1), 29–35 (2005)
14. Wu, Y., et al.: DB-Curve: a novel 2D method of DNA sequence visualization and representation. Chemical Physics Letters 367, 170–176 (2003)
15. Randic, M.: Novel 2-D graphical representation of DNA sequences and their numerical characterization. Chemical Physics Letters 368, 1–6 (2003)
16. Li, C., Wang, J.: A Simple Method for Characterization and Similarity Analysis of DNA sequences. Internet Electronic Journal of Molecular Design, 1–9 (2003)
17. Ji, M., Li, C.: TB-curve, a New 2-D Graphical Representation of DNA Sequences. Journal of Mathematical Chemistry 40(2), 185–193 (2006)
18. Randic, M., et al.: Four-color map representation of DNA or RNA sequences and their numerical characterization. Chemical Physics Letters 407(1-3), 205–208 (2005)
19. Song, J., Tang, H.: A new 2-D graphical representation of DNA sequences and their numerical characterization. Journal of Biochemical and Biophysical Methods 63(3), 228–239 (2005)
20. Liu, X., et al.: PNN-curve: A new 2D graphical representation of DNA sequences and its application. Journal of Theoretical Biology 243(4), 555–561 (2006)
21. Dai, Q., Liu, X., Wang, T.: A novel 2D graphical representation of DNA sequences and its application. Journal of Molecular Graphics and Modelling 25(3), 340–344 (2006)
22. Zhao, L., et al.: An S-Curve-Based Approach of Identifying Biological Sequences. Acta Biotheoretica 58(1), 1–14 (2009)

Similarity Analysis of DNA Barcodes Sequences Based on Compressed Feature Vectors

Hong-Jie Yu

School of Science, Anhui Science and Technology University,
Fengyang, Anhui 233100, China
Department of Automation, University of Science and Technology of China,
Hefei, Anhui 230027, China
Intelligent Computing Laboratory, Hefei Institute of Intelligent Machines,
Chinese Academy of Sciences, P.O. Box 1130, Hefei, Anhui 230031, China
yhj70@mail.ustc.edu.cn

Abstract. We provided a novel method for sequence analysis based on the compressed representation of sequences. In our work, we mapped DNA barcodes sequences into compressed feature vectors (CFV), which comprise 12 components. We used the Euclidean distance method (EMD) based on compressed feature vectors (CFV) to build dendrograms, which may have a biological interpretation and can be considered as a kind of phylogenetic tree. As a numeralization representation technique makes it easy to analyze the similarities between specimens in detail. The results show that CFV is a reasonable descriptor for DNA barcodes sequences.

Keywords: Comparison, Similarity analysis, Compressed feature vectors, DNA barcodes.

1   Introduction

The mitochondrial gene cytochrome c oxidase I (COI), extracted from a standardized region of the genome, can serve as the core for identifying species or specimens of animals [1-3]. The CO1 mitochondrial gene had been studied for more than two decades and sequenced in several thousands of animal species. Based on these data, the within- and among-species divergence of this gene was estimated, and it was proved to be an effective DNA barcode [4]. With the rapid growth of these sequences, coupled with their very broad taxonomic sampling, the CO1 mitochondrial gene could, potentially, provide insights into patterns of molecular evolution and population genetics [5]. Barcode quality index (BQI) is a novel, unified measure of sequence quality and contig overlap tailored to the needs of DNA barcoding. Re-analysis of published data demonstrates the utility of BQI [6]. Furthermore, how to deal with the large volume genomic DNA sequences is one of challenges with which bio-scientists are presently confronted.

Graphical representation provides a simple way of viewing, sorting and comparing various gene sequences with their intuitive pictures and pattern [7-10]. There have been several approaches using graphical representations of DNA sequences, many of

D.-S. Huang et al. (Eds.): ICIC 2011, LNBI 6840, pp. 470–477, 2012.

which have been covered in earlier reviews of the subject [11]. Part of the appeal of the graphical representation lies in the fact that relevant bits of information can be quickly obtained by visual inspection of the plot of a DNA sequence. There are several different techniques for plotting DNA sequences, ranging from a simple 2-D Cartesian method to complex 6-D methods [7, 8, 12, 13].

Similarity analysis is of great importance to bioinformatics and computational biology. The analysis of similarity of DNA sequences can be divided into two groups: the sequence alignment and the invariant-based comparison [14]. In the former, a distance function or a score function is used to represent insertion, deletion, and substitution of letters in the compared structures. Such approaches have been hitherto widely used. However, the computational complexity and the inherent ambiguity of the alignment cost criteria are still the bottleneck problems. The latter is based on the quantitative characterization of DNA sequences by ordered sets of invariants derived from the sequences, such as the leading eigenvalues of all kinds of matrices [15-17]. However, a trouble we must face is that the calculation of the matrix or eigenvalues will become more and more difficult with the length of the sequence longer.

In a recent study, the number of gene repetitions was shown to be a key aspect of gene expression and phenotype [18]. Apparently theses repetitions, not only at nucleotide level, might play a key role in genome organization and functionality of networks. The notions of entropy in DNA are unquestionably connected [19] and references therein–the degree of predictability of a sequence, which is closely related with its internal repetition and compression, can be measured by its entropy [20]. The major importance of these findings has provided evidence that is already too vast to fully account for. This creates the need for an efficient method to analyze, for different parameters sets, the degree or scale of each DNA region [21]. Liao et al. made a comparison for the first exon of beta-globin genes sequences belonging to eleven different species based on condensed matrices and information entropies to illustrate the utility of their approach [22].

In this study, we proposed a new compressed representation of DNA barcodes based on 12-tuple frequencies of nucleotide triplet. We applied this method in the phylogenetic tree constructing, the results of which are very close to those directly got from Neighbor-Joining method.

2   Materials and Methods

2.1   Dataset Preparation

In this study, 14 specimens of *Anostraca* are totally involved. DNA barcodes of these specimens are directly downloaded from http://www.boldsystems.org [23], and we choose aligned sequence model. In light of the required format, we must transform the original data file into the appropriate format with the specific file name extention, i.e. *.meg.

2.2   Nucleotide Frequencies Calculation of DNA Barcodes Sequence

Consequently, the nucleotide frequencies of each specimens' sequence are obtained by using software MEGA 4 [24, 25], which could provide each nucleotides' overall frequencies among all different triplet nucleotide, such as nucleotide frequencies of T_1, C_1, A_1, G_1, T_2, C_2, A_2, G_2, T_3, C_3, A_3, G_3. We can obtain a 12-tuple vector, which is mapped from 14 high dimensional sequences. As a result, the 12-tuple could serve as specimen's feature vector.

$$\overline{F} = (T_1, C_1, A_1, G_1, T_2, C_2, A_2, G_2, T_3, C_3, A_3, G_3) \cdot \tag{1}$$

Where T_p, C_p, A_p, G_p $(p=1,2,3)$ denote the frequencies of four types nucleotide at p-th position of triplet, respectively.

Furthermore, the dimension was drastically reduced, and the dimension is reduced into fixed number, i.e. twelve. Hence, we call these 12-tuple vectors $\overline{F}$ in (1) as compressed feature vectors (CFV).

2.3   A New Method for Computing the Distance of Two DNA Barcodes Sequences

Analysis for the similarity among DNA barcodes sequence comes down to calculation of their distances. The smaller value of the distance represents the more similar of two sequences. The Euclidean distances between the leading Eigen values of some matrices was used, that matrices was constructed from the points of the sequences and was used to represent the sequences, which was named as mathematical descriptors, such as E, D/D, M/M, L/L, k^L/k^L, k^M/k^M, CM matrices [26]. Further more, some researchers used the cumulative distance of every points or the last points distances (include the every component of points, or the angles distance) as the sequences' distance. Also many researchers had used these matrices to construct phylogenetic tree [26-28].

We can define Euclidean distance between the k-th and the l-th specimen as:

$$ED(k,l) = \sqrt{\sum_{j=1}^{12} (f_j^{(k)} - f_j^{(l)})^2} \cdot \tag{2}$$

Where $f_j^{(k)}$ and $f_j^{(l)}$ denotes the j-th element from the CFV of the k-th and the i-th specimen in (1).

Here we give a new distance computing method based on Euclidean distance. As in formula (2), this method can drastically compress the length of the sequences based on 12-tuple vector $\overline{F}$, so it is more effective. In formula (5), $ED(k, l)$ denotes the Euclidean distance between the k-th and l-th specimen's DNA barcodes sequence of *Anostraca*.

3   Results and Discussion

For better analyzing the sequences' similarity between specimens, we must map the primary sequences into numerical vectors.

3.1   Nucleotide Frequencies Matrix of DNA Barcodes Sequence

Using the software MEGA 4 to calculate triplet nucleotides' frequencies, we obtain the preprocessed data, which are show in Table 1. Each line stands for the compressed feature vector (CFV) of a specimen, while each column denotes one of the four type nucleotide frequencies at each position of the triplet. Hence, we have a total of fourteen CFVs in all, whose element is

$$\frac{\text{triplet nucleotide occurrence number}}{\text{sequence length}} \times 100 \cdot \tag{3}$$

Table 1. The 12-tuple vector value corresponding frequencies of 14 DNA barcodes sequences of Anostraca's specimens

No.s	T-1	C-1	A-1	G-1	T-2	C-2	A-2	G-2	T-3	C-3	A-3	G-3
01	23.1959	23.1959	25.7732	27.8351	43.8144	27.8351	12.3711	15.9794	36.5979	17.5258	35.0515	10.8247
02	21.6495	24.7423	25.7732	27.8351	43.8144	27.8351	12.3711	15.9794	35.5670	17.5258	36.5979	10.3093
03	22.6804	23.7113	25.7732	27.8351	43.8144	27.8351	12.3711	15.9794	36.0825	17.5258	37.1134	9.2784
04	22.6804	23.7113	25.7732	27.8351	43.8144	27.8351	12.3711	15.9794	35.5670	18.0412	38.1443	8.2474
05	23.1959	23.1959	25.7732	27.8351	43.8144	27.8351	12.3711	15.9794	36.0825	18.0412	37.1134	8.7629
06	23.7113	22.6804	25.7732	27.8351	43.8144	27.8351	12.3711	15.9794	35.0515	19.0722	37.1134	8.7629
07	22.6804	23.7113	25.7732	27.8351	43.8144	27.8351	12.3711	15.9794	37.1134	17.0103	36.5979	9.2784
08	22.6804	23.7113	25.7732	27.8351	43.8144	27.8351	12.3711	15.9794	37.1134	17.0103	36.5979	9.2784
09	22.6804	23.7113	26.2887	27.3196	43.8144	27.8351	12.3711	15.9794	37.1134	17.0103	36.5979	9.2784
10	22.6804	23.7113	25.7732	27.8351	43.8144	27.8351	12.3711	15.9794	37.6289	16.4948	35.5670	10.3093
11	22.1649	24.2268	25.7732	27.8351	43.8144	27.8351	12.3711	15.9794	36.5979	17.5258	33.5052	12.3711
12	22.1649	24.2268	25.7732	27.8351	43.8144	27.8351	12.3711	15.9794	37.1134	17.5258	34.5361	10.8247
13	22.1649	24.2268	25.7732	27.8351	43.8144	27.8351	12.3711	15.9794	36.5979	17.5258	33.5052	12.3711
14	22.6804	23.7113	25.7732	27.8351	43.8144	27.8351	12.3711	15.9794	35.0515	19.0722	35.0515	10.8247

*First column of Table 1 are the specimens' serial number GBFCL00 i -06lAY555252lAY555252lTanymasti, $i = 01,02,\cdots,14$

3.2   The Comparison Based on the Euclidean Distance of Two DNA Barcodes Sequences

To understand how close between two different specimens, we calculated their Euclidean distances between two different specimens' CFVs.

From Table 2, we can see that the 7-th and 8-th specimen are almost identical, because the distance is almost zero. Moreover, the 7-th and 8-th specimen are very close to the 9-th specimen, for the distance between No.7 and No.9 is as very small as that of distance between No.8 and No.9 (0.729). Similarly, the 11-th and 13-th specimen are far from the others, however, they are close to each other, because the distances between the other twelve specimens and 11-th specimen are identical to that of 13-th specimen. Simultaneously, their distance between 11-th and 13-th specimen is zero.

3.3   Relationship Analysis for Specimens

Using software MEGA4, we can construct the phylogenetic tree. Likewise, we can draw dendrogram through clustering. The compressed feature vectors (CFV) of each

Table 2. Euclidean Distance between different specimens based on 12-tuple nucleotide frequencies vector

No.	01	02	03	04	05	06	07	08	09	10	11	12	13	14
01	0	2.9159	2.7275	4.2506	3.0056	3.7171	2.4177	2.4177	2.5252	1.7857	2.6283	1.6301	2.6283	2.3052
02		0	1.9287	3.0057	2.8233	3.7171	2.4178	2.4178	2.5253	2.9160	3.9256	2.7275	3.9256	2.7276
03			0	1.6301	1.0309	2.4178	1.2626	1.2626	1.4580	2.6284	4.8354	3.2600	4.8354	3.1775
04				0	1.4579	2.1869	2.6284	2.6284	2.7276	4.1877	6.3550	4.7802	6.3550	4.1877
05					0	1.6301	1.7856	1.7856	1.9287	3.1775	5.3568	3.7878	5.3568	3.3406
06						0	3.3406	3.3406	3.4193	4.4938	5.9669	4.7243	5.9669	3.2601
07							0	0.0000	0.7290	1.6300	4.4936	2.7275	4.4936	3.6449
08								0	0.7290	1.6300	4.4936	2.7275	4.4936	3.6449
09									0	1.7856	4.5524	2.8233	4.5524	3.7171
10										0	3.3406	1.7857	3.3406	3.7172
11											0	1.9287	0.0000	3.1775
12												0	1.9287	2.7276
13													0	3.1775
14														0

*The specimens' serial number GBFCL00 i -06|AY555252|AY555252|Tanymasti, $i = 01,02,\cdots,14$

specimen's DNA barcode, can be used as a feature that identifies each specimen, thus CFV also applies to the comparison of species. These vectors are used to build dendrograms that show hierarchical clusters which could be interpreted as phylogenetic trees. The dendrograms are built using average linkage clustering and the similarity matrix was computed using the Euclidean distance [27].

The distance between pairs of specimens in 14-by-12 data matrix X. Each row of X corresponds to the specimen's CFV, which is a 12-tuple vector. Y is a row vector of length $n(n-1)/2$, corresponding to pairs of observations in X. The distances are arranged in the order $(2,1), (3,1), ..., (n,1), (3,2), ..., (n,2), ..., (n,n-1))$. Y is commonly used as a dissimilarity matrix in clustering or multidimensional scaling. To save space and computation time, Y is formatted as a vector. However, we can convert this vector into a square matrix using the squareform function so that element i, j in the matrix, where $i<j$, corresponds to the distance between specimen i and j in the original data set. $Y =$ pdist(X, metric) computes the distance between objects in the data matrix, X, using the method specified by metric, which is 'euclidean' (Euclidean distance) in this study.

3.4 Analysis for the Relationships within Intra-specimens

Evolutionary relationships can be seen via viewing Cladograms or Phylograms. Here software MEGA 4 is used to compute the similarity of sequences and construct the phylogenetic tree. These results are shown in Fig. 1.

Fig. 2 shows the phylogenetic tree for all the specimens in this study, where the original DNA barcodes sequences data were firstly converted to short vectors, i.e.

compressed feature vectors (CFV). The phylogenetic tree of our results (obtained with average linkage) displays a first branching between No. 1,11,12,13,14 specimens and the others. That is to say, at the first branching, these fourteen specimens fall into two categories, whose relationships are ((6, (4, (5, 3))), (2, (10, (9, (7, 8)))))) and (14, ((12, 1), (13, 11))), respectively. In contrast to Fig. 1, there is inconsistency with the relationship results between specimens in Fig. 2, where (a) No. 1 is classified into the opposite class during the first branching, (b) simultaneously, No. 3 is bracketed together with No. 5, rather than No. 4. According to the results shown in Fig. 1, all other branchings are correct. The results between Fig. 1 and Fig. 2 nearly makes no difference.

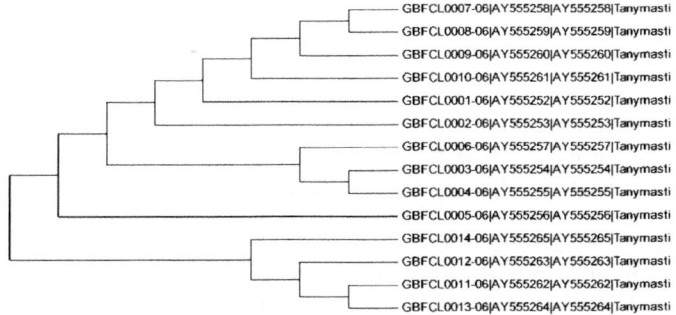

Fig. 1. Phylogenetic tree of the 14 specimens of *Anostraca* based on the Neighbor-Joining method

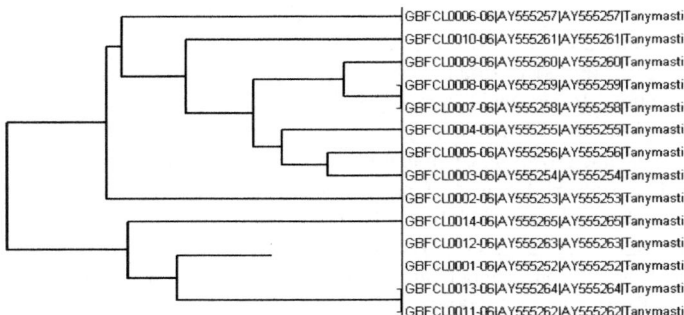

Fig. 2. Phylogenetic tree of the 14 specimens of *Anostraca* based on our new compressed mapping representation method

Understanding of phylogenetic information can provide the necessary assistance for us to conclude the biological relationship between DNA barcodes at the species level. Meanwhile, the phylogenetic information obtained by the proposed approach, can also provide the necessary information for other studies, such as analysis for amino acid sequence, evolution and so on.

4   Conclusion

In this study, a novel method is proposed to analyze DNA barcodes' similarity, based on compressed feature vectors (CFV) representation. The dendrograms may have a biological interpretation and may be considered as a kind of phylogenetic tree. The dendrograms is consistent with the results obtained directly from distance calculation of the original sequences. The results show that CFV is a suitable descriptor for DNA sequence.

Acknowledgement. This work was supported by the grants of the National Science Foundation of China, Nos. 31071168, 30900321, 60975005, 61005010, 60873012 & 60905023.

References

1. Hebert, P.D.N., Stoeckle, S.M., Zemlak, T.S., Francis, C.M.: Identification of birds through DNA barcodes. PLoS Biol. 2(10), 312 (2004)
2. Hebert, P.D.N., et al.: Biological identifications through DNA barcodes. Proceedings of the Royal Society B: Biological Sciences 270(1512), 313–321 (2003)
3. Hebert, P.D.N., Ratnasingham, S., de Waard, J.R.: Barcoding animal life: cytochrome c oxidase subunit 1 divergences among closely related species. Proceedings of the Royal Society B: Biological Sciences 270(suppl. 1), 96–99 (2003)
4. Shneyer, V.S.: DNA barcoding is a new approach in comparative genomics of plants. Russian Journal of Genetics 45(11), 1267–1278 (2009)
5. Hajibabaei, M., et al.: DNA barcoding: how it complements taxonomy, molecular phylogenetics and population genetics. Trends in Genetics 23(4), 167–172 (2007)
6. Little, D.P.: A unified index of sequence quality and contig overlap for DNA barcoding. Bioinformatics 26(21), 2780–2781 (2010)
7. Huang, G.: Similarity studies of DNA sequences based on a new 2D graphical representation. Biophysical Chemistry 143(1-2), 55–59 (2009)
8. Qi, Z.H., Fan, T.R.: PN-curve: A 3D graphical representation of DNA sequences and their numerical characterization. Chemical Physics Letters 442(4-6), 434–440 (2007)
9. Huang, G., et al.: H–L curve: A novel 2D graphical representation for DNA sequences. Chemical Physics Letters 462(1-3), 129–132 (2008)
10. Yu, J.F., Sun, X., Wang, J.H.: TN curve: A novel 3D graphical representation of DNA sequence based on trinucleotides and its applications. Journal of Theoretical Biology 261(3), 459–468 (2009)
11. Roy, A., Raychaudhury, C., Nandy, A.: Novel techniques of graphical representation and analysis of DNA sequences—a review. J. Biosci. 23(1), 55–71 (1998)
12. Cao, Z., Li, R., Chen, W.: A 3D graphical representation of DNA sequence based on numerical coding method. International Journal of Quantum Chemistry 110, 975–980 (2009)
13. Yao, Y.H., et al.: Similarity/dissimilarity studies of protein sequences based on a new 2D graphical representation. Journal of Computational Chemistry 31, 1045–1052 (2009)
14. Li, C., Wang, J.: Similarity analysis of DNA sequences based on the generalized LZ complexity of (0,1)-sequences. Journal of Mathematical Chemistry 43(1), 26–31 (2006)
15. Tang, X., Zhou, P., Qiu, W.: On the similarity/dissimilarity of DNA sequences based on 4D graphical representation. Chinese Science Bulletin 55(8), 701–704 (2010)

16. Liao, B., et al.: On the Similarity of DNA Primary Sequences Based on 5-D Representation. Journal of Mathematical Chemistry 42(1), 47–57 (2006)
17. Zhang, Z.J.: DV-Curve: a novel intuitive tool for visualizing and analyzing DNA sequences. Bioinformatics 25(9), 1112–1117 (2009)
18. Redon, R., et al.: Global variation in copy number in the human genome. Nature 444(7118), 444–454 (2006)
19. Herzel, H., Ebeling, W., Schmitt, A.O.: Entropies of biosequences: The role of repeats. Phys. Rev. E 50, 5061–5071 (1994)
20. Vinga, S., Almeida, J.: Renyi continuous entropy of DNA sequences. Journal of Theoretical Biology 231(3), 377–388 (2004)
21. Vinga, S., Almeida, J.S.: Local Renyi entropic profiles of DNA sequences. BMC Bioinformatics 8(1), 393 (2007)
22. Liao, B., Zhu, W.: Analysis of Similarity/Dissimilarity of DNA Primary Sequences Based on Condensed Matrices and Information Entropies. Current Computer-Aided Drug Design 2, 275–285 (2006)
23. Sujeevan, R., Hebert, P.D.N.: BOLD: The Barcode of Life Data System (www.barcodinglife.org). Molecular Ecology Notes (2007)
24. Stephen, F.A., Gish, W., Miller, W., Myers, E.W., Lipman, D.J.: Basic local alignment search tool. J. Mol. Biol. (215), 403–410 (1990)
25. Kimura, M.: A simple method for estimating evolutionary rates of base substitutions through comparative studies of nucleotide sequences. J. Mol. Evol. 16, 111–120 (1980)
26. Liao, B., Sun, X., Zeng, Q.: A Novel method for similarity analysis and protein sub-cellular localization prediction. Bioinformatics 26(21), 2678–2683 (2010)
27. Afreixo, V., Bastos, C.A.C., Pinho, A.J., Garcia, S.P., Ferreira, P.J.S.G.: Genome analysis with inter-nucleotide distances. Bioinformatics 25(23), 3064–3070 (2009)
28. Cai, S.J.J., Xia, D.K., Yuen, X.: Kwok-yung: MBEToolbox: a Matlab toolbox for sequence data analysis in molecular biology and evolution. BMC Bioinformatics 6(64), 1–8 (2005)

Indentifying Disease Genes Using Disease-Specific Amino Acid Usage

Fang Yuan[1], Jing Li[2], and Lun Li[3]

[1] Shenzhen Institute of Information Technology, Shenzhen 518029, China
[2] Huawei Technologies Corporation, Shenzhen 518129, China
[3] Hubei Bioinformatics and Molecular Imaging Key Laboratory,
Huazhong University of Science and Technology, Wuhan 430074, China
yuancopper@163.com

Abstract. The identification of disease genes from candidated regions is one of the most important tasks in bioinformatics research. Among all the approaches reported recently, methods based on sequence characteristics have the widest application range. However, their accuracies are usually low, because these methods take into account the overall differences between disease and non-disease gene, rather than specific characteristics. To tackle this problem, the statistical characteristics of the protein sequences between disease genes and non-disease genes have been analyzed. The analysis showed that the amino acids usage by a gene was similar to genes responsible for the same disease but remarkably different from others. An algorithm based on the amino acid usage characteristics was developed. And cross validation was performed for a set of 208 genes involved in 55 diseases with significant amino acid usage characteristics. The test demonstrated that, 15.4% target genes ranked first, and the target genes were in the top 5% with 44.2% chance. For those diseases with significant amino acid usage characteristics, this approach showed promising performance compared to other methods.

Keywords: disease gene prediction, disease phenotype, amino acid usage.

1   Introduction

The identification of disease genes is not only the critical to understand disease mechanisms, but also the essence for developing new diagnostics and therapeutics. Through complex-trait linkage studies, many disease genes are located within one or more specific chromosomal regions [1]. It is time-consuming and labor-intensive to perform random mutation analysis for the hundreds of genes in the regions of interest. Clearly, predicting the best candidate genes by computational approaches is necessary to facilitate the identification of disease-related genes for further study.

Many bioinformatics approaches for prioritizing candidate genes have been developed and released to public in recent years [2-12]. These approaches are based on data sources such as gene functional annotations [2, 3], gene sequence characteristics [4-7], expression profiles [8,9], protein-protein interactions [10-12], et al. The effectiveness and scope of these approaches depend heavily on the accuracy

D.-S. Huang et al. (Eds.): ICIC 2011, LNBI 6840, pp. 478–485, 2012.

and coverage of these data sources, which are typically incomplete and biased towards better-studied genes.

Differently, approaches based on gene sequence characteristics can be used for the genes lacking detailed annotations and avoid annotation bias. Thus, they have a broader range for disease gene prediction. For instance, researchers observed that, proteins involved in human diseases are longer than the rest of the proteins encoded in the human genome, and likely contain more alanine and glycine, and less histidines and lysines, etc. [5]. These characteristics can be used to distinguish the disease genes and the non-disease genes, but can not be used to identify specific disease genes.

To explore potential disease-specific features and help the identification of specific diseases, the statistical characteristics of the protein sequences between disease and non-disease genes were analyzed in this study. We found that amino acid usage characteristics of many disease-specific proteins are disease-specific. Based on the characteristics, we developed a novel approach to predict disease-specific genes.

2   Materials and Methods

2.1   Datasets

The disease gene data were derived from Morbid Map table in OMIM database (OMIM, http://www.ncbi.nlm.nih.gov/entrez/query.fcgi?db=OMIM). To assure the data reliability, we selected 2044 disease genes whose sequences are manually verified, by joining the two table "mim2gene" and "gene2refseq" to Morbid Map table.

To compare the characteristics between disease genes and non-disease genes, all of the manually verified human protein sequences were selected. And to ensure the data consistency, only the 12595 longest isoform sequences were derived for our comparison analysis.

2.2   Outline

Given a phenotypical definition of a disease (MIM number) and a chromosomal region where to search for causative genes associated with the disease of interest, the flowchart for prioritizing candidate genes is shown in Fig. 1.

The process includes the following two main phases:

1) Obtaining the disease-specific amino acids by analyzing amino acid usage characteristic of the known genes of the given disease.

2) Prioritizing the candidate genes in the chromosomal region of interest by similarity analysis based on amino acid usage. Firstly, we retrieve the amino acids of the coded proteins of the candidate genes in the given chromosomal region. Secondly, the degrees of amino acid usage similarity between the amino acids of candidate genes and the disease-specific amino acids are calculated. Finally, we prioritize the candidates according to the amino acid usage similarity degrees in descending order.

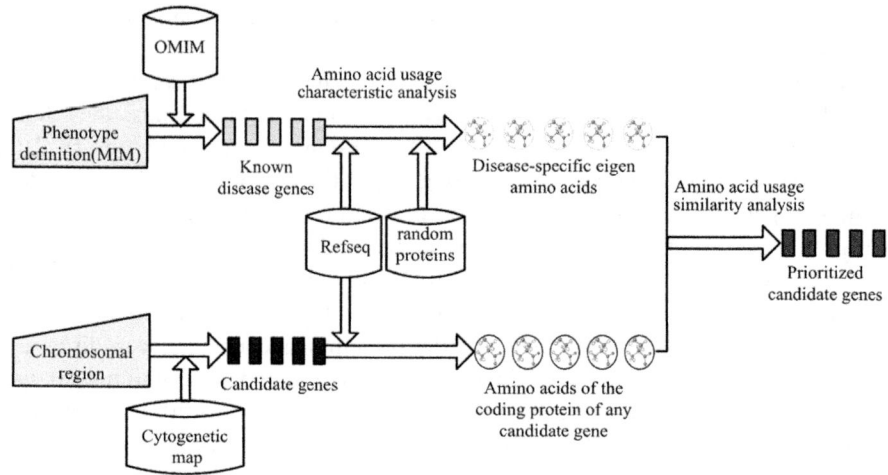

Fig. 1. The flowchart of disease gene prediction

2.3   Analysis of the Amino Acid Usage Characteristic

Here, we use the following models to analyze the disease-specific amino acid usage.

Given a set of proteins coded by genes involved in disease d, P_d (disease-specific proteins), and an amino acid a, we score the usage bias level of a in P_d as follows.

$$SA(a \mid P_d) = \ln \frac{f(a \mid P_d)}{f(a \mid P_h)} \qquad (1)$$

Where $f(a|P_d)$ and $f(a|P_h)$ are the average frequencies of a in P_d and in all human proteins P_h respectively. If the frequency of a in P_d is significantly different from that in P_h, $|SA(a|P_{Ri})|$, the absolute value of $SA(a|P_{Ri})$ will be very high.

To obtain the disease-specific amino acids, the levels of statistical significance (p-value) were calculated. Here, m (10000 in our analysis) sets of proteins (named as $P_R=\{P_{Ri} \mid_{i=1,2,\dots m}\}$), were randomly selected from all human proteins. Each set of P_{Ri} contains the same number of genes as the disease-specific protein set P_d. The level of statistical significance of amino acid a in disease-specific proteins is defined as follows.

$$PA(a \mid P_d) = \frac{Countif(P_R, |SA(a \mid P_{Ri})| > |SA(a \mid P_d)|)}{|P_R|} \qquad (2)$$

In the above formula, the numerator is the counter of P_R which counts when $|SA(a|P_{Ri})|$ is greater than $|SA(a|P_d)|$, and the denominator is the cardinality of P_R, which equals 10000 here. The lower $PA(a|P_d)$ is, the more remarkable the amino acid usage of a in the disease-specific proteins P_d is.

Similarly, given a protein set P and its corresponding specific amino acid set $A=\{a_i \mid i=1,2,\dots,n\}$, the composite score of amino acid usage bias level of P is defined as follows.

$$SP(P) = \sum_{i=1}^{n} SA(a_i \mid P) * f(a_i \mid P) \tag{3}$$

PP is defined to evaluate the statistical significance level of amino acid usage characteristic of the proteins involved in the disease **d** as follows.

$$PP(d) = \frac{Countif(P_R, |SP(P_{Ri})| > |SP(P_d)|)}{|P_R|} \tag{4}$$

The lower **PP(d)** is, the more remarkable the amino acid usage of disease **d** is. In this report, the diseases with a p-value lower than 0.1 are considered to be the ones with significant amino acid usage characteristics.

2.4   Prediction of Disease Genes by Amino Acid Usage Similarity Analysis

For diseases with significant amino acid usage bias, it is feasible to predict their potential genes by analyzing the similarity of amino acid usage between candidate genes and known genes.

Given P_g as the coded protein of the gene **g**, the relation of the candidate gene **g** to disease **d** is evaluated by **S(g|d)**.

$$S(g \mid d) = \sum_{i=1}^{n} SA(a_i \mid P_d) * f(a_i \mid P_g) \tag{5}$$

To avoid noise, only the amino acids with p-value **(PA)** lower than 0.1 are selected. Here, n is the number of the amino acids selected. **S** scores are used to rank all the candidates in the descending order. The top ones are considered as the candidate genes related to the given disease **d** with high risk.

2.5   Test Sets and Evaluation Methods

To assure the test validity, we picked 1022 from the selected 2044 disease genes as our base test set. The genes are involved in 295 OMIM disease phenotypes, and each disease phenotype has at least two known genes.

We calculated the amino acid usage bias level of all 295 groups of disease-specific proteins, using the algorithm mentioned above. The statistical analysis revealed that there were 55 groups of disease-specific proteins whose p-values **(PP)** were lower than 0.1. The corresponding 55 diseases which cover 208 known genes were considered to be the ones with significant amino acid usage characteristics, and were taken as our final test set.

To test the performance of our approach, we adopted the widely-used leave-one-out cross validation. Here, for each disease phenotype, we picked a fixed target gene from known genes, and selected 30 Mb around this gene as the target region, in which region all genes were considered as the candidates. To evaluate the performance of our approach, we calculated sensitivity **(Sn)** and specificity **(Sp)** [13].

3   Results and Discussion

3.1   Results of the Leave-One-Out Cross Validation

The disease gene prediction based on the amino acid usage characteristics was tested, using the 208 genes involved in 55 diseases. The results were shown as the solid curve in Fig.2. It indicated that, of the 208 genes, 15.4% targets rank first (Sn is 0.154 when 1-Sp is 0), 44.2% targets rank within the top 5% of the list, 57.2% targets rank within the top 10%, and 88.0% targets rank within the top 50%.

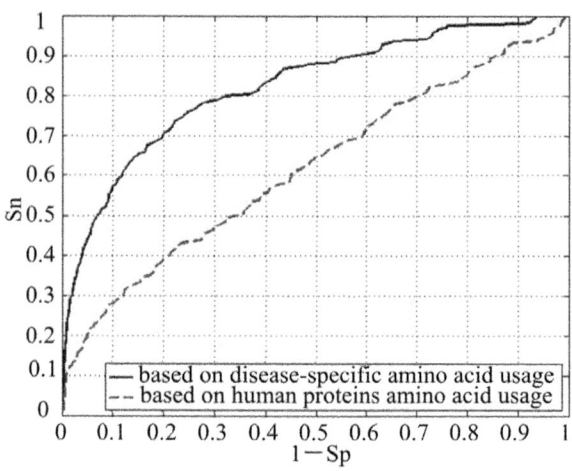

Fig. 2. Results of leave-one-out cross validation based on amino acid usage in disease-specific proteins and all human proteins

We have taken another test based on the amino acid usage characteristics in human proteins (the protein coded by the target gene in the target region is excluded). In this test, we used the same test dataset of the 208 genes. Shown as the dotted curve in Fig. 2, the results indicated that only 5.8% targets ranked first, 20.2% targets ranked within the top 5% of the prioritized list, 27.4% targets ranked within the top 10%, and 64.4% targets ranked within the top 50%. The solid curve was completely above the dotted one. Obviously, the performance of our approach was much higher than that based on the amino acid usage in all human proteins.

3.2   The Disease-Specific Amino Acid Usage Characteristics

Kondrashov et al. [4] have observed that disease-specific proteins contain more alanines and glycines, and less histidines, lysines and methionines. However, not all disease-specific proteins have the same amino acid usage characteristics. When the amino acid frequencies in Ehlers-Danlos ayndrome-specific proteins/Nemaline myopathy-specific proteins and in all human proteins were compared, a different scenario was observed (shown as Fig. 3). It was evident that the frequencies of glycine and proline in Ehlers-Danlos ayndrome-specific proteins were distinctly higher than their average frequencies in all human proteins, and the frequencies of lysine, serine, and threonine et al. in Ehlers-Danlos ayndrome-specific proteins were lower than that in all human proteins. Different from that in Ehlers-Danlos ayndrome-specific proteins, the specific amino acids with high frequencies[were glutamate, alanine and lysine et al. in Nemaline myopathy-specific proteins. This indicated that Ehlers-Danlos ayndrome and nemaline myopathy were both of amino acid usage bias, but the usage characteristics were disease-specific. These suggest that the amino acid usage characteristics of disease-specific proteins are also disease-specific.

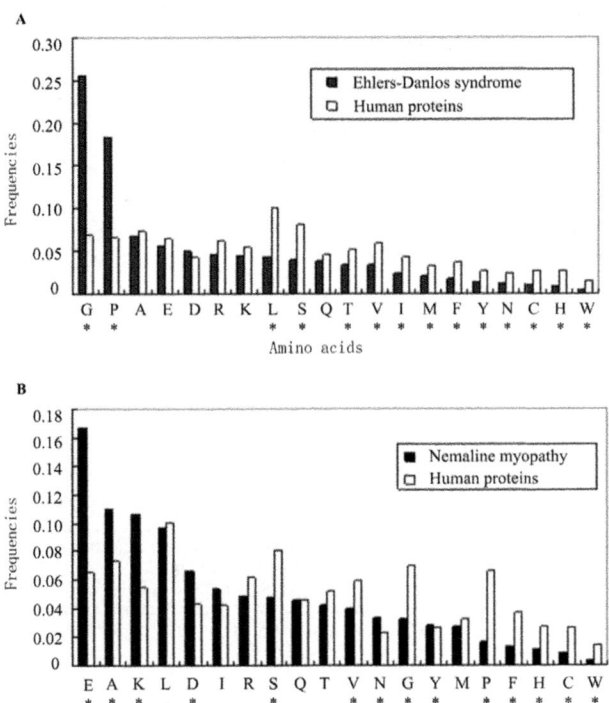

Fig. 3. Comparison of amino acid frequencies in disease-specific proteins and human proteins. The amino acids with * stand for disease-specific amino acids. A: Comparison of amino acid frequencies in Ehlers-Danlos syndrome and human proteins. B: Comparison of amino acid frequencies in Nemaline myopathy and human proteins.

3.3   Comparison with Other Similar Methods

Among similar methods that have been reported, PROSPECTR[6] and SUSPECTS [7] are also based on gene sequence characteristics. To compare the performance between these methods in a fair manner, we used the same test dataset. Of the 208 genes involved in 55 diseases, the ones that can not be identified by PROSPECTR or SUSPECTS were excluded, while the remaining 159 genes involved in 45 diseases were used as our test set. The corresponding ROC graphs were plotted (Fig. 4). When the false positive rate (1-Sp) was lower than 0.52, the true positive rate (Sn) of our method was higher than that of the other two, which suggest that the performance of our method was better. And the *AUC* of our method was 0.817, higher than that of PROSPECTR (0.703) and SUSPECTS (0.745). These results indicated that for diseases with significant amino acid usage characteristics, our method performed better than PROSPECTR and SUSPECTS.

The major difference between our method and PROSPECTR is that the latter combined 15 types of gene sequence characteristics of overall disease genes for disease gene prediction, but did not take the disease-specific characteristics into account. SUSPECTS is another tool based on gene sequences, and besides the above 15 characteristics, SUSPECTS also integrated disease-specific function, expression and InterPro domain data to identify disease genes. The performance of SUSPECTS was higher than that of PROSPECTR, indicating that disease-specific characteristics can improve the prediction performance. Our approach performed even better than SUSPECTS, indicating that the disease-specific amino acid usage is a more effective sequence characteristic for disease gene prediction.

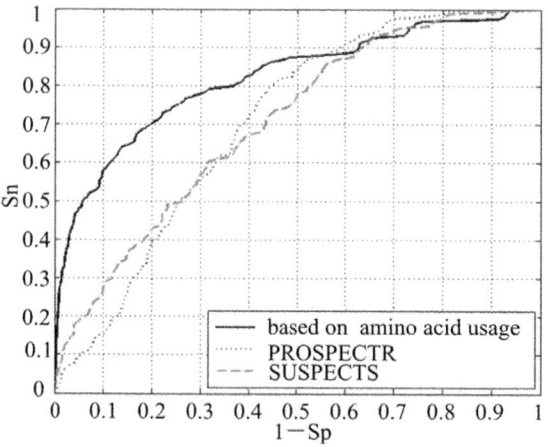

Fig. 4. Results of leave-one-out cross validation of our method based on amino acid usage, PROSPECTR and SUSPECTS

4   Conclusion and Prospect

We have presented a novel approach based on disease-specific amino acid usage bias to prioritize candidate disease genes located in a region of interest. Different from the methods that also base on gene sequences but only take into account the common characteristics of all disease genes, we focus on the disease-specific characteristics.

However, an obvious limitation of this approach is that diseases lacking significant amino acid usage bias can not be predicted effectively. And now data integration for improving gene identification is becoming one of the most important problems in biomedical research. In the near future, we plan to explore more other valuable disease-specific characteristics, such as the usage bias of amino acid repeats, to expand the scope of disease gene prediction and to improve the performances.

Acknowledgements. This work was supported by the Doctor Innovative Project of SZIIT (Grant No. BC2009011).

References

1. McCarthy, M.I., Smedley, D., Hide, W.: New methods for finding disease-susceptibility genes: impact and potential. Genome Biol. 4(10), 119 (2003)
2. Perez-Iratxeta, C., Bork, P., Andrade-Navarro, M.A.: Update of the G2D tool for prioritization of gene candidates to inherited diseases. Nucleic Acids Res. 35(Web Server issue), W212-W216 (2007)
3. Turner, F.S., Clutterbuck, D.R., Semple, C.A.: POCUS: mining genomic sequence annotation to predict disease genes. Genome Biol. 4(11), R75 (2003)
4. Kondrashov, F.A., Ogurtsov, A.Y., Kondrashov, A.S.: Bioinformatical assay of human gene morbidity. Nucleic Acids Res. 32(5), 1731–1737 (2004)
5. Lopez-Bigas, N., Ouzounis, C.A.: Genome-wide identification of genes likely to be involved in human genetic disease. Nucleic Acids Res. 32(10), 3108–3114 (2004)
6. Adie, E.A., et al.: Speeding disease gene discovery by sequence based candidate prioritization. BMC Bioinformatics 6, 55 (2005)
7. Adie, E.A., et al.: SUSPECTS: enabling fast and effective prioritization of positional candidates. Bioinformatics 22(6), 773–774 (2006)
8. van Driel, M.A., et al.: A new web-based data mining tool for the identification of candidate genes for human genetic disorders. Eur. J. Hum. Genet. 11(1), 57–63 (2003)
9. van Driel, M.A., et al.: GeneSeeker: extraction and integration of human disease-related information from web-based genetic databases. Nucleic Acids Res. 33(Web Server issue), W758–W761 (2005)
10. Oti, M., et al.: Predicting disease genes using protein-protein interactions. J. Med. Genet. 43(8), 691–698 (2006)
11. George, R.A., et al.: Analysis of protein sequence and interaction data for candidate disease gene prediction. Nucleic Acids Res. 34(19), e130 (2006)
12. Franke, L., et al.: Reconstruction of a functional human gene network, with an application for prioritizing positional candidate genes. Am J. Hum. Genet. 78(6), 1011–1025 (2006)
13. Aerts, S., et al.: Gene prioritization through genomic data fusion. Nat. Biotechnol. 24(5), 537–544 (2006)

DISCO2: A Comprehensive Peak Alignment Algorithm for Two-Dimensional Gas Chromatography Time-of-Flight Mass Spectrometry

Bing Wang[1], Aiqin Fang[1], Xue Shi[1], Seong Ho Kim[2], and Xiang Zhang[1]

[1] Department of Chemistry, University of Louisville, Louisville, KY 40292, USA
[2] Department of Biostatistics and Bioinformatics, University of Louisville,
Louisville, KY 40292, USA
wangbing@ustc.edu,
{a0fang01,x0shi013,s0kim023,xiang.zhang}@louisville.edu

Abstract. A novel two-stage peak alignment algorithm, DISCO2, was developed for analysis of comprehensive two-dimensional gas chromatography time-of-flight mass spectrometry (GC×GC-TOFMS) data. This algorithm uses a mixture similarity measure to determine whether two peaks were generated by the same type of compound in different experiments. In the first stage, the algorithm detects and merges multiple peak entries of the same compound into one peak entry in each input peak list. Landmark peaks are selected from all samples based on two-dimensional retention times and mass spectral similarity. In the second stage, the original retention time shifts in the two-dimensional GC are corrected using a local linear fitting method. A progressive retention time mapping method is then employed to align peaks in all samples based on the parameters optimized in the first alignment stage.

Keywords: GC×GC-TOFMS, peak alignment, retention time, mixture similarity.

1   Introduction

Peak alignment is one of key steps to deal with metabolomics ananlysis for data generated by comprehensive two-dimensional gas chromatography coupled with time-of-flight mass spectrometry (GC×GC-TOFMS) [1]. Fraga et al. developed a rank-based algorithm using the generalized rank annihilation method (GRAM) [2]. Mispelaar et al. developed a correlation-optimized shifting-based algorithm [3]. To correct the entire chromatogram in both GC dimensions, Pierce et al. proposed an indexing scheme together with a piecewise retention time alignment algorithm [4]. Zhang et al. developed a two-dimensional correlation optimized warping (2-D COW) method [5]. However, these methods align the GC×GC/TOF-MS data based on two-dimensional retention times alone, which may introduce a high rate of false alignment because some molecules with similar chemical functional groups have similar retention times in both GC dimensions. For this reason, two peak alignment methods, MSort [6] and DISCO [7], were developed. Both MSort and DISCO employ the two-dimensional retention times and the mass spectrum of metabolite fragment ions for peak alignment. MSort uses a user-defined retention time window with a fixed size in

D.-S. Huang et al. (Eds.): ICIC 2011, LNBI 6840, pp. 486–491, 2012.
© Springer-Verlag Berlin Heidelberg 2012

the two retention time dimensions to first filter the peak candidates. DISCO algorithm does not employ the two-dimensional retention window. The limitation of DISCO is that the sequential application of retention distance and spectral similarity may still generate false alignment. Kim et al. addressed this problem by using a mixture similarity to simultaneously evaluate the two-dimensional retention distance and spectrum similarity between two peaks of interest [8]. However, the variation of metabolite mass spectrum between experiments may result in small spectral similarity, and therefore affects the value of mixture similarity.

To overcome the limitations of current alignment algorithms, this paper reports a novel alignment algorithm for GCxGC-TOFMS, where the retention distance and spectral similarity are considered simultaneously during the alignment. Furthermore, the molecular spectral variability was considered by including the results of molecular identifications via mass spectral matching.

2   Algorithms

In the proposed algorithms, the instrument control software ChromaTOF was used to reduce the instrument data to peak lists. After peak merging and retention time transformation, a two-stage peak alignment approach, i.e., full alignment and partial alignment, is employed to align all peak lists. The full alignment finds peaks that were identified as the same molecule in all peak lists and aligns them together, while the partial alignment aligns the remaining peaks using a set of parameters derived from results of the full alignment.

2.1   Full Alignment

The purpose of full alignment is to find a list of peaks that present in all peak lists and each of these peaks must have the same identification in all samples. These peaks are termed as landmark peaks. By randomly selecting a sample from the sample set $S = \{s_1, s_2, ..., s_n\}$ as reference sample S_r, and considering the remaining samples $\overline{S} = \{s_1, s_2 .. s_{n-1}\}$ as target samples, a landmark peak pair $p_{r,k}$ and $p_{t,j}$ between the reference sample S_r and a target sample S_t can be found as follows:

$$d(p_{r,k}, p_{t,j} \mid p_{r,k}^{name} = p_{t,j}^{name}) = \min\{d_{(p_{r,k}, p_{t,g})}\} . \tag{1}$$

Here $p_{r,k}$ is the k-th peak in reference sample S_r, $p_{t,g}$ is the g-th peak in target sample S_t, $p_{r,k}^{name}$ is the identification result of spectral matching for a peak $p_{r,k}$, $p_{t,j}^{name}$ is the identification result of a peak $p_{t,j}$, $d(p_{r,k}, p_{t,g})$ is the Euclidean retention distance between two peaks $p_{r,k}$ and $p_{t,g}$, and m is the number of peaks in the t-th target sample.

2.2   Mixture Similarity

In order to simultaneously using the retention time and spectral similarity information for peak alignment, a mixture similarity measure is developed to determine whether two peaks in different peak lists were generated by the same metabolite. The mixture similarity is defined as follows:

$$S^m_{p_{r,k},p_{j,k}} = w_{r,t} \times [1 - (\frac{d_{p_{r,k},p_{t,k}}}{d_{max}})^2] + (1 - w_{r,t}) \times R_{p_{r,k},p_{t,k}} \quad .$$ (2)

Where $w_{r,t}$ is a weighting factor for the alignment of peaks in the reference sample S_r and peaks in a target sample S_t. $d_{p_{r,k},p_{t,j}}$ is the Euclidean distance between the k-th peak in the reference sample S_r and the j-th peak in the target sample S_t in the two-dimensional retention time space, d_{max} is the maximum distance in all samples, and $R_{p_{r,k},p_{t,j}}$ is Pearson's correlation coefficient between mass spectra of the k-th peak in the reference sample and the j-th peak in the target sample.

An optimal value of $w_{r,t}$ can be determined from a list of candidate values such as (0.05, 0.1, 0.2, 0.3, 0.4, 0.5, 0.6, 0.7, 0.8, 0.9, 0.95) by maximizing the sum of mixture similarity of all landmark peak pairs between the reference sample S_r and a target sample S_t:

$$w_{r,t} = \arg\max(\sum_{k=1}^{N_{lp}} S^m_{p_{r,k},p_{t,k}}) \quad .$$ (3)

Where N_{lp} is the number of landmark peak pairs in the sample pair (S_r, S_t). The optimal value of w is then used during the partial alignment to align the rest peaks in the same target sample and the reference sample.

2.3   Partial Alignment

The partial alignment is to align the remaining peaks which are not detected as the landmark peaks selected during the full alignment. Due to inaccuracy of identification, a high rate of false identifications may occur. It is possible that the peaks in different samples were identified as different molecules although they were actually generated by the same type of molecule. Therefore, the results of molecular identifications are not used during partial alignment. Several threshold values are first determined based on the results of full alignment after removing the outlier values using Grubbs' test. These thresholds include retention distance d_0, spectral similarity R_0 and mixture similarity S^m_0. The mixture similarity defined in equation (2) is then used to find the corresponding peak in a target sample for a peak in the reference sample with the following restrictions:

$$d_{p_{r,k},p_{j,k}} \leq d_0 \quad R_{p_{r,k},p_{t,k}} \geq R_0 \quad S^m_{p_{r,k},p_{j,k}} \geq S^m_0 \quad .$$ (4)

3   Results and Discussion

3.1   Experimental Data

A mixture of 76 compounds (8270 MegaMix, Restek Corp., Bellefonte, PA) and C7-C40 saturated n-alkanes (Sigma-Aldrich Corp., St. Louis, MO) were spiked with a deuterated six component semi-volatiles internal standard (ISTD) mixture (Restek Corp., Bellefonte, PA) at a concentration of 2.5 µg/mL prior to GC×GC/TOF-MS analysis.

Three replicate analyses of the same sample had been implemented using an identical two-dimensional GC configuration with different column temperature gradients, i.e., 5 °C/min, 7 °C/min, and 10 °C/min, respectively. The experiments ramped at different temperature gradients have been repeated different times, i.e., 10 replicate analyses for 5 °C/min, 3 replicate analyses for 7 °C/min, and 4 replicate analyses for 10 °C/min.

The raw instrument data were first reduced into peak list using ChromaTOF software version 4.21 equipped with the National Institutes of Standards and Technology (NIST) MS database (NIST MS Search 2.0, NIST/EPA/NIH Mass Spectral Library; NIST 2002). A total of 17 peak lists was generated. These peak lists are denoted as $S^{5C}(s_1^{5C}, s_2^{5C}, \cdots, s_{10}^{5C})$, $S^{7C}(s_1^{7C}, s_2^{7C}, s_3^{7C})$, and $S^{10C}(s_1^{10C}, s_2^{10C}, s_3^{10C}, s_4^{10C})$ for the temperature gradients of 5 °C/min, 7 °C/min, and 10 °C/min, respectively.

3.2   Selection of an Optimal w Value

The weight factor w in the mixture similarity weighs the contribution of retention distance and the spectrum similarity to the mixture similarity. An optimal value of $w_{r,t}$ can determined from a list of candidate values by maximizing the sum of mixture similarity of all landmark peak pairs, as specified in equation (3). By randomly chosen peak list S_1^{5C} as the reference sample, the other 16 samples ($S_2^{5C} \sim S_{10}^{5C}$, $S_1^{7C} \sim S_3^{7C}$, $S_1^{10C} \sim S_4^{10C}$) were considered as target samples. The full alignment between each of the target samples and the reference sample was performed. After elution order filtering, a list of landmark peaks was detected between each sample pair, i.e., a target sample and the reference sample. The optimal w value for the 16 sample pairs is 0.95, 0.95, 0.95, 0.95, 0.95, 0.95, 0.95, 0.95, 0.95, 0.05, 0.05, 0.05, 0.05, 0.05, 0.05, and 0.05, respectively. A larger value of w weighs more the contribution of retention distance.

3.3   Alignment Performance

By manually comparing the identified compounds in each peak list and the list of compound standards, there were 66, 72, 67 and 63 compound standards were identified in all samples of the four data sets S^{5C}, S^{7C}, S^{10C} and $S^{5C} + S^{7C} + S^{10C}$, respectively. These compounds are considered as true positive peak N_p and used to evaluate alignment performance. Table 1 lists the alignment results for the sample sets S^{5C}, S^{7C}, S^{10C} and $S^{5C} + S^{7C} + S^{10C}$.

It can be seen that the alignment performance decreases with the increase of sample size. The PPV values the proposed algorithm achieved 1.0 in all sample set, which means false-positive is zero. It should be noted that N_p was defined based on the name of molecular identifications. Due to the high rate of false identification, it is likely that some of the identifications may be false-positive because of very large retention time shift compared to the retention time shifts of other compound standards. For example,

Table 1. Overall performance of the proposed peak alignment method

	N_m	N_p	TPR	PPV	F1 score
S^{5C}	57	66	0.864	1.0	0.927
S^{7C}	64	72	0.889	1.0	0.941
S^{10C}	64	67	0.955	1.0	0.977
$S^{5C} + S^{7C} + S^{10C}$	43	63	0.68	1.0	0.811

Benzyl alcohol was identified in sample S_1^{5C} and S_5^{5C} with two-dimenstional retention times (744.43 s, 1.696 s) and (894.32 s, 1.399 s), respectively. The almost 150 s retention time shift in the first dimension retention time and 0.3 s shift in the second dimension retention time causes 0.538 shift of retention distance in the z-score transformed retention time space. The retention distance shift of the rest compound standards is only 0.017±0.012. Such a large retention distance shows that Benzyl alcohol is likely a false-positive identification in the sample S_1^{5C} and S_5^{5C}. These false identified compounds should be not counted as the true-positive compounds. That is the actual value of N_p should be smaller than the N_p value used here and therefore, the alignment performance should be better than the values listed in Table 1. However, it is difficult to further confirm the results of molecular identification.

3.4   Comparison with DISCO

Among the current peak alignment approaches, DISCO is the only one can deal with both the homogeneous and heterogeneous data. Table 2 lists the alignment results of DISCO on the same data set, where a spectral similarity threshold of 0.95 was set for peak merging and 0.80 for peak alignment. Comparing with the values of TPR, PPV and F1 listed in Table 1, it can be concluded that the proposed algorithm performs much better than DISCO. The F1 scores of the proposed algorithm ranges from 0.927 to 0.977 in analyzing homogenous data, while the F1 scores of DISCO ranges from 0.434 to 0.879. The proposed algorithm has a much better performance than DISCO in analyzing heterogeneous data with a F1 score of 0.811, while the F1 score of DISCO is only 0.121.

Table 2. Overall performance of DISCO

	N_m	N_p	TPR	PPV	F1 score
S^{5C}	24	66	0.354	0.561	0.434
S^{7C}	58	72	0.806	0.967	0.879
S^{10C}	37	67	0.561	0.902	0.692
$S^{5C} + S^{7C} + S^{10C}$	6	63	0.081	0.238	0.121

4   Conclusions

This paper proposed a new two-stage peak alignment algorithm, full alignment and partial alignment, to align the GCxGC-TOFMS data. The proposed algorithm incorporates the molecular spectral matching results into alignment to include the spectrum variations. It further eliminates any user-defined thresholds of retention time windows, as well as a threshold of spectrum similarity. The performance of the present algorithm was tested using experimental data, where a mixture of compound standards was analyzed under different experiment conditions. The results show that the proposed algorithm works well in comparison with literature reported method.

Acknowledgements. This work was supported by the National Science Foundation of China (Nos. 60803107), and National Institute of Health (NIH) grant 1RO1GM087735 through the National Institute of General Medical Sciences (NIGMS).

References

1. Ong, R.C., Marriott, P.J.: A review of basic concepts in comprehensive two-dimensional gas chromatography. J. Chromatogr Sci. 40(5), 276–291 (2002)
2. Fraga, C.G., Prazen, B.J., Synovec, R.E.: Objective data alignment and chemometric analysis of comprehensive two-dimensional separations with run-to-run peak shifting on both dimensions. Anal. Chem. 73(24), 5833–5840 (2001)
3. van Mispelaar, V.G., Tas, A.C., Smilde, A.K., Schoenmakers, P.J., van Asten, A.C.: Quantitative analysis of target components by comprehensive two-dimensional gas chromatography. J. Chromatogr. A 1019(1-2), 15–29 (2003)
4. Pierce, K.M., Wood, L.F., Wright, B.W., Synovec, R.E.: A comprehensive two-dimensional retention time alignment algorithm to enhance chemometric analysis of comprehensive two-dimensional separation data. Anal. Chem. 77(23), 7735–7743 (2005)
5. Zhang, D., Huang, X., Regnier, F.E., Zhang, M.: Two-dimensional correlation optimized warping algorithm for aligning GC x GC-MS data. Anal. Chem. 80(8), 2664–2671 (2008)
6. Oh, C., Huang, X., Regnier, F.E., Buck, C., Zhang, X.: Comprehensive two-dimensional gas chromatography/time-of-flight mass spectrometry peak sorting algorithm. Journal of Chromatography A 1179(2), 205–215 (2008)
7. Wang, B., Fang, A., Heim, J., Bogdanov, B., Pugh, S., Libardoni, M., Zhang, X.: DISCO: distance and spectrum correlation optimization alignment for two-dimensional gas chromatography time-of-flight mass spectrometry-based metabolomics. Analytical Chemistry 82(12), 5069–5081 (2010)
8. Kim, S., Fang, A., Wang, B., Jeong, J., Zhang, X.: An Optimal Peak Alignment For Comprehensive Two-Dimensional Gas Chromatography Mass Spectrometry Using Mixture Similarity Measure. Bioinformatics (2011)

Prediction of Human Proteins Interacting with Human Papillomavirus Proteins

Guangyu Cui, Chao Fang, and Kyungsook Han[*]

School of Computer Science and Engineering, Inha University, Incheon, South Korea
cuigy119@inhaian.net, fczqx@inha.edu, khan@inha.ac.kr

Abstract. Several computational methods have been developed for predicting protein-protein interactions, but most of these methods are intended for finding the protein-protein interactions within a species rather than for the interactions across different species. Methods for predicting the interactions between homogeneous proteins are not appropriate for predicting the interactions between heterogeneous proteins since they do not distinguish the interactions between proteins of the same species from those of different species. In this paper we present the development of a support vector machine (SVM) model that predicts the interactions between human papillomaviruses (HPV) proteins and human proteins using the sequence data. The average accuracy of the SVM model in predicting the interactions between HPV proteins and human proteins is 81.9%. Using the SVM model and the Gene Ontology (GO) annotations of proteins, we also predicted 130 new interactions between HPV and human proteins.

Keywords: protein-protein interaction, support vector machine, human papillomavirus.

1 Introduction

Recently a number of methods have been developed to predict protein-protein interactions in one species. The method developed by Bock and Gough [1] uses a support vector machine (SVM) classifier to identify protein-protein interactions from the primary structural and associated physiochemical properties. Predicting protein-protein interactions using protein's chemical information and signature products was proposed [2]. The Gene Ontology (GO) terms and other annotations of proteins and similarity measures were used to predict protein-protein interactions [3]. However, these methods are intended for the protein-protein interactions within a species rather than for the interactions across different species. Methods for predicting the interactions between homogeneous proteins are not appropriate for predicting the interactions between heterogeneous proteins since they do not distinguish the interactions between proteins of the same species from those of different species.

In this paper we propose a computational method to predict the interactions between human papillomaviruses (HPV) and human proteins. HPV is the primary

---

[*] Corresponding author.

D.-S. Huang et al. (Eds.): ICIC 2011, LNBI 6840, pp. 492–497, 2012.

cause of cervical cancer. Two vaccines against HPV were recently developed to prevent infection with HPV, but not effective for HPV infected people. Interactions between an HPV protein and a small number of host proteins have been studied previously. For example, it has been widely accepted that HPV E6 and E7 proteins function as the dominant oncoproteins of HPVs by altering the function of human proteins. However, the interactions between HPV proteins and a large number of human proteins targeted by HPV have not analyzed so far. Since recent studies have identified a number of additional human proteins with which HPV proteins interact during the cellular transformation, a systematic prediction of large-scale interactions between HPV proteins and human proteins would be interesting. Identifying more interactions between HPV and human proteins should help elucidate the interaction mechanism of HPV with host cells and can be helpful in the design of molecules that target the new interacting proteins.

2   Representation and Algorithm

2.1   Support Vector Machine

A support vector machine (SVM) has been applied to several biological problems such as homology detection [4], analysis of gene expression data [5] and prediction of protein-protein interactions [1]. Data examples labeled positive or negative are projected into a high-dimensional feature space using a kernel, and the hyper-plane in the feature space is optimized to maximize the margin between the positive and negative data examples. We implemented a SVM model using LIBSVM (http://www.csie.ntu.edu.tw/~cjlim/libsvm/) with the radial basis function (RBF) as a kernel function.

2.2   Data Set of Interacting Proteins

HPV is a member of the papillomavirus family of viruses that is capable of infecting humans. HPV types 16 and 18 cause 70% of cervical cancer [6, 7]. We extracted the interactions of HPV-16 and HPV-18 proteins with human proteins from NCBI. After removing redundancy, we identified a total of 252 interactions of HPV proteins with human proteins, and obtained Gene IDs from HPRD (http://www.hprd.org).

A positive data set for the SVM model consists of the 252 protein-protein interactions. To construct a negative data set, we randomly selected human proteins from HPRD, which are not included in the 252 interactions of the positive data set. We built a negative data set of 252 interactions for the balance with the positive data set.

We constructed a training set with 200 positive data and 200 negative data, which were selected from the 252 positive data and 252 negative data, respectively. The remaining 52 positive data and 52 negative data were used to construct a test set. To keep the same proportion of human proteins interacting with each HPV protein in both the training and test sets, we selected a training data by

$$N_i = N(Taining) \cdot \frac{N(T_i)}{N(Total)} \tag{1}$$

where T_i is the i-th HPV protein (i=1, 2, ..., 9), N(T_i) is the number of human proteins interacting with the i-th HPV protein, N(Training) is the total number of positive training data, and N(Total) is the total number of HPV-human protein interactions.

Table 1 shows the number of interactions in the training set. Negative data were also selected by equation 1. For example, there are 9 interactions between HPV E1 protein and human proteins. Since N(E1)=200×9/252=7, we selected 7 interactions as positive data and constructed the same number of interactions as negative data.

Table 1. The number of positive data in the training set

HPV protein	#interactions with human proteins	#interactions in the training set
E1	9	7
E2	36	29
E4	2	2
E5	13	10
E6	78	62
E7	76	60
E8	7	6
L1	20	16
L2	11	8
Total	252	200

2.3   Feature Vector

One of the main computational challenges in using an SVM model to predict PPIs is to find a suitable way to describe the important information of protein. We represent a protein sequence using three consecutive amino acids called *amino acid triplet*. For example, in the amino acid sequence TVAVTVA, there are four overlapping amino acid triplets: TVA, VAV, AVT and VTV, which appear 2, 1, 1, and 1 times in the sequence, respectively.

To reduce further the dimension of the vector space, we represent an amino acid sequence using the class of amino acids. Based on the biochemical similarity of amino acids, twenty amino acids were classified into six categories: {IVLM}, {FYW}, {HKR}, {DE}, {QNTP}, and {ACGS} [8, 9]. According to this classification, there are 6×6×6 = 216 possible triplets.

We use a binary space (V, F) to represent a protein sequence, in which V is a vector space of feature vectors with a fixed number of features and F is a vector space of frequency vectors. A protein sequence of variable length is first mapped to a feature vector v of fixed length. A feature vector v is then mapped to a frequency vector f_i (i=1, 2, ..., 216), which represents the frequency of each triplet type.

The difference between the frequency values of triplets can be too small to discriminate interacting sequences from others, so we modified the equation used in the study of Shen *et al.* [10] as follows:

$$d_i = \left\{ e^{\frac{f_i - \min\{f_1, f_2 \cdots f_{216}\}}{\max\{f_1, f_2 \cdots f_{216}\} - \min\{f_1, f_2 \cdots f_{216}\}}} \right\} - 1 \qquad (2)$$

where d_i is the relative frequency of the i-th triplet type in one sequence. The value of d_i ranges from 0 to e-1. The modified d_i has a value in a wider range than the frequency value used in the study of Shen *et al.* [10].

Two features were added at the end of a feature vector: index of an HPV protein (1 to 9 for 9 HPV proteins) and classification of the feature vector (1 for interaction and -1 for non-interaction). By encoding the index of an HPV protein, the SVM model can find a human protein interacting with the HPV protein.

3   Results

When the frequency of a triplet is low, the relative frequency value is also small in both methods. For example, for frequency of 0.3, the difference between the relative frequency values in both methods is only 0.05. However, as the frequency is increased, the difference between the relative frequency values becomes larger, which can improve the sensitivity of the SVM model.

We evaluated the performance of the SVM model using three measures: sensitivity, specificity and accuracy.

$$Sensitivity = \frac{TP}{TP + FN} \tag{3}$$

$$Specificity = \frac{TN}{TN + FP} \tag{4}$$

$$Accuracy = \frac{TP + TN}{TP + FP + TN + FN} \tag{5}$$

True positives (TP) are actual interacting proteins that are predicted correctly. True negatives (TN) are non-interacting proteins that are predicted correctly. False positives (FP) are non-interacting proteins that are predicted as interacting proteins. False negatives (FN) are interacting proteins that are missed by the SVM model.

Due to the randomness in drawing negative data from HPRD and positive data from the pool of HPV-human protein interactions for the training set, we prepared three test sets and training sets for the evaluation of the SVM model. As shown in Table 2, our method achieved a sensitivity of 85.6%, a specificity of 78.3% and an accuracy of 81.9% on average. Our method outperformed the method of Shen *et al.* [10], which showed a sensitivity of 81.7%, a specificity of 77.2% and an accuracy of 79.4% on average. In particular, our method showed the best performance in the second test set.

To find potentially new human proteins that interact with an HPV protein, we searched human proteins in NCBI that are similar to the 252 human proteins known to interact with an HPV protein. After running BLASTP (http://www.ncbi.nlm.nih.gov/BLAST/) with the E-value $\leq 10^{-20}$ and removing the redundant sequences with the 252 human proteins, we obtained a total of 560 human proteins. To predict reliable interactions, we selected human proteins that have a common cellular component gene ontology (GO) ID [11] with H_{HPV} (human proteins that are already known to interact with HPV proteins) for each HPV protein. For instance, the SVM model predicted 28 new human proteins as interacting partners of the HPV E2 protein, for which there are 36 H_{HPV} proteins. Twenty one out of the 28 new proteins had at least one common cellular component with the 36 H$_{HPV}$ proteins and were left as potential interacting partners of the HPV E2 protein. In this way, we predicted 130 new human proteins as interacting partners of HPV proteins.

Table 2. Comparison of our method and the method by Shen *et al.* [10] in terms of sensitivity, specificity and accuracy. SN: Sensitivity, SP: Specificity, AC: Accuracy.

Testing set	Our method			Method of Shen *et al.*		
	SN (%)	SP (%)	AC (%)	SN (%)	SP (%)	AC (%)
1	85.4	78.2	81.8	81.8	77.2	79.5
2	86.5	78.8	82.7	81.9	77.5	79.7
3	84.9	77.9	81.4	81.3	76.9	79.1
Average	85.6	78.3	81.9	81.7	77.2	79.4

4 Conclusion

Most methods for predicting protein-protein interactions focus on the interactions within a species rather than for the interactions across different species, such as the interactions between virus and host cell proteins. In this paper, we presented a support vector machine (SVM) model that predicts the interactions between human papillomavirus (HPV) proteins and human proteins.

We represented a protein sequence using three consecutive amino acids called amino acid triplet. We mapped a protein sequence of variable length to a feature vector of fixed length, and then mapped the feature vector to a frequency vector that represents the relative frequency of each triplet within the protein sequence. The SVM model showed an average accuracy of 81.9% in predicting the interactions between HPV proteins and human proteins, which is higher than the previous method by others [10]. Using the SVM model and the Gene Ontology (GO) annotations of proteins, we also predicted a total of 130 new human proteins that potentially interact with HPV proteins. As a future work, we plan to use additional biochemical properties of amino acids to improve the performance of the SVM model and perform a systematic analysis of the new interactions to identify specific human proteins that are biologically and clinically important in the HPV infection.

Acknowledgements. This research was supported by Basic Science Research Program through the National Research Foundation of Korea (NRF) funded by the Ministry of Education, Science and Technology (2010-0017213).

References

1. Bock, J.R., Gough, D.A.: Predicting protein-protein interactions from primary structure. Bioinformatics 17(5), 455–460 (2001)
2. Martin, S., Roe, D., Faulon, J.: Predicting protein-protein interactions using signature products. Bioinformatics 21(2), 218–226 (2005)

3. Wu, X.M., Zhu, L., Guo, J., Zhang, D.Y., Lin, K.: Prediction of yeast protein-protein interaction network: insights from the Gene Ontology and annotations. Nucleic Acids Res. 34(7), 2137–2150 (2006)
4. Leslie, C.S., Eskin, E., Cohen, A., Weston, J., Noble, W.S.: Mismatch string kernels for discriminative protein classification. Bioinformatics 20(4), 467–476 (2004)
5. Furey, T.S., Cristianini, N., Duffy, N., Bednarski, D.W., Schummer, M., Haussler, D.: Support vector machine classification and validation of cancer tissue samples using microarray expression data. Bioinformatics 16(10), 906–914 (2000)
6. Lowy, D.R., Schiller, J.T.: Prophylactic human papillomavirus vaccines. J. Clin. Invest. 116(5), 1167–1173 (2006)
7. Chaturvedi, A., Gillison, M.G.: Human Papillomavirus and Head and Neck Cancer. In: Olshan, A.F. (ed.) Epidemiology, Pathogenesis, and Prevention of Head and Neck Cancer. Springer, New York (2010)
8. Gomez, S.M., Noble, W.S., Rzhetsky, A.: Learning to predict protein-protein interactions from protein sequences. Bioinformatics 19(15), 1875–1881 (2003)
9. Taylor, W.R.: The classification of amino acid conservation. J. Theor. Biol. 119(2), 205–218 (1986)
10. Shen, J.W., Zhang, J., Luo, X., Zhu, W., Yu, K., Chen, K., Li, Y., Jiang, H.: Predicting protein-protein interactions based only on sequences information. Proc. Natl. Acad. Sci. USA 104(11), 4337–4341 (2007)
11. Ashburner, M., Ball, C.A., Blake, J.A., Botstein, D., Butler, H., Cherry, J.M., Davis, A.P., Dolinski, K., Dwight, S.S., Eppig, J.T., Harris, M.A., Hill, D.P., Issel-Tarver, L., Kasarskis, A., Lewis, S., Matese, J.C., Richardson, J.E., Ringwald, M., Rubin, G., Sherlock, G.: Gene Ontology: tool for the unification of biology. Nat. Genet. 25(1), 25–29 (2000)

Comparison of Data-Merging Methods with SVM Attribute Selection and Classification in Breast Cancer Gene Expression

Vitoantonio Bevilacqua[1], Paolo Pannarale[1], Mirko Abbrescia[1],
Claudia Cava[1], and Stefania Tommasi[2]

[1] Department of Electrical and Electronics, Polytechnic of Bari,
Via E. Orabona, 4 – 70125 Bari, Italy
[2] Istituto Oncologico "Giovanni Paolo II", I.R.C.C.S Ospedale Oncologico di Bari, Viale
Orazio Flacco 65 – 70124 Bari, Italy

Abstract. DNA microarray data are used to identify genes which could be considered prognostic markers. However, due to the limited sample size of each study, the signatures are unstable in terms of the composing genes and may be limited in terms of performances. Therefore, it is of great interest to integrate different studies thus increasing sample size. In the past, several studies explored the issue of microarray data merging, but the appearance of new techniques and a focus on SVM based classification needed further investigation. We used distant metastasis prediction based on SVM attribute selection and classification to three breast cancer data sets. The results showed that breast cancer classification does not take benefit of data merging, confirming the results found by other studies with different techniques.

Keywords: batch effect, gene expression, breast cancer, classification, SVM, pre-processing, ComBat, RMA, MBEI.

1 Introduction

DNA microarray technology and expression profiles are the most suitable tools to investigate gene activity with respect to the progress of disease. Furthermore, they are useful for molecular classification of tumor types [1] and to reveal complexity in the intrinsic cancer subtypes and for developing oncogenic pathway signatures as a guide to targeted therapies [2]. In particular, breast cancer has been extensively studied for gene expression in order to individualize a signature useful for molecular classification [3] and for prognostic purposes [4,5]. However, the sample size of each study is usually quite small with respect to the number of the genes in analysis, to permit an accurate statistical evaluation. Therefore, some authors used to analyze different data coming from different experiments with the goal of increasing sample size and thus increasing the power of the study. This could be done by meta-analysis [7,8], which means the statistical analysis of a large collection of results from individual studies for the purpose of combining their findings to reach a common result or by data merging, analyzing all together raw data coming from different

D.-S. Huang et al. (Eds.): ICIC 2011, LNBI 6840, pp. 498–507, 2012.
© Springer-Verlag Berlin Heidelberg 2012

studies with similar biological questions [6,9,10,11]. Typically, the first transformation applied to expression data, referred to as normalization and summarization, removes non-biological variability between arrays [12] and extracts gene level expression from probe intensities, respectively. However, transformation procedure cannot reduce completely the systematic differences from different data sets. When combining data sets from different experiments, non-biological experimental variation or "batch effects" are brought and therefore it is inappropriate to combine data sets without adjusting for batch effects [13,14].

The goal of this paper is to compare the performance of various data merging methods.[15] Our strategy for the biological comparison is to use microarray data with known phenotypes associated with specific gene sets (pathways).

In the literature, several techniques have been recently proposed for adjusting data for batch effects [10,13,16,17,18]. Many of these methods can be only applied to two batches at a time. In the previous studies, merging data sets were applied to develop a robust gene signature prognostic of survival outcome discretized into two [19] or more categorical values, or diagnostic of tumor subtypes, or predictive of treatment response [20]. A comparison between several techniques to merge different datasets like ComBat, Ratio_G, SVA, DWD, PAMR, has been done.

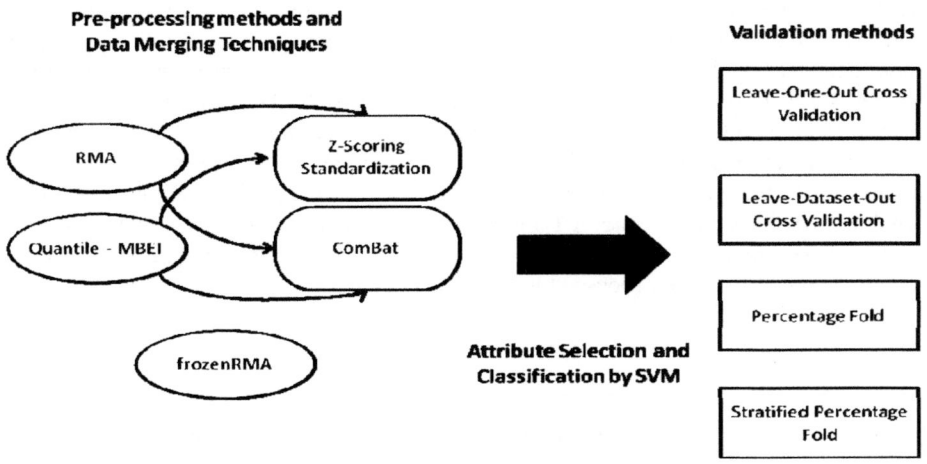

Fig. 1. Study Workflow

We used three breast cancer microarray data sets from three different studies in which all the samples come from lymph-node-negative patients who have not received adjuvant systemic treatment. We performed three pre-processing methods: Robust Multi-Chip Average [21], frozen RMA [22] and Quantile Normalization [12] with Model-Based Expression Index [23]; subsequently we applied two data merging approaches: ComBat [9] and z-scoring standardization procedure for each dataset [18]. frozenRMA is a recent method that performs batch effects removal inherently at summarization time and have not yet been compared to other methods in an independent study. ComBat is the so far best performing method [15], and z-score is one of the first methods used to remove batch effects. Chen and others [18] suggest that the data from two experiments could be integrated for the prognosis analysis after the data standardization. The methods were compared by a new point of view, i.e. in terms of SVM classification performances. Variation attributable to batch effects before and after batch adjustment were identified using principal variation component analysis (PVCA)[24]. Then, the three microarray datasets, differently processed, were examined for specific patterns of pathway deregulation with respect to clinical disease outcome. For this reason we used the most popular gene-set analysis method, Gene Set Enrichment Analysis (GSEA) [25].

2   Material and Methods

2.1   Dataset and Preprocessing Methods

We applied pre-processing and data merging techniques to three breast cancer data sets: GSE11121, GSE2990 (Gene Expression Omnibus) and a dataset used by Foekens et al. in [26], containing respectively 200, 125 and 180 samples from the same Affymetrix GeneChip Human Genome U133A platform. All of the patients in the data sets had lymphonode-negative tumors and did not receive adjuvant systemic treatment. The design of the study is reported in figure 1. Three pre-processing methods have been applied to microarray data to compute expression values from input CEL files: Robust Multi-array Average (RMA), Quantile normalization - Model-Based Expression Index (MBEI) and frozenRMA. The last method estimates batch effect directly from probe intensities. The other methods were integrated in independent manner with two data merging techniques: 1) ComBat 2) Z-score. All computations were done using the free statistical software package R and DNA-Chip Analyzer (dChip). Processing Affymetrix expression arrays usually consists of three steps: background adjustment, normalization and summarization [12] and, in this study, the data sets were processed in three ways (Table 1).

Table 1. Pre-processing methods

	Preprocessing		
Method	**Background correction**	**Normalization**	**Summarization**
RMA	Global correction from posterior mean given the observed PM	Quantile: intensities quantiles Average	Linear model including array and probe effects fittin by medianpolish
Quantile -MBEI	-	Quantile: baseline intensities	Model assuming multiplicative probe effect and additive error
FrozenRMA	RMA-like	Quantile: reference distribution intensity	RMA-like including batch effect in the model

Then, before proceeding to the next analysis, it is necessary to detect and remove the batch effects. Here, the term "batch effect" refers to experimental variations of dataset generated by different labs. An Empirical Bayes method, called Combating Batch Effects when Combining Batches of Gene Expression Microarray Data (ComBat) and Z-Score were used to adjust systematic difference of differently normalized data generated by the three different labs. Into the software package ComBat, including two empirical Bayes frameworks, a parametric and a non-parametric approaches, have been implemented the algorithm of Johnson et al where location and scale model parameters are specifically estimated by pooling information across genes in each batch to adjust for batch effect estimated toward the overall mean of the batch estimates. Each data set was alternatively standardized using a simple z-score transformation method and combined for analysis. In contrast to a classic pre processing and data merging techniques, we applied another pre-processing algorithm, Frozen Robust Multiarray Analysis (fRMA) which allows to analyze microarray in batches and then combine the data for analysis. fRMA without further adjustment for batch effects is similar to RMA when the data are analyzed as single batch.

2.2   Principal Variation Component Analysis (PVCA)

In this study we utilize PVCA to compute non biological experimental variation or "batch effects" bringing when we combined the three data sets from different experiments. The approach utilizes two data analysis methods: first, principal component analysis (PCA) is used to efficiently reduce data dimension with maintaining the majority of the variability in the data, and variance components analysis (VCA) fits a mixed linear model using factors of interest as random (or batch) effects and other variables (or covariates) to estimate and partition the total variability.

2.3   Classification

After these steps, above procedures were analyzed to extract a gene signature, for its association with distant-metastasis-free survival (dmfs) discretized into two values (0,1). The datasets where unbalanced towards class 0, but from a clinical stand point, the cost of false negatives is higher than false positive: patients that should receive treatment may be neglected not receiving treatment. Due to this the class 0 set where down-sampled until half spread. The training step applies a SVM based Recursive Feature Elimination (RFE) method [28] for the selection of genes by an iterative procedure of the features ranking and the worst one discarding, after sub sampling of the dataset to make equal he distribution size of the two classes. Then, this model, composed of 400 gene expression level, is used for the training of a linear support vector machine. The next step consists of testing on the samples of the dataset that has not been used to train the classifier. In this study, we used two performance measures of the classifier: recall (sensitivity) and precision.

2.4   Validation Methods

The models was validated using Leave-One-Out Cross Validation (LOOCV), Leave-Dataset-Out Cross Validation (LDOCV), Percentage Fold and Stratified Percentage Fold. The Leave-One-Out (or LOO) is a simple cross-validation. Each learning set is created by using all the samples except one, the test set being the sample left out.

Leave-one-out cross validation is used in the field of machine learning to determine how accurately a learning algorithm will be able to predict data that it was not trained on. It creates all the possible training/test sets by using a single observation from the original sample as the testing data, and the remaining observations as the training data. LDOCV is very similar to LOOCV but it considers the single dataset rather than samples. Percentage Fold randomly divides the merged data set into training and testing sets, 66% and 34% respectively; the Stratified Percentage Fold splits each datasets by preserving the same percentage for training and testing sets.

2.5 Pathway Analysis

For further validation of different data merging and processing technique comparison, we analyzed specific pathways associated with breast cancer progression: EGF, Stathmin, HER2, BRCA1, Homologous Recombination. The pathway database was compiled from the Kyoto Encyclopedia of Genes and Genomes (KEGG) and Biocarta. The GSEA algorithm [25] has been used, an established method in pathway enrichment analysis.

3   Results

3.1 Data Merging Validation

We directly merged the three microarray data sets, using the 22283 probe sets on Affymetrix HG-U133A microarray, to form an integrated data set in seven ways: RMA, Quantile/MBEI, RMA - ComBat, Quantile/MBEI-ComBat, RMA - Z-Score and Quantile/MBEI - Z-Score, frozenRMA. The integrated data set consists of 111 samples with distant metastases and 394 samples free of distant metastases, randomly divided into training and testing sets but with respect of proportions of complete dataset. We evaluate the classification performance using precision and recall of class 1 because of its clinical significance. The table 2 shows the results of the classifier which demonstrated less accuracy of data merging methods with respect to classification.

3.2 Data Merging Verification

The PVCA revealed that batch effects explained 22.4% of the overall variation in the RMA data (Fig. 2.a) and 32,3 % in the Quantile/MBEI   data (Fig. 2.d) .

After applying ComBat (Fig. 2.b - 2.e) and Z-Score (Fig. 2.c - 2.f) the variation were completely eliminated. The worst performance seemed to be of fRMA (Fig. 2.g) that shows 24,9 % threshold of variation of the batch effects.To assess the removal of microarray bias effect across data sets, Principal Component Analysis (PCA) were applied to the data sets after the application of data merging methods. The aim was to reveal intermixing of samples from different dataset before and after adjustment. The results of these approaches (Fig. 3) demonstrate that samples referring to the same dataset can't be grouped after using merging methods. This trend was respected in the RMA - ComBat data (Fig. 3.b), RMA - Z-Score data (Fig. 3.c), MBEI - ComBat data (Fig. 3.e) and MBEI - Z-Score data (Fig. 3.f), as shown in the graphs of the first three principal components. Conversely it's possible in the RMA data (Fig. 3.a), MBEI data (Fig. 3.d) and fRMA data (Fig. 3.g). Again, fRMA seemed to be unable to adjust the combination among the three datasets.

Table 2. Classification performance

RMA - Combat		Recall	Precision	RMA – Z-score		Recall	Precision
Validation		Recall	Precision	Validation		Recall	Precision
LDOCV	FK	0.568	0.263	LDOCV	FK	0.622	0.277
	GSE2990	0.5	0.286		GSE2990	0.536	0.3
	GSE11121	0.63	0.315		GSE11121	0.609	0.322
	Average	0.576	0.289		Average	0.596	0.301
LOOCV		0.631	0.370	LOOCV		0.512	0.317
Percentage Fold		0.531	0.273	Percentage Fold		0.563	0.228
Stratified Perc. Fold		0.677	0.198	Stratified Perc. Fold		0.613	0.25
RMA				**Quantile / MBEI - ComBat**			
Validation		Recall	Precision	Validation		Recall	Precision
LDOCV	FK	0.622	0.304	LDOCV	FK	0.649	0.282
	GSE2990	0.607	0.283		GSE2990	0.571	0.314
	GSE11121	0.609	0.280		GSE11121	0.652	0.341
	Average	0.613	0.289		Average	0.631	0.313
LOOCV		0.613	0.289	LOOCV		0.550	0.289
Percentage Fold		0.5	0.208	Percentage Fold		0.625	0.282
Stratified Perc. Fold		0.645	0.257	Stratified Perc. Fold		0.692	0.237
Quantile / MBEI - Z-Score				**Quantile / MBEI**			
Validation		Recall	Precision	Validation		Recall	Precision
LDOCV	FK	0.73	0.297	LDOCV	FK	0.486	0.290
	GSE2990	0.571	0.333		GSE2990	0.643	0.290
	GSE11121	0.696	0.36		GSE11121	0.522	0.414
	Average	0.677	0.331		Average	0.539	0.339
LOOCV		0.550	0.279	LOOCV		0.550	0.299
Percentage Fold		0.625	0.299	Percentage Fold		0.625	0.282
Stratified Perc. Fold		0.692	0.225	Stratified Perc. Fold		0.808	0.28
FrozenRMA							
Validation		Recall	Precision				
LDOCV	FK	0.649	0.279				
	GSE2990	0.464	0.283				
	GSE11121	0.5	0.319				
	Average	0.544	0.307				
LOOCV		0.512	0.308				
Percentage Fold		0.588	0.23				
Stratified Perc. Fold		0.581	0.265				

3.3   Patterns of Pathway Deregulation

We explored the performance of various pre-processing and data merging technique by using pathway and determining whether a group of differentially expressed genes is enriched for a particular set. We used several biological pathways: EGF, Stathmin, HER2, BRCA1, Homologous Recombination.

In a heat map, expression values are represented as colors, where the range of colors (red, pink, light blue, dark blue) shows the range of expression values (high, moderate, low, lowest). The light blue and dark blue bars reflect genes that are positively associated with DMFS (Disease Metastasis Free Survival), indicating a higher expression in tumors without metastatic capability. The red bars reflect genes that are negatively associated with DMFS, indicative of higher expression in tumors with metastatic capability.

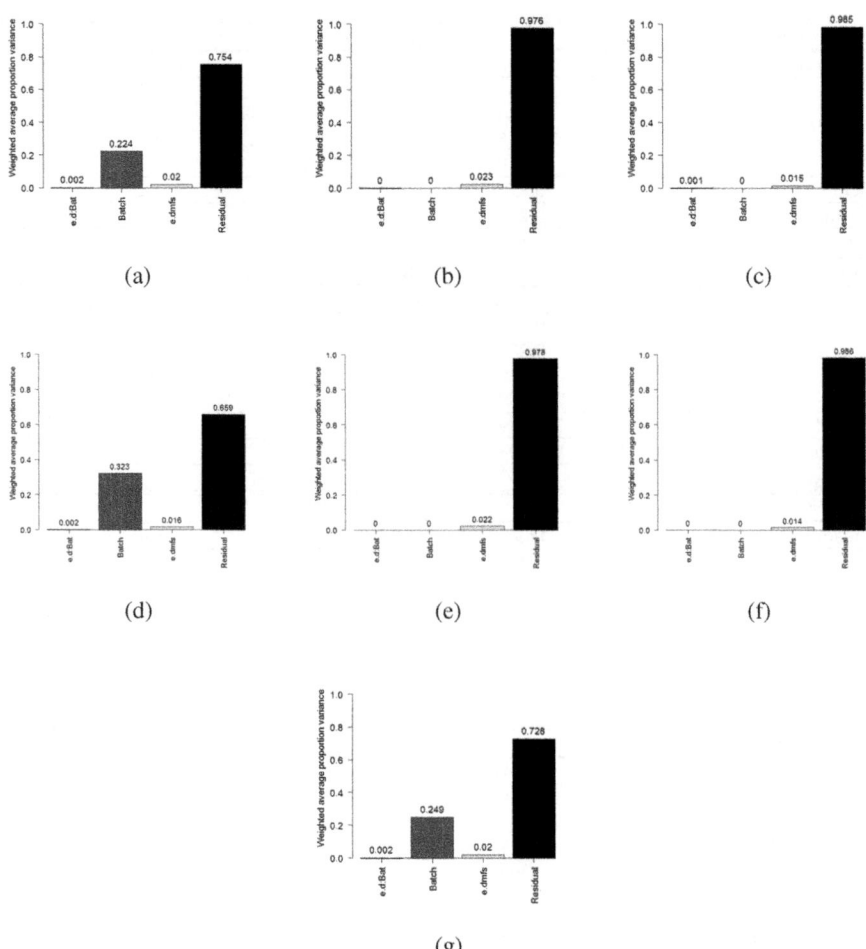

Fig. 2. PVCA results. RMA data (a), RMA - ComBat data (b), RMA - Z-Score data (c), MBEI data (d), MBEI - ComBat data (e), MBEI - Z-Score data (f), fRMA data (g).

The heat map of data sets that are pre-processed by robust multi-array average (RMA) and MBEI are available at [28]. Considering the expression level of the genes reported in each considered pathway, it can be notice that applying ComBat after RMA (Fig. 4b) or MBEI (Fig. 4e) lead to similar results than RMA (Fig. 4a) and MBEI (Fig. 4d) alone, respectively. Differently, Z-Score method reported activation of different genes in each pathway after both methods. (Fig. 4.c ; 4.f). All probe set were pre-ranked using SNR (signal to noise ratio) with respect to their correlation with distant metastasis free-survival, thereafter. The order probe set list was used as the GSEA input for pathway analysis.

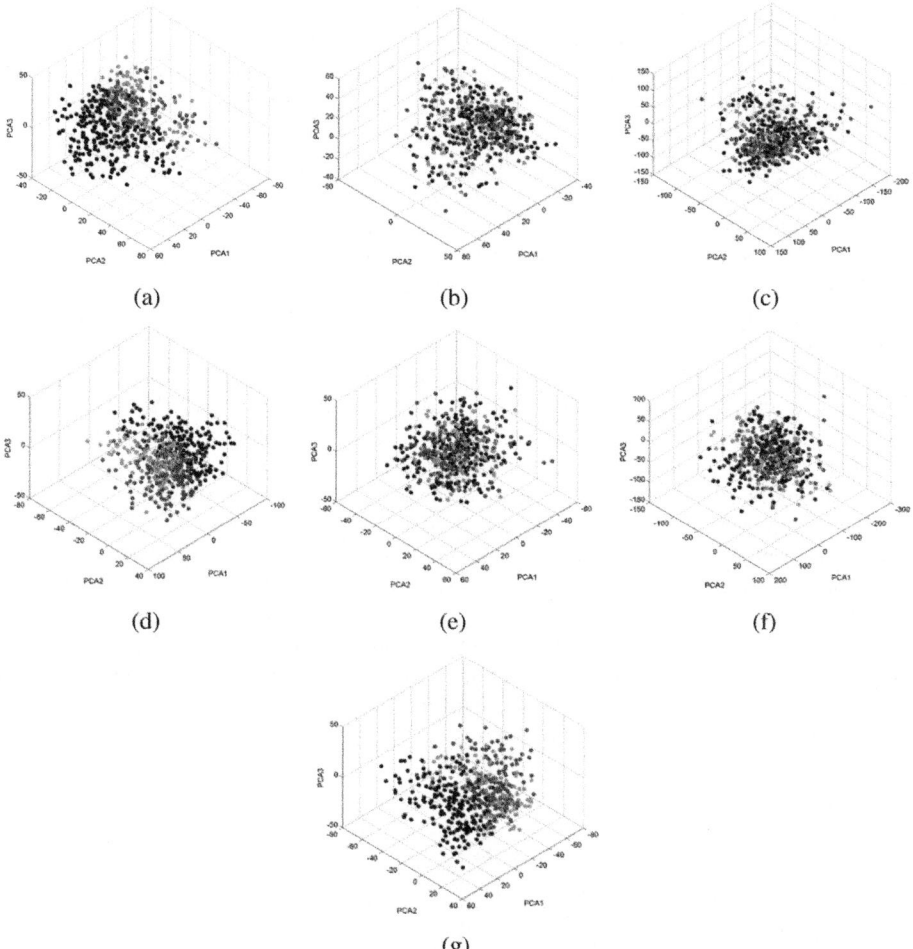

Fig. 3. PCA results. RMA data (a), RMA - ComBat data (b), RMA - Z-Score data (c), MBEI data (d), MBEI - ComBat data (e), MBEI - Z-Score data (f), fRMA data (g); FK - NCC samples in red, GSE2990 samples in green, GSE11121 samples in blue.

4   Discussion

In this study we showed the results of the classification of breast cancer microarray data, with respect to the event of distant metastasis free survival in terms of recall and precision. The results show very low classification performances. This is explained in part by heterogeneity of the data in terms of tumor grade, tumor size, histopathological tumor type, progesterone and estrogen receptor that might negatively influence the prediction. The previous works did not show benefits of survival prediction with merged dataset as compared to individual data sets using linear methods [11]. In this work we show that data merging may even show results worsening if applied to SVM based classification. Different pre-processing methods (MBEI and RMA) did not show significant variation giving rise to improvement of prediction.

506

The fact that the Stratified Percentage Fold validation method showed the highest performances indicates that the batch effect is markedly also after the batch effect removal by the three methods. Also the difference between Stratified Percentage Fold and weighted LDOCV average is very sharp in the MBEI pre-processed datasets showing that this method gets greater benefits from the merging algorithms. The comparison within the RMA processed datasets shows that the not merged dataset has even higher performances than the ComBat and Z-score merged ones.

A previous work making a comparison of merging techniques, not based on classification performances but on PVCA data analysis, showed that ComBat outperformed the other programs [15]. Also the analysis of gene expression level in specific pathways confirmed the better performance of ComBat with respect to Z-Score. Indeed, the datasets obtained by procedures based on the same pre-processing technique, RMA or MBEI, have almost similar GSEA results except for those merged by Z-Score transformation. Furthermore, also focusing on PVC Analysis fRMA resulted inaccurate to remove the batch effect from the data and this is evident by observing the graphical representation of data after PC Analysis.

In conclusion, the present data showed the difficulty to merge data from different datasets, even if coming from the same type of chip, due to the low accuracy even using different approaches. This can suggest that because of the low recall and precision of all methods, data merging does not seem the elective approach to implement samples in array analysis. Thus a better way to improve accurate signature from microarray datasets is to apply a meta-analysis more than to merge all raw data. However, apart frozenRMA which performs consistently worse, we showed that the other merging techniques performed with a good specificity in evidencing gene expression in different biological pathways.

References

1. Gatza, M.L., Lucas, J.E., Barry, W.T., Kim, J.W., et al.: A pathway-based classification of human breast cancer. PNAS 107(15), 6994–6999 (2010)
2. Bild, A.H., Yao, G., Chang, J.T., Wang, Q., et al.: Oncogenic pathway signatures in human cancers as a guide to targeted therapies, vol. 439 (January 19, 2006), doi:10.1038/nature04296
3. Sorlie, T., Perou, C.M., Tibshirani, R., et al.: Gene expression patterns of breast carcinomas distinguish tumor subclasses with clinical implications. Proc. Natl. Acad. Sci. U S A 98, 10869–10874 (2001)
4. Van de Vijver, M.J., He, Y.D., van 't Veer, L.J., Dai, H., et al.: A Gene- Expression Signature as a Predictor of Survival in Breast Cancer. N Engl. J. Med. 347(25), 1999–2009 (2002)
5. Van't Veer, L.J., Dai, H., van de Vijver, M.J., He, Y.D., et al.: Gene expression profiling predicts clinical outcome of breast cancer. Nature 415, 530–536 (2002)
6. Xu, L., Choon Tan, A., Winslow, R.L., Geman, D.: Merging microarray data from separate breast cancer studies provides a robust prognostic test. BMC Bioinformatics 9, 125 (2008)
7. Rhodes, D.R., Yu, J., Shanker, K., Deshpande, N., Varambally, R., et al.: Oncomine: a cancer microarray database and integrated data-mining platform. Neoplasia 6, 1–6 (2004)
8. Wirapati, P., Sotiriou, C., Kunkel, S., Farmer, P., Pradervand, S., et al.: Metaanalysis of gene-expression profiles in breast cancer: toward a unified understanding of breast cancer sub-typing and prognosis signatures. Breast Cancer Research 10, R65+ (2008)

9. Johnson, W.E., Li, C.: Adjusting batch effects in microarray expression data using empirical Bayes methods. Biostatistics (Oxford, England) 8(1), 118–127 (2007)
10. Warnat, P., Eils, R., Brors, B.: Cross-platform analysis of cancer microarray data improves gene expression based classification of phenotypes. BMC Bioinformatics 6, 265 (2005)
11. Yasrebi, H., Sperisen, P., Praz, V., Bucher, P.: Can Survival Prediction Be Improved By Merging Gene Expression Data Sets? PLoS One 4(10), e7431 (2009)
12. Bolstad, B.M., Irizarry, R.A., Åstrand, M., Speed, T.P.: A comparison of normalization methods for high density oligonucleotide array data based on variance and bias. Bioinformatics 19(2), 185–193 (2003)
13. Benito, M., Parker, J., Du, Q., Wu, J., Xiang, D., Perou, C.M., Marron, J.S.: Adjustment of systematic microarray data biases. Bioinformatics 20(1), 105–114 (2004)
14. Lander, E.S.: Array of hope. Nature Genetics 21, 3–4 (1999)
15. Chen, C., Grennan, K., Badner, J., Zhang, D., Gershon, E., Jin, L., Liu, C.: Removing Batch Effects in Analysis of Expression Microarray Data: An Evaluation of Six Batch Adjustment Methods. PLoS One 6(2), 17238 (2011)
16. Alter, O., Brown, P.O., Botstein, D.: Singular value decomposition for genome-wide expression data processing and modeling. Proceedings of the National Academy of Sciences of the United States of America 97, 10101–10106 (2000)
17. Jiang, H., Deng, Y., Chen, H.S., Tao, L., Sha, Q., et al.: Joint analysis of two microarray gene-expression data sets to select lung adenocarcinoma marker genes. BMC Bioinformatics 5, 81 (2004)
18. Chen, Q.R., Song, Y.K., Wei, J.S., Bilke, S., Asgharzadeh, S., et al.: An integrated cross-platform prognosis study on neuroblastoma patients. Genomics 92, 195–203 (2008)
19. Reyal, F., Van Vliet, M.H., Armstrong, N.J., Horlings, H.M., de Visser, K.E., et al.: A comprehensive analysis of prognostic signatures reveals the high predictive capacity of Proliferation, Immune response and RNA splicing modules in breast cancer. Breast Cancer Research 10, R93+ (2008)
20. Acharya, C.R., Hsu, D.S., Anders, C.K., Anguiano, A., Salter, K.H., et al.: Gene expression signatures, clinicopathological features, and individualized therapy in breast cancer. JAMA 299, 1574–1587 (2008)
21. Irizarry, R.A., Hobbs, B., Collin, F., Beazer-Barclay, Y.D., Antonellis, K.J., Scherf, U., Speed, T.P.: Exploration, normalization, and summaries of high density oligonucleotide array probe level data. Biostatistics 4(2), 249–264 (2003)
22. McCall, M.N., Bolstad, B.M., Irizarry, R.A.: Department of Biostatistics, Johns Hopkins University, Baltimore, MD 21205, USA Frozen robust multiarray analysis (fRMA)
23. Li, C., Wong, W.: Model-based analysis of oligonucleotide arrays: Expression index computation and outlier detection. Proceedings of the National Academy of Science U S A 98, 31–36 (2001)
24. Scherer, A. (ed.): Batch Effects and Noise in Microarray Experiments: Sources and Solutions. John Wiley & Sons, Chichester (2009)
25. Subramanian, A., Tamayoa, P., Mootha, V.K., Mukherje, S., et al.: Gene set enrichment analysis: A knowledge-based approach for interpreting genome-wide expression profiles. PNAS (August 2, 2005)
26. Foekens, J.A., Atkins, D., Zhang, Y., Sweep, F.C.G., et al.: Multicenter Validation of a Gene Expression–Based Prognostic Signature in Lymph Node–Negative Primary Breast Cancer. Journal of Clinical Oncology 24, 1665–1671 (2006)
27. Guyon, I., Weston, J., Barnhill, S.: Machine Learning Gene Selection for Cancer Classification using Support Vector Machines 46, 389–422 (2002)
28. http://www.vitoantoniobevilacqua.it/supplementarymaterials/ICIC2011_1945

Evolving Gene Regulatory Networks: A Sensitivity-Based Approach

Yu-Ting Hsiao and Wei-Po Lee

Department of Information Management
National Sun Yat-sen University
Kaohsiung, Taiwan

Abstract. Constructing genetic regulatory networks (GRNs) from expression data is one of the most important issues in systems biology research. To automate the procedure of network construction, we develop an evolution framework to infer the S-system network models. Our framework mainly includes a sensitivity analysis method and a hybrid GA-PSO method to infer appropriate network parameters. To validate the proposed methods, experiments have been conducted and the results show the promise of our approach.

Keywords: reverse engineering, gene regulatory network, parameter sensitivity, S-system model, GA-PSO.

1 Introduction

Gene regulatory networks (GRNs) are essential for cellular metabolism during the development of living organisms. To explore the system dynamics of GRNs, biologists and computational scientists have been working on creating predictive dynamical models from the experimentally measured time-series data. Traditionally, to reconstruct GRNs manually takes a considerable amount of time, therefore an automated procedure (i.e. reverse engineering) is now advocated [1][2]. This procedure involves altering the gene network in some ways, observing the outcome and using mathematics and logic (i.e. computational methods) to infer the underlying principles of the network. The present work is to establish a practical methodology for inferring gene networks.

In the study of GRN modeling, many models have been proposed in which the ordinary differential equations (ODEs) model has been widely used in order to capture biochemical interactions between genes and to explore the dynamic network behaviors. The most popular and well-researched ODEs model is the S-system model. It consists of a particular set of tightly coupled ODEs, in which the component processes are characterized by power law functions. This model has been extensively studied, and is considered suitable to characterize the gene regulations [3][4]. Therefore, in our work we adopt this model to represent GRNs. An S-system model has the following form:

D.-S. Huang et al. (Eds.): ICIC 2011, LNBI 6840, pp. 508–513, 2012.
© Springer-Verlag Berlin Heidelberg 2012

$$\frac{dx_i}{dt} = \alpha_i \prod_{j=1}^{N} x_j^{g_{i,j}} - \beta_i \prod_{j=1}^{N} x_j^{h_{i,j}} \qquad (1)$$

Here x_i is the expression level of gene i and N is the number of genes in a genetic network. The non-negative parameters α_i and β_i are rate constants that indicate the direction of mass flow. The real number exponents $g_{i,j}$ and $h_{i,j}$ are kinetic orders that reflect the intensity of interaction from gene j to i. The above set of parameters defines an S-system model. To infer an S-system model is, thus, to estimate all of the $2N(N+1)$ parameters simultaneously.

In this work, we develop an evolution framework to deal with the problems of scalability and network robustness in inferring gene networks. Our framework includes two major parts. The first part is a sensitivity analysis method responsible for selecting sensitive network parameters. The second is an evolution mechanism to infer the selected network parameters. To validate the proposed framework, experiments have been conducted. The results show that our approach can be used to infer robust networks successfully from the gene expression profiles.

2   The Proposed Approach

To deal with the scalability problem and to infer robust networks, we adopt the concept of incremental evolution with a parameter sensitivity analysis method. In the proposed incremental evolution framework, a sensitivity-based method is developed to iteratively determine the priorities of the parameters to be evolved and an evolution mechanism is constructed to search the selected parameter dimensions. Figure 1 shows the main flow of the proposed incremental evolution framework.

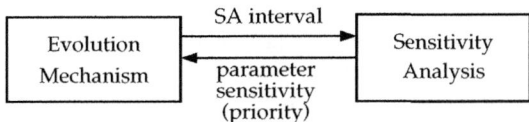

Fig. 1. The main flow of the proposed approach

The right-hand side block of the figure is the sensitivity analysis method mentioned above to select sensitive network parameters. Sensitivity analysis (SA) is an indispensable technique that helps researchers to investigate the property of robustness of an inferred network. In general, the sensitivity of a parameter is defined as [5]:

$$S_p^M = \frac{\partial M / M}{\partial P / P} = \frac{\text{percentage change in } M}{\text{percentage change in } P} \qquad (2)$$

where P represents the parameter that are varied in a given range, M is the mathematical function describing the system behavior, and ∂M means the change in M due to the value changed in ∂P with respect to P. In the work here, the above function means the fitness function to be optimized in the evolution algorithm, which is the mean squared error over the time course:

$$\sum_{i=1}^{N}\sum_{t=1}^{T}\left\{\frac{x_i^a(t)-x_i^d(t)}{x_i^d(t)}\right\}^2 \tag{3}$$

in which $x_i^d(t)$ is the desired expression level of gene i at time t, $x_i^a(t)$ is the actual value obtained from the inferred model, N is the number of genes in the network, and T is the number of time points in measuring gene expression data during the period.

In order to consider multiple network parameters simultaneously under the situation that the structure information is not available, we propose a new approach that is modified from the widely used SA technique, multi-parameter sensitivity analysis (MPSA) [5][6]. In our modified method m-MPSA, exactly one parameter is selected at a time. Then a repeated process is developed to calculate sensitivity for each parameter. Finally, the m-MPSA ranks the sensitivities for all parameters. In previous studies, researchers have shown that the most influential (sensitive) parameters may play key roles in modeling a GRN; that is, a model's behavior varies due largely to its sensitive parameters [7]. Our current study draws on the characteristic of sensitive parameters, and incorporates the sensitivity analysis method into the incremental evolution approach to infer a GRN.

As mentioned above, in addition to the sensitivity analysis method, our framework also includes an evolutionary mechanism for parameter optimization. Here, we take a hybrid GA-PSO method to exploit the characteristic of solution memory of PSO and the genetic operations of GA at the same time. Our main goal is to investigate how the parameter sensitivity can be used to derive robust networks. To concentrate on this issue, we thus choose to implement a popular GA-PSO hybrid method, Breeding Swarms ([8]), as the evolutionary mechanism and build our sensitivity analysis method on it.

In our current implementation, we take a direct encoding scheme to represent solutions for both GA and PSO parts, in which the network parameters (i.e., α_i, β_i, $g_{i,j}$, and $h_{i,j}$ in the S-system model) are arranged as a linear string chromosome of floating-point numbers. The goal here is to minimize the accumulated discrepancy between the gene expression data recorded in the dataset and the values produced by the inferred model. Therefore, the error (fitness) function defined above is used directly for performance measurement. For the PSO part, the equations for updating the particle's velocity and position are the same as the ones listed in [9], and for the GA part, the operations of crossover and mutation described in [8] are used.

3   Experiments and Results

To verify the proposed approach, we have conducted a series of experiments on two popular datasets. In the experiments, two different methods, including a traditional PSO method and a hybrid GA-PSO have been implemented to work with the SA method for performance comparison.

The first dataset was taken from [10]. It was an eight node gene network created manually by the popular GRN simulation software tool Genexp. To collect time series data, we started the network operations and continued the operations for thirty simulation time steps. The second dataset was a ten-node network described in [11]:

$$\dot{X}_1 = 5.0 X_4 X_6^{-2.0} - 10.0 X_1^{2.0}$$
$$\dot{X}_2 = 10.0 X_3 X_8^{1.0} - 10.0 X_2^{2.0}$$
$$\dot{X}_3 = 8.0 X_1^{-1.0} X_4^{-1.0} - 10.0 X_3^{2.0}$$
$$\dot{X}_4 = 10.0 X_5^{2.0} X_9 - 10.0 X_4^{2.0}$$
$$\dot{X}_5 = 10.0 X_2^{2.0} X_6^{-1.0} - 10.0 X_5^{2.0}$$
$$\dot{X}_6 = 5.0 X_9^{2.0} X_{10}^{-2.0} - 10.0 X_6^{2.0}$$
$$\dot{X}_7 = 10.0 X_6 X_{10}^{-1.0} - 10.0 X_7^{2.0}$$
$$\dot{X}_8 = 5.0 X_1 X_2^{-2.0} X_7 - 10.0 X_8^{2.0}$$
$$\dot{X}_9 = 10.0 X_3 X_8^{-2.0} - 10.0 X_9^{2.0}$$
$$\dot{X}_{10} = 8.0 X_1^{2.0} X_7^{-1.0} - 10.0 X_{10}^{2.0}$$

In the experiments, the population size used for the two datasets was 1000 and 1600, respectively. Tables 1-2 show the results, in which values are averaged over twenty independent runs. From these tables, we can see that the proposed approach outperforms (i.e., with lower fitness values) the other in inferring gene networks, no matter working with a traditional PSO method or a popular hybrid GA-PSO method. In addition, Figures 2-3 compare the behaviors of the networks with the inferred (left) and the original (right) networks by our method, in which the x-axis represents time step, and the y-axis, the concentration of different gene components. As can be observed, very similar network behaviors can be obtained, and which show the success of the proposed approach.

To evaluate the robustness of the evolved gene networks, we compare the sensitivity value (averaged over all parameters) of the best solutions obtained from the final generation. The results are listed in Tables 1-2. The values recorded in the table are the averaged results of the twenty independent runs performed for each setting. They show that the proposed approach is able to evolve networks with lower fitness values and lower sensitivity values simultaneously.

Table 1. Comparisons of the fitness and the relative sensitivity value of the best solutions collected from the final generations

	original PSO		PSO with SA	
	Fitness	sensitivity	fitness	Sensitivity
dataset 1	0.5789	0.7334	0.3606	0.7002
dataset 2	1.9133	0.6865	1.0983	0.6463

Table 2. Comparisons of the fitness and the relative sensitivity value of the best solutions collected from the final generations

	original GA-PSO		GA-PSO with SA	
	Fitness	sensitivity	fitness	Sensitivity
dataset 1	0.0595	0.8249	0.0193	0.7581
dataset 2	0.3590	0.8432	0.1406	0.7502

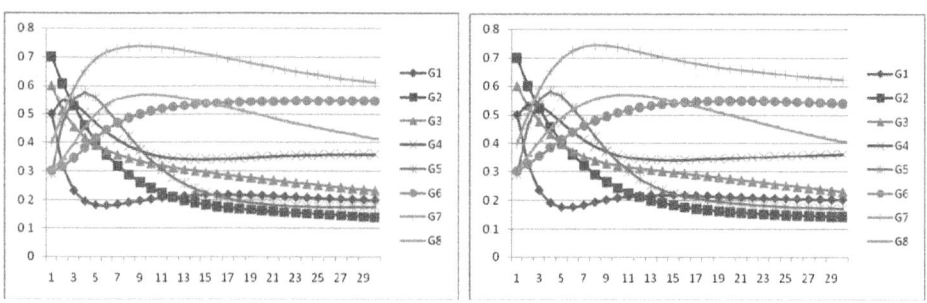

Fig. 2. Network behaviors of the evolved (left) and original (right) networks for dataset 1

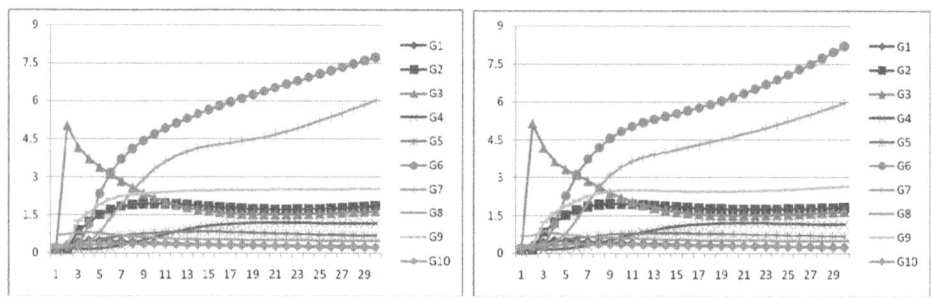

Fig. 3. Network behaviors of the evolved (left) and original (right) network for dataset 2

4   Conclusions

In this work, we indicated the importance of inferring gene networks from expression data. Among different computational models, the S-system model can work as a dynamical system as GRN, therefore our work adopted this model and developed an incremental evolution framework to infer networks. Our framework includes two parts. The first part involves a sensitivity analysis method to select sensitive parameters; and the second part, an evolution mechanism that takes the bounds for parameter optimization. In this way, the evolved networks can be robust and have desired behaviors at the same time.

Based on the work presented, we are currently investigating how to integrate gene clustering techniques to the proposed framework to infer networks with even more genes. In addition, we plan to investigate more advanced techniques that can automatically extract more useful information from various biological knowledge domains, and then integrate the knowledge into the sensitivity analysis method to identify critical parameters in a gene network.

References

1. Ingolia, N.T., Weissman, J.S.: Systems Biology: Reverse Engineering the Cell. Nature 454, 1059–1062 (2008)
2. Lee, W.-P., Tzou, W.-S.: Computational Methods for Discovering Gene Networks from Expression Data. Briefings in Bioinformatics 10(4), 408–423 (2009)

3. Kikuchi, S., Tominaga, D., Arita, M., et al.: Dynamic Modeling of Genetic Networks Using Genetic Algorithm and S-system. Bioinformatics 19(5), 643–650 (2003)
4. Ho, S.-Y., Hsieh, C.-H., Yu, F.-C., et al.: An Intelligent Two-Stage Evolutionary Algorithm for Dynamic Pathway Identification From Gene Expression Profiles. IEEE/ACM Trans. on Computational. Biology and Bioinformatics 4(4), 648–704 (2007)
5. Cho, K., Shin, S., Kolch, W., Wolkenhauer, O.: Experimental Design in Systems Biology, Based on Parameter Sensitivity Analysis Using a Monte Carlo Method: A Case Study for the TNFα-Mediated NF-kB Signal Transduction Pathway. Simulation 79(12), 726–729 (2003)
6. van Riel, N.A.W.: Dynamic Modeling and Analysis of Biochemical Networks: Mechanism-based Models And Model-based Experiments. Briefings in Bioinformatics 7(4), 364–374 (2006)
7. Fomekong-Nanfack, Y., Postma, M., Kaandorp, J.: Inferring Drosophila Gap Gene Regulatory Network: A Parameter Sensitivity and Perturbation Analysis. BMC Systems Biology 3(1), 94 (2009)
8. Settles, M., Soule, T.: Breeding Swarms: a GA/PSO Hybrid. In: Proceedings of Genetic and Evolutionary Computation Conference, pp. 161–168 (2005)
9. Kennedy, J., Eberhart, R.: Swarm Intelligence. Morgan Kaufmann Publishers, San Francisco (2001)
10. Lee, W.-P., Yang, K.-C.: A Clustering-Based Approach for Inferring Recurrent Neural Networks as Gene Regulatory Networks. Neurocomputing 71(4-6), 600–610 (2008)
11. Ho, S.-Y., Hsieh, C.-H., Yu, F.-C., et al.: An Intelligent Two-Stage Evolutionary Algorithm for Dynamic Pathway Identification From Gene Expression Profiles. Biology and Bioinformatics 4(4), 648–704 (2007)

Blind Testing of Quasi Brain Deaths Based on Analysis of EEG Energy

Wei Zhou[1], Gang Liu[1], Qiwei Shi[1], Shilei Cui[2],
Yina Zhou[2], Huili Zhu[2], Rubin Wang[4], and Jianting Cao[1,3,4]

[1] Saitama Institute of Technology
1690 Fusaiji, Fukaya-shi, Saitama 369-0293, Japan
[2] Huadong Hospital Affiliated to Fudan University
221 Yanan West Rd, Shanghai 200040, China
[3] Brain Science Institute, Riken
2-1 Hirosawa, Wako-shi, Saitama 351-0198, Japan
[4] East China University of Science and Technology
130 Meilong Rd, Shanghai 200237, China
cao@sit.ac.jp,zhuhuili@sh163b.sta.net.cn

Abstract. This paper presents a power spectral pattern analysis method for quasi-brain-death EEG based on Empirical Mode Decomposition (EMD) under the condition of unknowing the clinical symptoms of patients. EMD method is a time-frequency analysis method for analyzing the nonlinear and non-stationary data. In this paper,we decompose a single-channel recorded EEG data into a number of components with different frequencies, we calculate the power spectral or energy of the decomposed components in a suitable frequency band. Based on the EEG power spectral analysis, the patients are classified into two categories: existence of the brain activities or absence of the brain activities. The experimental results illustrate the effectiveness of our proposed method.

Keywords: Electroencephalography (EEG), empirical mode decomposition (EMD), power spectral pattern, coma, quasi-brain-death.

1   Introduction

The brain-death is defined as the irreversible and end of all brain activity including involuntary activity necessary to sustain life due to total necrosis of cerebral neurons following loss of blood flow and oxygenation [1], [2]. According to the definition, the Japanese criterion for diagnosing the brain-death includes the following items: 1) a prolonged coma: when the body stimuli painfully, there is no scowl response in a face or supra orbital; 2) pupil expansion: both of the pupils become expended, up to $4mm$; 3) absence of brain-stem reflexes: there is not any responses in brain-stem to stimulation of pillar light, corneal, vestibule-ocular, oculocephalic, gag and cough and etc; 4) horizontal brain waves: there is no electrical brain activity at least 30 minutes which are recorded by not less than 4 exploring electrode in the confirmatory EEG test, and the maximum

D.-S. Huang et al. (Eds.): ICIC 2011, LNBI 6840, pp. 514–520, 2012.
© Springer-Verlag Berlin Heidelberg 2012

sensitivity is $2\mu V/mm$; 5) cessation of spontaneous respiration: The cessation of spontaneous respiration must be confirmed for a patient by disconnecting the ventilator in the apnea test.

In the standard process of brain-death diagnosis, it often involves certain risks and takes a long time (e.g. shortly remove the breath machine in a spontaneous respiration test and the confirmatory EEG test which record the absence of electrical brain activity during at least 12 hours). In order to reduce the risks and to save invaluable time for the medical care for the patients, it is desirable to develop a practical, safe and reliable pre-test method in the diagnosis of brain death. As one of the criteria in the diagnosis of brain-death, Electroencephalography (EEG) plays a very important role in the analysis process [3], [4].

In this paper, we present a power spectral pattern analysis method for quasi-brain-death EEG based on Empirical Mode Decomposition (EMD) under the condition of unknowing the clinical symptoms of patients. In general, the power spectral of EEG to a comatose patient and a brain death is different. Therefore, we can establish a criterion to evaluate the power spectral pattern which originally created from a comatose patient or measure from a brain death. We decompose the recorded EEG data into a number of components with different frequencies by using EMD, and estimate the power spectral of the components by the Fast Fourier Transform (FFT). By the comparing, we can evaluate the power spectral patterns between comatose patients and quasi-brain-deaths. The experimental results illustrate the effectiveness and performance of our proposed method in the can evaluation of the power spectral patterns between comatose patients and quasi-brain-deaths.

2   Empirical Mode Decomposition

The EMD method is a necessary tool to reduce any given data into a collection of intrinsic mode functions (IMF) in which the Hilbert spectral analysis can be applied. The IMF components usually expressed as the standard Hilbert transforms, from which the instantaneous frequencies can be calculated. The local energy and the instantaneous frequency derived from the IMF components through the Hilbert transform can be given a full energy-frequency-time distribution of the data.

For an observed time domain signal $x(t)$, we can always obtain its Hilbert Transform $f(t)$, such as

$$f(t) = \frac{1}{\pi} P \int_{-\infty}^{\infty} \frac{x(\tau)}{t - \tau} d\tau. \tag{1}$$

It is impossible to calculate the Hilbert transform as an ordinary improper integral because of the pole at $\tau = t$. However, the P in front of the integral denotes the Cauchy principal value which expanding the class of functions for which the integral in Eq. (1) [5].

With this definition, $x(t)$ and $f(t)$ form the complex conjugate pair, so the complex signal $Z(t)$ can be formulated as

$$Z(t) = x(t) + if(t) = a(t)e^{i\theta(t)}, \tag{2}$$

in which an instantaneous amplitude $a(t)$ and an instantaneous phase $\theta(t)$ are presented by

$$a(t) = \sqrt{x^2(t) + f^2(t)}, \tag{3}$$

$$\theta(t) = \tan^{-1}\left(\frac{f(t)}{x(t)}\right). \tag{4}$$

The instantaneous frequency $w(t)$ of the signal $x(t)$ can be defined as

$$w(t) = \frac{d\theta(t)}{dt}. \tag{5}$$

In principle, it is necessary that one limitation is a narrow band signal for the instantaneous frequency by Eq. (5).

An IMF component as a narrow band signal is a function that satisfies two conditions [6]:

(i) In the whole data set, the number of extrema and the number of zero crossings must either equal or differ at most by one.

(ii) At any point, the mean value of the upper envelope with the lower envelope is zero.

The procedure to obtain the IMF components from an observed signal is called sifting [6] and it consists of the following steps:

(a) Identification of the extrema of an observed signal waveform $x(t)$.

(b) Generation of the waveform envelopes by connecting local maxims as the upper envelope, and connection of local minims as the lower envelope.

(c) Computation of the local mean $u_1(t)$ by averaging the upper and lower envelopes.

(d) Subtraction of the mean from the data for a primitive value of IMF component as $h_1(t) = x(t) - u_1(t)$.

(e) Repetition step (a)-(d) q times, until $h_q(t)$ is an IMF component, $h_{q-1}(t) - u_q(t) = h_q(t)$.

(f) Designation the first IMF component as $c_1(t) = h_q(t)$ from the data, so that the residue component is $r_1(t) = x(t) - c_1(t)$.

(g) Repetition step (a)-(f) n times, the residue component contains information about longer periods which will be further resifted to find additional IMF components, by $r_n(t) = x(t) - \sum_{i=1}^{n} c_i(t)$.

The sifting algorithm is applied to calculate the IMF components based on a criterion by limiting the size of the standard deviation (SD) computed from the two consecutive sifting results as

$$SD = \sum_{t=0}^{T}\left[\frac{(h_{k-1}(t) - h_k(t))^2}{h_{k-1}^2(t)}\right]. \tag{6}$$

Based on the sifting procedure for a channel of the recorded data, we finally obtain

$$x(t) = \sum_{i=1}^{n} c_i(t) + r(t). \tag{7}$$

In Eq. (7), $c_i(t)(i = 1, \cdots, n)$ represents n IMF components, and r represent one residue component. These components can be either the mean trend or a constant.

3   Experiments and Results

3.1   EEG Experiments

A portable EEG system (NEUROSCAN ESI) was used to record the patient's brain activity. In the EEG recording, only nine electrodes are chosen to apply to patients. Among these electrodes, six exploring electrodes (Fp1, Fp2, F3, F4, F7, F8) as well as GND were placed on the forehead, and two electrodes (A1, A2) as the reference were placed on the earlobes (Fig. 1). The sampling rate of EEG data was 1000 Hz and the resistances of electrodes were set under 10 $k\Omega$. Expecting the brain activities of a comatose patient dominate in lower frequency bands, in our experimental analysis, we focus on the δ-wave (0.1 $\sim$ 3 Hz), the θ-wave (4 $\sim$ 7 Hz) and α-wave (8 $\sim$ 13 Hz). In the following part, EEG were recorded from the patients who at that time were still under treatment and final clinical diagnosis reports were unknown.

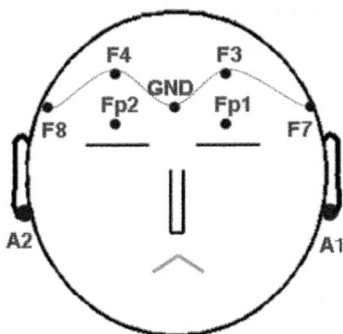

Fig. 1. The layout of six exploring electrodes, GND and two reference electrodes

3.2   Empirical Mode Decomposition of Two Patients

The EEG data was directly recorded at the bedside of the patients in the intensive care unit (ICU) at a hospital in China. In this paper, we applied the EMD method to 11 quasi-brain-death patients' data (11 cases) which were in the condition of unknowing the clinical symptoms from December 2009 to now. As an example , we have chosen Patient 9-1 and Patient 7-2's data. The EEG activity oscillates shown in the top of Fig. 2, we a randomly selected a channel, such as the sample of channel F3 from the beginning of 350 sec and F7 from 4600 sec. In time domain analysis, raw EEG signal F3 was decomposed to six IMF components ($C_1 \sim C_6$), and one residue component (r). In the right column, these components are transferred to in the frequency domain by applying

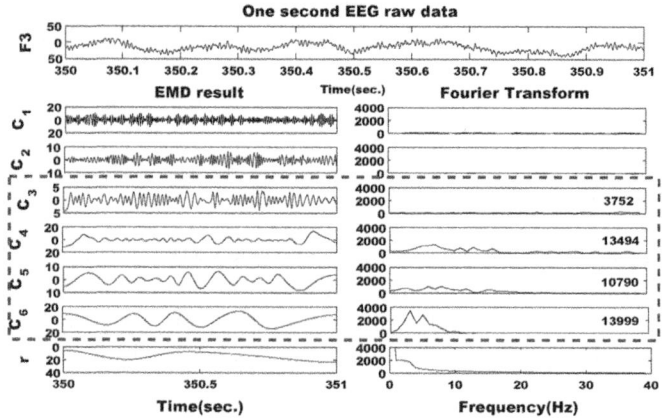

Fig. 2. EMD result of Patient 9-1

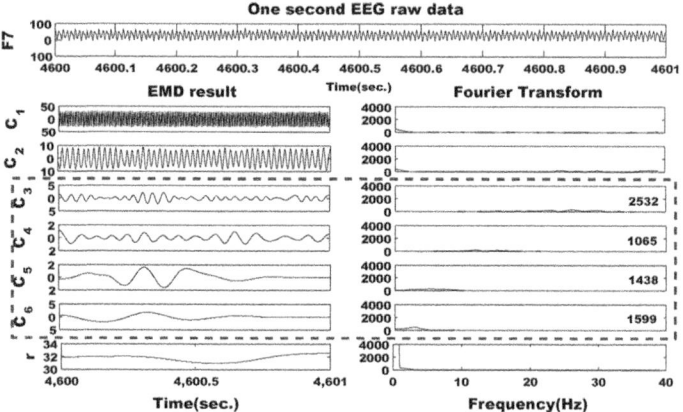

Fig. 3. EMD result of Patient 7-2

the Fast Fourier Transform (FFT). Since we focus on the brain activities with in the frequency band of α, β, θ waves, therefore, we chose four decomposed IMF components ($C_3 \sim C_6$), and calculated the sum of their power spectrum, receptively (see Fig. 2 right column),the value of the power spectrum is used to evaluate the states of patient. The result of applied the same EMD method to Patient 7-2 is shown in Fig. 3. Comparing the results shown in Fig. 2 and Fig. 3, we found that the wave form of the decomposed components ($C_3 \sim C_6$) are similar in the time domain, however, the power spectrum of components ($C_3 \sim C_6$) are different. In this case, the sum of power spectral for Patient 9-1 and Patient 7-2 in a channel are P9-1$_{F3}$=3752+13494+10790+13999=42035, and P7-2$_{F7}$=2532+1065+1438+1599=6636, respectively. These results imply that Patient 9-1 may has brain activities but Patient 7-2 may has not drain activities.

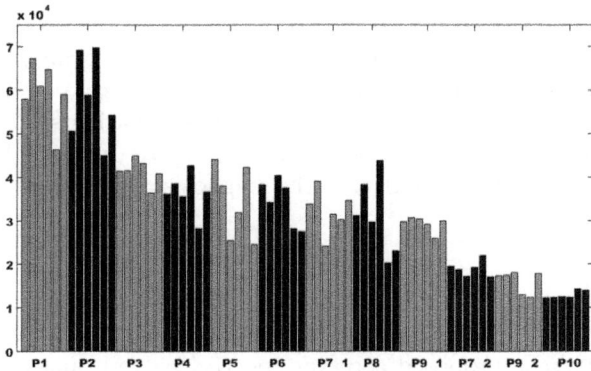

Fig. 4. The average power individual of each channel for 10 patients

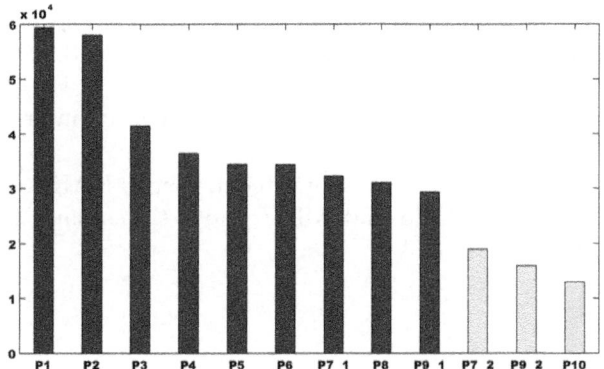

Fig. 5. The average power of total channels for 10 patients (12 cases)

3.3   The EEG Power Spectrum of Each Patient

In the previous sub-section, we have analyzed two patient's EEG data in one second and one channel. In this sub-section, we will further analyze 10 patients (12 cases), 6 channels EEG data in the recorded time.

Depending on the health condition of patient, we recorded 10 patient's EEG data. The duration of recording is different, the shortest one is about 14 minutes, the longest one is one hour and 15 minutes. Moreover, most of patients only recorded in one time, but one patient is recored for two times.

The averaged power spectrum of 50 seconds interval in each channel is shown in Fig. 4. In Fig. 4 the first bar to 6th bar corresponding to the channel Fp1, Fp2, F3, F4, F7, F8, in which it is result of a patient or a case of patient. As shown in Fig. 4, we found that the average power spectral of a patient or a case of patient is different each others.

In order to evaluate the power spectral of each patient or each case of patient, we calculated the average power spectral of each channel in each case. The result is shown in Fig. 5. As shown in Fig. 5, we found that two groups of patient are classified based on a big difference of calculated the average power spectral.

From this data analysis result, we concluded that patients P7-2, P9-2, P10 are brain death and other patients are in a coma state. In fact, this EEG data analysis result is completely identical to the diagnosis results achieved by clinical doctors.

4   Conclusions

In this paper, we have proposed a power spectral pattern analysis method for quasi-brain-death EEG based on EMD under the condition of unknowing the clinical symptoms of patient.

Significant feature differences in the power spectral pattern is used to evaluated the patient in a coma state or a brain-death state from data analysis point of view. Moreover, the data analyzed results are compared to the diagnosis results achieved by the clinical doctors. In this paper we conclude that the results are completely identical for 10 patients in 12 recorded cases. In the future studies, the extension of EMD and more patients EEG data collection are necessary.

Acknowledgments. This work was supported in part by KAKENHI (22560425) (JAPAN), and in part by JSPS and NSFC under the Japan-China Scientific Cooperation Program.

References

1. Mollaret, P., Goulon, M.: Le coma dépassé. Rev Neurol(Paris) 101, 3–15 (1959) (in French)
2. Taylor, R.M.: Reexamining the Definition and of Death. Semin Neurol. 17, 265–270 (1997)
3. Cao, J.: Analysis of the Quasi-Brain-Death EEG Data Based on a Robust ICA Approach. In: Gabrys, B., Howlett, R.J., Jain, L.C. (eds.) KES 2006. LNCS (LNAI), vol. 4253, pp. 1240–1247. Springer, Heidelberg (2006)
4. Chen, Z., Cao, J.: An Empirical Quantitative EEG Analysis for Evaluating Clinical Brain Death. In: Processings of the 2007 IEEE Engineering in Medicine and Biology 29th Annual Conference, Lyon, France, pp. 3880–3883 (2007)
5. Cardoso, J.F., Souloumiac, A.: Jacobi Angles for Simultaneous Diagonalization. SIAM J. Mat. Anal. Appl. 17(1), 161–164 (1996)
6. Huang, N.E., Shen, Z., Long, S.R., Wu, M.C., Shih, H.H., Zheng, Q., Yen, N.-C., Tung, C.C., Liu, H.H.: The Empirical Mode Decomposition and the Hilbert Spectrum for Nonlinear and Non-stationary Time Series Analysis. Proceedings of the Royal Society of London A 454, 903–995 (1998)

Connectivity Analysis of Hippocampus in Alzheimer's Brain Using Probabilistic Tractography

Md. Kamrul Hasan[1], Wook Lee[2], Byungkyu Park[1], and Kyungsook Han[2,*]

[1] Institute for Information and Electronics Research, Inha University, Incheon, South Korea
[2] School of Computer Science and Engineering, Inha University, Incheon, South Korea
{kamrul,wook,bpark,khan}@inha.ac.kr

Abstract. In recent years diffusion tensor imaging (DTI) has received increasing attention from several studies of Alzheimer's disease (AD) since it can reveal the microscopic tissue structure of brain white matter. In Alzheimer's brain both brain regions and inter-regional communications through the white matter are often hampered. In this study, we investigated the white matter tracts in the time series of the three dimensional DTI data obtained from 12 patients with AD and 24 normal control subjects. The probabilistic tractography revealed that the fiber paths of AD patients from the hippocampus toward other brain regions are more scattered and dispersed with less neurotransmitters than those of normal control subjects. Similar patterns were observed in the fiber paths in the reverse directions (i.e., the fiber paths from other brain regions toward the hippocampus). The analysis results can help diagnose AD or predict the prognosis of patients with mild AD.

Keywords: Brain Connectivity, Probabilistic Tractography, Alzheimer's Disease, Diffusion Tensor Imaging.

1   Introduction

Alzheimer's disease (AD) is one of the most devastating diseases for people in advanced age between 40 to 60 years. It affects parts of the brain that control thought, memory, and sometimes language. The most affected part is the memory which is primarily controlled by the hippocampus. Thus, a comprehensive study of the connectivity of the hippocampus with other brain regions in both AD patients and healthy subjects may reveal a new pattern specific to AD.

So far the connectivity of regions in Alzheimer's brain has been studied mostly at the functional level using functional MRI (fMRI) [1-3]. It is until recently that the importance of the physical connectivity of regions in Alzheimer's brain has become evident [4-7]. The cause of AD is now believed to be the excessive storage of Amyloid Beta (Aβ) plaque [8] into the white fiber tracts. So, the study of fiber tracts can reveal the actual degeneration process.

Recently diffusion tensor imaging (DTI) has received much attention from brain studies since it can capture the fine fiber orientations of brain white matter. Water

---

* Corresponding author.

D.-S. Huang et al. (Eds.): ICIC 2011, LNBI 6840, pp. 521–528, 2012.

molecules diffuse in the direction of the fiber more rapidly due to the myelination property of the tracts, and DTI records the diffusion. However, analyzing the fibers is not easy, in particular in crossing-fiber regions. Analyzing the inter-regional connectivity without tracking crossing-fiber regions often provides coarse and inaccurate results. In two recent studies [7, 10] a few deep brain fibers were analyzed but the inter-regional fiber connectivity was not analyzed in their studies. The inter-regional fiber connectivity analysis can provide comprehensive and measurable degradation of fiber tracts in AD patients' brains.

In this study, we tracked inter-region fibers tracts using a new method for detecting crossing fibers in voxels and streamlined the fibers by probabilistic tractography [9] to generate an inter-region physical connectivity map starting from the hippocampus. The primary focus of this study is to investigate the connectivity of the left and right hippocampus with other brain regions. We sampled DTI scans of 12 AD patients from the Alzheimer's Disease Neuroimaging Initiative (ADNI) database and DTI scans of 24 normal control (NC) subjects from the International Consortium for Brain Mapping (ICBM) database. We tracked fiber pathways from the left and right hippocampus to 13 other regions of interest (ROI) of brain. We built fiber tract maps of AD patients and NC subjects onto a standard space of MNI152 to compare them visually and quantitatively. The rest of this paper presents our method of tracking the fiber pathways and major results of the analysis.

2   Related Works

A recent study by Douaud *et al.* [10] has used tract-based spatial statistics (TBSS) [11] to detect region changes in AD and mild cognitive impaired (MCI) brains, and examined the changes in crossing fiber regions using probabilistic tractography [9]. Using voxelwise statistics they found that cingulum bundle, the uncinate fasciculus, the entire corpus callosum and the superior longitudinal fasciculus are the most affected white matter tracts in AD. Corpus callosum manifested significant changes from MCI to AD, which was confirmed by probabilistic tractography [9]. In this study, we also used the probabilistic tractography [9], but we used it for studying the whole brain rather than a restricted region of brain. Fiber tracking can picture the whole brain connectivity at the expense of lot of computing time if not parallelized. Fiber tracking of a whole brain usually takes several days to months in a single computer. Qiu *et al.* [7] have used the shape of few deep brain fiber tracts to design a classifier for AD patients and NC. Abnormal connectivity between posterior cingulate and hippocampus in AD and MCI was reported in other's work [6].

The hippocampus is in charge of storage of memory, and is the most damaged part of Alzheimer's brain. The functionality of other brain parts will also be affected which involve memory manipulation. Brain regions are interconnected through fiber paths, termed as neural communication pathways. So, inter-region fiber connections should be studied in details. In this paper we tracked fiber paths from the left and right hippocampus to 13 other regions. Our study was conducted to see how inter-region fiber tracts are affected in AD, which will in turn reveal the communication disruption to and from the hippocampus. We used probabilistic tractography to study the whole brain segmented in 15 regions of interest. The inter-region fiber statistics can also be used to design a classifier, which we consider as our future work.

3   Data Set and Analysis Method

We collected DTI scans of 36 subjects. 12 of them are AD patients and 24 are normal aged subjects (NC). AD patient data were obtained from the Alzheimer's Disease Neuroimaging Initiative (ADNI) database (http://adni.loni.ucla.edu) and NC data were obtained from the International Consortium for Brain Mapping (ICBM) database (http://www.loni.ucla.edu/ICBM). Table 1 shows the summary of the data collected.

Table 1. Group demography

	Alzheimer's disease patients	Normal controls
Number of subjects	12	24
Number of males (%)	6 (50%)	12 (50%)
Number of females (%)	6 (50%)	12 (50%)
Average age (SD)	74.9 (6.64)	52.9 (6.25)

First, we converted the data from the DICOM (.dcm) format to the FSL compatible NiFTi (.nii) format using the dcm2nii program (http://www.cabiatl.com/mricro). For each subject, we combined the 3D NiFTi files to construct a time series data file. During the process of data conversion, we also generated gradient values (bvals) and gradient vectors (bvecs).

After converting the data to the NiFTi format, we used the Oxford FMRIB Software Library's (FSL) Fiber Diffusion Toolbox (FDT) for the analysis of the data. FSL provides a complete set of tools for the analysis of fMRI and DTI scans [12].

The time series data were then corrected to remove distortions caused by eddy currents and simple head motion. From each time series data file, we generated a brain mask and extracted a frame with no diffusion in itself. The mask and frame were used as a reference for the DTIFIT process. The frame image must be pure brain without skull for later registration, so we discarded scalp boundaries from the reference frame using BET of FSL. DTIFIT of FSL was used to make diffusion tensors fit the data. Using the eddy corrected data along with the brain mask, bvals and bvecs for each subject in DTIFIT [13], we computed the fractional anisotropy (FA), and generated three principal eigenvector maps. DTIFIT is a diffusion tensor model at each voxel, and the output of DTIFIT was used in Bayesian probabilistic tractography (BEDPOSTX) [9] for crossing fiber tracking in principal diffusion directions (PDDs). We used the Bayesian tractography to detect two crossing fibers in each voxel. We then used the probabilistic tractography (PROBTRACKX) [9] to streamline the fibers from a seed region of interest (ROI) and target ROIs. The ROIs should be selected in a standard space for direct comparison and for generating statistics. The MNI152 T1 weighted image was used as a standard image in our study. The MNI152 T1 weighted image was then segmented into 15 ROIs using FIRST [14]. Since the ROIs are in standard space rather than in the diffusion space of the subjects, we computed non-linear transformation matrices using FMRIB's Non-linear Image Registration Tool (FNIRT) [15]. This is the registration process for the 36 data. The matrices were later used in PROBTRACKX for normalizing data with standard image. Due to the registration process, the data of AD patients can be compared with the data of NC although they were obtained from different databases.

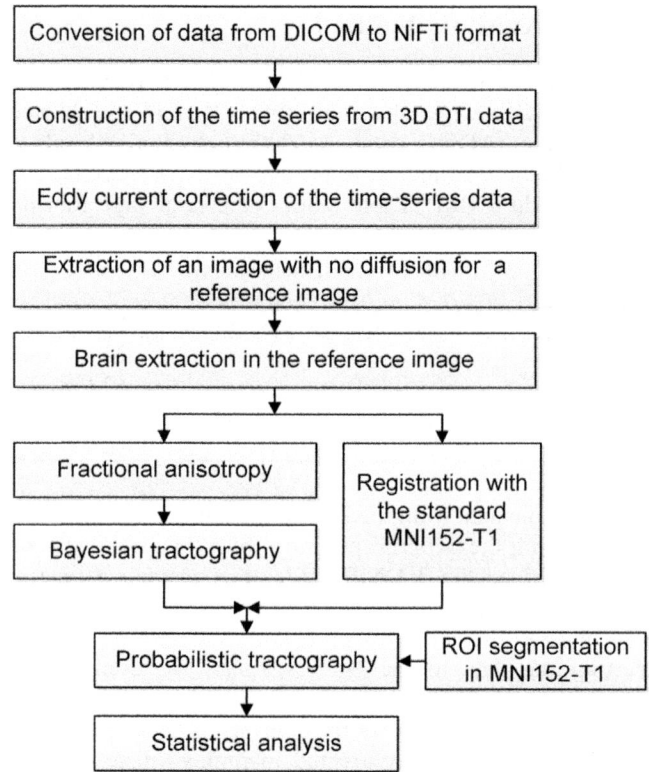

Fig. 1. Framework for analyzing fiber pathways in DTI data

Among the ROIs, the left and right hippocampus (Lhippo and Rhippo, respectively) were taken as seed masks, and the remaining 13 ROIs were used as target masks. The remaining 13 ROIs were used as target masks. In the probabilistic tractography (PROBTRACKX) [9], the left hippocampus (Lhippo) and the right hippocampus (Rhippo) were selected as seed masks. Each of the remaining 13 ROIs was used as a waypoint mask in the probabilistic tractography. The seed space that is not in diffusion and non-linear checkboxes was selected to feed the non-linear transformation matrices. The transformation plots all the paths to the standard MNI152 space.

4   Analysis Results

Since we tracked the fiber from left and right hippocampus to 13 other brains regions, we obtained a total of $2 \times 13 = 26$ fiber tracts for each of the 36 subjects. We analyzed the fiber paths after plotting them on the MNI152 standard space. Figs. 2 and 3 show an example of the dispersion of fiber paths in Alzheimer's disease patients and normal controls. Overall, fiber tracts were found scattered in AD patients' brain image.

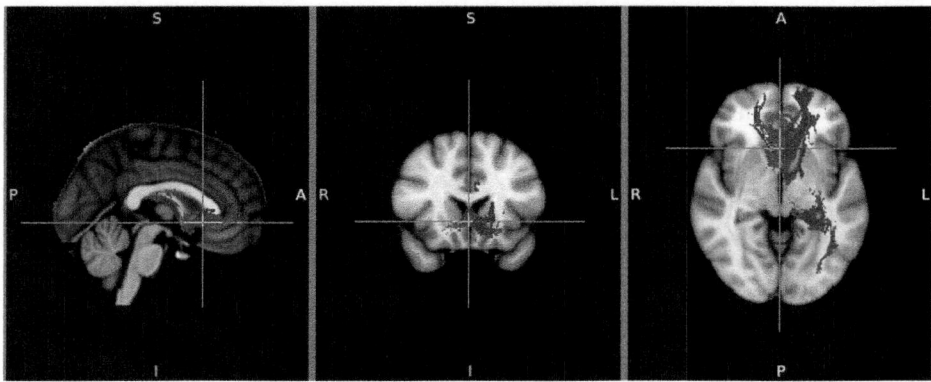

Fig. 2. Average fiber tracts from the left hippocampus to the left accumbens nucleus in Alzheimer's disease patients (red) and normal controls (green). *Left*: Sagittal view taken in planes parallel to the plane running through the nose keeping the eyes on both sides. *Middle*: coronal view taken in planes parallel to the plane intercepting both ears vertically. *Right*: axial views taken in planes perpendicular to spinal cord axis. P: posterior, A: anterior, R: right, L: left, S: superior, I: inferior.

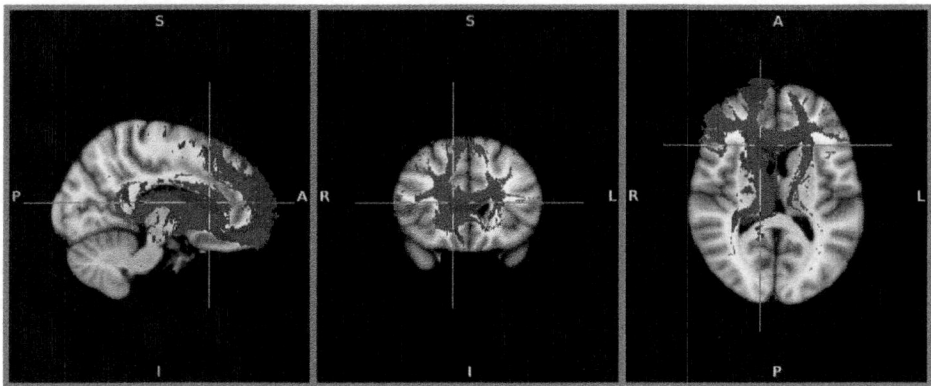

Fig. 3. Average fiber tracts from the right hippocampus to the right caudate in Alzheimer's disease patients (red) and normal controls (green). *Left*: Sagittal view taken in planes parallel to the plane running through the nose keeping the eyes on both sides. *Middle*: coronal view taken in planes parallel to the plane intercepting both ears vertically. *Right*: axial views taken in planes perpendicular to spinal cord axis. P: posterior, A: anterior, R: right, L: left, S: superior, I: inferior.

Alzheimer's brain has scattered and dispersed fiber paths, and its fiber paths have more non-zero voxels than those in NC (Fig. 4). In contrast, the average intensity of fiber paths in AD patients is much lower than that of NC (Fig. 5). The intensity of the fiber paths actually represents their freshness and the amount of neurotransmitters in them. More intense fiber paths contain more water molecules with neurotransmitters than less intense ones. In our data set, AD patients are 22 years older than NC on average. We performed further analysis to examine whether the differences in the number of non-zero voxels and intensity between AD patients and NC were caused by AD or simply by normal aging. The analysis results showed that the number of non-zero voxels and intensity are not affected by aging in both groups, and thus the differences are due to AD (data not shown).

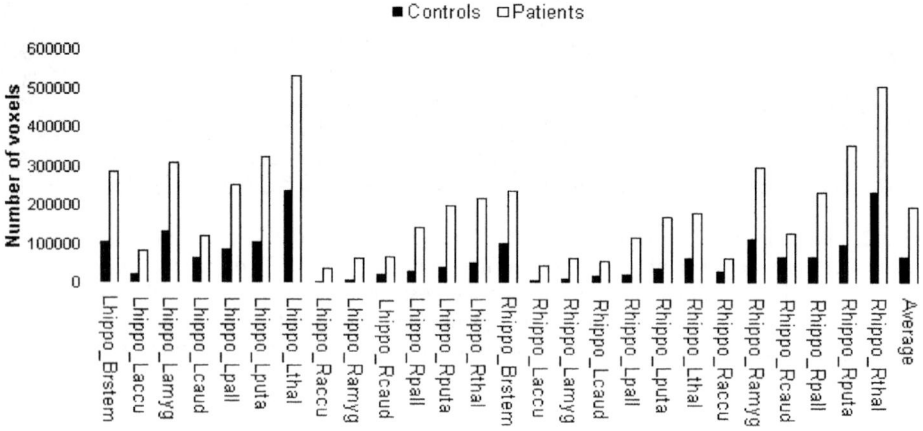

Fig. 4. The average number of non-zero voxels in AD patients and normal controls. The term x_y denotes the connection from x to y.

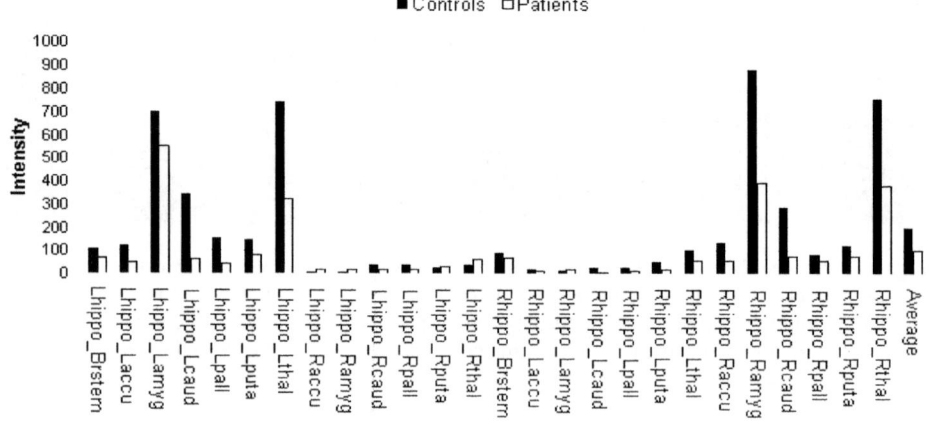

Fig. 5. The average intensity of fiber paths in AD patients and normal controls

In particular, the intensity of the fiber paths from the hippocampus (both left and right hippocampus) to the amygdale, thalamus and caudate showed a big difference between AD patients and NC. The brain stem is important because nerve connections of the motor and sensory systems from the main part of the brain to the rest of the body pass through it. It plays a pivotal role in maintaining consciousness and sleep cycle. The accumbens nucleus is thought to play important role in reward, pleasure, laughter, aggression and healing. The amygdala plays a significant role in processing of memory and emotional reaction. The caudate is important for memory and learning system, and the pallidum is responsible for planning. The main functionality of putamen is regulating movement and various types of learning. The functionality of the thalamus includes maintaining consciousness, sleep and alertness along with relaying sensation. Therefore, the damaged connection of the hippocampus to these regions will cause a significant impairment in planning, learning, and pleasure/aggression sense based on memory or past experience.

5   Conclusion and Future Works

We studied the fiber paths in 12 Alzheimer's disease (AD) patients and 24 normal controls (NC). The fiber paths from the left and right hippocampus toward other brain regions were traced to measure the damage of fibers. In both males and females, AD patients had more non-zero voxels than NC in their fiber paths, but the fibers in AD patients were more scattered and dispersed. The fiber paths of AD patients showed a much lower intensity than those of NC. Although not shown in this paper, similar patterns were observed when we tracked the fiber paths in the reverse directions (i.e., the fiber paths from other brain regions toward the hippocampus). The scattered fiber paths might have caused communication disruption from and to the hippocampus and impaired the memory related functionalities in all other brain regions. This is a preliminary result of the connectivity analysis of Alzheimer's brain, but may be used to diagnose AD or to predict the prognosis of patients with mild AD. We plan to perform a further connectivity analysis of Alzheimer's brain and find potential imaging markers that can differentiate Alzheimer patients from healthy elderly people.

Acknowledgments. This work was supported by the Key Research Institute Program through the National Research Foundation of Korea (NRF) funded by the Ministry of Education, Science and Technology (2011-0018394) and in part by the Basic Science Research Program through NRF funded by the Ministry of Education, Science and Technology (2010-0017213). Data used in the preparation of this article were obtained from the Alzheimer's Disease Neuroimaging Initiative (ADNI) database (http://adni.loni.ucla.edu).

References

1. Pihlajamäki, M., Sperling, R.A.: fMRI: Use In Early Alzheimer's Disease And In Clinical Trials. Future Neurology 3(4), 409–421 (2008)
2. Golby, A., Silverberg, G., Race, E., Gabrieli, S., O'Shea, J., Knierim, K., Stebbins, G., Gabrieli, J.: Memory encoding in Alzheimer's disease: an fMRI study of explicit and implicit memory. Brain 128(4), 773–787 (2005)
3. Petrella, J.R., Wang, L., Krishnan, S., Slavin, M.J., Prince, S.E., Tran, T.T., Doraiswamy, P.M.: Cortical Deactivation in Mild Cognitive Impairment: High-Field-Strength Functional MR Imaging. Radiology 245(1), 224–235 (2007)

4. Liu, Y., Spulber, G., Lehtimäki, K.K., Könönen, M., Hallikainen, I., Gröhn, H., Kivipelto, M., Hallikainen, M., Vanninen, R., Soininen, H.: Diffusion tensor imaging and Tract-Based Spatial Statistics in Alzheimer's disease and mild cognitive impairment. Neurobiology of Aging (2009)
5. Serra, L., Cercignani, M., Lenzi, D., Perri, R., Fadda, L., Caltagirone, C., Macaluso, E., Bozzali, M.: Grey and white matter changes at different stages of Alzheimer's disease. Journal of Alzheimer's Disease 19(1), 147–159 (2010)
6. Zhou, Y., Dougherty, J.H., Hubner, K.F., Bai, B., Cannon, R.L., Hutson, R.K.: Abnormal connectivity in the posterior cingulate and hippocampus in early Alzheimer's disease and mild cognitive impairment. Alzheimer's & Dementia 4(4), 265–270 (2008)
7. Qiu, A., Oishi, K., Miller, M.I., Lyketsos, C.G., Mori, S., Albert, M.: Surface-Based Analysis on Shape and Fractional Anisotropy of White Matter Tracts in Alzheimer's Disease. PLoS ONE 5(3), e9811 (2010)
8. Shankar, G.M., Li, S., Mehta, T.H., et al.: Amyloid-beta protein dimers isolated directly from Alzheimer's brains impair synaptic plasticity and memory. Nature Medicine 14(8), 837–842 (2008)
9. Behrens, T.E.J., Berg, H.J., Jbabdi, S., Rushworth, M.F.S., Woolrich, M.W.: Probabilistic diffusion tractography with multiple fibre orientations. What can we gain? NeuroImage 34(1), 144–155 (2007)
10. Douaud, G., Jbabdi, S., Behrens, T.E.J., Menke, R.A., Gass, A., Monsch, A.U., Rao, A., Whitcher, B., Kindlmann, G., Matthews, P.M., Smith, S.: DTI measures in crossing-fibre areas: Increased diffusion anisotropy reveals early white matter alteration in MCI and mild Alzheimer's disease. NeuroImage 55(3), 880–890 (2011)
11. Smith, S.M., Jenkinson, M., Johansen-Berg, H., Rueckert, D., Nichols, T.E., Mackay, C.E., Watkins, K.E., Ciccarelli, O., Cader, M.Z., Matthews, P.M., Behrens, T.E.J.: Tract-based spatial statistics: Voxelwise analysis of multi-subject diffusion data. NeuroImage 31(4), 1487–1505 (2006)
12. Smith, S.M., Jenkinson, M., Woolrich, M.W., et al.: Advances in functional and structural MR image analysis and implementation as FSL. NeuroImage 23(S1), 208–219 (2004)
13. Behrens, T.E.J., Woolrich, M.W., Jenkinson, M., Johansen-Berg, H., Nunes, R.G., Clare, S., Matthews, P.M., Brady, J.M., Smith, S.M.: Characterization and Propagation of Uncertainty in Diffusion Weighted MR images. Magnetic Resonance in Medicine 50(5), 1077–1088 (2003)
14. Patenaude, B., Smith, S.M., Kennedy, D.N., Jenkinson, M.: A Bayesian Model of Shape and Appearance for Subcortical Brain Segmentation. NeuroImage (in press, 2011)
15. Andersson, J., Smith, S., Jenkinson, M.: FNIRT–FMRIB's Non-linear Image Registration Tool. In: Fourteenth Annual Meeting of the Organization for Human Brain Mapping (2008)

A PACE Sensor System with Machine Learning-Based Energy Expenditure Regression Algorithm

Jeen-Shing Wang[1], Che-Wei Lin[1], Ya-Ting C. Yang[2], Tzu-Ping Kao[1],
Wei-Hsin Wang[1], and Yen-Shiun Chen[1]

[1] Department of Electrical Engineering,
[2] Institute of Education & Center for Teacher Education,
National Cheng Kung University
Tainan 701, Taiwan, R.O.C.
jeenshin@mail.ncku.edu.tw

Abstract. This paper presents a portable-accelerometer and electrocardiogram (PACE) sensor system and a machine learning-based energy expenditure regression algorithm. The PACE sensor system includes motion sensors and an electrocardiogram sensor, a MCU module (microcontroller), a wireless communication module (a RF transceiver and a Bluetooth® module), and a storage module (flash memory). A machine learning-based energy expenditure regression algorithm consisting of the procedures of data collection, data preprocessing, feature selection, and construction of energy expenditure regression model has been developed in this study. The sequential forward search and the sequential backward search were employed as the feature selection strategies, and a generalized regression neural network were employed as the energy expenditure regression models in this study. Our experimental results exhibited that the proposed machine learning-based energy expenditure regression algorithm can achieve satisfactory energy expenditure estimation by combing appropriate feature selection technique with machine learning-based regression models.

Keywords: Energy expenditure, accelerometer, electrocardiogram, feature selection, and generalized regression neural network.

1 Introduction

Inactive lifestyle is a key factor which causes chronic diseases such as diabetes, obesity, and cardiovascular diseases [1]. In order to improve people's lifestyle from inactive to active, providing necessary information such as energy expenditure of daily activities is essential. In the past decades, owing to the rapid development of MEMS and IC technology, physiological signal sensors such as accelerometers for motion detection and electrocardiogram (ECG) for cardiovascular diseases became cheaper and easier to use. Much literature aimed at using activity acceleration or heart rate data to develop energy expenditure regression algorithms [2,3,4,5]. Early approaches of energy expenditure regression models assume that there exists a linear relation between energy expenditure and a single parameter (e.g. count: a feature derived from acceleration signal) [2]. Using a single parameter to fit energy

D.-S. Huang et al. (Eds.): ICIC 2011, LNBI 6840, pp. 529–536, 2012.

expenditure by a linear regression is advantageous for low computation complexity, but the generalization property of the regression model is poor in general. To increase the accuracy of energy expenditure regression models, different approaches have been studied such as adding the number of sensors [3], combining multi-physiological signals with acceleration signal in energy expenditure regression problems [4]. Recently, artificial neural networks (ANN) have been widely adopted in the energy expenditure regression model construction [5]. Moreover, many well-developed theories of machine learning have been actively utilized in the energy expenditure regression problem.

To investigate the possible combinations of feature selection techniques and machine learning-based regression models for good energy expenditure estimation is one of the research focuses of this study. This paper proposed a PACE sensor system and a machine learning-based energy expenditure regression algorithm. In order to construct a regression model with the least number of features and good regression performance, the sequential forward search (SFS) and the sequential backward search (SBS) [6] were employed as the feature selection techniques, and a generalized regression neural network (GRNN) was employed as the regression model of energy expenditure.

The rest of the paper is organized as follow: In Section 2, we will introduce the hardware components of the PACE sensor system. The proposed machine learning-based energy expenditure regression algorithm is introduced in Section 3. Section 4 provides the experimental results of the proposed machine learning-based energy expenditure regression algorithm. Finally, conclusions are presented in Section 5.

2 PACE Sensor System

We have developed a PACE sensor system which is composed of motion sensors and an ECG sensor. The motion sensors are mounted on users' wrist, waist, and ankle and ECG sensor is attached on users' breast. The motion sensors are responsible for sensing acceleration signal from users' daily activities. The ECG sensor is responsible to collect users' ECG data during activities. The hardware components of the motion sensors and the ECG sensor are shown in Fig. 1. The hardware components of each sensor include a signal-sensing module, a MCU module, a wireless communication module, and a storage module. The major component in the motion sensor is a triaxial digital accelerometer for sensing motion/activity acceleration signal and the sensing module in the ECG sensor is an amplifying/filtering circuit. The accelerometer IC employed in the PACE sensor system is Freescale® MMA7455L triaxial digital output accelerometer. The Freescale® MMA7455L accelerometer possesses a user selectable full scale of $\pm 2g$, $\pm 4g$, and $\pm 8g$ and is able to measure acceleration signal over the bandwidth of 125 kHz for all axes. The sensitivity of accelerometers was set from $-8g$ to $8g$ in this study. The ECG amplifying/filtering circuit is composed of an instrument amplifier, a high-pass filter, a low-pass filter, and a notch filter as shown in Fig. 2. The instrument amplifier is responsible for amplifying differential input signals from the heart of a user, and rejecting common-mode noise such as electromyography (EMG) interference in ECG. The common mode rejection ratio (CMRR) of the instrument amplifier is higher than 90 dB in this study. The high-pass, low-pass, and notch filters in our design are responsible to reject the noises that are out of ECG's signal bandwidth. According to the nature of ECG, the filters are set to attenuate the signal that is under 0.05 Hz (by the high-pass filter), above 100 Hz (by the low-pass filter), and 60 Hz (power-line interference) (by the notch filter).

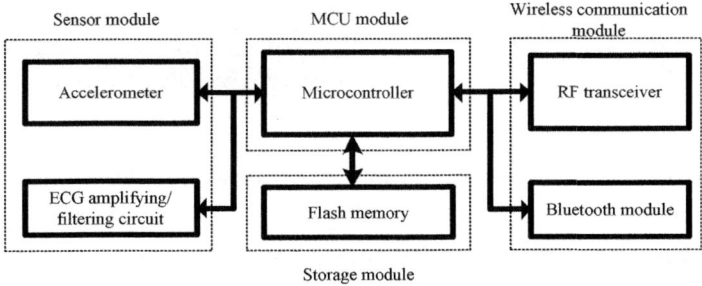

Fig. 1. Hardware components of the PACE sensor system

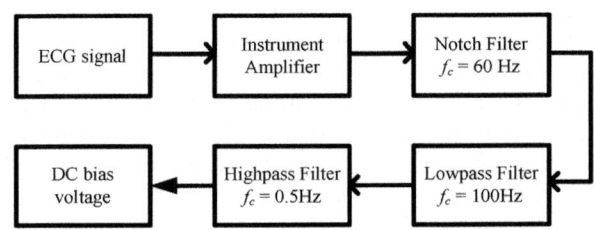

Fig. 2. The block diagrams of hardware components of the ECG amplifying/filtering circuit

The Microchip® PIC24FJ64GA002 was selected as the MCU module. The MCU module is responsible for the following tasks: 1) timing flow control (regularly retrieving data from the sensor module via I²C bus from accelerometers or from the 12-bit A/D converter embedded in the MCU from the ECG sensor), 2) wireless communication control, 3) peripheral component control. The wireless communication module includes a Nordic® nRF24L01+ wireless RF transceiver and a BTM-162 Bluetooth® module. The Nordic® nRF24L01+ wireless RF transceiver is employed to execute timing synchronizations and data transmission from the motion sensors and ECG sensor. The BTM-162 Bluetooth® module is served as the standard communication channel between the PACE sensor system and standard devices such as PCs or smart phones. The storage module employed in the PACE sensor system is MXIC® MX25L128 flash memory with 16 MB storage capacity. Considering high performance and low power consumption, we selected the Microchip® PIC24FJ64GA002 as the main controller. The sampling rates (f_s) of the motion sensors and ECG sensor are 30 Hz and 200 Hz with 8-bit and 12-bit data resolution, respectively. The power consumption of motion sensors and ECG sensor is 12.2 mA and 20 mA, respectively. The sizes of the motion sensors and the ECG sensor are 32mm × 30mm × 5mm and 35mm × 19mm × 6mm, respectively. The circuit boards of the motion sensor and ECG sensor are shown in Fig. 3.

      (a)   (b)

Fig. 3. The circuit boards of (a) motion sensors and (b) ECG sensor

3 Machine Learning-Based Energy Expenditure Regression Algorithm

The proposed machine learning-based energy expenditure regression algorithm includes the procedures of data collection, data preprocessing, feature selection, and construction of energy expenditure regression model. At the beginning, the data collection process collects users' activity acceleration signal and ECG data by the PACE sensor system. The collected data is stored in the storage module of the PACE sensor system, and can transmit to a PC via the Bluetooth® communication protocol. The processes of data preprocessing, feature selection and energy expenditure regression model construction are all executed on a PC. The data preprocessing process includes a noise removal process for acceleration signal, heart rate (HR) retrieval process for ECG signals, a windowing process, a feature generation process, and an activity type classification process. In the data preprocessing process, an elliptic high-pass filter (f_c = 0.05 Hz) and a Hanning low-pass filter (f_c = 10 Hz) were employed to filter the noise of the acceleration signals. A real-time QRS detection algorithm is utilized in the HR retrieval process to detect the R wave in the ECG to form the HR series [7]. After the noise removal and the HR retrieval process, the windowing process segments acceleration signals and HR data into a series of consecutive one-minute windows. In the feature generation process, a total of 12 features were derived from the motion sensors (4 features are generated from three motion sensors (wrist, waist, and ankle). Four features generated from three motion sensors are: 1) count, 2) mean of signal magnitude area (SMA), 3) standard deviation of SMA, and 4) median of SMA) [8]. In addition to the features derived from the acceleration signal, there are 9 features derived from the HR series: 1) mean, 2) standard deviation, 3) variance, 4) interquartile range, 5) skew, 6) kurtosis, 7) mean of HR difference series, 8) standard deviation of HR difference series, and 9) variance of HR difference series [8]. The detail procedures of the activity classification process can be found in [9]. The purpose for activity classification process is to construct energy expenditure regression models for different types of activities for better energy expenditure estimation.

SFS and SBS were employed in the feature selection process in this study. SFS and SBS are feature selection techniques that attempt to find a subset of n features out of m ($m > n$) features so that the selection of features can maximize the performance of a criterion function. SFS starts from an empty set. The features are added one by one, and sent to the criterion function to examine its effectiveness. The selected features are sent to construct an energy expenditure regression model and to employ the mean squared error (MSE) as the performance of the criterion function in our study. The search direction of SBS is opposite to that of SFS. SBS starts from the whole feature set and iteratively deletes features based on the combination performance of the criterion function. In this study, a GRNN were employed as the regression model. In addition to RFBN, GRNN is employed due to its good universal approximation property [10].

4   Experimental Design and Results

The effectiveness of the proposed PACE sensor system and the accuracy of the proposed energy expenditure regression algorithm were examined by an experiment. Twenty-six (20 men and 6 women) healthy college students with age 22.18 ± 2.94 joined our experiment. Participants were all non-smokers and free from diseases that may affect their metabolic rate. The demographics of participants are shown in Table 1.

Table 1. Demographics of participants

	All Subjects ($n = 26$)	Men ($n = 20$)	Women ($n = 6$)
Age, year	22.18 ± 2.94	23.55 ± 1.12	26.67 ± 2.13
Height, cm	170.92 ± 7.59	173.55 ± 0.06	162.17 ± 6.12
Weight, kg	65.04 ± 11.06	67.75 ± 10.74	56 ± 6.27
BMI, kg/m^2	22.18 ± 2.94	22.44 ± 3.08	21.29 ± 2.18

The experiment procedure of the study is shown in Fig. 4. In the beginning, the participants' personal data were collected (height, weight, age, and gender). Then the participants were instructed to wear the data collection instruments including Cosmed K4b^2 portable gas analyzer (to measure participants' energy expenditure) and the motion sensors. The Comsed K4b2 also measured users' HR data by the Polar heart rate electrode. The HR data measured by the Comsed K4b2 was used to get better data synchronization between the HR data and the energy expenditure. The participants all performed each activity for 6 minutes, and took a rest between activities for at least 5 minutes to ensure their heart rates were lower than 100bpm. The participants performed three types of activity; type 1 activity includes lying, desk working, sitting, and standing (sedentary activities). The average HR of Type 1 activity is lower than 95bpm, and the average MET consumption is lower than 1.5 METs. Type 2 activity includes sweep, mopping, walking (3mph/4mph), downstairs, and bicycling (50Watt) (moderate activities). The average energy expenditure of type 2 activity is between 3METs to 6METs and the HR is between 95bpm to 130bpm. The type 3 activity includes upstairs, running (6mph/7mph), and bicycling (100Watt) (vigorous activities); the average HR of Type 3 is higher than 130bpm and the average MET consumption is higher than 6 METs.

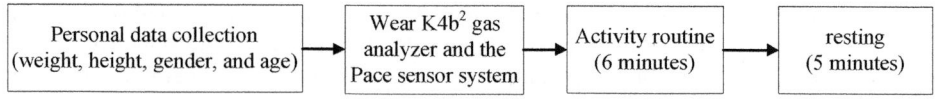

Fig. 4. Experimental procedures

4.1   Selected Features and Energy Expenditure Performance by GRNN

Twenty-three features (12 features derived from the acceleration signals, 9 features derived from the HR series, and 2 personal information features including weight and height) were generated as the feature candidates of the energy expenditure regression models. SFS and SBS were employed as the search strategies, and the reciprocal of MSE of the constructed energy expenditure regression model (GRNN) was chosen as the criterion function. The numbers of selected features in the SFS/SBS process were set from one to seven to examine the relations between the number of features and the performance of the energy expenditure regression model.

Table 2. Selected features based on SFS/SBS and GRNN as the criterion function

Ranking	Sequential Forward/Backward Search		
	Type 1	Type 2	Type 3
1	Varian of HR	HR	HR
2	Count of wrist	Count of wrist	Count of ankle
3	HR	Count of waist	Variance of HR
4	Weight	Difference of HR	Difference of HR
5	Height	Variance of HR	Count of waist
6	Count of ankle	Weight	Weight
7	Count of waist	Height	Kurtosis of HR

After the feature selection process, the performance of energy expenditure regression models was evaluated by constructing a GRNN using the selected features. The performance was evaluated by the standard error of the estimate (SEE) and the coefficient of determination (R^2) [11] between the desired energy expenditure and the estimated energy expenditure. The selected features based on SFS and SBS with GRNN as the criterion function are shown in Table 2. The selected features from SFS and SBS are identical in this experiment. From Table 2, most of the features selected for Type 1~3 are the frequently used features: count from wrist, wait, and ankle. The HR, difference of HR, and variance of HR are also selected in the energy expenditure regression model for Type 1~3. The SEE and R^2 of GRNN with the top 7 features are shown in Fig. 5 and Fig. 6. Fig. 5 shows that SEE is under 0.1 in all types of activities when the selected feature number is greater than 3. Fig. 6 shows that the R^2 is greater than 0.9 when the selected feature number are greater than 3. We found that using GRNN with few appropriate features can achieve satisfactory performance (SEE < 0.1 and R^2 > 0.9).

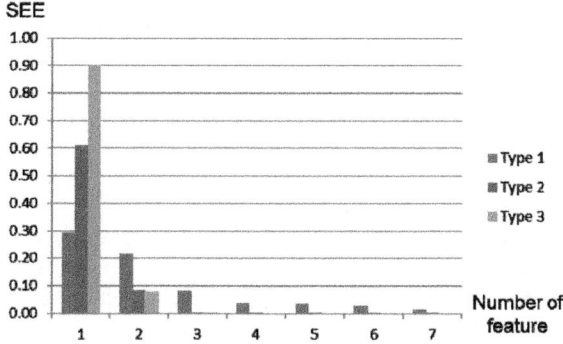

Fig. 5. SEE of energy expenditure regression model based on SFS and GRNN as the criterion function

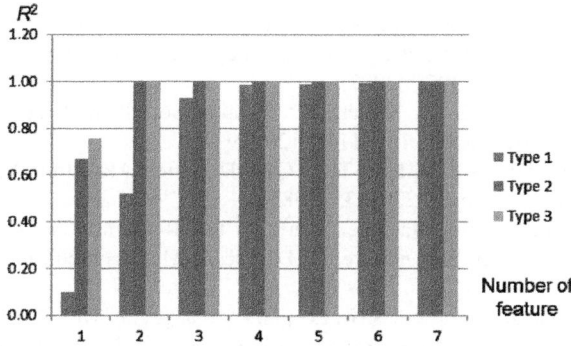

Fig. 6. R^2 of energy expenditure regression model based on SFS and GRNN as the criterion function

5   Conclusions

A realization of the PACE sensor system and the development of a machine learning-based energy expenditure regression algorithm have been presented in this paper. The PACE sensor system can be used to collect the acceleration signals of daily activities and ECG data simultaneously. In order to construct an efficient regression model with the least number of features and good regression performance, SFS and SBS were employed as the feature selection techniques. One representative neural network: GRNN was employed as the regression model of energy expenditure estimation. The SEE and R^2 in the GRNN regression model with the top 3 features can achieve 0.08/0/0 (SEE) and 0.93/1/1(R^2) for Type 1, 2, 3 activities, respectively. From these results, we have successfully validated the effectiveness of the proposed sensor system as well as the construction algorithm.

Acknowledgements. The funding for this research was provided by the National Science Council of Taiwan, under grants NSC 99-2628-S-006-001-MY3 and 99-2628-E-006-161-MY1.

References

1. Physical Inactivity and Cardiovascular Disease, http://www.health.state.ny.us/diseases/chronic/cvd.htm
2. Rothney, M.P., Schaefer, E.V., Neumann, M.M., Choi, L., Chen, K.Y.: Validity of Physical Activity Intensity Predictions by Actigraph, Actical, and RT3 Accelerometers. Obesity 16, 1946–1952 (2008)
3. Chen, K.Y., Sun, M.: Improving Energy Expenditure Estimation by using a Triaxial Accelerometer. J. Appl. Physiology 83, 2112–2122 (1997)
4. Brage, S., Brage, N., Franks, P.W., Ekelund, U., Wareham, N.J.: Reliability and Validity of the Combined Heart Rate and Movement Sensor Actiheart. Eur. J. Clin. Nutr. 59, 561–570 (2005)
5. Rothney, M.P., Neumann, M., Béziat, A., Chen, K.Y.: An Artificial Neural Network Model of Energy Expenditure Using Nonintegrated Acceleration Signals. J. of Applied Physiology 103, 1419–1427 (2007)
6. Pudil, P., Novovičovál, J., Kittlera, J.: Floating Search Methods in Feature Selection. Pattern Recognition Letters 15, 1119–1125 (1994)
7. Pan, J., Tompkins, W.J.: A Real-Time QRS Detection Algorithm. IEEE Transactions on Biomedical Engineering 3, 230–235 (1985)
8. Munguia, T.E.: Using Machine Learning for Real-time Activity Recognition and Estimation of Energy Expenditure. Ph.D. Thesis, Massachusetts Institute of Technology (2008)
9. Chen, Y.P., Yang, J.Y., Liou, S.N., Lee, G.Y., Wang, J.S.: Online Classifier Construction Algorithm for Human Motion Detection Using an Accelerometer. Applied Mathematics and Computation 205, 849–860 (2008)
10. Specht, D.F.: A General Regression Neural Network. Neural Networks 2, 568–576 (1991)
11. Engle, R.F., Granger, C.W.J.: Co-integration and Error Correction: Representation, Estimation and Testing. Econometrica 55, 251–276 (1987)

A Faster Haplotyping Algorithm Based on Block Partition, and Greedy Ligation Strategy

Xiaohui Yao[1,2] and Yun Xu[1,2,*], and Jiaoyun Yang[1,2]

[1] Department of Computer Science, University of Science and Technology of China, Hefei, Anhui 230026, China
[2] Anhui Province-MOST Co-Key Laboratory of High Performance Computing and Its Application, Hefei, Anhui 230027, China
xhyao@mail.ustc.edu.cn, xuyun@ustc.edu.cn

Abstract. Haplotype played a very important role in the study of some disease gene and drug response tests over the past years. However, it is both time consuming and very costly to obtain haplotypes by experimental way. Therefore haplotype inference was proposed which deduce haplotypes from the genotypes through computing methods. Some genetic models were presented to solve the haplotype inference problem, and Maximum Parsimony model was one of them, but at present the methods based on this principle are either simple greedy heuristic or exact ones, which are adequate only for moderate size instances. In this paper, we presented a faster greedy algorithm named FHBPGL applying partition and ligation strategy. Theoretical analysis shows that this strategy can reduce the running time for large scale dataset and following experiments demonstrated that our algorithm gained comparable accuracy compared to exact haplotyping algorithms with less time.

1 Introduction

Single nucleotide polymorphism (SNP), as the most typical genetic variation, has been widely studied in disease gene and drug responsiveness. However, many diseases such as asthma, cancer and other complex ones may be influenced by not only a single SNP, but multi-SNPs in a region[1]. These linked SNPs on a chromosome constitute a string named *haplotype*. Because of the fact that most vegetal and animal, genomes are diploid, it is very costly and time consuming that obtain the two separate haplotypes through biological experiments[2]. Actually, *genotype* is more economical and easier got in the laboratory. Genotype is the combination of two haplotypes from paired chromosomes, but we do not know the position of each allele.

In theory, there are exponential possible haplotype configurations for a given genotypes, however in reality the haplotype patterns' number is much small for a certain population, which makes it possible to derive haplotypes from genotypes directly. Therefore it is very necessary and important to develop efficient and

---

* Corresponding author.

D.-S. Huang et al. (Eds.): ICIC 2011, LNBI 6840, pp. 537–544, 2012.

accurate computational methods for inferring haplotypes from genotype data (i.e. *haplotyping* or *haplotype inference*).

There are two categories for the haplotype inference: statistical methods and rule based methods. Statistical methods[3]-[6] focus on estimating the haplotype frequencies according to certain statistical models. Although they can give the numerical assessments of the haplotype configurations and choose the most probable haplotype pairs, they are often time consuming and can not provide the exact haplotype patterns.

Compared with the statistical methods, the rule based methods[7]-[11] focus on finding the exact solution based on reasonable biological assumptions. Two models were considered: the perfect phylogeny and the maximum parsimony principle. In the fact of that the real number of the haplotype for a certain population was limited; therefore the smallest haplotype set was closest to the reality. Clark[7] described a simple greedy inference rule to compute a set of haplotypes which resolve a given genotypes set. Gusfield[8] formulated the problem and proposed an integer linear programming algorithm. Wang and Xu[9] proposed a branch and bound algorithm called HAPAR to find the optimal solution. Lancia et al.[10] proved the APX-hardness of the problem. Zhang et al.[11] presented an approximation algorithm combined the greedy algorithm and the branch-and-bound algorithm. However, because of the problem's NP-hard property, it is difficult to gain the optimal solution in acceptable time for most algorithms based on this model. This meant that most methods cannot handle large amount of SNPs.

In this article, we proposed a faster approximate algorithm based on block partition and greedy ligation to solve the problem. The rest of the paper is organized as follows. Section 2 formalizes the problem of haplotyping and the maximum parsimony model. Section 3 provides a detailed description of our algorithm. Section 4 analysis the algorithm's time complexity. The experimental results based on the real and simulation dataset are given in Section 5. At last, section 6 concludes the whole paper and brings forward some promising directions for future work.

2   Preliminary Definitions

Haplotype data is a sequence of string vector, each site of which corresponds to a SNP of the chromosome. For a m-long haplotype, there are two possible alleles for each site which can be denoted by '0' and '1'. Without loss of generality, we use '0' standing for wild-type and '1' standing for the mutant allele. Then a haplotype can be identified as a $[0,1]^m$ vector.

Genotype is the combination of two haplotypes h_1 and h_2 in which one comes from the father and the other comes from the mother. If we use '0' standing for (0,0), '1' standing for (1,1), and '2' standing for (0,1), then a genotype can be identified as a $[0,1,2]^m$ vector. If a locus of the genotype is '0' or '1', we call it *homozygous* locus, else *heterozygous* locus.

We put it down as $g = h_1 \oplus h_2$ if h_1 and h_2 can solve the g. A sample population of genotypes can be presented as a set of vectors $G = \{g_0, g_1, \ldots, g_{n-1}\}$. Each vector is a sequence associated with m sites. The purpose of haplotyping is to find out a haplotype set H to make sure that there always can be found an explanation for each vector g_i in G.

We formalize the problem of haplotyping and the maximum parsimony model as follows.

Definition 1. Haplotyping *Given a genotype set $G = \{g_0, g_1, \ldots, g_{n-1}\}$ with the length of m, where each component has a value of '0', '1' or '2', find out a haplotype set $H = \{h_0, h_1, \ldots, h_{t-1}\}$ (also with the length of m), so that for any genotype $g \in G$, there is a pair of haplotypes $h_i, h_j \in H$ so that $g = h_i \oplus h_j (i, j \in [0..t-1])$.*

We call that H explants G. Obviously, the number of possible haplotype pairs will increase exponentially 2^s as the number of heterozygous sites increase s. Fortunately large numbers of real data and experiments show that the number of haplotypes is limited and the solution which has small number of haplotypes is nearest to the reality. The observation and such idea is formalized as the model of haplotyping by maximum parsimony.

Definition 2. Haplotyping by Maximum Parsimony (HMP)[8] *Given a set $G = \{g_0, g_1, \ldots, g_{n-1}\}$ of genotypes with the length of m, where each component has a value '0', '1' or '2', find out a set $H = \{h_0, h_1, \ldots, h_{t-1}\}$ of haplotypes (also with the length of m) which can explain G such that $|H|$ is minimized.*

3   Methods

Although most exact algorithms based on maximum parsimony principle can work well to resolve small-scale dataset, they are exponential time algorithms and can not be applied to large-scale instances.

We proposed an approximate algorithm to solve the problem adopting the traditional uniform partition and adding the overlap strategy which could both reduce the time complexity and increase the accuracy.

3.1   Step 1: Block Partitioning with Overlap

We partition the G sequentially into blocks each of which has the same size s and with the overlap of o_num, then the block number is larger than m/s, where m is the length of G. Figure 1 shows an example for the partition.

So we can calculate the block number (noted b_num) as follows:

$$b\_num = \begin{cases} \dfrac{m-s}{s-o\_num} + 1, & \text{if } (m-s)\%(s-o\_num) = 0 \\[2ex] \dfrac{m-s}{s-o\_num} + 2, & \text{otherwise} \end{cases}$$

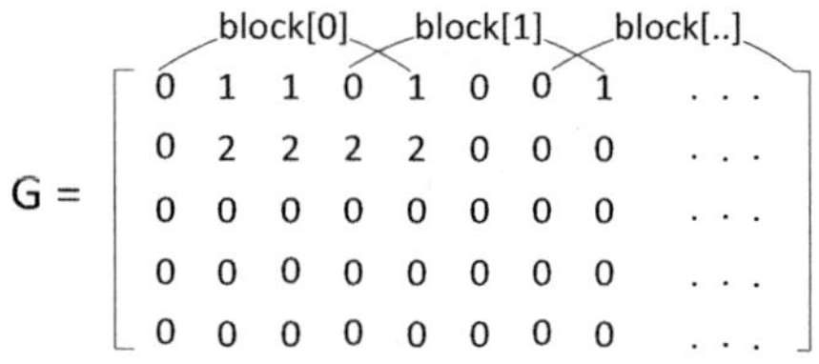

$$G = \begin{matrix} \overbrace{\quad}^{block[0]} \; \overbrace{\quad}^{block[1]} \; \overbrace{\quad}^{block[..]} \end{matrix}$$

$$G = \begin{bmatrix} 0 & 1 & 1 & 0 & 1 & 0 & 0 & 1 & \cdots \\ 0 & 2 & 2 & 2 & 2 & 0 & 0 & 0 & \cdots \\ 0 & 0 & 0 & 0 & 0 & 0 & 0 & 0 & \cdots \\ 0 & 0 & 0 & 0 & 0 & 0 & 0 & 0 & \cdots \\ 0 & 0 & 0 & 0 & 0 & 0 & 0 & 0 & \cdots \end{bmatrix}$$

Fig. 1. Block partition with overlap($s = 5, o\_num = 2$)

3.2 Step 2: Haplotyping in Each Block

We adopt maximum parsimony model to solve the haplotyping problem for each block. We also adopt some tricks to reduce the running time and improve the accuracy. When n is far larger than s, the diversity of the n genotype segments is most likely less than n, then we can merge the genotype segments which have the same configuration to reduce the block scale. There may be not only one best solution for a given genotype set, which one should be selected is not easy to decide. We stored all possible results in buffer and randomly selected one as the final solution to improve the chance that the right one be selected.

3.3 Step 3: Ligating among Blocks

The ligation of the blocks is sequential and for each step the next block[i..j] will be ligated with the former accomplished block[0..i-1].

Definition 3. Homology *In a block[i..j], if there are two (or more) genotype segments g_s and g_t share one haplotype h, then g_s and g_t are homology on h over [i..j].*

Definition 4. Compatibility and Incompatibility *For two haplotypes h_1 and h_2 which will be ligated, if the sequences on the overlap sites of them are equal, then h_1 and h_2 can be called compatibility; else can be called incompatibility.*

We put it down as $h_1 \cdot h_2$ if the haplotypes h_1 ligates to h_2 and they are compatibility. Considering the sth genotype, denote $g_{s,k-1}$ which is from block[0..i − 1 + o_num] and $g_{s,k}$ which is from block[i..j]. Correspondingly, their haplotype configurations are denoted as $(h_{s,k-1}, h'_{s,k-1})$ and $(h_{s,k}, h'_{s,k})$. While ligating, there have three cases as follows and Figure 2 illustrate case 3 which is more complex than others.

Case 1: $g_{s,k-1}$ and $g_{s,k}$ are both homozygous. There are $h_{s,k-1} = h'_{s,k-1}$ and $h_{s,k} = h'_{s,k}$. We can ligate $h_{s,k-1}$ and $h_{s,k}$, $h'_{s,k-1}$ and $h'_{s,k}$ directly. The result is $h_{s,k-1} \cdot h_{s,k}$ and $h'_{s,k-1} \cdot h'_{s,k}$.

Case 2: $g_{s,k-1}$ is homozygous and $g_{s,k}$ is heterozygous ($g_{s,k-1}$ is heterozygous and $g_{s,k}$ is homozygous, respectively). We can ligate $h_{s,k-1}$ and $h_{s,k}$, $h'_{s,k-1}$ and $h'_{s,k}$. The result is $h_{s,k-1} \cdot h_{s,k}$ and $h'_{s,k-1} \cdot h'_{s,k}$.

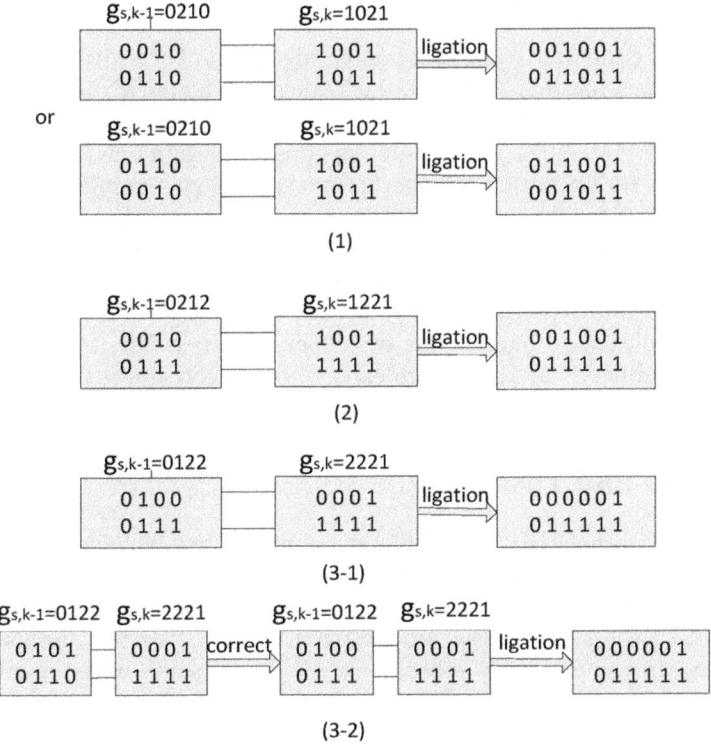

Fig. 2. Case3: $g_{s,k-1}$ and $g_{s,k}$ are both heterozygous with different heterozygous number of the overlap.(1)0 heterozygous loci. (2)1 heterozygous locus. (3-1) and (3-2) are 2 or more heterozygous loci with same configuration and different configurations.

Case 3: $g_{s,k-1}$ and $g_{s,k}$ are both heterozygous. For this case $h_{s,k-1} \neq h'_{s,k-1}$ and $h_{s,k} \neq h'_{s,k}$. There are three cases according to the heterozygous number of the overlap as follows.

1): 0 heterozygous loci. The four haplotypes are same on the overlap sites and we do not know how to select the suitable association of them if there is no additional information. Now we can ligate them according to the ligation situation of their homology haplotypes.

2): 1 heterozygous locus. On the overlap sites, the two haplotypes are different for each genotype and there must have $h_{s,k-1} = h_{s,k}$ and $h'_{s,k-1} = h'_{s,k}$, or $h_{s,k-1} = h'_{s,k}$ and $h'_{s,k-1} = h_{s,k}$. So we can ligate the two having equal configuration on the overlap.

3): 2 or more heterozygous loci. On the overlap sites, there may be two situations for the two haplotype configurations. First, $h_{s,k-1} = h_{s,k}$ and $h'_{s,k-1} = h'_{s,k}$, or $h_{s,k-1} = h'_{s,k}$ and $h'_{s,k-1} = h_{s,k}$, and we can ligate directly. Second, $h_{s,k-1} \neq h_{s,k} \neq h'_{s,k-1} \neq h'_{s,k}$, and this means one of the configurations must be error. We can search and correct the error one according to their homology haplotypes and then it will be turned to be the first situation.

4   Complexity Analysis

Now we analyze the time complexity of our algorithm. For a given genotype set G with n genotypes $\{g_0, g_1, \ldots, g_{n-1}\}$, the branch and bound algorithm searches all possible solutions and selects the best one. Then the search time complexity will be $O(2^{k_0+k_1++k_{n-1}-n})$ (k_i is the heterozygous locus number for each genotype g_i), if there is no limitation for the heterozygous locus number, there has the equation $O(2^{k_0+k_1++k_{n-1}-n}) = O(2^{mn})$ (m is the length of the genotype).

When the m, n is large enough, the time complexity can not be abided. While the length of the block is s and the overlap size is o_num, there are about $(m-s)/(s-o\_num)$ blocks. For each block, the time complexity is $O(2^{sn})$ and the time complexity of ligation for two blocks is $O(n)$. Due to the s and c are both smaller integer, the time complexity of our algorithm is $O(m2^{sn})$ which is much lower than the exact algorithm.

5   Results and Discussion

We ran our program FHBPGL for a large amount of real biological data as well as simulation data to demonstrate the performance of our program. We compared our program with four widely used existing programs including PLEM [14], HAPLOTYPER[4], HAPAR[9], and fastPHASE[15]. We will discuss different sets of data in the following subsection. All experiments were completed on a Windows server with 3.20 GHz CPU and 1GB RAM.

To evaluate the accuracy of our algorithm, we used third evaluative standards: the first is the individual error rate (IER), which was a widely used criteria to access the performance of individual haplotyping[16], the second is the precision which is the rate of the correct haplotypes deduced the number of haplotypes we inferred, and the last is recall which is the rate of correct haplotypes deduced with the number of real haplotypes.

5.1   Real Data

We ran the five programs PLEM, HAPLOTYPER, HAPAR, fastPHASE and FHBPGL on the human angiotensin converting enzyme dataset (ACE), which was provided by Rieder et al.[17]. It contained the genotypes of 11 unrelated individuals at 52 SNPs and 13 distinct haplotypes.

The block size of FHBPGL was set to 4 and the overlap size was set to be 2. The estimation of accuracy and running time can be described in Table 1. Among all the five algorithms our FHBPGL used least time and yielded better accuracy.

5.2   Simulated Data

To test the programs under more realistic data samples, we introduce a powerful model called the coalescent theory. Developed by R. Hudson[18], the program ms uses coalescent theory to generate simulating population samples of haplotypes.

Table 1. Comparison of performance of four programs on ACE data set

Program	IER	Precision	Recall	Running time(s)
fastPHASE	0.202	0.658	0.658	17.34
HAPLOTYPER	0.191	0.677	0.677	0.13
HAPAR	0.273	0.636	0.538	16.11
PLEM	0.218	0.631	0.631	0.15
FHBPGL	0.182	0.692	0.692	0.02

For each sample, 50-80 individuals were generated and the accuracies were estimated. The block size of FHBPGL was set to be 8. For other algorithms, we just chose their default parameter settings. To estimate the average performances of various algorithms, 10 independent runs were performed for each sample. Table 2 presented the accuracy and running time comparison of various algorithms.

Table 2. Comparison of performance of four programs on simulated data set

Program	IER	Precision	Recall	Running time(s)
fastPHASE	0.369	0.582	0.682	165.7
HAPLOTYPER	0.287	0.651	0.729	3.65
PLEM	0.566	0.411	0.441	0.63
FHBPGL	0.311	0.638	0.770	0.06

For the simulated data, HAPLOTYPER yielded the lowest IER and FHBPGL yielded the highest recall rate and comparable precision. PLEM and FHBPGL were much faster than other algorithms, and HAPAR failed to gain a solution in 10 hours so we did not list it here. FHBPGL used shortest time and yielded comparable accuracy.

6   Conclusion

In this article, we proposed a faster partition and ligation strategy. We partitioned the haplotype blocks using an overlap method, then applied maximum parsimony principle for each block and greedy ligated the adjacent blocks according the overlap and the homology of blocks. Compared with other algorithms, our algorithm gained comparable accuracy with the least time, as the sample is larger our algorithm also gained highest recall rate with least time.

Acknowledgment. We thank Bo Chen, Ying Wang and Pengyu Nie, who provided many helpful suggestions for our article. This work is supported by the National Natural Science Foundation of China (No.61033009 and No.60970085).

References

1. International HapMap Consortium: The international HapMap project. Nature 426 789–796 (2003)
2. Gusfield, D.: An Overview of Combinatorial Methods for Haplotype Inference. In: Istrail, S., Waterman, M.S., Clark, A. (eds.) DIMACS/RECOMB Satellite Workshop 2002. LNCS (LNBI), vol. 2983, pp. 9–25. Springer, Heidelberg (2004)
3. Excoffier, L., Slatkin, M.: Maximum-likelihood estimation of molecular haplotype frequencies in a diploid population. Molecular Biology and Evolution 12(5), 921–927 (1995)
4. Niu, T., Qin, Z.S., Xu, X., Liu, J.S.: Bayesian haplotyping interface for multiple linked single-nucleotide polymorphisms. Am J. Hum. Genet. 70(1), 157–169 (2002)
5. Xing, E.P., Jordan, M.I., Sharan, R.: Bayesian haplotype inference via the Dirichlet process. Journal of Computational Biology (JCB) 14(3), 267–284 (2007)
6. Zhao, Y.Z., Xu, Y., Yao, X.H., et al.: A better block partition and ligation strategy for individual haplotyping. Bioinformatics 24(23), 2720–2725 (2008)
7. Clark, A.: Inference of haplotypes from PCR-amplified samples of diploid populations. Molecular Biology and Evolution 7(2), 111–122 (1990)
8. Gusfield, D.: Haplotype inference by pure parsimony. In: Baeza-Yates, R., Chávez, E., Crochemore, M. (eds.) CPM 2003. LNCS, vol. 2676, pp. 144–155. Springer, Heidelberg (2003)
9. Wang, L.S., Xu, Y.: Haplotype inference by maximum parsimony. Bioinformatics 19(14), 1773–1780 (2003)
10. Lancia, G., Pinotti, C., Rizzi, R.: Haplotyping populations by pure parsimony: Complexity of exact and approximation algorithms. INFORMS J. Comp. 16, 348–359 (2004)
11. Zhang, Q., Che, H., Chen, G., Sun, G.: A Practical Algorithm for Haplotyping by Maximum Parsimony. Journal of Software 16(10), 1699–1707 (2005)
12. Daly, M.J., et al.: High-resolution haplotype structure in the human genome. Nat. Genet. 29, 229–232 (2001)
13. Gabriel, S.B., et al.: The structure of haplotype blocks in the human genome. Science 296, 2225–2229 (2002)
14. Qin, Z.S., et al.: Partition-Ligation EM algorithm for haplotype inference with single nucleotide polymorphisms. Am. J. Hum. Genet. 71, 1242–1247 (2002)
15. Scheet, P., Stephens, M.: A fast and flexible statistical model for largescale population genotype data: applications to inferring missing genotypes and haplotypic phase. Am. J. Hum. Genet. 78, 629–644 (2006)
16. Delaneau, O., et al.: ISHAPE: new rapid and accurate software for haplotyping. BMC Bioinformatics 8, 205 (2007)
17. Rieder, M.J., et al.: Sequence variation in the human angiotensin converting enzyme. Nat. Genet. 22, 59–62 (1999)
18. Hudson, R.R.: Generating samples under a wright-fisher neutral model of genetic variation. Bioinformatics 18(2), 337–338 (2002)

An Effective ECG Arrhythmia Classification Algorithm

Jeen-Shing Wang[1], Wei-Chun Chiang[1], Ya-Ting C. Yang[2], and Yu-Liang Hsu[1]

[1] Department of Electrical Engineering
[2] Institute of Education & Center for Teacher Education,
National Cheng Kung University,
Tainan 701, Taiwan, R.O.C.
jeenshin@mail.ncku.edu.tw

Abstract. This paper presents an effective electrocardiogram (ECG) arrhythmia classification scheme consisting of a feature reduction method combining principal component analysis (PCA) with linear discriminant analysis (LDA), and a probabilistic neural network (PNN) classifier to discriminate eight different types of arrhythmia from ECG beats. Each ECG beat sample composed of 200 sampling points at a 360 Hz sampling rate around an R peak is extracted from ECG signals. The feature reduction method is employed to find important features from ECG beats, and to improve the classification accuracy of the classifier. With the features, the PNN is then trained to serve as classifier for discriminating eight different types of ECG beats. The average classification accuracy of the proposed scheme is 99.71%. Our experimental results have successfully validated the integration of the PNN classifier with the proposed feature reduction method can achieve satisfactory classification accuracy.

Keywords: Arrhythmia classification, feature reduction, principal component analysis, linear discriminant analysis, probabilistic neural networks.

1 Introduction

Electrocardiograms (ECG) which reveal the rhythm and function of the heart is an important non-invasive clinical tool for cardiologists to diagnose various heart diseases. A successful ECG arrhythmia classification usually involves three important procedures: feature extraction, feature selection, and classifier construction. Feature extraction and reduction are two important procedures that usually influence the classification performance of any ECG arrhythmia classification system. Therefore, to extract sufficient features and reduce their dimensions for classifiers to achieve optimal classification results have become primary tasks for the ECG arrhythmia classification problems.

From a review of literature, we found that various ECG feature extraction methods have been successfully applied for arrhythmia classification. The feature extraction methods include 1) time-domain methods [11]; 2) frequency-domain methods [5]; 3) time-frequency domain analysis [1]; and 4) wavelet transform [4]. To name a few, Yu and Chou [11] extracted a total of 200 points before and after the R points to form the

D.-S. Huang et al. (Eds.): ICIC 2011, LNBI 6840, pp. 545–550, 2012.
© Springer-Verlag Berlin Heidelberg 2012

ECG samples. In addition, R-R time intervals were also extracted as the characteristic features for arrhythmia detection. Minami *et al.* [5] combined the Fourier transform to observe the changes in QRS complex, and used a neural network to discriminate three kinds of rhythms. Afonso and Tompkins [1] used the time-frequency distribution (TFD) of normal sinus rhythm, ventricular tachycardia, ventricular flutter, and ventricular fibrillation signals to detect ventricular fibrillation. Korürek and Nizam [4] extracted six time-domain features and PCA compressed wavelet coefficients from ECG signals for the ant colony optimization (ACO) classifier to discriminate six kinds of arrhythmias.

Furthermore, feature reduction methods have been widely utilized to reduce dimensions of input features and improve classification performances of classifiers. Recently, researchers have presented different feature reduction methods to reduce input dimensions of neural classifiers. To name a few, Yu and Chen [12] used the independent component analysis (ICA) method to reduce the input features from 200 to 17 to represent an ECG beat sample. Subsequently, the 17 ICA-based features and the R-R time interval were treated as the input features for the classifier.

Recently, neural networks have been treated as a powerful classifier to deal with ECG arrhythmia classification problems [7]. For example, Ceylan *et al.* [3] developed a type-2 fuzzy clustering neural network consisting of the type-2 fuzzy *c*-means clustering scheme and backpropagation learning to classify ten types of ECG arrhythmias. The accuracy of the proposed system achieved 99%. Özbay and Tezel [8] implemented novel neural networks with an adaptive activation function to classify EEG arrhythmias and the average accuracy achieved 98.19%.

In this paper, we propose an ECG arrhythmia classification scheme based on the feature reduction method and the PNN classifier to discriminate eight types of ECG arrhythmias, which are normal beat (NORMAL), premature ventricular contraction (PVC), paced beat (PACE), right bundle branch block beat (RBBB), left bundle branch block beat (LBBB), atrial premature contraction (APC), ventricular flutter wave (VLWAV), and ventricular escape beat (VESC) obtained from the MIT-BIH arrhythmia database [13]. The scheme includes the following steps: 1) data acquisition, 2) feature extraction and normalization, 3) feature reduction, and 4) classification. The rest of this paper is organized as follows. In Section 2, the whole database, feature extraction and normalization, feature reduction methods, and the PNN classifier for classifying arrhythmia are described detailedly. The results and discussion are presented in Section 3. Finally, conclusions are presented in Section 4.

2   Methods

The ECG arrhythmia classification scheme is composed of data acquisition, feature extraction and normalization, feature reduction, and classification. The block diagram of the proposed scheme is shown in Fig. 1. This paper randomly selects 9800 samples of eight ECG beat types from the MIT-BIH arrhythmia database in which half of the samples are selected as the training subset and the other half as the testing subset. We now introduce the ECG arrhythmia classification scheme in detail as follows.

2.1   Feature Extraction and Normalization

In the feature extraction procedure, a fraction of signal around the R peak is extracted as the time-domain features since the R peak of ECG signals is an important index for cardiac diseases. To ensure the important characteristic points of ECG like P, Q, R, S and T are included, a total of 200 sampling points at a 360 Hz sampling rate before and after the R peak are collected as one ECG beat sample. However, the signal amplitude biases of the waveforms of the ECG beat samples are inconsistent due to instrumental and human differences. Hence, we utilize the Z-score method to reduce the abovementioned differences of each waveform of each ECG beat sample. Through the Z-score method, the mean value of each ECG sample is firstly subtracted from each ECG sample to eliminate the offset effect, and then divided by its standard deviation. This procedure results in a normalized ECG beat sample with zero mean and unity standard deviation.

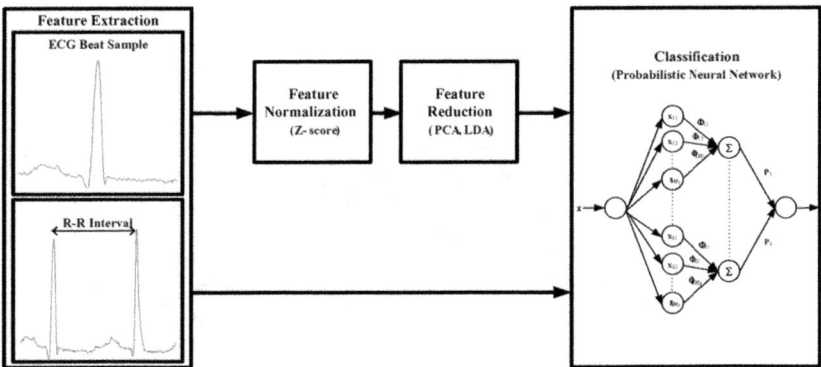

Fig. 1. The block diagram of the proposed scheme

2.2   Feature Reduction Method

2.2.1   Principal Component Analysis

The PCA method has been widely used in statistical data analysis, feature extraction, feature reduction, and data compression [2]. The goal of PCA is to transform one set of variables in $\Re^k$ space into another set in $\Re^p$ space containing the maximum amount of variance in the data where normally p is smaller than k. For details about the PCA, please refer to [2].

2.2.2   Linear Discriminant Analysis

The LDA method is an effective supervised dimension reduction method for pattern recognition problems [10]. The goal of LDA is to utilize a transformation matrix, which can maximizes the ratio of the between-class scatter matrix to the within-class scatter matrix, to transform the original feature vectors into a lower dimensional feature space by a linear transformation. For details about the LDA, please refer to [10].

In addition, we extract the R-R time interval (time elapse between successive R peaks) as the other feature, since it is also an important feature for classifying ECG arrhythmias. The R-R time interval is calculated as the time difference between the R points of the two successive beats. Fortunately, the information of the R-R time interval is available in the reference annotation files in the MIT-BIH arrhythmia database. After the feature reduction method, we integrate the PCA-based, LDA-based, or PCA+LDA-based features with the R-R time interval as the input features for the following neural classifier.

2.3   Probabilistic Neural Network Classifier

The PNN is proposed by Specht, which is guaranteed to converge to a Bayesian classifier with enough training data [6], [9]. According to the Bayes' strategy, the outputs of the PNN can be represented as the estimation of the probability of the class membership and the training rule of the PNN is based on the estimation of a probability density functions (pdf) of the classes. The PNN consists of an input layer, a pattern layer, a summation layer, and a competitive output layer. This architecture is illustrated in Fig. 2. For details about the PNN, please refer to [6], [9].

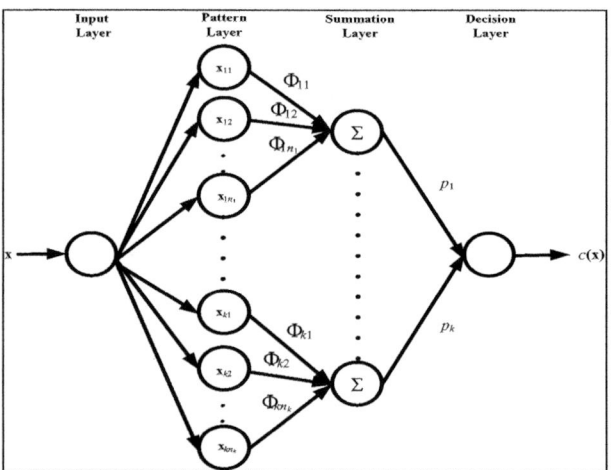

Fig. 2. The topology of a PNN classifier

3   Results and Discussion

The classification performance of the proposed ECG arrhythmia classification scheme was validated by a total number of 9800 samples of eight ECG beat types from the MIT-BIH arrhythmia database including 3600 NORMAL samples, 2460 PVC samples, 800 PACE samples, 800 RBBB samples, 800 LBBB samples, 764 APC samples, 472 VLWAV samples, and 104 VESC samples. The ECG beat samples were randomly separated into two subsets: One half of the samples were selected as the training subset and the other half as the testing subset. To evaluate the performance of the proposed classification scheme, three common measures of sensitivity, specificity, and accuracy are used.

3.1   Feature Dimension Reduction

The normalized ECG beat samples and the R-R time intervals of ECG signals were applied as the features for the proposed scheme in which only the dimensions of the normalized ECG beat samples were reduced by the proposed feature reduction methods, PCA and LDA. To estimate an optimal number of the principle components (PCs) of PCA, we varied the numbers of the PCs from 1 to 20. The experiments with the PNN classifier were repeated 10 times for different PC numbers. We found that the average accuracy reached a plateau at around 8 PCs and did not show a noticeable improvement since then. Similarly, the procedure to estimate the optimal dimensions of LDA was the same as PCA. However, the upper bound of the dimension of the features using LDA was 7 since there were eight types of ECG beats needed to be classified. Hence, the dimensions of LDA were varied from 1 to 7. The best average accuracy could be achieved when the dimension of LDA was 7. We combined PCA with LDA to reduce the dimensions of the normalized ECG beat samples. The dimension of each ECG beat sample was firstly reduced from 200 to 7 using the PCA method. Subsequently, the PCA-based features were treated as the inputs for LDA.

3.2   Feature Reduction Method Comparison

Once the optimal dimensions of each of the abovementioned feature reduction methods were estimated, we compared the classification performances of the PNN classifier between the three different feature reduction methods. The results of ECG arrhythmia classification using the PCA+PNN scheme, the LDA+PNN scheme, and the PCA+LDA+PNN scheme are shown in Table 1. The overall accuracy of the PCA+PNN scheme, the LDA+PNN scheme, and the PCA+LDA+PNN scheme were 99.61%, 98.26%, and 99.71%, respectively. Therefore, from the results, the PCA+LDA+PNN scheme is the best for the ECG arrhythmia classification task.

Table 1. Performance analysis of the proposed feature reduction methods with the PNN classifier

	PCA+PNN			LDA+PNN			PCA+LDA+PNN		
	Sensitivity (%)	Specificity (%)	Accuracy (%)	Sensitivity (%)	Specificity (%)	Accuracy (%)	Sensitivity (%)	Specificity (%)	Accuracy (%)
NORMAL	99.11	99.32	99.24	95.22	95.52	95.41	99.56	99.39	99.45
PVC	98.37	99.26	99.04	94.31	97.74	96.88	98.37	99.46	99.18
PACE	100.00	99.98	99.98	97.50	99.91	99.71	100.00	100.00	100.00
RBBB	98.75	99.87	99.78	96.50	99.58	99.33	99.25	100.00	99.94
LBBB	98.25	99.84	99.71	94.50	99.02	98.65	98.75	99.89	99.80
APC	98.17	99.85	99.71	81.68	99.27	97.90	97.91	99.91	99.76
VLWAV	92.37	95.41	99.51	73.31	96.05	98.33	95.76	95.19	99.61
VESC	92.31	98.98	99.88	86.54	99.06	99.84	94.23	99.00	99.94
Average	97.17	99.06	99.61	89.94	98.27	98.26	97.98	99.10	99.71

4   Conclusion

In this paper, an effective ECG arrhythmia classification scheme that consists of the PCA+LDA feature reduction method and the PNN classifier has been proposed. The 200-dimensional feature vector around each R peak of ECG signals is extracted. In order to reduce the training time and improve the classification accuracy of the classifier, the 200-dimensional features are reduced to 7 new features by the PCA+LDA feature reduction method. The reduced features and the R-R time interval are then treated as the inputs of the PNN classifier to discriminate eight different types of ECG arrhythmias. The average classification accuracy of the proposed PCA+LDA+PNN scheme is 99.71%. The effectiveness of the proposed PCA+LDA+PNN scheme for ECG arrhythmia classification has been successfully validated by the experiment. We believe that the proposed scheme can be served as an effective tool for cardiologists to diagnose heart diseases based on ECG signals.

Acknowledgements. The funding for this research was provided by the National Science Council of Taiwan, under grants NSC 99-2628-S-006-001-MY3 and 99-2628-E-006-161-MY1.

References

1. Afonso, V.X., Tompkins, W.J.: Detecting Ventricular Fibrillation: Selecting the Appropriate Time-Frequency Analysis Tool for the Application. IEEE Engineering in Medicine and Biology 14, 152–159 (1995)
2. Castells, F., Laguna, P., Sörnmo, L., Bollmann, A., Roig, J.M.: Principle Component Analysis in ECG Signal Processing. EURASIP Journal on Advances in Signal Processing, 1–21 (2007)
3. Ceylan, R., Özbay, Y., Karlik, B.: A Novel Approach for Classification of ECG Arrhythmias: Type-2 Fuzzy Clustering Neural Network. Expert Systems with Applications 36, 6721–6726 (2009)
4. Korürek, M., Nizam, A.: Clustering MIT-BIH Arrhythmias with Ant Colony Optimization Using Time Domain and PCA Compressed Wavelet Coefficients. Digital Signal Processing 20, 1050–1060 (2010)
5. Minami, K.I., Nakajima, H., Toyoshima, T.: Real-Time Discrimination of Ventricular Tachyarrhythmia with Fourier-Transform Neural Network. IEEE Trans. Biomedical Engineering 46, 179–185 (1999)
6. Mao, K.Z., Tan, K.C., Ser, W.: Probabilistic Neural-Network Structure Determination for Pattern Classification. IEEE Trans. Neural Networks 11, 1009–1016 (2000)
7. Moavenian, M., Khorrami, H.: A Qualitative Comparison of Artificial Neural Networks and Support Vector Machines in ECG Arrhythmias Classification. Expert Systems with Applications 37, 3088–3093 (2010)
8. Özbay, Y., Tezel, G.: A New Method for Classification of ECG Arrhythmias Using Neural Network with Adaptive Activation Function. Digital Signal Processing 20, 1040–1049 (2010)
9. Specht, D.F.: Probabilistic Neural Network. Neural Networks 3, 109–118 (1990)
10. Yeh, Y.C., Wang, W.J., Chiou, C.W.: Cardiac Arrhythmia Diagnosis Method Using Linear Discriminant Analysis on ECG Signals. Measurement 42, 778–789 (2009)
11. Yu, S.N., Chou, K.T.: Integration of Independent Component Analysis and Neural Networks for ECG Beat Classification. Expert Systems with Applications 34, 2841–2846 (2008)
12. Yu, S.N., Chou, K.T.: Selection of Significant Independent Components for ECG Beat Classification. Expert Systems with Applications 36, 2088–2096 (2009)
13. MIT-BIH Database Distribution, Massachusetts Institute of Technology, 77 MassachusettsAvenue, Cambridge, MA 02139 (1998),
   http://www.physionet.org/physiobank/database/mitdb

A New Subspace Approach for Face Recognition

Jian-Xun Mi

School of Electronics and Information Engineering
Tongji University, 4800 Caoan Road, Shanghai 201804, China
Bio-Computing Research Center, Shenzhen Graduate School
Harbin Institute of Technology, Shenzhen, China
mijianxun@gmail.com

Abstract. In this paper, we proposed a new linear regression-based approach for face recognition, called farthest subspace classification. In previous literatures, it was believed that the facial images from a specific object class tend to lie on a linear subspace, i.e. the class-specific subspace. Therefore a query image will be considered belonging to its nearest subspace (NS) of a class. The distance from a query image to each class-specific subspace is calculated simply by the linear regression. In this paper, we proposed a novel notion of face recognition that in the complete feature space spanned by all the gallery images, each class-specific subspace has not only common subspace shared by every class-specific subspace, but also its unique coordinate bases, which are available discriminative information. Based on this notion, we develop farthest subspace (FS) classifier to perform face recognition. The experimental results supported the proposed novel concept. Furthermore, we proposed nearest-farthest subspace (NFS) classification using both NS and FS rules, which outperform NS used alone.

Keywords: Face recognition, linear regression, nearest subspace classification, farthest subspace classification.

1 Introduction

In recent years, face recognition has become a popular investigated topic in biometrics. A lot of pattern recognition methods were proposed for addressing identification of facial images. To avoid the "curse of dimensionality", feature extraction is necessary in face recognition system in general [11]. Every image then will be represented in the feature space. Therefore, in this paper, we only discuss face recognition in feature space. After the discriminant features are extracted, a robust classifier is the remaining key point. One of the simplest is the nearest neighbor (NN) pattern classifier [1] which intelligibly assigns the facial image of query to the class which includes the nearest neighbor of the query among all the gallery images. An improved method, called nearest line (NL) classifier, was proposed by Li and Lu [2] which is more robust than the NN classifier. The NL classifier assigns the label of a query image to a class which includes a nearest feature line, passing through any two gallery images of a class. A systematical study of nearest feature classifiers was presented by Chien and Wu [3], who also proposed nearest plane (NP) classifier and

D.-S. Huang et al. (Eds.): ICIC 2011, LNBI 6840, pp. 551–557, 2012.

nearest space classifier, which is also known as nearest subspace (NS) classifier. In [3], it is reported that NS has the highest recognition rate among the mentioned nearest feature classifiers on IIS and ORL face databases. A recent ace recognition research paper [4] investigated the NS classification with downsampled face images. The experimental results indicated that such fairly simple linear regression classification method could compete with sparse representation classifier [5] which is more sophisticated and needs to solve $l_1 - norm$ optimization.

Previous studies suggested [10] that facial samples form a specific object class tend to lie on a linear subspace. However, the subspace formed by the gallery images of a person could be incomplete due to the small seize of gallery images in general face recognition systems. In this paper, we propose a new classification rule for the problem of face identification, farthest subspace classification, which is based on the premise that each class-specific subspace has its own coordinate bases. Our approach needs to construct a "leave-one-class-out" subspace of each class at first. The subspace without the gallery images of a certain person does not have its specific coordinate bases. Therefore the query images of that person will have a maximum distance to the subspace without his/her prototypes.

For NS and FS exploit different characteristics of the class-specific subspaces respectively, we then developed a new approach, NFS classification, using both NS and FS rules. The new classifier outperforms classifier using NS alone.

The rest of the paper is organized as follows: In Section 2, we introduce the NS classifier. The new proposed FS approach is described and discussed in Section 3, and followed by introducing nearest-farthest subspace classification. In Section 4, we verify the proposed classifiers on two popular face data sets and compare the proposed approaches with NS approach from recent published paper [4]. The paper concludes in Section 5.

2   Nearest Feature Classifiers

In 1967, Cover and Hart proposed a nonparametric classifier called nearest neighbor decision [1]. NN is the first nearest feature classifier which can produce acceptable recognition results without any parameter estimation and model training. The NN is feasible to face recognition, if there is an ideal face database which saves a large amount of images for each face covering a wide-range variation due to pose, illumination, expression deterioration, etc. However, such database with dense samples is practically unrealizable. It is, therefore, not robust to classify variational facial images via NN on databases with insufficient prototypes. To increase the robustness, the NL classifier was proposed in 1999 and the more robust methods, which are the NP classifier and NS classifier, were presented in 2002. It is easy to prove [3] that the NN, NL, NP are the particular cases of NS since they can always be written as the same form as NS. For a wider space can be extrapolated by NS, we are not surprised that NS can obtain highest recognition accuracy [3] among above mentioned nearest feature classifications.

The NS actually is based on the premise that facial samples from a specific object class tend to lie on a linear subspace. The decision process, hence, is to find which class specific subspace has minimum distance to the query image. The following part of this section describes the details of NS classification. First, we present the common face recognition scenario. There are N distinguished classes and p_i prototype images for ith class, $i = 1, 2, \cdots, N$. Each prototype will be represented in feature space as a vector, $a_i^{(m)} \in \mathbb{R}^{q \times 1}$, where $m = 1, 2, \cdots, p_i$ and q is the dimension of feature space. Then we stack the prototype vectors of a class to build a class-specific model $\mathbf{A}_i$ which defines class-specific subspace mentioned in previous section:

$$\mathbf{A}_i = [a_i^{(1)} a_i^{(2)} \cdots a_i^{(p_i)}] \tag{1}$$

Let y be the point of the unlabeled probe image in feature space. Then we calculate the distance between y and subspace spanned by $\mathbf{A}_i$, which is

$$d_i = \|y - \hat{y}_i\|_2 \tag{2}$$

where $\hat{y}_i$ is the best "explanation" of y that ith class can give, or, in other words, the projection of y onto the ith subspace. $\hat{y}_i$ can be obtained simply by using least-squares estimation,

$$\hat{y}_i = \mathbf{A}_i (\mathbf{A}_i^T \mathbf{A}_i)^{-1} \mathbf{A}_i^T y \tag{3}$$

where $\mathbf{A}_i^T \mathbf{A}_i$ should be well conditioned, otherwise we need to use principal component analysis (PCA) to make it invertible. Finally, we classify y to the class with minimum distance, d_i.

3 Face Recognition Based on FS Classification

In face recognition, approaches use various discriminant features of the face image. The concept that samples from a specific object class tend to lie on a particular linear subspace is a good discriminative criterion which has been used by NS classification. We discovered that the class-specific subspace of an object has some unique coordinate bases, which are available for face recognition. In this paper we propose a new approach of face recognition called farthest subspace classification. For FS classification, we first need to construct a "leave-one-class-out" subspace:

$$\mathbf{B}_i = [\mathbf{A}_1 \mathbf{A}_2 \cdots \mathbf{A}_{i-1} \mathbf{A}_{i+1} \cdots \mathbf{A}_N] \tag{4}$$

The subspace $\mathbf{B}_i$ does not have the unique coordinate bases of the ith class, but contains the common coordinate bases shared by all the class-specific subspaces. Therefore, the samples of the ith class will have maximum distance to $\mathbf{B}_i$ among all the "leave-one-class-out" subspaces.

Then we calculate the distance between y and subspace $\mathbf{B}_i$:

$$l_i = \|y - \overline{y}_i\|_2 \tag{5}$$

where $\overline{y}_i = \mathbf{B}_i (\mathbf{B}_i^T \mathbf{B}_i)^{-1} \mathbf{B}_i^T y$ which is the projection of y on to $\mathbf{B}_i$ and PCA is generally required to make $\mathbf{B}_i^T \mathbf{B}_i$ invertible. In general we should keep enough principal components so as to ensure the unique coordinate bases of every class are retained as many as possible. Regarding decision, we assign the query image to the class with the maximum distance, l_i.

Since the NS and FS classification use different characteristics of the class-specific subspace, a classification approach, NFS classification, are proposed which is regarded as extension for both NS and FS. For NFS classification, we need to calculate two distances, i.e., d_i and l_i. We then classify the query to the class that minimizes:

$$j_i = d_i / l_i \tag{6}$$

4   Experimental Results

In this section, we tested our proposed methods on two popular public face databases which are Extended Yale B [7] and Georgia Tech [9]. In [4], a state of the art NS classification approach is reported. Since this newly proposed approach outperforms several other classification methods, we mainly compared our proposed methods with the approach in [4] and hence followed its experimental protocols.

4.1   Extend Yale B

The Extended Yale B database consists of 2,414 frontal face images of 38 subjects (near 64 images for each subject). We divided the database into 5 subsets depending on the lighting conditions from moderate to extreme luminance [7]. We downsampled the images to an order of 20×20. The images in subset 1 were used as training data and others were used for validation. The results are shown in Table 1.

Table 1. Classification accuracies for FS and NFS compared with NS, PCA, ICA on the Extended Yale B database

Approach	Subset 2	Subset3	Subset 4	Subset 5
NS	100%	100%	83.27%	33.61%
FS	100%	99.46%	73.04%	20.20%
NFS	**100%**	**100%**	**87.10%**	**42.78%**

For subsets 2 and 3, accuracies of all three classification approaches were very high, reaching or close to 100%. For the rest subsets, although the FS could not compete with NS, NFS outperformed NS by 3.83% and 9.17% on subset 4 and subset 5, respectively.

As mentioned before, PCA was used in FS and NFS. In above experiments, to avoid the singularity, we kept the first 99 principal components (PCs) of $\mathbf{B}_i^T \mathbf{B}_i$ which remained more than 99.99% variance of the matrix. Next, we illustrated how number of PCs affected the recognition accuracy in Fig 1. From Fig.1 we see that the dimension of $\mathbf{B}_i^T \mathbf{B}_i$ has severe effects on FS, which implies that the unique coordinate bases for discrimination will lose if only keeping a small amount of PCs. We should point out that even when FS performed very poorly, the NFS can still outperform NS. With respect to an increasing dimensionality, that the almost synchronized increasing performance of NFS and FS indicates that the FS gives a reliable assistance to improve the recognition capability of NFS.

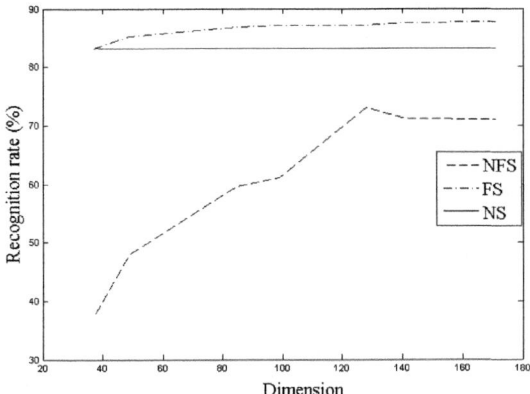

Fig. 1. The recognition accuracy versus dimension of $\mathbf{B}_i^T\mathbf{B}_i$ for subset 4 of Extended Yale B database using NFS and FS. As a comparison, a solid line indicating the recognition accuracy for NS was shown.

4.2   Georgia Tech (GT) Database

We also verified our proposed approaches on GT database which consists of 50 subjects with 15 images per subject. All images were downsampled to an order of 15×15. First, we used the same evaluation protocol as [4]. The first eight images of each subject were used for training, while the remaining 7 images were for testing. In this experimental scenario, NFS and NS achieved identical recognition accuracy, up to 92.57%.

Then we conducted a new evaluation experimental protocol which took a random subset with p ($p = 5, 10, 20, 30, 40, 50$) images of each subject to form a training data, and took the rest as testing images (each trial for a certain seize of training data was repeated for 20 times). The experimental results are shown in Table 2.

Table 2. Experimental results on GT database

Training Number	NS	FS	NFS
5	59.42 ± 0.02%	**78.59 ± 0.01%**	66.77 ± 0.02%
10	82.35 ± 0.01%	**88.03 ± 0.01%**	85.98 ± 0.01%
20	91.36 ± 0.01%	92.14 ± 0.01%	**93.31 ± 0.01%**
30	94.20 ± 0.01%	93.46 ± 0.01%	**95.52 ± 0.01%**
40	95.28 ± 0.01%	94.27 ± 0.01%	**96.33 ± 0.01%**
50	95.82 ± 0.01%	94.82 ± 0.01%	**96.91 ± 0.01%**

It was somewhat unexpected that FS outperforms NS and NFS by margins of 19.17% and 11.82% respectively when the training subsets were very small ($p = 5$). Unlike Extended Yale B, there are more variations incorporated in GT database. When using a small randomly selected subset of images as prototypes ($p = 5, 10$), the class-specific subspace is severely fragmentary. Nevertheless, for the FS, $\mathbf{B}_i$ consists of almost all variations except the unique coordinate bases of ith objective so that $\mathbf{B}_i$ has stronger discriminative power than $\mathbf{A}_i$ in this case. When we increased the training set, FS is still comparable to NS. In summary, the NFS outperforms NS verified on both used databases.

5   Conclusions

In this paper, we proposed a FS classification rule for face recognition. The FS classifier adopts a new notion of identification of face images: the farthest linear subspace. Intuitively the premise of FS is quite strong. However the experimental results verified that FS could attain acceptable recognition accuracy in general, and obtain even fairly competitive results in some cases. Further, we proposed NFS classifier which is still a simple linear regression-based classification approach. That our proposed NFS always outperformed NS verifies the discriminant power of FS. Therefore, our main theoretical contribution of this paper is the discovery that there are unique coordinate bases of each class-specific subspace.

Acknowledgements. The author would like to thank Can-Yi Lu and Dr. Jie Gui for suggestions. This work was supported by the grants of the National Science Foundation of China, Nos. 60975005, 61005010, 60873012, 60805021, 60905023, 31071168 & 30900321, and the grant of China Postdoctoral Science Foundation, No. 20100480708.

References

1. Cover, T.M., Hart, P.E.: Nearest Neighbor Pattern Classification. IEEE Trans. Information Theory 13, 21–27 (1967)
2. Li, S., Lu, J.: Face Recognition Using Nearest Feature Line. IEEE Trans. Neural Networks 10, 439–443 (1999)
3. Chien, J., Wu, C.: Discriminant Waveletfaces and Nearest Feature Classifiers for Face Recognition. IEEE Trans. Pattern Analysis and Machine Intelligence 24, 1644–1649 (2002)
4. Naseem, I., Togneri, R., Bennamoun, M.: Linear Regression for Face Recognition. IEEE Trans. Pattern Analysis and Machine Intelligence 32, 2106–2112 (2010)
5. Wright, J., Yang, A.Y., Ganesh, A., Sastry, S.S., Ma, Y.: Robust Face Recognition via Sparse Representation. IEEE Trans. Pattern Analysis and Machine Intelligence 31, 210–227 (2009)
6. Belhumeur, P., Hespanha, J., Kriegman, D.: Eigenfaces vs. Fisherfaces: Recognition Using Class Specific Linear Projection. IEEE Trans. Pattern Analysis and Machine Intelligence 19, 711–720 (1997)

7. Lee, K.C., Ho, J., Kriegman, D.: Acquiring Linear Subspaces for Face Recognition under Variable Lighting. IEEE Trans. Pattern Analysis and Machine Intelligence 27, 684–698 (2005)
8. Samaria, F., Harter, A.: Parameterisation of a Stochastic Model for Human Face Identification. In: Proc. Second IEEE Workshop Applications of Computer Vision (1994)
9. Georgia Tech Face Database, http://www.anefian.com/face_reco.htm
10. Barsi, R., Jacobs, D.: Lambertian Reflection and Linear Subspaces. IEEE Trans. Pattern Analysis and Machine Intelligence 25, 218–233 (2003)
11. Xu, Y., Zhang, D., Yang, J.Y.: A Feature Extraction Method for Use with Bimodal Biometics. Pattern Recognition 43, 1106–1115 (2010)

A Novel DE-ABC-Based Hybrid Algorithm for Global Optimization

Li Li[1], Fangmin Yao[1], Lijing Tan[2], Ben Niu[1,3,*], and Jun Xu[3]

[1] College of Management, Shenzhen University, Shenzhen 518060, China
[2] Management School, Jinan University, Guangzhou 510632, China
[3] e-Business Technology Institute, The University of Hongkong, Hongkong, China
drniuben@gmail.com

Abstract. A novel hybrid swarm intelligent algorithm DEABC, integrating differential evolution (DE) and artificial bee colony (ABC) algorithm, is proposed in this paper. By using global information obtained form DE population and bee colony, the exploration and exploitation abilities of DEABC algorithm are balanced. The DE population uses the global best to generate offspring every generation. The bee colony acquires the best individual after few generations. The experiments are performed on six benchmark functions to compare the efficiencies of DE, ABC, PSO and DEABC. The numerical results indicate the proposed algorithm outperforms other algorithms in terms of accuracy and convergence speed.

Keywords: Artificial bee colony algorithm, differential evolution, hybrid optimization methods.

1   Introduction

Swarm intelligence (SI) has become an innovative artificial intelligence technique for solving optimization problems during the last two decades. Artificial bee colony (ABC) algorithm, a recently developed SI algorithm, was first proposed by D. Karaboga in 2005[1]. Since ABC algorithm is robust and simple in concept, easy to use and has few control variables, it has successfully been used in science and engineering fields [2-3]. However, the convergence rate of ABC algorithm is poorer while dealing with constrained problems and composite functions.

In order to improve the ABC algorithm performance, various attempts have been made to enhance its optimization capability, including hybrid models. Haiyan Zhao et al [4] introduce a hybrid ABC with GA, in which exchanges information between bee colony and chromosome population. Xiaohu Shi et al [5] introduce hybrid model that two information exchanging processes to share information mutually between particle swarm and bee colony. B. Akay et al [6] propose modified versions of the ABC algorithm, which employ a control parameter and a scaling factor. Compared to other algorithms, it produces promising results on hybrid functions.

---

* Corresponding author.

D.-S. Huang et al. (Eds.): ICIC 2011, LNBI 6840, pp. 558–565, 2012.
© Springer-Verlag Berlin Heidelberg 2012

Differential evolution (DE), a stochastic parallel search method, was first introduced by Price in 1995[7]. It is one of the powerful evolutionary algorithms, which replaces the poorest performing population member through mutation, crossover, and selection. However, DE may fall into local optima and have a slow convergence speed in the last period of iterations.

In this paper, a novel algorithm DEABC is proposed by synthesizing differential evolution and artificial bee colony algorithm. DEABC algorithm enhances individuals by sharing information between DE population and bee colony. For every generation, the best individual is selected as parents for crossover operation. After few generations, the best may become a new food for bee colony to exploit. Offspring produced by DEABC algorithm are expected to perform better and the poor-performed individuals can be weeded out from generation to generation. Therefore the hybrid approach possesses both the merits of DE and ABC. The section 2 gives a briefly introduce to the ABC, DE and describes DEABC. Section 3 evaluates the performance of DEABC on the six benchmark test functions. Section 4 is experiment results. A few conclusions are given in Section 5.

2 DEABC Algorithm

2.1 Backgrounds of ABC and DE

Artificial bee colony algorithm (ABC) consists of three kinds of bees: employed bees, onlooker bees, and scouts. A food source position represents a possible solution. There is only one employed bee for every food source. Meanwhile, the number of employed bees is equal to onlooker bees. Employed bees exploit a food source and carry the information to onlooker bees. Onlooker bees wait in the hive and select good food sources through the information shared by employed bees. The employed and onlooker bees find new food source through a neighborhood search.ïAfter all the employed bees exploit a new food source and the onlooker bees are allocated a food source, if a food source is not improved further through a predetermined number called *limit*, then it is abandoned and the employed bee is converted to a scout. Scouts search the space for new food sources randomly.

DE algorithm is a population-based stochastic algorithm that creates new individuals through mutation, crossover and selection. As a greedy algorithm, it exploits potential solutions to effectively search the optima. *DE/best/1/exp* scheme [8] is adopted later on, which is detailed descried in references 8.

2.2 The Hybrid DEABC Algorithm

In this paper, we propose a new algorithm based on ABC and DE. Although DE algorithm maintains the diversity of population and the good local converging speed, it may encounter the premature convergence in optimizing multimodal problems. In ABC algorithm, owing to greedy selection schemes and the neighbor production mechanism, the most significant features are the self-improvement of solution and local search ability. But ABC algorithm has a slow convergence speed for unimodal function. However, ABC algorithm has no mechanism to use the global information in the search space, so it easily results in a waste of computing power and gets trapped in local optima. Share the differential information can be helpful for the search ability. A novel algorithm based on ABC and DE is proposed, in which objective is to get benefits form both approaches.

DEABC generates two populations, one generated by the ABC and the other by DE. When they are executed in parallel, information is exchanged between two populations. The information exchange among individuals of different populations will help the individuals to avoid misjudging information and becoming trapped by poor local minima. The ABC algorithm shares the global information through scouts. The scouts may obtain the information of best individual from global.

In DEABCïalgorithm, after all employed bees search and onlookers selection and search processes are performedïïf the food source is unable to be further improved through the predetermined number $limit$, it is abandoned. The scout makes a search using the equation (1) below:

$$x_i^j = \begin{cases} x_{min}^j + r(x_{max}^j - x_{min}^j), & \text{if } R < S \\ x_{best}^j, & \text{otherwise} \end{cases} \tag{1}$$

R is a random value between 0 and 1. S is a designated probability, which controls the balance of bee colony explorations. A higher value of S may cause reducing information exchange while a lower one may cause too little diversity. So S is set as 0.5 in Section 3 of this paper. x_{best}^j is the global best position at jth dimension from the two populations. According to equation (1), a 'food' position is randomly selected from the best individual as scout initial search, while random number is not less than designated probability S. Otherwise, scout makes a random search.

On the other hand, $\vec{x}_{best}^t$ is chosen according to the last global best positions from the two populations in mutation operator. The information exchanged between the two populations influences the exploration and maintains some diversity in the whole population. Meanwhile, it may reduce the risk of convergence to local sub-optima.

The pseudo code and flow chart of the proposed hybrid algorithm DEABC could be shown as follows:

1. Initialize DE and ABC sub-systems respectively. Evaluate the fitness value of each individual.
2. Compare the fitness value of DE and ABC, memorize the best solution.
3. Perform employed bees search and onlookers selection and search processes.
4. Execute scouts search process. The new search points should be determined according to a given probability, whether are randomly produced or obtained from the best positions.
5. Update $\vec{x}_{best}^t$ and execute mutation, crossover and selection operators.
6. If the termination criterion is not met, go to step 2. Otherwise, output the best solution and the global best fitness of the whole swarm.

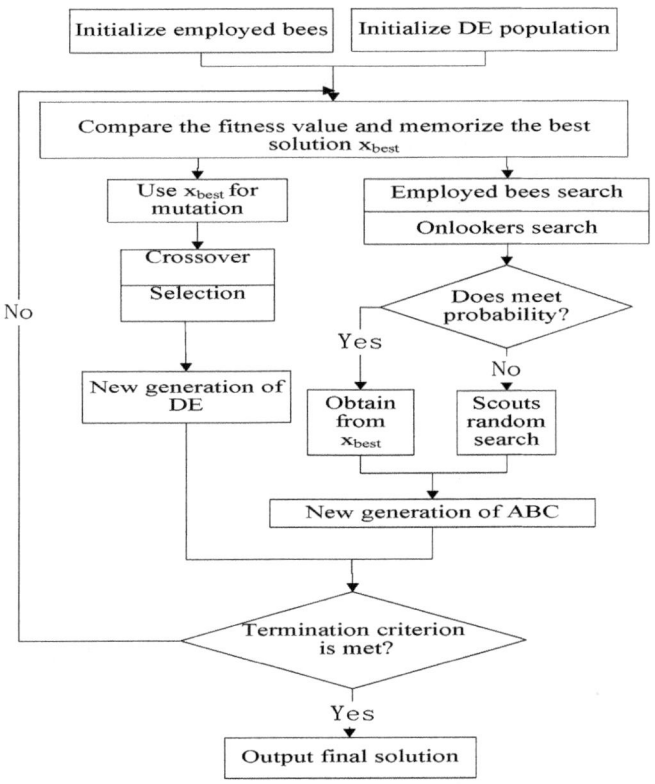

Fig. 1. Flow chart of DEABC

3   Experiment Setting and Benchmark Problems

To test the capability of DEABC, six benchmark optimization problems that are commonly used in literature [9-10] are used. Compare its results with improved PSO based on the linearly-decrease inertia weight (PSO-LIW) [11], DE and ABC. All functions are minimization problems, which present different difficulties to the optimization algorithms. They are Sphere function, Rosenbrock function, Rastrigin function, Noncontinuous Rastrigin function, Griewank function and Weierstrass function.

In every experiment, the w in PSO-LIW is during $[0.9, 0.4]$, that is $w_{start} = 0.9$, $w_{end} = 0.4$. Other parameters are $c_1 = c_2 = 2.0$, $t_{max} = 0.5$. The DE parameters used here are $F = 0.8$ and $CR = 0.5$. The parameters of ABC are as follows: limit is 100. For DEABC, the parameters $F, CR,$ limit are all the same with those defined in ABC and DE. The designated probability S of sharing information is 0.5. The population size is set as 40, the dimension of the functions is 10. A total of 20 runs for each experimental setting are conducted.

4   Experiment Results

PSO-LIW, ABC, DE, and DEABC were tested to optimize the six benchmark functions using the settings presented in the previous paragraph. Figs 2-5 show the convergence graphs of four functions. Table 1 lists the comparison results including the best values, the worst values, the mean values and standard deviations. All results below were reported as '0000e+000'.

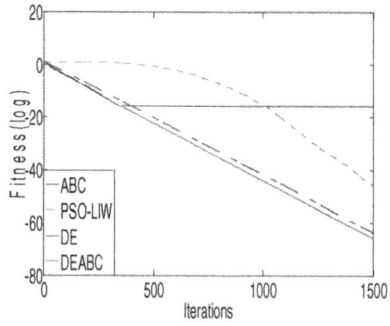

Fig. 2. Sphere function

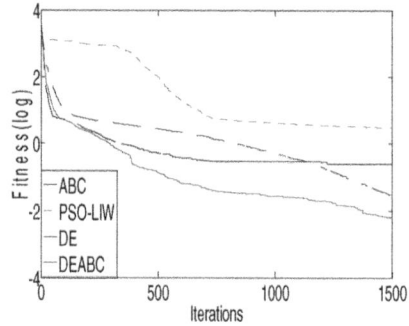

Fig. 3. Rosenbrock function

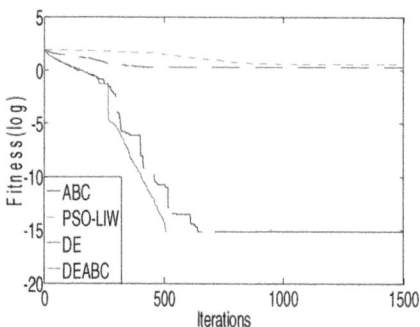

Fig. 4. Rastrigin fuction

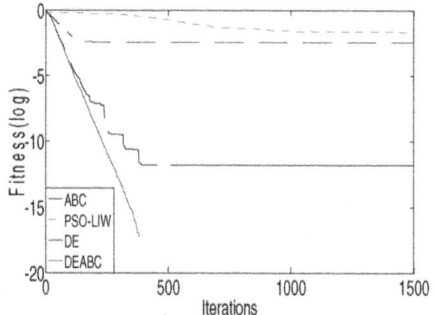

Fig. 5. Griewank function

For the continuous and unimodal Sphere function, except for ABC and PSO-LIW, DEABC and DE converge fast toward the fitness optimum. ABC has a fast convergence before 300 iterations, as can be seen from Fig.2. But after 300 iterations, ABC stagnates and can't converge toward the fitness optimum. In the Table 1 DEABC possesses the best results.

Rosenbrock function is a famous classic optimization problem, in which variables are strongly dependent, and the gradients generally mislead the direction of search. It can be seen from Fig.3 that DEABC algorithm outperforms DE, ABC, PSO-LIW.

Table 1. Results for all algorithms on benchmark problems

N=10		PSO-LIW	ABC	DE	DEABC
Sphere	Best	6.657e-050	7.462e-017	1.750e-066	5.286e-068
	Worst	1.635e-045	3.284e-016	1.464e-063	1.055e-065
	Mean	1.196e-046	1.814e-016	2.055e-064	**1.628e-066**
	Std	**1.378e-091**	1.029e-016	3.442e-064	2.712e-066
Rosenbrock	Best	1.200e-003	5.100e-003	2.400e-003	3.533e-007
	Worst	7.046e+000	1.821e+000	1.010e-001	3.220e-002
	Mean	2.930e+000	2.490e-001	2.860e-002	**6.400e-003**
	Std	3.768e+000	4.155e-001	3.000e-002	**9.500e-003**
Rastrigrin	Best	9.950e-001	0.000e+000	0.000e+000	0.000e+000
	Worst	5.969e+000	1.421e-014	4.974e+000	0.000e+000
	Mean	3.283e+000	7.105e-016	2.039e+000	**0.000e+000**
	Std	3.135e+000	3.177e-015	1.349e+000	**0.000e+000**
Noncontinuous	Best	1.125e+000	0.000e+000	0.000e+000	0.000e+000
Rastrigrin	Worst	4.751e+000	1.065e-014	2.250e+000	0.000e+000
	Mean	2.714e+000	5.329e-016	1.096e+000	**0.000e+000**
	Std	8.705e-001	2.383e-015	5.907e-001	**0.000e+000**
Griewank	Best	0.000e+000	0.000e+000	0.000e+000	0.000e+000
	Worst	1.106e-001	3.199e-011	2.220e-002	0.000e+000
	Mean	2.120e-002	1.600e-012	3.600e-003	**0.000e+000**
	Std	8.566e-004	7.154e-012	5.900e-003	**0.000e+000**
Weierstrass	Best	2.261e+000	0.000e+000	2.645e-001	0.000e+000
	Worst	9.244e+000	0.000e+000	4.100e+000	0.000e+000
	Mean	6.433e+000	**0.000e+000**	2.859e+000	**0.000e+000**
	Std	3.176e+000	**0.000e+000**	8.228e-001	**0.000e+000**

As Rastrigrin is a highly multi-modal function, the difficulty of this function is the optimization algorithm can easily be trapped in a local optimum when searches for the global optimum. DEABC converges very fast to good values near the optimum, while DE, ABC and PSO-LIW straggle with premature convergence. From the Table 1 and Fig.3, We can see DEABC is superior to DE, ABC and PSO-LIW.

For the Noncontinuous Rastrigin function which is based on the Rastrigin function, DEABC performs significantly better than DE, ABC and PSO-LIW in Table 1.

Griewank is a multi-modal function with significant interaction between its variables. The multimodality disappears for high dimensionalities and the problem becomes unimodal. DE and ABC perform worse than DEABC. On all of algorithms DEABC clearly performs best and gives consistently an optimum result.

Weierstrass function is famous because it is continuous but differentiable only on a set of points. In the Table 1, ABC and DEABC obtain the best results of four algorithms.

It is shown in Table 1 that the standard deviation of DEABC is significantly low compared with DE, ABC, PSO-LIW except for PSO-LIW in Sphere function. This illustrates that the results generated by DEABC is robust. As DEABC possesses the best results of six functions, so it performs best of four methods. Once the evolution process of ABC is in stagnation state, the mutation of DE may be introduced to eliminate stagnation and avoid premature. From the perspective of DE, DEABC make full use of the best individuals of DE population and bee colony into the evolution. Thus, the evolution of population is no longer restricted in the same generation, and offspring which produced by the better performed individuals can replace the worse-performed ones.

5   Conclusions and Future Work

Based on Differential Evolution and Artificial Bee Colony algorithm, an integrated swarm intelligent approach is presented. Six benchmark functions have been used to test DEABC in comparison with ABC and DE. In DEABC, information exchanges among different populations avoid the individuals trapped in poor local minima. Population uses the global best to generate offspring every generation. Numerical results show that DEABC performs better than ABC and DE on these six benchmark functions. It can be concluded that by integration of the two methods can produce a hybrid optimization method possessing the advantages of both methods.

DEABC algorithm is only used for solving single-objective optimization problems in this paper. The future work is focused on solving multi-objective questions and some other real-world problems. In addition, different hybrid models of ABC and DE algorithm will be studied.

Acknowledgements. This work is supported by National Natural Science Foundation of China (Grant No.71001072, 61005010), China Postdoctoral Science Foundation (Grant No. 20100480705, 20100480708), Science and Technology Project of Shenzhen (Grant No. JC201005280492A), The Natural Science Foundation of Guangdong Province (Grant no. 9451806001002294), Soft Science Research Project of Guangdong Province (Grant no. 2010B070300106).

References

1. Karaboga, D.: An Idea Based on Honey Bee Swarm for Numerical Optimization. Technical Report TR06 (2005)
2. Karaboga, D., Ozturk, C.: A Novel Clustering Approach: Artificial Bee Colony (ABC) Algorithm. Applied Soft Computing 11, 652–657 (2011)
3. Karaboga, D., Basturk, B.: Artificial Bee Colony (ABC) Optimization Algorithm for Solving Constrained Optimization Problems. In: Melin, P., Castillo, O., Aguilar, L.T., Kacprzyk, J., Pedrycz, W. (eds.) IFSA 2007. LNCS (LNAI), vol. 4529, pp. 789–798. Springer, Heidelberg (2007)

4. Zhao, H.Y., Pei, Z.L., Jiang, J.Q., Guan, R.C., Wang, C.Y., Shi, X.H.: A Hybrid Swarm Intelligent Method Based on Genetic Algorithm and Artificial Bee Colony. In: Tan, Y., Shi, Y., Tan, K.C. (eds.) ICSI 2010. LNCS, vol. 6145, pp. 558–565. Springer, Heidelberg (2010)
5. Shi, X.H., Li, Y.W., Li, H.J., Guan, R.C., Wang, L.P., Liang, Y.C.: An Integrated Algorithm Based on Artificial Bee Colony Optimization and Particle Swarm Optimization. In: 2010 Sixth International Conference on Natural Computation (ICNC), pp. 2586–2590. IEEE Press, Yantai (2010)
6. Akay, B., Karaboga, D.: A Modified Artificial Bee Colony Algorithm for Real-Parameter Optimization. Information Sciences, 1–23 (2010)
7. Price, K.V.: An Introduction to Differential Evolution. In: Corne, D., Dorigo, M., Glover, F., Dasgupta, D., Moscato, P., Poli, R., Price, K.V. (eds.) New Ideas in Optimization, pp. 79–108. McGraw-Hill Ltd., London (1999)
8. Storn, R.: On the Usage of Differential Evolution for Function Optimization. In: Biennial Conference of the North American on Fuzzy Information Processing Society, pp. 519–523. IEEE Press, Berkeley (1996)
9. Niu, B., Zhu, Y.L., He, X.X.: MCPSO: A Multi-Swarm Cooperative Particle Swarm Optimizer. Applied Mathematics and Computation 185, 1050–1062 (2007)
10. Karaboga, D., Akay, B.: A Comparative Study of Artificial Bee Colony Algorithm. Applied Mathematics and Computation 214, 108–132 (2009)
11. Shi, Y.H., Eberhart, R.: A Modified Particle Swarm Optimizer. In: The 1998 IEEE International Conference on Evolutionary Computation Proceedings, pp. 69–73. IEEE Press, Anchorage (1998)

A Discrete Artificial Bee Colony Algorithm for TSP Problem

Li Li[1], Yurong Cheng[1], Lijing Tan[2], and Ben Niu[1,*]

[1] College of Management, Shenzhen University,
Shenzhen 518060, China
[2] Management School, Jinan University
Guangzhou 510632, China
drniuben@gmail.com

Abstract. In this paper, a new discrete artificial bee colony algorithm is used to solve the symmetric traveling salesman problem (TSP). The concept of Swap Operator has been introduced to the original artificial bee colony (ABC) algorithm which can help the bees to generate a better candidate tour by greedy selection. By taken six typical TSP instances as examples, the proposed algorithm is compared with particle swarm optimization (PSO) algorithm to validate its performance. In the experimental study, we also analysis the important parameters in the artificial bee colony algorithm and their influence have been verified.

Keywords: Artificial bee colony, swap operator, traveling salesman problem.

1 Introduction

The Traveling Salesman Problem (TSP) is one of the most widely studied combinatorial optimization problems. It is a NP-hard problem whose computational complexity rises exponentially when the number of cities increasing.

Many exact and heuristic algorithms have been proposed to solve this problem in recent years. The exact methods include branch-and-bound and enumeration and cutting planes, while the heuristic algorithms consists of Particle Swarm Optimization [2], Neural Network [4], Ant Colony Optimization [5], Genetic Algorithms [6], Simulated Annealing [7] and so on. As swarm intelligence draws more and more interest to the scientists in related fields, new algorithms such as artificial bee colony (ABC) algorithm emerged. It was developed by Karaboga in 2005 in order to optimize multi-variable and multi-modal continuous functions [9] through imitating the specific intelligent foraging behaviors of bee swarms. The ABC algorithm was initially designed for numerical optimization problems. And numerical comparison results demonstrated that the performance of ABC algorithm is competitive to other swarm based algorithms. Because of its simplicity and easy implementation, the ABC algorithm has attracted much more attention and has been applied to many practical optimization problems. Till now, the ABC algorithm has been proven to succeed in continuous problems.

---

* Corresponding author.

D.-S. Huang et al. (Eds.): ICIC 2011, LNBI 6840, pp. 566–573, 2012.
© Springer-Verlag Berlin Heidelberg 2012

However, little work has been done to use ABC algorithm in some real-world applications for discrete problems. Based on this reason, we are presenting a new discrete ABC algorithm by applying Swap Operator [2] to solve the TSP problem and make a broad advancement to extend ABC to the domain of combinatorial problems. To validate the proposed algorithm, computational simulations were performed in comparison with Particle Swarm Optimization (PSO) algorithm. The results showed that the proposed discrete ABC algorithm outperforms the PSO algorithm for solving the TSP problem.

2   The Basic Artificial Bee Colony Algorithm

The Artificial Bee Colony (ABC) algorithm was proposed by Karaboga [1] based on the foraging behaviour of honey bees in 2005. In ABC algorithm, the colony of artificial bees contains three groups of bees: employed bees, onlookers and scouts. A bee visits the food source by itself is an employed bee while a bee waiting on the dance area is an onlooker. A scout carries out random search for a new food source. The colony is equally separated into employed bees and onlookers. The number of food sources is set equal to the number of employed bees. The employed bee whose food source is abandoned by the employed bees and onlookers becomes a scout and only one scout is allowed to occur in each cycle.

There are three important steps in basic ABC algorithm. In the first step, generate initial food source positions randomly. In order to update feasible solutions, all employed bees select a new candidate food source position, which is different from the previous one. The position of the new food source is calculated by the following equation:

$$v_{ij} = x_{ij} + r(x_{ij} - x_{kj})$$ (1)

In the equation (1), v_{ij} is a new feasible solution that is modified from its previous solution (x_{ij}) based on a comparison with its neighboring solution (x_{kj}). r is a random number between [-1,1]. $k \in \{1,2,3,\cdots,SN\}$, $j \in \{1,2,3,\cdots,D\}$ and SN is the number of food sources and D is the dimension of problem. The new source replaces the previous one in the employed bee's memory if it's better than previous position, otherwise keep the position of the previous one.

In the second step, each onlooker selects one of the proposed food sources obtained from the employed bees by using roulette wheel rule. The probability that a food source will be selected can be obtained from an equation below:

$$P_i = \frac{fit_i}{\sum fit_n}$$ (2)

where fit_i is the fitness value of the food source i . After selecting the food source, onlooker goes to the selected food source and selects a new candidate food source.

In the last step, *limit* is a predetermined number of cycles in ABC algorithm and it controls the times of a certain solution which has not updated. Any food source that does not improve over *limit* will be abandoned and replaced by a new position and the employed bee becomes a scout. The new random position chosen by the scout will be calculated by the equation below:

$$x_{ij} = x_{min}^j + rand(0,1)(x_{max}^j - x_{min}^j) \tag{3}$$

where x_{min}^i is the lower bound of the food source position in dimension j and x_{max}^j is the upper bound of the food source position in dimension j.

3   Discrete Artificial Bee Colony Algorithm for TSP Problem

This study illustrates a novel ABC algorithm for solving TSP problem. ABC algorithm was initially designed for solving numerical optimization problems which are continuous problems and had been proved successful. To make basic ABC algorithm available for discrete problems, the concepts of Swap Operator and Swap Sequence [2] are introduced here.

Let a normal solution sequence of a TSP problem with n cities, $S = (a_i), i = 1,2,\cdots,n$ be given. Now prescribe the meaning of Swap Operator, which was proposed by Kangping Wang, et al. firstly in 2003. $SO(i_1, i_2)$ is a process of exchanging city a_{i1} and city a_{i2} in solution S. Then define $S_{new} = S + SO(i_1, i_2)$ as a new solution on which operator $SO(i_1, i_2)$ acts.

A concrete example can be drawn as follows. Consider there is a TSP problem with ten cities, this is one solution: $S = (3,5,1,8,6,10,2,7,4,9)$ and the Swap Operator is $SO(4,9)$, then

$$S_{new} = S + SO(4,9) = (3,5,1,8,6,10,2,7,4,9) + SO(4,9) = (3,5,1,4,6,10,2,7,8,9)$$

And a Swap Sequence SS consists of one or more Swap Operators. That is $SS = (SO_1, SO_2, SO_3, \cdots, SO_n)$. Where $SO_1, SO_2, \cdots, SO_n$ are Swap Operators and the sequence of the Swap Operators in SS is very important. Swap Sequence working on a solution means all the Swap Operators in the Swap Sequence act on the solution one after another. This can be presented by:

$$S_{new} = S + SS = (SO_1, SO_2, SO_3, \cdots, SO_n) = ((S + SO_1) + SO_2) + \cdots + SO_n \tag{4}$$

Swap Operator is applied during the solution updating step in basic ABC algorithm, with the additional benefit of improved performance. The significance of Swap Operator is to help the bees to generate a better candidate tour by greedy selection. This method can improve the precision of the results. In the new discrete ABC algorithm for TSP problem, a solution refers to a tour consisting of all the cities which are arrayed differently and the fitness value of the solution refers to the length of the tour. The optimal value is the lowest length of all the possible tours and is called the optimal tour.

Having discussed the basic ABC algorithm and the concept of Swap Operator, the computational procedures of the proposed discrete ABC algorithm for TSP problem can be outlined as follows:

1. Initialize all the parameters: the number of colony size (*Colony*), *limit*, *Maxcycle*, *runtime*, initial tours and so on.

2. Perform Swap Operator processes to each employed bee: (1) Randomly create a Swap Sequence for each employed bee, that is, $SS_j = (SO_{j1}, SO_{j2}, SO_{j3}, \cdots, SO_{jm})$ where $j = 1, 2, \cdots, n$ and n stands for the number of employed bees and m is the number of Swap Operators in each Swap Sequence. (2) Create a new neighbour tour by the previous tour and *SS*. (3)Evaluate the new tour and update *SS* and the number of *trail*, which record the cycle number of all unimproved tours.

3. Perform Swap Operator processes to each onlooker bee: (1)Select tour from employed bees with a probability calculated by the equation (2) by using the common roulette wheel rule. (2)Create a new neighbor tour by the selected tour and updated *SS*. (3)Evaluate the new tour and update *SS* and the number of *trail*.

4. Calculate the number of *trail* and record the maximum number as *M*. If *M>limit*, then abandon that unimproved tour and randomly create a new tour for the scout. If *M<limit*, then the scout bee takes the previous tour.

5. Judge the stop criterion, namely, whether the iteration reaches the given number of *Maxcycle* or the best fitness reaches the designated value. If the criterion is satisfied then stop the program, else go to step 2.

4   Experimental Study

All the experiments are performed on a PC (Intel I3-370M CPU, 2 GB of memory, Win2007, Matlab 7.0). The algorithm described in the paper is tested on a set of benchmark problems taken from TSP library (TSPLIB), such as BURMA14, BAYS29, DANTZIG42, BERLIN52, KROA100 and CH130.

To confirm the importance of some parameters to the performance of ABC algorithm and make a comparison of proposed discrete ABC algorithm with other heuristic algorithms, two experiments have been done separately in this section.

4.1   Experiment I: The Importance of Parameters: *Nse* and *Limit*

In the first experiment, BURMA14 and DANTZIG42 are used as typical examples. The number of Swap Operators in a Swap Sequence, named as *Nse*, is important to the performance of the discrete ABC algorithm. In this experiment, the number of the colony is set equal to 2*N (N refers to the size of TSP problem) and the *Maxcycle* is 200 and *limit* is 50. The BURMA14 results with *Nse* set to be 1, 5, 7, 10 and 14 are shown in Fig. 2. Fig. 3 presents DANTZIG42 results with *Nse* set to be 5, 10, 21, 32 and 42.

In addition, *limit* is a special control parameter in ABC algorithm. In this part, *Nse* is equal to *N* in BURMA14 problem and *Nse* is set to *N/2* in DANTZIG42 problem. *Limit* has been set separately to be 10, 20, 50, 100 and 120 in both of the problems. And the results are presented in Fig. 4 and Fig. 5.

Figs 2-3 demonstrate that the results are worse and the best lengths are very long when *Nse* are too small. To BURMA14, it can obtain better tour when *Nse* is half of the problem size or above, which can be seen in Fig. 2.

However, when *Nse* is half of the problem size, optimal tour is obtained to DANTZIG42. It gets worse when *Nse* is higher and more time is needed to search in the space. This makes the performance not comparable to the result when *Nse* is 21. All these observations show that *Nse* should not be configured either too high or too low. To achieve a reasonable balance between quality of solution and computational cost, *Nse* should be set a little high to small-size problems and set almost half of scale of the problem to high-scale problems.

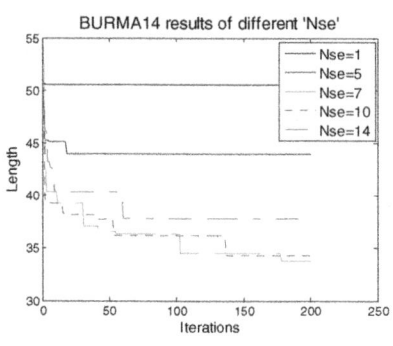

Fig. 2. BURMA14 results

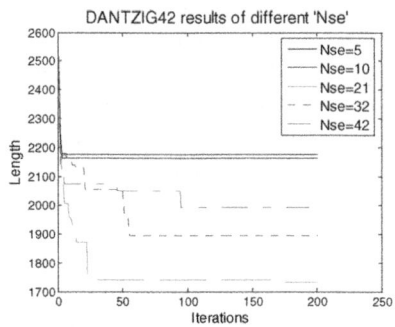

Fig. 3. DANTZIG42 results

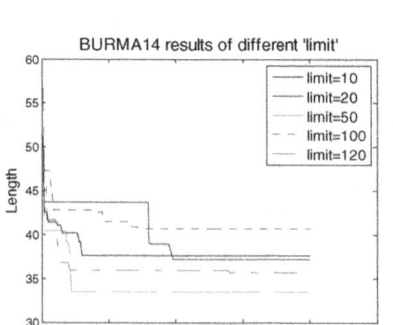

Fig. 4. BURMA14 results

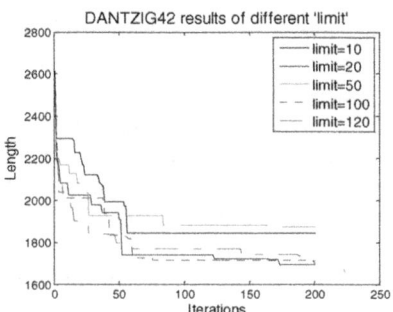

Fig. 5. DANTZIG42 results

Parameter *limit* presents the times of a certain unimproved solution. So by setting *limit* to a low value, it is expected that it can contribute to the optimal tours. However, the experiments do not show the expected results. In Fig. 4 and Fig. 5, the results of different *limit* are quite stochastic and the consequences at the same *limit* have too big difference to different problems. A conclusion can be drawn that *limit* has little influence on the performance of the algorithm. The reason may lie in that there is only one employed bee turn to be scout to carry out random search in each cycle.

4.2   Experiment II: The Comparison of DABC and PSO

This experiment compares the discrete ABC algorithm (abbreviated as DABC) with PSO algorithm which is a swarm based algorithm and performs well in numerical optimization problems. The performance of these two algorithms has been tested by taking BURMA14 and DANTZIG42 as examples. To make them comparable, initial parameters setting (table 1) were the same for both DABC algorithm and PSO algorithm [2]. *Colony* stands for the number of the population and *Runtime* is the repetition times of the program.

Table 2 presents the summary results of the computational experiments of the two algorithms. In the table, *Scale* refers to the size of different TSP problems, *Best* is the shortest tour length found in twenty runtimes and *Average* is the average closed tour length among twenty trials. *Worst* stands for the longest length found among twenty runtimes. *Error* is calculated by the equation: Error= (Average- Best)/ Best*100%. And it denotes the percent difference of the average tour length.

The experimental results show that the performance of ABC algorithm integrated with Swap Operator outperforms the PSO algorithm with the same discrete mechanism. The difference of the obtained average tour length of the algorithm is little if the indicator of *Error* is small. It is obviously that there are small differences between twenty optimal tours obtained respectively in twenty runtimes.

Table 1. Initialized parameters

Algorithms	Colony	Runtime	Maxcycle	Limit	Nse	α	β
DABC	2*N	20	1000	100	N/2	no	no
PSO	2*N	20	1000	no	N/2	0.25	0.25

Table 2. Summary results of the computational experiment

Problem	Scale	Algorithms	Best	Average	Worst	Error (%)
BURMA14	14	DABC	**34.4838**	34.7607	35.9549	0.80
		PSO	35.3788	36.9253	38.8274	4.37
BAYS29	29	DABC	**16024**	16751	17221	4.54
		PSO	16995	17953	18533	5.64
DANTZIG42	42	DABC	**1622.6**	1733.1	1835.2	6.41
		PSO	1933.8	1987.7	2038.6	2.79
BERLIN52	52	DABC	**21031**	21338	21768	1.46
		PSO	21274	21827	22532	2.60
KROA100	100	DABC	**129110**	129960	131720	0.66
		PSO	130210	132400	135450	1.68
CH130	130	DABC	**37039**	37243	38121	0.55
		PSO	37079	37982	38713	2.44

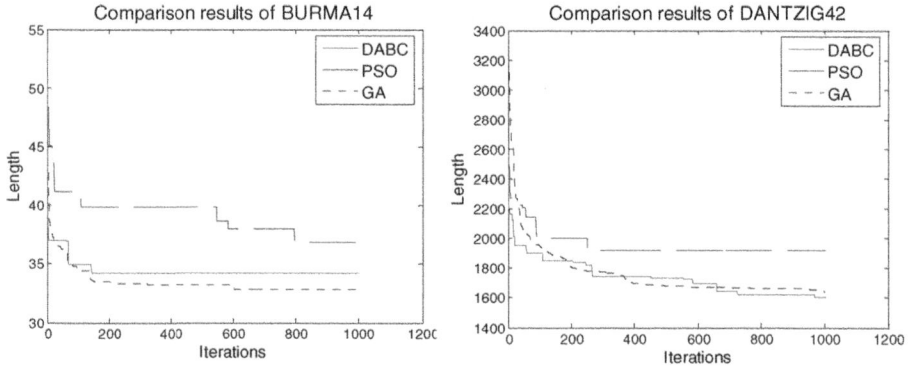

Fig. 6. BURMA14 results   **Fig. 7.** DANTZIG42 results

To confirm the efficiency of the proposed algorithm further, it is also compared with another traditional heuristic algorithm, Genetic Algorithm (GA) [6]. Fig. 6 presents the convergence efficiency of DABC, PSO and GA to the problem of BURMA14. Together with a problem of DANTZIG42, comparison results are shown in Fig. 7. They imply that the three algorithms converge to an optimal value quickly and the convergence rate of the new algorithm is high. In Fig. 6 and Fig. 7, the results of the DABC algorithm are worse than GA sometimes but the differences are small. So the new algorithm is competitive to GA.

5   Conclusions

In this paper, a new discrete ABC algorithm is proposed to solve the TSP problem. Swap Operator is applied to the basic ABC algorithm and six typical TSP problems are chosen to evaluate the performance of the new algorithm. The simulation results indicate that, in comparison to PSO algorithm, the new discrete ABC algorithm could provide more effective results for TSP problem.

Although the results are not the best known optimal tour costs for these instances, applying ABC algorithm to TSP problem is still a new attempt and is also applicable to other combinatorial problems. On the other hand, the discrete ABC algorithm performs well in the first period, but it fails to make further progress at the final stage and it reaches local optima easily. Future work should focus on enhancing population diversity and global search ability of the algorithm to improve its accuracy.

Acknowledgements. This work is supported by National Natural Science Foundation of China (Grant No.71001072, 61005010), China Postdoctoral Science Foundation (Grant No. 20100480705, 20100480708), Science and Technology Project of Shenzhen (Grant No. JC201005280492A), The Natural Science Foundation of Guangdong Province (Grant no. 9451806001002294), Soft Science Research Project of Guangdong Province (Grant no. 2010B070300106).

References

1. Karaboga, D.: An Idea Based on Honey Bee Swarm for Numerical Optimization. Technical Report (TR06), Computer Engineering Department, Erciyes University, Turkey (2005)
2. Wang, K.P., Huang, L., Zhou, C.G., Pang, W.: Particle Swarm Optimization for Traveling Salesman Problem. In: 2nd IEEE International Conference on Machine Learning and Cybernetics, pp. 1583–1585. IEEE Press, Xi'an (2003)
3. Marinakis, Y., Marinaki, M., Dounias, G.: A Hybrid Particle Swarm Optimization Algorithm for the Vehicle Routing Problem. Engineering Applications of Artificial Intelligence 23, 463–472 (2010)
4. Masutti, T.A.S., Castro, L.N.D.: A Self-Organizing Neural Network Using Ideas from the Immune System to Solve the Traveling Salesman Problem. Information Sciences 179, 1454–1468 (2009)
5. Tsai, C.F., Tsai, C.W., Tseng, C.C.: A New Hybrid Heuristic Approach for Solving Large Traveling Salesman Problem. Information Sciences 166, 67–81 (2004)
6. Heinrich, B.: On Solving Traveling Salesman Problems by Genetic Algorithms. In: Schwefel, H.-P., Männer, R. (eds.) PPSN 1990. LNCS, vol. 496, pp. 129–133. Springer, Heidelberg (1991)
7. Meer, K.: Simulated Annealing versus Metropolis for a TSP Instance. Information Processing Letters 104, 216–219 (2007)
8. Banharnsakun, A., Achalakul, T., Sirinaovakul, B.: ABC-GSX: A Hybrid Method for Solving the Traveling Salesman Problem. In: 2nd IEEE World Congress on Nature and Biologically Inspired Computing, pp. 7–12. IEEE Press, Fukuoka (2010)
9. Karaboga, D., Basturk, B.: A Powerful and Efficient Algorithm for Numerical Function Optimization: Artificial Bee Colony (ABC) Algorithm. Journal of Global Optimization 39, 459–471 (2007)
10. Karaboga, D., Basturk, B.: On the Performance of Artificial Bee Colony (ABC) Algorithm. Applied Soft Computing 8, 687–697 (2008)

Restoration of Epipolar Line Based on Multi-population Cooperative Particle Swarm Optimization

Hongwei Gao[1,4], Xiaofeng Liu[2], Jinguo Liu[4], Fuguo Chen[1], and Ben Niu[3,*]

[1] School of Information Science & Engineering, Shenyang Ligong University,
Shenyang, 110159, China
[2] Jilin Provincial Institute of Education, Changchu, 130022, China
[3] College of Management, ShenzhenUniversity, Shenzhen, 518060, China
[4] State Key Laboratory of Robotics, Shenyang Institute of Automation,
Chinese Academy of Sciences, Shenyang 110016, China
drniuben@gmail.com

Abstract. A high precision epipolar line restoration algorithm based on Multi-population Cooperative PSO (MCPSO) is proposed in this paper. It adopts Harris operator to extract corner point and finishes gray cross -correlation matching. Firstly, the fundamental matrix initial value between matches in two images is calculated by 8 pairs matches algorithm. And then the optimal value of this matrix is gotten by MCPSO and PSO respectively based on the object function which is the distance between the point and corresponding polar line. Finally, the experiment results prove the validity and practicability of the proposed method.

Keywords: Epipolar line, Fundamental matrix, PSO, MCPSO.

1 Introduction

For a pair of related views obtained from binocular vision system the epipolar geometry provides a complete description of relative camera geometry [1]. It can be computed from a certain number of point correspondences obtained from the pair of images independently of any other knowledge about the world. Epipolar geometry can be represented by a fundamental matrix (F matrix) in algebra, dimension of the matrix is defined as 3×3 and rank of which is 2 [2]. The simple method is to select more than 8 matches manually and uses 8 point algorithm to calculate F matrix [3, 4]. But this method is not accurate for the precision of corner point coordinate. The practical method is restoration online [5] which uses two cameras to acquire some pair of real scene images. Firstly, the corner points are detected, and then initial matching is finished by cross-correlation algorithm and relaxation algorithm, finally, nonlinear optimization algorithm is applied to get more accurate F matrix. Binocular vision can be divided into parallel binocular vision and non- parallel binocular vision according to the optical axis of two cameras, and the former is widely used such as the binocular vision on the Mars rover. The epipolar geometry restoration for binocular vision is

---

[*] Corresponding author.

D.-S. Huang et al. (Eds.): ICIC 2011, LNBI 6840, pp. 574–581, 2012.

investigated in this paper. The initial value of F matrix is gotten by 8 point algorithm, and then the object function is defined as the distance between the point and corresponding polar line, subsequently, the optimal value of this matrix is gotten by MCPSO and PSO respectively, finally, the experiment results prove the validity of the related algorithms.

2   Epipolar Geometry for Binocular Vision

The parallel binocular vision system is characterized by its parallel optical axis of two cameras. Figure 1 shows the configuration of the parallel binocular vision system. There exists only translation between two cameras along the x direction, where the translation b is the length of baseline. The image coordinate along the v direction of the matching points is equal while u coordinate is not same. $u_1 - u_2$ is defined to be disparity. The epipolar lines are parallel with the camera scanlines. Epipoles in two images are all at infinity. As shown in figure 1, E_{1i} and E_{2i} represent two epipolar lines, $P(x_{1i}, y_{1i}, z_{1i})$ is any single physical point in space, the corresponding projective points in two image are p_{1i} and p_{2i}.

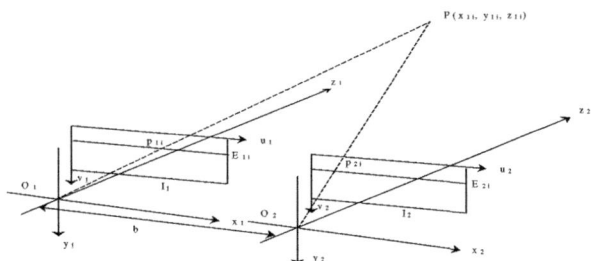

Fig. 1. Configuration of the parallel binocular vision system

3   Feature Extraction and Matching

Harris [5] and Stephens put forward a point feature extraction operator based on static image which is called Harris operator. According to the definition of Harris operator, for any point in the image, if its horizontal curvature value and vertical curvature value are bigger than the other points in the neighborhood, then this point is considered as a corner point. Generally speaking, the corner points are detected by Harris operator in pixel precision, then sub-pixel precision corner points are gotten by interpolation. According to corner points extracted in two images acquired by binocular vision, matching process can be completed by cross-correlation algorithm [6].

4   Estimation of F Matrix

4.1   Epipolar Line Equation and Initial Value for F Matrix

The epipolar line equation is as follows:

$$\begin{bmatrix} u_1 & v_1 & 1 \end{bmatrix} \begin{bmatrix} F_{11} & F_{12} & F_{13} \\ F_{21} & F_{22} & F_{23} \\ F_{31} & F_{32} & F_{33} \end{bmatrix} \begin{bmatrix} u_2 \\ v_2 \\ 1 \end{bmatrix} = 0 \tag{1}$$

Where F_{11}~F_{32} are the unknown variable in the F matrix, it can be calculated by at least 8 pairs of matching points. In order to promote the F matrix accuracy, we use more than 8 pairs of matching points and calculate it by least square algorithm.

4.2   Nonlinear Optimization Estimation for F Matrix

Nonlinear iterative optimization algorithm is adopted here in order to achieve higher estimation accuracy. In figure 1, m_i is defined as image coordinate of p_{1i} in left image, m_i' represents the matching point image coordinate of p_{2i} in right image of p_{1i}. $l_i' = Fm_i$ is the epipolar line equation in right image of point m_i, $l_i = F^T m_i'$ is the epipolar line equation in right image of point m_i'.

$$d_i'(m_i', l_i') = \frac{\left| m_i'^T F m_i \right|}{\sqrt{\alpha_i'^2 + \beta_i'^2}} \tag{2}$$

$$d_i(m_i, l_i) = \frac{\left| m_i^T F^T m_i' \right|}{\sqrt{\alpha_i^2 + \beta_i^2}} \tag{3}$$

$$r_i = \left| m_i'^T F m_i \right| \tag{4}$$

$$D = \min\left(\sum_{i=1}^{n} (d_i^2(m_i, l_i) + d_i'^2(m_i' + l_i'))\right) \tag{5}$$

Equation (2) and (3) represent the distance from the point m_i' and m_i to its corresponding epipolar line respectively. Equation (4) is the epipolar line equation residue. Then the estimation of fundamental matrix is converted to solve the minimal value of equation (5).

The corresponding nonlinear iterative optimization algorithm is as follows:

(1) Calculate the initial value of fundamental matrix by least square algorithm using all corners.

(2) Take the initial value of fundamental matrix into equation (5) to achieve optimal values by nonlinear optimization.

(3) Take the rank 2 constraint into consideration. Make singular value decomposition to fundamental matrix. Set $F = UDV^T$, where $D = \text{diag}(r_1, r_2, r_3)$, $r_1 \geq r_2 \geq r_3$, meanwhile, set $F' = U\text{diag}(r_1, r_2, 0)V^T$, then the new fundamental matrix F is replaced by F' .

(4) Compare the sequent two fundamental matrix, if $\|F'-F\| < \delta$ is satisfied, then finish the calculation, else turn to step (2).

Threshold δ is usually considered to be 0.05. The smaller δ is, the more the circular times of calculation is, but the calculation result will be more accurate.

4.3   F Matrix Optimization Based on PSO

The particle swarm optimization (PSO) algorithm stems from researching the predation behavior of birds; it is an iterative optimization tool which is similar with genetic algorithm (GA) [7]. Initialize PSO as a group of random particle (random solution), the particles update themselves by tracing two extreme value. The first is the best solution found by particle itself and named individual extreme point(its position is expressed by $pbest$).There is another extreme point, called global extreme point(its position is expressed by $gbest$) in the global version PSO, is the best solution which is found in the entire swarm at present. After the two best solution are found, the particles update equation (6) and (7) according to the following speed and position so that update their own speed and position.

$$v_{id}^{k+1} = v_{id}^k + c_1 rand_1^k \left(pbest_{id}^k - x_{id}^k\right) + c_2 rand_2^k \left(gbest_d^k - x_{id}^k\right) \tag{6}$$

$$x_{id}^{k+1} = x_{id}^k + v_{id}^{k+1} \tag{7}$$

4.4   Fmatrix Optimization Based on MCPSO

What have been proposed and set by MCPSO is a supposition which is shared in society based on the same kind of information. It has reflected the individual (fish, bird, insect) cooperation relation in a group (school, herd).It's obvious that it is not the relation of natural attribute. Many species evolve their surviving by collaborating (within their own species) and interacting with other species in natural ecosystem. The collaboration can be found in the organism from cell(for example, eukaryotic organisms produced by mutualistic symbiosis and interaction between the prokaryotic cells and other removed cells by them) to superior being(for example, African ticking bird gets stable food source by clearing the parasite on giraffe, zebras and other animals), the collaboration is also called co-evolution symbiosis [8].

The master-slave model is introduced to PSO in the paper which is aroused by symbiotic relationship of natural ecological system, then multi-swarm (species) cooperative optimization (MCPSO) will be generated [9]. In the method, the swarm consists of master swarm and slave swarm. The symbiotic relationship of master-slave will balance detection and development which is crucial for optimization.

Figure2 shows the relationship model between master and slave, and the model has assigned a suitable evaluation and maintained the algorithm synchronization.

The slave swarm is an independent swarm (species) from each other which connected by "node". Each node executes a sole PSO or its variant and it includes updating position and speed and the generation of the new local swarm. Each node sends the best individual in the local area to the master swarm node when all the nodes are ready for the new generation. Master swarm nodes choose the best individual of all the nodes being sent and evolve(evolution) according the following formula:

$$v_i^M(t+1) = wv_i^M(t) + R_1c_1(p_i^M - x_i^M(t)) + R_2c_2(p_g^M - x_i^M(t)) + R_3c_3(p_g^s - x_i^M(t)) \tag{8}$$

$$x_i^M(t+1) = x_i^M(t) +_i^M(t) \tag{9}$$

M is master swarm, c_3 is transport coefficient, R_3 is an uniform random sequence varying from [0,1]. The speed flag of master swarm is related to three factors:

i. p_{id}^M : Previous best position of the master swarm.

ii. p_{gd}^M : Best global position of the master swarm.

iii. p_{gd}^S : Previous best position of the slave swarms.

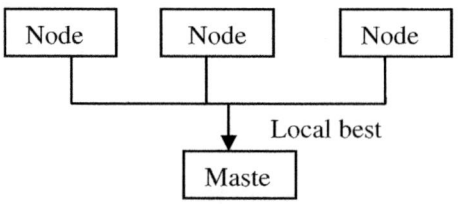

Fig. 2. The master-slave model

It is shown in equation (8) that the first term is the sum of inertia (the particles go on moving along their original direction), and the second term represents the memory (the optimal point of particle attracted to the track), the third represents collaboration (master swarm find all the particles that attracted to optimal position), the last term represents information transformation (slave swarm find all the particles that attracted to optimal position). Pseudo code of MCPSO algorithm can be found in literature [9].

5   Simulation and Experiment Results

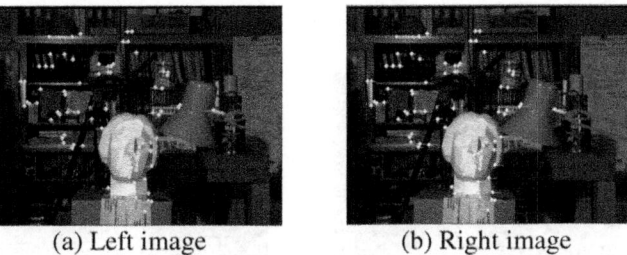

(a) Left image   (b) Right image

Fig. 3. The grey cross-correlation matching results

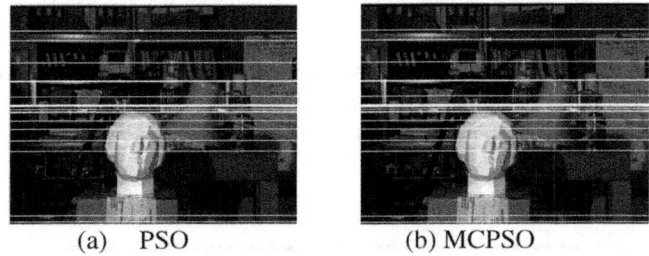

(a)   PSO   (b) MCPSO

Fig. 4. Epipolar line with fundamental optimized by PSO and MCPSO

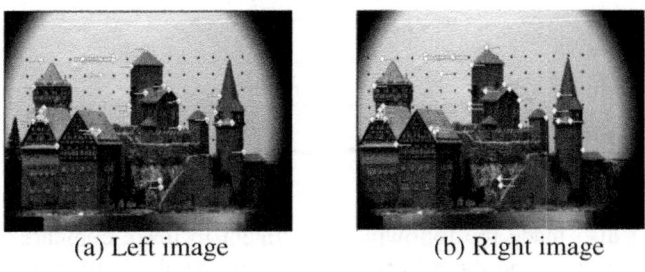

(a) Left image   (b) Right image

Fig. 5. The grey cross-correlation matching results

Two pairs of standard test stereo images are downloaded from internet to test the related algorithms. Take the Head portrait image as example, the feature points extracted by Harris operator distribute the whole image in uniformity which can be seen from the left and right images. There are 68 pairs of matches in the images, the line direction is the disparity direction and length of it represents the disparity value.

Some restored epipolar lines in the left image according to the F matrix optimized by PSO and MCPSO are shown in figure 4. We can see that the precision of F matrix optimized by tow algorithms is similar and all the epipolar lines are parallel with scanlines which illustrates the accuracy of 8 points algorithm and the validity of the optimal strategy. The number of epipolar lines in two images is almost equal.

Take the Castle image as example, the feature points extracted by Harris operator also distribute the whole image in uniformity and there are 33 pairs of matches in the images which can be seen form the images.

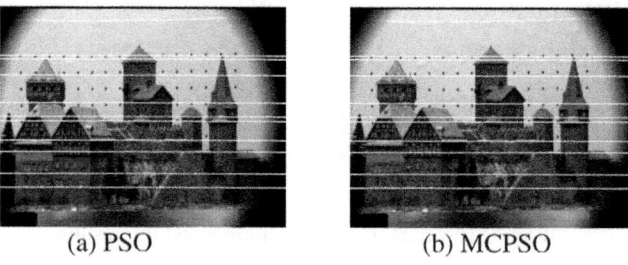

(a) PSO   (b) MCPSO

Fig. 6. Epipolar line with fundamental optimized by PSO and MCPSO

We can see from figure 6 that the precision of F matrix optimized by tow algorithms is similar and all the epipolar lines are parallel with scanlines. The number of epipolar lines in two images is also almost equal.

Table 1. The Comparison between PSO and MCPSO for different images

		Equation residual	Iteration
Head	PSO	16.52	30
portrait	MCPSO	15.86	50
Castle	PSO	18.27	20
image	MCPSO	18.01	40

The author also made the following experiment to make comparison on equation residual and iteration, and the results are shown in the above figures and table 1. For the Head portrait image, it shows that the descent speed of residual is slower in MCPSO than in PSO, and the residual of MCPSO tends to stable about 50th iteration, and the residual of PSO tends to stable about 30th iteration. The difference is determined by master-slave swarm structure so that the MCPSO is more time-consuming. However, the equation residual with MCPSO is 15.86 less than 16.52 with traditional PSO. This demonstrates that the capacity of searching optimal solution is better in MCPSO than in PSO. For the Castle image, the descent speed of residual with MCPSO is also slower than that with PSO, and the residual of MCPSO tends to stable about 40th iteration, and the residual of PSO tends to stable about 20th iteration. The equation residual with MCPSO is 18.01 less than 18.27 with traditional PSO.

6   Conclusions

On the basis of Harris corner point detection and matching, the initial value of F matrix is calculated by 8 points algorithm in this paper. Then the object function is formed by the distance between the point and corresponding polar line, and the optimization of F matrix is executed by PSO and MCPSO. The performance between the two optimal algorithms is compared in the experiment process and the results show that the proposed algorithm is effective and practical.

Acknowledgement. This work is supported by National Natural Science Foundation of China (Grant No.71001072), China Postdoctoral Science Foundation (Grant No. 20100480705), Science and Technology Project of Shenzhen (Grant No. JC201005280492A), Advanced Manufacturing Technology R&D Base Foundation of Chinese Academy of Sciences (No. Y0F7010701) and the CAS President's Award Winner Foundation.

References

1. Chen, Z.Z., Wu, C.K.: The Accurate Estimation of F Matrix. Pattern Recognition and Artificial Intelligence 13, 146–150 (2000) (in Chinese)
2. Ma, S.D., Zhang, Z.Z.: Computer Vision-calculation Theory and Algorithm. Science Press, Beijing (2003)
3. Longuet-Higgins, H.C.: A Computer Algorithm for Reconstructing a Scene from Two Projections. Nature 293, 133–135 (1981)
4. Hartley, R.: In defense of the eight-point algorithm. In: Proceedings of the 5th International Conference on Computer Vision, pp. 1064–1070. IEEE Computer Society Press, Boston (1995)
5. Harris, C.G., StePhens, M.: A Combined Corner and Edge Detector. In: Proceedings of the 4th Alvey Vision Conference, Manchester, pp. 147–151 (1988)
6. Gao, H.W., Wu, C.D., Li, B.: 3D Reconstruction Based on a Fast Area-based Stereo Matching Algorithm. In: Proceedings of the 6th World Congress on Control and Automation, Dalian, China, June 21 - 23, pp. 9542–9546 (2006)
7. Eberhart, R.C., Kennedy, J.: A new optimizer using particle swarm theory. In: Proceedings of the Sixth International Symposium on Micro Machine and Human, pp. 39–43 (1995)
8. Wiegand, R.P.: An analysis of cooperative co-evolutionary Algorithms. PhD thesis, George Mason University, Fairfax, Virginia, USA (2004)
9. Ben, N., Zhu, Y.L., He, X.X.: Construction of fuzzy models for dynamic systems using multi-population cooperative particle swarm optimizer. In: Wang, L., Jin, Y. (eds.) FSKD 2005. LNCS (LNAI), vol. 3613, pp. 987–1000. Springer, Heidelberg (2005)

Multi-objective Optimization Using BFO Algorithm

Ben Niu[1,2,4], Hong Wang[2], Lijing Tan[3], and Jun Xu[4]

[1] Hefei Institute of Intelligent Machines, Chinese Academy of Sciences, Hefei 230031, China
[2] College of Management, Shenzhen University, Shenzhen 518060, China
[3] Management School, Jinan University, Guangzhou 510632, China
[4] e-Business Technology Institute, The University of Hongkong, Hongkong, China
drniuben@gmail.com

Abstract. This paper describes a novel bacterial foraging optimization (BFO) approach to multi-objective optimization, called Multi-objective Bacterial Foraging Optimization (MBFO). The search for Pareto optimal set of multi-objective optimization problems is implemented. Compared with the proposed algorithm MOPSO and NSGAII, simulation results (measured by Diversity and Generational Distance metric) on test problems show that the proposed MBFO is able to find a much better spread of solutions and faster convergence to the true Pareto-optimal front. It suggests that the proposed MBFO is very promising in dealing with multi-objective optimization problems.

Keywords: Bacterial foraging algorithm, multi-objective optimization, pareto optimal.

1 Introduction

Multi-objective optimization (MO) problems [1] contain more than one objective that needs to be achieved simultaneously. Such problems arise in many applications, where two or more, sometimes competing and/or incommensurable objective functions have to be minimized concurrently.

There are several GA or evolutionary based methods for solving Multi-Objective Problems, such as Vector Evaluated Genetic Algorithm (VEGA) [2], Weight-based Genetic Algorithm (WBGA) [3], Non-dominated Sorting Genetic Algorithm (NSGA) [4], Strength Pareto Evolutionary Algorithm (SPEA) [5], Pareto Archive Evolution Strategy (PAES), and enhanced NSGA named NSGA-II [6]. As numerous GA based approaches were proposed, many researchers are interested in employing interesting in particles swarm intelligence (PSO) to solve multi-objective problems [7]. Although the GA has a wide searching ability of avoiding the local search and can increase the probability of finding the global best. It converges slower and spends more time to generating new offspring. On the other hand, the PSO has the ability to quickly converge to a reasonably acceptable solution, but it may be trapped in the local optimum while solving complex problems due to the searching behavior of the particles.

This paper proposes the multi-objective optimization based BFO for dealing with the problems mentioned above. The experimental results proved that the proposed

D.-S. Huang et al. (Eds.): ICIC 2011, LNBI 6840, pp. 582–587, 2012.

method can find better solutions when compared to other approaches. The proposed method has more capabilities to find more solutions located on/near the Pareto front for much kind of multi-objective problems.

The rest of the paper is organized as follows: Section 2 describes the original BFO methodology, Section 3 presents the proposed method, Section 4 presents the experimental results and Section 5 of the paper contains the conclusion

2   The Bacterial Foraging Algorithm

Bacterial Foraging Optimization (BFO) algorithm has been applied to model the E. coli bacteria foraging behavior for solving optimization problems. It is known that bacteria swim by rotating whip-like flagella driven by a reversible motor embedded in the cell wall. For E. coli have 8-10 flagella placed randomly on a cell body. When all flagella rotate counterclockwise, they form a compact, helically propelling the cell along a helical trajectory, which is called run. When the flagella rotate clockwise, they all pull on the bacterium in different directions, which causes the bacteria to tumble [8]. The cycle of optimization can be divided into three parts: Chemotaxis, Reproduction, Elimination and Dispersal. A brief introduction of these processes is given as follows:

Researchers have found that the movement of bacteria is accomplished by flagellum. The flagella can simultaneously alternate between moving clockwise and counter-clockwise so that the bacterium will alternatively tumble and swim. But between two modes will move the bacterium, in random directions, and this enables it to "search" for nutrients. The behavior of bacterial foraging process (such as swimming, tumbling) can be seen as the process of optimization called "Chemotaxis". After a period of chemotaxis, some bacteria distinctly have no advantage of searching for nutrition resulting departing from Pareto front. Therefore, it's essential to wash out some poor ones. The process of reproduction keeps the good individuals and deletes bad ones, which wildly increase the speed of searching for Pareto front. Bacteria will be greatly influenced as the environment of living life changed. The environment can't be invariable, so bacteria will be changed with the decrease of food. The dispersion operation occurs after a certain number of reproduction processes.

The bacterium is trying to swim from places with low concentrations of nutrients to places with high nutrients. It is never satisfied with the amount of surrounding food, so it always seeks higher concentrations of nutrients. Hence, the previous three parts: Chemotaxis, Reproduction, Elimination and Dispersal process are continuous and effective.

The following work is to apply the BFO to solve multiple objectives.

3   Multi-objective Bacterial Foraging Optimization

Since the BFO algorithms could solve single-objective optimization problems, the idea of solving multi-objective optimization problems with BFO algorithms was tested. However, the purpose of multi-objective optimization problems is to find all values which are possibly satisfied to all functions. Since different decision makers have different ideas about objective functions, it is not easy to choose a single solution for a multi-objective optimization problem without interaction with the decision makers. Thus, all we could do is to show the set of Pareto optimal solutions to decision makers. The main goal of multi-objective optimization problems is to obtain a non-dominated front which is close to the true Pareto front. The details of the new optimization algorithm based on BFO are given in the following sections.

In what follows we briefly outline the Multi-objective Bacterial Foraging Optimization (MBFO) step by step:

Table 1. Pseudocode for the MBFO algorithm

Algorithm MBFO
Begin
    Initialize all the parameters and positions
    While (a terminate-condition is met)
      For (Elimination-dispersal loop)
        For (Reproduction loop)
        For (Chemotaxis loop)
          Compute two fitness functions J_1 and J_2.
          Let $J_{last1} = J_1$, and $J_{last2} = J_2$
          Update of the positions
        End For (Chemotaxis)
          Compute two health values $J_{health1}$ and $J_{health2}$
          Sort bacteria based on health values
          Copy the best bacteria
        End For (Reproduction)
          Eliminate and disperse each bacterium with probability P_{ed}
      End For (Elimination-dispersal)
    End While
  End

---

4   Experiments and Results

To measure the performance of MBFO quantitatively, two performance metrics: Diversity (Δ) [9] and Generational Distance (*GD*) [10] were introduced to evaluate and compare the algorithms in this paper.

In order to prove the effectiveness in solving the multi-objective problems, two test problems as examples are used to compare the performance of NSGAII and MOPSO. Test problems are chosen from a number of significant past studies in this area, including Schaer's study (SCH) [11], Fonseca and Fleming's study (FON) [12].

The performance of our algorithm is compared against MOPSO and NSGAII. The two algorithms use a population of 100 individuals. In NSGA-II, the generation is setting 200. The solutions accepted after iteration processes are used to calculate the performance metrics. Tables 2~3 are used to keep the performance metrics results.

Test Problem 1 (SCH): This problem is a convex function with one variable. The goal is to minimize :

$$\text{minimize} \quad f_1(x) = x^2$$

$$\text{minimize} \quad f_2(x) = (x-2)^2$$

$$where \ : n = 1, x \in [-5,7]$$

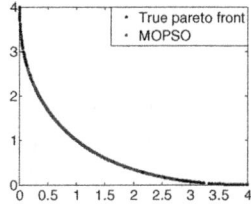 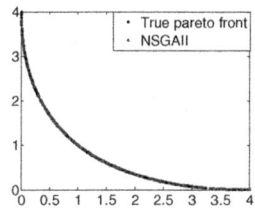 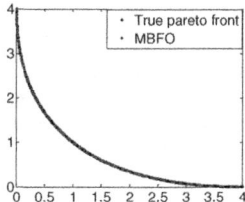

Fig. 1. Pareto front obtained by MBFO, MPSO, and NSGAII on Test Problem 1 (SCH1)

Table 2. Comparison of performance on Test Problem 1 (SCH1)

Problem function (SCH1)	Diversity Measure Δ			Generational Distance GD		
	MBFO	MOPSO	NSGAII	MBFO	MOPSO	NSGAII
Best	0.5582	0.6305	0.5730	0.0059	0.0054	0.0060
Worst	0.6485	0.6691	0.5910	0.0072	0.0060	0.0066
Mean	0.5771	0.6405	0.5821	0.0060	0.0052	0.0062

Fig. 1 shows all non-dominated solutions achieved by MBFO, MPSO, and NSGAII after the iteration process completed. True pareto front is also shown in the figure. The three algorithms all converge to the Pareto optimal front. MBFO is considerably better than MOPSO and NSGAII in terms of Diversity Measure, and there is little difference between MBFO and NSGAII in Generational Distance in this Test Problem (SCH) as shown in Table 2.

Test Problem 2 (FON): This problem has three variables and is a non-linear and non-convex issue. The goal is to minimize:

$$\text{Minimize} \ f_1 = 1 - \exp(-\sum_{i=1}^{3} (x_i + \frac{1}{\sqrt{3}})^2)$$

$$\text{Minimize} \ f_2 = 1 - \exp(-\sum_{i=1}^{3} (x_i - \frac{1}{\sqrt{3}})^2)$$

$$where \ : n = 3, x_i \in [-4,4], i = 1,2,3$$

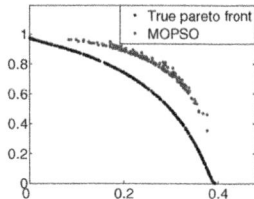 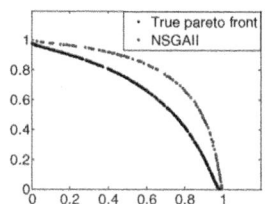 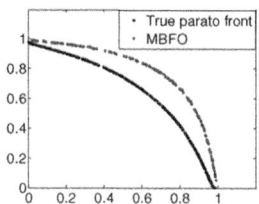

Fig. 2. Pareto front obtained by MBFO, MPSO, and NSGAII on Test Problem 3 (FON)

Table 3. Comparison of performance on Test Problem 3 (FON)

Problem function (FON)	Diversity Measure Δ			Generational Distance GD		
	MBFO	MOPSO	NSGAII	MBFO	MOPSO	NSGAII
Best	0.5864	2.2090	0.6072	0.0665	0.1390	0.0834
Worst	0.6718	3.8179	0.7924	0.1325	0.1460	0.0973
Mean	0.6205	2.9427	0.6941	0.0952	0.1427	0.0871

Fig. 2 shows the graphical results obtained by MBFO, MOPSO, and NSGAII on Test Problem 3. Similar to the results in the first two problems, MBFO has better performance in diversity and generational distance than MOPSO, and both the diversity and GD results are also closed to NSGAII shown in Table 3. Especially, MBFO have the better performance of diversity compared with two other algorithms.

5   Conclusions and Future Work

In this paper, a new algorithm MBFO based on Bacterial Foraging Algorithm (BFO) is proposed to solve multi-objective problems. Two performance metrics: Diversity (Δ) and Generational Distance (GD) were introduced to evaluate and compare the algorithms in this paper. The performed tests confirm the MBFO's ability in finding Pareto-optimal solutions. Through the comparison tests with the proposed MOPSO and NSGAII, the effectiveness of MBFO significantly outperforms MOPSO and competes with NSGAII. The results of the preliminary experiments presented in this paper are promising.

Such study is definitely helpful to choose and modify an algorithm for solving these problems. In future, more powerful and complicated MBFO variants like should be undertaken. In addition, we plan to study the ability of communication between the adjacent individuals to improve the speed of searching for Pareto optimal.

Acknowledgements. This work is supported by National Natural Science Foundation of China (Grant No.71001072, 61005010, 60805021, 31071168, 30900321), China Postdoctoral Science Foundation (Grant No. 20100480705, 20100480708), Science and Technology Project of Shenzhen (Grant No. JC201005280492A), The Natural Science Foundation of Guangdong Province (Grant no. 9451806001002294).

References

1. Deb, K.: Multi-Objective Optimization Using Evolutionary Algorithms. John Wiley & Sons, Chichester (2001)
2. Schaffer, J.D.: Multiple Objective Optimization with Vector Evaluated Genetic Algorithms. Ph.D. Thesis, Vanderbilt University (2004)
3. Haiela, P., Lin, C.Y.: Genetic Search Strategies in Multi-Criterion Optimal Design. Structural and Multidisciplinary Optimization 4(2), 99–107 (2002)
4. Srinivas, N., Deb, K.: Multi-Objective Optimization Using Non-Dominated Sorting in Genetic Algorithms. Evolutionary Computation 2(3), 221–248 (2001)
5. Zitzler, E., Thiele, L.: Multi-Objective Evolutionary Algorithms: A Comparative Case Study and the Strength Pareto Approach. IEEE Transactions on Evolutionary Computation 3(4), 257–271 (2005)
6. Deb, K., Pratap, A., Agarwal, S., Meyarivan, T.: A Fast and Elitist Multi-Objective Genetic Algorithm: NSGA-II. IEEE Transactions on Evolutionary Computation 6(2), 182–197 (2002)
7. Fieldsend, J.E., Singh, S.: A Multi-objective Algorithm Based Upon Particle Swarm Optimization, and Efficient Data Structure and Turbulence. In: Workshop on Computational Intelligence, pp. 34–44 (2002)
8. Passino, K.M.: Biomimicry of Bacterial Foraging for Distributed Optimization and Control. IEEE Control Systems Magazine 22(3), 52–67 (2002)
9. Deb, K., Pratap, A., Agarwal, S., Meyarivan, T.: A Fast and Elitist Multi-objective Genetic Algorithm: NSGA-II. IEEE Transactions on Evolutionary Computation 6(2), 182–197 (2002)
10. Van Veldhuizen, D.A., Lamont, G.B.: Evolutionary Computation and Convergence to Pareto Front// Late Breaking Papers at the Genetic Programming Conference. Stanford University Bookstore. Stanford, CA, USA (1998)
11. Schaffer, J.D.: Multiple Objective Optimization with Vector Evaluated Genetic Algorithms. In: 1st International Conference on Genetic Algorithms (ICGA), Hillsdale, NJ, USA, pp. 93–100 (1985)
12. Fonseca, C.M., Flemming, P.J.: Multi-objective Optimization and Multiple Constraint Handling with Evolutionary Algorithms-Part II: Application Example. IEEE Transactions on Systems, Man and Cybernetics 28, 38–47 (1998)

Using 2D Principal Component Analysis to Reduce Dimensionality of Gene Expression Profiles for Tumor Classification

Shu-Lin Wang[1], Min Li[1], and Hongqiang Wang[2]

[1] College of Information Science and Engineering,
Hunan University, Changsha, Hunan, 410082, China
jt_slwang@hotmail.com
[2] The Intelligent Computing Laboratory, Hefei Institute of Intelligent Machines,
Chinese Academy of Sciences, Hefei, Anhui, 230031, China

Abstract. In the last ten years, numerous methods have been proposed for accurate classification of tumor subtype based on gene expression profiles (GEP). Among these methods, feature extraction methods play an important role in constructing classification model. However, traditional methods view a gene expression sample as 1D vector, which does not sufficiently utilize the correlation and structure information among many genes. We, therefore, introduce 2D principal component analysis (2DPCA) to extract features for tumor classification by converting 1D sample vector into 2D sample matrix. To evaluate its performance, we perform a series of experiments on four tumor datasets. The experimental results indicate that the obtained performance by using 2DPCA is superior to the classic principal component analysis.

Keywords: Gene expression profiles, tumor classification, dimensionality reduction, 2D principal component analysis.

1 Introduction

More accurate diagnosis of tumor subtype is crucial for successful treatment. However, it is usually difficult to perform early tumor diagnosis by using traditional appearance-based diagnosis methods. Over the last decade, the advent of DNA microarray technique brings hope into the tumor diagnosis [1], and a great number of tumor classification methods based on gene expression profiles (GEP) have been proposed and extensively studied [2], but the challenges from GEP still exist due to the curse of dimensionality that the number of genes far exceeds the number of samples.

When facing these problems, adopting single dimensionality reduction method to extract efficient and powerful features to construct classification model is very difficult due to the complexity of GEP. Previous studies show that two-stage dimensionality reduction model is more efficient for the feature extraction of GEP than single method. Usually, the first stage is to select those differentially expressed genes from GEP dataset so as to enhance the discriminative information. The second stage is to extract few features from those selected genes that include importantly

D.-S. Huang et al. (Eds.): ICIC 2011, LNBI 6840, pp. 588–595, 2012.
© Springer-Verlag Berlin Heidelberg 2012

discriminative information. For example, based on two-stage method, Wang *et al* [3] designed 18 methods for the comparison of classification performance by combining two gene filter methods (Kruskal-Wallis rank sum test and Relief-F [4]) with three classical dimensionality reduction methods: principal component analysis (PCA), linear discriminative analysis (LDA), and non-linear multidimensional scaling (MDS). The comparison suggests that among these methods there is no optimal method that is consistently superior to other methods in performance on all the selected datasets. Therefore, it is very difficult to design the best dimensionality reduction approach to tumor classification due to what we do not know about the statistical distribution of measured gene expression levels.

Traditional dimensionality reduction methods such as PCA view a gene expression sample as a 1D vector, which does not sufficiently utilize the correlation and structure information among many genes. Recently, two-dimensional principal component analysis (2DPCA) [5] was successfully developed for image representation, which is superior to PCA in extracting features from 2D image. Similarly, 2DPCA can be introduced into the dimensionality reduction of 2D GEP sample matrix obtained by transforming from 1D sample vector into 2D matrix. So this paper proposes a 2DPCA-based feature extraction method that can sufficiently utilize 2D structure information to construct tumor classification model by using k-nearest neighbor (KNN) method.

2   Methods

2.1   Analysis Framework

Fig. 1 shows the analysis framework of GEP-based tumor dataset, which consists of four crucial steps. First, to avoid the bias of different ways on division of a whole dataset, we randomly split whole dataset into two parts: training set and test set. Second, to avoid the effects of tumor-unrelated genes, differentially expressed genes are selected by adopting the gene filters technique Relief-F, and 1D sample vector is further converted into 2D sample matrix. The main idea of Relief-F is to select samples at random, compute their nearest neighbors, and adjust a feature weighting vector to assign more weight to features that discriminate the sample from the neighbors belonging to different subclasses. Third, feature extraction technique 2DPCA is applied to the selected genes to extract discriminative features. Finally, a classification model is constructed and evaluated by KNN and the test set.

Obviously, our dimensionality reduction method is based on two-stage one. Usually, due to the fact that there are numerous tumor-unrelated genes in GEP, it is unnecessary to extract features of GEP by using all genes. In fact, two GEP samples are very similar when considering the expressions of too many genes, which can be likened to the similarity of two pictures. For example, the following pictures in Fig. 2 are very similar when we focus on the whole pictures, because only the upper-left corners of the two pictures are different. However, it is obvious that the two pictures are very different when we focus on only the upper-left corners of the two pictures. Selecting differentially expressed genes is also likened to selecting differential portion in the two pictures, which make us focus on the differential portion.

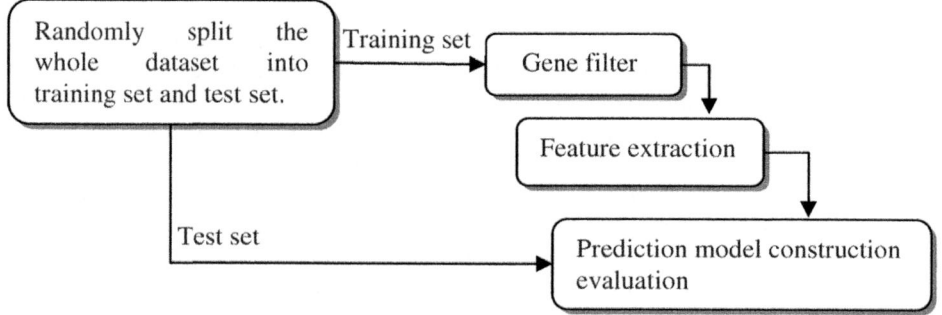

Fig. 1. The framework of our analysis method

Fig. 2. Two similar pictures containing two photos in the up-left corners

2.2   Representation of GEP

DNA microarray is composed of thousands of individual DNA sequences printed in a high density array on a glass microscope slide. Samples are generated under multiple conditions which may be a time series during a biological process or a collection of different tissue samples. Let $G = \{g_1, \cdots, g_n\}$ be a set of genes and $S = \{s_1, \cdots s_m\}$ be a set of samples with k subclasses $C = \{c_1, \cdots, c_k\}$. The corresponding gene expression matrix can be represented as $X = \{x_{i,j} | 1 \leq i \leq m, 1 \leq j \leq n\}$. The matrix X is composed of m row vectors. $x_i \in R^n, i = 1,2,\cdots,m$, m denotes the number of samples, and n denotes the number of genes measured.

Traditionally, a dimensionality reduction method attempts to find a matrix A so as to transform dataset $X = \{x_1, \cdots, x_m\}$ $(x_i \in R^n)$ into a new dataset $Y = \{y_1, \cdots, y_m\}$ $(y_i \in R^d)$ with dimensionality d, while retaining the geometry of dataset X as much as possible, where the intrinsic dimensionality d. That is $Y = A^T X$. However, if we convert 1D sample s_i into 2D square matrix V_i (suppose the dimensionality of s_i is a square number), 2DPCA can be applied to extract features from the 2D square matrix V_i.

2.3   Two-Dimensional Principal Component Analysis

Principal component analysis (PCA) attempts to find a linear basis of reduced dimensionality for dataset, in which the amount of variance in the dataset is maximal, to construct a low-dimensionality representation of the dataset. To achieve dimensionality reduction, PCA finds a linear transformation T that maximizes $T^T \text{cov}_{X-\bar{X}} \, T$, where $\text{cov}_{X-\bar{X}}$ is the covariance matrix of the dataset X. Hence PCA solves the eigenproblem.

$$\text{cov}_{X-\bar{X}} \, v = \lambda v \tag{1}$$

The eigenproblem is solved for the d principal eigenvalues λ. The corresponding eigenvector form the columns of the linear transform matrix T. The low-dimensionality representation Y of the original dataset X are computed by linear transform matrix T, i.e., $Y = (X - \bar{X})T$.

2DPCA [5-6] that can sufficiently utilize 2D structure information is a subspace learning algorithm, which is of great benefit to classification. Let V_i, $i = 1, \cdots, m$, represent m training sample matrices. The size of the matrix V_i is $h \times w$. 2DPCA aims to find a projection matrix $U = [u_1, u_2, \cdots, u_d] \in R^{w \times d}$ that minimizes the mean square reconstruction error. The matrix representation reduces the learning process of 2DPCA to an eigendecomposition problem of a small covariance matrix.

$$Cov = \sum_i^m (V_i - \bar{V})^T (V_i - \bar{V}) \in R^{w \times w} \tag{2}$$

where $\bar{V}$ is the mean sample matrix

$$\bar{V} = \frac{1}{m} \sum_i^m V_i. \tag{3}$$

The covariance matrix in 2DPCA is of size $w \times w$.

3   Experiments

3.1   Four Tumor Datasets

Four public available tumor datasets are applied to our experiments. They are Leukemia [7], high-grade gliomas (Gliomas) [8], Acute Lymphoblastic Leukemia (ALL) [9], and Small Round Blue Cell Tumor (SRBCT) [10], which are briefly described in Table 1.

The Leukemia dataset contains 72 samples with three subtypes or subclasses, i.e., MLL, AML and ALL. The Gliomas dataset consists of 50 samples with two subclasses, i.e., Glioblastomas and Anaplastic Oligodendrogliomas (AO). The ALL dataset totally contains 248 samples that belong to six tumor subtypes: BCR-ABL, E2A-PBX1, Hyperdip>50, MLL, T-ALL and TEL-AML1. The SRBCT dataset contains 83 samples with 2,308 genes in each sample. According to the original literature, there are 63 training samples and 20 test samples. The 63 training samples contain 23 Ewing family of tumors (EWS), 20 rhabdomyosarcoma (RMS), 12 neuroblastoma (NB), and eight Burkitt lymphomas (BL) samples. The test samples contain six EWSs, five RMSs, six NBs, and three BLs.

Table 1. The descriptions of four tumor datasets

No.	Datasets	Platform	#Samples	#Genes	#Subclasses
1	Leukemia	Affy HGU95a	72	12,582	3
2	Gliomas	Affy U95Av2	50	12,625	2
3	ALL	Affy HGU95Av2	248	12,626	6
4	SRBCT	cDNA	83	2,308	4

3.2   Experimental Methods and Parameter Setting

In order to exhibit the superiority of 2DPCA, we also adopt classical PCA to extract principal components from 1D sample vector to construct classification model by using KNN. The two methods are correspondently called 2DPCAKNN and PCAKNN, respectively.

However, to achieve honest and reliable results there are several parameters to be appropriately set for each method before evaluating test set. And all parameters must be determined within training set. Inappropriate parameter selection can easily lead to over-fitting and selection bias. For example, if gene selection is performed on whole dataset including training set and test set, the obtained prediction accuracy on the test set will be upwardly biased [11]. If we set the parameters according to the prediction accuracies, the bias of results can also be caused. If we repeatedly optimize the parameters on training set, over-fitting might be occurred.

Therefore, some parameters should be appropriately set in advance. For Relief-F, the number of the selected genes is set to the square number of natural number, such as $15^2, 18^2$, etc.. For PCA, we fixedly extract 5 principal components for each dataset, and for 2DPCA we extract the eigenvectors that correspond to the top 5 eigenvalues. For KNN classifier, 5 nearest neighbors are used for label decision. At last, each result is obtained by performing the construction process of classification model shown in Fig. 1 200 times to obtain average accuracies for each method and dataset. In each time, 50% samples in whole dataset are randomly selected and used for training set and the remains are used for the test set.

3.3   Experimental Results

Table 2 shows the comparison of prediction accuracies obtained by PCAKNN and 2DPCAKNN methods on the four tumor datasets under the condition of the dataset division that 50% of total dataset is randomly divided into training set. The comparison indicates that 2DPCAKNN obviously outperforms PCAKNN in mean prediction accuracies on the four datasets.

Table 2. The comparison of prediction accuracies obtained by both PCAKNN and 2DPCAKNN methods

Datasets	PCAKNN	2DPCAKNN
Leukemia	89.58±4.65	**91.38±3.60**
Gliomas	72.87±8.68	**73.09±8.56**
ALL	89.09±2.45	**94.49±1.61**
SRBCT	88.31±6.76	**91.13±5.09**

Fig. 3 shows the prediction accuracies varying with different proportion of training samples on the selected four tumor datasets, in which the X-axis denotes the proportion of training set in whole dataset and the Y-axis denotes the prediction axis. From Fig. 3 we can see that the prediction accuracies obtained by PCAKNN and 2DPCAKNN increase generally with the increase of the number of training set and 2DPCAKNN always outperforms PCAKNN in accuracies. Strangely, for the Leukemia dataset we do not explain why the prediction accuracies with 40% of total dataset as training set are lower than the prediction accuracies with 35 of total dataset as training set.

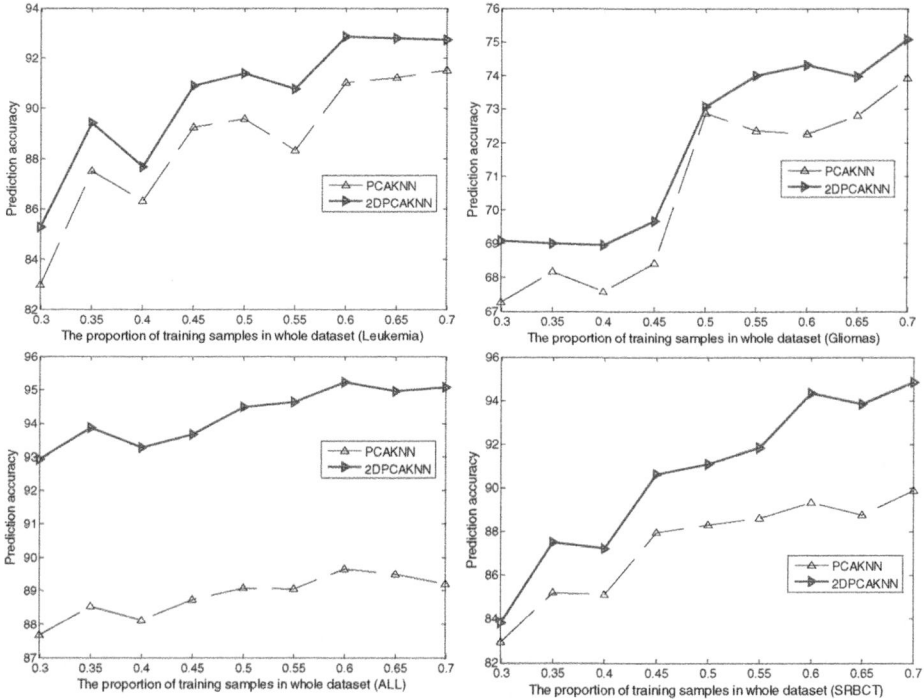

Fig. 3. The prediction accuracies varying with different proportion of training samples on four tumor datasets, respectively

Different number of the selected genes can also lead to different results and in fact for a certain dataset we do not know how many genes are related to tumor. So we select different number of top-ranked genes ranked by Relief-F to evaluate the classification performance. Fig. 4 shows the prediction accuracies varying with different number of the selected genes on the four tumor datasets, in which the X-axis denotes the square root of the number of the selected genes (the number of genes ranges from 15^2 to 35^2) and the Y-axis denotes the prediction accuracies. Generally, the prediction accuracies decrease slowly with the increase of the number of the selected genes on the four datasets.

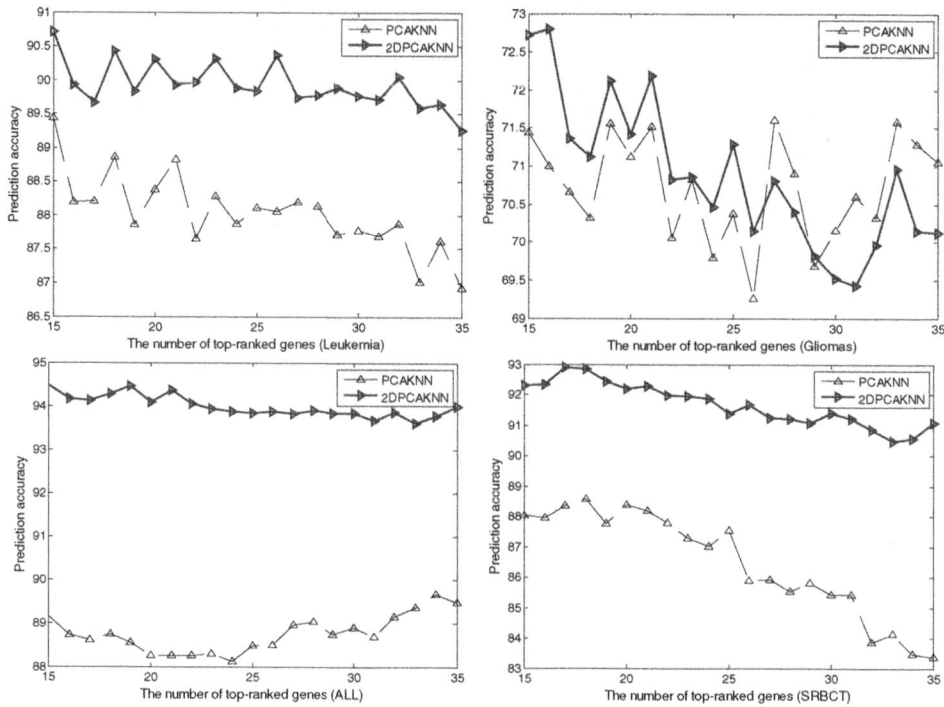

Fig. 4. The prediction accuracies varying with different number of top-ranked genes ranked by Relief-F on four datasets, respectively

4   Conclusions

Designing better feature extraction methods for tumor classification based on gene expression profiles to achieve better performance is always one of our important goals, which is of great benefit to clinical tumor diagnosing. In this paper, a new dimensionality reduction method is designed for tumor classification by introducing 2DPCA method, which can sufficiently utilize 2D structure information in the process of extracting features. In fact, the new method is based on two-stage dimensionality reduction model. The first stage is to adopt Relief-F method to rank genes and select a certain number of top-ranked genes. The second stage is to convert 1D sample vector into 2D sample matrix and apply 2DPCA to extract features from the 2D sample matrix. Our experiments on four public available tumor datasets have proved the effectiveness of the proposed methods. Compared with classical PCA method, 2DPCA can obtain better classification performance on the four tumor datasets.

Acknowledgments. This work was supported by the National Science Foundation of China (grant nos. 60973153 and 30900321), the China Postdoctoral Science Foundation (grant no. 20090450825).

References

1. Golub, T.R., et al.: Molecular Classification of Cancer: Class Discovery and Class Prediction by Gene Expression Monitoring. Science 286, 531–537 (1999)
2. Asyali, M.H., et al.: Gene Expression Profile Classification: A review. Current Bioinformatics 1, 55–73 (2006)
3. Wang, S.L., et al.: Performance Comparison of Tumor Classification Based on Linear and Non-linear Dimensionality Reduction Methods. Advanced Intelligent Computing Theories and Applications 6215, 291–300 (2010)
4. Kononenko, I.: Estimating attributes: Analysis and Extensions of Relief. In: Bergadano, F., De Raedt, L. (eds.) ECML 1994. LNCS, vol. 784, pp. 171–182. Springer, Heidelberg (1994)
5. Yang, J., Zhang, D., Frangi, A.F., Yang, J.Y.: Two-dimensional PCA: A new Approach to Appearance-based Face Representation and Recognition. IEEE Transactions on Pattern Analysis and Machine Intelligence 26, 131–137 (2004)
6. Li, X.L., Pang, Y.W., Yuan, Y.A.: L1-Norm-Based 2DPCA. IEEE T. Syst. Man Cy. B 40, 1170–1175 (2010)
7. Armstrong, S.A., et al.: MLL Translocations Specify a Distinct Gene Expression Profile that Distinguishes a Unique Leukemia. Nature Genetics 30, 41–47 (2002)
8. Nutt, C.L., et al.: Gene Expression-based Classification of Malignant Gliomas Correlates Better with Survival than Tistological Classification. Cancer Research 63, 1602–1607 (2003)
9. Yeoh, E.J., et al.: Classification, Subtype Discovery, and Prediction of Outcome in Pediatric Acute Lymphoblastic Leukemia by Gene Expression Profiling. Cancer Cell 1, 133–143 (2002)
10. Khan, J., et al.: Classification and Diagnostic Prediction of Cancers Using Gene Expression Profiling and Artificial Neural Networks. Nature Medicine 7, 673–679 (2001)
11. Ambroise, C., McLachlan, G.J.: Selection Bias in Gene Extraction on the Basis of Microarray Gene-expression Data. Proceedings of the National Academy of Sciences of the United States of America 99, 6562–6566 (2002)

3D Virtual Colonoscopy for Polyps Detection by Supervised Artificial Neural Networks

Vitoantonio Bevilacqua[1,2,*], Domenico De Fano[1], Silvia Giannini[1],
Giuseppe Mastronardi[1,2], Valerio Paradiso[1], Marcello Pennini[1], Michele Piccinni[1],
Giuseppe Angelelli[3], and Marco Moschetta[3]

[1] Dipartimento di Elettrotecnica ed Elettronica, Politecnico di Bari, Bari, Italy
[2] e.B.I.S. s.r.l., Politecnico di Bari Spin-Off, Bari, Italy
[3] Dipartimento di Medicina Interna e Medicina Pubblica (Di.M.I.M.P.),
Sezione di Diagnostica per Immagini, Università degli Studi di Bari, Bari, Italy
bevilacqua@poliba.it

Abstract. The occurrence of false-positives (FPs) is still an important concern and source of unreliability in computer-aided diagnosis systems developed for 3D virtual colonoscopy. This work presents three different supervised approaches, based on supervised artificial neural networks (ANNs) architectures tested on 16 rows helical multi-slice computer tomography. The performance of the best ANN architecture developed, by using the volumes belonging to only 4 of 7 available nodules diagnosed by expert radiologists as polyps and non-polyps were evaluated in terms of FPs and false-negatives. It revealed good performance in terms of generalization and FPs reduction, correctly detecting all 7 polyps.

Keywords: Computer-aided diagnosis, 3D virtual colonoscopy, supervised artificial neural network, colonic polyps detection.

1 Introduction: Materials and Methods

The colon and rectal cancers are estimated to be the third carcinoma death cause in western countries. Every year approximately 678.000 new cases are diagnosed in the world and 150.000 in Europe. Although this form of cancer is more curable than other forms of digestive apparatus carcinoma, the possibilities of 5 years surviving from the diagnosis stands at 40-50%, reaching 80-90% in early cases. These statistics show how important is to detect colorectal neoplasia at an early stage in order to ensure the effectiveness of the therapies and reduce the risk of death. Screening programs are, in this perspective, fundamental instruments of diagnosis. Computed tomography colonography (CTC), also known as virtual colonoscopy, is one of the most recent screening test techniques. Although many computer-aided diagnosis (CAD) architectures have been investigated, the occurrence of false-positives (FPs) is still a problem that can lead to less confidence of behalf of technicians in the system and to the eventuality of non-distinction. The aim of this work is to develop a CAD system for CTC that could automatically detect polyps and, in the future, interact with the 3D

---

* Corresponding author.

D.-S. Huang et al. (Eds.): ICIC 2011, LNBI 6840, pp. 596–603, 2012.
© Springer-Verlag Berlin Heidelberg 2012

reconstruction and rendering of the colon lumen, in order to display polyps and fasten radiologists reviews. The data set available, obtained by using a 16-row helical-CT multi-slice scanner with a 1 mm resolution, consists of 10 volumetric regions diagnosed as polyps by expert radiologists in 6 different patients, and of a number of several regions belonging to the same patients correctly detected as colon folds and used as samples of non-polyps. Only 7 polyps were useful for the analysis, excluding 2 tumors, with a diameter bigger than 1 cm, and one polyp hardly recognizable due to fecal stool. The DICOM data have been provided by the operative unit called "Sezione di diagnostica per immagini" of "Dipartimento di medicina interna e medicina pubblica" of Policlinico of Bari. The principal techniques developed for colon polyps detection are based on characteristics, such as tissue density, shape and edges, and involve the analysis of volumetric and surface data, geometric and texture features or intensity values distribution. All of these can be led to a specific pattern recognition problem, solvable by ANNs [2,4]. This study focused on assessing the opportunity of using supervised ANNs approach to detect polyps and reduce FPs in 3D virtual colonoscopy after an opportune preprocessing phase. Starting from this assumption, three methods based on three different architectures have been investigated. At the end of this work, we can state that ANNs appear to provide robust performance in terms of classification, sensitivity and specificity. Moreover, the reduction of FPs findings has been achieved, from a theoretical point of view, thanks to the supervised ANNs ability of generalising their knowledge, acquired through a proper training set sampled by two radiologists among a restricted number of cases previously collected. In particular, with the limited set now available, our approach seems to show its better performance, in terms of correct detection of true polyps and no misclassification of any colon folds. Although our method appears promising, an extension to a larger database, retrospectively and prospectively, will be needed to confirm the usefulness of the method. In the following sections the tree methods are explained. The first uses a cascade of two ANNs working both with 3D input data with particular attention to the evaluation of the effectiveness of the shape feature for the recognition task. The second method uses a sequence of two ANNs: a 2D one scans CT slices in order to find possible polyps centers, and a 3D one processes the spheres centered where stated by the first ANN. For the realization of this system, the massive-training artificial neural network (MTANN) approach has been investigated to reducing FPs, based on the construction of a polyp model through a 3D gaussian density distribution, introduced by [3]. Finally the third method implemented works exclusively on 3D images, using a single ANN trained to recognize polyps and discard all the other structures. The results of the third method, being the better performing, are presented quantitatively at the end of the paper.

2.1 First Method: 3D ANN Approach

The aim of this approach was to test the performance of using a cascade of two ANNs, working on 3D data, especially focusing on the shape recognition skill of ANNs. Considering the 7x7x7 pixels cubic volume containing a polyp, it is reasonable to think that the volume useful to detect the nodule is the inner one. In order to decrease the computational load, the three innermost slices of the cube are convoluted. The result is a matrix analyzed by means of its entropy variation. This approach is justified by the empirical verification that polyps, having denser nuclei, tend to originate a greater entropy than normal colon walls. Those matrices, whose entropy is greater than a 0.5 threshold value, are organized in a vector of 343 elements and then put as input of the ANN.

2.1.1 Implementation of the First Volumetric ANN

The first ANN was trained using vectors extracted from real polyps and FPs. Its architecture consists of three layers (see tab. 1). A positive result of such an ANN is that it successfully recognizes all the polyps, including those not included in the training set, returning no false-negatives and showing good generalization skills.

2.1.2 Implementation of the Second Volumetric ANN

The second ANN implementation tested if the only shape properties of a polyp can well perform the recognition task, using the sole shape as a training parameter for the ANN. An OR function has then been applied to the sequences of 7 slices, passed through a Sobel edge filter. The final image exalts the polyp shape and its variations considering the entire volume. The input for the second ANN is a 49 elements vector (fig. 1), representing shape variations in a 7x7x7 volume, selected on the basis of the first ANN results. The scan of an entire exam, with the architecture explained in tab. 1, was performed in a short time. As a final stage, centers recognized by both the first and the second ANN were filtered to eliminate replicas.

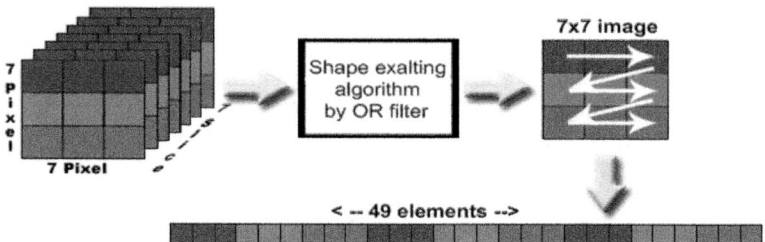

Fig. 1. Image processing for the second ANN implementation

2.2   Second Method: 3D ANN Approach

2.2.1 Implementation of the Second Volumetric ANN

The first ANN works on single CTC preprocessed slices. A 7x7 pixels square window moves along the borders of the lumen, assuming as central pixel those returned by the segmentation process. The 49 pixels contained in each window are linearized by columns and stored in a vector constituting the input for the ANN. Aiming at creating the training set, the polyps diagnosed in the whole dataset have been manually divided into four classes on the basis of their shape and position. For each class, the pattern having the most defined shape has been chosen and it has been extracted the 7x7 pixels square window that best showed the polyp shape (the same technique has been used for the FPs). We use area dimension according to the idea that a nodule is easily recognizable when its diameter is larger than 6 mm. The fundamental analysis of polyp gray-scale values has been focused at this stage on the experimental observation that 7x7 pixels polyps matrices include only a specific range of gray-scale values, experimentally estimated to be a mean value floating in the range [0.1, 0.45]. The assignment parameters about the ANN structure, layers transfer function, the training algorithm and the best training set have been chosen upon evaluation of the ANN performance changes, gradually varying the parameters. As a final model, it was selected a feed forward multi-layer perceptron (see tab. 1). A positive result of such a ANN is that it successfully recognizes all the polyps in the dataset, including those not included in the training set, returning no false-negatives and showing good generalization skills.

Table 1. Projected ANNs

		First method 3D ANN	First method "OR" ANN	Second method 2D ANN	Second method 3D ANN	Third method 3D ANN
First hidden layer	**neurons**	45	15	49	171	65
	activation function	logsig	logsig	tansig	tansig	logsig
Second hidden layer	**neurons**	9	9	3	/	37
	activation function	logsig	logsig	tansig	/	logsig
Output layer	**neurons**	1	1	1	171	1
	activation function	tansig	tansig	purelin	purelin	purelin
Learning algorithm		Resilient back-propagation	Resilient back-propagation	Resilient back-propagation	Resilient back-propagation	Resilient back-propagation
Output threshold		0.9	0.9	0.7	mse = 0.04 avg = 0.45	0.98
Output training value	**Polyp**	1	1	1	See fig.4	1
	Lumen	0	0	0	See fig.4	-1
	FPs	0	0	0.5	See fig.4	-0.5

2.2.2   Design of a Volumetric ANN Architecture

A 3D MTANN is a supervised volume-processing technique capable of directly operate on image data. The process involves the comparison between an input volume and a teaching volume, in order to facilitate the task of distinguishing a specific opacity from other opacities in medical images. In our study, starting from the hypothesis brought forward by [3], the teaching volume has been modelled using a 3D gaussian distribution for polyps and 0 values for FPs. The second ANN worked on volumetric data obtained considering the entire slices sequence, analysed through a window containing 7 slices at a time and moving with a step of 1 slice. In detail, spheres having a diameter of 7 pixels have been extracted for each window (fig. 2),

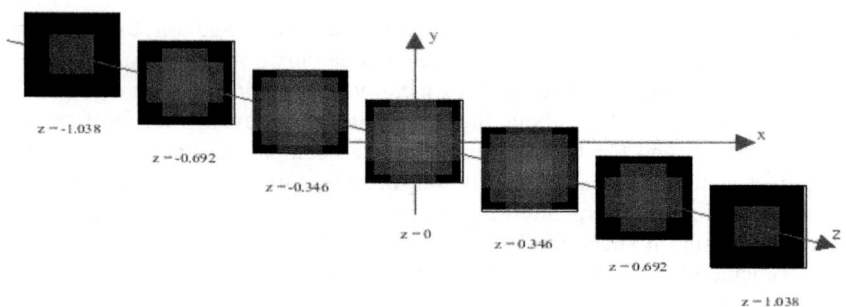

Fig. 2. The teaching model for the MTANN

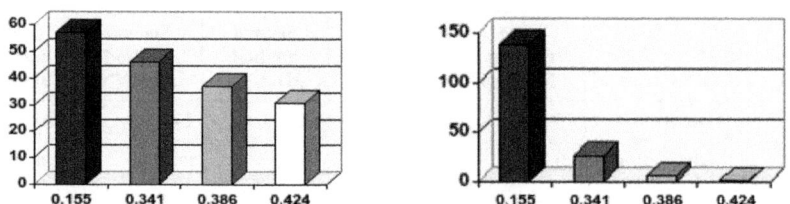

Fig. 3. Number of pixels falling in each of the 4 intensity values classes considering all the real polyps in the dataset. On the right graph it is shown the same distribution calculated for the model sphere confirming the hypothesis that polyps gray-scale values distribution can be associated to a 3D gaussian distribution. On the abscissa mean values are reported for the corresponding intervals.

in correspondence to the candidate polyps centers returned by the previous analysis. For each sphere, the pixels were rearranged in a vector containing 171 elements. It has been preferred to use a spherical volume instead of a cubic one since the sphere better reproduces a polyp 3D shape, allowing at the same time a 48% reduction of computational costs. The MTANN returns 171 output values, a volume reproducing the input forced to be more similar to the correspondent teaching volume. The result is a neural filter that exalts the polyps shape by making them brighter while at the same time darkening the FPs.

The polyp teaching volume has to correctly reproduce a generic polyp physical conformation and its density distribution. To achieve this, intensity histograms have been analysed for each polyp in the dataset, grouping all the gray-scale values in 4 intensity classes ([0, 0.31], [0.31, 0.372], [0.375, 0.398] and [0.398, 0.45]). Following [3], the model sphere has been built using the of a normalized 3D gaussian distribution. The brightest pixel is the central pixel of the central square and darker gray-scale values are assigned to external pixels (fig. 2), according to the fact that polyps are thicker in their center and degrading towards the borders. This assumption allowed the computation of a 0.969 value for the standard deviation σ and the absolute value of 0.346 that had to be assigned to x, y or z in one pixel shifts from the central one. Further tests led to a multiplication of all the values in the model by a factor of 2.5 in order to move the gray-scale values in a range between 0.5 and 1. This operation allowed a better definition of the spherical shape and at the same time forced the MTANN to assume higher output values for polyps, increasing the filtering efficiency. Experimental tests on the ANN architecture, varying layers transfer functions, allowed the selection, as a final model, of a feed-forward multi-layer perceptron, with the parameters shown in tab. 1. It has been applied an output threshold based on the mean square error calculated between the model and the output volume. Low error values indicate a greater probability of being in presence of a polyp. Another threshold calculating the mean value of the entire output volume has then been added.

2.3 Third Method: ANN Approach

Unlike the previous method, the following has been developed with the purpose of understanding if a single ANN can be capable of correctly identify polyps. The implementation of a 3D ANN has been chosen, in order to provide a bigger amount of distribution-related information, while at the same time reducing the computational complexity that would have been generated from the analysis of single 2D images using a 2D ANN.

2.3.1 ANN Development
This time the attention has been focused on the density values composition of cubes containing polyps. Such an analysis led to the finding that, as the size of the cube extracted is increased, starting from the central position, the distribution of densitometry values becomes more similar to that of the lumen. Following this strategy, the slices of the cubes are ordered on the basis of their position (fig. 4). Then the vector creation process considers each slice of the cube as a matrix composed by the centre and three concentric circumferences. The linearisation of the cubic structure is realized through an algorithm that, by making use of the standard circumference formula, orders the elements, placing the ones at shorter distance from the centre at the beginning of the vector. The final result is a linear structure of 343 elements (fig. 4). All the other information useful to correctly train the ANN (lumen and other intestinal structures) have been obtained in the same way as mentioned above. In this case too the ANN structure that proved to be the most efficient for the accomplishment of the task was a 3-layered supervised ANN. Representative parameters used for the building process of the final ANN are shown in tab. 1. By experimental results on the available dataset, the finale ANN developed was able to correctly detect all the polyps while returning no FPs.

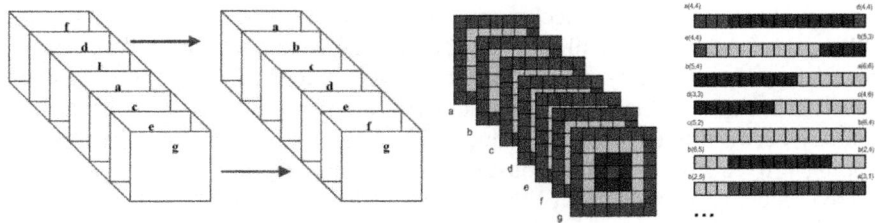

Fig. 4. Graphical representation of the matrices and the vector

3 Results and Conclusions

Finally we can conclude that the first method resulted to be the fastest. A complete scan took about 50 minutes, mostly due to the fact that the second ANN only examined a single image for each volume. However it also returned the worst results, since even though it has been capable of detecting all the polyps in the dataset, it returned also an average number of 25 FPs per exam. It has been then concluded that even though shape information is fundamental to the pattern recognition task, it is still important to consider other factors such as the density. Results and comparisons between the second and third method tested have been evaluated in terms of sensitivity, false positive rate (FPR), accuracy (ACC) and specificity (SPC), defined by receiver operating characteristic (ROC) analysis.

Table 2. Performance indexes extracted for the ANN architecture by the scan of the entire dataset

	TPR	FPR	ACC	SPC
2D ANN	100%	0.4%	99.6%	99.6%
MTANN	100%	0.08%	99.2%	99.2%
Third method ANN	100%	0%	100%	100%

Aiming at demonstrating the efficiency of the MTANN even with the limited dataset available, random noise has been introduced in real polyps to generate new examples, and the MTANN has been tested on these new ones. The MTANN resulted to be robust, always detecting the polyps. In one case the polyp was discarded by the output filter, though it had a diameter smaller than 4 mm, and therefore not included in the considered range. The performance of the system, involves a 90 minutes complete scan time on a desktop PC equipped with a Pentium dual core CPU running at a frequency of 2 GHz and 384 MB RAM. The simulation output of the third method described, as previously reported, gave no FPs and detected all the polyps. The execution time was slightly longer than the first method, employing 110 minutes on the same PC. In this paper three different supervised ANNs approaches have been presented for automatic polyp detection in 3D virtual colonoscopy. We are limited in explaining from a practical point of view the achieved generalisation about the detection accuracy of our CAD schemes, due to the small test set available since now. Anyway both 2D and 3D techniques proved to be efficient and complementary, although trained with a small number of cases; the generalization skill still remains to be confirmed, especially regarding the second and third methods developed. An interesting aspect of the work is that it allows the confirmation of the existence of an analogy between polyps density values and those of a 3D gaussian distribution. This finding proved useful to the robustness and generalization skills of the system and helped reducing the number of FPs, together with other CAD algorithms that operated by identifying the shape and CT numbers variation to classify voxels into polypoid or non-polypoid areas.

References

1. O'Connor, S.D., Summers, R.M., Yao, J., Pickhardt, P.J., Choi, J.R.: CT Colonography with Computer-aided Polyp Detection: Volume and Attenuation Thresholds to Reduce False-Positive Findings Owing to the Ileocecal Valve. Radiology 241, 426–432 (2006)
2. Wang, Z., Liang, Z., Li, L., Li, X., Li, B., Anderson, J., et al.: Reduction of False Positives by Internal Features for Polyp Detection in CT-Based Virtual Colonoscopy. Medical Physics 32(12), 3602–3616 (2005)
3. Suzuki, K., Yoshida, H., Näppi, J., Dachman, A.H.: Massive-training Artificial Neural Network (MTANN) for Reduction of False-positives in Computer-aided Detection of Polyps: Suppression of Rectal Tubes. Medical Physics 33(10), 3814–3824 (2006)
4. Suzuki, K., Horiba, I., Sugie, N.: A Simple Neural Network Pruning Algorithm with Application to Filter Synthesis. Neural Processing Letters 13(1), 43–53 (2001)

Face Recognition System Robust to Occlusion

Mohit Sharma, Surya Prakash, and Phalguni Gupta

Department of Computer Science and Engineering,
Indian Institute of Technology Kanpur
Kanpur 208016, India
{msharma,psurya,pg}@cse.iitk.ac.in

Abstract. This paper presents an efficient face recognition system which can handle partial occlusion in both training and test image sets. We hybridize Gabor filter with Eigen faces which make use of localization effect of Gabor filter and whole appearance effect of Eigen faces. It has been tested on AR database which contains naturally occluded face images.

Keywords: Principal Component Analysis, Eigen Faces, Gabor Filter, City Block Distance.

1 Introduction

Most of the face recognition systems mainly deal with problems like variation in illumination, expression, pose etc but all of them have their own limitations. One of the most challenging problems in face recognition is partially occluded face. A face recognition system should be able to handle faces occluded by accessories and objects, such as scarf or sunglasses, hands on the face, and external sources that partially occlude the camera view. Therefore, the system has to be robust to deliver facial recognition over partial occlusion with high probability. This paper attempts to solve the problem occurred in a face recognition system due to occlusion in face image. It makes use of Principal component analysis along with Gabor filter. Any well known approach for recognition system considers the face with some feature points which could be used to describe the face. Features are extracted out of the facial image in such a manner that it is common to all individuals but they are sufficiently different for each.

This paper is organized as follows. Section 2 describes the proposed algorithm. Experimental results have been analyzed in Section 3. Conclusion are given in the last section.

2 Proposed System

An image I is convolved with 8 different orientations ($\frac{u\pi}{8}$ $\forall u \in [1,8]$) of Gabor filter (Fig. 1) to get 8 different response images I_1 to I_8 (Fig. 2). Each of these 8 response images is a part of 8 different subspaces. Each training image is convolved in a similar manner to get corresponding response images for each

D.-S. Huang et al. (Eds.): ICIC 2011, LNBI 6840, pp. 604–609, 2012.

subspaces separately. Training image set matrix is made in each of these subspaces individually. Eigen faces in each subspace is obtained by calculating Eigen vectors of training matrix of corresponding subspace. Therefore, 8 different sets of Eigen faces is obtained and call them *Gabor-Eigen-faces*. Depending upon the distribution of training response images set in each subspace, *best* Eigen faces (Fig. 3) of each subspace are taken. So, finally one gets top Eigen faces from each subspace.

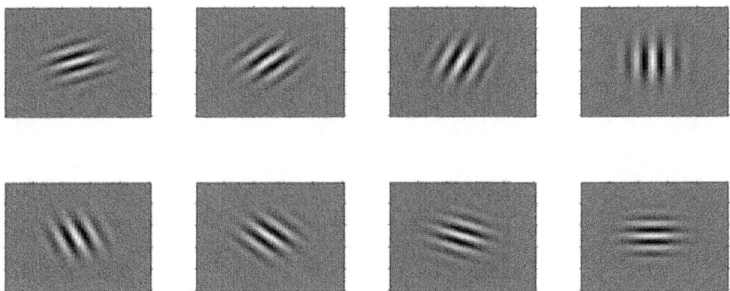

Fig. 1. 8 different Gabor filters used for convolution

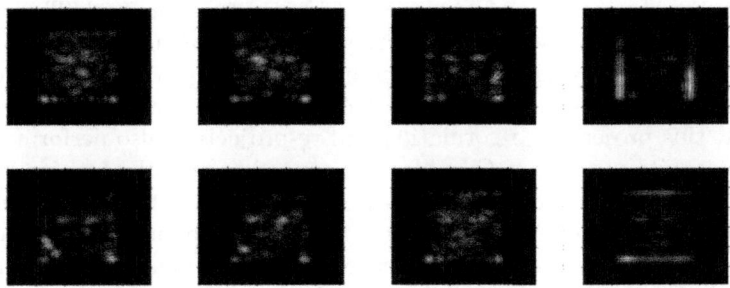

Fig. 2. 8 Different response images for one subject for a image

Each image response in each subspace is projected onto the Gabor-Eigen faces of that subspace and their weight is being recorded. These weights for each subspace of each image are stored as the features of that image in corresponding subspace. Therefore, for each training image, 8 different subsets of feature vectors are obtained.

Test image is convolved with Gabor filters and response images in each subspaces is obtained. Then, projection of response images to the chosen Gabor-Eigen faces of each subspace are noted. Because the original test image may contain the occlusion, these projection weights are distracted. Therefore, the image is projected back to the original image subspace and reformed image is obtained. By doing this, reformed image have almost same projection for

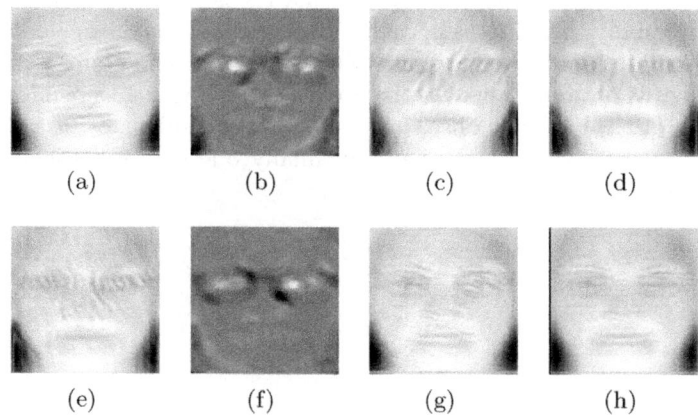

Fig. 3. *Best* Gabor-Eigen face of each subspace

non-occluded part as of original but it is not the case for occluded area, it contains contribution of same area of all training images thereby decreasing the effect of occlusion. The reformed image is projected to the Gabor-Eigen faces of corresponding subspace and the weights associated are calculated. Thereby, 8 different sets of weights (or features) for test image is obtained. This projection reformation sets the test image into the framework of the training images and projection to Gabor-Eigen faces gives more matching for non-similar parts of face. Because of this, when training images contain occlusion and test image does not, this projection, reformation and re-projection also perform better.

For recognition purpose, City-block distance between the test image and the training image features of each subspace is calculated. If two images are of same person then the difference of features in each of the subspace would be less. For two different person, the difference would would not be negligible in at least one of the subspaces. So, the *net distance* between 2 distinct images is the maximum distance between their features in all subspace. For verification purpose, the person is verified if the *net distance* between the test image and the training image is less than the verification threshold. Mathematically, distance D_l between test image and gallery image in l^{th} Gabor sub-space can be calculated as:

$$\text{City Block Distance } (D_l) = \sum_{i=1}^{m_l} (\, ||FT_i^l - FG_i^l|| \,)$$

where FT_i^l and FG_i^l are i^{th} features of test image and gallery image for l^{th} Gabor sub-space and m_l is number of *top* eigen vectors considered for l^{th} Gabor Space. *Net Distance* can be calculated as:

$$\text{Net Distance} = \text{maximum} \, \{D_1, D_2, \ldots, D_8\}$$

3   Experimental Results

This section discusses the experimental results of the proposed algorithm. It has been tested on AR face database [4] which contains frontal face images of 100 subjects taken under different conditions. For each subject, it has 13 images per session and comprises of 2 different session taken at 2 different times. We have used same labeling for images as done in [3]. Let {a, b, c, d, e, f} be the set of images for each subject of Session-I while {a', b', c', d', e', f'} be set of images obtained in Session-II.

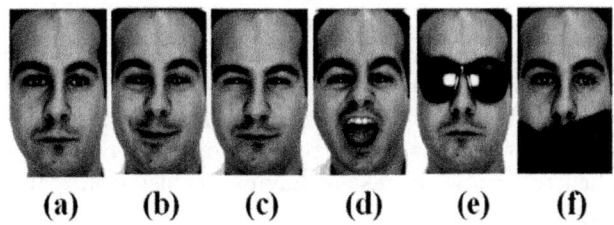

(a)   **(b)**   **(c)**   **(d)**   **(e)**   **(f)**

Fig. 4. 6 Images for one subject of Session-I of AR database

In one set of experiment, we use the same setup for experiment as used in [3]. In this paper, the set of images {a, b, c, a', b', c'} is used for training set while the set of images {e, e', f, f'} is used for testing. The results are compared with the results of [3] and [1] and are shown in Table 1. The algorithm results in recognition rate of over 95% for the whole test set.

Table 1. Recognition Results

Training Set Images	Test Set Images	[3]	[1]	Proposed Algorithm
{a, b, c, a', b', c'}	{e, e'}	93%	95%	96%
{a, b, c, a', b', c'}	{f, f'}	87%	93%	94.5%
{a, b, c, a', b', c'}	{e, e', f, f'}	90%	94%	95.3%

Any face recognition system designed for solving problems like occlusion should also give good result with neutral images. We have test the algorithm on non-occluded images and compared the results of [3] and [1]. In one experiment, the set of images {a, b, c} is used for training and the set of image {d} which are for screaming faces, is used for testing. In this case, the algorithm results in improvement recognition rate of 9% over [3]. In another experiment, the set of images {a, b, c, d} is used for training while the set of images {a', b', c', d'} is used for testing. The results are compared in Table 2. Highest recognition rate reached is 98.1% while testing for the set {a'} of image.

Table 2. Recognition Results

Training Set Images	Test Set Images	[3]	Proposed Algorithm
{a, b, c}	{d}	85%	94%
{a, b, c, d}	{a'}	91%	98.1%
{a, b, c, d}	{b'}	91%	96.7%
{a, b, c, d}	{c'}	89%	96.1%
{a, b, c, d}	{d'}	78%	86.2%

Table 3 provides the results of simple nearest neighbor approach with 2-Norm $\{NN_1\}$ and 1-Norm$\{NN_2\}$. In this set, there is occlusion in training set while there is no occlusion in test set. Table 3 presents some of the most difficult cases including up to $\approx 50\%$ occlusion in both training and testing image sets. The algorithm results in as high as 93.7% recognition rate. Highest recognition rate has been reached to 98.9% for case of having occlusion in training images and no occlusion in testing image.

Table 3. Recognition Results

Training Set Images	Test Set Images	[3]	[2]	N_2	N_1	Proposed Algorithm
{e,f}	{a}	96.0%	89.0%	45.0%	79.0%	98.9%
{e,f}	{a'}	79.4%	71.0%	31.0%	50.0%	92.9%
{e,f}	{b,c,d}	80.0%	72.0%	31.7%	59.7%	85.0%
{e,f}	{b', c', d'}	58.7%	47.3%	20.3%	32.7%	79.8%
{e,f}	{e', f'}	57.0%	55.0%	25.5%	29.0%	93.7%

In another experiment, the algorithm is tested with much closer scenario to the real world. We may not have many images per subject in database. Considering this, we have trained with only one neutral image {a} of each subject and tested the system against occluded set {e, e', f, f'} of images. The experimental results are shown in Table 4. Algorithm reports recognition rate up-to 98.6%.

4   Conclusion

We have used Gabor Eigen face algorithm for face recognition with partial occlusion. This algorithm allows to have occlusion in both training and test set images. It has achieved high recognition rate pertaining to the use of projection-reformation of test image which converts the test image into the frame work of the training images. Because of this key idea, the proposed algorithm provides better recognition rate compared to other well known algorithms.

Table 4. Recognition Results

Training Set Images	Test Set Images	Proposed Algorithm
{ a }	{ e }	96.2%
{ a }	{ e'}	98.6%
{ a }	{ f }	95.9%
{ a }	{ f'}	90.2%

Acknowledgments. This work has been supported by the Department of Information Technology, Government of India, New Delhi, INDIA. Authors are thankful to all anonymous reviewers for their valuable comments and suggestions to improve the quality of the paper.

References

1. Fidler, S., Skocaj, D., Leonardis, A.: Combining reconstructive and discriminative subspace methods for robust classification and regression by subsampling. IEEE Transactions on Pattern Analysis and Machine Intelligence 28(3), 337–350 (2006)
2. Jia, H., Martinez, A.M.: Face recognition with occlusions in the training and testing sets. In: IEEE International Conference on Automatic Face and Gesture Recognition, pp. 1–6 (2008)
3. Jia, H., Martinez, A.M.: Support vector machines in face recognition with occlusions. In: Computer Vision and Pattern Recognition, pp. 136–141 (2009)
4. Martinez, A., Benavente, R.: The ar face database. Tech. Rep. 24, Computer Vision Center, Bellatera (1998), http://www.cat.uab.cat/Publications/1998/MB98

A Pathway-Based Classification Method That Can Improve Microarray-Based Colorectal Cancer Diagnosis

Hong-Qiang Wang[1,*], Xin-Ping Xie[2], and Chun-Hou Zheng[3.]

[1] Intelligent Computing Lab, Hefei Institute of Physical Science,
CAS, P.O.1130, 230031, Hefei, China
hqwang126@126.com
[2] Department of Mathematics and physics,
Anhui University of Architecture, 230022, Hefei, China
[3] College of Information and Communication Technology,
Qufu Normal University, 276826, Rizhao, China

Abstract. Colorectal cancer is the third most commonly diagnosed cancer in the world. Microarray-based colorectal cancer diagnosis is increasingly paid more and more attentions. In view of a number of pathway information available in the KEGG database, this paper proposes to model pathways for colorectal cancer diagnosis, and as a result, a pathway-based classification method is developed. The proposed method can extract pathway information through modeling gene associations in a pathway via regression. Experimental results on six pathways show that the proposed method remarkably improves the performance of microarray-based colorectal cancer diagnosis.

Keywords: DNA microarray, colorectal cancer, pathways, cancer diagnosis.

1 Introduction

Colorectal cancer is a cancer characterized by neoplasia in colon, rectum, or vermiform appendix. Medically, it originates from the epithelial cells lining the colon or rectum of the gastrointestinal tract, most frequently as a result of mutations in the Wnt signaling pathway that artificially increase signaling activity. The mutations can be inherited or are acquired, and must probably occur in the intestinal crypt stem cell.

Clinically, colorectal cancer can take many years to develop, and early detection of colorectal cancer can greatly improve the chances of a cure. Traditionally, diagnostics of colorectal cancer is based on the morphological appearance of colon tissues. These methods of such kind include digital rectal exam (DRE), Fecal occult blood (FOBT) and Endoscopy and so on. However, their diagnostic performance is dissatisfying due to the genetic essence of colorectal cancer. The recently developed DNA microarray technology can simultaneously measure the expression of tens of thousands of genes in cells, making it possible to early predict cancer in terms of a genome-wide gene expression profile. Biologically, microarray-based cancer diagnosis will be more reliable and more accurate compared with a histomorphology-based method [1-3]. As the first trial, using the technology, Alon et al measured and analyzed gene expression profiles

D.-S. Huang et al. (Eds.): ICIC 2011, LNBI 6840, pp. 610–617, 2012.

in tens of tumor/normal colorectal tissues [4]. This study and thereafter ones show that gene expression profile can help characterize the differences between normal and cancerous colon tissues, and thus lead to a more reliable diagnosis of colorectal cancer.

However, microarray data have many non-typical properties for traditional data analysis tools [5]. Clinically, it remains challenging to efficiently apply microarray technology to the diagnosis of colorectal cancer. Over the past decade, a number of machine learning and statistical methods, such as artificial neural networks [6, 7], support vector machines [8, 9], have been applied to microarray-based cancer diagnosis. A primary challenge in applying these methods is the "small N, large P" problem, where the number P of variables (genes) is typically much larger than the number N of available samples. So far, there are still no methods that can efficiently deal with this problem [10]. Generally, one has to pre-select a small set of significantly differentially expressed genes and then build a classifier using them [11]. Such kind of methods limits the accuracy of cancer diagnosis due to the loss of association information between genes.

Microarray-based cancer research is shifting from single gene expression signature-based ones to pathway signature-based ones [12-16]. It is critical and necessary to model pathway activity to understand genetically complex cancer and to further classify cancer. Gatza et al have proposed a pathway-based method for human breast cancer classification[13]. Briefly speaking, the authors first used principal component analysis to identify principle signatures hidden in pathways, and then built a probit regression model to classify breast cancer. Rapaport et al proposed to apply spectrum graph theory to analyzing a pathway, and as a result, a classification model was built to classify irradiated and non-irradiated yeast strains [20]. One main disadvantage of these methods is that these methods did not model a pathway in terms of gene associations and can not essentially represent the activity of the pathway in a sample.

In this paper, we propose a novel method for pathway modeling and use it for colorectal cancer diagnosis. We assume that the expression of a gene in a pathway can be predicted by the other genes. In particular, we use the sigmoid function to fit non-linear associations in a pathway and the regularized least squares regression algorithm to extract pathway information. In evaluation experiment, we apply the proposed method to the Alon's data and compare the method with several conventional classification methods.

2   Materials and Methods

2.1   Data Sets

Using Affymetrix Oligonucleotide Array, named Hum6000, containing about 65,000 features each containing '107 strands of a DNA 25-mer oligonucleotide, Alon et al measured gene expression profiles in 40 colon adenocarcinoma specimens and 22 paired normal colon tissues [4]. In data preprocessing, the authors used a neighbor-based outlier filter to keep 2000 of the 65,000 features in the final data set. In addition, to compensate for possible variations of measurement between arrays, the expression value of each gene was standardized by dividing its original intensity by the average of all genes on that array and multiplying by a nominal average intensity of 50. We used the data set for method evaluation in this paper. The data set can be downloaded online from http://www.molbio.princeton.edu/colondata.

2.2   Classification Method

Modeling Pathway

Assume a pathway consisting of p genes and let $\mathbf{x} = [x_1, x_2, \cdots x_p]^T$ denote their measured expression levels in a sample. In view of the complex relationships between genes in the pathway [23, 24], we non-linearly transform gene expression data via a sigmoid transformation as follows:

$$\mathbf{v} = [v_1, v_2, \cdots, v_p]^T, v_i = \left(1 + e^{-\beta(\frac{x_i - \mu_i}{\sigma_i})^2}\right)^{-1}, i = 1, 2, \cdots, p \qquad (1)$$

where $\beta \in (0,1]$ is a constant, referred to as the sigmoid parameter, and μ_i and σ_i are the location and width parameters respectively, which can be estimated as the mean and standard deviation of gene expression levels. Note that the transformed expression levels are used as the expression levels below.

To model a pathway, we assume that the expression of a gene g in the pathway is correlated to and can be predicted by those of the other p-1 genes. Let y denote the expression level of gene g and $\breve{\mathbf{v}}$ the expression vector of the rest p-1 genes, mathematically, we have the following equations:

$$\begin{cases} \hat{y} = f(\breve{\mathbf{v}}) \\ \epsilon = (y - \hat{y}) \sim N(0,1) \\ E(y, \breve{\mathbf{v}}) = 0.5\epsilon^2 \end{cases} \qquad (2)$$

where $f(\breve{\mathbf{v}})$ is a prediction function of y by $\breve{\mathbf{v}}$, which specifies the internal relationship in the pathway, and E defines a cost function whose value measures the disagreement between the observed expression of gene g and the predicted one by other p-1 genes. For simplicity, we take $f(\breve{\mathbf{v}}) = A\breve{\mathbf{v}} + b$, where b is a constant and will be equal to zero for data centered to be of zero mean, and A is referred to as the association coefficient vector, whose element a_i measures the association of gene i to gene g: a positive value denotes an expression promotion on gene g while a negative one an expression repression.

We solve the model by using the regularized least squares (RLS) algorithm. Without loss of generality, consider a microarray data consisting of l samples: the expression levels of p-1 genes $\breve{V} = [\breve{\mathbf{v}}^1, \breve{\mathbf{v}}^2, \breve{\mathbf{v}}^3, \cdots, \breve{\mathbf{v}}^l]$ and the expression levels of gene g $\mathbf{y} = [y_1, y_2, \cdots y_l]$. According to the RLS algorithm, the A can be solved as

$$A = (\breve{V}^T \breve{V} - \alpha I)^{-1} \breve{V}^T \mathbf{y} \qquad (3)$$

where α represents the regularization parameter. For the detail of the RLS solution, readers can refer to [25].

A Pathway-Based Cancer Classifier

The model in Eq.(2) reflects gene association between one gene in a pathway and the others, and for each gene g_i ($i=1,2,..., p$) in the pathway, a cost function E_i can be obtained via Eq.(3). Because not all genes in a pathway are connected to each other, not every gene can be exactly predicted by the left. Therefore, we only consider k genes with highest signal-to-noise ratio (SNR) values (refer to [3]) to model a pathway for constructing a pathway-based colorectal classifier. In particular, let E_i^1 and E_i^2 ($i=1,2,...,k$) represent two sets of cost functions for the two tissue classes, normal and tumor. Given a test sample s, its class will be predicted as follows:

$$c = \arg \min_{j \in \{1,2\}} \left\{ \sum_{i=1}^{k} E_i^j(t) \right\} \tag{4}$$

where $j=1$ and 2 mean the normal and tumor classes, respectively. An interpretation for the diagnosis rule can be put in this way: for a sample, only when it matches the internal relationship of a model will the corresponding cost value approach zero or be small enough.

Performance Evaluation

We employ leave-one-out cross validation (LOOCV) for performance evaluation. Briefly speaking, in LOOCV, each sample in the data set is used as a test set in turn and the classifier is constructed using the rest sample, and finally, the mean correct classification rate, referred to as LOOCV accuracy, is calculated and used as a performance indicator. The LOOCV can overcome the evaluation bias caused when using the alternative random training/test set split strategy, and is especially suitable for small sample problem [26].

3   Experimental Results

We applied the proposed method to the Alon's data set. In the application, six pathways known to be related to colorectal cancer were collected from the KEGG pathway database [27], as listed in Table 1. Each of these pathways was modeled and used to build a classifier for colorectal cancer diagnosis. Since only a part of the genes in these pathways are present in the Alon's data, we used these genes for colorectal cancer diagnosis in our experiment. Table 1 lists the total number of genes in each pathway and the corresponding number of genes also present in the Alon's data. To obtain the best performance for the data, for each pathway setting, we optimized the three parameters of the proposed method using a three-dimensional grid search technique, where β varies in the range [5,1,0.5,0.1,0.05,0.01,0.001], α in the range [10,1,0.5,0.1,0.05,0.01,0.001,0.0001], and k in the range [1, 2,..., p], respectively.

As a result, Table 2 reports the LOOCV accuracies obtained in each pathway case. From Table 2, it can be seen that all six LOOCV accuracies are larger than 77%, and in particular, based on the pathway called path:hsa04010:MAPK signaling pathway, our method achieved the accuracy of up to 86.36%. Such a high accuracy should be related to the large number (25) of genes available in the Alon data, because more genes available in modeling a pathway allow extracting more pathway information for cancer classification.

For performance comparison, several conventional methods, including two support vector machines (SVMs) with linear (linear-SVM) and radial basis function (rbf-SVM) kernels (http://sourceforge.net/projects/svm/), k-nearest neighborhood (KNN) and Fisher linear discriminant (FLD), were implemented in the experiment, and were used to re-classify the Alon's data based on the same sets of genes for each pathway case. To obtain the optimal LOOCV accuracies for these methods, their parameters were optimally chosen. For the linear-SVM, its regularization parameter was varied in a range of $\{2^{12}, 2^{11}, \cdots, 2^{-1}, 2^{-2}\}$, and for the rbf-SVM, its regularization and kernel width parameters were optimized based on a two-dimensional grid search technique within the ranges, $\{2^{12}, 2^{11}, \cdots, 2^{-1}, 2^{-2}\}$ and $\{2^{4}, 2^{3}, \cdots, 2^{-9}, 2^{-10}\}$. Table 2 compares the LOOCV accuracies between the conventional methods and our method. From Table 2, it can be seen that our classifier outperformed all the previous methods for three of the six pathways, and achieved the highest accuracy of 86% among these methods.

To explore why the proposed method performs very well for colorectal cancer diagnosis, we examined the distribution of cost values of samples with respect to the models of normal/tumor classes, as the boxplots shown in Fig.1. In this figure, the top and bottom of each box indicate the 25th and 75%, respectively, and the red horizontal line within each box indicates the median cost values. From Fig.1, it can be seen that normal/tumor samples have a very low cost value with respective to the corresponding models of normal/tumor classes, but a large one with respect to the models of the opposite classes. These results show that the models learned by our method efficiently captured classification information hidden in the pathways, and thus led to the better classification performance than the conventional methods. Generally, SVMs are good at dealing with the high-dimensional problem, and KNN at the multi-class classification problem. In addition, these conventional methods are a general classification method, and are unable to capture the specific pathway information as our method does.

Table 1. Related information of pathways used in the experiment

Names of Pathways	Short name	No. of total genes/genes present in the colon data
path:hsa04350:TGF-beta signaling pathway	hsa04350	11/85
path:hsa05210:Colorectal cancer	hsa05210	8/62
path:hsa04310:Wnt signaling pathway	hsa04310	18/151
path:hsa04210:Apoptosis	hsa04210	10/88
path:hsa04110:Cell cycle	hsa04110	20/128
path:hsa04010:MAPK signaling pathway	hsa04010	25/268

Table 2. Comparison of cross-validation accuracies with previous methods for each pathway

Pathway	Our method	Linear SVM	non-linear SVM	KNN	FLD
hsa04350	0.8159	0.7204	0.825	0.6795	0.7023
hsa05210	**0.7727**	0.6295	0.7295	0.684	0.5795
hsa04310	**0.8136**	0.6273	0.8	0.7136	0.6159
hsa04210	0.7727	0.7477	0.7863	0.7045	0.7227
hsa04110	0.8318	0.7268	0.8404	0.7804	0.6963
hsa04010	**0.8636**	0.7273	0.8409	0.7273	0.6363

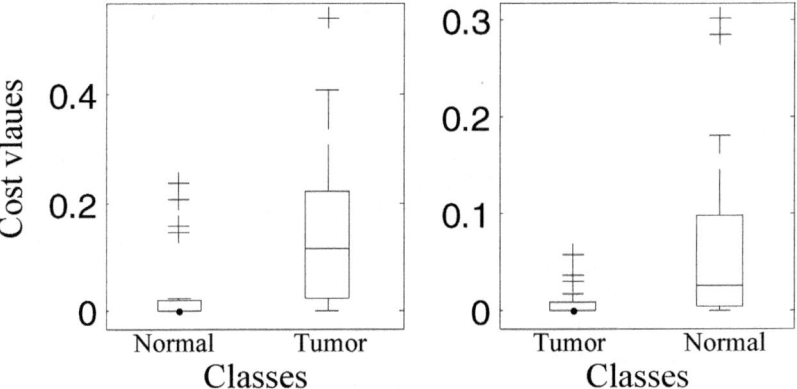

Fig. 1. Boxplots of the cost values of normal/tumor samples with respect to the models for normal class (left) /tumor class (right)

4 Conclusions

Despite recent great advances in microarray data analysis, how to conveniently apply microarray technology to routine clinical diagnostic is still a challenge. For this problem, there are two main issues needing to be deal with: biomarker discovery and the classification performance. In this paper, we have proposed a pathway-based classification method for colorectal cancer diagnosis. The proposed method can extract pathway information through modeling gene associations for cancer classification. Comparison experiments showed the remarkable advantage of the method in classification accuracy over several conventional methods.

Acknowledgement. This work was supported by the grants of the National Science Foundation of China, Nos. 31071168, 30900321, 60975005, 61005010, 60873012, 60973153 and 60905023.

References

1. Liao, J.G., Chin, K.-V.: Logistic regression for disease classification using microarray data: model selection in a large p and small n case. Bioinformatics 23(15), 1945–1951 (2007)
2. Shipp, M., et al.: Diffuse large B-cell lymphoma outcome prediction by gene-expression profiling and supervised machine learning. Nat. Med. 8, 68–74 (2002)
3. Golub, T.R., et al.: Molecular classification of cancer: class discovery and class prediction by gene expression monitoring. Science 286, 531–537 (1999)
4. Alon, U., et al.: Broad patterns of gene expression revealed by clustering analysis of tumor and normal colon tissues probed by oligonucleotide arrays. Proc. Natl. Acad. Sci. USA 96, 6745–6750 (1999)
5. Leek, J.T., Storey, J.D.: Capturing Heterogeneity in Gene Expression Studies by Surrogate Variable Analysis. PLoS Genet. 3(9), e161 (2007)
6. Khan, J., et al.: Classification and diagnostic prediction of cancers using gene expression profiling and artificial neural networks. Nature Medicine 7(6), 673–679 (2001)
7. Wang, Z., Palade, V., Xu, Y.: Neuro-Fuzzy Ensemble Approach for Microarray Cancer Gene Expression Data Analysis. In: 2006 International Symposium on Evolving Fuzzy Systems (2006)
8. Duan, K.-B., et al.: Multiple SVM-RFE for gene selection in cancer classification with expression data. IEEE Transactions on Nanobioscience 4(3), 228 (2005)
9. Yousef, M., et al.: Classification and biomarker identification using gene network modules and support vector machines. BMC Bioinformatics 10(1), 337 (2009)
10. Qiu, P., Wang, Z.J., Liu, K.J.R.: Genomic processing for cancer classification and prediction - Abroad review of the recent advances in model-based genomoric and proteomic signal processing for cancer detection. IEEE Transaction on Signal Processing Magazine 24(1), 100–110 (2007)
11. Zeng, X.-Q., et al.: Dimension reduction with redundant gene elimination for tumor classification. BMC Bioinformatics 9(suppl. 6), 8 (2008)
12. Khosravi-Far, R.: Oncogenic Ras activation of Raf/mitogen-activated protein kinase-independent pathways is sufficient to cause tumorigenic transformation. Mol. Cell. Biol. 16, 3923–3933 (1996)
13. Gatza, M.L., et al.: A pathway-based classification of human breast cancer. Proceedings of the National Academy of Sciences 107(15), 6994–6999 (2010)
14. Huang, E., et al.: Gene expression phenotypic models that predict the activity of oncogenic pathways. Nat. Genet. 34(2), 226–230 (2003)
15. Tomfohr, J., Lu, J., Kepler, T.: Pathway level analysis of gene expression using singular value decomposition. BMC Bioinformatics 6(1), 225 (2005)
16. West, M., et al.: Predicting the clinical status of human breast cancer by using gene expression profiles. Proc. Natl. Acad. Sci. U S A 98(20), 11462–11467 (2001)
17. Segal, E., et al.: From signatures to models: understanding cancer using microar-rays. Nat. Genet. 37, S38–S45 (2005)
18. Tlsty, T.: Cancer: Whispering sweet somethings. Nature 453(7195), 604–605 (2008)
19. Lee, E., et al.: Inferring Pathway Activity toward Precise Disease Classification. PLoS Comput. Biol. 4(11), e1000217 (2008)
20. Rapaport, F., et al.: Classification of microarray data using gene networks. BMC Bioinformatics 8(1), 35 (2007)
21. Basso, K., et al.: Reverse engineering of regulatory networks in human B cells. Nature Genetics 37(4), 382–390 (2005)

22. Calvano, S.E., et al.: A network-based analysis of systemic inflammation in hu-mans. Nature 437(7061), 1032–1037 (2005)
23. Shawe-Taylor, J., Cristianini, N.: Kernel methods for pattern analysis. Cambridge University Press, Cambridge (2004)
24. Du, K.-L., Swamy, M.N.S.: Neural networks in a soft-computing framework. Springer-Verlag London Limited, London (2006)
25. Myers, R.H., Montgomery, D.C., Vining, G.G.: Generalized Linear Models, with Applications in Engineering and the Sciences. John Wiley & Sons, Chichester (2002)
26. McLachlan, G., Do, K.A., Ambroise, C.: Analyzing microarray gene expres-sion data. Wiley, Chichester (2004)
27. Kanehisa, M., et al.: KEGG for representation and analysis of molecular networks involving diseases and drugs. Nucleic Acids Research 38(suppl. 1), 355–360 (2010)

Structure-Function Relationship in Olfactory Receptors

M. Michael Gromiha[1,*], R. Sowdhamini[2], and K. Fukui[3]

[1] Department of Biotechnology, Indian Institute of Technology Madras,
Chennai 600 036, Tamilnadu, India
gromiha@iitm.ac.in
[2] National Center for Biological Sciences, Bangalore, India
[3] Computational Biology Research Center, National Institute of Advanced Industrial Science
and Technology, 2-4-7 Aomi, Koto-ku, Tokyo 135-0064, Japan

Abstract. Olfactory receptors are key components in signal transduction. The sequence and structural analysis of olfactory receptors provides deep insights to understand their function. In this work, we have systematically analyzed the relationship between various physical, chemical, energetic and conformational properties of amino acid residues, and the change of half maximal effective concentration (EC50) due to amino acid substitutions. We observed that the odorant molecule (lignad) as well as amino acid properties are important for EC50. The inclusion of neighboring residues information of the mutants enhanced the correlation. Further, amino acid properties have been combined systematically and we obtained a correlation of 0.90-0.98 with functional data for different (goldfish, mouse and human) olfactory receptors.

Keywords: membrane protein, olfactory receptor, protein function, amino acid properties.

1 Introduction

Membrane proteins perform several functions, including the transport of ions and molecules across the membrane, bind to small molecules at the extra cellular space, recognize the immune system and energy transducers. Olfactory receptors (OR) are membrane proteins, belonging to the G Protein-Coupled Receptor superfamily, which are characterized by the presence of hydrophobic transmembrane domains. The odorant response of an organism by ORs to its environment forms the basis for our understanding in intra-species interactions, host-pathogen interactions, balance of chemicals, cell-cell interactions and other fundamental processes. Further, ORs have been analyzed to understand the mechanism of chloride uptake [1], modulation of signaling [2], functional architecture [3], unitary response [4], structural and functional plasticity at binding pocket [5] and conversion of chemical information into electronic signals in olfactory sensory neurons [6,7].

The importance of specific amino acid residues in ORs and other membrane proteins has been demonstrated through site directed mutagenesis experiments. The experimental data on EC50, maximal velocity of transport, odorant response, percentage uptake of compounds, affinity and specificity has been accumulated in the database for functional residues in membrane proteins [8]. Luu et al. [9] elucidated the features of olfactory receptors for determining ligand specificity using different amino

D.-S. Huang et al. (Eds.): ICIC 2011, LNBI 6840, pp. 618–623, 2012.
© Springer-Verlag Berlin Heidelberg 2012

acid agonists. The structural basis for mouse and human ORs to EC50 data has been analyzed by systematically substituting amino acid residues at different positions [10,11]. On the other hand, computational methods have been proposed to understand the binding affinity of ligands with ORs using the template structure of rhodopsin [12].

In spite of these studies, the role of amino acid properties for the change of EC50 has not yet explored. In this work, we have constructed different datasets for goldfish, mouse and human ORs as well as the mutants with change in EC50 values. The difference in experimental functional data has been related with physical, chemical, energetic and conformational properties of amino acid residues. The important amino acid properties have been brought out. The combinations of amino acid properties and the influence of neighboring residues have been successfully used to relate the experimental functional data.

2   Materials and Methods

2.1   Datasets

We have constructed different datasets using the experimental data, EC50 for goldfish, mouse and human ORs. The final dataset contains 47 data with the following categories: (i) goldfish OR: 12; mouse OR: 28 and human OR: 7.

2.2   Amino Acid Properties

We used a set of 49 diverse amino acid properties [13], which represent physical, chemical, energetic and conformational features, in the present study. The amino acid properties were normalized between 0 and 1. The numerical and normalized values for all the 49 properties used in this study along with their brief descriptions are available on the web at http://www.cbrc.jp/~gromiha/fold_rate/property.html. These properties have been successfully used to understand the folding and stability of proteins [14-18].

2.3   Computational Procedure

The mutation induced changes in property values $\Delta P(i)$ was computed using the equation [14]:

$$\Delta P(i) = P_{mut}(i) - P_{wild}(i) \tag{1}$$

where $P_{mut}(i) - P_{wild}(i)$ are, respectively, the property value of the ith mutant and wild type residues, and i varies from 1 to N; total number of mutants. The computed difference in property values ΔP was related with experimental EC50 or odorant response using single correlation coefficient.

2.4   Local Sequence Effects

The effect of local sequence, $P_{seq}(i)$, was included using the equation [14]:

$$P_{seq}(i) = [\sum_{j=i-k}^{j=i+k} P_j(i)] - P_{mut}(i) \tag{2}$$

where, P_{mut} (i) is the property value of the ith mutant residue and $\Sigma P_j(i)$ is the total property value of the segment of (2k+1) residues ranging from i-k to i+k about the ith residue of wild type.

2.5   Multiple Regression Analysis

We have combined the amino acid properties using multiple regression technique: multiple correlation coefficients and regression equations were determined using standard procedures [19].

3   Results and Discussion

3.1   Relationship between Amino Acid Properties and Change in EC50 upon Mutation: Goldfish OR with Lys Potency

We have computed the changes in amino acid properties using Eqn.1 and related with change in EC50 upon mutation using correlation coefficient. The accessible surface area of unfolding showed a negative correlation of -0.86 with goldfish OR, and other physical properties, bulkiness, volume etc. showed appreciable negative correlation with ΔEC50. Figure 1a shows the relationship between ΔASA and ΔEC50. On the other hand, entropy change has high positive correlation (r = 0.85) with ΔEC50.

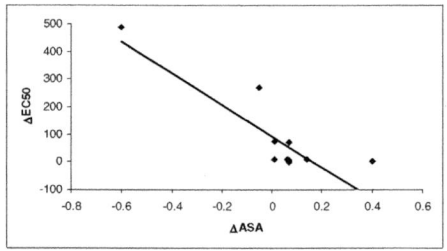

Fig. 1a. Relationship between ΔASA and ΔEC50

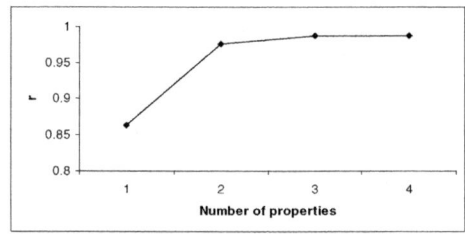

Fig. 1b. Variation of correlation coefficient with number of properties

We have combined different amino acid properties and related with ΔEC50 values. The variation of correlation coefficient with number of properties is shown in Figure 1b. We noticed that the combination of four properties raised the correlation up to 0.988.

3.2 Relationship between Amino Acid Properties and Change in EC50: Mouse OR

Katada et al. [10] measured the EC50 values for the mutants at various position in the transmembrane helices of mouse OR. Figure 2 shows a model for mouse OR and the information about mutated residues.

We have computed the difference in amino acid properties and related with difference in EC50 values. We observed a maximum correlation of just 0.38 and the combination of five properties raised the correlation only up to 0.56. We have included the information on neighboring residues (Eqn. 2), which increased the correlation up to 0.76.

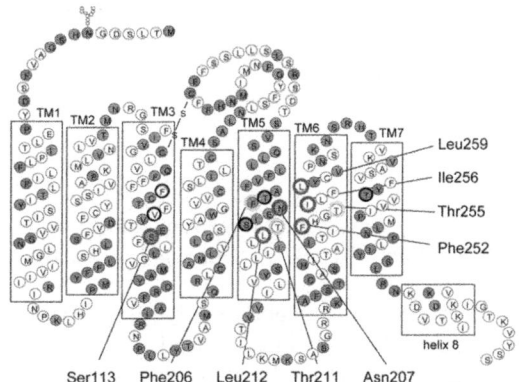

Fig. 2. A model for mouse OR Figure was adapted from [10]

3.3 Relationship between Amino Acid Properties and Change in EC50 upon Mutation: Human OR

Schmiedeberg et al. [17] measured the EC50 values for seven mutants in human OR. We found that bulkiness has a correlation of 0.71 and the information on three neighboring residues increased the correlation up to 0.92 (Figure 3). This analysis reveals the importance of neighboring residues for determining the change in EC50. Further, the combination of just two properties increased the correlation up to 0.999.

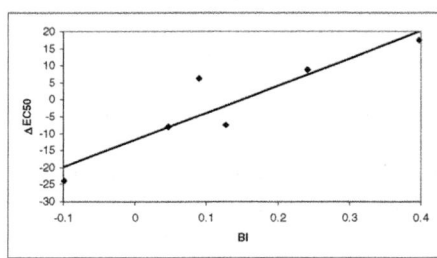

Fig. 3. Relationship between bulkiness and ΔEC50 in human OR

4   Conclusions

We have constructed different datasets for mouse, goldfish and human ORs and the experimental data, EC50. The EC50 values have been systematically analyzed with physical, chemical, energetic and conformational properties of amino acid residues and important properties have been brought out. Further, we have combined the amino acid properties using multiple regression analysis, which relates experimental EC50 very well. The results obtained in the present work would help to understand the importance of amino acid properties to the functions of ORs.

Acknowledgments. This research was partially supported by Indian Institute of Technology Madras research grant (BIO/10-11/540/NFSC/MICH), Department of Biotechnology, Government of India and National Institute of Advanced Industrial Science and Technology, Japan.

References

1. Jaén, C., Ozdener, M.H., Reisert, J.: Mechanisms of chloride uptake in frog olfactory receptor neurons. J. Comp. Physiol. A Neuroethol. Sens. Neural Behav. Physiol. 197, 339–349 (2011)
2. Hall, R.A.: Autonomic modulation of olfactory signaling. Sci. Signal 4, pe1 (2011)
3. Abuin, L., Bargeton, B., Ulbrich, M.H., Isacoff, E.Y., Kellenberger, S., Benton, R.: Functional architecture of olfactory ionotropic glutamate receptors. Neuron 69, 44–60 (2011)
4. Ben-Chaim, Y., Cheng, M.M., Yau, K.W.: Unitary response of mouse olfactory receptor neurons. Proc. Natl. Acad. Sci. U S A 108, 822–827 (2011)
5. Baud, O., Etter, S., Spreafico, M., Bordoli, L., Schwede, T., Vogel, H., Pick, H.: The mouse eugenol odorant receptor: structural and functional plasticity of a broadly tuned odorant binding pocket. Biochemistry 50, 843–853 (2011)
6. Buck, L.B.: Information coding in the vertebrate olfactory system. Annu. Rev. Neurosci. 19, 517–544 (1996)
7. Mombaerts, P.: Genes and ligands for odorant, vomeronasal and taste receptors. Nat. Rev. Neurosci. 5, 263–278 (2004)
8. Gromiha, M.M., Yabuki, Y., Suresh, M.X., Thangakani, A.M., Suwa, M., Fukui, K.: TMFunction: database for functional residues in membrane proteins. Nucleic Acids Res. 37(Database issue), D201–D204 (2009)
9. Luu, P., Acher, F., Bertrand, H.O., Fan, J., Ngai, J.: Molecular determinants of ligand selectivity in a vertebrate odorant receptor. J. Neurosci. 24, 10128–10137 (2004)
10. Katada, S., Hirokawa, T., Oka, Y., Suwa, M., Touhara, K.: Structural basis for a broad but selective ligand spectrum of a mouse olfactory receptor: mapping the odorant-binding site. J. Neurosci. 25, 1806–1815 (2005)
11. Schmiedeberg, K., Shirokova, E., Weber, H.P., Schilling, B., Meyerhof, W., Krautwurst, D.: Structural determinants of odorant recognition by the human olfactory receptors OR1A1 and OR1A2. J. Struct. Biol. 159, 400–412 (2007)
12. Man, O., Gilad, Y., Lancet, D.: Prediction of the odorant binding site of olfactory receptor proteins by human-mouse comparisons. Protein Sci. 13, 240–254 (2004)

13. Gromiha, M.M., Oobatake, M., Sarai, A.: Important amino acid properties for enhanced thermostability from mesophilic to thermophilic proteins. Biophys. Chem. 82, 51–67 (1999)
14. Gromiha, M.M., Oobatake, M., Kono, H., Uedaira, H., Sarai, A.: Role of structural and sequence information in the prediction of protein stability changes: comparison between buried and partially buried mutations. Protein Eng. 12, 549–555 (1999)
15. Huang, L.T., Gromiha, M.M.: Reliable prediction of protein thermostability change upon double mutation from amino acid sequence. Bioinformatics 25, 2181–2187 (2009)
16. Ou, Y.Y., Chen, S.A., Gromiha, M.M.: Classification of transporters using efficient radial basis function networks with position-specific scoring matrices and biochemical properties. Proteins 78, 1789–1797 (2010)
17. Gromiha, M.M., Selvaraj, S.: Inter-residue interactions in protein folding and stability. Prog. Biophys. Mol. Biol. 86, 235–277 (2004)
18. Gromiha, M.M.: Influence of long-range contacts and surrounding residues on the transition state structures of proteins. Anal. Biochem. 408, 32–36 (2011)
19. Grewal, P.S.: Numerical Methods of Statistical Analysis. Sterling Publ., New Delhi (1987)

First Report of Knowledge Discovery in Predicting Protein Folding Rate Change upon Single Mutation

Lien-Fu Lai[1], Chao-Chin Wu[1], and Liang-Tsung Huang[2,*]

[1] Department of Computer Science and Information Engineering,
National Changhua University of Education, Changhua 500, Taiwan
[2] Department of Biotechnology, Mingdao University, Changhua 523, Taiwan
larry@mdu.edu.tw

Abstract. To explore the mechanism of protein folding is one of the important topics in protein research. The accurate prediction of protein folding rate change is helpful and useful in protein design. In earlier study, we have firstly analyzed the prediction of folding rate change upon single point mutation and constructed a non-redundant dataset of F467. F467 consists of 467 mutants with various features and widely distributed on secondary structure, solvent accessibility, conservation score and long-range contacts. In this work, we therefore focused on effectively developing the knowledge in F467 dataset. We have systematically analyzed the dataset and presented several representative data mining techniques, including decision tree, decision table and association rule algorithms. Furthermore, we have interpreted, evaluated, and compared the knowledge obtained from different techniques. The experimental results showed that the present approach can effectively develop the knowledge in the dataset and the outcomes can increase the understanding of predicting protein folding rate change upon single mutation. We have also created a website with related information about this work and it is freely available at http://bioinformatics.myweb.hinet.net/kdfreedom.htm.

Keywords: Knowledge discovery, data mining, protein folding rate, single point mutation.

1 Introduction

In protein folding process, a polypeptide chain of amino acid residues folds into a three-dimensional structure. The protein folding rate measures the tendency of folding (slow/fast) from its unfolded state to its native state.

Currently, several methods have been proposed for predicting protein folding rates from amino acid sequence and/or structural information [1-7]. Further, in order to understand and predict protein folding rates due to amino acid replacements, we have developed a method based on quadratic regression models for discriminating the accelerating and decelerating mutants [8], which is the first report about the issue.

---

* Corresponding author.

D.-S. Huang et al. (Eds.): ICIC 2011, LNBI 6840, pp. 624–631, 2012.
© Springer-Verlag Berlin Heidelberg 2012

Knowledge discovery in bioinformatics is driven massively by available biological experimental data [9]. To discover novel knowledge or generate novel hypothesis from the existing data is extremely useful in life science [10]. In this work, we adopted a knowledge discovery approach and focused on effectively developing the knowledge for increasing the understanding of predicting protein folding rate change upon single mutation.

2   Materials

2.1   Dataset Construction and Preprocessing

In earlier work, we have originally constructed a dataset (F467) of protein mutants with experimental k_f values and relevant attributes [8]. F467 is obtained from published reports in the literature as well as from two kinetic databases, PFD [11] and kineticDB [12]. The collection conditions are: (i) all single mutants, (ii) kinetic type two, and (iii) k_f values are extrapolated to zero concentration (i.e. water). The folding rate change upon single mutation is calculated by $\Delta k_f = k_f^{mutant} - k_f^{wild}$, where k_f^{mutant} and k_f^{wild} are k_f values for mutant and wild-type residues, respectively. F467 consists of 467 unique mutants from 15 different proteins and the number of accelerating and decelerating mutants is 79 and 338, respectively.

2.2   Accessible Surface Area (ASA)

F467 consists of 467 mutants with various attributes, such as secondary structure, solvent accessibility, conservation score and long-range contacts etc. Earlier literatures have revealed that solvent accessibility is an important parameter to predict protein mutant stability [13]. In this study, we have utilized accessible surface area (ASA) to quantify solvent accessibility as the leading attribute. We obtained the information of accessible surface area (ASA) for all the wild type residues from DSSP, Dictionary of Secondary Structure of Proteins [14].

3   Methods

In this work, we focused on effectively developing the knowledge in predicting protein folding rate change upon single mutation. We presented a knowledge discovery approach, which systematically analyzes a dataset, implements different data mining techniques, and then interprets, evaluates, compares and integrates the mined knowledge.

A rule-based knowledge representation (KR) can provide more intuitive interpretation than many other types of representation. Therefore, we implemented three rule-based mining techniques, i.e. association rule [15], decision tree [16], and decision table [17] algorithms. The first one can generate association rules and the others generate classification rules.

3.1   Decision Tree Mining Technique

The PART algorithm [18] is based on partial decision trees [19] for generating accurate rules. It adopts the separate-and-conquer strategy, where there are three main steps: (i) builds a rule; (ii) removes the instances it covers; and (iii) continues creating rules recursively for the remaining instances until none are left.

3.2   Decision Table Mining Technique

In this study, the inducer of decision table majority algorithm [17] was implemented to build decision tables. A decision table consists of two components: a schema and a body. The schema is a set of attributes that are included into the table; and the body is a list of instances with values defined by the attributes in the schema. The set of instances with the same values for attributes in the schema is named a cell and the cell can be presented by if-then rules.

Once a decision table is available, the outcome of a query instance can be predicted by the following criteria. Let L be the set of instances in the cell that matches the query instance over the attributes in the schema.

(i) If $L \neq \emptyset$, return the majority class of instances in L for the nominal class (or the average class for numeric classes).

(ii) Otherwise ($L = \emptyset$), return the majority (or average) class of instances in the decision table.

3.3   Association Rule Mining Technique

The Apriori [15] is a classic algorithm for learning association rules. The algorithm does not consider the transactions (instances) in the database. Instead, it generates iteratively the candidate itemsets (attribute sets) by using only the itemsets that are found large in the previous pass. Therefore, the candidate itemsets having k items can be generated by joining large itemsets having k-1 items, and deleting those that contain any subset that is not large. This procedure results in the generation of a much smaller number of candidate itemsets.

3.4   Performance Measure Scores

Various measure scores were used to evaluate the performance of predicting the decreasing and accelerating mutants. Given a rule: IF A THEN B, where A is the antecedent and B is the consequent, support is calculated by SU $= N_A / N \times 100\%,$ where N_A is the number of instances with antecedent A of the rule. Similarly, confidence (CO) is calculated by CO$= N_{A \cup B} / N_A, \times 100\%$ where $N_{A \cup B}$ is the number of instances with both antecedent A and consequent B. Support (SU) can quantify the coverage of the rule on a dataset and confidence (CO) can indicate the correctness of the rule.

4   Results and Discussions

4.1   Direct Effects of Wild and Mutant Residues on Knowledge Discovery

The constructed F467 dataset comprised various properties of attributes. Firstly, we have mined F467 dataset with two basic ones, wild and mutant residues. As shown in

Figure 1, we analyzed the distribution of residues based on different amino acid types. The significant difference appears at residue types D and I. We compared the means of two types of instances (decelerating and accelerating) and performed a paired t–test with a level of significance of $\alpha = 0.05$, showing the means are not statistically significant (p-value is $0.5 > 0.05$).

Further, we have implemented three mining techniques on F467 dataset with the attributes of wild and mutant residues. Table 1 lists the mined rules with higher confidence and the rules take the form: IF A THEN B, where A is called the antecedent or premise and B is the consequent or conclusion. The rule ID is named according to the mining strategies and attributes. The corresponding relationships between them are listed at the Knowledge Base page of the website.

Interestingly, we found that rules A1-1 and B1-2 are obtained from association and decision tree techniques, respectively but both the rules have the same antecedent "WILD = I" and consequent "Decreasing", which means 'If the wild residue of the protein is Isoleucine then the folding rate of the mutant may be decreasing". For the rule A1-2, the antecedent "MUTANT = G" has a consequent "Decreasing". These results agree with the observations in Figure 1.

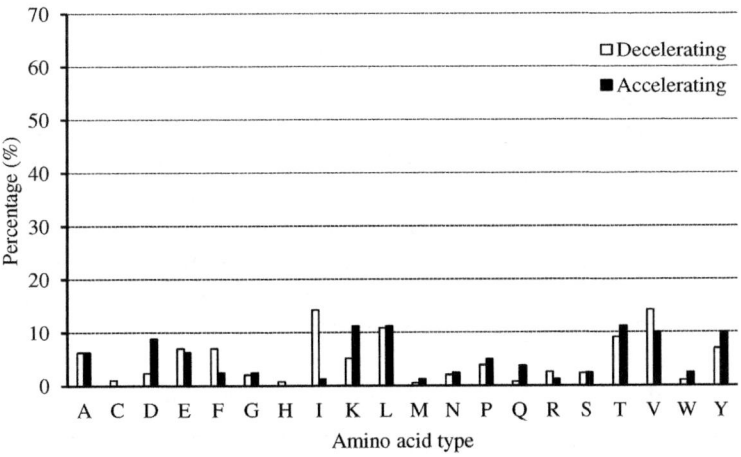

Fig. 1. Distribution of amino acid types for wild residues. It exhibits the percentage difference of the wild residue between decelerating and accelerating instances

Table 1. Antecedent and consequent of mined rules with confidence > 90% and support > 5% from the attributes of wild and mutant residues

Rule ID	Antecedent	Consequent	Support (%)	Confidence (%)	Mining technique
A1-1	WILD = I	Decreasing	12.0	98.2	Association
A1-2	MUTANT = G	Decreasing	17.1	90.0	Association
B1-2	WILD = I	Decreasing	12.0	98.2	Decision tree

4.2   Environmental Impacts of Neighboring Residues on Improving Knowledge Discovery

To observe the environmental impacts of neighboring residues, we have further mining F467 dataset by appending the attributes of neighboring residues. Table 2 lists additional rules that are mined after including 3 neighboring residues toward both the N- and C- terminus. N0i/C0i denotes the i-th residue along N-/C-terminus at mutation position. We observed that the decision table technique significantly mines more rules with high confidence, showing the attributes of neighboring residues may increase the performance of the knowledge discovery.

Table 2. Antecedent and consequent of additional rules with confidence > 90% and support > 5% after including the atrributes of neighboring residues

Rule ID	Antecedent	Consequent	Support (%)	Confidence (%)	Mining technique
A2-2	N03 = E	Decreasing	11.3	90.6	Association
C2-5	N02 = G	Decreasing	6.2	93.1	Decision table
C2-6	N02 = D	Decreasing	5.8	92.6	Decision table
C2-8	N02 = E	Decreasing	9.2	90.7	Decision table

4.3   Comparison between Different Mining Techniques

We have further compared the performance among three mining techniques. In Table 3, three techniques have similar rule length (1) when using 8 attributes. Association and decision tree give less rules (3 and 2, respectively) than decision table (21) but higher support (13.5 and 12.7, respectively) and confidence (92.9% and 92.8%, respectively). The results indicate that the association and decision tree techniques perform better than decision table technique on the data mining of F467 dataset. We analyzed the overlap of rules mined from three techniques, showing that only 1 rule overlaps between association and decision tree.

Table 3. Summary of mined rules with different mining techniques

Mining technique	Number of attributes	Number of rules	Average length	Average support (%)	Average confidence (%)
Association	8	3	1	13.5	92.9
Decision tree	8	2	1	12.7	92.8
Decision table	8	21	1	4.8	84.4

4.4   Persistent Strong Rules for Discriminating Folding Rate Change

A rule common to more than one data mining method is called persistent rule. In addition, a rule meeting a defined confidence level is called strong rule. The persistent strong rule discovery suggests these rules that may be the most robust, consistent, and noteworthy among the candidate rules [20].

Accordingly, we have examined all the mined rules by the criterion of the persistent strong rule with a confidence level of higher than 90%. For F467 dataset, the persistent strong rule we found is rule A1-1 mined by association technique (or B1-2 by decision tree technique) in Table 1. The concept of persistent rules can improve the usefulness of rules.

4.5   Knowledge Usefulness for the Instances with Different ASA Ranges

Furthermore, we have tested the persistent strong rule on F467 dataset with different ASA ranges. The instances of the dataset were divided into four groups according to the ASA of the mutation position. As shown in Figure 2, the instances are generally distributed among different ranges, i.e. the buried/partially buried region (ASA<20%), partially exposed (20% < ASA < 50%) and exposed (ASA > 50%) regions.

Next, the persistent strong rule A1-1 was tested on each group. In Figure 2, the results show that the rule has highest support within the buried regions and performs good confidence (>90%) for the instances with different ASA ranges.

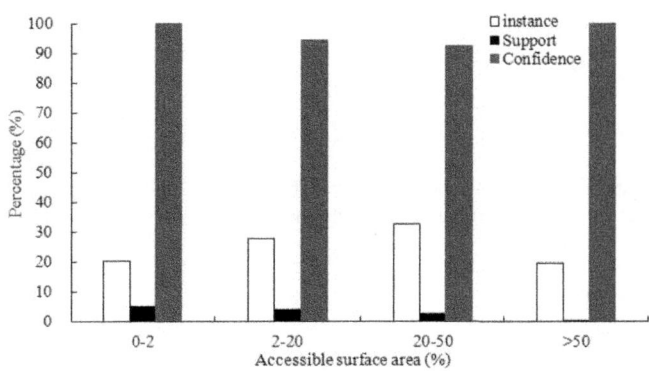

Fig. 2. Instance distribution based on different accessible surface area and the performance of the persistent strong rule.

4.6   A Web Knowledge Base

We have constructed a website for developing the knowledge base for predicting stability change upon single mutation. The website is freely available at http://bioinformatics.myweb.hinet.net/kdfreedom.htm. The main page of the website is showed in Figure 3. It provides the information about the work, including Introduction, Dataset, Knowledge Base and References items.

Fig. 3. Combination screenshot of multiple web pages. The website provides the mined rules for predicting the folding rate change upon single mutation.

5 Conclusions

In conclusion, we have proposed a knowledge discovery approach on developing the rules of predicting folding rate change for protein mutants. The approach integrates three different data mining techniques and the results show that the mined rules perform well on the tests. We have created a website with the information about this work and it is available at http://bioinformatics.myweb.hinet.net/kdfreedom.htm.

This study is a first but preliminary report about the knowledge discovery in predicting protein folding rate change up single mutation. Meanwhile, we are working on including and analyzing more attributes, covering the secondary structure, conservation score and contacts. We suggest that the present approach can be adopted to develop the potential knowledge and increase the understanding of predicting protein folding rate change upon single mutation.

Acknowledgments. The authors would like to thank Dr. M. Michael Gromiha for valuable comments. Dr. Wu acknowledges the support from National Science Council (Grant Numbers NSC98-2221-E-018-008-MY2), Taiwan.

References

1. Gromiha, M.M.: Importance of native-state topology for determining the folding rate of two-state proteins. J. Chem. Inf. Comput. Sci. 43, 1481–1485 (2003)
2. Gromiha, M.M.: A statistical model for predicting protein folding rates from amino acid sequence with structural class information. J. Chem. Inf. Model 45, 494–501 (2005)
3. Huang, J.T., Tian, J.: Amino acid sequence predicts folding rate for middle-size two-state proteins. Proteins 63, 551–554 (2006)
4. Gromiha, M.M., Thangakani, A.M., Selvaraj, S.: FOLD-RATE: prediction of protein folding rates from amino acid sequence. Nucleic Acids Res. 34, W70–W74 (2006)
5. Huang, J.T., Cheng, J.P., Chen, H.: Secondary structure length as a determinant of folding rate of proteins with two- and three-state kinetics. Proteins 67, 12–17 (2007)
6. Gromiha, M.M.: Multiple contact network is a key determinant to protein folding rates. J. Chem. Inf. Model 49, 1130–1135 (2009)
7. Micheletti, C.: Prediction of folding rates and transition-state placement from native-state geometry. Proteins 51, 74–84 (2003)
8. Huang, L.T., Gromiha, M.M.: First insight into the prediction of protein folding rate change upon point mutation. Bioinformatics 26, 2121–2127 (2010)
9. Hu, X., Pan, Y.: Knowledge discovery in bioinformatics: techniques, methods, and applications. Wiley Interscience, Hoboken (2007)
10. Fred, A., Dietz, J.L.G., Liu, K., Filipe, J.: IC3K 2009. Communications in Computer and Information Science, vol. 128. Springer, Heidelberg (2011) (revised selected papers)
11. Fulton, K.F., Devlin, G.L., Jodun, R.A., Silvestri, L., Bottomley, S.P., Fersht, A.R., Buckle, A.M.: PFD: a database for the investigation of protein folding kinetics and stability. Nucleic Acids Res. 33, D279–D283 (2005)
12. Bogatyreva, N.S., Osypov, A.A., Ivankov, D.N.: KineticDB: a database of protein folding kinetics. Nucleic Acids Res. 37, D342–D346 (2009)
13. Gromiha, M.M., Oobatake, M., Kono, H., Uedaira, H., Sarai, A.: Role of structural and sequence information in the prediction of protein stability changes: comparison between buried and partially buried mutations. Protein Eng. 12, 549–555 (1999)
14. Kabsch, W., Sander, C.: Dictionary of protein secondary structure: pattern recognition of hydrogen-bonded and geometrical features. Biopolymers 22, 2577–2637 (1983)
15. Agrawal, R., Srikant, R.: Fast Algorithms for Mining Association Rules in Large Databases. In: VLDB 1994: Proceedings of the 20th International Conference on Very Large Data Bases, pp. 487–499. Morgan Kaufmann Publishers Inc., San Francisco (1994)
16. Quinlan, J.R.: Induction of decision trees. Machine Learning 1, 81–106 (1986)
17. Kohavi, R.: The Power of Decision Tables. In: Lavrač, N., Wrobel, S. (eds.) ECML 1995. LNCS, vol. 912, pp. 174–189. Springer, Heidelberg (1995)
18. Frank, E., Witten, I.: Generating accurate rule sets without global optimization. In: Proc. 15th International Conf. on Machine Learning, pp. 144–151. Morgan Kaufmann, San Francisco (1998)
19. Quinlan, J.R.: C4.5: programs for machine learning. Morgan Kaufmann Publishers, San Mateo (1993)
20. Senthil kumar, A.: Knowledge discovery practices and emerging applications of data mining: trends and new domains. Information Science Reference, Hershey, Pa (2011)

Detection of Protein Spots from Complex Region on Real Gel Image

Cheng-li Sun[1,2], Yong Xu[1,*], Jie Jia[2], and Yu He[2]

[1] Bio-Computing Research Center, Shenzhen Graduate School,
Harbin Institute of Technology, Shenzhen, 518055, China
[2] Institute of Space Information and Security Technology,
Nanchang Hangkong University, Nanchang, 330063, China
laterfall2@yahoo.com.cn

Abstract. Protein spot detection is an important step in gel image analysis. The results of spot detection may substantially influence gel image analysis. One of the most challenges of spot detection is separation of spots from complex region, where spots are overlapped and saturated. In this paper, we propose a spot detection algorithm which is capable of detection of spots from complex region. The proposed approach is realized by using European distance transform and centroid marker based watershed. Firstly, we apply a standard marker-controlled watershed algorithm to roughly segment the gel image into multiple regions. For those complex region containing several overlapped spots, we estimate spot centroids through a European distance transform. The estimated centroids are then served as spot markers in marker-controlled watershed algorithm to make a fine spot segmentation. Experimental results show the proposed method can obtain high detection performance in real gel images, and can accurately identify overlapped spots in complex region, when compared to other popular spot detection methods.

Keywords: gel electrophoresis, spot detection, watershed transform.

1 Introduction

Proteomics is the field that studies multi-protein systems, focusing on the interplay of multiple proteins as functional components in a biological system. The leading technique for protein analysis is 2DGE (two-dimensional gel electrophoresis), this technique sorts proteins according to two independent properties: the isoelectric point and molecular weight. Once the complex protein pattern in 2D gel is visualized, the two dimensional gel is then scanned by an image capture device. After gel scanning, protein species are depicted as spots of varying size and positions in the resulting gel image, which can be further analyzed in order to quantitate the relative amount of each protein in the gel sample or to compare the samples from a gel database [1].

The most common approaches in spot detection are spot modeling and spot segmentation. Model based approaches try to model a spot's intensity as a Gaussian distribution or diffusion function [2-4], the spot's quantity and boundaries are then

---

* Corresponding author.

D.-S. Huang et al. (Eds.): ICIC 2011, LNBI 6840, pp. 632–640, 2012.

derived from the model. The model-based method works well in protein spots with standard shape, but is unsuitable for irregular or clustered spots since those spots cannot be modeled accurately by any function. Segmentation is the process of segmenting the gel into small regions, each of which may contain one spot. Watershed is one of popular spot segmentation techniques. However, over-segmentation is a well-known problem in watershed [5]. Cutler et al. proposed pixel value collection approach [6], they try to segment protein spot through stepwise shareholding the gel density. In the pixel-based approach, unwanted background variation outside the protein spots are not removed prior to the analysis, making it sensitive to noise and artifacts [7]. Liu propose spot detection for a 2-DE gel image using a slice tree with confidence evaluation [8]. The proposed method takes slices of a gel image in the gray level direction, and builds them into a slice tree, which in turn is adopted to perform spot detection and confidence evaluation. A comprehensive review of published spot detection methods and commercial software for 2-D gel images analysis can be referred to [9, 10].

One of the main challenges of spot detection is segmentation of spots from complex region. In the complex region overlapped with many spots, the peak area becomes flat and the shape can be irregular, therefore it is hard to prediction with the traditional method. Efrat et al. formulate the problem of segmentation for complex region as a Linear Programming formulation [11]. They use linear programming method to conform the elliptic equation to the edge of complex region. However, their approach is time-consuming and inaccuracy because only two-dimensional information adopted. Savelonas proposed an active contour-based spot segmentation approach [12], their approach is based on the observation that the spot boundaries within the overlapped region are associated with local intensity minima, but this method is fail in the case of spot over-saturation. Srinark proposed a suite of spot detection algorithms [13], which employs the watershed segmentation, k-means analysis and distance transform to perform spot segmentation. Their method can obtain good performance on the complex region but suffering from discontinuity of segmented results.

In this paper, we present a new spot detection algorithm which is capable of detection of spots from complex region. The presented algorithm is realized based on a two-phase procedure. In the first phase, we apply a standard marker-controlled watershed algorithm to roughly segment the gel image into multiple regions. While the second phase is to identify and perform fine segmentation on the complex regions, which are under-segmented regions after the first phase segmentation and should be further analyzed. We specially address the issue of detection of spots from complex region. The proposed complex region segmentation algorithm consists of following steps: Firstly, we estimate the spot centroids within complex region through a European distance transform. The estimated spot centroids are then combined with their corresponding outer marker to form a modified marker map. Finally, we make a fine segmentation to the complex region using watershed transform with the modified markers. Experimental results demonstrate that the proposed approach can obtain high performance on real gel image containing lots of overlapped spots, and less sensitive to the influence of noise and background.

2   Marker-Controlled Watershed

Watershed transform technique is a region-growing algorithm that analyzes an image as a topographic surface. It detects minima of the gradient image and grows these minima according to the gradient values. However, direct application of the watershed segmentation algorithm by flooding from the regional minimums generally leads to over-segmentation due to noise, and much over-segmentation will render the result of watershed transform useless. To alleviate the over-segmentation effects, marker controlled watershed transform are usually applied [14, 15]. The marker image used for watershed segmentation is a binary image consisting of either single marker points or larger marker regions, where each connected marker is placed inside an object of interest. Marker-controlled watershed floods from the markers instead of flooding from the regional minimums. The standard marker-controlled watershed algorithm can be divided into following steps.

(1) Find the inner markers in gel image using H-minima transform.
(2) Find the outer markers.
(3) Construct marker image using inner markers and outer markers.
(4) Compute the gradient image G.
(5) Minima superimpose markers on the gradient image G.
(6) Apply watershed algorithm on the modified gradient image to do spot segmentation.

In the gel image analysis, because the protein spots which we want to find out are dark against to the background, we can use this feature to mark them. The inner marker is very important because it represents the spot object which we want to segment. It can be obtained through H-minima transform. Let $I(x, y)$ denotes the pixel intensity of gel image I. The H-minima transform is defined as:

$$H_h(x, y) = R_I^\varepsilon (I(x, y) + h) \ . \tag{1}$$

Where h is a constant and represents a given depth. R and ε represent the morphological reconstruction and erosion operator. By using the H-minima transform, all local minima whose depth is lower than the given h-value are suppressed, and the regional minima of source image is obtained by the following formula:

$$M(x, y) = H_h(x, y) - I(x, y) = R_I^\varepsilon (I(x, y) + h) - I(x, y) \ . \tag{2}$$

The regional minima correspond to inner markers of I, and the inner marker image is a binary image in which pixels are set to 1 identify regional minima and all other pixels are set to 0.

When obtain the inner markers, the next step is to find outer markers. A outer marker defines the scope of the corresponding inner marker. So we can use the watershed transform of inner markers to find their corresponding outer markers. Inner markers and outer markers are then combined to form the marker image M. The pixel value of M is defined by

$$M(x, y) = \begin{cases} 1 & \text{if } (x,y) \text{ belongs to marker} \\ 0 & \text{otherwise} \end{cases} \ . \tag{3}$$

The third step is to remove undesired minima by minima imposition. Here the gradient image instead of the source image is used, and we can use marker image modify the gradient image so that it has regional minima only occur at the pixels belong to markers, It is achieved by

$$G' = R^\varepsilon_{(G+1)\wedge M}(G) \quad .$$ (4)

Where G is gradient of I, and G' is its reconstructed image. the symbol '$\wedge$' stands for point-wise minimum of pixels between two images.

After minima imposition, we can reserve the desired minima and remove the irrelevant minima. In this way, the over-segmentation problem can be solved because the unnecessary minima are removed. Finally, watershed algorithm can be used on the modified gradient image to segment the protein spot contours. Fig.1 (a) shows the standard watershed segmented results. It can see that most of isolated spots were correctly detected after the first phase of segmentation. However some issues remain unsolved. One is spurious spots, and the other is that spots were not correctly detected from complex region. In our previous work [14], we have proposed a spot verification technique to reject the spurious spots. In this work, we will emphasize address the latter issue.

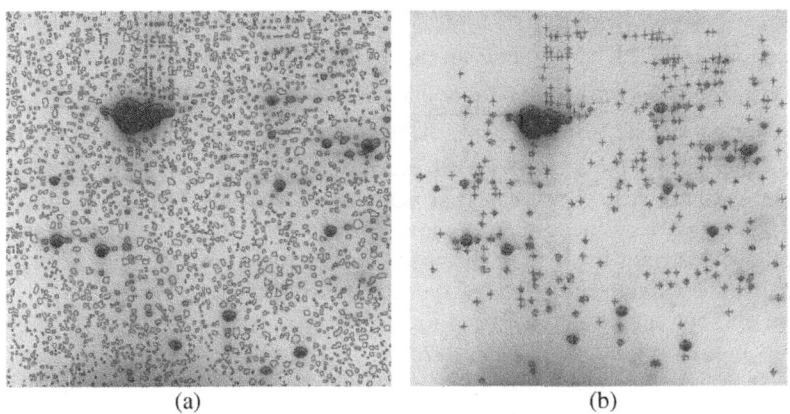

(a)   (b)

Fig. 1. Comparison of segmented results from different phases
(a) Segmented results of first phase (b) Segmented results of second phase

3   Proposed Spot Detection Method

Although marker-controlled watershed methods could effectively handle over-segmentation problem, it only works on the premise that the extracted markers really represent the true objects. When perform inner marker exaction on the complex region, it is difficult to exact corrected markers using H-minima transform because the peaks of overlapped spots may be flat or even disappear. In this paper, we propose a spot detection algorithm which is capable of detection of overlapped spots from complex region. We firstly apply a standard marker-controlled watershed algorithm to roughly segment the gel image into multiple regions. Then identify the complex regions and estimate the spot centroids inside complex region through a distance transform. Finally, the estimated centroids are served as spot inner markers in marker-controlled watershed algorithm to make a fine spot segmentation.

Suppose R_i (i=1,2...N) is a segmented region produced by the initial watershed segmentation, where N is the total number of segmented regions. We compute the shape factor of the segmented region by fitting a 2-D Gaussian function to the spot foreground. The 2-D Gaussian function is defined as

$$G(x, y) = \frac{1}{2\pi\sigma_x\sigma_y} e^{-[(x-\mu_x)^2/2\sigma_x^2+(y-\mu_y)^2/2\sigma_y^2]} .$$

(5)

Where σ_x and σ_y are standard deviations, (μ_x, μ_y) is the centroid location of each segmented region. We consider the region whose the ratio between σ_x and σ_y is less than λ as a complex region, or namely $\min(\sigma_x/\sigma_y, \sigma_y/\sigma_x) \leq \lambda$, where $\lambda = 0.25$.

For each complex region R_i, we estimate spot centroids using a European distance matrix, which is computed through a distance transform of binary image where "1" represents the pixel belong to R_i and "0" represents otherwise case. In the European distance matrix, the distance between that pixel and the nearest non-zero pixel of the image is assigned. All elements of the distance matrix corresponding to the R_i pixels are non-zero, and elements that out of R_i pixels are zero. The elements that have high distance values are identified as the centroid pixels, which represent spot centers. From the distance matrix, we can find many spot centroids that are local maxima of the distance matrix. Since some complex region may have multiple centroids, we merge those centroids the nearest distance between them are less than a threshold, and replace by a new one from the average of the old centroids.

Once all spot centroids have been identified, spot boundaries can be inferred. if a complex region contain multiple centroids, it means that region should be split to several spots. To accurately draw the underlying spot boundaries in complex region, a modified marker-controlled watershed algorithm was proposed. We construct a new marker map for guidance of watershed segmentation, where inner marker was constructed by the estimated spot centroids and outer marker was constructed by the regional boundaries of R_i. Finally, the new marker map was used for minima imposition on the gradient image and followed by watershed transform to make fine segmentation.

In summary, the main steps of complex region segmentation algorithm are as follows:

Suppose s_i is the spot foreground of the complex region R_i, and c_i is the estimated centroid of s_i.

Step 1. For each complex region R_i which has no centroid, calculate the European distance transform $D_i(x) = \inf_{y \in R_i} D(x, y)$,

Where $D(x, y)$ is the European distance between points x and y, and $D_i(x)$ corresponds to the minimum distance between x and s_i.

Step 2. Estimate all the spot centroids by finding the regional maxima in the European distance transform matrix.

Step 3. Choice two nearest centroids from the estimated centroid set. If the distance between these centroids is less than a threshold T_c, compute a new centroid from the average of the old centroids, iterate this process until all the centroid distances are greater than the threshold T_c.

Step 4. Construct a new marker map by the estimated spot centroids and boundaries of R_i.

Step 5. Perform gel image segmentation by the modified marker controlled watershed algorithm.

Fig. 1(b) shows the segmented results of second phase. Compared with that of the first phase, we can see that the overlapped spots are correctly detected, and the segmented results are comparable with human sensing, thus illustrates the effects of our approach.

4   Experiments and Results

The data set we tested is 10 real gel images collected by the gel electrophoresis experiment, which digitized at 1250×1000 pixels at 16-bit grey level depth, and each gel contains approximate 1200 spots. We compared our two-phase marker-controlled watershed (TPMCW) based spot detection method with two well-known 2D-GE image analyze software: PDQuest (version 8.0.1) and GELLAB-II. PDQuest is a state-of-art commercial protein analysis software where spots are detected by a Gaussian model (GM). While GELLAB-II is an open-source protein analysis software and its detection algorithm is based on Laplacian of Gaussian (LOG). The default turning parameters h=4, T_c=8 were adopted for the proposed method. While the parameters of PDQuest and GELLAB-II were manual adjusted according the actually maximal and minimal size of spots.

The comparison of three methods on a real gel sub-image example was shown in Fig.2. It can be observed that the detection results of GM and LOG have many spurious spots, and merely detect overlapping spots without accurately identifying their boundaries. Whereas our proposed method have the least number of spurious spots. Moreover, our method also show high detection effects on complex region, as shown in Fig. 2(b), where GM always tends to false detection and LOG tends to missing detection, as shown in Fig. 2(c) and Fig. 2(d) respectively. The main reason is that in the complex region, many spots are overlapped and the foreground intensities may be of over-saturation, it is hard to prediction with the traditional model based method or peak finding based method.

For exactly comparison the performance of proposed method, we assess the detection results using the following measures: true-positives rate(TP), false-positives rate(FP), and false-negatives rate(FN). Table 1 shows the TP, FP and FN of spot detection by the three methods. The results show our method have slightly higher TP over the other two methods. Compared the FP results we can see that the detection methods of GM and LoG is mostly sensitive to noise with high FP value, our method obtain best performance with 3.1% on gel image dataset. This result indeed indicates that our proposed method has greater immunity to the noise since our method is based on homogeneity of intensity.

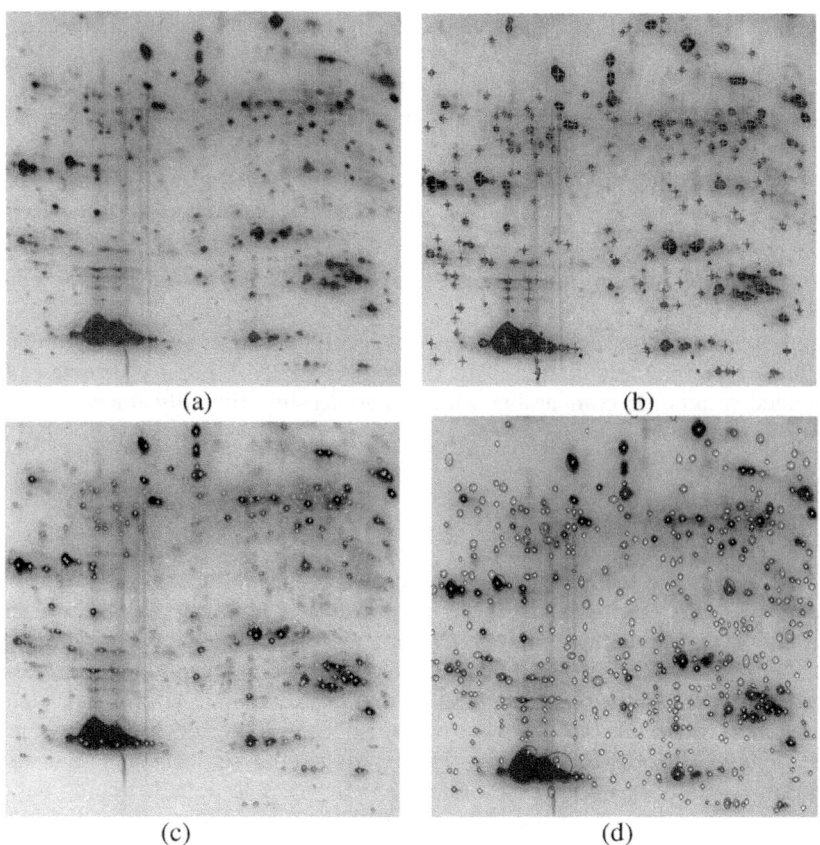

Fig. 2. Comparison of spot detection methods with real gel image
(a) Source gel image (b) TPMCW method (c) GM method (d) LOG method

Table 1. Summary of spot detection results

Methods	TP(%)	FP(%)	FN(%)
Gaussian model	97.5	4.6	5.8
LOG	96.5	3.9	6.7
TPMCW	97.9	3.1	3.0

5   Conclusions

In this work, we present a new spot detection algorithm which is capable of detection of spots from complex region. The presented algorithm is realized based on a two-phase procedure. In the first phase, we roughly segment gel image into a lot of spot regions using standard marker-controlled watershed method. The segmented complex regions are then identified and performed a fine segmentation in the second phase by a modified marker-controlled watershed transform. Unlike traditional approaches, our method does not depend on spot models and is less sensitive to the artifacts. It can steadily obtain high detection performance in real gel images, compared with other two popular methods: GM and LOG.

Acknowledgments. This article is partly supported by National Nature Science Committee of China under Grant Nos. 61071179, 60973098, 60803090 and 60902099, 60973048 the Fundamental Research Funds, and the Foundation of Jiangxi Provincial Department of Education (No. GJJ11517). foundation of Nanchang Hangkong university (No. 2009ZC56).

References

1. Rabillouda, T., Chevalletb, M., Lucheb, S.: Two-dimensional gel electrophoresis in proteomics: Past, present and future. Journal of Proteomics 73, 2064–2077 (2010)
2. Pleissner, K.P., Hoffmann, F., Kriegel, K., Wenk, C., Wegner, C., Sahlstrohm, A., Oswald, H., Alt, H., Fleck, E.: New algorithmic approaches to protein spot detection and pattern matching in two dimensional electrophoresis gels. Electrophoresis 20, 755–765 (1999)
3. Iakovidis, D.K., Maroulis, D., Zacharia, E., Kossida, S.: A Genetic Approach to Spot Detection in two-Dimensional Gel Electrophoresis images. In: Proceedings of International Conference on Information Technology in Biomedicine (ITAB), Ioannina, Greece (2006)
4. He, F., Xiong, B., Sun, C., Xia, X.: A laplacian of gaussian-based approach for spot detection in two-dimensional gel electrophoresis images. In: Li, D., Liu, Y., Chen, Y. (eds.) CCTA 2010, Part IV. IFIP Advances in Information and Communication Technology, vol. 347, pp. 8–15. Springer, Heidelberg (2011)
5. Malpica, N., et al.: Applying watershed algorithms to the segmentation of clustered nuclei. Cytometry 28, 289–297 (1997)
6. Cutler, P., Heald, G., White, I., Ruan, J.: A novel approach to spot detection for two-dimensional gel electrophoresis images using pixel value collection. Journal of Proteomics 2, 392–401 (2003)
7. Rye, M.B., Færgestad, E.M., Martens, H.: An Improved Pixel-Based Approach for Analyzing Images in Two-Dimensional Gel Electrophoresis. Electrophoresis 29, 1382–1393 (2008)
8. Liu, Y., Chena, S., Liu, R.: Spot detection for a 2-DE gel image using a slice tree with confidence evaluation. Mathematical and Computer Modelling 50, 1–14 (2009)
9. Berth, M., Moser, F., Kolbe, M., Bernhardt, J.: The state of the art in the analysis of two-dimensional gel electrophoresis images. Microbiol. Biotechnol. 76, 1223–1243 (2007)
10. Dowsey, A.W., Dunn, M.J., Yang, G.Z.: The role of bioinformatics in two-dimensional gel electrophoresis. Proteomics 3(8), 1567–1596 (2003)
11. Efrat et, A., Hoffmann, F., Kriegel, K., et al.: Geometric Algorithms for the Analysis of 2D electrophoresis Gels. Journal of Computational Biology 9(2), 299–315 (2002)

12. Savelonas, M., Maroulis, D., Mylona, E.: Segmentation of Two-Dimensional Gel Electrophoresis Images containing Overlapping Spots. In: Proceedings of the 9th International Conference on Information Technology and Applications, Larnaca, Cyprus (2009)
13. Srinark, T., Kambhamettu, C.: An image analysis suite for spot detection and spot matching in two-dimensional electrophoresis gels. Electrophoresis 29, 706–715 (2008)
14. Sun, C., Wang, X.: Spot segmentation and verification based on improved marker based watershed transform. In: International Conference on Computer Science and Information Technology, vol. 8, pp. 63–66 (2010)
15. Tsai, M.-H., Hsu, H.-H., Cheng, C.C.: Watershed-based Protein Spot Detection in 2D-GE Images. In: Proceeding of the International Computer Symposium, pp. 1334–1338 (2006)

A Novel Approach to Clustering and Assembly of Large-Scale Roche 454 Transcriptome Data for Gene Validation and Alternative Splicing Analysis

Vitoantonio Bevilacqua[1,3,*], Fabio Stroppa[1],
Stefano Saladino[1], and Ernesto Picardi[2,4]

[1] Dipartimento di Elettrotecnica ed Elettronica
Politecnico di Bari - Via Orabona, 4, Bari, 70125, Italy
[2] Dipartimento di Biochimica e Biologia Molecolare "E. Quagliariello"
University of Bari, Bari
[3] e.B.I.S. s.r.l. (electronic Business in Security), Spin-Off of Polytechnic of Bari
Via Pavoncelli, 139 Bari, Italy
[4] Istituto di Biomembrane e Bioenergetica del Consiglio Nazionale delle Ricerche, Bari
bevilacqua@poliba.it

Abstract. In this paper we propose a new implementation of EasyCluster, a robust software developed to generate reliable clusters by expressed sequence tags (EST). Such clusters can be used to infer and improve gene structures as well as discover potential alternative splicing events. The new version of EasyCluster software is able to manage genome scale transcriptome data produced by massive sequencing using Roche 454 sequencers. Moreover it can speed up the creation of gene-oriented clusters and facilitate downstream analyses as the assembly of full-length transcripts. Finally available annotations can now be employed to improve the overall clustering procedure. The new EasyCluster implementation embeds also a graphical browser to provide an overview of results at genome level, simplifying the interpretation of findings to researchers with no specific skills in bioinformatics.

Keywords: EasyCluster, expressed sequence tags, 454 reads, alternative splicing.

1 Introduction

Expressed sequence tags and full-length cDNAs represent an invaluable source of evidence for inferring reliable gene structures and discovering potential alternative splicing events (1). However, to fully exploit their biological potential, correct and reliable EST clusters are required. To fill this gap we developed the program EasyCluster that resulted the most accurate when compared to software representing the state of the art in this field (2). Recent technological advances are dramatically increasing the number of available transcriptome reads. Indeed, EST-like sequences can now be generated by pyrosequencing using Roche 454 platforms which generate approximately one million reads per run. Handling such huge amount of EST-like

---

* Corresponding author.

D.-S. Huang et al. (Eds.): ICIC 2011, LNBI 6840, pp. 641–648, 2012.

data is basic to detect alternative splicing events, improve gene annotations or simply create gene-oriented clusters for expression studies. Sometimes EST-like data provide a fragmented overview of their genomic loci of origin and, thus, transcript assembly may be an optimal solution to annotate user-produced sequences. For these reasons we propose here a new implementation of EasyCluster able to manage genome scale transcriptome data and generate reliable gene-oriented clusters from 454 reads.

The new version of EasyCluster software can facilitate downstream analyses allowing the assembly of full-length transcripts per cluster, improving the clustering procedure by available annotations (if any) and embedding a graphical browser to inspect results at genome level.   Differently from other existing clustering methods based on BLAST or Blat, EasyCluster takes advantage from the well-known EST-to-genome mapping program GMAP (3). The main benefit of using GMAP is that it can perform a very quick mapping of whatever expressed sequence onto a genomic sequence attended by an alignment optimization. In particular, GMAP can detect splicing sites according to a so defined "sandwich" dynamic programming that is organism independent. Given a genomic sequence and a pool of EST-like data from Roche 454 sequencers, EasyCluster initially builds a GMAP database of the genomic sequence to speed-up the mapping and a local EST database storing all provided expressed sequences. Subsequently, it runs GMAP program and parses results in order to create an initial collection of pseudo-clusters. Each pseudo-cluster is obtained by grouping EST-like reads according to the overlap of their genomic coordinates on the same strand. In the next step, EasyCluster refines the EST grouping by including in each cluster only expressed sequences sharing at least one splice site. Unspliced ESTs are finally added to each refined cluster and an *ad hoc* procedure is used to correct potential GMAP errors occurring near splice sites. Refined clusters are used to assemble full-length transcripts, valuate the alternative splicing extent and provide gene expression levels according to user supplied annotations. EST groupings produced by EasyCluster can be graphically inspected by means of an embedded browser. The main clustering procedure is a two steps methodology:

- a pseudo-cluster is identified according to overlapping genome coordinates;
- each pseudo-cluster is refined identifying sub-clusters sharing at least one splice site. This represents a real novelty over previous genome-based clustering methods.

In summary, the main goal of this work is to develop and re-implement a new algorithm with the following features:

- Optimization of the procedure to refine clusters reducing biases due to erroneous GMAP mapping. This improvement is expected to reduce the overall number of gene-oriented clusters and increase the reliability of generated EST groupings.
- Implementation of an ad hoc algorithm to check EST-to-genome alignments near splice sites in order to correct mapping errors mainly due to indels in homopolymeric regions (generally introduced by pyrosequencing).

2   Materials and Methods

Materials

EasyCluster reads in mapping info in the standard GFF3 (Generic Feature Format Version 3) format as generated by GMAP. GFF3 format extends the most common GFF format maintaining the compatibility. In particular, GFF3 format: 1) adds a mechanism for representing more than one level of hierarchical grouping of features and sub features; 2) separates the ideas of group membership and feature name/id 3) constrains the feature type field to be taken from a controlled vocabulary; 4) allows a single feature, such as an exon, to belong to more than one group at a time; 5) provides an explicit convention for pairwise alignments; 6) provides an explicit convention for features that occupy disjoint regions.

EasyCluster input data are represented by mapped ESTs and the related GFF3 file reports individual exons and their corresponding genome coordinates. Each EST is considered as an individual mRNA transcript. GFF3 files are obtained by GMAP using as input a multi-fasta file containing raw EST data. Such EST-like sequences have been generated by simulation of RefSeq sequences employing a 454 error model. All simulations have been carried out by *Metasim* software [5].

Methods and Work-Flow

The software workflow is shown below:

- A single GFF3 file is used as input;
- The file is parsed in memory exploiting JAVA classes of a specific developed library;
- ESTs are grouped according to their features (mostly the 'exon' feature of GFF3 file);
- Initial clusters are generated according to EST overlaps (only genomic coordinates are used in this step);
- Refined Clusters are created according to the biological criterion of splice site sharing;
- Potential GMAP errors are corrected using an appropriate re-aligning strategy.

The clustering is done for each genomic region found in the GFF3 file and separately for each chromosome.Transcript orientation (mapping strand) has also been taken into account.

ESTs in the GFF3 file have been classified in two types: Unique and Mixed:

- Unique ESTs are those mapping in unique genome locations and thus only one alignment can be retrieved;
- Mixed ESTs instead show multiple paths and thus more than one alignment is generated.

ESTs have also been divided into Spliced and Unspliced:

- Spliced ESTs are those having more than one exon;
- Unspliced ESTs are those defined intronless (only one exon is available).

Not all ESTs are used in the first clustering procedure but only Unique and Spliced sequences. In summary, EasyCluster groups each EST found in the GFF3 file according to the specific genomic region and mapping coordinates other than strand. The first algorithm phase concerns data acquisition and creation of appropriately designed dedicated data structures. Moreover, ESTs are sorted by increasing coordinates along the acquisition step.

First Clustering

The first clustering procedure begins by instantiating an object representing the cluster. Such object is part of our library and its start and end coordinates are set-up to ones of the first EST included into the cluster.

For each EST belonging to the EST set to be clustered, we verify if its start coordinate is smaller than the cluster end coordinate. If this condition is satisfied the EST is added to the corresponding cluster.

Furthermore, if the EST belongs to the cluster, it is necessary to verify whether or not its end coordinate is smaller than the cluster end coordinate. If the condition is true, the end coordinate of the new Cluster is updated accordingly. The above procedure is performed on the overall set of ESTs sorted by coordinates in ascending order.

Second Clustering

After the generation of pseudo-Clusters, EasyCluster refines each group to create biologically useful sub-groups. In this phase ESTs are assigned to the same group according to the biological criterion of splicing site sharing. During this second clustering step, EasyCluster keeps track of all ESTs having common coordinates at exon level, excluding start or end coordinates of initial or terminal exons since they do not mark splice site boundaries.

Figure 3 shows an example of a refined Cluster. Some ESTs share the same Splicing Site, i.e. they have same donor and/or acceptor coordinates.

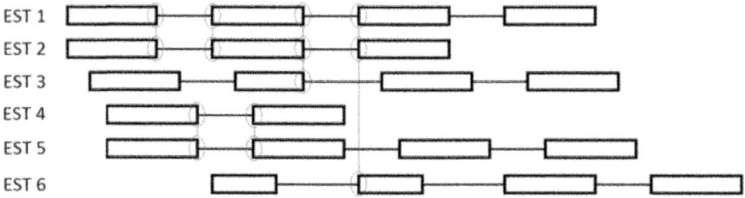

Fig. 1. Example of Clusters with sharing Splicing Sites

The Second algorithmic procedure to Cluster ESTs is more complex than the First one. The most important aspect is the high computational complexity. Indeed, the algorithm has to check for each exon of each EST if a particular donor/acceptor coordinate is present in other ESTs belonging to the same pseudo-Cluster. At computational level the search can be optimized because ESTs are already pre-ordered by their start coordinates avoiding unavailing loops. In addition, it is useful to consider the following particular case:

• The splicing site search is done on a particular exon and scrolling through the subsequent exon list checking whether the coordinate of each of them is the same or not; the search will always least up to the last exon of the last EST because more then one EST may have a splicing site in common, and all information so obtained are being stored for future utilization. When it is necessary to analyze an exon splice site which has already been analyzed by a previous iteration, it means that all the next ESTs, with the same splicing site, will be treated as already considered; for this reason, it is useless and inefficient to continue the research for this coordinate because it has already been done and the information sought are already been provided.

Finally, the schema shown below represents the information to be found according to the algorithm:

- A Second Cluster vector in which storing effective Clusters is established. Each vector element will be a list of EST names and two dictionaries structures (key-value). In these dictionaries keys are donors and acceptors coordinates, while the corresponding values will be a list of all ESTs sharing the splicing site under investigation.
- We create another dictionary structure in which each key is the EST name and the value the element index of the vector defined above after the searching procedure. In other words, this dictionary contains all ESTs of each pseudo-Cluster. Moreover, Java has a simple procedure to verify the presence or the absence of a given EST in one of the newly created cluster during the iteration.

New improvements over the first EasyCluster clustering method

Starting point is that input GFF3 file may contain errors due to 454 sequencer or GMAP mapping. For example an EST could be assigned to a singleton (a cluster with only one EST) even though according to its real mapping strand it should be included in other overlapping cluster but in opposite orientation. Such errors could be caused by a variety of factors leading to wrong Clusters in overlap with real Clusters including a significant number of ESTs (called here Master Clusters). If these Clusters, called here Satellites Clusters, are due to mapping errors they could be corrected trying to realign erroneous ESTs to the Master Cluster.

A specific master cluster corresponds to the cluster with the highest number of ESTs among every cluster of the same set. In this way we take advantage of deep coverage by high-throughput sequencing. In contrast, a Satellite Cluster is a set of two or more ESTs in which start and end coordinates fall within the end and start coordinates of a longer master Cluster (but with opposite orientation). Clusters having only one coordinate inside of the two coordinates of another Cluster are not considered in overlap (as in the case of First Clustering procedure).

Specifically, the new strategy to fix potential errors of reads is based on the creation of a *Cluster's Profile*, containing the ordered list of all the donors and acceptors of every EST of a given Cluster, considering also their occurrences.

Pre-processing

The first step includes an algorithm to identify the *Satellites Clusters* over a *Master Cluster*, which could be reconciled in a unique Cluster. Consider that this approach will be applied to all *Satellites Clusters* independently to the stand of the *Master* Cluster.In this specific step EasyCluster verifies if every EST belongs to overlapping Clusters and compares such EST with the Master Cluster Profile. This comparison is carried out by checking if in the region surrounding each   splicing site of the considered EST there is a correspondence of the same type in the *Cluster Profile* (taking into account also site occurrence). If the correspondence is found, then EasyCluster tries to realign the EST using the Smith-Waterman algorithm [4].

Fixing methods

This kind of method is used in two main occasions:

- every ESTs of the *Satellite Clusters* is compared to the *Master Cluster Profile*
- every EST of a Cluster is compared with the own *Cluster Profile*

According to our strategy, corrections will be applied to overlapping clusters as well as the cluster itself. In this procedure the Smith-Waterman algorithm is used. Sequences are provided by 454 reads in FASTA format (file extension .fasta or .fna) from which it is possible to extract nucleotides for every single splicing site. By using the Smith-Waterman algorithm on each splicing site it is possible to provide a score for each alignment according to a simple scoring table. If there is a high alignment score between the erroneous EST and the expected correct genomic region, the EST alignment will be fixed according to this finding. The region subjected to Smith-Waterman alignment is the one surrounding each splice site.

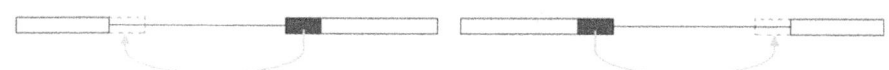

Fig. 2. On the left, donor splicing site on the first exon is in line with a donor of the Profile of Cluster belonging to the right neighbourhood searched: part of the next exon is shifted to the first exon. On the right, donor splicing site on the first exon is in line with a donor of the Profile of Cluster belonging to the left neighbourhood searched: part of this exon is shifted to the next exon.

The goal of this part of the algorithm is to check the shift of an EST substring (or rather a portion of its Exons) considering all Splicing Sites of the *Cluster Profile*. To this aim EasyCluster works   as follow:

- For all donor sites of each EST (as thus, for each Exon) the algorithm verifies a potential correspondence; in other words, it checks if a specific coordinate of each EST Exon under analysis is present in the *Profile of Cluster*.
- This kind of check is done in a region surrounding the splicing site consisting in 10 nucleotides, 5upstream and 5 downstream the selected site.
- Then the algorithm verifies if there is a correspondence of coordinates in the *Profile* for each coordinate of nucleotide in neighborhood, testing if coordinates *"splicing_site_+1"*, *"splicing_site_+2"*, *"splicing_site_+3"* (and so on) belong to the *Profile*.
- If a correspondence is found, the Smith-Waterman algorithm is used to verify the quality of the alignment onto the corresponding genomic region . The exon under investigation is then cut according to Smith-Waterman result and   ready to be shifted to the next exon (or to the previous).

Figures below graphically explain our fixing algorithm.

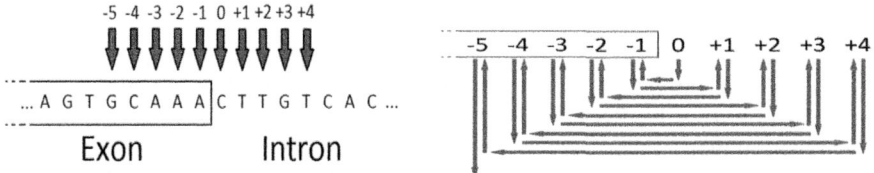

Fig. 3a. Research in neighborhood   **Fig. 3b.** Search Algorithm

3   Results

We tested our method in diverse executions of MetaSim and, thus, using different multifasta files containing EST-like sequences generated according to the 454 error model. Table 1 summarizes preliminary results.

The algorithm recognizes potential errors included in the GFF3 file and proposes a correction. Consider that the number of clusters decreases by an average of 28%, making more reliable the final result. Most importantly, our implementation does not include unspliced ESTs (at least in this preliminary release) and, thus, not all gene loci are completely covered. For this reason the final number of clusters per starting RefSeq is higher than the expected one. The inclusion of Unspliced ESTs will fix this aspect. However, each final cluster covers a specific region of the corresponding gene locus and thus all predicted clusters are correctly part of the same RefSeq genomic region.

Table 1. Preliminary Results

RefSeq	ESTs	Exons	Elaborated Clusters from Second Clustering	Fixed Exons in overlapping Clusters	Fixed Exons in each Cluster	Final number of Clusters	% of fixed Clusters
NM_000824	4918	7998	10	3	7	7	30,00%
NM_000827	4930	7547	14	8	21	11	21,43%
NM_000828	4939	7751	27	13	48	19	29,63%
NM_000829	4943	7770	16	6	23	11	31,25%

Finally, it been realized a graphical brwoser in order to facilitate the analysis of the processing results. EasyCluster draws PNG files during the clustering procedure which represents all the clusters created (either first cluster than second).

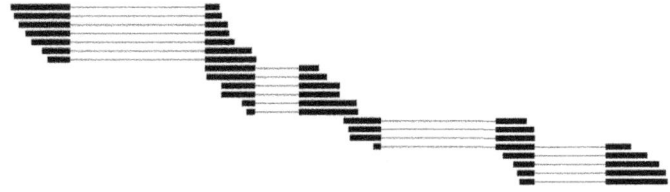

Fig. 4. This is an example of clusters: rectangles represent exons and lines represents introns

4   Conclusions

The EasyCluster application provides clusters of 454 sequences ready to be used in gene prediction pipelines, to detect alternative splicing events or expression profiles. Our new implementation has been mainly conceived to handle huge amount of EST-like reads produced by Roche 454 sequencers and supplies a unique tool to cluster such transcriptome reads (very soon we will also able to implement an assembly algorithm).   Differently to the previous version, EasyCluster can includ unspliced reads and takes benefit from available annotations. For each cluster, the program can valuate the alternative splicing pattern and will reconstruct the most likely full-length transcript sequences according to read-to-genome alignments. During the clustering, EST alignments are refined near exon-intron boundaries in order to reduce potential mapping errors due to GMAP. The new EasyCluster program is in Java programming language and, thus, platform independent. Performances and accuracy have been extensively tested on simulated 454 reads by MetaSim software. Our preliminary results indicate that the new EasyCluster implementation is highly efficient to manage and analyze deep transcriptome data from Roche 454 technology.

References

1. Nagaraj, S.H., Gasser, R.B., Ranganathan, S.: A Hitchhiker's Guide to Expressed Sequence tag (EST) Analysis. Briefings in Bioinformatics 8, 6–21 (2007)
2. Picardi, E., Mignone, F., Pesole, G.: EasyCluster: A Fast and Efficient Gene-oriented Clustering Tool for Large-scale Transcriptome Data. BMC Bioinformatics 10(suppl. 6), 10 (2009)
3. Wu, T.D., Watanabe, C.K.: GMAP: A Genomic Mapping and Alignment Program for mRNA and EST Sequences. Bioinformatics (Oxford, England) 21, 1859–1875 (2005)
4. Smith, T.F., Waterman, M.S.: Identification of Common Molecular Subsequences. Journal of Molecular Biology 147, 195–197 (1981)
5. Richter, D.C., Ott, F., Auch, A.F., Schmid, R., Huson, D.H.: MetaSim—A Sequencing Simulator for Genomics and Metagenomics. PLoS ONE 3(10), e3373 (2008)

Sparse Maximum Margin Discriminant Analysis for Gene Selection

Yan Cui[1], Jian Yang[1], and Chun-Hou Zheng[2]

[1] School of Computer Science and Technology, Nanjing University of Science and Technology,
Nanjing, Jiangsu, China
yancui128@gmail.com, csjyang@mail.njust.edu.cn
[2] College of Information and Communication Technology,
Qufu Normal University, Rizhao, Shandong, China
zhengch99@126.com

Abstract. Dimensionality reduction is necessary for gene expression data classification. In this paper, based on sparse representation, we propose a sparse maximum margin discriminant analysis (SMMDA) method for reducing the dimensionality of gene expression data. It could find the one dimension projection in the most separable direction of gene expression data, thus one can use sparse representation technique to regress the projection to obtain the relevance vector for the gene set and select genes according to the vector. Extensive experiments on publicly available gene expression datasets show that SMMDA is efficient for gene selection.

Keywords: Sparse maximum margin, gene selection, gene Expression Data.

1   Introduction

The rapid development of microarray technologies, which can simultaneously assess the expression level of thousands of genes, makes the precise, objective, and systematic analyses, and diagnoses of human cancers possible. Gene expression data from DNA microarray can be characterized by many variables (genes), but with only a few observations (experiments). Mathematically, they can be expressed as a gene expression matrix, where each row represents a gene, while each column represents a sample or a patient for tumor diagnosis.

Statistically, the fact of the very large number of variables (genes) with only a small number of observations (samples) makes most of the classical classification methods infeasible. Fortunately, this problem can be avoided by selecting only the relevant features or extracting the essential features from the original data, where the former methodology is belonging to feature selection or subset selection and the latter is just feature extraction. For feature selection, i.e. selecting a subset of genes from the original data, many related works on tumor classification have been reported [1,2].

In recent years, many nonparametric discriminant analysis methods have been developed, such as maximum margin criterion (MMC) [3]. It measures between-class scatter based on marginal information. In this paper, we propose a new maximum margin characterization method by virtue of sparse representation using for gene

D.-S. Huang et al. (Eds.): ICIC 2011, LNBI 6840, pp. 649–656, 2012.

selection. The presented method, named sparse maximum margin discriminant analysis (SMMDA), could find the one dimension projection in the most separable direction of gene expression data. Thus one can use sparse representation technique to regress the projection to obtain the relevance vector for the gene set, and select genes according to the vector.

1.1   Maximum Margin Criterion

Given samples $(x_1, y_1), (x_2, y_2), ..., (x_n, y_n) \in \mathbb{R}^m \times \{C_1, ..., C_l\}$, where pattern x_i is a m-dimensional vector, y_i is the corresponding class label for sample x_i $(i = 1, 2, ..., n)$. Maximum margin criterion aims to maximize the distances between classes after the transformation, and the criterion is [3]

$$J = \frac{1}{2} \sum_{i=1}^{l} \sum_{j=1}^{l} p_i p_j d(C_i, C_j) \tag{1}$$

Then we obtain

$$J = tr(\mathbf{S}_b - \mathbf{S}_w) \tag{2}$$

Since $tr(\mathbf{S}_b)$ measures the overall variance of the class mean vectors, a large $tr(\mathbf{S}_b)$ implies that the class mean vectors scatter in a large space. On the other hand, a small $tr(\mathbf{S}_w)$ implies that every class has a small spread. Thus, a large J indicates that patterns are close to each other if they are from the same class but are far from each other if they are from different classes. The derivation is listed in literature [3].

Considering a mapping W, we would like to maximize

$$J(W) = tr(S_b^W - S_w^W) = tr(W^T(S_b - S_w)W) \tag{3}$$

The optimal projection W_{opt} is chosen as the matrix with orthogonal columns $[w_1, ..., w_n]$ following the criterion

$$W_{opt} = \arg\max_{W} tr(W^T(S_b - S_w)W) = \arg\max_{W} \sum_{w_i}^{n} w_i^T(S_b - S_w)w_i \tag{4}$$

Here $\{w_i | i = 1, ..., n\}$ are the generalized eigenvectors of the equation $(S_b - S_w)w = \lambda w$ corresponding the n largest generalized eigenvalues $[\lambda_1, ..., \lambda_n]$. Compute the d largest generalized eigenvalues and the corresponding eigenvector matrix W of $(S_b - S_w)$, and obtain the d-dimensional data $Y = W^T X$.

1.2   Sparse Maximum Margin Discriminant Analysis for Gene Selection

We denote the gene expression matrix as $X = (x_{ij})_{m \times n}$, where each row represents a gene, while each column represents a sample. s is the projection of the most separable direction of gene expression data for two-class cases, which can be obtained by MMC, i.e., $s = w_1^T X$, where w_1 is the first eigenvector of the equation $(S_b - S_w)w = \lambda w$.

We construct the model as follows:

$$s = \alpha^T X \qquad (5)$$

In this equation, s can be seen as the signal and X is the dictionary. However, α might not be sparse. We would like to find the vector α as sparse as possible i.e., $\|\alpha\|_0$ as small as possible. A sparse α indicates the relevance of the corresponding genes to s. In the vector α, the nonzero elements correspond to the genes that have contributions to s, whereas the zero ones correspond to the genes that have no contribution to s. Based on this idea, we can select genes according to the nonzero elements of α. To this end, we build the following model

$$\min\|\alpha\|_0 \quad \text{subject to } \alpha^T X = s \qquad (6)$$

It is an NP-hard problem, but if the solution is sparse enough, the solution of Eq.(6) is equivalent to the following L_1-minimization problem [4]:

$$\min\|\alpha\|_1 \quad \text{subject to } \alpha^T X = s \qquad (7)$$

$$\tilde{\alpha} = abs(\alpha) \qquad (8)$$

Here $\tilde{\alpha}$ is the relevance vector for the gene set, it is absolute score of sparse vector α. $\tilde{\alpha}_j, j \in \{1, 2, ..., m\}$ indicates the relevance of j th gene to s. The larger the $\tilde{\alpha}_j$ is, the higher classification ability the j th gene has. The genes then sorted in descending order according to the $\tilde{\alpha}_j$. Finally, we select the first M genes for sample classification.

It is also easy to extend the gene selection algorithm to multi-class cases. Suppose there are l pattern classes $C_1, ..., C_l$. We convert the multi-class cases into two-class cases in the following way: C_i is viewed as one class and the remaining is viewed as the other class, s^i is the projection in the most separable direction of gene expression data for the i th two-class case. Then we can have

$$\min\|\alpha_i\|_1 \quad \text{subject to } \alpha_i^T X = s^i \qquad (9)$$

$$\tilde{\alpha} = \sum_{i=1}^{l} abs(\alpha_i) \qquad (10)$$

The algorithm can be summarized as follows:

Step 1: For the i th two-class cases $(i = 1, 2, ..., l)$, compute the projection s^i $(i = 1, 2, ..., l)$ by using MMC;

Step   2:   For each s^i $(i = 1, 2, ..., l)$, calculate α_i by solving Eq.(9);

Step   3:   Obtaining the absolute score $\tilde{\alpha}$ using Eq.(10);

Step   4:   The m genes are sorted in descending order according to the $\tilde{\alpha}_j$, then select the first M genes.

Since it is based sparse representation, we named it sparse maximum margin discriminant analysis (SMMDA).

2   Experimental Results

After achieving the lower dimensional data by gene selection, we adopt a suitable classifier to classify the data. In this paper, we apply K-NN classifier for its simplicity. Moreover, another classifier, the sparse representation classifier [5], was also adopted for comparison in the experiments. The performance of SMMDA is evaluated in comparison with other gene selection methods, such as the ratio of genes between-group to within-group sums of squares (BW) [6] and one-way analysis of variance (ANOVA) [7].

2.1   Two-Class Classification Experiments

Table 1. Description of the data sets

No.	Data sets	Samples	Genes	Classes
1	Colon	62	2000	2
2	Acute leukemia	72	7192	2
3	Breast	97	24188	2
4	Glioma	50	12625	2
5	Prostate	136	12600	2

To evaluate the performance of the proposed method, we apply it to five publicly available datasets, i.e., colon dataset [8], acute leukemia dataset[9], breast dataset [10], glioma dataset[11] and prostate dataset [12]. These datasets used in the experiments are summarized in Table 1.

To obtain reliable experimental results with comparability and repeatability for different numerical experiments, we not only used the original division of each data set for training and testing, but also reshuffled all data sets randomly in the experiments. In total, the numerical experiments were performed with 20 random splitting of the five original data sets. In addition, the randomized training and test sets contain the same amount of samples of each class, as those of the original training and test sets. We built the classification models using the training samples and estimated the correct classification rates using the test set.

Table 2. The mean classification accuracies and standard deviations on the five data sets with gene selection

No.	Methods	Datasets				
		Colon	Acute leukemia	Breast	Glioma	Prostate
1	SMMDA+KNN	91.25±2.61	96.88±3.13	70.61±10.63	75.71±8.67	93.21±3.28
2	BW+K-NN	88.28±2.99	92.19±5.85	69.12±10.05	73.86±7.92	90.65±3.62
3	ANOVA+KNN	86.98±3.65	91.73±4.64	70.32±11.27	74.02±7.85	90.92±4.08

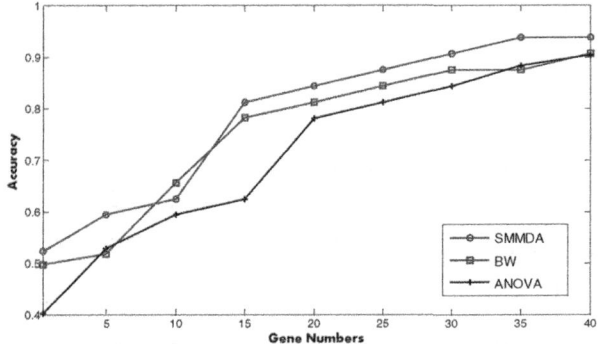

Fig. 1. The mean classification accuracy on the test set of the Colon data set with gene selection

For gene selection, the highest classification results using BW, ANOVA and SMMDA are listed in Table 2. The corresponding gene numbers of five datasets are selected 40, 90, 100, 50 and 100, respectively. For each classification problem, the experimental results were reported by the mean value and the corresponding standard deviation of the accuracy on the original data set and the 20 randomly partitioned sets.

Method 2 in Table 2 performed a preliminary selection of genes on the basis of the ratio of their between-groups to within-groups sum of squares, and then used K-NN for classification. Method 3 is non-parametric one-way Analysis of Variance (ANOVA). The experimental results of the three methods listed in Table 2 are their best classification results, and Fig.1 shows the accuracies of these methods on colon data. In Fig.1, the X-axis represents the M genes, i.e., the number of selected genes used for classification, and the Y-axis represents the mean accuracy of the 21 experiments. We added 5 genes in each step. It is noted that the accuracy is descent when the gene numbers larger than 55.

From Table 2 and Fig. 1 it can be seen that, SMMDA could achieve good classification results. Though both BW and ANOVA are also efficient, yet they are not stable than SMMDA.

We also adopted SRC as classifier in the gene selection experiments. The corresponding experimental results are listed in Table 3. From this experiment, we can achieve the similar results with using K-NN.

Table 3. The classification results on the five data sets with gene selection using SRC

No.	Methods	Datasets				
		Colon	Acute leukemia	Breast	Glioma	Prostate
1	SMMDA+SRC	90.00±2.62	95.31±4.03	70.21±10.05	75.00±7.91	92.24±3.00
2	BW+SRC	88.03±2.74	90.63±4.94	68.42±10.53	73.63±7.62	90.26±4.13
3	ANOVA+SRC	86.88±4.08	91.47±3.86	70.45±10.67	73.50±7.94	90.65±3.87

2.2   Multi-class Classification Experiments

In the multi-class classification experiments, we used another two data sets to further evaluate the performance of the proposed method. One is the Leukemia data set [13], which has become a benchmark in cancer classification. Another data set is the Brain tumor [2].

Table 4. The multi-class classification results by different methods with gene selection

No.	Methods	Datasets	
		Leukemia2	Brain tumor2
1	SMMDA+K-NN	96.55±3.15	86.69±3.36
2	BW+K-NN	93.90±3.55	83.47±3.55
3	ANOVA+K-NN	92.09±4.03	82.83±3.58

The experiment was run 21 times. Table 4 and Fig. 2 are gene selection experimental results. The corresponding selected gene numbers of two datasets are both 500. It should be noted that the accuracy will descent when the gene numbers larger than 800. In Fig.2, we select additional 50 genes in each step. Moreover we apply SRC classifier to the above experiments. Tables 5 lists the experimental results using SRC classifier. From the experiments, we can see that, compared with in two-class classification, the advantage of our method is more obvious in the multi-class classification task.

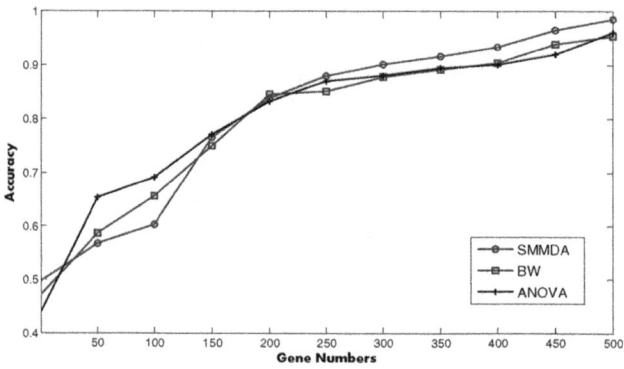

Fig. 2. The mean classification accuracy on the test set of the Leukemia data set with gene selection

Table 5. The multi-class classification results with gene selection using SRC

No.	Methods	Datasets	
		Leukemia	Brain tumor
1	SMMDA+SRC	95.89±3.43	84.36±4.18
2	BW+SRC	92.83±4.02	82.95±3.69
3	ANOVA+SRC	92.00±4.33	80.79±3.61

From the above experiments, it can be found that the proposed method achieves much higher and more stable accuracy than BW, ANOVA for gene selection. It should be noted that standard deviations of the proposed method are not smaller than other methods when it is used on the high dimensional data sets. Of course, this is just some primary conclusion from our experiments. In the future, more experiments will be done to verify the conclusion.

3   Conclusions

This article presented a new maximum margin discriminant analysis method based on sparse representation (SMMDA). For gene selection, SMMDA finds the one dimension projection in the most separable direction of gene expression data, and then uses sparse representation technique to regress the projection to obtain the relevance vector for the gene set and select genes according to the vector. Compared with other widely used gene selection algorithms, the experimental results demonstrated that it is very effective in selecting genes. In future, we will study new algorithm to extend our method and try to achieve better results.

Acknowledgements. This work was partially supported by the Foundation for Young Scientist of Shandong Province, China under Grant No. 2008BS01010, the Program for New Century Excellent Talents in University of China, the NUST Outstanding Scholar Supporting Program, the National Science Foundation of China under Grants No. 60973098.

References

1. Liao, J.G., Chin, K.V.: Logistic Regression for Disease Classification Using Microarray Data: Model Selection In Large p and Small n Case. Bioinformatics 23, 1945–1951 (2007)
2. Zheng, C.H., Huang, D.S., Zhang, L., Kong, X.Z.: Tumor Clustering Using Non-Negative Matrix Factorization with Gene Selection. IEEE Transactions on Information Technology in Biomedicine 13(4), 599–607 (2009)
3. Li, H.F., Jiang, T., Zhang, K.S.: Efficient and Robust Feature Extraction by Maximum Margin Criterion. IEEE Transactions on Neural Networks 17, 157–165 (2006)
4. Donoho, D.: For Most Large Underdetermined Systems of Linear Equations the Minimal l_1-Norm Solution is Also the Sparsest Solution. Comm. Pure and Applied Math. 59(6), 797–829 (2006)

5. Wright, J., Yang, A.Y., Ganesh, A., Sastry, S.S., Ma, Y.: Robust Face Recognition via Sparse Representation. IEEE Transactions on Pattern Analysis and Machine Intelligence 31(2), 210–227 (2009)

6. Dudoit, S., Fridlyand, J., Speed, T.P.: Comparison of Discrimination Methods for the Classification of Tumor Using Gene Expression Data. Journal of the American Statistical Association 97, 77–87 (2002)

7. Jones, B.: Matlab Statistics Toolbox. The MathWorks, Inc., Natick (1997)

8. Alon, A., Barkai, N., Notterman, D.A., et al.: Broad Patterns of Gene Expression Revealed by Clustering Analysis of Tumor and Normal Colon Tissues Probed by Oligonucleotide Arrays. Proc. Natl. Acad. Sci. USA 96, 6745–6750 (1999)

9. Golub, T.R., Slonim, D.K., Tamayo, P., et al.: Molecular Classification of Cancer: Class Discovery and Class Prediction by Gene Expression Monitoring. Science 286, 531–537 (1999)

10. Van't Veer, L.J., Dai, H., et al.: Gene Expression Profiling Predicts Clinical Outcome of Breast Cancer. Nature 415, 530–536 (2002)

11. Nutt, C.L., Mani, D.R., Betensky, R.A., et al.: Gene expression-Based Classification of Malignant Gliomas Correlates Better with Survival than Histological Classification. Cancer Research 63(7), 1602–1607 (2003)

12. Singh, D., Febbo, P.G., Ros, S.K., et al.: Gene Expression Correlates of Clinical Prostate Cancer Behavior. Cancer Cell 1, 203–209 (2002)

13. Armstrong, S.A., Staunton, J.E., Silverman, L.B., et al.: MLL Translocations Specify a Distinct Gene Expression Profile That Distinguishes a Unique Leukemia. Nature Genetics 1, 41–47 (2002)

The Models of Dominance–Based Multigranulation Rough Sets

Xibei Yang*

Jiangsu Sunboon Information Technology Co., Ltd., Wuxi,
Jiangsu, 214072, P.R. China
yangxibei@hotmail.com

Abstract. In this paper, the dominance–based rough set approach is introduced into multigranulation environment. Two different dominance–based multigranulation rough sets models: dominance–based optimistic multigranulation rough set and dominance–based pessimistic multigranulation rough set are constructed, respectively. Not only the properties of these two dominance–based multigranulation rough sets are discussed, but also the relationships among dominance–based optimistic multigranulation rough set, dominance–based pessimistic multigranulation rough set and the classical dominance–based rough set are investigated.

Keywords: Dominance relation, dominance–based rough set, dominance–based multigranulation rough set.

1 Introduction

The theory of rough set [1–4], which was firstly proposed by Pawlak, has been demonstrated in the fields such as data mining, pattern recognition, knowledge discovering, medical diagnose and so on.

It is well known that Pawlak's rough set, is constructed on the basis of an indiscernibility relation, which is the intersection of some equivalence relations. However, such indiscernibility relation cannot be used to describe the inconsistencies coming from the criteria, such as product quality, market share, and debt ratio. To solve this problem, Greco et al. have proposed an extension of the Classic Rough Sets Approach, which is called the Dominance-based Rough Sets Approach (DRSA) [5–7]. This innovation is mainly based on substitution of the indiscernibility relation by a dominance relation. Thus, the knowledge to be approximated are upward and downward unions of decision classes and the basic information granules are sets of objects defined by using a dominance relation.

From discussions above, we can see that the dominance–based rough set was constructed on the basis of one and only one dominance relation. However, it should be noticed that in Ref. [8], Qian et al. argued that we often need to describe concurrently a target concept through multi binary relations (e.g. equivalence relation, tolerance relation, reflexive relation and neighborhood relation)

---

* Corresponding author.

D.-S. Huang et al. (Eds.): ICIC 2011, LNBI 6840, pp. 657–664, 2012.

on the universe according to a user's requirements or targets of problem solving. Therefore, they proposed the concept of multigranulation rough set model. The first multigranulation rough set model was proposed by Qian et al. in Ref. [8]. Following such work, Qian classified his multigranulation rough set theory into two parts: one is the optimistic multigranulation rough set [9, 10] and the other is pessimistic multigranulation rough set [11].

The purpose of this paper is to further introduce the dominance principle into multigranulation rough set theory. This is mainly because Qian's multi-granulation rough sets are constructed on the basis of a family of the indiscerni-bility relations, these indiscernibility relations cannot be used to describe the inconsistencies coming from the criteria. Therefore, by using the basic idea of multigranulation rough set, we present the dominance–based multigranulation rough sets. dominance–based multigranulation rough sets also has two cases, i.e. dominance–based optimistic multigranulation rough set and dominance–based pessimistic multigranulation rough set. It is proved that the basic properties of dominance–based rough set are still satisfied with these two dominance–based multigranulation rough sets.

2   Preliminary Knowledge on Rough Sets

2.1   Pawlak's Rough Set

Formally, an information system can be considered as a pair $I =< U, AT >$, where

- U is a non–empty finite set of objects, it is called the universe;
- AT is a non–empty finite set of attributes, such that $\forall a \in AT$, V_a is the domain of attribute a.

$\forall x \in U$, let us denote by $a(x)$ the value that x holds on $a(a \in AT)$. For an information system I, one then can describe the relationship between objects through their attributes values. With respect to a subset of attributes such that $A \subseteq AT$, an indiscernibility relation $IND(A)$ may be defined as

$$IND(A) = \{(x, y) \in U^2 : a(x) = a(y), \forall a \in A\}. \tag{1}$$

The relation $IND(A)$ is reflexive, symmetric and transitive, then $IND(A)$ is an equivalence relation. By the indiscernibility relation $IND(A)$, one can derive the lower and upper approximations of an arbitrary subset X of U. They are defined as

$$\underline{A}(X) = \{x \in U : [x]_A \subseteq X\} \text{ and } \overline{A}(X) = \{x \in U : [x]_A \cap X \neq \emptyset\} \tag{2}$$

respectively, where $[x]_A = \{y \in U : (x, y) \in IND(A)\}$ is the A–equivalence class containing x. The pair $[\underline{A}(X), \overline{A}(X)]$ is referred to as the Pawlak's rough set of X with respect to the set of attributes A.

2.2 Multigranulation Rough Set

Optimistic Multigranulation Rough Set

Definition 1. *[9] Let I be an information system in which $A_1, A_2, \cdots, A_m \subseteq AT$, then $\forall X \subseteq U$, the optimistic multigranulation lower and upper approximations are denoted by $\underline{\sum_{i=1}^{m} A_i}^{O}(X)$ and $\overline{\sum_{i=1}^{m} A_i}^{O}(X)$, respectively,*

$$\underline{\sum_{i=1}^{m} A_i}^{O}(X) = \{x \in U : [x]_{A_1} \subseteq X \vee \cdots \vee [x]_{A_m} \subseteq X\}; \tag{3}$$

$$\overline{\sum_{i=1}^{m} A_i}^{O}(X) = \sim \underline{\sum_{i=1}^{m} A_i}^{O}(\sim X); \tag{4}$$

where $[x]_{A_i}$ ($1 \leq i \leq m$) is the equivalence class of x in terms of set of attributes A_i, $\sim X$ is the complement of set X.

Theorem 1. *Let I be an information system in which $A_1, A_2, \cdots, A_m \subseteq AT$, then $\forall X \subseteq U$, we have*

$$\overline{\sum_{i=1}^{m} A_i}^{O}(X) = \{x \in U : [x]_{A_1} \cap X \neq \emptyset \wedge \cdots \wedge [x]_{A_m} \cap X \neq \emptyset\}. \tag{5}$$

Pessimistic Multigranulation Rough Set

Definition 2. *[11] Let I be an information system in which $A_1, A_2, \cdots, A_m \subseteq AT$, then $\forall X \subseteq U$, the pessimistic multigranulation lower and upper approximations are denoted by $\underline{\sum_{i=1}^{m} A_i}^{P}(X)$ and $\overline{\sum_{i=1}^{m} A_i}^{P}(X)$, respectively,*

$$\underline{\sum_{i=1}^{m} A_i}^{P}(X) = \{x \in U : [x]_{A_1} \subseteq X \wedge \cdots \wedge [x]_{A_m} \subseteq X\}; \tag{6}$$

$$\overline{\sum_{i=1}^{m} A_i}^{P}(X) = \sim \underline{\sum_{i=1}^{m} A_i}^{P}(\sim X). \tag{7}$$

Theorem 2. *Let I be an information system in which $A_1, A_2, \cdots, A_m \subseteq AT$, then $\forall X \subseteq U$, we have*

$$\overline{\sum_{i=1}^{m} A_i}^{P}(X) = \{x \in U : [x]_{A_1} \cap X \neq \emptyset \vee \cdots \vee [x]_{A_m} \cap X \neq \emptyset\}. \tag{8}$$

2.3   Dominance–Based Rough Set

Given an information system I, it is assumed here that the domain of a criterion $a \in AT$ is completely preordered by an outranking relation $\succeq_a$; $x \succeq_a y$ means "x is at least as good as (outranks) y with respect to criterion a". In the following, without any loss of generality, we consider a condition criterion having a numerical domain, that is, $V_a \in \mathbb{R}$ ($\mathbb{R}$ denotes the set of real numbers) and being of type gain, that is, $x \succeq_a y \Leftrightarrow f(x, a) \geq f(y, a)$ (according to increasing preference) or $x \preceq_a y \Leftrightarrow f(x, a) \leq f(y, a)$ (according to decreasing preference), where $a \in AT$, $x, y \in U$. For a subset of attributes $A \subseteq AT$, we define $x \succeq_A y \Leftrightarrow \forall a \in A, x \succeq_a y$. Formally, the dominance relation is showed in Definition 3.

Definition 3. *Let I be an information system in which $A \subseteq AT$, then the dominance relation in terms of A is:*

$$D_A = \{(x, y) \in U^2 : \forall a \in A, a(x) \geq a(y)\}. \tag{9}$$

Moreover, given a decision system I, assume that the decision attribute d makes a partition of U into a finite number of classes; let $\mathbf{CL} = \{CL_t, t \in T\}$, $T = \{1, 2, \cdots, l\}$, be a set of these classes that are ordered, that is, $\forall r, s \in T$ such that $r > s$, the objects from CL_r are preferred to the objects from CL_s. In dominance–based rough set approach, the sets to be approximated are upward unions and downward unions of classes, which are defined respectively as $CL_t^{\geq} = \bigcup_{s \geq t} CL_s$, $CL_t^{\leq} = \bigcup_{s \leq t} CL_s$, $t \in T$. The statement $x \in CL_t^{\geq}$ means "x belongs to at least class CL_t", where $x \in CL_t^{\leq}$ means "x belongs to at most class CL_t".

Definition 4. *[5–7] Let I be a decision system in which $A \subseteq AT$, then $\forall CL_t^{\geq}$ ($1 \leq t \leq l$), the lower and upper approximations of $CL_t^{\geq}$ are defined as:*

$$\underline{A_D}(CL_t^{\geq}) = \{x \in U : D_A^+(x) \subseteq CL_t^{\geq}\}, \tag{10}$$

$$\overline{A_D}(CL_t^{\geq}) = \{x \in U : D_A^-(x) \cap CL_t^{\geq} \neq \emptyset\}; \tag{11}$$

$\forall CL_t^{\leq}$ ($1 \leq t \leq l$), the lower and upper approximations of $CL_t^{\leq}$ are defined as:

$$\underline{A_D}(CL_t^{\leq}) = \{x \in U : D_A^-(x) \subseteq CL_t^{\leq}\}, \tag{12}$$

$$\overline{A_D}(CL_t^{\leq}) = \{x \in U : D_A^+(x) \cap CL_t^{\leq} \neq \emptyset\}; \tag{13}$$

where $D_A^+(x) = \{y \in U : (y, x) \in D_A\}$ is the set of objects, which are dominating x in terms of A; $D_A^-(x) = \{y \in U : (x, y) \in D_A\}$ is the set of objects, which are dominated by x in terms of A.

More details about the dominance–based rough set can be found in Ref. [5–7].

3   Dominance–Based Multigranulation Rough Set

Obviously, Greco's dominance–based rough set is constructed on the basis of one and only one dominance relation. However, following the basic idea of multigranulation rough set approach, it is natural to introduce Greco's dominance–based rough set model into multigranulation environment. This is what will be discussed in the following.

3.1   Dominance–Based Optimistic Multigranulation Rough Set

Definition 5. *Let I be a decision system in which $A_1, A_2, \cdots, A_m \subseteq AT$, then $\forall CL_t^{\geq} (1 \leq t \leq l)$, the optimistic multigranulation lower and upper approximations of $CL_t^{\geq}$ in terms of the dominance relations are denoted by $\sum_{i=1}^{m} A_i^{O}(CL_t^{\geq})$ and $\overline{\sum_{i=1}^{m} A_i}^{O}(CL_t^{\geq})$, respectively,*

$$\sum_{i=1}^{m} A_i^{O}(CL_t^{\geq}) = \{x \in U : D_{A_1}^{+}(x) \subseteq CL_t^{\geq} \vee \cdots \vee D_{A_m}^{+}(x) \subseteq CL_t^{\geq}\}; \quad (14)$$

$$\overline{\sum_{i=1}^{m} A_i}^{O}(CL_t^{\geq}) = \{x \in U : D_{A_1}^{-}(x) \cap CL_t^{\geq} \neq \emptyset \wedge \cdots \wedge D_{A_m}^{-}(x) \cap CL_t^{\geq} \neq \emptyset\}; (15)$$

$\forall CL_t^{\leq} (1 \leq t \leq l)$, the optimistic multigranulation lower and upper approximations of $CL_t^{\leq}$ in terms of the dominance relations are denoted by $\sum_{i=1}^{m} A_i^{O}(CL_t^{\leq})$ and $\overline{\sum_{i=1}^{m} A_i}^{O}(CL_t^{\leq})$, respectively,

$$\sum_{i=1}^{m} A_i^{O}(CL_t^{\leq}) = \{x \in U : D_{A_1}^{-}(x) \subseteq CL_t^{\leq} \vee \cdots \vee D_{A_m}^{-}(x) \subseteq CL_t^{\leq}\}; \quad (16)$$

$$\overline{\sum_{i=1}^{m} A_i}^{O}(CL_t^{\leq}) = \{x \in U : D_{A_1}^{+}(x) \cap CL_t^{\leq} \neq \emptyset \wedge \cdots \wedge D_{A_m}^{+}(x) \cap CL_t^{\leq} \neq \emptyset\}. (17)$$

By the above definition, the optimistic multigranulation boundary regions of $CL_t^{\geq}$ and $CL_t^{\leq}$ in terms of the dominance relations are

$$BN_{\sum_{i=1}^{m} A_i}^{O}(CL_t^{\geq}) = \overline{\sum_{i=1}^{m} A_i}^{O}(CL_t^{\geq}) - \sum_{i=1}^{m} A_i^{O}(CL_t^{\geq}); \quad (18)$$

$$BN_{\sum_{i=1}^{m} A_i}^{O}(CL_t^{\leq}) = \overline{\sum_{i=1}^{m} A_i}^{O}(CL_t^{\leq}) - \sum_{i=1}^{m} A_i^{O}(CL_t^{\leq}). \quad (19)$$

Theorem 3. *Let I be a decision system in which $A_1, A_2, \cdots, A_m \subseteq AT$, then $\forall t \in T$, the dominance–based optimistic multigranulation rough sets have the following properties:*

1. $\underline{\sum_{i=1}^{m} A_i}^{O}(CL_t^{\geq}) \subseteq CL_t^{\geq} \subseteq \overline{\sum_{i=1}^{m} A_i}^{O}(CL_t^{\geq})$,

   $\underline{\sum_{i=1}^{m} A_i}^{O}(CL_t^{\leq}) \subseteq CL_t^{\leq} \subseteq \overline{\sum_{i=1}^{m} A_i}^{O}(CL_t^{\leq})$;

2. $\overline{\sum_{i=1}^{m} A_i}^{O}(CL_t^{\geq}) = \bigcap_{i=1}^{m} \left(\bigcup_{y \in CL_t^{\geq}} D_{A_i}^{+}(y) \right)$,

   $\overline{\sum_{i=1}^{m} A_i}^{O}(CL_t^{\leq}) = \bigcap_{i=1}^{m} \left(\bigcup_{y \in CL_t^{\leq}} D_{A_i}^{-}(y) \right)$;

3. (a) $\underline{\sum_{i=1}^{m} A_i}^{O}(CL_t^{\geq}) = U - \overline{\sum_{i=1}^{m} A_i}^{O}(CL_{t-1}^{\leq})$ for $t = 2, \cdots, l$,

   (b) $\underline{\sum_{i=1}^{m} A_i}^{O}(CL_t^{\leq}) = U - \overline{\sum_{i=1}^{m} A_i}^{O}(CL_{t+1}^{\geq})$ for $t = 1, \cdots, l-1$,

   (c) $\overline{\sum_{i=1}^{m} A_i}^{O}(CL_t^{\geq}) = U - \underline{\sum_{i=1}^{m} A_i}^{O}(CL_{t-1}^{\leq})$ for $t = 2, \cdots, l$,

   (d) $\overline{\sum_{i=1}^{m} A_i}^{O}(CL_t^{\leq}) = U - \underline{\sum_{i=1}^{m} A_i}^{O}(CL_{t+1}^{\geq})$ for $t = 1, \cdots, l-1$,

   (e) $BN^{O}_{\sum_{i=1}^{m} A_i}(CL_t^{\geq}) = BN^{O}_{\sum_{i=1}^{m} A_i}(CL_{t-1}^{\leq})$ for $t = 2, \cdots, l$.

3.2   Dominance–Based Pessimistic Multigranulation Rough Set

Definition 6. *Let I be a decision system in which $A_1, A_2, \cdots, A_m \subseteq AT$, then $\forall CL_t^{\geq} (1 \leq t \leq l)$, the pessimistic multigranulation lower and upper approximations of $CL_t^{\geq}$ in terms of the dominance relations are denoted by $\underline{\sum_{i=1}^{m} A_i}^{P}(CL_t^{\geq})$ and $\overline{\sum_{i=1}^{m} A_i}^{P}(CL_t^{\geq})$, respectively,*

$$\underline{\sum_{i=1}^{m} A_i}^{P}(CL_t^{\geq}) = \{x \in U : D_{A_1}^{+}(x) \subseteq CL_t^{\geq} \wedge \cdots \wedge D_{A_m}^{+}(x) \subseteq CL_t^{\geq}\}; \qquad (20)$$

$$\overline{\sum_{i=1}^{m} A_i}^{P}(CL_t^{\geq}) = \{x \in U : D_{A_1}^{-}(x) \cap CL_t^{\geq} \neq \emptyset \vee \cdots \vee D_{A_m}^{-}(x) \cap CL_t^{\geq} \neq \emptyset\} \quad (21)$$

$\forall CL_t^{\leq} (1 \leq t \leq l)$, the pessimistic multigranulation lower and upper approximations of $CL_t^{\leq}$ in terms of the dominance relations are denoted by $\underline{\sum_{i=1}^{m} A_i}^{P}(CL_t^{\leq})$ and $\overline{\sum_{i=1}^{m} A_i}^{P}(CL_t^{\leq})$, respectively,

$$\underline{\sum_{i=1}^{m} A_i}^{P}(CL_t^{\leq}) = \{x \in U : D_{A_1}^{-}(x) \subseteq CL_t^{\leq} \wedge \cdots \wedge D_{A_m}^{-}(x) \subseteq CL_t^{\leq}\}; \qquad (22)$$

$$\overline{\sum_{i=1}^{m} A_i}^{P}(CL_t^{\leq}) = \{x \in U : D_{A_1}^{+}(x) \cap CL_t^{\leq} \neq \emptyset \vee \cdots \vee D_{A_m}^{+}(x) \cap CL_t^{\leq} \neq \emptyset\}. \quad (23)$$

By the above definition, the pessimistic multigranulation boundary regions of $CL_t^{\geq}$ and $CL_t^{\leq}$ in terms of the dominance relations are

$$BN^{P}_{\sum_{i=1}^{m} A_i}(CL_t^{\geq}) = \overline{\sum_{i=1}^{m} A_i}^{P}(CL_t^{\geq}) - \underline{\sum_{i=1}^{m} A_i}^{P}(CL_t^{\geq}); \qquad (24)$$

$$BN^P_{\sum^m_{i=1} A_i}(CL^{\leq}_t) = \sum^m_{i=1} \overline{A_i}^P (CL^{\leq}_t) - \sum^m_{i=1} \underline{A_i}^P (CL^{\leq}_t). \tag{25}$$

Theorem 4. *Let I be a decision system in which $A_1, A_2, \cdots, A_m \subseteq AT$, then $\forall t \in T$, the dominance–based pessimistic multigranulation rough sets have the following properties:*

1. $\sum^m_{i=1} \underline{A_i}^P (CL^{\geq}_t) \subseteq CL^{\geq}_t \subseteq \overline{\sum^m_{i=1} A_i}^P (CL^{\geq}_t),$
   $\sum^m_{i=1} \underline{A_i}^P (CL^{\leq}_t) \subseteq CL^{\leq}_t \subseteq \overline{\sum^m_{i=1} A_i}^P (CL^{\leq}_t);$

2. $\overline{\sum^m_{i=1} A_i}^P (CL^{\geq}_t) = \bigcup^m_{i=1} \left(\bigcup_{y \in CL^{\geq}_t} D^+_{A_i}(y) \right),$
   $\overline{\sum^m_{i=1} A_i}^P (CL^{\leq}_t) = \bigcup^m_{i=1} \left(\bigcup_{y \in CL^{\leq}_t} D^-_{A_i}(y) \right);$

3. (a) $\sum^m_{i=1} \underline{A_i}^P (CL^{\geq}_t) = U - \overline{\sum^m_{i=1} A_i}^P (CL^{\leq}_{t-1})$ *for* $t = 2, \cdots, l,$
   (b) $\sum^m_{i=1} \underline{A_i}^P (CL^{\leq}_t) = U - \overline{\sum^m_{i=1} A_i}^P (CL^{\geq}_{t+1})$ *for* $t = 1, \cdots, l-1,$
   (c) $\overline{\sum^m_{i=1} A_i}^P (CL^{\geq}_t) = U - \sum^m_{i=1} \underline{A_i}^P (CL^{\leq}_{t-1})$ *for* $t = 2, \cdots, l,$
   (d) $\overline{\sum^m_{i=1} A_i}^P (CL^{\leq}_t) = U - \sum^m_{i=1} \underline{A_i}^P (CL^{\geq}_{t+1})$ *for* $t = 1, \cdots, l-1,$
   (e) $BN^P_{\sum^m_{i=1} A_i}(CL^{\geq}_t) = BN^P_{\sum^m_{i=1} A_i}(CL^{\leq}_{t-1})$ *for* $t = 2, \cdots, l.$

Similar to Theorem 3, the properties of dominance–based pessimistic multi-granulation rough set are still consistent to the properties of classical dominance–based rough set.

Theorem 5. *Let I be a decision system in which $A = \{a_1, a_2, \cdots, a_m\} \subseteq AT$, then $\forall t \in T$, we have*

$$\sum^m_{i=1} \underline{a_i}^P (CL^{\geq}_t) \subseteq \sum^m_{i=1} \underline{a_i}^O (CL^{\geq}_t) \subseteq \underline{A_D}(CL^{\geq}_t); \tag{26}$$

$$\overline{A_D}(CL^{\geq}_t) \subseteq \sum^m_{i=1} \overline{a_i}^O (CL^{\geq}_t) \subseteq \sum^m_{i=1} \overline{a_i}^P (CL^{\geq}_t); \tag{27}$$

$$\sum^m_{i=1} \underline{a_i}^P (CL^{\leq}_t) \subseteq \sum^m_{i=1} \underline{a_i}^O (CL^{\leq}_t) \subseteq \underline{A_D}(CL^{\leq}_t); \tag{28}$$

$$\overline{A_D}(CL^{\leq}_t) \subseteq \sum^m_{i=1} \overline{a_i}^O (CL^{\leq}_t) \subseteq \sum^m_{i=1} \overline{a_i}^P (CL^{\leq}_t). \tag{29}$$

4   Conclusions

In this paper, we have further generalized the multigranulation rough set. The models of dominance–based optimistic and pessimistic multigranulation rough sets are presented respectively. These two rough set approaches are different from the Qian's multigranulation rough set since we use the dominance relations instead of the indiscernibility relations for target approximation; they are also different from Greco's dominance–based rough set since we use a family of dominance relations instead of a single dominance relation for target approximation.

Furthermore, the reductions in dominance–based optimistic and pessimistic multigranulation rough sets is an interesting topic to be addressed.

Acknowledgments. This work is supported by the Natural Science Foundation of China (Nos.60632050, 60903110) and Postdoctoral Science Foundation of China (No.20100481149).

References

1. Pawlak, Z.: Rough Sets–Theoretical Aspects of Reasoning About Data. Kluwer Academic Publishers, Dordrecht (1991)
2. Pawlak, Z., Skowron, A.: Rudiments of Rough Sets. Inform. Sci. 177, 3–27 (2007)
3. Pawlak, Z., Skowron, A.: Rough Sets: Some Extensions. Inform. Sci. 177, 28–40 (2007)
4. Pawlak, Z., Skowron, A.: Rough Sets and Boolean Reasoning. Inform. Sci. 177, 41–73 (2007)
5. Greco, S., Inuiguchi, M., Słowiński, R.: Fuzzy Rough Sets and Multiple–Premise Gradual Decision Rules. Int. J. Approx. Reason. 41, 179–211 (2006)
6. Greco, S., Matarazzo, B., Słowiński, R.: Rough Approximation by Dominance Relations. Int. J. Intell. Syst. 17, 153–171 (2002)
7. Greco, S., Matarazzo, B., Słowiński, R.: Rough Sets Theory for Multicriteria Decision Analysis. Euro. J. Oper. Res. 129, 1–47 (2002)
8. Qian, Y.H., Dang, C.Y., Liang, J.Y.: MGRS in Incomplete Information Systems. In: 2007 IEEE International Conference on Granular Computing, pp. 163–168 (2007)
9. Qian, Y.H., Liang, J.Y., Yao, Y.Y., Dang, C.Y.: MGRS: A Multi–Granulation Rough set. Inform. Sci. 180, 949–970 (2010)
10. Qian, Y.H., Liang, J.Y., Dang, C.Y.: Incomplete Multigranulation Rough set, IEEE T. Syst. Man Cy. B. 20, 420–431 (2010)
11. Qian, Y.H., Liang, J.Y., Wei, W.: Pessimistic Rough Decision. In: Second International Workshop on Rough Sets Theory, pp. 440–449 (2010)

An Intuitionistic Fuzzy Dominance–Based Rough Set

Yanqin Zhang[1,*] and Xibei Yang[2]

[1] School of Economics, Xuzhou Institute of Technology, Xuzhou, 221000, P.R. China
[2] School of Computer Science and Engineering, Jiangsu University of Science and Technology, Zhenjiang, Jiangsu, 212003, P.R. China
zyqxuzhou@163.com,
yangxibei@hotmail.com

Abstract. The dominance–based rough set approach plays an important role in the development of the rough set theory. It can be used to describe the inconsistencies coming from consideration of the preference–ordered domains of the attributes. The purpose of this paper is to further generalize the dominance–based rough set model to fuzzy environment. The constructive approach is used to define the intuitionistic fuzzy dominance–based lower and upper approximations, respectively. Basic properties of the intuitionistic fuzzy dominance–based rough approximations are then examined.

Keywords: Dominance–based rough set, dominance–based fuzzy rough set, intuitionistic fuzzy dominance relation, intuitionistic fuzzy dominance–based rough set.

1 Introduction

Though the rough set [1–4] has been demonstrated to be useful in the fields of knowledge discovery, decision analysis, pattern recognition and so on, it is not able, however, to discover inconsistencies coming from consideration of criteria, that is, attributes with preference–ordered domains, such as product quality, market share and debt ratio. To solve this problem, Greco et al. have proposed an extension of Pawlak's rough set approach, which is called the Dominance–based Rough Set Approach (DRSA) [5–7]. This innovation is mainly based on substitution of the indiscernibility relation by a dominance relation. Presently, work on dominance–based rough set model also progressing rapidly. For example, by considering two different types of semantic explanations of unknown values, Shao et al. and Yang et al. generalized the DRSA to incomplete environments in Ref. [8] and Ref. [9] respectively. With introduction of the concept of variable precision rough set into DRSA, Błszczyńki et al. proposed the variable consistency dominance–based rough set approach [10, 11], Inuiguchi et al. proposed the variable precision dominance–based rough set [12]. Kotłowski [13] introduced

---

* Corresponding author.

D.-S. Huang et al. (Eds.): ICIC 2011, LNBI 6840, pp. 665–672, 2012.
© Springer-Verlag Berlin Heidelberg 2012

a new approximation of DRSA which is based on the probabilistic model for the ordinal classification problems. Greco et al. generalized the DRSA to fuzzy environment and then presented the model of dominance–based rough fuzzy set in Ref. [5]. By using a fuzzy dominance relation, the same authors also presented dominance–based fuzzy rough set [14] in their literatures.

As a generalization of the Zadeh fuzzy set, the notion of intuitionistic fuzzy set was suggested for the first time by Atanassov [15, 16]. An intuitionistic fuzzy set allocates to each element both a degree of membership and one of non–membership, and it was applied to the fields of approximate inference, signal transmission and controller, etc. In this paper, the intuitionistic fuzzy set will be combined with the DRSA and then the model of Intuitionistic Fuzzy Dominance–based Rough Set(IFDRS) is presented.

2   Intuitionistic Fuzzy Dominance–Based Rough Set

A decision system is a pair $\mathscr{I} =< U, AT \cup \{d\} >$, where U is a non–empty finite set of objects, it is called the universe; AT is a non–empty finite set of attributes; d is the decision attribute where $AT \cap \{d\} = \emptyset$.

$\forall a \in AT$, V_a is used to represent the domain of attribute a and then $V = V_{AT} = \bigcup_{a \in AT} V_a$ is the domain of all attributes. Moreover, for each $x \in U$, let us denote by $a(x)$ the value that x holds on a ($a \in AT$).

By considering the preference–ordered domains of attributes (criteria), Greco et al. have proposed an extension of the classical rough set that is able to deal with inconsistencies typical to exemplary decisions in Multi–Criteria Decision Making (MCDM) problems, which is called the Dominance–based Rough Set Approach (DRSA). let $\succeq_a$ be a weak preference relation on U (often called outranking) representing a preference on the set of objects with respect to criterion a ($a \in AT$); $x \succeq_a y$ means "x is at least as good as y with respect to criterion a". We say that x dominates y with respect to $A \subseteq AT$, if $x \succeq_a y$ for each $a \in A$.

In the traditional DRSA, we assume here that the decision attribute d determines a partition of U into a finite number of classes; let $\mathbf{CL} = \{CL_n, n \in N\}$, $N = \{1, 2, \cdots, m\}$, be a set of these classes that are ordered. Different from Pawlak's rough approximation, in DRSA, the sets to be approximated are an upward union and a downward union of decision classes, which are defined respectively as $CL_n^{\geq} = \bigcup_{n' \geq n} CL_{n'}$, $CL_n^{\leq} = \bigcup_{n' \leq n} CL_{n'}$, $n, n' \in N$.

An intuitionistic fuzzy set $\mathscr{F}$ in U is given by $\mathscr{F} = \{< x, u_{\mathscr{F}}(x), v_{\mathscr{F}}(x) >: x \in U\}$ where $u_{\mathscr{F}} : U \to [0,1]$ and $v_{\mathscr{F}} : U \to [0,1]$ with the condition such that $0 \leq u_{\mathscr{F}}(x) + v_{\mathscr{F}}(x) \leq 1$. The numbers $u_{\mathscr{F}}(x), v_{\mathscr{F}}(x) \in [0,1]$ denote the degree of membership and non–membership of x to $\mathscr{F}$, respectively. Obviously, when $u_{\mathscr{F}}(x) + v_{\mathscr{F}}(x) = 1$, for all elements of the universe, the traditional fuzzy set concept is recovered. The family of all intuitionistic fuzzy subsets in U is denoted by $\mathscr{IF}(U)$. Let us review some basic operations on $\mathscr{IF}(U)$ as follows:

By the definition of intuitionistic fuzzy set, we know that an intuitionistic fuzzy relation $\mathscr{R}$ on U is an intuitionistic fuzzy subset of $U \times U$, namely, $\mathscr{R}$ is given by $\mathscr{R} = \{< (x,y), u_{\mathscr{R}}(x,y), v_{\mathscr{R}}(x,y) >: (x,y) \in U \times U >\}$, where

$u_{\mathscr{R}} : U \times U \rightarrow [0,1]$ and $v_{\mathscr{R}} : U \times U \rightarrow [0,1]$ satisfy with the condition $0 \leq u_{\mathscr{R}}(x,y) + v_{\mathscr{R}}(x,y) \leq 1$ for each $(x,y) \in U \times U$. The set of all intuitionistic fuzzy relation on U is denoted by $\mathscr{IFR}(U \times U)$.

Definition 1. *Let U be the universe of discourse, $\forall \mathscr{R} \in \mathscr{IFR}(U \times U)$, if*

1. *$u_{\mathscr{R}}(x,y)$ represents the credibility of the proposition "x is at least as good as y in $\mathscr{R}$";*
2. *$v_{\mathscr{R}}(x,y)$ represents the non–credibility of the proposition "x is at least as good as y in $\mathscr{R}$";*

then $\mathscr{R}$ is referred to as an intuitionistic fuzzy dominance relation.

By the above definition, we can see that the intuitionistic fuzzy dominance relation can describe not only the credibility of dominance principle between pairs of objects, but also the non–credibility of dominance principle between pairs of objects. In a decision system, suppose that for each $a \in AT$, we have an intuitionistic fuzzy dominance relation $\mathscr{R}_a$ which is related to a, then the intuitionistic fuzzy dominance relation which is related to AT is denoted by $\mathscr{R}_{AT}$ such that $\mathscr{R}_{AT}(x,y) = < u_{\mathscr{R}_A}(x,y), v_{\mathscr{R}_A}(x,y) > = \big\langle \wedge \{u_{\mathscr{R}_a}(x,y) : a \in AT\}, \vee\{v_{\mathscr{R}_a}(x,y) : a \in AT\}\big\rangle$. where $(x,y) \in U \times U$. To simplify our discussion, the intuitionistic fuzzy dominance relation we used in this paper is always reflexive, i.e. $\mathscr{R}_a(x,x) = 1$ $(u_{\mathscr{R}_a}(x,x) = 1, v_{\mathscr{R}_a}(x,x) = 0)$ for each $x \in U$ and each $a \in AT$.

Definition 2. *Let $\mathscr{I}$ be a decision system in which $A \subseteq AT$, $\mathscr{R}_A$ is a intuitionistic fuzzy dominance relation with respect to A, $\forall n \in N$, the A–lower approximation and A–upper approximation of $CL_n^{\geq}$ with respect to intuitionistic fuzzy dominance relation $\mathscr{R}_A$ are denoted by $\underline{A_{\mathscr{R}}}(CL_n^{\geq})$ and $\overline{A_{\mathscr{R}}}(CL_n^{\geq})$ respectively such that*

$$\underline{A_{\mathscr{R}}}(CL_n^{\geq}) = \{< x, u_{\underline{A_{\mathscr{R}}}(CL_n^{\geq})}(x), v_{\underline{A_{\mathscr{R}}}(CL_n^{\geq})}(x) >: x \in U\}, \tag{1}$$

$$\overline{A_{\mathscr{R}}}(CL_n^{\geq}) = \{< x, u_{\overline{A_{\mathscr{R}}}(CL_n^{\geq})}(x), v_{\overline{A_{\mathscr{R}}}(CL_n^{\geq})}(x) >: x \in U\}, \tag{2}$$

where

$$u_{\underline{A_{\mathscr{R}}}(CL_n^{\geq})}(x) = \wedge_{y \in U}\big(u_{CL_n^{\geq}}(y) \vee v_{\mathscr{R}_A}(y,x)\big);$$

$$v_{\underline{A_{\mathscr{R}}}(CL_n^{\geq})}(x) = \vee_{y \in U}\big(v_{CL_n^{\geq}}(y) \wedge u_{\mathscr{R}_A}(y,x)\big);$$

$$u_{\overline{A_{\mathscr{R}}}(CL_n^{\geq})}(x) = \vee_{y \in U}\big(u_{CL_n^{\geq}}(y) \wedge u_{\mathscr{R}_A}(x,y)\big);$$

$$v_{\overline{A_{\mathscr{R}}}(CL_n^{\geq})}(x) = \wedge_{y \in U}\big(v_{CL_n^{\geq}}(y) \vee v_{\mathscr{R}_A}(x,y)\big);$$

the A–lower approximation and A–upper approximation of $CL_n^{\leq}$ with respect to intuitionistic fuzzy dominance relation $\mathscr{R}_A$ are denoted by $\underline{A_{\mathscr{R}}}(CL_n^{\leq})$ and $\overline{A_{\mathscr{R}}}(CL_n^{\leq})$ respectively such that

$$\underline{A_{\mathscr{R}}}(CL_n^{\leq}) = \{< x, u_{\underline{A_{\mathscr{R}}}(CL_n^{\leq})}(x), v_{\underline{A_{\mathscr{R}}}(CL_n^{\leq})}(x) >: x \in U\}, \tag{3}$$

$$\overline{A_{\mathscr{R}}}(CL_n^{\leq}) = \{< x, u_{\overline{A_{\mathscr{R}}}(CL_n^{\leq})}(x), v_{\overline{A_{\mathscr{R}}}(CL_n^{\leq})}(x) >: x \in U\}, \tag{4}$$

where

$$u_{\underline{A_{\mathscr{R}}}(CL_{\tilde{n}}^{\leq})}(x) = \wedge_{y \in U} \left(u_{CL_{\tilde{n}}^{\leq}}(y) \vee v_{\mathscr{R}_A}(x, y) \right);$$

$$v_{\underline{A_{\mathscr{R}}}(CL_{\tilde{n}}^{\leq})}(x) = \vee_{y \in U} \left(v_{CL_{\tilde{n}}^{\leq}}(y) \wedge u_{\mathscr{R}_A}(x, y) \right);$$

$$u_{\overline{A_{\mathscr{R}}}(CL_{\tilde{n}}^{\leq})}(x) = \vee_{y \in U} \left(u_{CL_{\tilde{n}}^{\leq}}(y) \wedge u_{\mathscr{R}_A}(y, x) \right);$$

$$v_{\overline{A_{\mathscr{R}}}(CL_{\tilde{n}}^{\leq})}(x) = \wedge_{y \in U} \left(v_{CL_{\tilde{n}}^{\leq}}(y) \vee v_{\mathscr{R}_A}(y, x) \right).$$

Theorem 1. *Let $\mathscr{I}$ be a decision system in which $A \subseteq AT$, the intuitionistic fuzzy dominance–based rough approximations have the following properties:*

1. *Contraction and extension:*

$$\underline{A_{\mathscr{R}}}(CL_{\tilde{n}}^{\geq}) \subseteq CL_{\tilde{n}}^{\geq} \subseteq \overline{A_{\mathscr{R}}}(CL_{\tilde{n}}^{\geq});$$
$$\underline{A_{\mathscr{R}}}(CL_{\tilde{n}}^{\leq}) \subseteq CL_{\tilde{n}}^{\leq} \subseteq \overline{A_{\mathscr{R}}}(CL_{\tilde{n}}^{\leq});$$

2. *Complements:*

$$\underline{A_{\mathscr{R}}}(CL_{\tilde{n}}^{\geq}) = U - \overline{A_{\mathscr{R}}}(CL_{\tilde{n}-1}^{\leq}), n = 2 \cdots m$$
$$\underline{A_{\mathscr{R}}}(CL_{\tilde{n}}^{\leq}) = U - \overline{A_{\mathscr{R}}}(CL_{\tilde{n}+1}^{\geq}), n = 1 \cdots m-1$$
$$\overline{A_{\mathscr{R}}}(CL_{\tilde{n}}^{\geq}) = U - \underline{A_{\mathscr{R}}}(CL_{\tilde{n}-1}^{\leq}), n = 2 \cdots m$$
$$\overline{A_{\mathscr{R}}}(CL_{\tilde{n}}^{\leq}) = U - \underline{A_{\mathscr{R}}}(CL_{\tilde{n}+1}^{\leq}), n = 1 \cdots m-1$$

3. *Monotones with attributes:*

$$\underline{A_{\mathscr{R}}}(CL_{\tilde{n}}^{\geq}) \subseteq \underline{AT_{\mathscr{R}}}(CL_{\tilde{n}}^{\geq}); \overline{A_{\mathscr{R}}}(CL_{\tilde{n}}^{\geq}) \supseteq \overline{AT_{\mathscr{R}}}(CL_{\tilde{n}}^{\geq});$$
$$\underline{A_{\mathscr{R}}}(CL_{\tilde{n}}^{\leq}) \subseteq \underline{AT_{\mathscr{R}}}(CL_{\tilde{n}}^{\leq}); \overline{A_{\mathscr{R}}}(CL_{\tilde{n}}^{\leq}) \supseteq \overline{AT_{\mathscr{R}}}(CL_{\tilde{n}}^{\leq});$$

4. *Monotones with decision classes:*
   $n_1, n_2 \in N$ *such that* $n_1 \leq n_2$

$$\underline{A_{\mathscr{R}}}(CL_{\tilde{n_1}}^{\geq}) \supseteq \underline{A_{\mathscr{R}}}(CL_{\tilde{n_2}}^{\geq}); \overline{A_{\mathscr{R}}}(CL_{\tilde{n_1}}^{\geq}) \supseteq \overline{A_{\mathscr{R}}}(CL_{\tilde{n_2}}^{\geq});$$
$$\underline{A_{\mathscr{R}}}(CL_{\tilde{n_1}}^{\leq}) \subseteq \underline{A_{\mathscr{R}}}(CL_{\tilde{n_2}}^{\leq}); \overline{A_{\mathscr{R}}}(CL_{\tilde{n_1}}^{\leq}) \subseteq \overline{A_{\mathscr{R}}}(CL_{\tilde{n_2}}^{\leq}).$$

3   Intuitionistic Fuzzy Dominance–Based Rough Set in Decision System with Probabilistic Interpretation

It is well known that Greco's traditional DRSA was firstly proposed for dealing with complete system with preference–ordered domains of the attributes. In this section, we will illustrate how the proposed intuitionistic fuzzy dominance–based rough set can be used in the decision system with probabilistic interpretation.

For a decision system $\mathscr{I}$, if $\forall x \in U$ and $\forall a \in AT$, $a(x) \subseteq V_a$ instead of $a(x) \in V_a$, i.e.

$$a : U \to P(V_a)$$

where $P(V_a)$ is the collection of all nonempty subsets of V_a, then such system is referred to as a set–valued decision system. Obviously, in a set–valued decision system $\mathscr{I}$, x holds a set of values instead of a single value on each attribute.

Furthermore, in a set–valued decision system with probabilistic interpretation, $\forall v \in V_a$, $a(x)(v) \in [0,1]$ represents the possibility of state v. $\forall x \in U$, $\forall a \in AT$, we assume here that $\sum_{v \in V_a} a(x)(v) = 1$

It is clear that every set value is expressed in a probability distribution over the elements contained in such set. This leads to that the set value can be expressed in terms of a probability distribution such that $a(x) = \{v_1/a(x)(v_1), v_2/a(x)(v_2), \cdots, v_k/a(x)(v_k)\}$ where $v_1, v_2, \cdots, v_k \in V_a$.

Actually, the set–valued decision system with probabilistic interpretation has been analyzed by rough set technique. For example, in valued tolerance relation based rough set for dealing with incomplete information systems, each unknown value is expressed in a uniform probability distribution over the elements contained in the domain of the corresponding attribute. Suppose that $V_a = \{a_1, a_2, a_3, a_4\}$, if $a(x) = *$ where $*$ denotes the "do not care" unknown value, then the probability distribution can be written such that $a(x) = \{a_1/0.25, a_2/0.25, a_3/0.25, a_4/0.25\}$. This tells us that if the value that x holds on a is unknown, then x may hold any one of the values in V_a. Moreover, the probabilistic degrees that x holds each value are equal. However, valued tolerance and dominance relations only consider the memberships of tolerance degree and dominance degree, they do not take the non–memberships into account. To overcome this limitation, the intuitionistic fuzzy rough technique has become a necessity.

Let us consider Table 1, it is a set–valued decision system with probabilistic interpretation. In Table 1,

- $U = \{x_1, x_2, \cdots, x_{10}\}$ is the universe of discourse;
- $AT = \{a, b, c, d, e\}$ denotes the set of condition attributes;
- $V_a = \{a_0, a_1, a_2\}$, $V_b = \{b_0, b_1, b_2\}$, $V_c = \{c_0, c_1, c_2\}$, $V_d = \{d_0, d_1, d_2\}$, $V_e = \{e_0, e_1, e_2\}$, $a_0 < a_1 < a_2$, $b_0 < b_1 < b_2$, $c_0 < c_1 < c_2$, $d_0 < d_1 < d_2$, $e_0 < e_1 < e_2$;
- f is the decision attribute where $V_f = \{1, 2\}$

$\forall (x, y) \in U \times U$, let us denote the intuitionistic fuzzy dominance relation as following:

$$\mathscr{R}_{AT}(x, y) = \begin{cases} [1, 0] & : \quad x = y \\ < u_{\mathscr{R}_{AT}}(x, y), v_{\mathscr{R}_{AT}}(x, y) > & : \quad \text{otherwise} \end{cases}$$

where $\forall a \in AT$,

$$u_{\mathscr{R}_a}(x, y) = \sum_{v_1 > v_2, v_1, v_2 \in V_a} a(x)(v_1) \cdot a(x)(v_2)$$

$$v_{\mathscr{R}_a}(x, y) = \sum_{v_1 < v_2, v_1, v_2 \in V_a} a(x)(v_1) \cdot a(x)(v_2)$$

Table 1. A decision system with probabilistic interpretation

U	a	b	c	d	e	f
x_1	$\{a_1/1\}$	$\{b_0/0.7, b_1/0.3\}$	$\{c_0/1\}$	$\{d_1/0/4, d_2/0.6\}$	$\{e_2/1\}$	2
x_2	$\{a_0/0.3, a_1/0.7\}$	$\{b_2/1\}$	$\{c_1/0.5, c_2/0.5\}$	$\{d_0/1\}$	$\{e_0/1\}$	1
x_3	$\{a_0/1\}$	$\{b_1/0.4, b_2/0.6\}$	$\{c_1/1\}$	$\{d_0/0.3, d_1/0.7\}$	$\{e_0/1\}$	1
x_4	$\{a_0/0.9, a_1/0.1\}$	$\{b_1/1\}$	$\{c_1/1\}$	$\{d_1/1\}$	$\{e_0/0.2, e_2/0.8\}$	1
x_5	$\{a_1/0.8, a_2/0.2\}$	$\{b_1/1\}$	$\{c_0/0.6, c_1/0.4\}$	$\{d_0/1\}$	$\{e_1/1\}$	2
x_6	$\{a_0/0.5, a_2/0.5\}$	$\{b_1/1\}$	$\{c_0/0.3, c_1/0.7\}$	$\{d_0/1\}$	$\{e_1/1\}$	1
x_7	$\{a_1/1\}$	$\{b_0/0.2, b_2/0.8\}$	$\{c_0/0.1, c_1/0.9\}$	$\{d_1/1\}$	$\{e_2/1\}$	2
x_8	$\{a_0/1\}$	$\{b_2/1\}$	$\{c_1/1\}$	$\{d_0/1\}$	$\{e_0/0.9, e_1/0.1\}$	1
x_9	$\{a_1/1\}$	$\{b_0/0.8, b_1/0.2\}$	$\{c_0/0.5, c_2/0.5\}$	$\{d_1/1\}$	$\{e_2/1\}$	2
x_{10}	$\{a_1/1\}$	$\{b_1/1\}$	$\{c_2/1\}$	$\{d_0/0.8, d_1/0.2\}$	$\{e_2/1\}$	2

In the above definition, $u_{\mathscr{R}_{AT}}(x, y)$ denotes the degree of *dominance principle* in terms of the set of attributes AT while $v_{\mathscr{R}_{AT}}(x, y)$ denotes the degree of *non-dominance principle* in terms of the set of attributes AT. For instance,

$$u_{\mathscr{R}_b}(x_2, x_1) = \sum_{v_1 \geq v_2, v_1, v_2 \in V_b} b(x_2)(v_1) \cdot b(x_1)(v_2)$$
$$= b(x_2)(b_2) \cdot b(x_1)(b_0) + b(x_2)(b_2) \cdot b(x_1)(b_1)$$
$$= 1$$
$$v_{\mathscr{R}_b}(x_2, x_1) = \sum_{v_1 < v_2, v_1, v_2 \in V_b} b(x_2)(v_1) \cdot b(x_1)(v_2)$$
$$= 0$$

Similarly, the intuitionistic fuzzy dominance relation in Table 1 is showed in Table 2.

According to the above intuitionistic fuzzy dominance relation, we can obtain the corresponding rough approximate memberships and non–memberships by Definition 2.

Table 2. Intuitionistic fuzzy dominance relation in Table 1

$x\backslash y$	x_1	x_2	x_3	x_4	x_5	x_6	x_7	x_8	x_9	x_{10}
x_1	[1.0,0.0]	[0.0,1.0]	[0.0,1.0]	[0.0,1.0]	[0.0,0.7]	[0.0,0.7]	[0.0,0.9]	[0.0,1.0]	[0.0,0.5]	[0.0,1.0]
x_2	[0.0,1.0]	[1.0,0.0]	[0.0,0.7]	[0.0,1.0]	[0.0,1.0]	[0.0,1.0]	[0.0,1.0]	[0.0,0.1]	[0.0,1.0]	[0.0,1.0]
x_3	[0.0,1.0]	[0.0,0.7]	[1.0,0.0]	[0.0,0.8]	[0.0,1.0]	[0.0,1.0]	[0.0,1.0]	[0.0,0.4]	[0.0,1.0]	[0.0,1.0]
x_4	[0.0,0.9]	[0.0,1.0]	[0.0,0.6]	[1.0,0.0]	[0.0,0.92]	[0.0,0.5]	[0.0,0.9]	[0.0,1.0]	[0.0,0.9]	[0.0,1.0]
x_5	[0.0,1.0]	[0.0,1.0]	[0.0,0.7]	[0.0,1.0]	[1.0,0.0]	[0.0,0.42]	[0.0,1.0]	[0.0,1.0]	[0.0,1.0]	[0.0,1.0]
x_6	[0.0,1.0]	[0.0,1.0]	[0.0,0.7]	[0.0,1.0]	[0.0,0.5]	[1.0,0.0]	[0.0,1.0]	[0.0,1.0]	[0.0,1.0]	[0.0,1.0]
x_7	[0.0,0.6]	[0.0,0.55]	[0.0,0.2]	[0.0,0.2]	[0.0,0.2]	[0.27,0.5]	[1.0,0.0]	[0.0,0.2]	[0.0,0.5]	[0.0,1.0]
x_8	[0.0,1.0]	[0.0,0.7]	[0.0,0.7]	[0.0,1.0]	[0.0,1.0]	[0.0,0.9]	[0.0,1.0]	[1.0,0.0]	[0.0,1.0]	[0.0,1.0]
x_9	[0.0,0.6]	[0.0,1.0]	[0.0,0.92]	[0.0,0.8]	[0.0,0.8]	[0.0,0.8]	[0.0,0.8]	[0.0,1.0]	[1.0,0.0]	[0.0,0.8]
x_{10}	[0.0,0.92]	[0.0,1.0]	[0.0,0.6]	[0.0,0.8]	[0.0,0.2]	[0.0,0.5]	[0.0,0.8]	[0.0,1.0]	[0.0,0.8]	[1.0,0.0]

By the decision attribute f, the universe can be partitioned into decision classes such that $\mathbf{CL} = \{CL_1, CL_2\} = \{\{x_2, x_3, x_4, x_6, x_8\}, \{x_1, x_5, x_7, x_9, x_{10}\}\}$. The results of intuitionistic fuzzy dominance–based rough approximations in Table 1 are showed in Table 3.

Table 3. Intuitionistic fuzzy dominance–based rough approximations in Table 1

$x \backslash y$	x_1	x_2	x_3	x_4	x_5	x_6	x_7	x_8	x_9	x_{10}
$u_{\underline{AT_{\mathscr{R}}}(CL_1^{\geq})}(x)$	1	1	1	1	1	1	1	1	1	1
$u_{\underline{AT_{\mathscr{R}}}(CL_2^{\geq})}(x)$	0.9	0	0	0	0.5	0	0.9	0	0.9	1
$v_{\underline{AT_{\mathscr{R}}}(CL_1^{\geq})}(x)$	0	0	0	0	0	0	0	0	0	0
$v_{\underline{AT_{\mathscr{R}}}(CL_2^{\geq})}(x)$	0	1	1	1	0	1	0	1	0	0
$u_{\overline{AT_{\mathscr{R}}}(CL_1^{\geq})}(x)$	1	1	1	1	1	1	1	1	1	1
$u_{\overline{AT_{\mathscr{R}}}(CL_2^{\geq})}(x)$	1	0	0	0	1	0	1	0	1	1
$v_{\overline{AT_{\mathscr{R}}}(CL_1^{\geq})}(x)$	0	0	0	0	0	0	0	0	0	0
$v_{\overline{AT_{\mathscr{R}}}(CL_2^{\geq})}(x)$	0	1	1	0.9	0	0.5	0	1	0	0
$u_{\underline{AT_{\mathscr{R}}}(CL_1^{\leq})}(x)$	0	1	1	0.9	0	0.5	0	1	0	0
$u_{\underline{AT_{\mathscr{R}}}(CL_2^{\leq})}(x)$	1	1	1	1	1	1	1	1	1	1
$v_{\underline{AT_{\mathscr{R}}}(CL_1^{\leq})}(x)$	1	0	0	0	1	0	1	0	1	1
$v_{\underline{AT_{\mathscr{R}}}(CL_2^{\leq})}(x)$	0	0	0	0	0	0	0	0	0	0
$u_{\overline{AT_{\mathscr{R}}}(CL_1^{\leq})}(x)$	0	1	1	1	0	1	0	1	0	0
$u_{\overline{AT_{\mathscr{R}}}(CL_2^{\leq})}(x)$	1	1	1	1	1	1	1	1	1	1
$v_{\overline{AT_{\mathscr{R}}}(CL_1^{\leq})}(x)$	0.9	0	0	0	0.5	0	0.9	0	0.9	1
$v_{\overline{AT_{\mathscr{R}}}(CL_2^{\leq})}(x)$	0	0	0	0	0	0	0	0	0	0

4   Conclusions

In this paper, we have developed a general framework for the generalization of dominance–based rough set. In our approach, the concept of intuitionistic fuzzy set is combined with the DRSA and then the intuitionistic fuzzy dominance–based rough set is defined. Different from the previous DRSA, we use an intuitionistic fuzzy dominance relation instead of the crisp or fuzzy dominance relation to defined dominance–based rough set model.

Acknowledgments. This work is supported by Postdoctoral Science subsubsection Foundation of China (No.20100481149).

References

1. Pawlak, Z.: Rough Sets–Theoretical Aspects of Reasoning About Data. Kluwer Academic Publishers, Dordrecht (1991)
2. Pawlak, Z., Skowron, A.: Rudiments of Rough Sets. Inform. Sci. 177, 3–27 (2007)

3. Pawlak, Z., Skowron, A.: Rough Sets: Some Extensions. Inform. Sci. 177, 28–40 (2007)
4. Pawlak, Z., Skowron, A.: Rough Sets and Boolean Reasoning. Inform. Sci. 177, 41–73 (2007)
5. Greco, S., Inuiguchi, M., Słowiński, R.: Fuzzy Rough Sets and Multiple–Premise Gradual Decision Rules. Int. J. Approx. Reason. 41, 179–211 (2006)
6. Greco, S., Matarazzo, B., Słowiński, R.: Rough Approximation by Dominance Relations. Int. J. Intell. Syst. 17, 153–171 (2002)
7. Greco, S., Matarazzo, B., Słowiński, R.: Rough Sets Theory for Multicriteria Decision Analysis. Euro. J. Oper. Res. 129, 1–47 (2002)
8. Shao, M.W., Zhang, W.X.: Dominance Relation and Rules in an Incomplete Ordered Information System. Int. J. Intell. Syst. 20, 13–27 (2005)
9. Yang, X.B., Yang, J.Y., Wu, C., Yu, D.J.: Dominance–Based Rough Set Approach and Knowledge Reductions in Incomplete Ordered Information System. Inform. Sci. 178, 1219–1234 (2008)
10. Błaszczyński, J., Greco, S., Słowiński, R., Szeląg, M.: On Variable Consistency Dominance-Based Rough Set Approaches. In: Greco, S., Hata, Y., Hirano, S., Inuiguchi, M., Miyamoto, S., Nguyen, H.S., Słowiński, R. (eds.) RSCTC 2006. LNCS (LNAI), vol. 4259, pp. 191–202. Springer, Heidelberg (2006)
11. Błaszczyński, J., Greco, S., Słowiński, R., Szeląg, M.: Monotonic Variable Consistency Rough Set Approaches. In: Yao, J., Lingras, P., Wu, W.-Z., Szczuka, M.S., Cercone, N.J., Ślęzak, D. (eds.) RSKT 2007. LNCS (LNAI), vol. 4481, pp. 126–133. Springer, Heidelberg (2007)
12. Inuiguchi, M., Yoshioka, Y., Kusunoki, Y.: Variable–precision Dominance-based Rough Set Approach and Attribute Reduction. Int. J. Approx. Reason. 20, 1199–1214 (2009)
13. Kotłowski, W., Dembczyński, K., Greco, S., Słowiński, R.: Stochastic Dominance-based Rough Set Model for Ordinal classification. Inform. Sci. 178, 4019–4037 (2008)
14. Greco, S., Matarazzo, B., Słowiński, R.: Fuzzy Set Extensions of the Dominance–based Rough Set Approach. In: Bustince, H., et al. (eds.) Fuzzy Sets and Their Extensions: Representation, Aggregation and Models, pp. 239–261. Springer, Heidelberg (2008)
15. Atanassov, K.: Intuitionistic Fuzzy Sets. Fuzzy Set. Syst. 20, 87–96 (1986)
16. Atanassov, K.: Intuitionistic Fuzzy Sets: Theory and Applications. Physica-Verlag, Heidelberg (1999)

A Covering-Based Pessimistic Multigranulation Rough Set

Guoping Lin* and Jinjin Li

Department of Mathematics and Information Science,
Zhangzhou Normal University, Zhangzhou, 363000, Fujian, China

Abstract. In view of granular computing, the classical optimistic and pessimistic multigranulation rough set models are both primarily based on simple granules among multiple granular structures, namely multiple partitions of the universe in MGRS. This correspondence paper presents a new rough set model where set approximations are defined by using multiple coverings on the universe. In order to distinguish Qian's covering-based optimistic multigranulation rough set model, we call the new rough set model as covering-based pessimistic multigranulation rough set model. The key distinction between covering-based pessimistic multigranulation rough set model and Qian's covering-based optimistic multigranulation rough set model is set approximation descriptions. Then some properties are proposed for covering-based pessimistic multigranulation rough set model.

Keywords: Covering, Multigranulation, Pessimistic, Rough sets.

1 Introduction

Rough set theory, proposed by Pawlak [1, 2], is a well-established mechanism for vagueness and uncertainty in data analysis. It has been applied to a variety of problems, such as feather selection [3, 4], knowledge reduction [5, 6, 7], rule extraction [8, 9], uncertainty reasoning [2, 10, 11, 12] and granular computing [13, 14, 15]. The theory of rough set is originally constructed on the basis of a indiscernibility relation(equivalence relation)or partition. However, it is restrictive for many applications. To address this issue, there are two meaningful methods to generalize the traditional rough set. One is to relax the equivalence relation, for example, many expanded rough set models have been proposed which based on similarity relation or tolerance relation [16, 17] and so on. Another method is the relaxation of partition to covering [18, 19, 20, 21]. In fact, the covering generalized rough sets are improvement of traditional rough set models to handle more complex practical problems. Chen et al. [21] first proposed several covering backgrounds and presented a new definition of covering to construct the upper and lower approximation of an arbitrary set. Synchronously, many researchers considered a covering as a granular space in view of granular computing.

---

* Corresponding author.

D.-S. Huang et al. (Eds.): ICIC 2011, LNBI 6840, pp. 673–680, 2012.

From the above discussion, we can see that all of extensional rough set models are concerned with approximations of set described only by a single binary relation or single covering on the given universe. With the view of granular computing, the above rough set models are only based on a single granulation. These rough set models are still have some restrictive for some special practical applied background [22]. Recently, Qian et al. first took multiple binary relations into account and proposed an optimistic and pessimistic multigranulation rough models [22, 23], in which a target concept was described through multiple binary relations on the universe according to user different requirements or targets of problem solving. Later, Qian et al. also proposed an optimistic covering-based multigranulation rough model so that MGRS theory can be used to wider applicable range[24]. In this paper, we present a new multigranulation rough set based on multiple coverings. In order to distinguish Qian's covering-based optimistic multigranulation rough set model, we call the new rough set model as covering-based pessimistic multigranulation rough set model. The key distinction between covering-based pessimistic multigranulation rough set model and Qian's covering-based optimistic multigranulation rough set model is set approximation descriptions.

The main objective of this correspondence paper is to establish a rough set model based on multiple coverings in the practical applied background. The rest of this paper is organized as follows. Some basic concepts in complete and incomplete MGRS respectively are briefly reviewed in Section 2. In Section 3, a covering-based pessimistic multigranulation rough set is presented, some of its important properties are investigated. Finally, Section 4 concludes this paper.

2   Preliminary

In this section, we recall some basic concepts on covering rough set [22, 25, 28].

Definition 2.1[18]. Let U be a finite nonempty universe, C a family of subsets of U. If none of the subsets in C is empty and $\bigcup C = U$, C is called a cover of U, then the ordered pair $< U, C >$ is called a covering approximation space.

Definition 2.3[18]. For any $X \subseteq U$, $\underline{C}X = \bigcup\{K \in C | K \subseteq X\}$ is called the lower approximation of X.

For the mutual related concepts of extension and intension, there are many different presentations for upper approximation of X on the universe. In this paper, we use following definition of upper approximation of X.

Definition 2.4[25]. For any $X \subseteq U$, $\overline{C}X = \bigcup\{K | K \in C, K \cap X \neq \emptyset\}$ is called the upper approximation of X.

Definition 2.5[21]. Let $C = \{K_1, K_2 \cdots, K_m\}$ be a covering of U. For every $x \in U$, let $K_x = \bigcap\{K_j | K_j \in C, x \in K_j\}$, $Cov(C) = \{K_x | x \in U\}$ is then also a cover of U.

Definition 2.6[21]. Let $\Delta = \{C_1, C_2 \cdots, C_m\}$ be a family of covers of U. For $X \subseteq U$, let $\Delta_x = \bigcap\{K_{ix} \in Cov(C_i), x \in K_{ix}\}$, then $Cov(\Delta) = \{\Delta_x : x \in U\}$ is also a covering of U.

Definition 2.7[21]. For $X \subseteq U$, the lower and upper approximation of X with respect to $Cov(\Delta)$ are defined as follows

$$\underline{\Delta}(X) = \bigcup\{\Delta_x : \Delta_x \subset X\}, \tag{1}$$

$$\overline{\Delta}(X) = \bigcup\{\Delta_x : \Delta_x \cap X \neq \emptyset\}. \tag{2}$$

3   Covering-Based Multigranulation Rough Set Model

In this Section, we present a new multigranulation rough set based on multiple coverings, where approximation descriptions of a target concept are different from those in Qian's covering-based optimistic multigranulation rough set model. In the following, we recall Qian's covering-based optimistic multigranulation rough set model and then present a new covering-based pessimistic multigranulation rough set model which is based on "Concept description" strategy.

Definition 3.1[24]. Let (U, Δ) be a covering approximation space, $\Delta = \{C_1, \cdots, C_m\}$ a family of coverings of U, where $C_i = \{K_{i1}, K_{i2}, \cdots, K_{it_i}\}$, and $X \subseteq U$, an optimistic lower approximation and upper approximation of X with respect to Δ are denoted by $\sum_{i=1}^{m} C_i^O X$ and $\overline{\sum_{i=1}^{m} C_i}^O X$, respectively, where

$$\sum_{i=1}^{m} C_i^O X = \bigcup\{K_{it_i} \in C_i \mid K_{it_i} \subseteq X, t_i = 1, 2, \cdots, |C_i|\}, \tag{3}$$

$$\overline{\sum_{i=1}^{m} C_i}^O X = \bigcup\{K_{it_i} \in C_i : K_{it_i} \cap X \neq \emptyset, t_i = 1, 2, \cdots, |C_i|\}. \tag{4}$$

Where $C_i = \{K_{i1}, K_{i2}, \cdots, K_{i|C_i|}\}$.

And the *area of uncertainty* or *boundary region* of X relative to Δ in an optimistic multigranulation rough set model is

$$Bn_{\sum_{i=1}^{m} C_i}^O(X) = \overline{\sum_{i=1}^{m} C_i}^O X \backslash \sum_{i=1}^{m} C_i^O X.$$

Definition 3.2. Let U be a finite universe of discourse, C_1 and C_2 two different covers of U, and $K(x)$ is a subset including x. For any $K(x) \in C_1$, if any $L(x) \in C_2$ such that $K(x) \subseteq L(x)$, then we call C_1 is uniform finer than C_2, called uniform partial relation between C_1 and C_2, denoted $C_1 \preceq^c C_2$.

Especially, let C_1 and C_2 be two different partitions of U, and $K(x)$ is a subset including x. If any $K(x) \in C_1$, there exists $L(x) \in C_2$ such that $K(x) \subseteq L(x)$, then we call C_1 is finer than C_2, denoted $C_1 \preceq C_2$.

Theorem 3.1. The partial relation $\preceq$ is a special case of uniform partial relation $\preceq^c$.

Proof. It can be proof easily by Definition 3.2.

Definition 3.3. Let U be a finite universe of discourse, $C_1 = \{K_{11}, K_{12}, \cdots, K_{1t_1}\}$, $C_2 = \{K_{21}, K_{22}, \cdots, K_{2t_2}\}$ two coverings of U, then join operation between C_1 and C_2 is defined as follows:

$$C_1 \cap C_2 = \{K_{1t_1} \cap K_{2t_2} : K_{1t_1} \in C_1, K_{2t_2} \in C_2, 1 \le t_1 \le |C_1|, 1 \le t_2 \le |C_2|\}.$$

Definition 3.4. Let (U, Δ) be a covering approximation space, $\Delta = \{C_1, C_2, \cdots, C_m\}$ a family of coverings of U, where $C_i = \{K_{i1}, K_{i2}, \cdots, K_{it_i}\}$, and $X \subseteq U$, a pessimistic lower approximation and upper approximation of X with respect to Δ are denoted by $\underline{\sum_{i=1}^{m} C_i}^{P} X$ and $\overline{\sum_{i=1}^{m} C_i}^{P} X$, respectively, where

$$\underline{\sum_{i=1}^{m} C_i}^{P} X = \bigcup \{K_{it_i} \in C_i \mid \wedge_{i=1}^{m}(K_{it_i} \subseteq X), t_i = 1, 2, \cdots, |C_i|\}, \qquad (5)$$

$$\overline{\sum_{i=1}^{m} C_i}^{P} X = \bigcap_{i=1}^{m}(\overline{C_i}X). \qquad (6)$$

Where $C_i = \{K_{i1}, K_{i2}, \cdots, K_{it_i}\}$. And the *area of uncertainty* or *boundary region* of X relative to Δ in CMGRS is

$$Bn_{\sum_{i=1}^{m} C_i}^{P}(X) = \overline{\sum_{i=1}^{m} C_i}^{P} X \backslash \underline{\sum_{i=1}^{m} C_i}^{P} X.$$

Example 3.1. Let us consider an evaluation problem of credit card applicant. Suppose $U = \{x_1, x_2, x_3, x_4, x_5, x_6, x_7, x_8, x_9\}$ is a set of nine applicants, $E = \{education; salary\}$ is a set of two condition attributes, the values of ‘*education*‘are $\{best; better; good\}$, and the values of ‘*salary*‘ are $\{high; middle; low\}$. We have three specialist $\{A, B, C\}$ to evaluate the attributes of these applicants. It is possible that their evaluation results to the same attribute values are not the to one another. The evaluation results are listed below as Table 1. In following Table 1, y_i, n_i, $(i = 1, 2, 3)$ denotes the evaluation result by specialist A, B, C, respectively, where y_i means "yes"and n_i means "no".

Example 3.2 (Continued from Example 3.1). From Table 1, we can get two coverings of U, such as $C_1 = \{\{x_1, x_2, x_4, x_5, x_7, x_8\}, \{x_2, x_5, x_8\}, \{x_3, x_6, x_5, x_9\}\}$, $C_2 = \{\{x_1, x_2, x_3\}, \{x_4, x_5, x_6, x_7, x_8\}, \{x_7, x_8, x_9\}\}$. For a target concept $X = \{x_1, x_2, x_5, x_8\} \subseteq U$, by Definition 3.4, we have $(\underline{C_1 + C_2})^P X = \{x_2, x_5, x_8\} \cap \emptyset = \emptyset$; $(\overline{C_1 + C_2})^P X = \overline{C_1}X \cap \overline{C_2}X = \{x_1, x_2, x_3, x_4, x_5, x_6, x_7, x_8, x_9\}$. And $C_1 \cap C_2 = \{\{x_1, x_2\}, \{x_2\}, \{x_3\}, \{x_4, x_5, x_7, x_8\}, \{x_5, x_8\}, \{x_5, x_6\}, \{x_2, x_8\}, \{x_8\}, \{x_9\}\}$,

Table 1. Evaluation information system

U \ A	education									salary								
attribute value	best			better			good			hign			middle			low		
	A	B	C	A	B	C	A	B	C	A	B	C	A	B	C	A	B	C
x_1	y_1	y_2	y_3	n_1	n_2	n_3	n_1	n_2	n_3	y_1	y_2	y_3	n_1	n_2	n_3	n_1	n_2	n_3
x_2	n_1	y_2	n_3	y_1	n_2	y_3	n_1	n_2	n_3	y_1	y_2	y_3	n_1	n_2	n_3	n_1	n_2	n_3
x_3	n_1	n_2	n_3	n_1	n_2	n_3	y_1	y_2	y_3	y_1	y_2	y_3	n_1	n_2	n_3	n_1	n_2	n_3
x_4	y_1	y_2	y_3	n_1	n_2	n_3	n_1	n_2	n_3	n_1	n_2	n_3	y_1	y_2	y_3	n_1	n_2	n_3
x_5	y_1	n_2	n_3	n_1	y_2	n_3	n_1	n_2	y_3	n_1	n_2	n_3	y_1	y_2	y_3	n_1	n_2	n_3
x_6	n_1	n_2	n_3	n_1	n_2	n_3	y_1	y_2	y_3	n_1	n_2	n_3	y_1	y_2	y_3	n_1	n_2	n_3
x_7	y_1	y_2	y_3	n_1	n_2	n_3	n_1	n_2	n_3	n_1	n_2	n_3	y_1	y_2	n_3	n_1	n_2	y_3
x_8	n_1	y_2	n_3	y_1	n_2	y_3	n_1	n_2	n_3	n_1	n_2	n_3	y_1	n_2	y_3	n_1	y_2	n_3
x_9	n_1	n_2	n_3	n_1	n_2	n_3	y_1	y_2	y_3	n_1	n_2	n_3	n_1	n_2	n_3	y_1	y_2	y_3

then we have $\underline{C_1 \cap C_2}(X) = \{x_1, x_2, x_5, x_8\}$; $\overline{C_1 \cap C_2}(X) = \{x_1, x_2, x_4, x_5, x_6, x_7, x_8\}$. As a result of this example, we can see that the pessimistic lower approximation of X induced by $C_1 + C_2$ is not bigger than that induced by $C_1 \cap C_2$, then we have the following propositions.

Proposition 3.1. Let (U, Δ) be a covering approximation space, $\Delta = \{C_1, C_2, \cdots, C_m\}$ a family of coverings of U, and $X \subseteq U$, then, the following properties hold:

$$\underline{\sum_{i=1}^{m} C_i}^P X \subseteq \bigcap_{i=1}^{m} \underline{C_i X}; \tag{7}$$

$$\overline{\sum_{i=1}^{m} C_i}^P X \supseteq \bigcap_{i=1}^{m} \overline{C_i X}. \tag{8}$$

Proof. (1) For any $x \in \sum_{i=1}^{m} C_i^P X$, from Definition 3.4, it follows that there must exist $K_{1t_1}(x) \in C_1, K_{2t_2}(x) \in C_2 \cdots, K_{mt_m}(x) \in C_m$, such that $x \in K_{it_i}(x), i = 1, 2, \cdots, m$; where $K_{it_i}(x)$ consists x. Hence, $x \in \bigcap_{i=1}^{m} K_{it_i}(x)$, But $\bigcap_{i=1}^{m} K_{it_i}(x) \subseteq \Delta_x$, for any $x \in U$, and $\bigcap_{i=1}^{m} \underline{C_i X} = \bigcup\{\Delta_x : \Delta_x \subseteq X\}$, therefore $x \in \bigcap_{i=1}^{m} \underline{C_i X}$.

(2) For any $x \in \bigcap_{i=1}^{m} \overline{C_i X}$, then there exists Δ_x, such that $x \in \Delta_x$, and $\Delta_x \cap X \neq \emptyset$, but $\Delta_x \subseteq K_{it_i}(x), (i = 1, 2, \cdots, m)$. Hence, $K_{it_i}(x) \cap X \neq \emptyset$, i.e., $x \in \overline{\sum_{i=1}^{m} C_i}^P X$, therefore $\overline{\sum_{i=1}^{m} C_i}^P X \supseteq \bigcap_{i=1}^{m} \overline{C_i X}$.

Proposition 3.2. Let $\Delta = \{C_1, C_2, \cdots, C_m\}$ be a family of coverings of U, and $X \subseteq U$, then following properties hold:

$$\underline{\sum_{i=1}^{m} C_i}^P U = \overline{\sum_{i=1}^{m} C_i}^P U = U; \tag{9}$$

$$\sum_{i=1}^{m} C_i^{\underline{P}}\ \emptyset = \sum_{i=1}^{m} C_i^{\overline{P}}\ \emptyset = \emptyset;$$ (10)

$$\sum_{i=1}^{m} C_i^{\underline{P}}\ X \subseteq X \subseteq \sum_{i=1}^{m} C_i^{\overline{P}}\ X;$$ (11)

$$\sum_{i=1}^{m} C_i^{\overline{P}}\ (X \cup Y) = \sum_{i=1}^{m} C_i^{\overline{P}}\ X \cup \sum_{i=1}^{m} C_i^{\overline{p}}\ Y;$$ (12)

$$\sum_{i=1}^{m} C_i^{\underline{P}}\ X = \sum_{i=1}^{m} C_i^{\underline{P}}\ X.$$ (13)

Proof. These can be proved easily by Definition 3.1.

However, in covering-based pessimistic multigranulation rough set model, there are some propositions which have not been hold as follows

$$\sum_{i=1}^{m} C_i^{\underline{P}}\ (X \cap Y) = \sum_{i=1}^{m} C_i^{\underline{P}}\ X \cap \sum_{i=1}^{m} C_i^{\underline{P}}\ Y;$$ (14)

$$\sum_{i=1}^{m} C_i^{\overline{P}}\ X = \sum_{i=1}^{m} C_i^{\overline{P}}\ X;$$ (15)

$$\sum_{i=1}^{m} C_i^{\underline{P}}\ X = \wr \sum_{i=1}^{m} C_i^{\overline{P}}\ (\wr X).$$ (16)

Theorem 3.2. Let $\Delta = \{C_1, C_2, \cdots, C_m\}$ be a family of coverings of U, and $X_1 \subseteq X_2 \subseteq \cdots \subseteq X_n$, and $X_i \subseteq U, i = 1, 2, \cdots, n$, then we have

$$\sum_{i=1}^{m} C_i^{\underline{P}}\ X_1 \subseteq \sum_{i=1}^{m} C_i^{\underline{P}}\ X_2 \subseteq \cdots \subseteq \sum_{i=1}^{m} C_i^{\underline{P}}\ X_n;$$ (17)

$$\sum_{i=1}^{m} C_i^{\overline{P}}\ X_1 \subseteq \sum_{i=1}^{m} C_i^{\overline{P}}\ X_2 \subseteq \cdots \subseteq \sum_{i=1}^{m} C_i^{\overline{P}}\ X_n.$$ (18)

Proof. These can be proved by Definition 3.4.

Theorem 3.3. Let $\Delta = \{C_1, C_2, \cdots, C_m\}$ be a family of coverings of U, and $X \subseteq U$, suppose that $C_1 \preceq^c C_2 \preceq^c \cdots \preceq^c C_m$, then we have

$$\sum_{i=1}^{m} C_i^{\underline{P}}\ X = \underline{C_m}X;$$ (19)

$$\overline{\sum_{i=1}^{m} C_i}^{P} X = \overline{C_m}X. \tag{20}$$

Proof. (1) For any $x \in \overline{\sum_{i=1}^{m} C_i}^{P} X$, then we have $K_{it_i}(x) \subseteq X$, for $i = 1, 2, \cdots, m$, it follows that $x \in \overline{C_m}X$; for any $x \in \overline{C_m}X$, we have $K_{mt_m} \subseteq X$. Moreover, since $C_1 \preceq^c C_2 \preceq^c \cdots \preceq^c C_m$, then we have $K_{1t_1}(x) \subseteq K_{2t_2}(x) \subseteq \cdots \subseteq K_{mt_m}(x) \subseteq X$, then by Definition 3.4, we have $x \in \overline{\sum_{i=1}^{m} C_i}X$. hence, we can conclude that $\overline{\sum_{i=1}^{m} C_i}^{P} X = \overline{C_m}X$. Similarly, (2) can be proved.

4 Conclusion

The contribution of this paper has proposed a covering-based pessimistic multi-granulation rough set model which is different from Qian's optimistic covering multigranulation model. Under the covering-based pessimistic multigranulation, we have given some important properties. Accordingly, covering-based multi-granulation rough set theory can be used to wider application fields.

Acknowledgments. This work was supported by National Natural Science Fund of China (No. 60903110, 10971186, 10671173, 11061004), Natural Science Fund of Fujian Province (No.2010J01018), and Education Committee of Fujian Province(No. JA09167).

References

1. Pawlak, Z.: Rough sets. International Journal of computer and Information Sciences 11, 341–365 (1982)
2. Pawlak, Z.: Rough sets, Theoretical aspects of reasoning about data. Kluwer Academic Publishers, Dordrecht (1991)
3. Liang, J.Y., Li, D.Y.: Uncertainty and knowledge acquisition in information systems. Science Press, Beijing (2005)
4. Swiniarski, W., Skowron, A.: Rough set methods in feature selection and recognition. Pattern Recognition Letters 24(6), 849–883 (2003)
5. Li, D., Zhang, B., Leung, Y.: On knowledge reduction in inconsistent decision information systems. International Journal of Uncertainty, Fuzziness Knowledge-Based Systems 12(5), 651–672 (2004)
6. Wang, G., Yu, H., Yang, D.: Decision table reduction based on conditional information entropy. Chinese Journal of Computers 25(7), 1–9 (2002)
7. Qian, Y.H., Liang, J.Y., Witold, P.: Positive approximation: an accelerator for attribute reduction in rough set theory. Artificial Intelligence 174, 597–618 (2010)
8. Apolloni, B., Brega, A., Malchiodi, D., Palmas, G., Zanaboni, A.M.: Learning rule representations from data. IEEE Transactions Systems Man and CYbernetics Part B 36(5), 1010–1028 (2006)
9. Tsumoto, S.: Automated extraction of hierarchical decision rules from clinical databases using rough set model. Expert Systems and Application 24(2), 189–197 (2003)

10. Liang, J.Y., Qian, Y.H.: Information granules and entropy theory in information systems. Science in China-Series F: Information Sciences 51(9), 1–18 (2008)
11. Qian, Y.H., Liang, J.Y., Li, D.Y.: Measures for evaluating the decision performance of a decision table in rough set theory. Informations Science (178), 181–202 (2008)
12. Qian, Y.H., Liang, J.Y., Dang, C.Y.: Consistency measure, inclusion degree and fuzzy measure in decision tables. Fuzzy Sets and Systems (159), 2353–2377 (2008)
13. Liang, J., Qian, Y.: Axiomatic approach of knowledge granulation in information system. In: Sattar, A., Kang, B.-h. (eds.) AI 2006. LNCS (LNAI), vol. 4304, pp. 1074–1078. Springer, Heidelberg (2006)
14. Lin, T.Y.: Granular Computing on binary relations I: data mining and neighourhood systems. Rough sets and Knowledge Discovery, 107–121 (1998)
15. Yao, Y.Y.: Information granulation and rough set approximation. International Journal of Intelligence Systems 16(1), 87–104 (2001)
16. Slowinski, R., Vanderpooten, D.: A generalized definition of rough approximations based on similarity. IEEE Transactions on Knowledge and Data Engineering 12, 331–336 (2000)
17. Kryszkiewicz, M.: Rough set approach to incomplete information systems. Information Sciences 112, 39–49 (1998)
18. Bonikowski, Z., Brynirski, E., Wybraniec, U.: Extensions and intentions in the rough set theory. Information Sciences 107, 149–167 (1998)
19. ZaKowski, W.: Approximations in the space (U, Π). Demonstration Mathematica 16, 761–769 (1983)
20. Zhu, W., Wang, F.Y.: On three types of covering rough sets. IEEE Transactions on Knowledge Data Engineering 19(8), 1131–1144 (2007)
21. Chen, D.G., Wang, C.Z., Hu, Q.H.: A new approach to attribute reduction of consistent and inconsistent covering decision systems with covering rough sets. Information Sciences 177, 3500–3518 (2007)
22. Qian, Y.H., Liang, J.Y., Yao, Y.Y., Dang, C.Y.: MGRS: A multigranulation rough set. Information Sciences 180, 949–970 (2010)
23. Qian, Y.H., Liang, J.Y., Wei, W.: Pessimistic rough decision. In: 2nd International Workshop on Rough Sets Theory, pp. 19–21. Zhoushan (2010)
24. Qian, Y. H., Liang, J. Y.: Granulation mechanism and Data Modeling for Complex Data. Phd thesis, Shanxi University (2011)
25. Pomykala, J.A.: Approximation operations in approximation space. Bulletin of the Polish Academy of Sciences 9-10, 653–662 (1987)

A Generalized Multi-granulation Rough Set Approach

Weihua Xu*, Xiantao Zhang, and Qiaorong Wang

School of Mathematics and Statistics,
Chongqing University of Technology, 400054 Chongqing, P.R. China
chxuwh@gmail.com

Abstract. A generalized multi-granulation rough set is proposed in this paper. In the new model, supporting characteristic function is defined and a parameter called information level is introduced to investigate that an object supports a concept precisely under majority granulations. Moreover, some important properties are discussed on the new multi-granulation rough set. And it can be found that the proposed model is more valid than old multiple granulation rough set models and Pawlak rough set model.

Keywords: Information level, Lower and upper approximation sets, Multi-granulation rough set, Supporting characteristic function, Majority granulations.

1 Introduction

The rough set theory proposed by Z. Pawlak ([2]) in 1980's is a useful soft computing tool for reasoning from data and a new mathematical approach to handle imprecision, vagueness and uncertainty in data analysis. As this theory has been applied to various fields such as medicine, engineering, management, economy, finance and security, many generalized rough set models are developed and studied ([1, 3, 7–12]).

Let $I = (U, A, V, f)$ be an information system. $B \subseteq A$ is an attribute subset. The equivalence relation corresponding to the attribute subset B is still denoted by itself. For an arbitrary set $X \subseteq U$, it may be impossible to describe X precisely using the equivalence classes $[x]_B = \{y \in U | f(x, a) = f(y, a), \forall a \in A\}$, that is, X can't be equal to the combination of some equivalence classes. In this case, one can depict the concept X by a pair of sets so called lower and upper approximation sets which are precise with respect to B. And the pair of sets can be defined as

$$\underline{B}(X) = \{x \in U | [x]_B \subseteq X\}, \quad \overline{B}(X) = \{x \in U | [x]_B \cap X \neq \varnothing\}.$$

X is fine if and only if $\underline{B}(X) = \overline{B}(X)$, otherwise X is rough if and only if $\underline{B}(X) \neq \overline{B}(X)$. The sets $\underline{B}(X)$ and $\overline{B}(X)$ are called, respectively, the lower approximation set and upper approximation set of X([1–3, 7, 12]).

---

* W.H. Xu is a Ph. D and a Prof. of Chongqing University of Technology. His main research fields are rough set, fuzzy set and artificial intelligence.

D.-S. Huang et al. (Eds.): ICIC 2011, LNBI 6840, pp. 681–689, 2012.

2   Multi-granulation Rough Set

Multi-granulation rough set model (MGRS) was studied as an expanding of Pawlak rough set model in references [4–6, 10]. An equivalence class of an object with respect to an attribute subset is a granularity in the view of granular computing. And a partition of the universe is a granular space. Then the classical rough set model is a single granulation rough set model (SGRS) and the granular space in this model is induced by the indiscernibility relation of attribute set. In cases referred in referrence [6](Case1, Case2 and Case3), there are limitations in SGRS for dealing with practical problems, and MGRS now can be used to solve these problems.

In MGRS, unlike SGRS, a concept is approximated through multiple partitions of U induced by multiple equivalence relations. And we have a brief introduction of MGRS in this section. As we have studied multi-granulation rough set further, we now in this section illustrate the two forms of multi-granulation rough set in our submitted paper [10].

Definition 2.1.([6, 10]) Let $I = (U, A, V, f)$ be an information system, $X \subseteq U$ and $P = \{P_i \subseteq A | P_i \cap P_j = \varnothing (i \neq j), i, j \leq l\}$. The lower and upper approximation sets of X with respect to P can be defined by following.

$$\underline{OM}(X) = \{x \in U | \vee ([x]_{P_i} \subseteq X), i \leq l\},$$

$$\overline{OM}(X) = \{x \in U | \wedge ([x]_{P_i} \cap X \neq \varnothing), i \leq l\}.$$

where "$\vee$" means the logical operator "OR" and "$\wedge$" means the logical operator "AND".

X is definable if and only if $\underline{OM}(X) = \overline{OM}(X)$; otherwise X is rough if and only if $\underline{OM}(X) \neq \overline{OM}(X)$. This model can be called the optimistic multi-granulation rough set model, denoted by OMGRS. And $\underline{OM}(X)$ and $\overline{OM}(X)$ are called, respectively, optimistic lower and upper approximation sets.

From the above definition, the operators "$\vee$" and "$\wedge$" can be exchanged between the lower approximation set and the upper approximation set. Corresponding to OMGRS, the pessimistic multi-granulation rough set model, denoted by PMGRS, can be defined in the following .

Definition2.2.([10]) Let $I = (U, A, V, f)$ be an information system, $X \subseteq U$ and $P = \{P_i \subseteq A | P_i \cap P_j = \varnothing (i \neq j), i, j \leq l\}$. The pessimistic lower and upper approximation sets of X with respect to P can be defined as follows.

$$\underline{PM}(X) = \{x \in U | \wedge ([x]_{P_i} \subseteq X), i \leq l\},$$

$$\overline{PM}(X) = \{x \in U | \vee ([x]_{P_i} \cap X \neq \varnothing), i \leq l\}.$$

X is definable if and only if $\underline{PM}(X) = \overline{PM}(X)$, otherwise X is rough if and only if $\underline{PM}(X) \neq \overline{PM}(X)$. $\underline{PM}(X)$ and $\overline{PM}(X)$ are called,respectively, pessimistic lower and upper approximation sets.

As generalizations of Pawlak rough set model, we merely show the relations of OMGRS, PMGRS and SGRS in the next proposition. Other descriptions on multi-granulation rough set can be reviewed in references [4–6, 10].

Proposition 2.1.([6, 10]) Let $I = (U, A, V, f)$ be an information system, $X \subseteq U$ and $P = \{P_i \subseteq A | P_i \cap P_j = \varnothing (i \neq j), i, j \leq l\}$. The following propositions hold.

(1) $\underline{OM}(X) = \overset{l}{\underset{i=1}{\cup}} \underline{P_i}(X);$

(2) $\overline{OM}(X) = \overset{l}{\underset{i=1}{\cap}} \overline{P_i}(X);$

(3) $\underline{PM}(X) = \overset{l}{\underset{i=1}{\cap}} \underline{P_i}(X);$

(4) $\overline{PM}(X) = \overset{l}{\underset{i=1}{\cup}} \overline{P_i}(X);$

(5) $\underline{PM}(X) \subseteq \underline{OM}(X);$

(6) $\overline{OM}(X) \subseteq \overline{PM}(X).$

3   A Generalized Multi-Granulation Rough Set

The two forms of multi-granulation rough set model illustrated in Section 2 are only special ones . From these two forms, we propose a more generalized and logical one in this section.

In order to present the generalized multi-granulation rough set model, we define a function called supporting characteristic function firstly.

Definition 3.1. Let $I = (U, A, V, f)$ be an information system, $X \subseteq U$ and $P = \{P_i \subseteq A | P_i \cap P_j = \varnothing (i \neq j), i, j \leq l\}$. Characteristic function $S_X^{P_i}(x)$, describing the inclusion relation between the class $[x]_{P_i}$ and the concept X, is defined as follows.

$$S_X^{P_i}(x) = \begin{cases} 1, & [x]_{P_i} \subseteq X \\ 0, & else \end{cases} \quad (i \leq l).$$

We call $S_X^{P_i}(x)$ supporting characteristic function of $x \in U$. It shows the object x supports the concept X precisely or not with respect to P_i.

Proposition 3.1. For any $x \in U$ and $P_i \in P$, the following properties of $S_X^{P_i}(x)$ hold.

(1) $S_{\sim X}^{P_i}(x) = \begin{cases} 1, & [x]_{P_i} \cap X = \varnothing \\ 0, & [x]_{P_i} \cap X \neq \varnothing \end{cases};$

(2) $S_\varnothing^{P_i}(x) = 0, \ S_U^{P_i}(x) = 1;$

(3) $S_{X \cup Y}^{P_i}(x) \geq S_X^{P_i}(x) \vee S_Y^{P_i}(x);$

(4) $S_{X \cap Y}^{P_i}(x) = S_X^{P_i}(x) \wedge S_Y^{P_i}(x);$

(5) $X \subseteq Y \Rightarrow S_X^{P_i}(x) \leq S_Y^{P_i}(x);$

(6) $X \subseteq Y \Rightarrow S_{\sim X}^{P_i}(x) \geq S_{\sim Y}^{P_i}(x).$

where, "$\wedge$" and "$\vee$" are, respectively, operations "minimum" and "maximum" in this proposition.

Proof. (1) Since $[x]_{P_i} \subseteq \sim X \Leftrightarrow [x]_{P_i} \cap X = \varnothing$ and $[x]_{P_i} \nsubseteq \sim X \Leftrightarrow [x]_{P_i} \cap X \neq \varnothing$. So this proposition is obvious.

(2) According to Definition 3.1, one can have that

$\forall x \in U \Rightarrow [x]_{P_i} \nsubseteq \varnothing$, i.e., $S_{\varnothing}^{P_i}(x) = 0$; $\forall x \in U \Rightarrow [x]_{P_i} \subseteq U$, i.e., $S_U^{P_i}(x) = 1$. This item is proved.

(3) As is known, "$Z \subseteq X$ or $Z \subseteq Y \Rightarrow (\nLeftarrow) Z \subseteq X \cup Y$" holds for any set Z. Thus, we have

$$S_X^{P_i}(x) \vee S_Y^{P_i}(x) = 1 \Leftrightarrow S_X^{P_i}(x) = 1 \text{ or } S_Y^{P_i}(x) = 1 \Leftrightarrow [x]_{P_i} \subseteq X \text{ or } [x]_{P_i} \subseteq Y$$
$$\Rightarrow (\nLeftarrow)[x]_{P_i} \subseteq X \cup Y \Leftrightarrow S_{X \cup Y}^{P_i}(x) = 1.$$

If $X \cup Y = U$, then $S_{X \cup Y}^{P_i}(x) = S_U^{P_i}(x) = 1$ is obvious. If $X \cup Y \neq U$, then $\sim (X \cup Y) = \sim X \cap \sim Y \neq \varnothing$. So, we have that

$$S_{X \cup Y}^{P_i}(x) = 0 \Leftrightarrow [x]_{P_i} \cap \sim (X \cup Y) \neq \varnothing \Leftrightarrow [x]_{P_i} \cap \sim X \cap \sim Y \neq \varnothing$$
$$\Rightarrow (\nLeftarrow)[x]_{P_i} \cap \sim X \neq \varnothing \text{ and } [x]_{P_i} \cap \sim Y \neq \varnothing$$
$$\Leftrightarrow S_X^{P_i}(x) = 0 \text{ and } S_Y^{P_i}(x) = 0 \Leftrightarrow S_X^{P_i}(x) \vee S_Y^{P_i}(x) = 0.$$

That is to say that $S_{X \cup Y}^{P_i}(x) \geq S_X^{P_i}(x) \vee S_Y^{P_i}(x)$ holds for any $x \in U$.

(4) Since "$Z \subseteq X$ and $Z \subseteq Y \Leftrightarrow Z \subseteq X \cap Y$" holds for any set Z, we can obviously have

$$S_{X \cap Y}^{P_i}(x) = 0 \Leftrightarrow [x]_{P_i} \cap \sim (X \cap Y) \neq \varnothing \Leftrightarrow [x]_{P_i} \cap (\sim X \cup \sim Y) \neq \varnothing$$
$$\Leftrightarrow ([x]_{P_i} \cap \sim X) \cup ([x]_{P_i} \cap \sim Y) \neq \varnothing$$
$$\Leftrightarrow [x]_{P_i} \cap \sim X \neq \varnothing \text{ or } [x]_{P_i} \cap \sim Y \neq \varnothing$$
$$\Leftrightarrow S_X^{P_i}(x) = 0 \text{ or } S_Y^{P_i}(x) = 0 \Leftrightarrow S_X^{P_i}(x) \wedge S_Y^{P_i}(x) = 0;$$

and

$$S_{X \cap Y}^{P_i}(x) = 1 \Leftrightarrow [x]_{P_i} \subseteq X \cap Y \Leftrightarrow [x]_{P_i} \subseteq X \text{ and } [x]_{P_i} \subseteq Y$$
$$\Leftrightarrow S_X^{P_i}(x) = 1 \text{ and } S_Y^{P_i}(x) = 1 \Leftrightarrow S_X^{P_i}(x) \wedge S_Y^{P_i}(x) = 1.$$

Then, $S_{X \cap Y}^{P_i}(x) = S_X^{P_i}(x) \wedge S_Y^{P_i}(x)$ holds for any $x \in U$. This item is proved.

(5) The case is obvious if $[x]_{P_i} \nsubseteq X$ by Definition 3.1 and $S_X^{P_i}(x) = 0 \leq S_Y^{P_i}(x)$. If $[x]_{P_i} \subseteq X$, one can have that $[x]_{P_i} \subseteq X \subseteq Y$, i.e., $S_X^{P_i}(x) = 1 = S_Y^{P_i}(x)$. Then this item is proved.

(6) From (1), this item can be proved similarly as (5).

For any $x \in U$ and $X \subseteq U$, the number of equivalence classes $[x]_{P_i}$ satisfying $[x]_{P_i} \subseteq X$ can be represented as $\sum_{i=1}^{l} S_X^{P_i}(x)$ by supporting characteristic function and the number of equivalence classes $[x]_{P_i}$ satisfying $[x]_{P_i} \cap X \neq \varnothing$ can be represented as $\sum_{i=1}^{l} (1 - S_{\sim X}^{P_i}(x))$. Moreover, we have the following proposition.

Proposition 3.2. By supporting characteristic function, the lower and upper approximation sets in OMGRS and PMGRS can be represented, respectively, in the following form.

$$(1)\ \underline{OM}(X) = \{x \in U | \frac{\sum\limits_{i=1}^{l} S_X^{P_i}(x)}{l} > 0\};$$

$$\overline{OM}(X) = \{x \in U | \frac{\sum\limits_{i=1}^{l} (1 - S_{\sim X}^{P_i}(x))}{l} \geq 1\}.$$

$$(2)\ \underline{PM}(X) = \{x \in U | \frac{\sum\limits_{i=1}^{l} S_X^{P_i}(x)}{l} \geq 1\};$$

$$\overline{PM}(X) = \{x \in U | \frac{\sum\limits_{i=1}^{l} (1 - S_{\sim X}^{P_i}(x))}{l} > 0\}.$$

Proof. It can be proved easily from Definition 2.1, 2.2 and 3.1.

In the view of granular computing, models in the above proposition may be not always effective in practice. OMGRS may be so loose that the approximation sets can't describe concepts as precisely as possible. And PMGRS may be too strict to depict concepts on universe.

As a generalization of OMGRS and PMGRS, we will propose a new multi-granulation rough set model with a parameter $\beta \in (0.5, 1]$. We introduce this parameter to realize that the objects supporting a concept in majority granulations are included and the ones possibly describing the concept below the corresponding level are ignored. This model is presented in the definition below.

Definition 3.2. Let $I = (U, A, V, f)$ be an information system, $X \subseteq U$ and $P = \{P_i \subseteq A | P_i \cap P_j = \varnothing (i \neq j), i, j \leq l\}$. $S_X^{P_i}(x)$ is supporting characteristic function of x. For any $\beta \in (0.5, 1]$, the lower and upper approximation sets of X with respect to P are defined as follows.

$$\underline{P}(X)_\beta = \{x \in U | \frac{\sum\limits_{i=1}^{l} S_X^{P_i}(x)}{l} \geq \beta\},$$

$$\overline{P}(X)_\beta = \{x \in U | \frac{\sum\limits_{i=1}^{l} (1 - S_{\sim X}^{P_i}(x))}{l} > 1 - \beta\}.$$

X is called definable if and only if $\underline{P}(X)_\beta = \overline{P}(X)_\beta$, otherwise X is rough if and only if $\underline{P}(X)_\beta \neq \overline{P}(X)_\beta$. We denote this generalized multi-granulation rough set model by GMGRS and call β the information level with respect to P.

GMGRS is a generalization of OMGRS and PMGRS. The approximations in these models can reflect this. In the following proposition, we present the relations between GMGRS and OMGRS (PMGRS).

Proposition 3.3. Let $I = (U, A, V, F)$ be an information system, $X \subseteq U$, $\beta \in (0.5, 1]$ and $P = \{P_i \subseteq A | P_i \cap P_j = \varnothing (i \neq j), i, j \leq l\}$. The lower and upper approximations in GMGRS have the following relation with those in OMGRS and PMGRS.

(1) $\underline{PM}(X) \subseteq \underline{P}(X)_\beta \subseteq \underline{OM}(X)$;
(2) $\overline{OM}(X) \subseteq \overline{P}(X)_\beta \subseteq \overline{PM}(X)$.

This proposition can be proved easily by Definition3.2 and Proposition3.2. Details will not be illustrated on these two properties.

Remark 1. The relation of inclusion between $\underline{P}(X)_\beta$ and an arbitrary $\underline{P_i}(X)$ is uncertain. And so does it between $\overline{P}(X)_\beta$ and an arbitrary $\overline{P_i}(X)$.

Propositions of approximations in rough set theory are useful and important in theoretical research and practice. Thus, we will investigate some important properties as Pawlak rough set in the following.

Proposition 3.4. Let $I = (U, A, V, F)$ be an information system, $X \subseteq U$ and $P = \{P_i \subseteq A | P_i \cap P_j = \varnothing (i \neq j), i, j \leq l\}$. For any $\beta \in (0.5, 1]$, we have that

(1a) $\underline{P}(\sim X)_\beta =\sim \overline{P}(X)_\beta$;
(1b) $\overline{P}(\sim X)_\beta =\sim \underline{P}(X)_\beta$;
(2a) $\underline{P}(X)_\beta \subseteq X$;
(2b) $X \subseteq \overline{P}(X)_\beta$;
(3a) $\underline{P}(\varnothing)_\beta = \overline{P}(\varnothing)_\beta = \varnothing$;
(3b) $\underline{P}(U)_\beta = \overline{P}(U)_\beta = U$;
(4a) $X \subseteq Y \Rightarrow \underline{P}(X)_\beta \subseteq \underline{P}(Y)_\beta$;
(4b) $X \subseteq Y \Rightarrow \overline{P}(X)_\beta \subseteq \overline{P}(Y)_\beta$;
(5a) $\underline{P}(X \cap Y)_\beta \subseteq \underline{P}(X)_\beta \cap \underline{P}(Y)_\beta$;
(5b) $\overline{P}(X \cup Y)_\beta \supseteq \overline{P}(X)_\beta \cup \overline{P}(Y)_\beta$;
(6a) $\underline{P}(X \cup Y)_\beta \supseteq \underline{P}(X)_\beta \cup \underline{P}(Y)_\beta$;
(6b) $\overline{P}(X \cap Y)_\beta \subseteq \overline{P}(X)_\beta \cap \overline{P}(Y)_\beta$.
where, "$\sim$" is means the complementary operation of cantor sets.

Proof. (1a) Since $x \in \overline{P}(X)_\beta \Leftrightarrow \dfrac{\sum\limits_{i=1}^{l} (1-S_{\sim X}^{P_i}(x))}{l} > 1 - \beta$. Then,

$$x \in\sim \overline{P}(X)_\beta \Leftrightarrow \frac{\sum\limits_{i=1}^{l} (1-S_{\sim X}^{P_i}(x))}{l} \leq 1 - \beta \Leftrightarrow \frac{\sum\limits_{i=1}^{l} S_{\sim X}^{P_i}(x)}{l} \geq \beta \Leftrightarrow x \in \underline{P}(\sim X)_\beta.$$

This item is proved. Item (1b) can be proved similarly as (1a).

(2a) For any $x \in \underline{P}(X)_\beta$, we have that $\dfrac{\sum\limits_{i=1}^{l} S_X^{P_i}(x)}{l} \geq \beta > 0$. Then, there must exist $i \leq l$ such that $[x]_{P_i} \subseteq X$. Thus $x \in X$. $\underline{P}(X)_\beta \subseteq X$ is proved.

(2b) By $\sim \overline{P}(X)_\beta = \underline{P}(\sim X)_\beta \subseteq\sim X$, we can have $X \subseteq \overline{P}(X)_\beta$ directly.

(3a)(3b) From Proposition 3.1 $S_\varnothing^{P_i}(x) = 0$ and $S_U^{P_i}(x) = 1$ ($\forall x \in U$), we have that

$$\underline{P}(\varnothing)_\beta = \{x \in U | \frac{\sum\limits_{i=1}^{l} S_\varnothing^{P_i}(x)}{l} = \frac{\sum\limits_{i=1}^{l} 0}{l} = 0 \geq \beta\} = \varnothing,$$

$$\underline{P}(U)_\beta = \{x \in U | \frac{\sum\limits_{i=1}^{l} S_U^{P_i}(x)}{l} = \frac{\sum\limits_{i=1}^{l} 1}{l} = 1 \geq \beta\} = U.$$

From the duality (1a)(1b), we can easily have $\overline{P}(\varnothing)_\beta = \varnothing$ and $\overline{P}(U)_\beta = U$.

(4a) For any $x \in \underline{P}(X)_\beta$, we have $\frac{\sum\limits_{i=1}^{l} S_X^{P_i}(x)}{l} \geq \beta$. Since $X \subseteq Y$, one can have $S_X^{P_i}(x) \leq S_Y^{P_i}(x)$. Then,

$$\frac{\sum\limits_{i=1}^{l} S_Y^{P_i}(x)}{l} \geq \frac{\sum\limits_{i=1}^{l} S_X^{P_i}(x)}{l} \geq \beta.$$

So $x \in \underline{P}(Y)_\beta$ is obtained. Thus, this item is proved and item (4b) can be proved similarly.

(5a) From the propositions of $S_X^{P_i}(x)$, for any $x \in \underline{P}(X \cap Y)_\beta$, we have that

$$x \in \underline{P}(X \cap Y)_\beta \Leftrightarrow \frac{\sum\limits_{i=1}^{l} S_{X \cap Y}^{P_i}(x)}{l} = \frac{\sum\limits_{i=1}^{l} S_X^{P_i}(x) \wedge \sum\limits_{i=1}^{l} S_Y^{P_i}(x)}{l} \geq \beta$$

$$\Leftrightarrow \frac{\sum\limits_{i=1}^{l} S_X^{P_i}(x)}{l} \geq \beta \text{ and } \frac{\sum\limits_{i=1}^{l} S_Y^{P_i}(x)}{l} \geq \beta$$

$$\Leftrightarrow x \in \underline{P}(X)_\beta \text{ and } x \in \underline{P}(Y)_\beta$$

$$\Leftrightarrow x \in \underline{P}(X)_\beta \cap \underline{P}(Y)_\beta.$$

(5b) From the duality property, this item can be proved easily by (5a).
(6a)(6b) can be proved directly by properties (4a) and (4b).

Remark 2. Propositions (4),(5) in Proposition 3.4 are not the same as SGRS, OMGRS and PMGRS. And the properties $\underline{P}(\underline{P}(X)_\beta)_\beta = \underline{P}(X)_\beta = \overline{P}(\underline{P}(X)_\beta)_\beta$ and $\overline{P}(\overline{P}(X)_\beta)_\beta = \overline{P}(X)_\beta = \underline{P}(\overline{P}(X)_\beta)_\beta$ don't hold in GMGRS.

For different information levels, one can consider the difference between approximations and the following properties hold.

Proposition 3.5. Let $I = (U, A, V, F)$ be an information system, $X \subseteq U$ and $P = \{P_i \subseteq A | P_i \cap P_j = \varnothing (i \neq j), i, j \leq l\}$. For any $\alpha \leq \beta$ and $\alpha, \beta \in (0.5, 1]$, the following propositions hold.
  (1) $\underline{P}(X)_\beta \subseteq \underline{P}(X)_\alpha$,
  (2) $\overline{P}(X)_\alpha \subseteq \overline{P}(X)_\beta$.

Proof. It can be proved easily by Definition 3.1 and Proposition 3.4.

In this section, we proposed a generalized multi-granulation rough set model and studied some important propositions. From these propositions, we can easily have that GMGRS is a more generalized and logical multi-granulation rough set model than ones refereed in Section 2. By the information level $\beta \in (0.5, 1]$, GMGRS has the ability to discover more affirmative information than PMGRS and leave out some useless possible knowledge in information systems. Furthermore, GMGRS popularize OMGRS and discover information more precise than OMGRS. Propositions studied in this section have important effect in practice and make it convenient to solve problems using the new model we propose.

4 Conclusions

We proposed a generalized multi-granulation rough set model denoted by GM-GRS in this paper. And some important propositions were discussed in detail. GMGRS is a generalization of OMGRS and PMGRS. More useful information and descriptions can be employed in GMGRS to represent the knowledge precisely for supportting the concept in majority granulations. Correspondingly, many useless information and descriptions can be thrown off since they have so less effect on possible knowledge representation that they can be ignored in sense of multi-granulation.

From the paper, one can find that GMGRS is more valid than PMGRS and OMGRS in the view of multi-granulation. This model is a complement of multi-granulation rough set theory and may make great effect in practice.

Acknowledgement. This paper is supported by National Natural Science Foundation of China (No. 11001227, 71071124).

References

1. Komorowski, J., Polkowski, L., Skowron, A.: Rough sets: A Tutorial. Rough Fuzzy Hybridization. In: A New Trend in Decision Making. Springer-Verlag New York, Inc., New York (1999)
2. Pawlak, Z.: Rough sets. International Journal of Computer and Information Science 11, 341–356 (1982)
3. Pawlak, Z., Skowron, A.: Rough sets: Some Extensions. Information Sciences 177, 28–40 (2007)
4. Qian, Y.H., Liang, J.Y.: Rough Set Method Based on Multi-granulations: The 5th IEEE International Conference on Cognitive Informatics, Beijing, China (2006)
5. Qian, Y.H., Liang, J.Y., Dang, C.Y.: Incomplete Multigranulation Rough Set. IEEE Transactions on Systems, Man and Sybernetics-Part A: Systems and Human 40(2), 420–431 (2010)
6. Qian, Y.H., Liang, J.Y., Yao, Y.Y., et al.: MGRS: A multi-granulation rough set. Information Sciences 180(6), 949–970 (2010)
7. Skowron, A., Peters, J.F.: Rough Sets Trends and Challenges Extended Abstract. In: Wang, G., Liu, Q., Yao, Y., Skowron, A. (eds.) RSFDGrC 2003. LNCS (LNAI), vol. 2639, pp. 25–34. Springer, Heidelberg (2003)

8. Xu, W.H., Zhang, X.Y., Zhong, J.M., Zhang, W.X.: Attribute Reduction in Ordered Information Systems Based on Evidence Theory. Knowledge and Information Systems 25(1), 169–184 (2010)
9. Xu, W.H., Zhang, X.Y., Zhang, W.X.: Knowledge Granulation, Knowledge Entropy and Knowledge Uncertainty Measure in Ordered Information Systems. Applied Soft Computing (9), 1244–1251 (2009)
10. Xu, W.H., Zhang, X.Y., Zhang, W.X.: Two New Types of Multiple Granulation rough set. Information Science (submitted)
11. Yao, Y.Y.: Information Granulation and Rough Set. International Journal of Intelligent Systems 16(1), 87–104 (2001)
12. Zhang, W.X., Leuang, Y., Wu, W.Z.: Information System and Knowledge Discovery. Science Press, Beijing (2003)

Closed-Label Concept Lattice Based Rule Extraction Approach

Junhong Wang, Jiye Liang, and Yuhua Qian

[1] School of Computer and Information Technology, Shanxi University, Taiyuan, 030006, Shanxi, China
[2] Key Laboratory of Computational Intelligence and Chinese Information Processing of Ministry of Education, Taiyuan, 030006, China
{wjhwjh,ljy}@sxu.edu.cn, jinchengqyh@126.com

Abstract. Concept lattice is an effective tool for data analysis and extracting classification rules. However, the classical concept lattice often produce a lot of redundant rules. Closed-label concept lattice realizes reduction of concept intention, which can be used to extract fewer rules than the classical concept lattice. This paper presents a method for classification rules extraction based on the closed-label concept lattice. Examples show that the proposed method is effective for extracting more concise classification rules.

Keywords: Concept lattice, Closed-label concept lattice, Classification rules, ID3.

1 Introduction

Concept lattice theory was proposed by Wille[1] in 1982. In the concept lattice theory, the data for analysis are described by a formal context (U, A, I), which consists of universe U, attributes set A, and relation $I \in U \times A$. Based on the formal context, we can construct some formal concepts and the set of all the above formal concepts forms a concept lattice. The concepts are constituted by two parts: intension, which comprises all attributes shared by the objects, and extension, which consists of all objects belonging to the concept. The concept lattice reflects the relationship of generalization and specialization among concepts. It is an effective and intuitive way to represent, design and discover knowledge structures.

Concept lattice theory is a kind of important mathematical tool for conceptual knowledge processing and data analysis. It provides a theoretical framework for the discovery and design of concept hierarchies from relational information systems. Most of the researches on concept lattice focus on such topics as: construction of concept lattice[2,3], extended model of concept lattice[6], acquisition of rules[4,5], relationship with rough set[7-10], and attribute reduction[11-14]. To date, concept lattice has been applied to digital library, information retrieval, software engineering and other aspects[15-17].

D.-S. Huang et al. (Eds.): ICIC 2011, LNBI 6840, pp. 690–698, 2012.

Classification rules mining is an important data mining task. The typical classification rule mining methods include decision tree, neural networks, rough sets and so on. Since concept lattice reflects the relationship between the formal concepts, it can be seen as a natural platform for rule extraction. Therefore the framework of the concept lattice is meaningful to discuss the issue of classification rules mining[4,5]. However, the classical concept lattice often produce a lot of redundant rules. In [6], we introduced a new lattice structure called closed-label concept lattice. In the new structure, all concepts are depicted by its intension reduction. Closed-label concept lattice realizes reduction of concept intention, which can be used to extract fewer rules than the classical concept lattice. The paper aims to present a method for extracting classification rules based on the closed-label concept lattice.

This paper is organized as follows. Basic definitions of concept lattice and closed label lattice are recalled in Section 2 and 3. In Section 4, closed-label concept lattice based rule extraction method is introduced. In Section 5, the proposed method compared with ID3 algorithm is given. Finally, some conclusions are given in Section 6.

2   Preliminaries

In this section, we review some basic concepts of concept lattice [1,18].

Definition 1. A formal context is a triplet (U, A, I), where $U = \{x_1, x_2, \ldots, x_n\}$ is a non-empty, finite set of objects called the universe of discourse, $A = \{a_1, a_2, \ldots, a_m\}$ is a non-empty, finite set of attributes, and $I \subseteq U \times A$ is a binary relation between U and A, where $(x, a) \in I$ means that object x has attribute a.

In this paper, we assume that the binary relation I is regular, that is, it satisfies the following conditions: for any $(x, a) \in U \times A$,

(1) there exist $a_1, a_2 \in A$ such that $(x, a_1) \in I$ and $(x, a_2) \notin I$,

(2) there exist $x_1, x_2 \in U$ such that $(x_1, a) \in I$ and $(x_2, a) \notin I$.

For $X \subseteq U$ and $B \subseteq A$, we define

$$X^* = \{a \in A : \forall x \in X, (x, a) \in I\},$$
$$B' = \{x \in U : \forall a \in B, (x, a) \in I\}. \tag{1}$$

X^* is the maximal set of attributes shared by all objects in X. Similarly, B' is the maximal set of objects that have all attributes in B. For $x \in U$ and $a \in A$, we denote $x^* = \{x\}^*$ and $a' = \{a\}'$. Thus x^* is the set of attributes possessed by x, and a' is the set of objects having attribute a.

Definition 2. Let (U, A, I) be a formal context. A pair (X, B), with $X \subseteq U$ and $B \subseteq A$, is called a formal concept of the context (U, A, I) if $X^* = B$ and $B' = X$. The set of objects X and the set of attributes B are respectively called the extension and the intension of the formal concept (X, B).

In the paper, for a formal concept C, the extension noted as extension(C) and the intension noted as intension(C).

Table 1. A formal context

U	a	b	c	d	e
1	0	1	0	1	0
2	1	0	1	0	1
3	1	1	0	0	1
4	0	1	1	1	0
5	1	0	0	0	1

The set of all formal concepts forms a complete lattice called a concept lattice and is denoted by $L(U, A, I)$. The meet and join of the lattice are given by:

$$(X_1, B_1) \wedge (X_2, B_2) = (X_1 \cap X_2, (B_1 \cup B_2)'^*),$$
$$(X_1, B_1) \vee (X_2, B_2) = ((X_1 \cup X_2)^{*'}, B_1 \cap B_2). \tag{2}$$

The corresponding partial order relation $\leq$ in the concept lattice $L(U, A, I)$ is given as follows: for $(X_1, B_1), (X_2, B_2) \in L(U, A, I)$,

$$(X_1, B_1) \leq (X_2, B_2) \Longleftrightarrow X_1 \subseteq X_2 \Longleftrightarrow B_1 \supseteq B_2. \tag{3}$$

A formal context can be represented by a table the rows of which are headed by the object names and the columns headed by the attribute names. A value 1 in row x and column a means that the object x has the attribute a.

In this paper, we use the table in [12] for example.

Example 1. Table 1 depicts an example of formal context $F = (U, A, I)$, where $U = \{1, 2, 3, 4, 5\}$, $A = \{a, b, c, d, e\}$, and for each $(x_i, a_j) \in U \times A$, $(x_i, a_j) \in I$ iff object x_i has value 1 in attribute a_j, i.e., $a_j(x_i) = 1$. Fig. 1 is the Hasse diagram of the concept lattice derived from Table 1.

3   Closed-Label Concept Lattice

In [6], we introduced a new lattice structure called closed-label concept lattice. In the new structure, all concepts are depicted with its intension reduction. In the following, we briefly introduce some concepts of closed-label concept lattice.

Let (U, A, I) be a formal context, $Y \subseteq A$, then Y is a closed itemset iff $Y'^* = Y$. closed itemset is the maximal set of attributes that shared by some object set. Thus, (1) Every intension of concept in concept lattice is a closed itemset. (2) The set of intension for all concepts in concept lattice equals to the set of closed itemset generated from formal context.

Definition 3. Let (U, A, I) be a formal context. The closed-label of C, expressed as $Closedlabel(C)$. For $\forall Y \in Closedlabel(C)$, satisfied:
(1) $Y'^* \neq Y$
(2) $Y'^* = intension(C)$
(3) $\forall Z \subset Y, Z'^* \subset Y'^*$

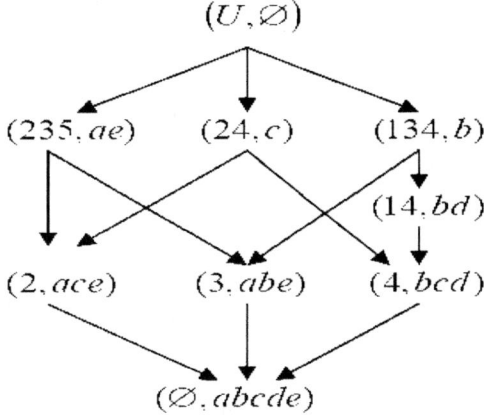

Fig. 1. $L(U, \{a, b, c, d, e\}, I)$

Definition 4. Let (U, A, I) be a formal context. C called a closed-label concept, expressed as $(extension(C), Closedlabel(C), intension(C))$.

The set of all closed-label concepts forms a closed-label concept lattice and is denoted by $CL(U, A, I)$.

Example 2. The closed-label concept lattice of Table 1 is shown in Fig. 2.

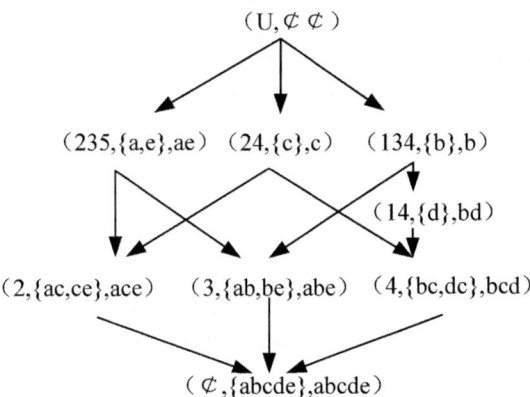

Fig. 2. $CL(U, \{a, b, c, d, e\}, I)$

4 Closed-Label Concept Lattice Based Rule Extraction Approach

In this section, a classification rule extration method based on closed-label concept lattice is given.

Theorem 1. Let $K = (U, A, I)$ be a formal context and $L(U, A, I)$ the associated concept lattice. $L_U(U, A, I)$ is the set of extension of all concepts: $L_U(U, A, I) = \{X|(X, B) \in L(U, A, I)\}$. Then, $L_U(U, A, I) = U$. $L_U(U, A, I)$ forms a cover of the domain.

Proof. The conclusion can be obtained directly.

Let $K' = (U, A \bigcup d, I)$ be a decision table, A the set of condition attributes and d the decision attribute. $\pi_d = \{d_i|i = 1, 2, \cdots, n\}$ denote that d divided U into d class.

Definition 5. Let $K' = (U, A \bigcup d, I)$ be a decision table, $L(U, A, I)$ the associated concept lattice induced by (U, A, I), if $\bigcup\{extention(C)|extension(C) \subseteq d_i, extension(C) \in L_U(U, A, I), i = 1, 2, \cdots, n\} = U$, then $K' = (U, A \bigcup d, I)$ is consistent.

In the following, a classification rule mining method based on closed-label concept lattice is proposed.

Algorithm 1: A classification rule mining method based on closed-label concept lattice

Input: A decision table $K' = (U, A \bigcup d, I)$, where $U = \{u_1, u_2, \cdots, u_{|U|}\}$, $A = \{a_1, a_2, \cdots, a_{|A|}\}$, and associated closed-label concept lattice $CL(U, A, I)$, $\pi_d = \{d_i|i = 1, 2, \cdots, n\}$

Output: A classification rule set R

01 Initialize $R = \emptyset, \Omega = \emptyset$
02   For every decision class d_i begin
03   Sort the concept as $|extention(C)|$ descending
04   for every $extention(C) \in L_U(U, A, I)$
05   if $extention(C) \subseteq d_i$, and $extention(C) \not\subseteq \Omega$
06   then $R = R \bigcup \{Closedlable(C) \Longrightarrow d_i\}, \Omega = \Omega \bigcup extention(C)$
07   End
08   End
09 Output R
10 End

As follows, we analyze the complexity for Algorithm 1. We suppose $|C|$ is the number of concepts in closed label lattice, $|d|$ is the number of decision class. Then, without considering the complexity of lattice construction. The time complexity of this algorithm is $O(|C||d|)$.

Example 3. Table 2 depicts an example of decision table $K' = (U, A \bigcup class, I)$, where $U = \{1, 2, 3, 4, 5, 6, 7, 8\}$, $A = \{Height, Hair, Eyes\}$. The corresponding formal context is shown as Table 3. where a denote $Height = short$, b denote $Height = tall$, c denote $Hair = blond$, d denote $Hair = red$, e denote $Hair = dark$, f denote $Eyes = brown$, g denote $Eyes = blue$. Fig. 3 is the Hasse diagram of the closed-label concept lattice derived from Table 2.

Then, four classification rules can be induced based on Algorithm 1: $f \Longrightarrow class = -$, $e \Longrightarrow class = -$, $cg \Longrightarrow class = +$, $d \Longrightarrow class = +$, i.e.,

Table 2. Decision table

	Height	Hair	Eyes	Class
1	short	blond	blue	−
2	short	blond	brown	-
3	tall	red	blue	−
4	tall	dark	blue	-
5	tall	dark	blue	-
6	tall	blond	blue	−
7	tall	dark	brown	-
8	short	blond	brown	-

Table 3. Formal context of Table2

U	a	b	c	d	e	f	g
1	1	0	1	0	0	0	1
2	1	0	1	0	0	1	0
3	0	1	0	1	0	0	1
4	0	1	0	0	1	0	1
5	0	1	0	0	1	0	1
6	0	1	1	0	0	0	1
7	0	1	0	0	1	1	0
8	1	0	1	0	0	1	0

$r_1 : Eyes = brown \Longrightarrow class = -;$
$r_2 : Hair = dark \Longrightarrow class = -;$
$r_3 : Hair = blond \wedge Eyes = blue \Longrightarrow class = +;$
$r_4 : Hair = red \Longrightarrow class = +.$

5   Comparison with ID3 Algorithm

Decision tree is an instance-based inductive learning algorithm, which induce classification rules from a group of objects. ID3 algorithm [19]proposed by Quinlan is a class of decision tree algorithm, which use information gain as the selection criteria of attribute.

Fig. 4 is the decision tree derived from Table 2 using ID3 Algorithm.

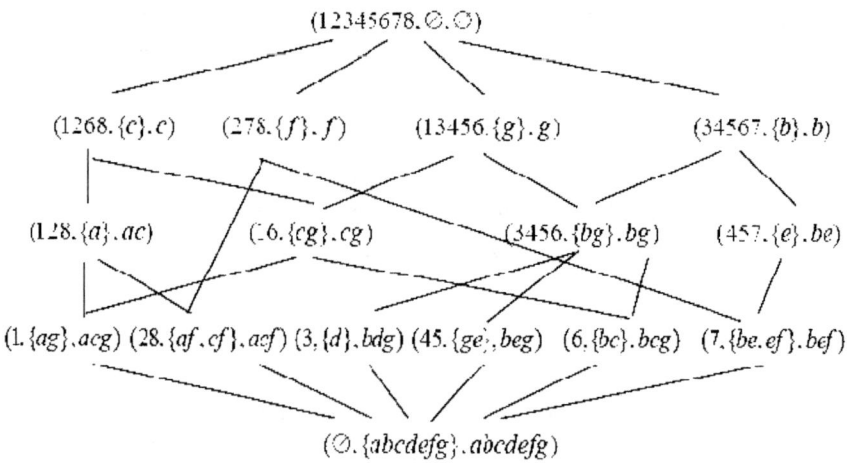

Fig. 3. Hasse diagram

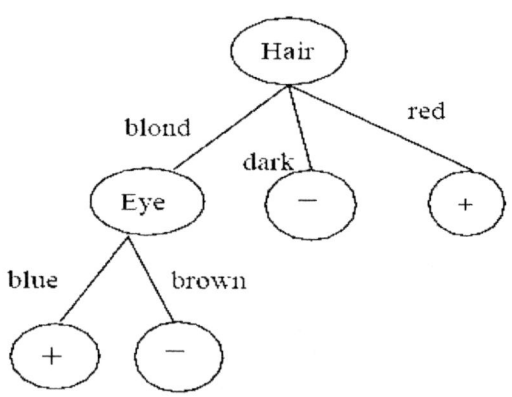

Fig. 4. Decision tree

Then, four classification rules can be induced from Fig. 4:

r'_1 : $Eyes = brown \wedge Eyes = brown \Longrightarrow class = -$;

r_2 : $Hair = dark \Longrightarrow class = -$;

r_3 : $Hair = blond \wedge Eyes = blue \Longrightarrow class = +$;

r_4 : $Hair = red \Longrightarrow class = +$;

From the two groups of rules, we can see that r_2, r_3, r_4, and r'_2, r'_3, r'_4 has same form, but r_1 has more concise form than r'_1.

Also, Hasse diagram provides a description of the whole information table, and can give users a more intuitive choice. In addition, when the test set increasing, concept lattice can be dynamically generated, which avoid the re-construction.

6   Conclusions

Classification is one of typical machine learning issues. As a novel soft computation method, the concept lattice show some advantages for data analysis and extracting rules. This paper has presented a method for classification rules extraction based on one generalization of concept lattice, called closed-label concept lattice. The rules extracted by the proposed method have much simpler forms.

Acknowledgements. This work was supported by the National Natural Science Foundation of China (Nos. 71031006, 70971080, 60903110), the Doctoral Program Foundation of Higher Education Research(No. 20101401110002), the Natural Science Foundation of Shanxi Province, China (Nos. 2010021017-3, 2009021017-1) and Technology Development Project of University of Shanxi Province, China.

References

1. Wille, R.: Restructuring Lattice Theory: An Approach Based on Hierarchies of Concepts. In: Rival, I. (ed.) Ordered Sets, pp. 445–470. Reidel, Dordrecht-Boston (1982)
2. Godin, R.: Incremental concept formation algorithm based on Galois (concept) lattice. Computational Intelligence 11(2), 246–267 (1995)
3. Ho, T.B.: An approach to concept formation based on formal concept analysis. IEICE Transaction on Information and Systems E78-D(5), 553–559 (1995)
4. Godin, R., Missaoui, R.: An incremental concept formation approach for learning from databases. Theoretical Computer Science, Special Issue on Formal Metthods in Databases and Software Engineering 133, 387–419 (1994)
5. Ho, T.B.: Discovering and using knowledge from unsupervised data, Decision Support System. Decision Support System 21(1), 27–41 (1997)
6. Liang, J.Y., Wang, J.H.: A new lattice structure and methor for extracting association rules based on concept lattcice. International Journal of Computer Science and Network Security 6(11), 107–114 (2006)
7. Yao, Y.Y.: A Comparative Study of Formal Concept Analysis and Rough Set Theory in Data Analysis. In: Tsumoto, S., Słowiński, R., Komorowski, J., Grzymała-Busse, J.W. (eds.) RSCTC 2004. LNCS (LNAI), vol. 3066, pp. 59–68. Springer, Heidelberg (2004)
8. Yao, Y.Y., Chen, Y.H.: Rough Set Approximations in Formal Concept Analysis. In: Peters, J.F., Skowron, A. (eds.) Transactions on Rough Sets V. LNCS, vol. 4100, pp. 285–305. Springer, Heidelberg (2006)
9. Shao, M.W., Liu, M., Zhang, W.X.: Set Approximations in Fuzzy Formal Concept Analysis. Fuzzy Sets and Systems 158(23), 2627–2640 (2007)
10. Hu, K., Sui, Y., Lu, Y.-c., Wang, J., Shi, C.-Y.: Concept Approximation in Concept Lattice. In: Cheung, D., Williams, G.J., Li, Q. (eds.) PAKDD 2001. LNCS (LNAI), vol. 2035, pp. 167–173. Springer, Heidelberg (2001)
11. Qian, Y.H., Liang, J.Y., Pedrycz, W., Dang, C.Y.: Positive approximation: An accelerator for attribute reduction in rough set theory. Artificial Intelligence 174(9-10), 597–618 (2010)

12. Wu, W.Z., Leung, Y., Mi, J.S.: Granular Computing and Knowledge Reduction in Formal Contexts. IEEE Transactions on Knowledge and Data Engineering 21(10), 1461–1474 (2009)
13. Zhang, W.X., Wei, L., Qi, J.J.: Attribute Reduction Theory and Approach to Concept Lattice, Science in China: Ser. F Information Sciences 48(6), 713–726 (2005)
14. Liu, M., Shao, M.W., Zhang, W.X., Wu, C.: Reduction Method for Concept Lattices Based on Rough Set Theory and Its Application. Computers and Mathematics with Applications 53(9), 1390–1410 (2007)
15. Jiang, G.Q., Chute, C.G.: Auditing the Semantic Completeness of SNOMED CT Using Formal Concept Analysis. Journal of the American Medical Informatics Association 16(1), 89–102 (2009)
16. Tonella, P.: Using a concept lattice of decomposition slices for program understanding and impact analysis. IEEE Transactions on Software Engineering 29(6), 495–509 (2003)
17. Yao, J.T., Yao, Y.Y.: A granular computing approach to machine learning. In: Proceedings of the 1st International Conference on Fuzzy Systems and Knowledge Discovery (FSKD 2002), Singapore, pp. 732–736 (2002)
18. Ganter, B., Wille, R.: Formal Concept Analysis, Mathematical Foundations. Springer, Berlin (1999)
19. Quinlan, J.R.: Induction of decision trees. Machine Learning 1, 81–106 (1986)

An Improved Extreme Learning Machine Based on Particle Swarm Optimization

Fei Han[1,*], Hai-Fen Yao[1], and Qing-Hua Ling[2]

[1] School of Computer Science and Telecommunication Engineering,
Jiangsu University, Zhenjiang, Jiangsu, 212013, China
[2] School of Computer Science and Engineering, Jiangsu University of Science and
Technology, Zhenjiang, Jiangsu, 212003, China
hanfei@ujs.edu.cn, {yao_hf0611,lingee_2000}@163.com

Abstract. Traditional extreme learning machine (ELM) may require high number of hidden neurons and lead to ill-condition problem due to the random determination of the input weights and hidden biases. In this paper, we use a modified particle swarm optimization (PSO) algorithm to select the input weights and hidden biases of single-hidden-layer feedforward neural networks (SLFN) and Moore–Penrose (MP) generalized inverse to analytically determine the output weights. The modified PSO optimizes the input weights and hidden biases according to not only the root mean squared error on validation set but also the norm of the output weights. The proposed algorithm has better generalization performance than other ELMs and its conditioning is also improved.

Keywords: Extreme learning machine, particle swarm optimization, generalization performance.

1 Introduction

In order to overcome the drawbacks of gradient-based methods, extreme learning machine (ELM) for single-hidden-layer feedforward neural networks (SLFN) was proposed [1]. ELM randomly chooses the input weights and hidden biases and analytically determines the output weights of SLFN through simple generalized inverse operation of the hidden layer output matrices. ELM not only learns much faster with higher generalization performance than the traditional gradient-based learning algorithms but also avoids many difficulties faced by gradient-based learning methods such as stopping criteria, learning rate, learning epochs, and local minima [2]. However, it is also found that ELM tends to require more hidden neurons than traditional gradient-based learning algorithms as well as result in ill-condition problem due to randomly selecting input weights and hidden biases [2,3,4].

In the literature [2], evolutionary ELM (E-ELM) was proposed which used the differential evolutionary algorithm to select the input weights and Moore–Penrose (MP) generalized inverse to analytically determine the output weights. The evolutionary ELM was able to achieve good generalization performance with much more compact

---

* Corresponding author.

D.-S. Huang et al. (Eds.): ICIC 2011, LNBI 6840, pp. 699–704, 2012.

networks. In the literature [5], particle swarm optimization (PSO) [6,7] was used to optimize the input weights and hidden biases of the SLFN to solve some prediction problems, which mainly encoded the boundary conditions into PSO to improve the performance of ELM. In the literature [3], in order to improve the conditioning of traditional ELM, an improved ELM was proposed by selecting input weights for an ELM with linear hidden neurons.

In this paper, PSO is also used to select the input weights and hidden biases of the SLFN, and the MP generalized inverse is used to analytically calculate the output weights. Unlike the hybrid algorithm in the literature [5], the proposed algorithm in this study optimizes the input weights and hidden biases according to the root mean squared error on validation set and the norm of the output weights. Thus, the proposed algorithm can obtain good performance with more compact and well-conditioned networks than other ELMs.

2   Particle Swarm Optimization

Particle swarm optimization (PSO) is a population-based stochastic optimization technique developed by Eberhart and Kennedy [6,7]. PSO simulates the social behavior of organisms, such as birds in a flock or fish in a school, and can be described as an automatically evolving system.

PSO works by initializing a flock of birds randomly over the searching space, where every bird is called as a "particle". These "particles" fly with a certain velocity and find the global best position after some iteration. At each iteration, each particle adjusts its velocity vector, based on its momentum and the influence of its best position (P_b) as well as the best position of its neighbors (P_g), and then a new position that the "particle" is to fly to is computed. Supposing the dimension of searching space is D, the total number of particles is n, the position of the i-th particle can be expressed as vector $X_i = (x_{i1}, x_{i2}, \cdots, x_{iD})$; the best position of the i-th particle searching until now is denoted as $P_{ib} = (p_{i1}, p_{i2}, \cdots, p_{iD})$, and the best position of all particles searching until now is denoted as vector $P_g = (p_{g1}, p_{g2}, \cdots, p_{gD})$; the velocity of the i-th particle is represented as vector $V_i = (v_{i1}, v_{i2}, \cdots, v_{iD})$. Then the original PSO [6,7] is described as:

$$v_{id}(t+1) = v_{id}(t) + c_1 * rand() * [p_{id}(t) - x_{id}(t)] + c_2 * rand() * [p_{gd}(t) - x_{id}(t)] \qquad (1)$$

$$x_{id}(t+1) = x_{id}(t) + v_{id}(t+1) \ 1 \le i \le n, \quad 1 \le d \le D \qquad (2)$$

where $c1$, $c2$ are the acceleration constants with positive values; $rand()$ is a random number between 0 and 1. In addition to the $c1$ and $c2$ parameters, the implementation of the original algorithm also requires to place a limit on the velocity (v_{max}).

The adaptive particle swarm optimization (APSO) algorithm is based on the original PSO algorithm, proposed by Shi & Eberhart [8]. The APSO can be described as follows:

$$v_{id}(t+1) = w*v_{id}(t) + c_1*rand()*[p_{id}(t) - x_{id}(t)] + c_2*rand()*[p_{gd}(t) - x_{id}(t)] \quad (3)$$

where w is a new inertial weight. The parameter can reduce gradually as the generation increases according to $w(t)=w_{max}-t*(w_{max}-w_{min})/itermax$ where w_{max}, w_{min} and $itermax$ are the initial inertial weight, the final inertial weight and the maximum searching generations, respectively. The APSO is more effective than original PSO, because the searching space reduces step by step, not linearly.

3   The Improved Extreme Learning Machine (IPSO-ELM)

In this paper, an approach named IPSO-ELM combining an improved PSO with ELM is proposed. This new ELM uses the improved PSO to select the input weights to improve the generalization performance and the conditioning of the SLFN. The detailed steps of the proposed method are as follows:

Firstly, the swarm is randomly generated. Each particle in the swarm is composed of a set of input weights and hidden biases: $P_i=[wh_{11},wh_{12},...,wh_{1n},..., wh_{21}, wh_{22},...,wh_{2n},..., wh_{H1},wh_{H2},...,wh_{Hn}, b_1,b_2,...,b_H]$. All components in the particle are randomly initialized within the range of [-1,1].

Secondly, for each particle, the corresponding output weights are computed according to ELM. Then the fitness of each particle is evaluated. In order to avoid overfitting of the SLFN, the fitness of each particle is adopted as the root mean squared error (RMSE) on the validation set only instead of the whole training set as used in [2,9].

Thirdly, with the fitness of all particles, the P_bs for all particles and the P_g for the swarm are computed. As analyzed by Bartlett and Huang [2,10], neural networks tend to have better generalization performance with the weights of smaller norm. Therefore, in order to further improve generalization performance, the norm of output weights along with the RMSE on the validation set are considered for determine the P_bs for all particles and the P_g for the swarm. The corresponding details are described as follows:

$$P_{ib} = \begin{cases} P_i & (f(P_{ib})-f(P_i) > \eta f(P_{ib})) \ or \ (|f(P_{ib})-f(P_i)| < \eta f(P_{ib}) \ and \ \|wo_{P_i}\| < \|wo_{P_{ib}}\|) \\ P_{ib} & else \end{cases} \quad (4)$$

$$P_g = \begin{cases} P_i & (f(P_g)-f(P_i) > \eta f(P_g)) \ or \ (|f(P_g)-f(P_i)| < \eta f(P_g) \ and \ \|wo_{P_i}\| < \|wo_{P_g}\|) \\ P_g & else \end{cases} \quad (5)$$

where $f(P_i)$, $f(P_{ib})$ and $f(P_g)$ are the corresponding fitness for the i-th particle, the best position of the i-th particle and global best position of all particles, respectively. wo_{P_i}, $wo_{P_{ib}}$ and wo_{P_g} are the corresponding output weights obtained by MP generalized inverse when the input weights are set as the i-th particle, the best position of the i-th particle and global best position of all particles, respectively. The parameter $\eta > 0$ is a tolerance rate.

Fourthly, each particle updates its position, and the new population is generated.

The above optimization process is repeated until the goal is met or the maximum optimization epochs are completed.

4   Experiment Results and Discussion

In this section, the IPSO-ELM are compared with E-ELM, LM (one of the fastest implementation of BP algorithms and is provided in the neural networks tools box of MATLAB.) and ELM. For the optimization algorithm in IPSO-ELM and E-ELM, the maximum optimization epochs both are 20, and the population size both are 200. All the results shown in this paper are the mean values of 50 trails. We conduct simulations on function approximation and benchmark classification problems. All the programs are run in MATLAB 7.0 environment.

To begin with, all the four algorithms are used to approximate the 'SinC' function:

$$y = \begin{cases} \sin(x) / x & x \neq 0 \\ 1 & x = 0 \end{cases} \qquad (6)$$

A training set (x_i, y_i) and testing set (x_i, y_i) with 1000 data, respectively, are created where x_is are uniformly randomly distributed on the interval (-10,10). Moreover, large uniform noise distributed in [-0.2,0.2] has been added to all the training samples while testing data remain noise-free. The corresponding results are shown in Table 1.

Table 1. The performance of four learning algorithms on approximating SinC function

Algorithms	Training RMSE	Testing RMSE	Dev	Cpu time Training	Hidden nodes	Condition	Norm
LM	0.1137	0.0223	0.0039	0.3079s	15	/	/
ELM	0.1138	0.0207	0.0038	0.0078s	30	2.9462e+14	3.2015e+9
E-ELM	0.1156	0.0168	0.0034	9.4652s	10	2.1416e+6	3.9983e+4
IPSO-ELM	0.1155	0.0163	0.0028	9.5808s	10	1.7569e+6	2.3355e+4

From Table 1, the following conclusions can be drawn as follows:

First, all ELMs have smaller testing RMSE than LM. Only ten hidden nodes both are assigned for E-ELM and IPSO-ELM, while 15 and 30 hidden nodes are used in LM and ELM, respectively.

Second, compared with ELM, the E-ELM and IPSO-ELM obtained smaller testing RMSE with less hidden nodes. This indicates that the E-ELM and IPSO-ELM can achieve better generalization performance with much more compact networks. Training Cpu time for the E-ELM and IPSO-ELM is more than that of the other algorithms, which is mainly spent on the selection of input weights.

Thirdly, the condition value of the hidden matrix H in the E-ELM and IPSO-ELM is much less than those of ELM. This shows that the networks trained by the E-ELM and IPSO-ELM are more well-conditioned that that of ELM. From the norm value of the output weights, it can be found that the E-ELM and IPSO-ELM have better generalization than ELM.

Finally, compare with E-ELM, IPSO-ELM has smaller RMSE, Deviation, condition and norm value, which shows that the IPSO-ELM is superior to the E-ELM on approximating SinC function.

The performance of IPSO-ELM is also tested on real benchmark classification problem-Diabetes classification. The training, testing and validation datasets are randomly regenerated at each trial of simulations for all the algorithms. The corresponding performance of four algorithms on the classification problem is listed in Table 2.

Table 2. Performance of four algorithms on Diabetes classification problem

Algorithm	Cpu time Training	Accuracy(%) Training	Accuracy(%) Testing+Dev.	Hidden neurons	Norm	Condition
LM	1.7386s	95.25	68.49+0.0312	20	/	/
ELM	0.0061s	81.45	75.19+0.0282	30	382.5409	7.6265e+3
E-ELM	12.2287s	76.98	75.51+0.0273	10	62.5096	1.2576e+3
IPSO-ELM	12.1810s	77.70	76.40+0.0241	10	72.4941	1.0556e+3

The similar conclusions to those in the experiment of the function approximation can be drawn from Table 2, which shows that the proposed method is an effective and efficient technique for training SLFN.

5 Conclusions

In this paper, we proposed an improved learning algorithm named IPSO-ELM which makes use of the advantages of both ELM and PSO. In the process of selecting the input weights, the modified PSO consider not only the RMSE on validation set but also the norm of the output weights. The proposed algorithm has better generalization performance than the E-ELM, ELM and LM algorithms. The system trained by the IPSO-ELM is well-conditioned. The simulation results also verified the effectiveness and efficiency of the proposed algorithm. Future research works will include how to apply the new learning algorithm to resolve more complex problem such as data mining on gene expression data.

Acknowledgments. This work was supported by the National Natural Science Foundation of China (Nos.60702056, 61003183), Natural Science Foundation of Jiangsu Province (No.BK2009197) and the Initial Foundation of Science Research of Jiangsu University (No.07JDG033).

References

1. Huang, G.-B., Zhu, Q.-Y., Siew, C.-K.: Extreme Learning Machine: A New Learning Scheme of Feedforward Neural Networks. In: 2004 International Joint Conference on Neural Networks (IJCNN 2004), Budapest, Hungary, July 25-29, pp. 985–990 (2004)
2. Zhu, Q.-Y., Qin, A.K., Suganthan, P.N., Huang, G.-B.: Evolutionary Extreme Learning Machine. Pattern Recognition 38(10), 1759–1763 (2005)
3. Zhao, G.P., Shen, Z.Q., Miao, C.Y., Man, Z.H.: On Improving the Conditioning of Extreme Learning Machine: A Linear Case. In: The 7th International Conference on Information, Communications and Signal Processing, ICICS 2009, Maucu, China, December 8-10, pp. 1–5 (2009)

4. Miche, Y., Sorjamaa, A., Bas, P., Simula, O., Jutten, C., Lendasse, A.: OP-ELM: Optimally Pruned Extreme Learning Machine. IEEE Transactions on Neural Networks 21(1), 158–162 (2010)
5. Xu, Y., Shu, Y.: Evolutionary Extreme Learning Machine – Based on Particle Swarm Optimization. In: Wang, J., Yi, Z., Żurada, J.M., Lu, B.-L., Yin, H. (eds.) ISNN 2006. LNCS, vol. 3971, pp. 644–652. Springer, Heidelberg (2006)
6. Eberhart, R.C., Kennedy, J.: A new optimizer using particles swarm theory. In: Proceeding of Sixth International Symposium on Micro Machine and Human Science, Nagoya, Japan, pp. 39–43 (1995)
7. Eberhart, R.C., Kennedy, J.: Particle swarm optimization. In: Proceeding of IEEE International Conference on Neural Network, Perth, Australia, pp. 1942–1948 (1995)
8. Shi, Y., Eberhart, R.C.: A modified particle swarm optimizer. In: Proceeding of IEEE World Conference on Computation Intelligence, pp. 69–73 (1998)
9. Ghosh, R., Verma, B.: A hierarchical method for finding optimal architecture and weights using evolutionary least square based learning. International Journal of Neural Systems 12(1), 13–24 (2003)
10. Bartlett, P.L.: The sample complexity of pattern classification with neural networks: the size of the weights is more important than the size of the network. IEEE Trans. Inform. Theory 44(2), 525–536 (1998)

Neural Network Generalized Inverse of Multi-motor Synchronous System Working on Vector Control Mode

Ding Jin-lin and Wang Feng

Department of Electronic Information Engineering, Suzhou vocational university,
Suzhou 215104, Jiangsu, China
{djinlin,wf}@jssvc.edu.cn

Abstract. Multi-motor synchronous system is a multi-input multi-output, non-linear and high coupling control system. The neural network generalized inverse system can realize the linearization and decoupling of the nonlinear control. Local minima, irrationality learning rate and over learning easily occur in traditional feedforward neural networks. To overcome the problems, it put forward a single hidden-layer feedforward neural network system (SLFNs) and constructed the two-motor synchronous combined system based on the SLFNs generalized inverse. The experimental results prove that the method can realize decoupling and the decoupling linearization subsystems are open-loop stability.

Keywords: SLFNs, multi-motor, neural network, generalized inverse.

1 Introduction

Multi-motor synchronous control is a nonlinear, strong-coupling, multi-input and multi-output(MIMO) system. Traditional methods [1,2] require accurate mathematical models and all state variables. The neural network generalized inverse system(NNGI) [3,4], which combined the linearization control theory of nonlinear system and the self-learning ability of neural network, can realize the linearization decoupling control of the unknown nonlinear system model.

Local extremum, irrationality learning rate and over learning easily occur in traditional feedback NNs[5]. The single hidden-layer feedforward neural network system (SLFNs) overcomes the problems. Weights of input layer and threshold of hidden layer can be arbitrarily assigned. The activation function can be arbitrary nonvanishing function. SLFNs can obtain the least training variance, the smallest weight norm and the best generalization. In this paper put forward the SLFNs, and constructed the two-motor synchronous combined system [6] base on SLFNs generalized inverse.

2 The Mathematical Model of Two-Motor Synchronous System and Its Reversibility

The control diagram of two-motor synchronous system is shown in Figure 1.

According to Hooke's law, considering the amount of forward slip, tension is described by the following equation

D.-S. Huang et al. (Eds.): ICIC 2011, LNBI 6840, pp. 705–710, 2012.
© Springer-Verlag Berlin Heidelberg 2012

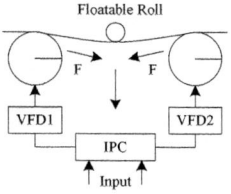

Fig. 1. The control diagram of two-motor synchronous system

$$\dot{F} = \frac{AE}{L_0}(\frac{1}{n_{p1}}r_1k_1\omega_{r1} - \frac{1}{n_{p2}}r_2k_2\omega_{r2}) - \frac{AV}{L_0}F \quad = \frac{K}{T}(\frac{1}{n_{p1}}r_1k_1\omega_{r1} - \frac{1}{n_{p2}}r_2k_2\omega_{r2}) - \frac{F}{T} \quad (1)$$

Under vector control mode, the linkage of rotor flux remained constant approximately, the mathematical model of the system can be simplified as follow

$$\begin{cases} \dot{x} = f(x,u) = \begin{bmatrix} \frac{n_{p1}}{J_1}[(u_1 - x_1)\frac{n_{p1}T_{r1}}{L_{r1}}\psi_{r1}^2 - (T_{L1} + r_1x_3)] \\ \frac{n_{p2}}{J_2}[(u_2 - x_2)\frac{n_{p2}T_{r2}}{L_{r2}}\psi_{r2}^2 - (T_{L2} - r_2x_3)] \\ \frac{K}{T}(\frac{1}{n_{p1}}r_1k_1x_1 - \frac{1}{n_{p2}}r_2k_2x_2) - \frac{x_3}{T} \end{bmatrix} \\ y = h(x) = [y_1, y_2]^T = [x_1, x_3]^T = [\omega_{r1}, F]^T \end{cases} \quad (2)$$

Where the state variable $x = [x_1, x_2, x_3]^T = [\omega_{r1}, \omega_{r2}, F]^T$, and control variable $u = [u_1, u_2]^T = [\omega_1, \omega_2]^T$. By analyzed the reversibility of the system, the relative degree is $\alpha = (1,2)$ and $\alpha_1 + \alpha_2 = 3 = n$. The two outputs of the original system are y_1 and y_2, whose the highest order are 1 and 2 respectively, the highest order of u_1 and u_2 are all zero. So inverse is $u = \xi(y_1, \dot{y}_1, y_2, \dot{y}_2, \ddot{y}_2)$. Connected with the NNGI, the original system became linear and unrelated. The decoupled system is a pseudo-linear composite system, which has good static and dynamic characteristics. Transfer function is expressed by the equation $G(s) = diag\{G_1(s), G_2(s)\}$, where $G_1(s) = (a_{11}s + a_{10})^{-1}$, $G_2(s) = (a_{22}s^2 + a_{21}s + a_{20})^{-1}$.

The generalized inverse of the system is $u = \bar{\phi}(y_1, \eta_1, y_2, \dot{y}_2, \eta_2)$, appendix PID adjuster, the control block diagram is shown in Figure 2.

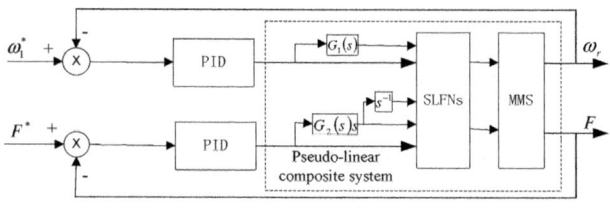

Fig. 2. Control block diagram of the NNGI

3   The Single Hidden-Layer Feedforward Neural Network System

The SLFNs has $\tilde{N}$ hidden layer neurons. Linear neuron is used in output layer. When the activation function is $g(x)$, the system can approach the sample in zero-error. The equation is $H\beta = T$, where

$$H\left(w_1,\cdots,w_{\tilde{N}},b_1,\cdots,b_{\tilde{N}},x_1,\cdots x_{\tilde{N}}\right)=\begin{bmatrix} g\left(w_1\cdot x_1+b_1\right) & \cdots & g\left(w_{\tilde{N}}\cdot x_1+b_{\tilde{N}}\right) \\ g\left(w_1\cdot x_N+b_1\right) & \cdots & g\left(w_{\tilde{N}}\cdot x_N+b_{\tilde{N}}\right) \end{bmatrix}_{N\times\tilde{N}} \quad (3)$$

$$\beta=\begin{bmatrix}\beta_1^T & \cdots & \beta_{\tilde{N}}^T\end{bmatrix}_{\tilde{N}\times m}^T, \quad T=\begin{bmatrix}t_1^T & \cdots & t_N^T\end{bmatrix}_{N\times m}^T$$

The conventional gradient descent learning algorithm has defects, such as difficulty of selecting learning rate, local minimum problem, over learning easily occur and lengthy. SLFNs is a novel learning algorithm. Training SLFNs is just to find a least squares solution β of the equation

$$\left\|H\left(w_1,\cdots,w_{\tilde{N}},b_1,\cdots,b_{\tilde{N}}\right)\hat{\beta}-T\right\|=\min_{\beta}\left\|H\left(w_1,\cdots,w_{\tilde{N}},b_1,\cdots,b_{\tilde{N}}\right)\beta-T\right\| \quad (4)$$

The minimum norm least squares solution of the equation (4) is $\hat{\beta}=H^+T$. This particular solution is uniquely determined, it is one of the least squares solutions, so it can obtain the least training mean square shown as equation (5).

$$\left\|H\hat{\beta}-T\right\|=\left\|HH^+T-T\right\|=\min_{\beta}\left\|H\beta-T\right\| \quad (5)$$

The particular solution has the minimum norm as equation (6)

$$\left\|\hat{\beta}\right\|=\left\|H^+T\right\|\le\left\|\beta\right\|,\forall\beta\in\left\{\beta:\left\|H\beta-T\right\|\le\left\|Hz-T\right\|,\forall z\in R^{\tilde{N}\times N}\right\} \quad (6)$$

The activation function of traditional learning algorithm must be differentiable, but SLFNs can adopt non-differentiable function. Routine learning algorithm has the problems of easily getting in the local minimum, irrational learning rate, over learning, SLFNs solve the problems and it is more simple and faster. The algorithm of gradient descent can be used to train multi hidden layer NN. SLFNs is applied to train a single hidden layer NN. Here chose SLFNs learning algorithm

$$\Psi=\left\{\left(x_i,t_i\right)\middle| x_i\in R^n,t_i\in R^m,i=1,\cdots,N\right\} \quad (7)$$

The activation function $g(x)$ can choose non-differentiable function. The number of hidden layer neurons is $\tilde{N}$. Initial values of input layer weight w_i and hidden layer threshold b_i is assigned arbitrarily, calculate the output matrix H of hidden NN, calculate the output weight $\beta:\hat{\beta}=H^+T$.

4   System Synthesis and Implementation of Controller

The NNGI series multi-motor system form composite controlled system, which is a linear decoupling special system, then speed and tension are relatively independent pseudo-linear subsystems. Design its controller based on linear control theory.

Pole of pseudo-linear subsystems is assigned again. The PI regulator is added in speed pseudo-linear subsystem. The parameter $K_p = 0.62, K_i = 30$. The PD regulator is added in tension pseudo-linear subsystem. The parameter $K_p = 20, K_d = 120$. Construct $\eta_1 = y_1 + \dot{y}_1, \eta_2 = y_2 + 1.414\dot{y}_2 + \ddot{y}_2$, the determine matrix A_1 and A_2 as follows

$$A_1 = \begin{bmatrix} 1 & 0 \\ 1 & 1 \end{bmatrix}, \quad A_2 = \begin{bmatrix} 1 & 0 & 0 \\ 0 & 1 & 0 \\ 1 & 1.414 & 1 \end{bmatrix} \tag{8}$$

The NNGI adopt a 3-layer feedforward NN of 5-12-2 structure and three linear elements. The three linear elements are listed in turn as equation (9)

$$\frac{1}{s+1}, \quad \frac{s}{s^2+1.414s+1}, \quad \frac{1}{s} \tag{9}$$

Chose 1500 groups training samples, the input and output samples of the system $\{u_1, u_2, y_1, y_2\}$ calculate off-line $\{\dot{y}_1, \dot{y}_2, \ddot{y}_2\}$.

Let $[z_{10} \ z_{11}]^T = A_1[y_1 \ \dot{y}_1]^T$, $[z_{20} \ z_{21} \ z_{22}]^T = A_2[y_2 \ \dot{y}_2 \ \ddot{y}_2]^T$, then calculate input data set of NN $\{z_{11}, z_{10}, z_{22}, z_{21}, z_{20}\}$ and unitary processing. The activation function of hidden-layer selects hyperbolic tangent function. The activation function of output layer is linear function. Assign arbitrary initial values to input layer weight and hidden layer threshold, training neural network off-line let it approximate ideal output sample set $\{u_1, u_2\}$, after training 200 times obtain the weight vector $\hat{\beta}$. $\varphi_1 = z_{11}, \varphi_2 = z_{22}$ were the input, connect the NNGI with its original MIMO system, then a linearized and decoupled composite system is formed. Speed and tension are decoupled relatively independent pseudo-linear subsystems.

5   Testing

The block diagram used in testing is shown in figure 3.

When speed remained constant, the tension increased at 120 second, the graph of speed response and tension response are shown in figure 4. Where (a) is the graph based on PID, (b) is the graph based on NNGI, the upward curve is the speed wave, the downward curve is the tension wave. The results indicated that NNGI can completely realize the decoupling between speed and tension, and NNGI achieve better results than PID.

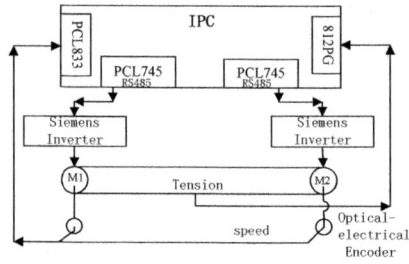

Fig. 3. The block diagram of the multi-motor synchronous system experimental installation

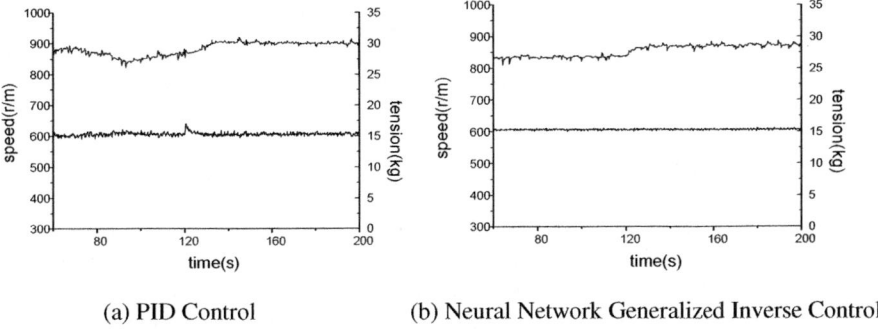

(a) PID Control   (b) Neural Network Generalized Inverse Control

Fig. 4. The graph of response when increase the tension suddenly

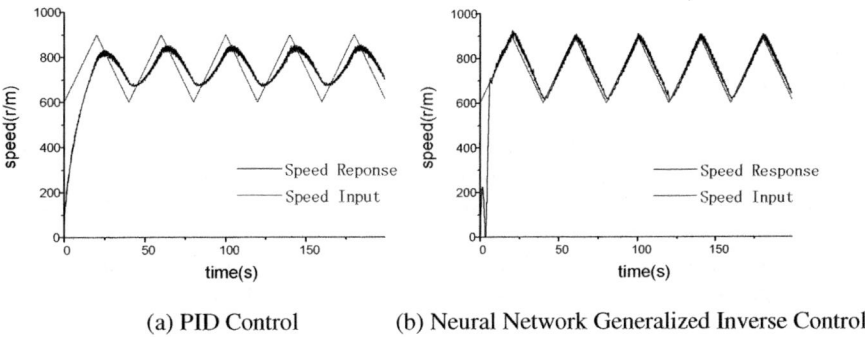

(a) PID Control   (b) Neural Network Generalized Inverse Control

Fig. 5. The graph of response when input is triangular-wave

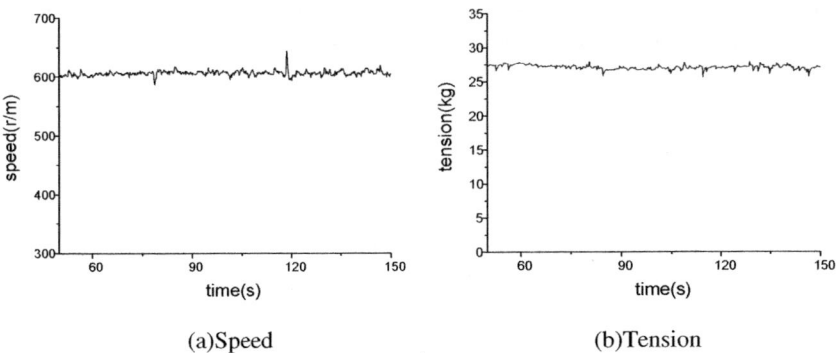

(a)Speed   (b)Tension

Fig. 6. The graph of response when increase or decrease the load suddenly

When the input speed is triangular-wave, the graph of speed response is shown in figure 5. Among which (a) is the graph based on PID, (b) is the graph based on NNGI. The solid line is the response of speed and the dotted line is the input of speed. The results showed that NNGI has better tracking ability, and has no distortion and hysteresis.

Under the mode of vector control, when the load of the two motors is increased or decreased suddenly at the same time, the graph of response is shown in figure 6, where (a) is the speed graph and (b) is the tension graph. The results showed that the speed and tension remain unchanged and would not be affected by abrupt load.

6   Conclusions

The two-motor synchronous system is a nonlinear coupling system, which connected with the NNGI and constructed the pseudo-linear combined system. The decoupling linearization subsystems are open-loop stability. The NNGI method has excellent control, compared with traditional PID method, the influence of increase or decrease the load suddenly is smaller, and has better tracking ability. It proved the superiority and advantage of NNGI. Furthermore, it can apply to general invertible nonlinear MIMO system.

References

1. Godbole, D.N., Sastry, S.S.: Approximate Decoupling and Asymptotic Tracking for MIMO Systems. IEEE Trans. on Automatic Control 40, 441–450 (1995)
2. Li, C., Feng, Y.: Inverse System Method of Multi-Variable Nonlinear Control. Tsinghua University Press, China (1991)
3. Dai, X., He, D., Zhang, X., Zhang, T.: MIMO system invertibility and decoupling control strategies based on ANN (th-order inversion). IEE Proc.-Control Theory Appl. 148, 125–136 (2001)
4. Huang, G.B., Zhu, Q.Y., Siew, C.K.: Real-time learning capability of neural networks. IEEE Transactions on Neural Networks 17, 863–878 (2006)
5. Huang, G.B., Chen, Y.Q., Babri, H.A.: Classification Ability of Single Hidden Layer Feed-Forward Neural Networks. IEEE Trans. on Neural Networks 11(3), 799–801 (2000)
6. Liu, G., Dai, X.: Decoupling control of an induction motor speeding system. Transactions of China Electrotechnical Society 16(5), 30–34 (2001)

Author Index